APPLIED OPTICS
AND
OPTICAL DESIGN

A. E. CONRADY

IN TWO PARTS

PART ONE

DOVER PUBLICATIONS, INC.
NEW YORK

Published in Canada by General Publishing Company, Ltd., 30 Lesmill Road, Don Mills, Toronto, Ontario.

Published in the United Kingdom by Constable and Company, Ltd., 3 The Lanchesters, 162–164 Fulham Palace Road, London W6 9ER.

This Dover edition, first published in 1992, is a reissue of the edition first published by Dover in 1957 (Part I) and 1960 (Part II). The Dover edition of Part I was an unabridged and corrected republication of the work originally published by the Oxford University Press, London, in 1929. Part II was originally published by Dover Publications, Inc., in 1960; this part was a posthumous publication edited and completed by Dr. Rudolf Kingslake (then Director of Optical Design, Eastman Kodak Company, Rochester, New York) with the assistance of Dr. Fred H. Perrin of the Eastman Kodak Company Research Laboratories.

Manufactured in the United States of America
Dover Publications, Inc., 31 East 2nd Street, Mineola, N.Y. 11501

Library of Congress Cataloging in Publication Data

Conrady, A. E. (Alexander Eugen)
Applied optics and optical design / A. E. Conrady ; edited and completed by Rudolf Kingslake.
p. cm.
Includes bibliographical references and index.
ISBN 0-486-67007-4 (pbk. : v. 1). — ISBN 0-486-67008-2 (pbk. : v. 2)
1. Optics. 2. Optical instruments. I. Kingslake, Rudolf. II. Title.
QC371.C62 1991
535—dc20 91-33198
CIP

PREFACE

THIS book is the final result of about thirty-five years of continuous devotion to the study and practice of applied optics, beginning about 1893 with first attempts, purely as a scientific hobby, to fathom the mysteries of the design of telescope and microscope objectives. The hobby developed into a profession some five years later, and a large number of new types of telescopic, microscopic, and photographic lens systems were the result, followed during the great war by the design of most of the new forms of submarine periscopes and of some other Service instruments. It was probably the uniform success of this work which caused me to be invited to the principal teaching position in the new Department of Technical Optics founded in 1917 at the Imperial College of Science and Technology. The increased opportunities for research thus provided, and especially the close contact with students, following upon a long period of widely varied practical experience, have been invaluable in determining the form in which the subject of applied optics is presented in the present work.

Every effort has been made to limit the subjects dealt with and the methods employed to what the late Silvanus P. Thomson called ' real Optics ' and to exclude the purely mathematical acrobatics, which he called ' examination optics '.

Progress in applied optics has been retarded to a most deplorable extent by the widely accepted belief that the two principal methods of attacking individual problems, namely the elegant but approximate algebraical method and the more empirical but rigorously exact method of trigonometrical ray-tracing, must necessarily be antagonistic and mutually exclusive. It is hoped that this book will permanently establish the opposite doctrine that the two methods are ideally fitted to be applied in closest conjunction, the analytical method readily finding a rough solution (which, however, is hardly ever close enough to admit of actual execution), while the trigonometrical method quickly and systematically adds the necessary finishing touches.

In order to raise this co-operation of the two methods to the highest possible efficiency, the algebraical expressions have been put into such a form and the adopted variables have been so chosen as to render the progression from the rough analytical solution to the exact trigonometrical calculations as smooth and as simple as possible, the aim throughout being to reach the exact final result with the least *total* expenditure of time and trouble. In some cases this last demand has led to the final adoption of equations which obviously could be algebraically transformed into apparently simpler expressions, the reason being, as is explicitly shown in one or two instances, that the algebraically simplest formula is by no means always that which can be most quickly evaluated numerically.

Considerable attention is devoted throughout the book to a subject which is hardly mentioned elsewhere. As the final prescription for a new optical system can only be executed within certain limits of precision, and since, moreover, in many cases—probably the majority—it is impossible to correct all the aberrations, the important question always arises in practice : At what magnitude does any one aberration become a serious menace to the proper performance of a given optical system ? It is this question which is answered in this book in the case of all the ordinary aberrations, chiefly on the basis of the important quarter-wave limit laid down by the third Lord Rayleigh in 1878, but strangely neglected until comparatively recent years.

The present volume includes all the ordinary ray-tracing methods, the general theory of perfect optical systems, the complete theory of the primary aberrations, and as much of the higher aberrations as is required for the design of all types of telescopes, of low-power microscopes, and of the simplest photographic objectives. A second volume will give the necessary additions to the theory, largely on the principle of equal optical paths, to extend the scope of the complete work to the systematic study and design of practically all types of optical systems, with special attention to high-power microscope objectives and to anastigmatic photographic objectives.

A. E. C.

CONTENTS

INTRODUCTION

ALTHOUGH even the simplest properties of lenses are proved in this book, the proofs and their discussion are brief, being included chiefly for the purpose of establishing a uniform and consistent system of nomenclature and of sign conventions. For that reason some acquaintance with elementary general optics will be distinctly helpful to the beginner, and the optical section of a good modern text-book on Physics will supply the desirable information. As nomenclature and sign conventions are almost sure to clash with those adopted in this book, this preliminary study should be, as far as possible, limited to the descriptive parts, keeping an open mind with regard to signs and symbols.

The mathematical knowledge which is assumed in the book does not, as a rule, go beyond ordinary geometry, algebra, and trigonometry, but analytical geometry and elements of the calculus are also employed in certain sections.

Serious students should bear in mind that the book is arranged in strict logical order and that it should therefore be worked through systematically from beginning to end. Moreover the present volume is almost entirely devoted to the development of general methods for the solution of optical problems, with very little specialization for isolated cases of restricted interest; hence there are very few sections which could be safely omitted. Owing to the large amount of ground to be covered repetitions or needless elaborations have been avoided, and even facts or conclusions of the highest importance may be found stated only once and in the fewest words compatible with clearness and completeness.

There is a deplorable tendency among students to concentrate on the mathematical equations and their proofs. These, whilst necessary and highly useful, are merely the dry bones of applied optics. The soul of the subject will be found in the numerous pages of plain letterpress with hardly any mathematical intermezzos, where the more important equations are discussed in order to discover their true optical significance and the best methods of applying them in the solution of practical problems. Students who intend to take up optical design as a profession should also realize from the very beginning that skill and ingenuity in numerical calculations will play a prominent part in determining their value to an optical establishment; they should therefore devote a considerable part of their available time to actual numerical work and should master as soon and as completely as possible the numerous hints and suggestions on this much neglected subject which will be found in almost every chapter of the book.

Whilst everything included in the text will find valuable applications in solving optical problems, it is not necessary, and would indeed be quite impossible for the vast majority of people, to remember the whole contents of the book. To assist the student in singling out the principal fundamental facts which should be firmly fixed upon the memory so as to be instantly available in thinking out or discussing optical phenomena or problems when books are not at hand, brief 'special memoranda' have been added at the end of each chapter. These should prove useful when their stated purpose is borne in mind ; but they should on no account be looked upon as representing everything that is of real value or importance in the particular chapter.

CHAPTER I

FUNDAMENTAL EQUATIONS

[1] SPHERICAL surfaces (including the plane as part of a spherical surface of infinite radius) are the only ones which can be produced by the optical grinding and polishing process with a sufficient approach to the necessary accuracy. Very slight departures from the spherical form, amounting at most to a few wave-lengths in depth, can indeed be secured by the process of 'figuring'; such small amounts can be taken care of in our computations by allowing for the slight departure from strictly spherical form by approximate corrections. We therefore limit ourselves in our general formulae to exact spherical surfaces.

A spherical surface is the simplest of all possible curved surfaces, for it is perfectly defined by its radius and by the location of its centre. A radius drawn from the centre through any point of a spherical surface stands at right angles to the tangent-plane in that point and is therefore the normal of the surface. As by the laws of reflection and refraction both the reflected and the refracted ray lie in the plane defined by the incident ray and the normal at the point of incidence ('incidence-normal') we can at once conclude that the plane of reflection and refraction in the case of spherical surfaces always contains the centre of the sphere and can thus be determined with the greatest ease.

In all ordinary optical instruments we have another vast simplification by reason of the centring of all the surfaces. It is intended, and usually achieved with sufficient accuracy, that the centres of curvature of all the component spherical surfaces shall lie on one and the same straight line, *the optical axis of the instrument.* That evidently means that any ray which originally cut the optical axis and therefore entered the system in a plane containing the optical axis will permanently remain in the same plane, and can therefore be traced right through the whole system by plane geometry or trigonometry. As this saves us the decidedly considerable trouble of determining a new incidence plane separately for each successive surface it will be one of the chief aims of our theoretical discussions to develop computing methods which avoid as far as possible the complication of tracing 'skew-rays'.

LAWS OF REFLECTION AND REFRACTION

[2] By the law of reflection the reflected ray lies in the plane defined by the incident ray and the incidence-normal and forms the same angle with the latter as the incident ray; but as the two rays lie on opposite sides of the normal, the angles have the opposite clock-sense and we express this by giving them the opposite sign: we therefore state the law of reflection as

$$I' = -I. \qquad \text{(I)}$$

By the law of refraction the refracted ray also lies in the plane defined by the incident ray and the normal of the refracting surface, but on the other side of both the surface and the normal: hence the angle of refraction has the same clock-sense or sign as the angle of incidence, and if N is the refractive index of the medium

[2] containing the incident ray, N' that of the medium containing the refracted ray, the law of refraction states that

(I)* $$N' \sin I' = N \, . \sin I.$$

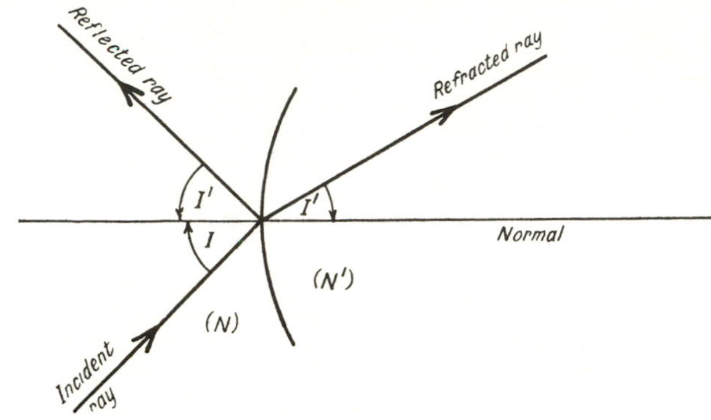

Fig. 1.

Small corresponding changes of I and I' are frequently of interest, and are found with sufficient accuracy by differentiating (I)* with the result

$$N' \cos I' . dI' = N \cos I \, . \, dI$$

or $$dI' = dI \, . \, N \cos I / N' \cos I',$$

and this equation will be included under (I)*.

When angles become very small their sines become equal to the angles themselves expressed in radians, hence the law of refraction for 'paraxial' rays which enter a refracting surface at very small angles with the incidence-normal becomes, using small letters for 'paraxial' angles,

(I)*$_p$ $$N'i' = Ni \, ; \; di' = di \, . \, N/N'.$$

Comparing the two fundamental laws, we at once see that we may treat the law of reflection mathematically as a particular case of the law of refraction, resulting from putting $N = -N'$, for this gives

$$N' \sin I' = -N' \sin I, \text{ or } \sin I' = -\sin I,$$

which with the necessarily acute angles can only be if $I' = -I$. This is a very important deduction because it enables us to apply practically all our refraction-formulae to problems of reflection by simply putting $N' = -N$ or $N'/N = -1$.

The Fundamental Formulae and Sign-Conventions

[3] It has been shown above that any problem of refraction at a spherical surface can be reduced to one of plane trigonometry by first finding the plane containing the ray to be traced and the centre of curvature.

In the diagram, Fig. 2, let the paper represent this incidence-plane, OP the ray to be traced through the refracting surface, and let the latter cut the incidence-plane in the circle AP with C as centre. Let ACB be a straight line passing through [3] the centre C to which it is convenient to refer the ray in the particular problem: very often it will be the optical axis of a complete system of lenses, but we do not

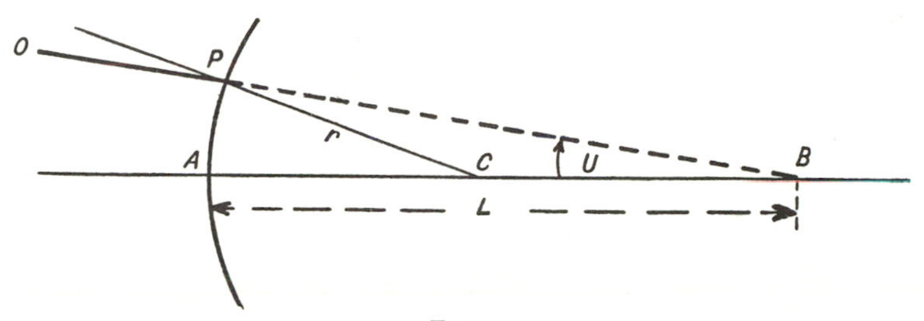

FIG. 2.

restrict ourselves at all to that assumption. A reference-axis ACB which does not coincide with the principal optical axis of a centred system will be referred to in future as an *auxiliary* optical axis.

We then define the position of the ray by the distance $AB = L$ from the pole or vertex of the surface to the point where the ray (produced if necessary) intersects the adopted axis ACB, and by the angle of obliquity or convergence U under which the ray meets that axis. It will be seen at once that both these quantities leave an ambiguity. We can find a point to the left of the pole A which is also at the given distance L from it, and we may in either of these intersection-points apply the angle U either above the axis AB as shown in Fig. 2 or below that axis. To remove these ambiguities we stipulate that the intersection-length AB shall be given the positive sign when it falls to the right of the pole A, and the negative sign when it falls to the left of the pole. In the case of the angle of obliquity U we stipulate that it shall always be an acute angle, and that it shall have the *positive* sign when a *clockwise* turn will bring a ruler from the direction of the adopted axis into that of the ray: and the negative sign if the turn is counter-clockwise. The only additional datum which enters into the calculation is the radius of curvature r of the refracting surface. Obviously this also requires a sign-convention to distinguish concave from convex surfaces. In agreement with the convention for the intersection-length L we stipulate that r shall have the positive sign when the centre C lies to the right of the pole A, and that r shall be negative when C lies to the left of A. Fig. 2 shows all the quantities in their positive position. The formulae to be deduced are such as will invariably bring out the *new* data for the *refracted* ray in accordance with the above sign-conventions.

The sign-conventions as stated are optically the most convenient, and therefore almost universally adopted. It should be noted at once that whilst the convention with regard to L and r agrees with that of co-ordinate geometry, the convention as to the sign of angles is the reverse of that usual in co-ordinate geometry. This occasionally will call for adjustments of the sign in equations derived in the first instance by the methods of analytical geometry.

The nomenclature employed in all our computing formulae has been carefully

[3] selected so as to be easily memorized, and also to be within the capacity of the ordinary typewriting machine. It is based on the following simple principles :

(1) Only English letters are employed, capitals for the data of rays at finite angles with, or at finite distances from, the optical axis, small letters for rays so close to the optical axis or to a principal ray as to allow of the use of simpler formulae.

(2) Vowels are invariably used for angles ; Consonants for lengths. Y is treated as a consonant.

(3) Quantities which are changed in value by the refraction at a surface are distinguished by the use of 'plain' and 'dashed' letters respectively.

(4) Suffixes such as I_1, U_4, &c., are only used when really necessary. Plain letters are usually retained for the surface actually under consideration, and the suffix (-1) then applies to the preceding, the suffix 1 to the following surface. In general formulae for a whole series of surfaces the latter are numbered successively 1, 2, 3, &c.

(5) In *lens-systems* the surfaces will always be numbered in the order from left to right. The refractive index† of the medium to the left of any surface will be denoted by plain N, that of the medium to the right by N', and plain L and U will be used for the ray in the left medium, L' and U' for the ray in the medium to the right.

It should be clearly realized and borne in mind that the use of 'plain' or 'dashed' letters is determined by the location of the actually existing ray, and not by that of the intersecting point B or B'. Thus in Fig. 2 the ray OP only exists in the medium to the left of the refracting surface, and its co-ordinates therefore are plain L and plain U, although the actual intersecting point B, found by producing the ray beyond the surface which really intercepts it and alters its subsequent course, lies to the right of the surface.

[4] In Fig. 2 (a) (which again shows all the quantities in their positive sense) we have the arriving ray OP, when produced, meeting the axis at B at the distance

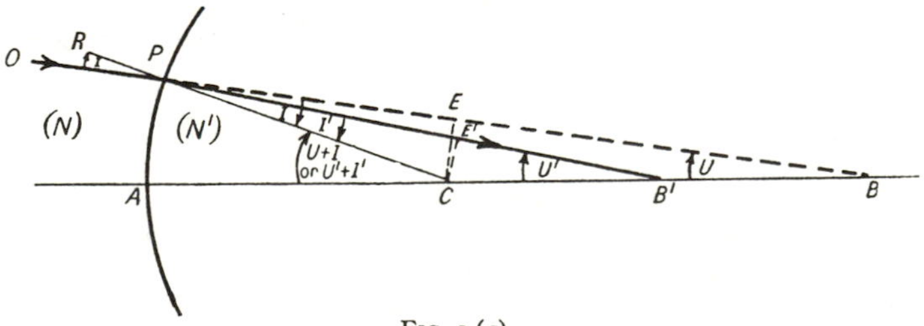

FIG. 2 (a).

$AB = L$ from the vertex A, and making with the axis the angle U. Drawing the radius CP through the point of incidence we have the angle of incidence OPR, for

† Capital N in preference to small n was originally adopted in the typed lecture-notes issued at the Imperial College because it was found to be more distinctive and to look better. Small Greek μ is inadmissible in practical applications of optics because it is the internationally agreed symbol for 0·001 mm., and in that meaning frequently enters into calculations which also involve refractive indices.

which we adopt the symbol I, also appearing at CPB in the triangle CPB. In this [4] triangle we know the side $CP = r$ and the side $CB = AB - AC = L - r$, also the angle $CBP = U$; dropping a perpendicular CE from C upon BP, we have

$$\sin I = \frac{CE}{r} \quad \text{and} \quad \frac{CE}{L-r} = \sin U,$$

and introducing the value of CE by the second equation into the first we obtain equation

$$\text{(1)} \qquad \sin I = \sin U \cdot \frac{L-r}{r},$$

by which we calculate the angle of incidence. A transposition of equation (I)* next gives

$$\text{(2)} \qquad \sin I' = \frac{N}{N'} \sin I$$

by the law of refraction, and we thus determine the direction of the refracted ray PB' in Fig. 2 (*a*).

We now have to determine L' and U' for the refracted ray. Referring to the diagram we see that the angle PCA at the centre of curvature is external angle to the triangle PCB and therefore $= (U+I)$, and is also external angle to the triangle PCB', corresponding to the refracted ray and therefore $= (U'+I')$. Consequently we have the important relation (which should be remembered as it is very frequently employed subsequently):

$$U+I = U'+I'$$

and by transposition of this we obtain the value of U', namely

$$\text{(3)} \qquad U' = U+I-I'.$$

In the triangle PCB' we now know the side $PC = r$ and the two angles I' and U': we therefore can determine the side $CB' = L' - r$. Employing the same reasoning already applied to the triangle PCB, we easily obtain

$$\text{(4)} \qquad L' - r = \sin I' \frac{r}{\sin U'} \quad \text{and from this}$$

$$\text{(5)} \qquad L' = (L' - r) + r$$

which completes the work for the surface under consideration.†

In accordance with a generally accepted custom the formulae have been deduced for a ray travelling from left to right. The formulae are not, however, limited in validity to this usual direction, for our definition of the meaning and of the signs of the L, U, and r is quite free from any reference to the direction in which the light is travelling. If we required to trace the same ray in the reverse direction we should have as given quantities its intersection-length $AB' = L'$ and its inclination U' to the optical axis in the medium of index N' to the right of the

† These extremely convenient computing formulae are quite ancient and probably date back to the eighteenth century. The German astronomer, Bessel, used them in his investigations of the Königsberg Heliometer (1841). Many attempts to find still better methods must be regarded as failing in attaining their object, and are therefore not included in the present book.

[4] surface, and we could compute its course by a simple transposition of the above formulae taken in inverse order. Equation (4) transposed would give

$$\text{(1) right-to-left:} \quad \sin I' = \sin U' \frac{L'-r}{r}.$$

By transposing equation (2) we should then have

$$\text{(2) right-to-left:} \quad \sin I = \frac{N'}{N} \sin I'.$$

Then by transposing (3)

$$\text{(3) right-to-left:} \quad U = U' + I' - I,$$

and finally by transposing (1)

$$\text{(4) right-to-left:} \quad L - r = \sin I \frac{r}{\sin U},$$

giving

$$\text{(5) right-to-left:} \quad L = (L-r) + r.$$

It will be noticed that these equations are mathematically identical with those for left-to-right calculations, from which they are obtainable by simply exchanging 'plain' and 'dashed' symbols.

This is very important, for many calculations of complicated lens-systems can be considerably shortened and simplified by using left-to-right and right-to-left calculations alternately.

[5] As has already been pointed out, the fundamental computing formulae are applicable to every case of refraction or reflection at spherical surfaces as soon as the incidence-plane has been determined. In the general case of a so-called skew-ray the determination of the incidence plane calls for a calculation of some difficulty at each successive surface. To avoid this, one of our principal aims will be to avoid the tracing of skew-rays as far as possible, and to derive the required results from rays proceeding in the plane of the optical axis of a centred lens-system. For such rays the transfer of the data from surface to surface is of the simplest kind.

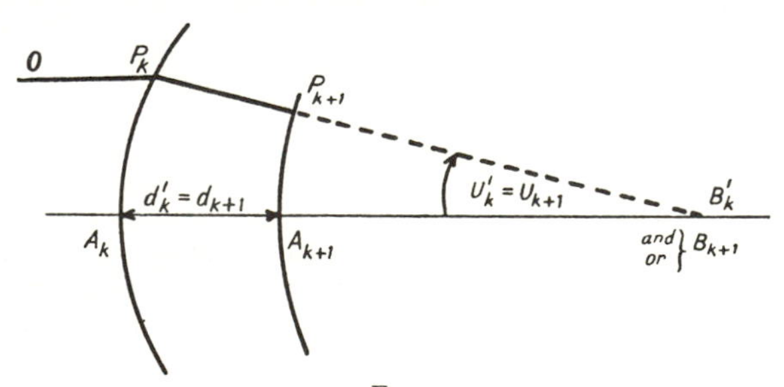

FIG. 3.

Assume that A_kP_k represents the kth surface of a centred system of lenses and that a ray OP_k in the plane of the optical axis of the system has been traced through the k surfaces. We shall know its intersection-length $A_kB'_k = L'_k$ and its angle of convergence U'_k. Let $A_{k+1}P_{k+1}$ be the next, $(k+1)$st, surface at an axial separation

$A_k A_{k+1} = d'_k$ from the kth surface. In accordance with our last stipulation as to [5] signs and symbols d'_k will necessarily lie to the right of surface $A_k P_k$ and is therefore a positive quantity. We shall usually employ the separation in this sense of a 'dashed' quantity. But we must note that for a right-to-left ray-tracing the same separation will more naturally be looked upon as the distance at which the kth surface lies to the *left* of the $(k+1)$st, and that it should then be called d_{k+1} (without a dash) and be treated as intrinsically negative. In the case of general theoretical formulae this alternative aspect of the separation is sometimes preferable, and the double name, implying a reversal of sign, of any given separation or thickness should therefore be clearly realized and well remembered. It means that in any formula we may replace d'_k by $-d_{k+1}$ and vice versa.

The refractive index of the medium between the two successive surfaces also has two names: it is N'_k when thought of in connexion with the kth surface, but N_{k+1} when associated with the $(k+1)$st surface. But as refractive indices are absolute positive numbers we have in this case $N'_k = N_{k+1}$.

We can now read the transfer-formulae directly from the diagram. The convergence-angle U'_k of the ray from or to the kth surface is obviously identical with that of the ray to or from the $(k+1)$st surface, therefore $U'_k = U_{k+1}$. But the intersection-lengths measured from the two surfaces are different by the separation. We thus arrive at the transfer formulae

(5)* $$L_{k+1} = L'_k - d'_k \; ; \; U_{k+1} = U'_k \; ; \; N_{k+1} = N'_k,$$

or for right-to-left calculations the transposed formulae

(5)* right-to-left $$L'_k = L_{k+1} + d'_k = L_{k+1} - d_{k+1} \; ; \; U'_k = U_{k+1} \; ; \; N'_k = N_{k+1}.$$

It should again be noted, as a rule which is in fact universally applicable, that in the equations for right-to-left ray-tracing 'plain' and 'dash' are exchanged and also k and $(k+1)$, the latter on account of the reversed sequence in which the surfaces are met by the light.

Finally we must discuss the most specialized of all applications of our standard [6] formulae, namely that to an *object-point* situated on the optical axis of a centred

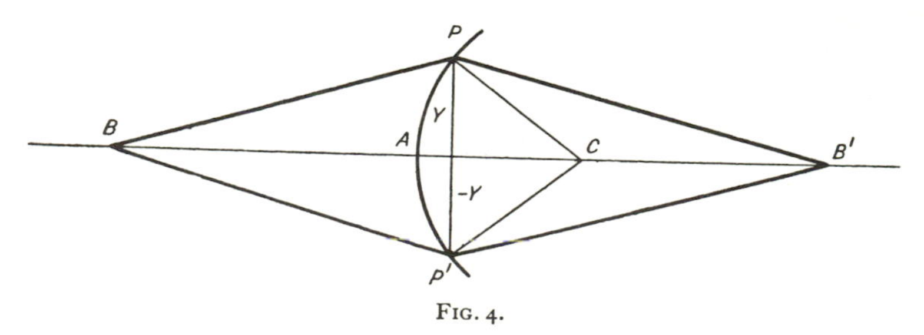

FIG. 4.

system. If B is such an object-point and P the point of incidence of a ray, we shall find the intersecting point B' of the refracted ray by the formulae in the ordinary way. But in this case we can draw a most important conclusion. By reason of the perfect symmetry of a spherical surface with reference to its centre

[6] it is at once obvious that if we calculated a ray from B through P', at the same distance Y *below* the optical axis as P is *above* it, we should get a diagram for the new ray exactly congruent to that for the first ray, and therefore, that we should find precisely the same intersecting point B'. This argument we can at once extend, for obviously the same reasoning applies to any section of our spherical surface laid through the optical axis AC, at any angle whatever with the plane of the diagram. We thus conclude that the image-point B' found in the first calculation must apply to a whole cone of rays which has its apex at B before refraction

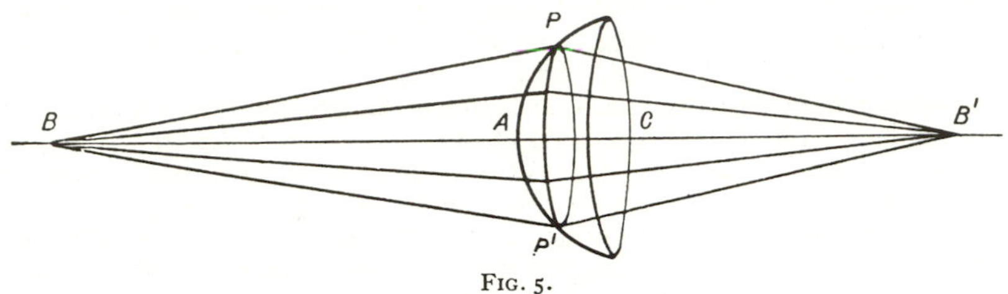

FIG. 5.

and at B' after refraction, and which passes through the refracting surface at the distance Y from the optical axis. The one calculation thus answers for a complete circular zone of the surface : and this evidently holds for any number of successive surfaces provided their centres of curvature lie on the original axis AC.

Important Note.

It should be observed that the change in the direction of a ray which is produced by refraction always depends on the *relative* refractive index N'/N. The indices determined by the spectrometer and communicated by the glassmakers are the relative indices of glass to ordinary air, and therefore allow automatically for the refractive index of air (about 1·00029) in glass-air refraction.

This fact justifies the universal optical practice of treating air as having a refractive index equal to 'one' exactly, at least in so far as the air may be treated as an unvarying medium. Really its index is approximately proportional to its density, and also varies with the humidity of the atmosphere. These disturbing causes may render the refractive indices of glass relative to air inaccurate by several units in the fifth decimal place, and are alone sufficient to demonstrate the futility of excessive accuracy in optical calculations.

NUMERICAL CALCULATIONS

A. Mathematical Tables and their Use

[7] Optical calculations can be carried out by a good calculating machine, and this method may sometimes save time and always reduces the mental strain of the work owing to the infallible accuracy with which the machine will produce correct results, provided the data put into it are correct.

For the present we will assume, however, that the calculations are to be carried out on paper with the aid of suitable mathematical tables. A good table of logarithms is then the primary requisite. For the majority of optical calculations the [7] accuracy afforded by logarithms to five places of decimals is both desirable and sufficient. Among the innumerable tables of this type first place for optical calculations must probably be given to one at present almost unobtainable in England, viz. Bremiker's table of logarithms and trigonometrical functions to five decimals. The strong points of the table are that it employs decimal division of the degree instead of the usual sexagesimal minutes and seconds. This greatly simplifies interpolation and reduces the risk of error in additions and subtractions of angles: moreover the table gives the trigonometrical functions at closer intervals than those of sexagesimal tables: the log sines of the first five degrees are given at intervals of 0·001° and the logs of all the functions of larger angles at intervals of 0·01°. The best five-figure table with sexagesimal division of the degree is Albrecht's; its strong points are the excellent and complete arrangements for interpolation and the inclusion of secants and cosecants in the trigonometrical part. The numerous small five-figure tables with big intervals and average differences for interpolation should be avoided on account of the great loss of accuracy and the waste of time in the inconvenient interpolations. Worst of all are those among this type of table which give sines, cosines, &c., in separate tables, for in optical (as in many physical and astronomical) calculations we frequently require several functions of the same angle almost simultaneously, and it is a gross waste of time to have to hunt these up on different pages instead of finding them all in one horizontal line.

The table used by optical students at the Imperial College and employed in calculating all the examples given in this treatise is Bremiker's six-figure table of logs and trigonometrical functions (English publisher: D. Nutt, London). The drawback of a bulk of about 600 pages is amply compensated by the closer intervals: the logs of numbers are given directly for five significant figures, and the logs of trigonometrical functions at intervals of 10 seconds of arc for the whole quadrant, with a most useful additional table of log sines and tans for every second up to 5°. It is employed in our optical practice so as to obtain rather more than the usual five-figure accuracy by roughly interpolating numbers beginning with 4 or a lower figure for a sixth significant figure and by interpolating angles to the nearest second. After making these interpolations the sixth decimal place of the logs may be either rounded off to the nearest five-figure number (that practice is followed in the examples given here), or it may be retained with a slight further gain in precision.

The three tables specifically mentioned are printed from stereotype-plates, and may with considerable confidence be regarded as absolutely free from misprints or errors.

In calculations by approximate analytical formulae, reciprocals, roots, and powers of natural numbers are of frequent occurrence; for that reason every optical computer should have close at hand *Barlow's Tables of squares, cubes, square roots, cube roots and reciprocals of all integer numbers up to 10,000* (London: E. & F. N. Spon, Ltd.), which is also a stereotype-edition and free from errors.

A few golden rules in the use of tables may be impressed on computers. Although good tables are almost sure to be free from error, some of the figures are apt to be indistinct or damaged. Errors arising from that source are avoided by acquiring the habit of glancing at the two neighbouring values and noting that the tabular

[7] number to be actually employed is the mean of its two neighbours within at most a few units of the last decimal place. As an example of the wisdom of thus checking a doubtful figure it may be related that a student in an examination required the reciprocal of 1·683, the true value of which is 0·5941771. In the specimen of Barlow's table employed the 9 had a damaged tail, and the student used 0·5042, and so spoilt an otherwise faultless calculation. As all the reciprocals in that neighbourhood begin with 59 he should have noticed this; moreover he should (in the absence of examination-fever) have realized that 0·50, &c., could not be the reciprocal of a number so far short of 2·0 as the one looked up! The most common errors in using tables arise in the interpolations. The amount resulting from the interpolation is apt to be added when it ought to be subtracted, or vice versa. The simplest and usually sufficient safeguard against this kind of error is to note that the interpolated value must necessarily fall between, and cannot lie outside, the tabular numbers between which interpolation was carried out. Thus, supposing log cos $48° 1' 34''$ to be required, the table gives for $48° 1' 30''$, the log cos = 9·825300, and for $48° 1' 40''$ 9·825277. Therefore we have a difference for $10''$ = 23 Units of the 6th decimal, and for $4''$ $4 \times 2{\cdot}3$ or 9 Units. This must be subtracted and gives 9·825291 as the correct value. If—from the habit acquired through working so largely with sines—the difference were added it would give 9·825309, and would be rejected mechanically if the common-sense rule given above is borne in mind. A determined effort should be made to learn to do the entire interpolation mentally, and not to make a written sum of it; if this easy trick cannot be acquired the working out should be done on a separate piece of paper, not on the margin or in vacant spots of the actual calculation. Gross errors quite commonly arise from the turning up of the correct minutes and seconds, but a wrong degree, or the degree at the top of the page when that at the bottom should be taken. Another error particularly common in optical calculations on account of the vast predominance of sines in them is that the occasional tan is apt to be fetched out of the accustomed sine-column. All these warnings may appear trite and superfluous, but long experience has shown that most of the mistakes in calculations arise from such elementary blunders.

A minor worry in numerical calculations presents itself in rounding off superfluous decimals if the numbers to be cut off are 5, either alone or followed by zeros. If the 5 is merely dropped without raising the last retained figure, the numbers actually employed will on the grand average be too small. If on the other hand the last figure is always raised when the dropped figure is 5, then on the average our numbers will be too large. A widely adopted rule which avoids this onesidedness and has everything in its favour is that in such cases we always round off to the nearest *even* number. Thus, supposing we wanted log tan $41° 41' 10''$ to four decimals, the 6-figure table giving 9·949650, we should employ 9·9496 because 6 is even, and not 9·9497, for 7 is odd. But if log sin $41° 44' 40''$ were required to four decimals, the 6-figure value being 9·823350, we should take 9·8234 because 4 is even. It would be equally justifiable to round off always to the odd figure, but the even one has the advantage that it leads to no further worry if one-half of the number turned up is really required, as is frequently the case. In some mathematical tables terminal fives are marked so as to indicate whether further decimals would lead to a value either larger or smaller than an exact 5 followed by endless zeros. In such cases the rounding off would of course be carried out correctly in accordance with the indication of a 'small' or a 'large' five. [7]

With regard to the characteristic of logarithms, which determines the position of the decimal point in the corresponding number, it will be found convenient and to lead to fewer mistakes to adopt the universal practice of the tables of logs of trigonometrical functions; that is to increase the negative characteristic of *all* proper fractions by 10 so as to operate always with purely positive logs. Thus, for the log of 0·0273, which is really $\bar{2}\cdot43616$, we write 8·43616. Strictly this requires the addition of − 10 to make it correct, but in practice it is not necessary to worry about this because it is unthinkable that the result of a calculation, at any rate of an optical system, should be in doubt to the extent of 10^{10} or 10^{-10} times the real value, and that is the only doubt that can arise. Thus, in the example, if the log of result 8·43616 had been found with utter disregard of positive or negative multiples of 10 in the characteristic we should know that the numerical result must be either

0·0273 (which is reasonable),

or 273,000,000

or 0·000000000000273,

or another 10, 20, 30, &c., zeros either before or after the decimal point, and all these numbers would be plainly absurd, being for practical purposes equivalent to either infinite or zero values. Doubts of that order could only arise in the occasional calculation of numbers beyond human conception, such as the number of vibrations which light performs in coming from a distant star to the earth; or the ratio of the mass of a milligramme to that of the earth.

The rules to be observed in working with these logs of proper fractions with positive characteristic are:

When determining the log of a proper decimal fraction the characteristic is = 9 − number of zeros after the decimal point.

When determining the number, known to be a proper fraction, to a given log, the number of zeros after the decimal point is = 9 − the positive characteristic of the log. Note that this rule is merely an algebraical transposition of the preceding one, so that really only one needs remembering.

In adding up a number of logs of this kind we proceed by the usual rule, but put down only the unit-place of the characteristic:

$$\begin{array}{rl} 9\cdot79437 & (-10) \\ +8\cdot96213 & (-10) \\ +9\cdot45927 & (-10) \\ \hline (2)8\cdot21577 & (-30), \end{array}$$

the 2 of 28 being thrown away, being in fact cancelled by two of the omitted (− 10)s.

In subtracting a log of a fraction borrow, if necessary, a ten in the minuendus; it is really the omitted (− 10) of the subtrahendus:

$$\begin{array}{rl} 1\cdot31216 & \\ -9\cdot72611 & (-(-10)) \\ \hline 1\cdot58605 & \end{array}$$

[7] Integer powers of proper fractions are calculated by the usual rule; supposing $(0{\cdot}0273)^3$ to be required, we put down log $(0{\cdot}0273)^3 = 3 \times \log.\ 0{\cdot}0273 = 3 \times (8{\cdot}43616(-10)) = (2)5{\cdot}30848(-30)$;

$$\therefore (0{\cdot}0273)^3 \quad = 0{\cdot}000020346.$$

To calculate complicated powers of proper fractions, the quickest and safest logarithmic method consists in using the true negative log of the fraction. Assuming $(0{\cdot}0273)^{1{\cdot}719}$ to be required, we use $\log 0{\cdot}0273 = 8{\cdot}43616 - 10 = -1{\cdot}56384$, then $\log (0{\cdot}0273)^{1{\cdot}719} = -1{\cdot}56384 \times 1{\cdot}719 = -2{\cdot}68825 = 7{\cdot}31175(-10)$;

$$\therefore (0{\cdot}0273)^{1{\cdot}719} = 0{\cdot}00204998.$$

When integer roots of proper fractions are to be evaluated by logs, the omitted -10 of the logs of proper fractions must be considered; in order to obtain the log of the root again with -10 as the omitted part, the -10 of the given log must be increased to $-10 \,.\, k$ if k is the index of the root, and the initial positive characteristic must, in compensation, be increased by $10(k-1)$ before dividing by k. Supposing $\sqrt[6]{0{\cdot}0273}$ to be required, the procedure is

$$\begin{aligned} \log \ 0{\cdot}0273 &= 8{\cdot}43616(-10) = 58{\cdot}43616(-60), \text{ then} \\ \log \sqrt[6]{0{\cdot}0273} &= \tfrac{1}{6} \log 0{\cdot}0273 \quad = \quad 9{\cdot}73936(-10); \\ \therefore \sqrt[6]{0{\cdot}0273} &= 0{\cdot}54873. \end{aligned}$$

In working out formulae of a number of terms connected by multiplication and division a practised computer always makes a straightforward addition-sum of the logarithmic work by using the logs of the reciprocals of denominator-terms, or their '*co*logs'. As a rule these cologs = log 1—the tabular log are not obtainable directly. By noting that log 1 is zero and may be written 9·9999(10), (i. e. all nines, but a ten, treated as a unit-number, in the last place), the subtraction of the tabular log from zero may be done without any borrowing and the resulting colog may be written down from left to right practically as quickly and with as little risk of error as the log itself. Thus supposing cos $31^\circ\ 5'\ 17''$ to occur in the denominator of a formula and no secant-table to be available. Turn up the log cos, which is 9·93266, *imagine* written above it 9·9999(10) and subtract from left to right: 9 from 9 = zero, 9 from 9 = zero, 3 from 9 = 6, 2 from 9 = 7, 6 from 9 = 3, and in the last place 6 from 10 = 4. These straightforward subtractions can be done as quickly as the results can be written down, and 0·06734 is on the computing paper almost as quickly as if the tabular log itself had been copied. Should the log to be subtracted from nothing end with one or several zeros, then the terminal (10) of the imagined sequence of nines must of course be applied to the last *significant* figure of the log and the colog is finished with as many zeros as are at the end of the log.

The accuracy of practically all *approximate* optical formulae—such as the *TL* ones in the theoretical part—does not exceed that obtainable from a 10-inch slide-rule. Much time and paper may be saved without real loss of precision by becoming reasonably expert in the use of this most convenient tool, which will be frequently referred to subsequently. The now universal arrangement with uppermost or *a* scale from 1 to 100, second or *b* scale of the same extent on the upper edge of the slide, and c and d scales from 1 to 10 at the lower meeting edge of slide and rule, will always be assumed. With reasonable care in setting and estimating fractions of the actual divisions the probable error of a simple calculation by slide-rule is about $\frac{1}{5}$ of 1 per cent. [7]

B. GENERAL HINTS ON NUMERICAL CALCULATIONS

[8] A neat and orderly arrangement of each calculation is of primary importance; it saves time and precludes mistakes even at the time when the work is done, but the chief advantage will be reaped subsequently when an old calculation is referred to as a guide in a new or modified design. Whilst it is not possible for most people to write with the regularity and precision of copperplate, no trouble should be spared in acquiring the habit of writing each figure so that it cannot be mistaken for another. 6 and 9 should have long tails so as to distinguish them clearly from 0; 2, 3 and 5 should have their characteristic differences well displayed, &c. Mistakes are made over and over again by a computer misreading his own 2 as a 3, even within a few minutes of writing it down, or a 7 with deficient upper works is read as 1, and so forth. Paper with impressed squares (about $\frac{1}{4}$ inch) will assist greatly in keeping numbers to be added or subtracted correctly alined and so to avoid confusion of the proper decimal places—another fruitful source of errors. It is a decided advantage to become used to writing rather small figures as it makes it possible to complete most calculations on a single sheet. Columns of $\frac{3}{4}$-inch width are then sufficient for 5-figure work. A definite scheme should be adopted for each type of calculation so that each quantity is always found in a definite part of the column. Practically all optical formulae embody some algebraical sign-convention; therefore $a+b$ does not necessarily mean that the numbers expressing a and b are to be added together. To avoid mistakes from this source, which are very prevalent, it is best to get rid entirely of the idea that $a+b$ means the sum and $a-b$ the difference and to substitute the infallible rule that a quantity which appears in an algebraical formula with the *positive* sign is to be used with *its own* sign, whilst one which appears in the formula with the *negative* sign is to be introduced with *reversed* sign. Budding computers cannot be too pedantic in watching the signs of all the quantities introduced into a calculation.

In logarithmic calculations a difficulty arises because logarithms can only be found for positive numbers; in order to avoid a separate investigation in each calculation as to the sign of the result, a mark should be put to each logarithm which belongs to a negative number and a small n is used for that purpose in the specimen calculations here given. When any one logarithmic formula (that is one containing only factors and divisors) has had all the logarithms extracted from the table and the n added to such as represent negative numbers, the sign of the result is at once found by counting the number of n's occurring in the sum. If their number is even the numerical result must be positive, whilst if there is an odd number of n's in the sum then the numerical result must be negative: this should be invariably indicated by putting the terminal n to the logarithm of the result, which will automatically lead to a minus-sign being put to the corresponding number found in the table of logs.

C. ACTUAL OPTICAL CALCULATIONS

[9] The ray-tracing formulae given in the theoretical part are extensively used in practically every optical designing problem : they should therefore be thoroughly mastered, and they should be employed in working out numerical examples until the process becomes almost automatic. A few progressive examples will be given in extenso for use as patterns in further work, for which plentiful material will be supplied subsequently. The examples actually given have been selected so as to be suitable for the testing of theoretical conclusions in later chapters.

Example 1

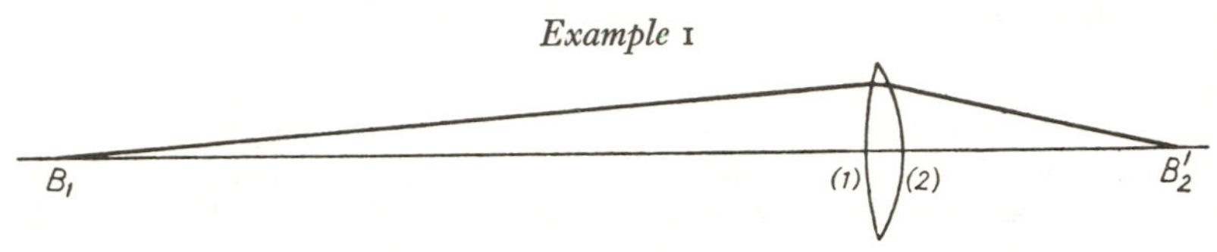

A simple biconvex lens of crown glass of the specification

$$r_1 = 10{\cdot}000'' \quad d_1' = 0''{\cdot}600$$
$$r_2 = -5{\cdot}000''$$

receives light from a real object-point B_1 at 24″ from the vertex of the shallower surface. For the usual left-to-right direction of ray-tracing we therefore have $L_1 = -24$. The refractive index of the glass is to be taken as $= 1{\cdot}5180$. Rays are to be traced which leave B_1 at angles of 1, 2, and 3 degrees with the optical axis. If the directly computed rays are to pass through the lens *above* the optical axis (another widely adopted custom) we must put $U_1 = -1^\circ, -2^\circ, -3^\circ$ respectively, because a ruler would have to be turned counter-clockwise to carry it from the direction of the optical axis into that of the ray. When several rays are to be traced through the same system it is best to do the work in parallel columns. The complete working out is then as given on page 17.

The two surfaces are here arranged side-by-side to fill the page. In actual practice it would usually be preferred to place the second surface calculation below that of the first ; the transferring of the last figures in numerical columns 1 to 3 to the heads of columns 4 to 6 would thus be saved.

Detailed Explanation of the Calculation

The calculation by standard schedule begins with equation (1) $\sin I = \sin U\,(L-r)/r$. For this we require $L-r$, and this is worked out in the first 3 horizontal lines, putting down L with its true sign as -24 and r (given as 10) with reversed sign as $-r = -10$, giving $L-r = -34$.

As, at the first surface, all three rays come from the same point B, this calculation of $(L-r)$ is only written down once ; at the second surface the rays have different L-values, and $(L-r)$ has to be calculated separately for each one.

The logarithmic work then begins. Log sin U to the given angles is turned up and as the angles, and therefore also their sines, are negative, the logs found have the n added. Log $(L-r)$ is similarly turned up and as $(L-r)$ is also negative its log is also followed by an n. Addition of the two logs then gives log $(L-r)\sin U$,

[9]

and as the logs added carry 2—an even number—of n's the product will be positive and its log does *not* receive an n. Division by r means subtraction of its log, and this completes the calculation of equation (1). We then compute equation (2) $\sin I' = \sin I \, . \, N/N'$. Log sin I is already on the paper. We therefore form log $N/N' = \log N - \log N'$, and as our rays pass from air (N taken as 1 exactly) into glass of index $N' = 1{\cdot}5180$, the log of which is 0·18127, we have

$$\log N/N' = 0{\cdot}00000 - 0{\cdot}18127 = -0{\cdot}18127\,\dagger :$$

and as this has, by the formula, to be algebraically added to log sin I, we put it down, as found, namely $-0{\cdot}18127$ and find log sin I'.

Original U	$-1°$	$-2°$	$-3°$	$-1°$	$-2°$	$-3°$
	First Surface			*Second Surface.*		
L	−24·000	same	same	147·938	144·551	139·046
$-r$	−10·000			+5·000	5·000	5·000
$L-r$	−34·000	same	same	152·938	149·551	144·046
log sin U	8·24186n	8·54282n	8·71880n	7·45050	7·76221	7·95625
+log $(L-r)$	1·53148n	1·53148n	1·53148n	2·18452	2·17479	2·15850
log $(L-r)$ sin U	9·77334	0·07430	0·25028	9·63502	9·93700	0·11475
−log r	−1·00000	−1·00000	−1·00000	−0·69897n	−0·69897n	−0·69897n
log sin I	8·77334	9·07430	9·25028	8·93605n	9·23803n	9·41578n
+log N/N'	−0·18127	−0·18127	−0·18127	+0·18127	+0·18127	+0·18127
log sin I'	8·59207	8·89303	9·06901	9·11732n	9·41930n	9·59705n
+log r	1·00000	1·00000	1·00000	0·60897n	0·69897n	0·69897n
log r sin I'	9·59207	9·89303	0·06901	9·81629	0·11827	0·29602
−log sin U'	−7·45050	−7·76221	−7·95625	−8·67929	−8·98890	−9·18026
log $(L'-r)$	2·14157	2·13082	2·11276	1·13700	1·12937	1·11576
U	−1- 0- 0	−2- 0- 0	− 3- 0- 0	0- 9-42	0-19-53	0-31- 5
$+I$	3-24- 7	6-48-53	10-15- 0	−4-57- 4	− 9-57-43	−15- 5-56
$U+I$	2-24- 7	4-48-53	7-15- 0	−4-47-22	− 9-37-50	−14-34-51
$-I'$	−2-14-25	−4-29- 0	− 6-43-55	7-31-42	15-13-28	23-17-30
U'	0- 9-42	0-19-53	0-31- 5	2-44-20	5-35-38	8-42-39
$L'-r$	138·538	135·151	129·646	13·7088	13·4701	13·0545
$+r$	10·000	10·000	10·000	− 5·0000	− 5·0000	− 5·0000
L'	148·538	145·151	139·646	8·7088	8·4701	8·0545
$-d'$	− 0·600	− 0·600	− 0·600			
L_2	147·938	144·551	139·046			

We then proceed to evaluating equation (3) $U' = U+I-I'$. Leaving 4 blank lines for the subsequent completion of the logarithmic work, we build up the angle-register, beginning by putting down the given U of each ray. We then turn up the 3 values of log sin I in the 5th line of the logarithmic work and enter the angles in a second line of the angle-register. In the present case sin I, and therefore I itself, is positive throughout. If log sin I had an n at the end we should have to put a minus-sign to the angles of incidence, as happens at the second surface. By

† In working out the standard ray-tracing formulae with a fractional N/N' the true negative log is preferable to its positive equivalent $(+9{\cdot}81873)$, because it has one figure less, and frequently turns up again with the positive sign at the next surface.

[9] equation (3) U and I have next to be algebraically added, but as the U are negative whilst the I are positive, the arithmetical operation amounts to subtracting U from I. We, however, must call the result by its algebraical name $(U+I)$. To complete the working out of (3) we next turn up the angles to the log sin I' found, and as the equation calls for $-I'$ we put down the angles found from the table with the reverse of their own sign, in this case with a minus-sign. We thus obtain U', the angle which the refracted ray forms with the optical axis.

We now return to the logarithmic work and compute equation (4) $L'-r = \sin I' . r/\sin U'$. Log sin I' is already down, hence we copy log r from the fourth line of the logarithmic work, but give it the + sign as it is now a factor. Addition gives log $r \sin I'$. To complete (4) we have to divide by sin U' or to subtract log sin U'. The latter is therefore turned up for each column and log $(L'-r)$ found by subtraction. The ray-tracing is then completed by (5) as shown under the angle-register, and as the rays are to be taken through a second surface we must prepare for this by (5)*, by deducting (algebraically) the given thickness $d' = 0{\cdot}600$ from the L' found. We thus arrive at the intersection-lengths of the three refracted rays referred to the vertex of the second surface, i. e. the correct L-values to be used in tracing these rays through that surface. In accordance with the second of (5)* the U' of the first surface are simply transferred as the U of the second surface, and similarly the log sin U' in the last-but-one line of the logarithmic work of the first surface is copied to serve as log sin U for the second surface. The calculation through this surface is then an exact repetition of that described for the first, but instructive variations in signs should be noted. The final results are seen in the last line of the angle-register and the last line of the whole calculation :

The first ray cuts the optical axis at a distance $L' = 8{\cdot}7088$ to the right of the vertex of the second surface and forms a clockwise angle $U' = 2^\circ\ 44'\ 20''$ with it.

For the second ray $L' = 8{\cdot}4701$ $U' = 5\text{-}35\text{-}38$.
For the third ray $L' = 8{\cdot}0545$ $U' = 8\text{-}42\text{-}39$.

The writing of angles with hyphens separating degrees, minutes, and seconds will be found less likely to lead to mistakes (such as misreading 2° as 20 or $4'$ as 41). It also acts as a constant reminder of the fact that in borrowing or carrying degrees, minutes and seconds the usual arithmetical rule does not hold, inasmuch as a degree is equal to *sixty* minutes and a minute equal to *sixty* seconds.

GENERAL VALIDITY OF THE STANDARD EQUATIONS

[10] The standard equations have been directly proved in [4] and [5] for the case in which all the quantities are positive. We must now prove that they will give the correct result in every conceivable case if the sign-conventions are observed. For this purpose we must examine the geometrical relations between the quantities in the various cases and must show that our formulae, under the sign-conventions, lead to the same result.

Referring back to Fig. 2 *a* it is evident that all the quantities referring to the incident ray will be positive and that no essential changes can occur as long as B lies to the right of the centre of curvature C. But if we imagine point B moving from an original great distance towards the refracting surface, there will occur a very special case when B coincides with C; the entering ray will then coincide with the radius PC which is the incidence-normal and will go straight on because the angle of incidence is zero. No calculation of any kind is therefore required and no question can arise as all quantities remain unchanged. If point B approaches the surface still more, it will fall to the left of C and we are confronted by a new case, Fig. 6. [10]

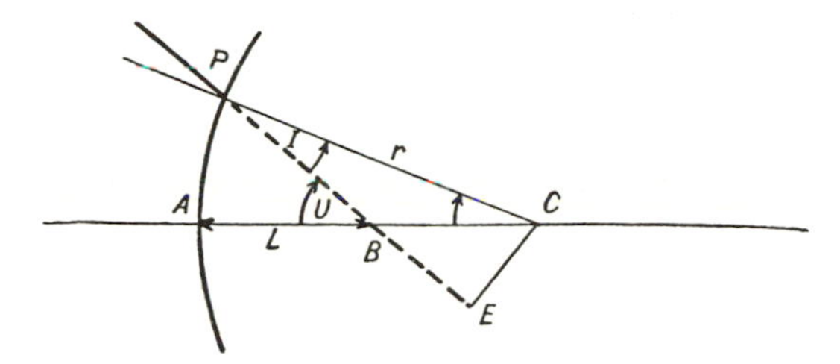

FIG. 6.

In considering it geometrically we treat all the quantities as positive or absolute; to distinguish these absolute values from the quantities in our trigonometrical computing formulae embodying our sign-conventions, we will enclose the symbols in a bracket, such as (L), in working out the geometrical result. Fig. 6 then shows that in the triangle PCB we have side $PC = (r)$, side $BC = (r)-(L)$ (because L is now smaller than r), and by the arguments used in deducing the computing formulae we find

$$\sin (I) = \sin (U) \frac{(r)-(L)}{(r)}.$$

For the important angle at C between the axis AC and the incidence normal PC we find, seeing that in the present case U is external angle to the triangle PCB,

$$\text{angle at } C = (U)-(I).$$

According to our computing formulae we have

$$\sin I = \sin U \frac{L-r}{r} \text{ and angle at } C = U+I,$$

both of which seem to contradict the geometrical result. But if we examine first the trigonometrical formulae for $\sin I$, we can write it in absolute values of the given quantities, which in this case are all positive,

$$\sin I = \sin (U) \frac{(L)-(r)}{(r)} = -\sin (U) \frac{(r)-(L)}{(r)};$$

and this shows that our computing formulae will give the same numerical value of $\sin I$ as that above deduced geometrically, but will attach a negative sign to $\sin I$ and therefore also to the angle I itself: or mathematically put, we shall find

[10] $I = -(I)$. If we then put this computed value into the computing formulae for the angle at $C = U+I$, we obtain in absolute value

$$\text{angle at } C = (U)-(I)$$

in precise agreement with the geometrical result above. Hence the entire result obtained from the computing formulae is in accordance with the geometrical deductions, simply because the angle of incidence is brought out by the formulae as a negative one.

These relations obviously hold for any position of B between the refracting surface and the centre of curvature.

A third class of cases arises when point B falls to the left of the refracting surface (Fig. 7).

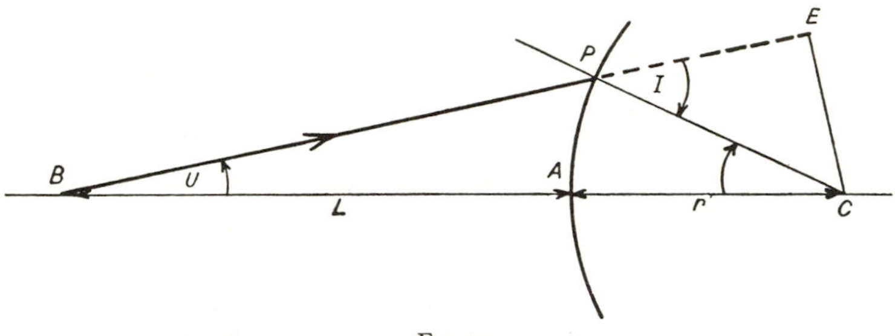

FIG. 7

Geometrically we shall then find

$$\sin(I) = \sin(U)\frac{(L)+(r)}{(r)} \text{ and angle at } C = (I)-(U).$$

The computing formulae will again be in the invariable form

$$\sin I = \sin U\frac{L-r}{r} \text{ and angle at } C = U+I,$$

but as B lies to the left of the surface, L will be negative under the sign-conventions, or $L = -(L)$, and as U is a counter-clockwise angle it is also negative or $U = -(U)$. Therefore the working out of the computing formulae for $\sin I$ will give

$$\sin I = \sin[-(U)]\frac{-(L)-(r)}{(r)} = -\sin(U)\frac{-[(L)+(r)]}{(r)} = \sin(U)\frac{(L)+(r)}{(r)}$$

in precise agreement with the geometrical result. We shall therefore obtain a positive angle of incidence or $I = (I)$, and if we now work out the angle at the centre of curvature $= U+I$ we shall find, as $U = -(U)$ and $I = (I)$,

$$\text{angle at } C = -(U)+(I) = (I)-(U),$$

also in agreement with the geometrical result. This case obviously covers every position of B between the refracting surface and negative infinity, and therefore completes the study for a surface of positive radius pierced at a point above the axis by a ray from any point of that axis. [10]

Geometrically we have exhausted the possible relations between the quantities, for the other nine cases which may occur lead to the same diagrams merely turned into various positions. Thus the turning of the three diagrams in their own plane through 180° (looking at them upside down) gives three cases for a concave surface passed by rays below the axis (Fig. 8).

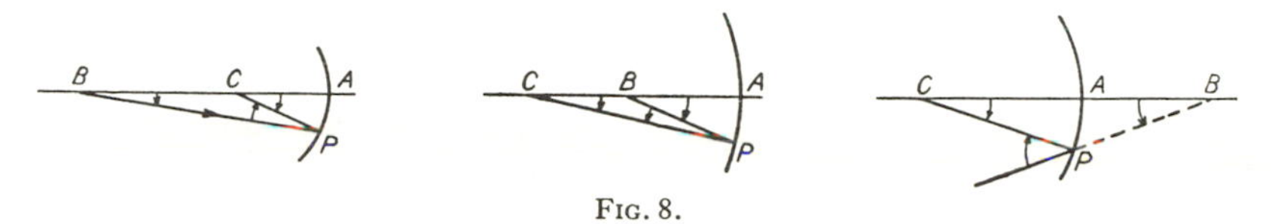

FIG. 8.

By our sign-conventions these cases have the same sign of the angles but the opposite sign of L and r as the corresponding first cases. Turning the original three diagrams over like the page of a book and looking at them through the paper gives three more cases for a concave surface with the ray passing above the axis; and these three diagrams viewed upside down (still through the paper) yield the last three possibilities of a convex surface passed by a ray below the axis. In the third set of three cases all signs are reversed as compared with the original cases. In the fourth set the signs of angles are reversed, but those of L and r are the same as in the original cases. Some of these additional cases may profitably be discussed by students in the manner employed above for cases 2 and 3 in order to verify still further the claim that the fundamental formulae may be applied with absolute confidence to every case that can possibly arise.

We have applied the proof to the point B; obviously it applies equally to point B'. And as the connexion between the two points is derived from the simple and unequivocal fact that the angle at C is the same for the ray after refraction as it is before refraction : $(U+I) = (U'+I')$, we may regard the proof as complete. It is important to be perfectly convinced of the universal validity of the fundamental formulae, as it enables us to accept as equally valid all deductions from them which depend only on the ordinary rules of algebra and trigonometry.

When the relations of the quantities before and after refraction are considered the total number of distinctive cases is raised from 12 to 20, for in the original cases 1 and 3, as well as in the inversion of their diagrams, L and L' may either fall both on the same side of the surface or they may fall on opposite sides, according to the values of N and N'. If we wanted to follow the practice of some elementary treatises in navigation and avoid the bother of watching plus or minus signs we should therefore be compelled to draw up separate instructions for twenty different possibilities, all of which are completely covered by our single set of computing formulae if the sign-conventions are adopted.

We have only *stipulated* the sign of the angles U or U' because that of all the other angles will follow from these. It is, however, useful at times to be able to tell the signs of the other angles directly from a diagram. Adhering to the method of defining the sign of angles by the clock-sense of the turn which will carry a ruler

[10] from the direction of one arm into that of the other arm of a given angle, and calling a clockwise angle positive, a counter-clockwise angle negative, the rules are :

The angles I and I' are positive if a clockwise turn will carry a ruler from the direction of the ray into the direction of the radius or incidence-normal.

The angles $U+I$ or $U'+I'$ are positive if a clockwise turn will carry a ruler from the direction of the adopted axis into that of the incidence-normal.

These rules are easily verified by the diagrams already given, in all of which the clock-sense of the various angles has been marked according to the preceding rules.

In the actual practical use of the computing formulae all the angles are, as a rule, less than 90°, and the angles I and I' necessarily must always be acute. An exception does, however, occur in high-power microscope objectives with regard to the angle $U+I$ at the centre of curvature, which may reach 100° or thereabout in the hyperhemispherical front-lenses of oil-immersion lenses. Theoretically, it is also thinkable that U' or U might reach more than 90° ; but a ray at such an excessive angle could not emerge from any lens-system and is therefore impossible in practical optics. If it occurred in the course of a new design it would indicate the imperative necessity of reducing the working aperture.

In developing new designs or in working out carelessly framed numerical exercises two other exceptional cases may be met with, and it is desirable to become acquainted with them and to know how our formulae indicate them. It may happen that a given ray misses a surface altogether on account of an excessive

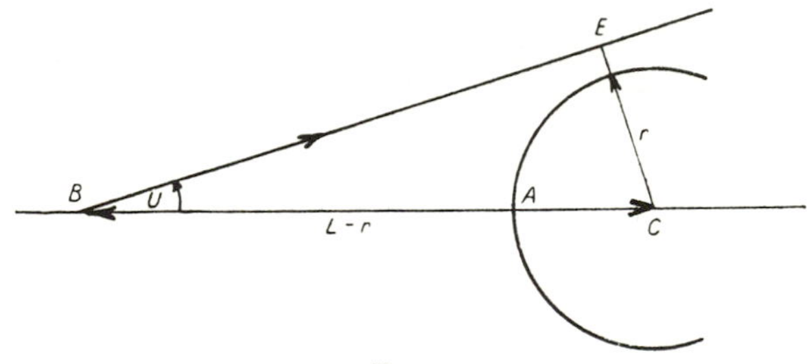

FIG. 9.

value of U or L. In the absence of a correctly drawn diagram the computer may not be aware of this and may apply the formulae in the usual way. He will work out by formula (1) $\sin I = \sin U \dfrac{L-r}{r}$. Now $\sin U(L-r)$ represents the perpendicular CE from the centre of curvature upon the given ray, and as Fig. 9 shows, this perpendicular will be greater than r if the ray misses the surface. Consequently the division of $\sin U(L-r)$ by r will give a result exceeding unity as the value of $\sin I$, and as the sine of any real angle is a proper fraction, the existence of an impossible case is pointed out to the computer. The other impossible case may easily turn up at refractions from a denser into a lighter medium. If the angle I is large, the multiplication of $\sin I$ by N/N'—which in such a case will be >1—may give a value of $\sin I'$ which is in excess of unity and therefore impossible. The ray consequently cannot pass into the lighter medium; as will be [10] known from elementary optics, the ray is then totally reflected at the surface and must be dealt with by the law of reflection, if at all. Our computing formulae therefore indicate total reflection by a value of the sine of the angle of refraction which exceeds unity and is therefore impossible.

Accuracy of the Standard Equations

[11] We must now discuss the formulae from the practical computer's point of view. The discussion of their validity has shown that they will always give the correct result if worked out *accurately*. In practice this working out is done most usually by logarithms with a limited number of decimal places—five places being the usual number in optical calculations. On account of the rounding off of the following decimal places such a five-figure logarithm, if taken directly—without interpolation—from the table, may be wrong by not more than half a unit of the fifth decimal place. Hence we shall have at the worst

$$\text{Tabular log} = \text{true log} \pm 0{\cdot}000005\,;$$

and if we turned up the corresponding numbers, let us say in Vega's ten-figure table, we should find by the usual rules of logarithmic calculations

Number really corresponding to the tabular log =
true number times, or divided by, 1·00001151;

and as the vulgar fraction corresponding to 0·00001151 is, nearly enough, 1/87,000th, it follows that any five-figure log taken directly from the table may really correspond to a number greater or smaller by 1/87,000th part than the number assigned to it by the table. When a log calls for interpolation, then the rounding off of the interpolated correction to five figures introduces an additional maximum uncertainty of half a unit, and as this may, at worst, become added to the half-unit error of the tabular log, the maximum error in the number corresponding to an interpolated five-figure log amounts to 1/43,500th part. In the case of four- or six-figure logs the uncertainty will of course be 10 times or 1/10th times the above fractions.

The estimates at which we have arrived are extreme values, but on the average we may take it that the number corresponding to any five-figure logarithm has a 'probable' error of about 1/100,000th part of its value as derived from the table. When a result is obtained by adding or subtracting a number of such logarithms the uncertainty necessarily increases, and in calculations such as are here considered we may estimate the *probable* uncertainty of the number resulting from the combination of about six logarithms as 1 part in 40000, but with a *possibility*—in unfavourable cases of coincidence of a number of one-sided errors of the combined logarithms—of an uncertainty up to perhaps 1 part in 10000.† Even this

† It may be pointed out that the inherent and inevitable inaccuracy of logarithmic calculations demonstrated above also renders futile any attempt to secure greater precision by carrying the interpolations beyond about 1/100,000th part of the number in the case of five-figure work. Beginners are usually very loth to accept this conclusion. It should also be realized that discussions like the one which next follows in the text are of fundamental importance and that they will frequently disclose the probability of grave inaccuracy in a logarithmic calculation even when based upon equations which are absolutely rigorous from the purely mathematical point of view.

[11] uncertainty would be permissible in the vast majority of optical calculations, as will be proved in later chapters. But we must now show that the simple computing formulae are of a type which in certain cases of frequent occurrence leads to a further great and frequently fatal increase of the uncertainty of the result.

We find the intersection-length L' after refraction by equation (4) $L'-r = \sin I'.r/\sin U'$, in which $r.\sin I'$ has been obtained by adding and subtracting a few logarithms of given quantities, and has therefore the average uncertainty of about 1 part in 40000. If we refer to the diagram, Fig. 10, we see that $r \sin I'$

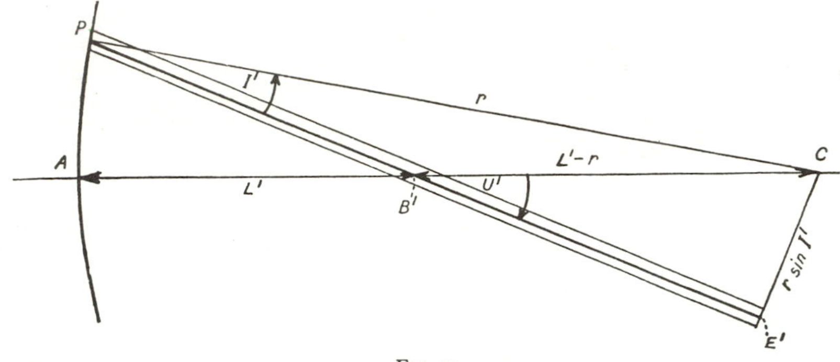

FIG. 10.

represents the perpendicular CE' dropped from the centre of curvature upon the refracted ray. Consequently the small errors of the logarithms employed render the position of E' uncertain, say between the two neighbouring points marked (with huge exaggeration) in the diagram. The division of $r \sin I'$ by $\sin U'$ then obviously amounts to laying a line under angle U' with the optical axis through the calculated (instead of the *exact*) point E', and even assuming that we knew the exact value of U', the calculated location of the emerging ray might be anywhere between the two parallels to the true PE' marked in the diagram. This shows that not only the calculated L' but also the point at which the refracted ray leaves the surface are misrepresented, and it is immediately apparent that if $L'-r$ is greater than L' itself, then the percentage error in the values of L' *and* of AP derived from the calculation will be greater and therefore more serious than that of CE'.

Cases are comparatively common in which CE' is up to ten and more times AP, and it will be realized that the estimated uncertainty of one part in 40000 of CE' then would lead to an uncertainty of one part in 4000 or less of L' and of the height AP of the point of emergence of the computed refracted ray ; this would be a serious error, especially when it is remembered that an unfavourable accumulation of the small individual errors of the several logarithms might raise it to three or more times the average value. Moreover, there is a further uncertainty, for the finding of U' by $U+I-I'$ involves the errors in the interpolation of the angles I and I' from the values of their sines, hence the calculated U' used in determining the direction of PE' will itself be slightly wrong, occasionally by a second or two. Consequently we shall obtain from the calculation an emerging ray which not only passes through a point slightly away from the true E' but also takes a slightly wrong direction, with an obvious further uncertainty in the values of AP and L' as derived from the calculation. [11]

It follows, firstly, that the tracing of a ray by the standard formulae *always* implies a *slight* discontinuity at each surface, inasmuch as the calculated point of emergence does not coincide with the point of incidence; secondly, that the formulae give a *seriously* inaccurate result in the case of a radius which is long compared to L' or even merely long compared to the other radii of a system. Evidently the difficulty may be met by calculating through the occasional surface of long radius with logarithms of one or two more decimal places, of course with correspondingly closer interpolation of all the numbers and angles. But the desired increased accuracy may be secured by logarithms of the usual number of places by using supplementary formulae which remove the *source* of the uncertainty. The first method of this description was apparently due to Seidel, and is given in Steinheil's *Handbook*. A decidedly simpler method,† however, will be given first place here because it simultaneously acts as an extremely reliable check against numerical errors in carrying out the main calculation by the standard formulae.

[12] The method consists in using the given data and the angle I directly derived from them to determine the distance PA from the point of incidence to the pole of the refracting surface (Fig. 11), and in laying the refracted ray through the point of incidence so determined at the angle U' obtained from the standard formulae. The discontinuity of the ray-tracing mentioned above is thus avoided and the accuracy of the resulting L' becomes quite independent of the length of the radius.

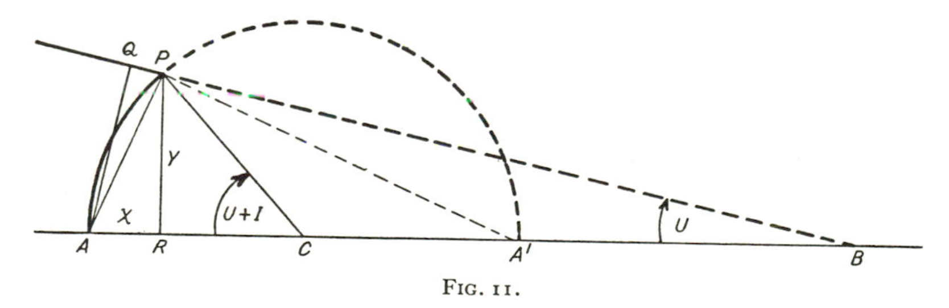

FIG. 11.

Referring to Fig. 11, we have $AB = L$, hence the perpendicular AQ dropped from A upon the incident ray is $= L \sin U$. The chord PA to be determined is the hypotenuse of the right-angled triangle AQP, therefore

$$PA = L \sin U \,.\, \sec \widehat{QAP}.$$

We then take from the diagram

$$\text{angle } QAP = \text{angle } QAB - \text{angle } PAC.$$

† The method was first published in my lectures at the Imperial College in the autumn of 1918.

[12] Now triangle BAQ is right-angled at Q, hence

$$\text{angle } QAB = 90^\circ - U.$$

Triangle PCA is isosceles because two sides are radii of the refracting surface; therefore, as the angle at C is $= U+I$,

$$\text{angle } PAC = 90^\circ - \frac{U+I}{2}.$$

Consequently $$\text{angle } QAP = 90^\circ - U - \left(90^\circ - \frac{U+I}{2}\right) = \frac{I-U}{2};$$

and putting this into the formula for PA we find

$$PA = L \sin U \sec \frac{I-U}{2}.$$

Evidently the refracted ray must lead to a formula of the same type with mere substitution of 'dashed' for plain letters; therefore we also have

$$PA = L' \sin U' \sec \frac{I'-U'}{2}$$

and can transpose this to solve for L'. The computing formulae therefore are

$$(5)^{**} \quad \begin{cases} PA = L \sin U \sec \dfrac{I-U}{2} \\ L' = PA \operatorname{cosec} U' \cos \dfrac{I'-U'}{2}. \end{cases}$$

For a ray from right-to-left the 'dashed' data are given and the plain ones are to be determined, hence we only have to invert the sequence to

$$(5)^{**} \text{ right-to-left} \quad \begin{cases} PA = L' \sin U' \sec \dfrac{I'-U'}{2} \\ L = PA \operatorname{cosec} U \cos \dfrac{I-U}{2}. \end{cases}$$

As the angle QAP is necessarily acute we need not trouble about its sign, for the cosine and secant of an acute angle are always positive, no matter what the sign of the angle may be. We may therefore use $\frac{1}{2}(U-I)$ instead of $\frac{1}{2}(I-U)$ according to convenience, the same applying to the 'dashed' angle $\frac{1}{2}(I'-U')$.

If increased accuracy is the only object of using (5)**, the standard formulae need only be calculated to (3) inclusive. But as the closing of the standard calculation implies very little work it should never be omitted, for the virtually independent values of L' by the standard equations and by (5)** respectively then supply an almost absolute safeguard against numerical errors, as they will only agree if both calculations have been carried through correctly. The L' (or L in right-to-left work) from (5)** is the more trustworthy one, and should therefore *always* be given the preference in carrying forward to the next surface. The PA-formulae should be tested for universal validity in the way exemplified in [10] for the standard formulae in order that they may be used with absolute confidence in all cases. All that is required is to show that the angle QAP is always either $\frac{1}{2}(I-U)$ or $\frac{1}{2}(U-I)$ no matter where C and B may be located. It will be found that PA always comes out positive when P lies above the axis, and negative for a P below the axis, in accordance with the usual signs of analytical geometry. [12]

In the case of a very long radius in which the check-formula (5)** *must* be used to obtain an accurate L' with logarithms of the usual number of places, the check-formula loses its full value as a safeguard against all numerical errors, and special care must then be taken in working it out, as the inaccurate L' from the standard equations only affords a rough control of the accurate value given by the check-formula.

[13] One source of error in optical calculations which is not very obvious, and therefore the more dangerous, is not covered by the PA or any other usual check. It is self-evident that any equations of whatever kind can only give the correct result if correct data are introduced into them. Every computer therefore should be extremely careful in putting in the given quantities. Our optical formulae provide an ugly trap in this respect in the introduction of the refractive index which enters only once, namely in obtaining $\sin I'$ from $\sin I$ by equation (2). No one would doubt for a moment that the introduction of a wrong value of N/N' would necessarily lead to a wrong result which would nevertheless satisfy any check that could be applied to it, simply because the result so obtained *is* the right one for the indices actually used. What is apt to be overlooked is that if the right N/N' is used but a mistake made in the multiplication, then the effect is precisely the same as if a wrong N/N' had been correctly applied ; therefore such an error will also escape detection by all usual checks. The safeguard against this particularly insidious error given in a subsequent practical section of this chapter really depends on an additional equation to be derived from the standard set.

Introduction of the value of $\sin I$ by (1) into (2) gives

$$\sin I' = \frac{N}{N'} \sin U \frac{L-r}{r},$$

and introducing this into (4) we obtain

$$L' - r = \frac{N}{N'} \sin U \frac{L-r}{r} \frac{r}{\sin U'}$$

which can be transposed immediately into

$$(L'-r)\, N' \sin U' = (L-r)\, N \sin U. \tag{6}$$

This equation is valuable in itself on account of important deductions to be obtained from it. It also forms an extremely safe check upon the correct application of the refractive indices if the two sides are calculated just as they stand in the equation and compared for *absolute* equality. The much simpler form of this check already referred to has however been found to answer so well that the separate calculation will only be advisable for computers who are particularly liable to make mistakes.

Serious inaccuracy in calculations by the standard formulae arises occasionally from the way in which the direction of the ray after refraction is derived from that

[13] before refraction by (3). The *change* in direction is obviously represented by $(I-I')$ which in the calculation is obtained by determining I and I' separately. If I and I' are large (and they occasionally reach 50 and more degrees) they cannot be found by the five-figure logarithm of their sines to the usually desirable accuracy within one second of arc, and as a consequence the precision of the whole calculation is seriously lowered. In such cases of exceptionally large angles, which frequently occur at cemented contact surfaces, the desired precision of the U'-values may be restored by calculating $I-I'$ directly from the values of the separate angles obtained by the standard formulae.

Equation (2) may be written in the form of a proportion

$$\frac{\sin I}{\sin I'} = \frac{N'}{N}$$

which yields by a well-known algebraical transformation

$$\frac{\sin I - \sin I'}{\sin I + \sin I'} = \frac{N'-N}{N'+N};$$

and applying the trigonometrical formula for the sums and differences of these sines of two angles this leads to

$$\frac{N'-N}{N'+N} = \frac{2\sin\frac{I-I'}{2}\cos\frac{I+I'}{2}}{2\sin\frac{I+I'}{2}\cos\frac{I-I'}{2}} = \frac{\tan\frac{I-I'}{2}}{\tan\frac{I+I'}{2}},$$

which we transpose into the computing formula

$$\tan\frac{I-I'}{2} = \tan\frac{I+I'}{2}\cdot\frac{N'-N}{N'+N}. \tag{7}$$

As soon as I and I' have been found by the standard formulae the right side of (7) can be calculated, and as $\frac{1}{2}(I-I')$ is a comparatively small angle, it is always obtained accurately, even to fractions of a second if required. For particularly erratic computers equation (7) supplies also by far the safest and surest check against errors caused by applying the refractive indices wrongly in the main calculation; such computers (who may be excellent designers in spite of their liability to numerical mistakes) will therefore be wise to adopt (7) instead of the simpler check depending on (6) as a regular part of their computing formulae. When (7) is used the U' for closing the standard calculation and for computing (5)** is of course determined as the starting value of U plus twice the value of $\frac{1}{2}(I-I')$ obtained from (7); as in all optical calculations, the sign of all the quantities must be carefully watched.

ADDITIONAL COMPUTING FORMULAE

[14] Frequently the rectangular co-ordinates X and Y, Fig. 11, of the point of incidence P of a ray are required in calculations. The simplest exact formula for X is obtained by continuing the trace AP of the refracting surface to its second intersection A' with the axis AC and joining PA'. The triangles APR and $AA'P$ are then right-angled, and similar because they have the angle at A in common. And as $AR = X$ and $AA' = 2r$, we obtain the proportion

$$X/PA = PA/2r$$

or

(8)a $$X = \tfrac{1}{2}(PA)^2/r.$$

Convenient formulae for X and Y in terms of PA and the angle $(U+I)$ are also obtainable from the triangle PAR in which the angle at A has already been determined in [12] as $= 90° - \tfrac{1}{2}(U+I)$, for this angle gives

(8)b $X = PA \sin \tfrac{1}{2}(U+I)$. (8)a* $Y = PA \cos \tfrac{1}{2}(U+I)$.

The last gives $PA = Y \sec \tfrac{1}{2}(U+I)$, and putting this into (8)a we obtain another good formula, viz.

(8)c $$X = \frac{Y^2}{2r} \cdot \sec^2 \tfrac{1}{2}(U+I).$$

The triangle PAR also gives X in terms of Y as

(8)d $$X = Y \,.\, \tan \tfrac{1}{2}(U+I).$$

The triangle PRC gives at sight

(8)b* $$Y = r \sin (U+I),$$

and the same triangle gives

(8)e $$X = AC - RC = r - r\cos(U+I) = 2r \sin^2 \tfrac{1}{2}(U+I).$$

It is easily shown that all these equations are universally valid if the sign-conventions are observed. As we have always and unconditionally $U+I = U'+I'$, the latter may be substituted for the former in any equation.

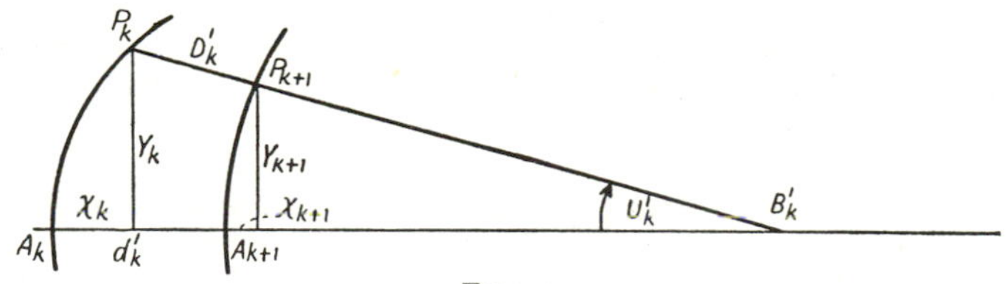

FIG. 12.

The values of X are most frequently required in the calculation of another important datum, namely, the length of the path (P_kP_{k+1} in Fig. 12) of a ray between successive refracting surfaces. By our conventions as to symbols D'_k is the proper name for this distance from the kth to the $(k+1)$st surface. By projecting

[14] it upon the axis we read off the length of the projection as $= d'_k + X_{k+1} - X_k$, and as D'_k forms an angle U'_k with its projection we obtain the computing formula

$$D'_k = (d'_k + X_{k+1} - X_k) \sec U'_k. \tag{9}$$

A check may be considered desirable. It is obtainable from the diagram which gives

$$P_k B'_k = Y_k / \sin U'_k \quad \text{and} \quad P_{k+1} B'_k = Y_{k+1} / \sin U'_k,$$

and as D'_k is the difference of these two lengths we obtain

(9) Check : $$D'_k = Y_k / \sin U'_k - Y_{k+1} / \sin U'_k.$$

This check gives a less accurate value than (9), especially when U'_k is a small angle; it should therefore not be substituted for (9) as a computing formula, but only used as a safeguard against gross errors. The X and Y in these equations must of course be calculated by (8) and (8)*, (8)a and (8)a* being the most accurate and convenient ones if the PA-check is used.

It is best always to calculate the necessarily positive D'_k by the above equations, and not to use the alternative D_{k+1} which would be a negative quantity. If ever D'_k is found to come out negative, which can only happen for a lens or space of the form of a collective lens, i.e., one getting thinner above and below the optical axis, it indicates that an insufficient axial thickness has been assigned to that lens or space, and that it comes to a sharp edge at a smaller diameter than that called for by the ray supposed to pass through it. Such an underestimate of the required axial d' explains some of the most insidious errors in optical calculations, and should be carefully guarded against by making an accurate scale-drawing of any doubtful case. Two precious months were lost in the early days of the war owing to an error of this kind in a complicated instrument computed by analytical methods without the precaution—particularly advisable in analytical designs—of making an accurate drawing.

PLANE SURFACES

[15] Plane surfaces occur more frequently in optical systems than they would do if designs were always carried to the utmost limit in finding the best radius for each surface, for if this were done it would be infinitely improbable that a true plane should ever be arrived at for a lens surface.

In any case it is impossible in actual practice, with the usual sizes of lenses, to distinguish a plane from a surface of a radius of plus or minus say 10 miles, and therefore it would be quite legitimate in the calculations to substitute for a nominal plane a surface of a finite radius of this order of magnitude, in which case the standard formulae, with the check given, would be available, and would give a perfectly accurate and reliable result. In the working instructions for the glass-shop the plane would of course be specified. But as the plane surface obviously represents a specially simple case, it is preferable and saves much time to deduce special formulae. In a centred lens system a plane surface necessarily stands at exact right-angles with the optical axis, and all its incidence normals are therefore parallels to the optical axis. Referring to Fig. 13 we see that with a plane PA and Y become identical, and by using the instruction for the sign of I and I'

(positive if clockwise from ray to radius or normal) we read off a first deduction of [15] considerable importance in theoretical discussions

Pl. (1) $$I = -U; \quad I' = -U'.$$

The same result is obtained from the general equation (1) by writing it $\sin I = \sin U(L/r - 1)$ and then letting r grow beyond all measure, which brings L/r down

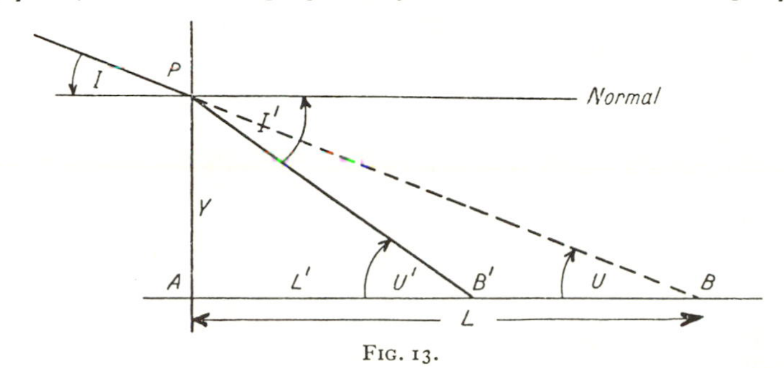

FIG. 13.

to zero for any finite L. Therefore, for a plane $\sin I = -\sin U$ or, as the angles are necessarily acute, $I = -U$. Equation (4) gives the corresponding result $I' = -U'$. The case, which mathematically invalidates this deduction, when L is also infinitely large, can only happen if the ray is parallel to the optical axis, therefore coincides with the normal at the point of incidence and suffers no refraction at all. Hence, in such a case, we ignore the plane and go directly to the next surface.

By putting $I = -U$ and $I' = -U'$ into standard equation (2) we obtain the first computing formula

Pl. (2) $$\sin U' = \sin U \,.\, N/N',$$

the minus-sign cancelling out because it occurs on both sides of the equation.

To find L' the diagram gives at sight

Pl. (2)* $$Y = L \tan U = L' \tan U'$$

which in this form serves to find the semi-aperture of a plane surface. A transposition of the second and third items of the equation gives one form of the computing equation for L', namely

Pl. (3) $$L' \doteq L \tan U \operatorname{cotan} U'.\dagger$$

If the practice usually followed in our numerical calculations is adopted of taking out the angles only to the nearest second this equation becomes inexact for small angles U and U', and it becomes necessary to use a modified form of Pl. (3), which in practice should be given preference in all cases because it is more easily and safely computed. Putting sin/cos in place of tan the equation becomes

$$L' = L\,(\sin U/\sin U') \,.\, (\cos U'/\cos U),$$

† In this book the cotangent is denoted by the symbol cotan instead of the more usual cot to avoid the well-known danger of confusion with cos.

[15] and as by Pl. (2) $\sin U/\sin U' = N'/N$ the computing equation becomes

Pl. (3)* $$L' = L \cdot \frac{N'}{N} \cdot \frac{\cos U'}{\cos U} = L \cdot \frac{N'}{N} \cdot \frac{\sec U}{\sec U'}.$$

The greater convenience and accuracy of this form is explained in the practical section of this chapter, section [22].

The formulae for a plane are so simple that an independent check may be considered unnecessary ; but accidents may happen, and it is therefore preferable to take no risks. (5)** cannot be used, for on putting $I = -U$ and $I' = -U'$ it is turned simply into our two equations for the plane-Y. But the triangle BPB', Fig. 13, gives at sight

$$BP = L \sec U\,; \quad BB' = L - L'\,; \quad \text{angle } BPB' = U' - U:$$

hence by the trigonometrical sine-relation

Pl. Check : $$L - L' = L \sec U \sin (U' - U)/\sin U'.$$

The result obtained by this equation should be compared with the value of $(L - L')$ obtained from Pl. (3) or (3)* ; and if there is agreement within the accuracy of the computation the result by the main calculation should be used for the carrying-forward because the check depends on sines of frequently small angles, and thus gives a less accurate result than Pl. (3) or (3)*. This check-formula may be rendered exceedingly searching and highly accurate by a transformation. By using $\sin a = 2 \sin \tfrac{1}{2}a \cos \tfrac{1}{2}a$ it becomes

$$L - L' = L \sec U \frac{2 \sin \frac{1}{2}(U' - U) \cos \frac{1}{2}(U' - U)}{\sin U'};$$

or, as $\sec \tfrac{1}{2}(U' + U) \,.\, \cos \tfrac{1}{2}(U' + U) = 1$ can be introduced without changing the value,

$$L - L' = L \sec U \cos \tfrac{1}{2}(U' - U) \sec \tfrac{1}{2}(U' + U) \frac{2 \sin \frac{1}{2}(U' - U) \cos \frac{1}{2}(U' + U)}{\sin U'}$$

$$= L \sec U \cos \tfrac{1}{2}(U' - U) \sec \tfrac{1}{2}(U' + U) \frac{\sin U' - \sin U}{\sin U'}.$$

From Pl. (2): $\sin U/\sin U' = N'/N$ we then deduce

$$(\sin U' - \sin U)/\sin U' = (N - N')/N$$

and obtain the modified check-formula

Pl. Check* $$L' = L - L \sec U \cos \tfrac{1}{2}(U' - U) \sec \tfrac{1}{2}(U' + U) \,.\, (N - N')/N.$$

It is not so laborious in use as it might seem at first sight because it contains only cosines and secants of usually small angles which call for little if any interpolation.

Plano-parallel Plates

A special case of the application of the formulae for a plane refracting surface is that of a parallel plate at right angles to the optical axis. Such plates often form parts of instruments in the obvious form of scale-plates or 'graticules' : still more frequently they are present in a hidden form which at first sight suggests formidable complications in a calculation, viz. in the form of reflecting prisms. It is easily [15] shown that a correctly formed reflecting prism always produces the same refraction-effects as a plano-parallel plate; if the prism is to be traversed by converging or diverging pencils of rays it is moreover necessary that the optical axis of the instrument should enter and leave it at exact right angles to the refracting surfaces. For these reasons we can in our ray-tracing always substitute a parallel plate at right angles to the optical axis for such prisms, the thickness of the plate being equal to the total geometrical length of the path of an axial ray through the prism. With reference to Fig. 13(a), the calculation of the effect of this plate then assumes a particularly simple form. Let an arriving ray be defined by L_1 and U_1 as usual. Let the thickness of the plate be d', and its relative refractive index with reference to the medium in which it stands (nearly always air) $= N$.

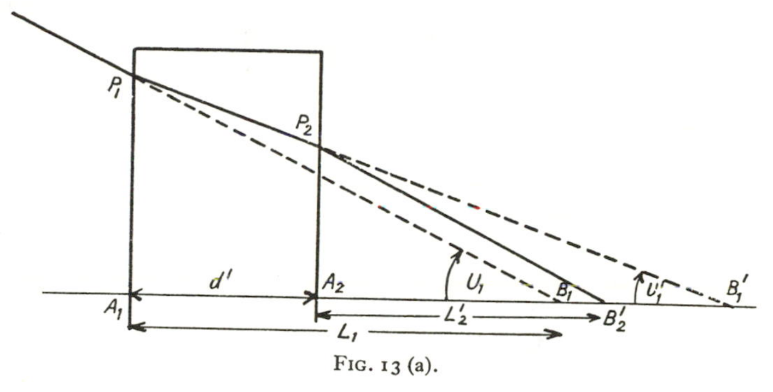

FIG. 13 (a).

We shall have U_1' defined by Pl. (2)

$$\sin U_1' = \sin U_1/N.$$

Next we have $U_2 = U_1'$ and then

$$\sin U_2' = N \sin U_1' = \sin U_1.$$

That is, the angle of convergence is not altered at all by the passage through the plate. For the intersection-lengths we find

$$L_1' = L_1 \tan U_1 \operatorname{cotan} U_1';$$

then

$$L_2 = L_1' - d',$$

and with this, on account of $U_1' = U_2$ and $U_2' = U_1$,

$$L_2' = (L_1' - d') \tan U_1' \operatorname{cotan} U_1 = L_1 - d' \tan U_1' \operatorname{cotan} U_1.$$

The effect of a parallel plate is therefore defined by

Pl. (4)

$$U_2' = U_1; \quad \sin U_1' = \sin U_1/N$$

$$L_2' = L_1 - d' \tan U_1' \operatorname{cotan} U_1.$$

[15] The term containing d' may be transformed to advantage in accordance with the process applied in Pl. (3)* and the formulae then become

Pl. (4)* $$U_2' = U_1\,; \quad \sin U_1' = \sin U_1/N$$

$$L_2' = L_1 - d' \cos U_1 \sec U_1'/N.$$

A further conclusion from the last equations is of considerable importance and practical utility. In the diagram the intersecting point of the incident ray is marked B_1, that of the final emerging ray as B_2'. The intersecting point is therefore shifted along the axis by the amount B_1B_2', in the direction of the ray. By following the same ray backwards (from right to left) it is easily seen that the following rule holds : A plano-parallel plate of higher index than the surrounding medium always shifts the intersecting point in the direction in which the light is travelling. The amount of the shift is easily read off the diagram, for we see that

$$B_1B_2' = d' + L_2' - L_1$$

or, introducing the value of L_2' by Pl. (4)*,

Pl. (5) $$B_1B_2' = d' - d' \cos U_1 \sec U_1'/N.$$

The shift is therefore entirely independent of the location of the focus, for Pl. (5) does not contain either L_1 or L_2' ; the shift depends solely on the thickness and refractive index of the plate and on the angles of obliquity. The simplest way of dealing with plano-parallel plates and reflecting prisms is therefore to calculate the lengthening of the intersection lengths by Pl. (5). Moreover, if there are several plates or prisms of the same index in the same air-space we can add their thicknesses and allow for the whole lot in a single calculation.

All the formulae for plane surfaces may be adapted to right-to-left calculations by turning plain into dashed letters and vice versa. In the equations for plano-parallel plates this process will turn the positive thickness d' into the negative plain d ; the more convenient positive d' may be restored by reversing the sign of the second part of Pl. (4) and (4)* and of Pl. (5).

OPENING A CALCULATION

[16] In beginning a calculation through a lens system it usually happens that the rays to be traced are not specified in the form assumed by the general computing formulae, that is by L and U, and the opening must then be modified. One of the most common cases is that the original object-point is at so great a distance that L may be treated as infinite (telescope and photographic objectives). The rays from the object-point on the optical axis will then be a parallel bundle and we have $L = \infty$ and $U = 0$ with which we cannot do anything. Such a parallel ray is therefore always specified by its distance from the optical axis, which evidently represents the Y of our general equations and of Fig. 14. On drawing and producing the radius CP we see that angle ACP is equal to the angle of incidence I, and the triangle QCP, with a right angle at Q, gives us at once

(1) for // light: $$\sin I = \frac{Y}{r}$$ [16]

which enables us to go on with the calculation by the standard schedule.

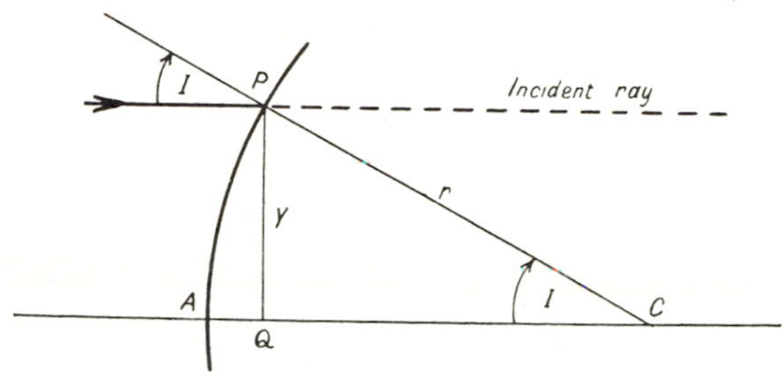

FIG. 14.

The check-formula also breaks down in this case because it contains $L \sin U$ which becomes infinity times zero. It is, however, easily adapted by noticing what becomes of $L \sin U$ in the formula when the object-point moves very far away from the first lens-surface.

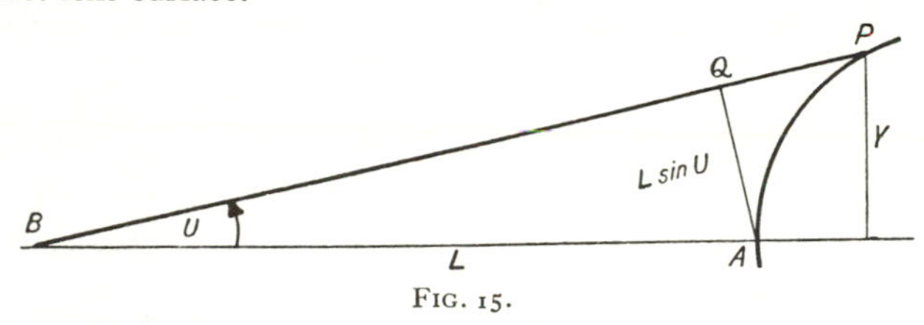

FIG. 15.

Referring to Fig. 15 we see that the perpendicular AQ dropped from the Pole A upon the entering ray is $= L \sin U$. If we now imagine L becoming longer and longer and U correspondingly smaller and smaller, AQ will become more and more nearly vertical to the optical axis, and simultaneously the ray BP will become more and more nearly parallel to the optical axis: hence, at the limit, AQ will become parallel to Y and QP simultaneously becomes parallel to the optical axis, and therefore for an infinitely distant point B we have rigorously $L \sin U = Y$. Hence the check-formula for parallel light is

(5)** Check for // light $$\begin{cases} PA = Y \sec \frac{1}{2} I \\ L' = PA \operatorname{cosec} U' \cos \frac{1}{2}(I' - U'); \end{cases}$$

for in the formula for PA $\sec \frac{1}{2}(I - U)$ becomes simply $\sec \frac{1}{2} I$ because for a parallel ray $U = 0$.

It may here be added that a ray may conceivably be found to emerge parallel to the optical axis at some subsequent surface of a system, which will be indicated by U' by (3) coming out as zero.† It is easily seen by drawing a diagram that in

† The case has never occurred in my extensive practice of about thirty years, the nearest approach experienced being a U' of one second, which could be and was dealt with by the ordinary formulae.

[16] this case $r \sin I'$ would represent the distance Y of the emerging parallel ray from the optical axis : hence the calculation would be broken off at that point, and the value $Y = r \sin I'$ used for a fresh start at the next surface by the above special formulae.

When the original object-point is at a finite distance, and L therefore known and given, the direction of the ray to be traced is usually defined by the semi-aperture Y at which it is to enter the first surface and a special opening is again required.

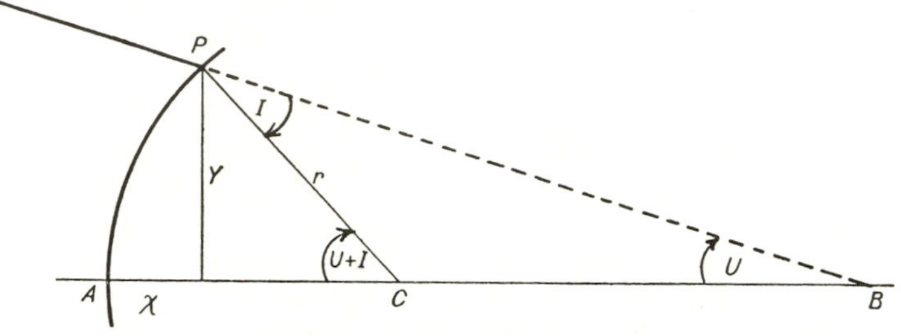

FIG. 16.

Referring to Fig. 16 we shall have given $AB = L$ and the ordinate Y of the point of incidence. Very frequently L is large compared to Y and still more so when compared to X, and as great accuracy in realizing the desired semi-aperture Y is rarely necessary, it frequently is sufficient to use

$$\sin U \text{ nearly} = Y/L$$

and to open the calculation with this approximate value. If it is desired to attain full accuracy, two methods are available.

First Solution : By transposing (8)b* we can calculate from the given data

$$\sin (U+I) = Y/r :$$

Then with the value of $(U+I)$ so secured, by (8)d

$$X = Y \tan \tfrac{1}{2} (U+I)$$

and we can then read off the diagram

$$\tan U = Y/(L-X).$$

Second Solution : Again we begin with

$$\sin (U+I) = Y/r.$$

From equation (1) we then derive algebraically

$$\frac{L-2r}{L} = \frac{\sin I - \sin U}{\sin I + \sin U} = \frac{\tan \frac{1}{2}(I-U)}{\tan \frac{1}{2}(I+U)}$$

from which we can (using the above value of $(U+I)$) compute

$$\tan \tfrac{1}{2} (I-U) = \tan \tfrac{1}{2} (I+U)\,(L-2r)/L.$$

We then know both $\frac{1}{2}(I+U)$ and $\frac{1}{2}(I-U)$, the sum of which gives I, whilst the difference gives U. If we want to economize time we can turn up $\sin I$ and begin [16] the usual calculation with equation (2). It will be safer to turn up $\sin U$ and to calculate the full standard schedule, for the finding of values of I and $(U+I)$ by this calculation which agree with the above will check the whole work.

PARAXIAL RAYS

The fundamental equations can be greatly simplified when they are restricted to [17] rays which make small angles with the optical axis and with the normals of the refracting surfaces. If an angle is expressed in 'radians' or 'circular measure', then

$$\sin A = A - \frac{1}{6}A^3 + \frac{1}{120}A^5 \text{ \&c.,}$$

and this shows that if A is a small fraction the higher terms of the series become extremely small, and we shall have with close approximation $\sin A = A$. For sufficiently small angles the sines in the fundamental formulae may therefore be replaced by the angles themselves. From the purely mathematical point of view we should limit this statement to infinitely small angles, and would then be restricted to the 'thread-like space around the optical axis' of books on examination-optics as the only legitimate home of paraxial rays. In reality we hardly ever are interested in the direction of rays within narrower limits than 1 second of arc or 0·0000048 radian, for in big astronomical telescopes, in which fractions of a second become important, the want of homogeneity of all optical glass renders the attainment of precision to even a whole second almost completely illusory by *calculation* of the object-glass ; it always depends on patient *empirical figuring* of the surfaces. If we take 1 second as the limit of admissible inaccuracy, the second approximation for the value of a sine

$$\sin A = A - \frac{1}{6}A^3$$

shows that we may treat $\sin A$ as $= A$ up to the value of A, which makes

$$\frac{1}{6}A^3 = 0{\cdot}0000048$$

or

$$A = 0{\cdot}031 = \text{nearly } 2^\circ.$$

As the largest angles occurring in lens-systems for the extreme marginal ray usually are less than 30°, this shows that the region within which the paraxial formulae may be regarded as sufficiently valid exceeds one twentieth of the full aperture of any ordinary lens system, and is thus decidedly finite and useful.

For these small angles cosines and secants become sensibly equal to 1, for the series developments of these functions in terms of the angle expressed in radians are

$$\cos A = 1 - \tfrac{1}{2}A^2 + \frac{1}{24}A^4, \text{ \&c.}$$

$$\sec A = 1 + \tfrac{1}{2}A^2 + \frac{5}{24}A^4, \text{ \&c.}$$

[17] We can therefore turn all the formulae deduced for rays at finite angles into their paraxial equivalents by replacing all sines by the angles and all cosines and secants by 'one'. Using the corresponding small letters throughout, we thus obtain the standard paraxial formulae

(1p) $i = u(l-r)/r$;
(2p) $i' = i \,.\, N/N'$;
(3p) $u' = u+i-i'$;
(4p) $l'-r = i' \,.\, r/u'$;
(5p) $l' = (l'-r)+r$;
(5p)* $l_{k+1} = l'_k - d'_k$; $u_{k+1}=u'_k$; $N_{k+1}=N'_k$.

In the check-formula PA becomes indistinguishable from the ordinate y of the point of incidence, therefore the check becomes

(5p)** $y = lu$; $l' = y/u'$; which give $lu = l'u' = y$.

The following formulae become

(6p) $(l'-r)\, N' \,.\, u' = (l-r)\, N \,.\, u$;

(7p) $i-i' = (i+i')\,(N'-N)/(N'+N)$.

Of the formulae for X and Y the only ones of interest and importance are

(8p) $x = \frac{1}{2}\, y^2/r = \frac{1}{2}\, y(u+i)$; (8p)* $y = r(u+i) = l \,.\, u = l' \,.\, u'$.

The conversion of equation (9) for D'_k requires care, for as (8p) shows, a paraxial x is a small quantity of the *second* order because it depends on the square of the small quantity of the first order y ; hence the paraxial D'_k differs from the axial d'_k only by small quantities of the second order. In first approximation the paraxial D'_k may therefore be taken as $= d'_k$, but as d'_k must be treated as a large (compared to y) quantity it follows that, if a second approximation is required, the factor sec U'_k cannot in the case of d'_k be put equal to 1, but must be put $= 1 + \frac{1}{2}\, u'^2_k$. Therefore the formula becomes

(9p) Paraxial $D'_k = d'_k\,(1 + \frac{1}{2}\, u'^2_k) + x_{k+1} - x_k$.

(9) Check is not suitable for conversion into a paraxial equation correct within second order terms.

The paraxial equations (8p), (8p)*, and (9p) are chiefly required in advanced work based on physical optics and the beginner need not worry about them till later.

The formulae for plane surfaces also are easily converted into their paraxial equivalents on the principles laid down, and are found as

Pl. (1p) $i = -u$; $i' = -u'$;
Pl. (2p) $u' = u \,.\, N/N'$;
Pl. (2p)* $y = l \,.\, u = l' \,.\, u'$;
Pl. (3p) $l' = l \,.\, N'/N$;
Pl. (p) Check* $l' = l - l(N-N')/N$;
Pl. (4p)* $\left\{ \begin{array}{l} u'_2 = u_1 \text{ ; } u'_1 = u_1/N \text{ ;} \\ l'_2 = l_1 - d'/N \text{ ;} \end{array} \right.$
Pl. (5p) $B_1B'_2 = d'(N-1)/N$.

The opening formulae become

(1p) for // light : $i = y/r$; (5p) Check for // light : $l' = y/u'$.

When y and l are given : $u = y/l$.

These opening formulae are only to be used when a paraxial ray is to be traced by itself without reference to a corresponding marginal ray from the same object-point. In the latter case, which is the usual one, the convenient and labour-saving [17] special instructions in a subsequent practical part of this chapter should be followed.

The important distinction between the exact trigonometrical formulae for rays at any angle and the paraxial ones for rays at very small angles is that the paraxial formulae are simple linear equations of the angles entering into them, and so lend themselves to the simplest possible algebraical treatment, whereas the exact formulae are transcendental and therefore incapable of ordinary algebraical discussion.

The linear nature of the paraxial ray-tracing equations immediately leads to a general conclusion of very great importance. Let us suppose that we had traced a paraxial ray by the equations (1p) to (4p) with a certain value of the initial u, and that we next repeated the calculation with the initial angle $= k.u$, k being any positive or negative number, but such that all the angles of the ray-tracing still remain of paraxial magnitude. By (1p) we should find

$$\text{new } i = k \,.\, u(l-r)/r = k \text{ (original } i).$$

Using this to calculate (2p) we should obtain

$$\text{new } i' = (\text{new } i) \,.\, N/N' = k \,.\, i \,.\, N/N' = k \text{ (original } i').$$

Equation (3p) would next give

$$\text{new } u' = \text{new } u + \text{new } i - \text{new } i' = k \,.\, u + k \,.\, i - k \,.\, i' = k \text{ (original } u')$$

and with this equation (4p) would produce

$$\text{new } (l'-r) = (\text{new } i') \,.\, r/(\text{new } u') = k \,.\, i' \,.\, r/k \,.\, u' = i' \,.\, r/u' = \text{original } (l'-r).$$

That is to say, within the paraxial region all rays from a given object-point at any distance l are refracted so as to pass through one and the same image-point at a perfectly definite conjugate distance l'. Although we have only proved this for one surface, it is evident that the proof could be repeated for any number of successive refractions at *centred* surfaces, and we can therefore state the first important general property of centred lens-systems in this form :

'Every centred lens-system brings all paraxial rays of any one colour sent out by any axial object-point to a sharp focus at a conjugate axial image-point.'

It is necessary to specify a definite colour on account of the variation of refractive indices with the colour and wavelength of the light, the effect of which we shall study in detail in Chapter IV.

From the practical computer's point of view the proof given above (that l' comes out unchanged regardless of the value of the initial u, and that the only change is a strictly proportional variation of *all* the paraxial angles) means that in our numerical calculations we can use any nominal finite value we please for the initial u of any paraxial ray-tracing ; how we take advantage of this will be shown in a following section on numerical calculations.

A very useful alternative solution of the paraxial ray-tracing problem is easily deduced from equation

$$\text{(6p)} \qquad (l'-r) \,.\, N' \,.\, u' = (l-r) \,.\, N \,.\, u.$$

If we introduce into this from (5p)** $u = y/l$ and $u' = y/l'$, then the y occurs

[17] on both sides and cancels out (thus directly demonstrating that within the paraxial region l' is independent of the aperture) and we obtain

(6p)* $$N'(l'-r)/l' = N(l-r)/l,$$

or $$N'-N'.r/l' = N-N.r/l;$$

and if this is divided throughout by r and rearranged we arrive at the equation

(6p)** $$\frac{N'}{l'} = \frac{N'-N}{r} + \frac{N}{l}$$

which is convenient for finding l' by slide-rule or calculating machine, but is chiefly valuable as a fundamental equation for the deduction of paraxial lens-formulae. Strange as it may seem, the apparently simple formula (6p)** has no advantage over the seemingly far more complicated standard paraxial equations (1p) to (5p) in logarithmic calculations, for (6p)** requires more references to the tables and more interpolations to arrive at the result. Moreover the standard equations give us the *angles* of the traced ray from which many valuable results are obtainable in addition to the determination of l', which is the only datum obtained directly by (6p)**. For this reason the standard equations very frequently deserve preference even in work by slide-rule or calculating machine.

Equation (6p)** becomes available for right-to-left calculations by treating l' as the given and l as the sought quantity.

Two special cases are of interest, namely when either l or l' becomes infinitely large and the corresponding rays become parallel to the optical axis. The resulting intersection-length after refraction is then called the principal focal length, with symbol f or f'. Putting $l = \infty$ in (6p)** we find

$$N'/f' = (N'-N)/r \quad \text{or} \quad f' = rN'/(N'-N)$$

in the medium to the right of the surface. Putting $l' = \infty$ we find for the medium to the left of the surface

$$0 = (N'-N)/r + N/f \quad \text{or} \quad f = -rN/(N'-N),$$

and comparing f and f' we find the important relation

$$f'/f = -N'/N.$$

THEOREM OF LAGRANGE

[18] A highly important deduction from equation (6p) is known as the 'theorem of Lagrange'; † it determines the linear magnification of the image of a small object which is produced by the refraction of paraxial rays at a spherical surface. In Fig. 17, $BB_o = h$ represents a small (in this case virtual) object, assumed for the present to lie on a spherical surface with C as centre. Let the paraxial image of point B be formed at B' in accordance with the paraxial equations (1p) to (5p). Then CB will $= (l-r)$ and $CB' = (l'-r)$ and we shall have the relation

(6p) $$(l'-r)\,N'.u' = (l-r)\,N.u.$$

† The theorem is probably most widely known by the name adopted in the text. Its full history is, as usual, complicated and possibly still uncertain. It has been traced back to the famous Huygens. The modern form chiefly used in the text is due to Helmholtz.

If we now draw an auxiliary optical axis by joining C to B_o we shall have CB_o [18] $= CB = (l-r)$, because BB_o is an arc of a circle having C as its centre. Hence any rays aiming at B_o within paraxial distance of the auxiliary axis will be under

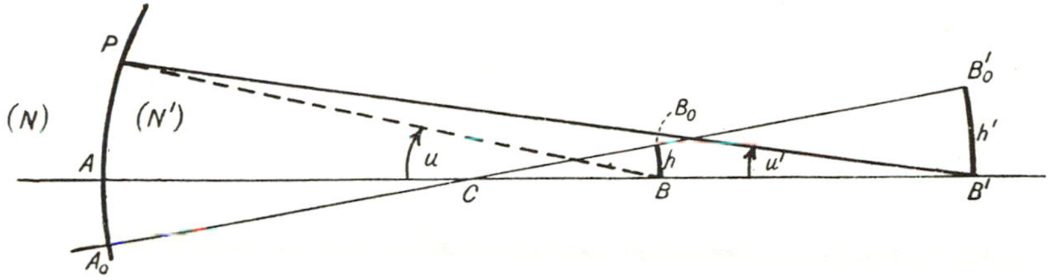

FIG. 17.

precisely the same conditions of refraction with reference to the auxiliary axis as rays aiming at B are with reference to the principal axis. Consequently B_o will be depicted at B_o' on the auxiliary axis and so that $CB_o' = CB' = (l'-r)$. As the same arguments obviously hold for all intermediate points of the object BB_o, it follows that the image $B'B_o' = h'$ lies on a sphere with C as centre. Therefore the sectors CBB_o and $CB'B_o'$ are similar and lead to the proportion

$$h'/(l'-r) = h/(l-r).$$

If we multiply (6 p) by this equation then $(l'-r)$ and $(l-r)$ cancel out and we obtain Lagrange's Theorem :

(10p) $$h'.N'.u' = h.N.u.$$

In order to make the proof a strictly Euclidian one we had to assume an object of the specified curvature. From the practical optical point of view this restriction is unimportant, for as all paraxial equations are limited to very small angles—including in this case the angle between the principal and the auxiliary axis—the difference in size and the departure from coincidence of the short arc BB_o and say a straight perpendicular to the principal optical axis are both negligibly small. We may therefore claim our equation as an exact paraxial one even for objects which are plane, or curved to some finite radius differing from the special values assumed in the geometrical proof.

The symbol h for the linear size of an object at right angles to the optical axis will be retained in future work, but in accordance with our principles of nomenclature it will be replaced by capital H for objects too large to be dealt with by paraxial equations. The symbol may be remembered by thinking of the *height* of an object or image.

The significance of Lagrange's theorem lies in the fact that h'/h, the ratio of the size of the image to that of the object, obviously represents the *linear magnification* for which we use the symbol m for paraxial rays, M for rays at finite angles. As in other cases we shall use the term magnification in a wider sense than the word strictly implies ; we retain both name and symbol when the image is smaller than the object, and when it would be really proper to speak of diminution, reduction, or minification ; in other words m and M simply stand for the ratio of the size

[18] of the image to that of the object, no matter which is the larger ; real magnification leads to a value of m or M greater than one, minification to fractional values. Moreover, we apply a sign to the magnification ; for erect images (h and h' on the same side of the axis) we use the positive sign, for inverted images (h and h' on opposite sides of the axis) the negative sign. Our equations automatically bring out the correct signs.

In order to distinguish magnification in the two possible directions of tracing rays, we again use the dash for the magnification in the medium to the right of a surface, the plain letter for the medium to the left, and therefore use the definitions

$$m' = h'/h \,;\; m = h/h' \,;\; \text{whence } m \,.\, m' = 1 \quad \text{or } m = 1/m'.$$

The two magnifications are therefore reciprocals.

For a single surface we now have by Lagrange's theorem $m' = h'/h = Nu/N'u'$, and if we add simple transformations of this by either (6p) which gives $Nu/N'u' = (l'-r)/(l-r)$, or by (5p)** which gives $u = y/l$ and $u' = y/l'$, we can collect the formulae

$$\text{(10p)*} \qquad m' = Nu/N'u' = (l'-r)/(l-r) = Nl'/N'l$$

$$\text{or} \qquad m = N'u'/Nu = (l-r)/(l'-r) = N'l/Nl',$$

the latter by the reciprocal relation just proved.

The most valuable application of Lagrange's theorem is its extension to any number of centred surfaces. The equation $h'N'u' = hNu$ is of course applicable to every refracting surface. If we write it down for the first surface of a centred system in the form

$$h_1'N_1'u_1' = h_1N_1u_1,$$

then for the second surface it will be

$$h_2'N_2'u_2' = h_2N_2u_2,$$

in which h_2 is the size of the object presented to the second surface. But this is the image h_1' produced by the first surface ; therefore $h_2 = h_1'$. In discussing the carrying on of a paraxial calculation from surface to surface we noted that $u_2 = u_1'$ and $N_2 = N_1'$. Therefore the right of the second equation is identically equal to the left of the first. Therefore for two surfaces

$$h_2'N_2'u_2' = h_1N_1u_1$$

quite regardless of the separation between the two surfaces. Evidently we can continue this process of reasoning for the third and any subsequent surfaces, with the result that no matter how many surfaces there may be in a centred optical system and what their axial separations may be, the product $h \,.\, N \,.\, u$ retains an absolutely fixed value throughout and we must have for a system of k surfaces

$$h_k'N_k'u_k' = h_1N_1u_1 \,;$$

and as h_k' (= size of final image) $/h_1$ (= size of original object) again is the linear magnification produced by the complete system we have, for any centred system,

$$\text{(10p)**} \quad \begin{cases} h_k'/h_1 = m' = N_1u_1/N_k'u_k' \\ \text{or } h_1/h_k' = m = N_k'u_k'/N_1u_1 \text{ for right-to-left work.} \end{cases}$$

The second form results by solving the equation [18]

$$h'_k N'_k u'_k = h_1 N_1 u_1 \text{ for } h_1/h'_k = m.$$

It should be carefully noted that in these equations for the paraxial magnification N_1 stands for the refractive index of the medium to the left of the first surface of the system and N'_k for the index of the medium to the right of the last surface. In the vast majority of actual lens systems both these media are the same, practically always air. Hence for all systems having object and image surrounded by the same medium, N_1 and N'_k cancel each other in the above equation and the magnification-formulae become simply

$$h'_k/h_1 = m' = u_1/u'_k \; ; \; h_1/h'_k = m = u'_k/u_1.$$

By the standard paraxial equations we obtain all the angles u in finding the l', hence this theorem of Lagrange enables us to find the linear magnification of the final or of any intermediate image by a simple division of known quantities. We shall subsequently become acquainted with the optical sine-condition which solves the same problem of linear magnification in an equally simple manner for the image produced by rays at finite angles, the paraxial angles u_1 and u'_k being merely replaced by sines of the corresponding finite angles U_1 and U'_k.

As Lagrange's theorem is valid for any centred lens system, we can make an important addition to the general *paraxial* properties of such systems noted in the previous section, namely,

> 'Every centred lens system sharply depicts a small object placed at right angles with its optical axis and at any distance from the system at a linear magnification which can be determined by Lagrange's theorem.'

The restriction to paraxial rays of one definite colour must of course be maintained.

The absolute constancy of the product $h \, . \, N \, . \, u$ for any given centred optical system when the size h_1 of the original object and the angle u_1 of the entering paraxial ray have been fixed is a fact of great importance and should be remembered. Such a quantity is usually called an 'invariant', and we shall refer to $h \, . \, N \, . \, u$ as the 'Lagrange Invariant'.

LONGITUDINAL MAGNIFICATION

The theorem of Lagrange enables us to calculate the *linear* magnification of the image of any object at right angles with the optical axis. Frequently we shall be interested in two neighbouring points *on* the optical axis which are simultaneously depicted by a surface or system. We can look upon two such points as the ends of a short object lying along the optical axis (instead of at right angles to it) and shall want to know the length of the image of such an axial distance or object. [19]

This problem is easily solved for any one surface by equation (6p)**. Let one of the two neighbouring points on the axis be at distance l from the surface, the other at distance $l + dl$, where for the present dl shall be treated as a finite quantity.

[19] Calling the corresponding intersection-distances after refraction l' and $l'+dl'$, (6p)** gives

$$N'/l' = (N'-N)/r + N/l$$
$$N'/(l'+dl') = (N'-N)/r + N/(l+dl);$$

or subtracting the second from the first

$$N'[1/l' - 1/(l'+dl')] = N[1/l - 1/(l+dl)]$$

and bringing the square brackets to a common denominator

$$N'dl'/l'(l'+dl') = N\,.\,dl/l(l+dl),$$

or transposing N' and l'

$$dl'/(l'+dl') = N\,.\,l'.\,dl/N'l(l+dl).$$

We now use the rule : if $a/b = c/d$ then also $a/(b-a) = c/(d-c)$

and find $\quad dl'/l' = N\,.\,l'.\,dl/[N'.\,l(l+dl) - N\,.\,l'.\,dl]$

or $\quad dl'/dl = N.\,l'^2/[N'l^2 + dl(N.'l - N\,.\,l')].$

This is an exact expression for the paraxial magnification of an axial object of any length. If we now restrict ourselves to a very small value of dl, the second term of the denominator on the right will become negligible beside the finite first term, with the exception of the case when l' is very large compared to l, which we will merely note. As a rule the axial or longitudinal magnification of a short length dl is therefore determined sufficiently accurately as

$$\frac{dl'}{dl} = \bar{m}' = \frac{N\,.\,l'^2}{N'\,.\,l^2}$$

where we have introduced $\bar{m}$ (*m-bar*) as a general symbol for this longitudinal magnification. If we introduce into this equation from (5p)** $l' = y/u'$ and $l = y/u$, it becomes

$$dl'/dl = Nu^2/N'u'^2,$$

the y^2 cancelling out. Transposed, this equation gives

$$dl'\,.\,N'\,.\,u'^2 = dl\,.\,N\,.\,u^2$$

and this form of the equation enables us to determine the longitudinal magnification for any centred system, in close analogy with the same stage in the development of the theorem of Lagrange. For a first surface we shall have

$$dl_1'\,.\,N_1'\,.\,u_1'^2 = dl_1\,.\,N_1\,.\,u_1^2,$$

and for a second surface

$$dl_2'\,.\,N_2'\,.\,u_2'^2 = dl_2\,.\,N_2\,.\,u_2^2.$$

Obviously the axial image dl_1' by the first surface acts as the axial object for the second, therefore $dl_1' \equiv dl_2$; and as $N_1' \equiv N_2$ and $u_1' \equiv u_2$ by the fundamental conventions, the left of the first equation is identical with the right of the second ; hence

$$dl_2'\,.\,N_2'\,.\,u_2'^2 = dl_1\,.\,N_1\,.\,u_1^2,$$

and it is evident that these arguments can be repeated any number of times, thus [19] leading to the general theorem of longitudinal magnification

$$dl'_k \,.\, N'_k \,.\, u'^2_k = dl_1 \,.\, N_1 \,.\, u_1^2,$$

which gives us another optical invariant : $dl \,.\, N \,.\, u^2$. We must bear in mind, however, that the present deductions are only *strictly* true if the dl are of differential smallness, but they give very good approximations as long as the dl are decidedly small fractions of the corresponding l-values, especially when the original dl is taken with its *centre* at the point B from which the paraxial ray-tracing starts.

The general equation by which the longitudinal magnification can be determined is now

$$dl'_k/dl_1 = \bar{m}' = N_1 u_1^2/N'_k u'^2_k.$$

If we compare this with the Lagrange equation for the linear magnification at the same conjugate points, namely

$$m' = N_1 u_1/N'_k u'_k,$$

we easily see that we have unconditionally

$$\bar{m}' = m'^2 \,.\, N'_k/N_1,$$

and also, that the longitudinal magnification can only become numerically equal to the linear magnification if $m' = \pm N_1/N'_k$, for this is the only value which gives $\bar{m}'$ also $= N_1/N'_k$. This relation discloses a remarkable and absolutely incurable defect in the images given by that vast majority of optical instruments which yield a linear magnification differing from our very special value N_1/N'_k. The axial dimension of any three-dimensional object is magnified at a different rate to the cross-dimensions. Thus, supposing that we looked at a cube with sides of 1 mm. through a magnifying glass giving a linear magnification of ten times, we should *not* see a *cube* with sides of 10 mm., but, supposing that one of the square faces were towards us, a *square stick* with endfaces of 10 mm. square, and a length, seen end on, of 10^2 or 100 mm. We will point out at once that in the case of low magnifications we usually fail to realize the enormity of the misrepresentation of axial dimensions because the single eye used on most optical instruments has very little power of correctly estimating distances ; but at the high magnifications of the compound microscope the terrific exaggeration of axial dimensions leads to the greatest difficulty in practical microscopy, namely, the almost infinitesimal depth of focus, which forces microscopists to form their judgement of the axial sequence of appearances not by direct vision but by the use of a delicate micrometrical 'fine adjustment' of the focus which gives a succession of 'optical cross-sections' of the object which the observer's mind must correlate.

The effects resulting from the difference between $\bar{m}'$ and m' become of the reversed type when the image observed or photographed is a minified rendering of the object. A cube of say 1 cm. side when depicted at a linear minification of 0·1 times would not be rendered as a cube of 1 mm. side but as a square flake 1 mm. square but only 0·1 mm. thick. This effect of minified images explains the remarkable depth of object-distances covered practically sharply by photographic cameras.

[19] The law of longitudinal magnification finds extensive applications in the theory of aberrations, and we will therefore collect the most useful formulae. By equations already used in the present and the preceding section we have

(10p)*** For any centred system:

$$\bar{m}' = N_1 u_1^2 / N'_k u'^2_k = m' \,.\, u_1 / u'_k = m'^2 \,.\, N'_k / N_1.$$

For a single surface:

$$\bar{m}' = Nu^2 / N'u'^2 = m' \,.\, u/u' = Nl'^2 / N'l^2 = \frac{l'-r}{u'} \cdot \frac{u}{l-r} \cdot$$

The last form given for a single surface is the most readily calculated one when a standard paraxial calculation is available, for, as will be seen in the next section, the logs of $l'-r$ and of u', of $l-r$ and of u, stand next to each other ready for the subtraction.

NUMERICAL CALCULATIONS (*continued*)

The Normal Type of an Optical Calculation

[20] The tracing of several rays at finite angles as exemplified in [9] is not the most usual type of an optical calculation. As a rule a good judgement of the state of correction of a lens or system with regard to the axial object-point can be based on the tracing of only two rays, of which one passes through the extreme margin of the aperture and is traced by the trigonometrical formulae (1) to (5), whilst the other is a paraxial ray traced by the formulae (1p) to (5p). We will therefore apply this method to the biconvex lens of [9] on the assumption that the lens is so mounted that the extreme ray which can enter from B_1 makes an angle of $-4°$ with the optical axis. With regard to the computation for the paraxial ray it was shown in the theoretical part that whilst the formulae are mathematically correct only for rays at minute angles, their linear character causes them to give precisely the same intersection-lengths, no matter what value may be given to the first u, and that the latter may therefore be arbitrarily chosen. We therefore choose a nominal value of the first u which leads to the maximum convenience both in the actual computing and in subsequent deductions from the paraxial results. This choice falls upon a nominal or fictitious value of the first u which is precisely equal to the first sin U of the marginal ray with which the paraxial one is associated. It is, however, important to note:

(1) That this rule is applied only at the *first* surface of any system. In passing from the first to the second and subsequent surfaces the u' found for the preceding surface must be carried forward as the u of the next surface in accordance with the second of (5p)*. At the surfaces following the first there is then nearly always a notable difference between u and sin U, and this difference has an important significance.

(2) That the use of these large nominal values of the paraxial angles does not imply a claim that the results are correct for such large angles: they are correct for angles which are a minute fraction of those carried on the register; for practical purposes we may generally take one hundredth of the nominal angles as being

small enough to be treated as paraxial quantities to which the resulting intersection lengths may safely be assigned. [20]

As the paraxial formulae are of precisely the same form as those for rays at finite angles, the calculations can and ought to be made in parallel columns. The only difference is that whilst the angles of the marginal ray are obtained from the log sin table, the paraxial angles (being in terms of the radian) are looked up in the table of the logs of natural numbers.

	First Surface.			*Second Surface.*	
	Marginal.	Paraxial.		Marginal.	Paraxial.
L	−24·000	same	l	131·893	149·211
$-r$	−10·000	same	$-r$	+ 5·000	+ 5·000
$L-r$	−34·000	same	$l-r$	136·893	154·211
log sin U	8·84358*n*	same	log (nominal u)	8·10568	8·04825
+log $(L-r)$	1·53148*n*	same	+log $(l-r)$	2·13638	2·18812
log $(L-r)$ sin U	0·37506	same	log $(l-r)\,u$	0·24206	0·23637
−log r	−1·00000	same	−log r	−0·69897*n*	−0·69897*n*
log sin I	9·37506	same	log i	9·54309*n*	9·53740*n*
+log N/N'	−0·18127	same	+log N/N'	0·18127	0·18127
log sin I'	9·19379	same	log i'	9·72436*n*	9·71867*n*
+log r	1·00000	same	+log r	0·69897*n*	0·69897*n*
log r sin I'	0·19379	0·19379	log $r\,.\,i'$	0·42333	0·41764
−log sin U'	−8·10568	−8·04825	−log u'	−9·32859	−9·27809
log $(L'-r)$	2·08811	2·14554	log $(l'-r)$	1·09474	1·13955
U	− 4- 0- 0	−0·069756	u	0-43-51	0·011175
$+I$	13-43-10	0·237170	$+i$	−20-26-21	−0·344668
$U+I$	9-43-10	0·167414	$u+i$	−19-42-30	−0·333493
$-I'$	− 8-59-19	−0·156239	$-i'$	32- 0-45	0·523203
U'	0-43-51	0·011175	u'	12-18-15	0·189710
$L'-r$	122·493	139·811	$l'-r$	12·4377	13·7895
$+r$	10·000	10·000	$+r$	− 5·0000	− 5·0000
L'	132·493	149·811	l'	7·4377	8·7895
$-d'$	−0·6	−0·6	$-d'$	log first u	8·84358*n*
L_1	131·893	149·211	l_1	−log last u'	−9·27809
				log m'	9·56549*n*
				m' =	−0·36770

The calculation begins with the determination of $(L-r)$ as before, and the logarithmic work for the marginal ray begins with log sin of $-4°$ also as before. By the convention stated the logarithmic work for the paraxial ray begins with $u = \sin U$, and therefore also with $\log u = \log \sin U$. The same number would therefore appear in the two columns. At a first surface receiving light from a real object-point this holds equally for $\log (l-r) = \log (L-r)$, and it always holds for the radius and the refractive index. The consequence is that there is in all cases less work connected with the tracing of a paraxial ray than with that of the associated ray at finite angles, and at a first surface the first nine lines would contain identical numbers throughout, and these numbers need therefore not be written twice. But on looking at the work for the second surface, where most of the

[20] numbers are different, it is seen that the differences are moderate. Hence the advantages of the employment of initial u = initial sin U from the computer's point of view are that there is a substantial saving at the very beginning, and that the parallel columns check each other throughout against gross errors.

With regard to the paraxial angles in the angle-register, u of the first surface is the natural number found in the first part of the log-table to the log sin U at the beginning of the logarithmic work ; given a table of natural trigonometrical functions, it could also be turned up (at a first surface only !) as the natural sine to $-4°$. The angles i and i' for the first surface are the natural numbers to log sin I and log sin I' in the first column ; this also applies only to a first surface. For a second or subsequent surface i and i' are the natural numbers to their logs as found in the paraxial column.

At the end of the paraxial calculation for the second surface an important further result is derived, namely the magnification of the paraxial image of a small object placed at B_1 at right angles with the optical axis, by the theorem of Lagrange. Equation (10p)** is applied, simplified by omission of N_1 and N'_k, because both object and image are in air of index 'one'. The result, $m' = -0{\cdot}36770$ means :

(1) On account of the negative sign of m' the image is an inverted one.

(2) m' being a proper fraction means that the image is smaller than the object in the proportion of 0·3677 to Unity. Roughly the image is a little larger than one-third natural size.

The magnification so determined only applies to paraxial rays, and would be closely realized if a small stop were applied to reduce the aperture of the lens. It will be shown subsequently that the magnification produced by a cone of rays of finite aperture can be determined in an equally simple manner (as first sin U divided by last sin U') by the optical sine-condition.

Example 1a. Trace the same five rays through the same lens for a more refrangible colour, employing the refractive index 1·527. This index will approximately correspond to the photographically most active rays if the index (1·518) first employed is taken to be that for the visually brightest yellow-green rays.

Example 1b. Carry out similar calculations for the original lens in the reverse position, with the surface of 5 inches radius next the object. Note once for all that when a lens or lens system is reversed all the radii and thicknesses will naturally be met in inverse order, and that all the radii will reverse their sign, the latter because a centre of curvature which was to the right of the surface in the first position will be to the left in the reversed position, and vice versa. The formula of the lens will therefore be

$$\begin{aligned} r_1 &= 5{\cdot}000 \quad d'_1 = 0{\cdot}600 \\ r_2 &= -10{\cdot}000 \end{aligned}$$

L remaining at -24 as before.

CHECKS

[21] It will have been noted that ray-tracing by the exact standard equations is a very simple and rapid operation. After some practice, when no time is wasted in thinking about each successive step, five or six minutes of time will be found to be

required to trace one ray through one surface—rather more for a marginal ray, but less for a paraxial one, as the logs of r and of N/N' are simply copied from the adjoining marginal column. The time stated includes the usual checking of additions and subtractions of logs, numbers, and angles, which should never be omitted if no independent check is to be employed.

The best method for this purely arithmetical checking is to check an addition of say $a+b = c$ by subtracting b from c and seeing that 'a' results, and a subtraction, say $a-b = c$ by adding b and c and again seeing that 'a' is obtained. This is far safer than a mere repetition of the original operation, for the reason that the subconscious mind seems to tend to repeat the same error. If we have in the original addition said $3+8 = 12$ (and that is the kind of thing everybody does at times !) we are apt to say it again when the eye is turned upon the same figures within a few minutes. The reversal of each operation in checking defeats this psychological perversity.

In the checking of the logarithmic part of a standard calculation we can save time and carry the independence of the check still further by applying the above method only to the first three and again to the last three logs. The rest of the arithmetical work can then be checked in one operation by ascertaining that the third log (of $(L-r)$ sin U) together with the sixth log (of N/N') gives the ninth log (of r sin I', but also of $(L'-r)$ sin U'). This check must always be satisfied because the skipped intermediate operations consist of first a subtraction and then an addition of the same log r, which obviously cancel each other. The trick is easily acquired of doing this little check sum 'in the air' by ignoring the intervening figures : but if this is found worrying, a piece of cardboard may be cut with two projecting prongs which cover the two pairs of logs to be ignored and leave visible the three logs to be tested. This check really verifies that

$$(L-r) \sin U \,.\, N/N' = (L'-r) \sin U',$$

and is the modified use of (6) as a check referred to in the theoretical part, section [13].

As an alternative and highly effective arithmetical check we may compute (6) independently on a separate piece of paper. In the case of the first column of the last complete example we should write out, mostly copied directly :

log sin U	8·84358n	log sin U'	8·10568
+log $(L-r)$	1·53148n	+log $(L'-r)$	2·08811
+log N	0·00000	+log N'	0·18127
log $(L-r)$ sin $U \,.\, N$ =	0·37506	log $(L'-r)$ sin $U' \,.\, N'$ =	0·37506

and the obtaining of *precisely* the same result would prove the whole of the additions and subtractions in the logarithmic work to be correct.

On account of the impossibility of detecting an error in the application of the refractive index by the PA-check, either the simple third-sixth-ninth log test or the complete test by (6) should never be omitted, even when the PA-check is adopted throughout. But in the latter case it is not necessary to check the first three and the last three logs, nor the angle-register nor the numerical work on $(L-r)$ and $(L'-r)$, as the check will detect all these errors, except in the case of a long radius.

[21] As an example of a complete calculation including the *PA*-check we will choose a cemented achromatic telescope objective of a very prevalent type, with the specification

Example 2.

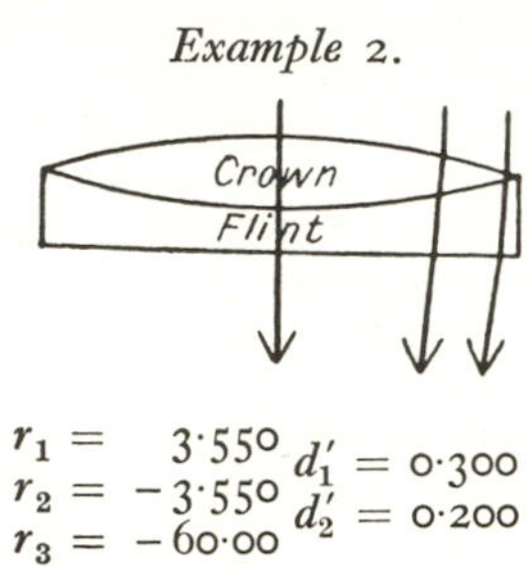

$$\begin{aligned} r_1 &= 3{\cdot}550 \quad d_1' = 0{\cdot}300 \\ r_2 &= -3{\cdot}550 \quad d_2' = 0{\cdot}200 \\ r_3 &= -60{\cdot}00 \end{aligned}$$

Clear aperture 2·00; refractive indices, for yellow-green light, of Crown = 1·5166, of Flint = 1·6256.

It may be briefly mentioned that the cementing of compound-lenses is done by a film of Canada-balsam (a variety of turpentine exuded by a Canadian tree) which is rendered hard by heating the cemented lenses for some time to a temperature of 100° – 120° Centigrade. From the strictly mathematical point of view the calculation ought to be made for the two surfaces of the film, but as the thickness of the latter is minute (a few ten thousandths of an inch) the calculation is always carried out as if the two glass-surfaces were in absolute contact.

Usually it is sufficient to trace through telescope-objectives only an extreme marginal ray—for which in this case we should have $Y = 1{\cdot}00$, i. e. half the clear full aperture—and the associated paraxial ray. But as this object-glass has very short radii compared to its aperture, and will therefore have some of the angles of incidence decidedly large, we will add an intermediate or 'zonal' ray passing at $\sqrt{0{\cdot}5} = 0{\cdot}7071$ of the full aperture. For this zonal ray we therefore have $Y = 0{\cdot}7071$. The reason for using this apparently odd value for the intermediate ray will become apparent in the later chapters. The complete calculation will then be as shown in the table on page 51.

The standard calculation† is carried out exactly as in example 1, but in the case of the first surface which receives parallel rays from the (assumed infinitely-) distant object-point, the special opening by $Y/r = \sin I$ is used. As Y takes the place of $(L-r) \sin U$, it is best to adopt the method shown in the example of entering $\log Y$ in what would usually be the third line of the logarithmic work; the standard pattern of a calculation is thus maintained as far as is possible. The same trick has been employed in the check-calculations shown side by side for a // ray and for an ordinary one. The third surface represents a case of a long radius in which only a decidedly uncertain value of the L' is obtained by the standard calculation, therefore the tracing of the rays through this surface must either be done by six-figure logs or—far preferably—it *must* be completed by the *PA*-check. Beginners

† There is one very small error in the above ray-tracing, too small to be detected by the *PA*-check and also too small to make any real difference in the final results.

especially will find it advisable to adopt this check *always*, as in the present example, for it will prevent disappointment by wrong results from unchecked calculations, provided that any disagreement between the check result and the one from the standard equations is investigated and the error located and corrected before going on to the next surface. [21]

The discussion of any such disagreement is rendered simple and definite by bearing in mind the practically constant *percentage* of uncertainty of numbers derived from logs, which was shown in section [11] to amount on an average to 1/40,000th part of the number, with a possible maximum of perhaps 1/10,000th part. Taking 1/20,000th part as reasonably possible and also as usually immaterial.

Example 2.

3—6—9 check

	First Surface.			*Second Surface.*			*Indices.*
	Marginal.	Paraxial.	Zonal.	Marginal.	Paraxial.	Zonal.	Crown.
L				9·9393	10·1215	10·0312	1·5166
$-r$				3·55	3·55	3·55	log
$L-r$				13·4893	13·6715	13·5812	0·18087
		Parallel rays.					Flint.
							1·6256
$\log \sin U$	$Y = 1$	Nominal	$Y = 0·7071$	8·99375	8·98207	8·83731	log
$+\log (L-r)$		$y = 1$		1·12999	1·13582	1·13294	0·21101
$\log (L-r) \sin U$							log
or Y	0·00000	same	9·84948	0·12374	0·11789	9·97025	relative N
$-\log r$	−0·55023	same	−0·55023	−0·55023n	−0·55023n	−0·55023n	0·03014
$\log \sin I$	9·44977	same	9·29925	9·57351n	9·56766n	9·42002n	
$+\log (N/N')$	−0·18087	same	−0·18087	−0·03014	−0·03014	−0·03014	
$\log \sin I'$	9·26890	same	9·11838	9·54337n	9·53752n	9·38988n	
$+\log r$	0·55023	same	0·55023	0·55023n	0·55023n	0·55023n	
$\log r \sin I'$	9·81913	9·81913	9·66861	0·09360	0·08775	9·94011	
$-\log \sin U'$	−8·99375	−8·98207	−8·83731	−8·85570	−8·85235	−8·70372	
$\log (L'-r)$	0·82538	0·83706	0·83130	1·23790	1·23540	1·23639	
U	0- 0- 0	0·000000	0- 0- 0	5-39-25	0·095956	3-56-33	
$+I$	16-21-40	0·281689	11-29-21	−21-59-48	−0·369539	−15-15- 2	
$U+I$	16-21-40	0·281689	11-29-21	−16-20-23	−0·273583	−11-18-29	
$-I'$	−10-42-15	−0·185738	−7-32-48	20-27-11	0·344762	14-12-20	
U'	5-39-25	0·095951	3-56-33	4- 6-48	0·071179	2-53-51	
$L'-r$	6·6893	6·8716	6·7811	17·2942	17·1949	17·2342	
$+r$	3·55	3·55	3·55	−3·55	−3·55	−3·55	
L'	10·2393	10·4216	10·3311	13·7442	13·6449	13·6842	
$-\frac{1}{2} U$	0- 0- 0		0- 0- 0	− 2-49-42		−1-58-16	
$\frac{1}{2} I$	8-10-50		5-44-40	−10-59-54		−7-37-31	
$\frac{1}{2} (I-U)$	8-10-50		5-44-40	−13-49-36		−9-35-47	
$\frac{1}{2} I'$	5-21 -8		3-46-24	−10-13-36		−7- 6-10	
$-\frac{1}{2} U'$	−2-49-43		−1-58-16	− 2- 3-24		−1-26-56	
$\frac{1}{2} (I'-U')$	2-31-25		1-48- 8	−12-17- 0		−8-33- 6	
$\log L$				0·99736	1·00524	1·00135	
$+\log \sin U$				8·99375	8·98207	8·83731	
or $\log Y$ (// light)	0·00000	0·00000	9·84948				
$+\log \sec \frac{1}{2} (I-U)$	0·00444		0·00219	0·01277		0·00612	
$\log PA$	0·00444		9·85167	0·00388		9·84478	
$+\log \operatorname{cosec} U'$	1·00625	1·01793	1·16269	1·14430	1·14765	1·29628	
$+\log \cos \frac{1}{2}(I'-U')$	9·99958		9·99979	9·98994		9·99514	
$\log L'$	1·01027	1·01793	1·01415	1·13812	1·13496	1·13620	
L'	10·2393 !	10·4215 !	10·3312 !	13·7442 !	13·6446 !	13·6836!	
$-d'$	−0·3	−0·3	−0·3	−0·2	−0·2	−0·2	
L_1	9·9393	10·1215	10·0312	13·5442	13·4446	13·4836	

[21]

Third Surface.

	Marginal.	Paraxial.	Zonal.
L	13·5442	13·4446	13·4836
$-r$	60·0000	60·0000	60·0000
$L-r$	73·5442	73·4446	73·4836
log sin U	8·85570	8·85235	8·70372
+log $(L-r)$	1·86655	1·86596	1·86619
log $(L-r)$ sin U	0·72225	0·71831	0·56991
−log r	−1·77815*n*	−1·77815*n*	−1·77815*n*
log sin I	8·94410*n*	8·94016*n*	8·79176*n*
+log N	0·21101	0·21101	0·21101
log sin I'	9·15511*n*	9·15117*n*	9·00277*n*
+log r	1·77815*n*	1·77815*n*	1·77815*n*
log r sin I'	0·93326	0·92932	0·78092
−log sin U'	−9·10322	−9·09929	−8·95092
log $(L'-r)$	1·83004	1·83003	1·83000
U	4- 6-48	0·071179	2-53-51
$+I$	−5- 2-39	−0·087128	−3-32-58
$U+I$	−0-55-51	−0·015949	−0-39- 7
$-I'$	8-13- 2	0·141635	5-46-34
U'	7-17-11	0·125686	5- 7-27
$L'-r$	67·614	67·613	67·608
$+r$	−60·000	−60·000	−60·000
L'	7·614	7·613	7·608

Checks.	Marginal.	Paraxial.	Zonal.
$-\frac{1}{2}U$	−2- 3-24		−1-26-56
$\frac{1}{2}I$	−2-31-20		−1-46-29
$\frac{1}{2}(I-U)$	−4-34-44		−3-13-25
$\frac{1}{2}I'$	−4- 6-31		−2-53-17
$-\frac{1}{2}U'$	−3-38-36		−2-33-44
$\frac{1}{2}(I'-U')$	−7-45- 7		−5-27- 1
log L	1·13175	1·12855	1·12981
+log sin U	8·85570	8·85235	8·70372
+log sec $\frac{1}{2}(I-U)$	139		69
log PA	9·98884		9·83422
+log cosec U'	0·89678	0·90071	1·04908
+log cos $\frac{1}{2}(I'-U')$	9·99601		9·99803
log L'	0·88163	0·88161	0·88133
L'	7·6143	7·6140	7·6090
$l'-L'$ marginal =		−0·0003	
$l'-L'$ zonal =		+0·0050	
log nominal y		0·00000	
−log final u'		−9·09929	
log f'		0·90071	
Equivalent focal length f' =		7·9563	

we can discuss any disagreement thus : In the standard calculation $(L'-r)$ is the number directly obtained from the logarithmic calculation. Therefore, if the disagreement in the two L' values is less than 1/20,000th part of the $(L'-r)$ of the standard calculation the latter may be accepted as correct. By the check we obtain L' itself as the direct result, and if the disagreement is also less than 1/20,000th part of L', the latter may also be accepted. A definite decision can thus nearly always be arrived at for the standard calculation ; but if the radius is long compared to L' and therefore $(L'-r)$ large compared to L', then a disagreement amounting to 1/20,000th part of $(L'-r)$ may be a very much larger fraction of L' without proving the check L' either right or wrong. The inherent greater precision of the check L' then depends absolutely on the numerical accuracy of the check calculation which in these cases of a long radius therefore requires all the care that can be bestowed upon it in order to avoid computing slips.

The *PA*-check should *invariably* be *preceded* by the 3rd-6th-9th log check as previously described and as indicated to the left of the calculation for the first surface ; otherwise a wrong result might still slip through.

Special care is also required in doing the apparently simple operation $(L_k'-d_k')=L_{k+1}$ by paying close attention to signs and ascertaining that the correct thickness of the particular lens or space is introduced, for an error in this transfer is not discovered by any of the checks already referred to. A practically infallible check of the transfer consists in *also* calculating $(L_{k+1}-r_{k+1})$ directly from $(L_k'-r_k)$ by the easily proved relation (really only an altered sequence of operations)

$$L_{k+1}-r_{k+1} = (r_k-r_{k+1}-d_k') + (L_k'-r_k),$$

which may also be employed regularly in unchecked calculations because it saves a few lines in the standard scheme.

At the end of the calculation for the third surface of our telescope objective the [21] paraxial equivalent focal length is determined by the equation $f' = \text{initial } y/\text{final } u'$, which will be proved in a later section of this chapter.

Example 2a. In preparation for the study of chromatic aberration repeat the calculation of the objective specified above for red light (*C*-line of the spectrum) and blue light (F-line of the spectrum), the indices to be used being

	C-line.	*F*-line.
Crown	1·51263	1·52080
Flint	1·61746	1·63466

The same three rays as used for yellow-green light in the specimen-calculation should be traced.

The results of these calculations will be discussed and added to in following chapters.

ALTERNATIVE CHECK-METHOD

The results from the standard computing formulae may be checked and simultaneously freed from the uncertainty which arises when a long radius occurs by the following method which is more laborious but, by way of compensation, even more certain than the *PA*-check. It depends on the independent determination of the incidence-height Y of a ray at each surface, and on obtaining L' from this.

At a first surface Y is very usually given or else can easily be computed, and the depth of curvature X is obtainable with equal facility. Reference to Fig. 12 then shows that for the first surface we have

$$L'_k = Y_k \,.\, \text{cotan}\, U'_k + X_k$$

and this is the check-value of L' on which the alternative method depends.

For the second and all following surfaces we determine the new Y from the preceding one by the distance $P_k P_{k+1} = D'_k$ and the slope $U'_k = U_{k+1}$ of the ray. Having completed the tracing of the ray through the new surface by the standard formulae, we calculate for the check

$$X_{k+1} = 2\, r_{k+1} \,.\, \sin^2 \tfrac{1}{2}\,(U_{k+1} + I_{k+1}) \qquad \text{by (8)e ;}$$

then

$$D'_k = (d'_k + X_{k+1} - X_k)\, \sec U_{k+1} \qquad \text{by (9) ;}$$

from this, by Fig. 12,

$$Y_{k+1} = Y_k - D'_k \,.\, \sin U_{k+1}$$

and then

$$\text{check : } \; L'_{k+1} = Y_{k+1} \,.\, \text{cotan}\, U'_{k+1} + X_{k+1}.$$

This should agree within the precision of the calculations with the L' from the standard equations. As in the case of the *PA*-check, the check-L' should be used in carrying the calculation forward to the next surface. The alternative check is superior to the simpler *PA*-check because it also checks the correctness of the allowance for the axial thickness d' in the usual transfer from surface to surface: any mistake in this will cause the check-L' to disagree with the L' from the standard formulae. Moreover, we obtain the D' of the marginal ray. If insufficient thickness

[21] has been assigned to any convex lens or airspace, this D' will come out negative, and this being obviously inadmissible, forcible attention is called by the alternative check to such a fatal oversight.

The alternative check is directly applicable to plane surfaces, at which it becomes simplified because the X of such a surface is obviously zero.

For paraxial rays the x must be treated as infinitely small and the sec u as exactly one. The companion paraxial formulae therefore are

$$\text{Alternative paraxial check:} \quad \begin{cases} y_{k+1} = y_k - d'_k \,.\, u_{k+1} \\ l'_{k+1} = y_{k+1}/u'_{k+1}. \end{cases}$$

In the simpler types of optical calculations the alternative check causes more work than the PA-check, chiefly on account of the X and D' which have to be determined in its course.

In Part II a highly convenient and accurate method of determining the chromatic aberration will be given which depends on the D' of the marginal ray. When that method is adopted, as it is almost sure to be, then the alternative check-method will actually save time in addition to being even more dependable than the PA-check.

The same remarks apply to certain calculations referring to the astigmatism of photographic objectives, and when these more advanced studies are reached it will be well worth while seriously to consider the adoption of the alternative check.

As an introduction to a numerical example we will collect formulae and special instructions as follows:

For a ray at finite angles:

$$\begin{aligned} X_{k+1} &= 2\, r_{k+1} \,.\, \sin^2 \tfrac{1}{2}\,(U_{k+1} + I_{k+1}) \\ D'_k &= (d'_k + X_{k+1} - X_k) \sec U_{k+1} \\ Y_{k+1} &= Y_k - D'_k \,.\, \sin U_{k+1} \\ L'_{k+1} &= Y_{k+1} \,.\, \text{cotan}\, U'_{k+1} + X_{k+1}. \end{aligned}$$

For a paraxial ray:

$$\begin{aligned} y_{k+1} &= y_k - d'_k \,.\, u_{k+1} \\ l'_{k+1} &= y_{k+1}/u'_{k+1}. \end{aligned}$$

At a first surface ($k = 0$) there is no D'_k to be computed, and the initial Y and y are either given or are computed as

$$\text{initial } Y = (L - X) \tan U\,; \quad \text{initial } y = l \,.\, u \text{ with } u = \sin U.$$

Note as highly important that when a marginal and associated paraxial ray are traced from an axial object-point at finite distance with the usual $u = \sin U$, the y of the paraxial ray will, with highly improbable exceptions, come out different from the Y of the marginal ray, on account of the tangent of U and the presence of X in the formula for initial Y. The initial y must be put equal to the initial Y only when the entering pencil consists of parallel rays from an infinitely distant axial object-point. In cases when the initial Y is given, but when the object-point is located at a finite distance, the U of the marginal ray must first be computed by one of the solutions in section [16], the first being the best for the present purpose, and then u must be fixed as $=\sin U$ and y calculated by $y = l \,.\, u$.

As a short example we will apply this method of checking to the calculation of the biconvex lens for associated marginal and paraxial rays in section [20], with use of the data of that calculation.

First Surface :	*Marginal.*		*Paraxial.*
$\log 2$	0·30103		
$+\log r_1$	1·00000		
$+2\log\sin\frac{1}{2}(U_1+I_1)$	7·85594		
$\log X_1$	9·15697	all x = *Zero.*	
$L_1 =$	-24·0000	l_1	-24·000
$-X_1 =$	$-\;$0·1435		
$L_1 - X_1 =$	-24·1435		
$\log(L_1 - X_1) =$	1·38280n	$\log l_1$	1·38021n
$+\log\tan U_1 =$	8·84464n	$+\log u_1$	8·84358n
$\log Y_1 =$	0·22744	$\log y_1$	0·22379
$Y_1 =$	1·68826	$y_1 =$	1·67413
$\log Y_1 =$	0·22744	$\log y_1 =$	0·22379
$+\log\cot\mathrm{an}\, U_1' =$	1·89428	$+\mathrm{colog}\, u_1' =$	1·95175
$\log Y_1\cot\mathrm{an}\, U_1' =$	2·12172	$\log l_1' =$	2·17554
Number : $=$	132·349		
$+X_1 =$	0·144		
$L_1' =$	132·493	$l_1' =$	149·810

It will be noticed that both values are to all intents and purposes identical with those found before. Paraxial quantities have been alined with corresponding marginal ones as far as possible. In actual practice the L_1' and l_1' by the check would be used in transferring to the second surface.

The latter will represent the normal case of a calculation by the alternative check-method :

Second Surface. $U_2 + I_2 = -19-42-30.$

	Marginal.	*Paraxial.*
$\log 2$	0·30103	
$+\log r_2$	0·69897n	
$+\;2\log\sin(U_2+I_2)$	8·46671·	
$\log X_2 =$	9·46671n	
$d_1' =$	0·6000	
$+X_2 =$	$-$0·2929	
$d_1' + X_2 =$	0·3071	
$-X_1 =$	$-$0·1435	
$d_1' + X_2 - X_1 =$	0·1636	

[21]

Marginal.		Paraxial.	
$\log (d_1' + X_2 - X_1) =$	9·21378		
$+\log \sec U_2 =$	0·00004		
$\log D_1' =$	9·21382	$\log d_1' =$	9·77815
$+\log \sin U_2 =$	8·10568	$+\log u_2 =$	8·04825
$\log D_1' . \sin U_2 =$	7·31950	$\log d_1' . u_2 =$	7·82640
$Y_1 =$	1·68826	$y_1 =$	1·67413
$-D_1' . \sin U_2 =$	−0·00209	$-d_1' . u_2 =$	−0·00671
$Y_2 =$	1·68617	$y_2 =$	1·66742
$\log Y_2 =$	0·22690	$\log y_2 =$	0·22205
$+\log \text{cotan } U_2' =$	0·66132	$+\text{colog } u_2' =$	0·72191
$\log Y_2 \text{ cotan } U_2' =$	0·88822	$\log l_2' =$	0·94396
$Y_2 \text{ cotan } U_2' =$	7·7307		
$+X_2 =$	−0·2929		
$L_2' =$	7·4378	$l_2' =$	8·7894

Again both results agree within 0·0001 with those by the standard formulae and prove the correctness of both calculations.

In calculating the X, note once for all that X always has the sign of r, for as r is only multiplied by 2 and by the square of a sine, both of which are necessarily positive, the sign is ruled by r. Errors in the sign of X introduce particularly awkward small discrepancies into the calculation, and should therefore be carefully avoided. Always make sure that D' comes out positive; a negative sign can only result for a convex lens or space and then means that the two surfaces form a sharp edge at less than the required aperture.

Like the *PA*-check, the present one does not and cannot disclose an error in the application of the refractive index in finding I' from I and i' from i. Great care must therefore always be taken to ascertain that the correct index has been put in and also, *never to be forgotten*, the third-sixth-ninth log check should be used to guard against errors in the arithmetical work.

[22] As an example of the employment of the special formulae for plane surfaces we will take the telescope objective in the previous section on the assumption that the slightly convex last surface were replaced by a plane, the thickness of the flint lens remaining unchanged. We can then take over from the previous calculation

	Marginal.	Paraxial.	Zonal.
L_3	13·5442	13·4446	13·4836
U_3	4–6–48	·071179	2–53–51
$\log \sin U_3$	8·85570	8·85235	8·70372

We have to compute

Pl. (2) $\sin U' = \sin U . N/N'$ $u' = u . N/N'$

and by the first method

Pl. (3) $L' = L . \tan U . \text{cotan } U'$ $l' = l . N'/N$

to which we will add as the simplest check [22]

Pl. Check : $L' = L - L \cdot \sec U \cdot \sin(U' - U)/\sin U'$; $l' = l - l(N - N')/N$.

With a refraction from flint glass into air we then have $N' = 1$ and $N = 1{\cdot}6256$, or $\log N/N' = 0{\cdot}21101$, and the calculation is :

	Marginal.		Paraxial.	Zonal.
U_3	4–6–48		0·071179	2–53–51
$\log \sin U_3$	8·85570		8·85235	8·70732
$+\log N_3/N_3'$	0·21101		0·21101	0·21101
$\log \sin U_3'$	9·06671		9·06336	8·91473
U_3'	6–41–46		0·115707	4–42–49
$\log L_3$	1·13175		1·12855	1·12981
$+\log \tan U_3$	8·85682	[$-\log N_3/N_3'$ –	0·21101]	8·70427
$+\log \operatorname{cotan} U_3'$	0·93032			1·08379
$\log L_3'$	0·91889		0·91754	0·91787
L_3'	8·2964		8·2707	8·2769

Naturally, all the final L' are longer than for the previous convex last surface. For the check we form

	Marginal		Paraxial	Zonal
$U_3' - U_3$	2–34–58			1–48–58
and		$N - N' =$	0·6256, and calculate :	
$\log L_3$	1·13175	$\log l_3$	1·12855	1·12981
$+\log \sec U_3$	0·00112	$+\log (N - N')$	9·79630	0·00056
$+\log \sin (U' - U)$	8·65382	$+\operatorname{colog} N$	9·78899	8·50095
$+\operatorname{colog} \sin U_3'$	0·93329			1·08527
$\log (L - L')$	0·71998		0·71384	0·71659
$L - L'$	5·2478		5·1742	5·2070
L	13·5442		13·4446	13·4836
L'	8·2964		8·2704	8·2766

The check-L' agrees in every case within 1/20,000th part with the direct calculation and the latter may therefore be accepted as correct.

As was stated in [15], it is quicker, safer, and more accurate to use Pl. (3)* for calculations of rays at finite angles through plane surfaces. The formula is

$$L' = L(N'/N) \cdot (\sec U/\sec U')$$

and by comparison with the paraxial formula $l' = l \cdot N'/N$ it is obvious that the final factor may be looked upon as a *correction of the relative index*, and as by the law of refraction the greater index corresponds to the smaller angle, it is easily seen that the correction by the secant ratio always increases the relative index provided the latter is taken, as advised in [9], always as greater divided by smaller index. For these reasons we can make an absolutely mechanical operation of the working out of Pl. (3)* by taking the logs of the secants of U and U', forming their difference and adding this to log (relative index) ; moreover, the equations show that the log of the index is always applied with opposite sign in the U and the L calculations respectively.

[22] In our present example we work thus :

	Marginal.	*Zonal.*
log sec U_3	0·00112	0·00056
log sec U_3'	0·00297	0·00147
Difference	0·00185	0·00091
(log (relative index)	0·21101	0·21101
log (index for Pl. (3)*)	0·21286	0·21192
log L_3	1·13175	1·12981
log L_3'	0·91889	0·91789
L_3'	8·2964	8·2773

The greater simplicity is self-evident. The considerable disagreement of the zonal result with the previous check is due to accidental cumulative error in the two in opposite directions ; the highly precise check given in [15] leads to 8·2769 as the true result.

We will choose a final example which brings into use most of the formulae and processes which have not yet been demonstrated. In many cases our usual five-figure logs give an entirely unnecessary precision ; it is for instance almost inconceivable that such precise results could be of any *practical* value in the case of the simple bi-convex lens of example (1). In such cases four-figure logs are indicated and save much time while diminishing the risk of numerical errors. The most efficient way of carrying out this type of calculation is to take the four-figure logs from the six-figure table and to avoid all interpolations by taking all natural numbers with the five significant figures given by the table, and all angles in the same way to the nearest 10-second value. This method is used in the example which follows. As another new feature we will choose a right-to-left ray-tracing through the bi-convex lens of example (1), stipulating that the object-point shall be placed at the previously found paraxial focus having $l_2' = 8{\cdot}7895$, so that we now also have $L_2' = 8{\cdot}7895$. The marginal ray is to enter the adjacent second surface of the lens at the incidence height $Y_2 = 1{\cdot}259$, and is to have a paraxial ray associated with it. As incidence height and object-distance are the direct data, we must begin by finding the initial U_2', and as we have a right-to-left tracing, we must transcribe the solutions given in section [16] by exchanging 'plain' and 'dashed' symbols. By the approximate method, which will be decidedly inaccurate in the present case because Y_2 is so large and at the same time r_2, decidedly short, we find $\sin U_2' = Y_2/L_2'$ or

$$\begin{aligned} \log Y_2 &= 0{\cdot}1000 \\ -\log L_2' &= -0{\cdot}9440n \\ \log \sin U_2' &= 9{\cdot}1560n \text{ (approximately).} \end{aligned}$$

By the first exact solution we have to calculate

$$\sin (U_2'+I_2') = Y_2/r_2\,;\;\; X_2 = Y_2 \,.\, \tan \tfrac{1}{2}\,(U_2'+I_2')\,;$$

$$\text{then } \tan U_2' = Y_2/(L_2'-X_2),$$

and as $r_2 = -5{\cdot}000$, $Y_2 = 1{\cdot}259$, $L_2' = 8{\cdot}7895$, we obtain : [22]

$\log Y_2$	$= 0{\cdot}1000$	$\log Y_2$	$= 0{\cdot}1000$	L_2'	$= 8{\cdot}7895$	$\log Y_2$	$= 0{\cdot}1000$
$-\log r_2$	$= -0{\cdot}6990n$	$+\log \tan \frac{1}{2}(U_2' + I_2')$	$= 9{\cdot}1071n$	$-X_2$	$= +0{\cdot}1611$	$-\log (L_2' - X_2)$	$= -0{\cdot}9519$
$\log \sin (U_2' + I_2')$	$= 9{\cdot}4010n$	$\log X_2$	$= 9{\cdot}2071n$	$L_2' - X_2$	$= 8{\cdot}9506$	$\log \tan U_2'$	$9{\cdot}1481$
$U_2' + I_2'$	$= -14-35-0$					U_2'	$= 8-0-20$

By the second exact solution we calculate $\sin (U_2' + I_2')$ as above. Then $\tan \frac{1}{2}(I_2' - U_2') = \tan \frac{1}{2}(I_2' + U_2') \,.\, (L_2' - 2\,r_2)/L_2'$ and

$$U_2' = \tfrac{1}{2}(I_2' + U_2') - \tfrac{1}{2}(I_2' - U_2').$$

$L_2' =$	$8{\cdot}7895$	$\log \tan \frac{1}{2}(I_2' + U_2')$	$= 9{\cdot}1071n$	$\frac{1}{2}(I_2' + U_2')$	$= -\ 7-17-30$
$-2\,r_2$	$= +10{\cdot}0000$	$+\log (L_2' - 2\,r_2)$	$= 1{\cdot}2739$	$-\frac{1}{2}(I_2' - U_2')$	$= 15-17-50$
$L_2' - 2\,r_2 =$	$18{\cdot}7895$	$+\text{colog}\, L_2'$	$= 9{\cdot}0560$	U_2'	$= 8-0-20$
		$\log \tan \frac{1}{2}(I_2' - U_2')$	$= 9{\cdot}4370n$	in agreement with the first result.	

There is not much to choose between these two solutions. For either of them the subsequent ray-tracing supplies a check, as the $U_2' + I_2'$ of the angle register must agree closely with that found in the course of the solution.

With $l_2' = L_2' = 8{\cdot}7895$, $U_2' = 8–0–20$, and $u_2' = \sin U_2'$ the calculation now is as follows :

(3–6–9 check)

	Second surface, into *denser* medium.		First surface, into *lighter* medium.	
	Marginal.	Paraxial.	Marginal.	Paraxial.
L	8·7895		962·10	149·70
$-r$	+5·0000		−10·00	−10·00
$L' - r$	13·7895	same	952·10	139·70
$\log \sin U'$	9·1439		7·1169	7·9143
$+\log (L' - r)$	1·1395		2·9787	2·1452
$\log (L' - r) \sin U'$	0·2834		0·0956	0·0595
$-\log r$	−0·6990*n*		−1·0000	−1·0000
$\log \sin I'$	9·5844*n*	same	9·0956	9·0595
$+\log N'/N$	−0·1813		+0·1813	+0·1813
$\log \sin I$	9·4031*n*		9·2769	9·2408
$+\log r$	0·6990*n*		1·0000	1·0000
$\log r \sin I$	0·1021	0·1021	0·2769	0·2408
$-\log \sin U$	−7·1169	−7·9143	−8·8065*n*	−8·7094*n*
$\log (L - r)$	2·9852	2·1878	1·4704*n*	1·5314*n*
U'	8- 0-20	0·13928	0- 4-30	0·00821
$+I'$	−22-35-10	−0·38406	7- 9-30	0·11468
$U' + I'$	−14-34-50	−0·24478	7-14- 0	0·12289
$-I$	+14-39-20	+0·25299	−10-54-20	−0·17410
U	+0- 4-30	+0·00821	−3-40-20	−0·05121
$L - r$	966·50	154·10	−29·539	−33·994
$+r$	−5·00	−5·00	+10·000	+10·000
L_2	961·50	149·10	−19·539	−23·994
$+d_1'$	+0·60	+0·60		
L_1'	962·10	149·70		

In the ray-tracing we have now to employ the usual formulae with exchange of 'plain' and 'dashed' letters, so that we have to calculate in succession

$$\sin I' = \sin U'(L' - r)/r\ ;\ \ \sin I = \sin I' \,.\, N'/N\ ;\ \ U = U' + I' - I\ ;$$
$$L - r = \sin I' \,.\, r/\sin U\ ;\ \ L = (L - r) + r\ ;$$

[22] and in the transfer to the next surface we must *add* the usual positive thickness d' because the new surface lies to the left.

We will here add and use a rule which can be adhered to in all ray-tracing, both in the left-to-right and in the right-to-left direction. It follows from the law of refraction that the angle of the ray with the normal is diminished on entering a denser medium and increased on entering a lighter medium. Hence an infallible rule that, if we always use the relative index as advised in section [9], namely greater index divided by smaller index, then the (positive) log of the relative index must be *added* if the refraction is *into a lighter medium* and the log of the relative index must be subtracted for a refraction into a denser medium.

We should expect as the paraxial l_1 the former value -24. The small discrepancy represents the accumulated rounding off errors of the four-figure calculation, and as it amounts to only 1/5600th of the $(l_1 - r_1)$ no question of a possible numerical error can arise. The accumulated error might easily reach 1/2000th of $(l_1 - r_1)$ with a four-figure calculation.

SIMPLE LENSES

[23] The exact ray-tracing formulae with which we have dealt up to now find their chief application in the later stages of the evolution of new optical designs. A very large amount of practically useless labour would be required if we attempted to test every rough preliminary scheme by these formulae and thus to feel our way, largely in a more or less blindfold manner, to the finished instrument. Progress towards the ultimate perfection will be quickest if we begin with a rough first approximation and gradually apply more refined methods as the work proceeds. In the majority of cases we can arrive at such a simple scheme most quickly and with very little trouble by determining what combination of thin simple lenses would produce the intended effect. Therefore we must now deduce suitable formulae which enable us to calculate the properties of simple lenses.

Equation (6p)** can be applied to the two surfaces of a simple lens so as to yield a single expression for its focal effects.

Referring to Fig. 18, we shall have by the transfer-equations (5p)* $l_2 = l_1' - d_1'$ and $N_1' = N_2$, therefore

for the first surface $\quad \dfrac{N_1'}{l_1'} = \dfrac{N_1' - N_1}{r_1} + \dfrac{N_1}{l_1} \quad \Big| \quad \times \dfrac{l_1'}{l_1' - d_1'}$

for the second surface $\quad \dfrac{N_2'}{l_2'} = \dfrac{N_2' - N_1'}{r_2} + \dfrac{N_1'}{l_1' - d_1'}.$

If the first equation is multiplied throughout by the stated factor its left side becomes equal to the last term on the right of the second equation ; these two terms therefore cancel each other when the second equation is added to the extended first, with the result

$$\frac{N_2'}{l_2'} = \frac{N_1' - N_1}{r_1} \cdot \frac{l_1'}{l_1' - d_1'} + \frac{N_2' - N_1'}{r_2} + \frac{N_1}{l_1} \cdot \frac{l_1'}{l_1' - d_1'}.$$

Taken together with the equation for the first surface, this is the most general, but decidedly inconvenient, solution of the problem of a thick lens. The equation becomes simpler if we assume that $N_1 = N_2'$, that is, that the lens is employed immersed in a medium of index N_1. [23]

We then have

$$\frac{N_1}{l_2'} = \frac{N_1' - N_1}{r_1} \cdot \frac{l_1'}{l_1' - d_1'} + \frac{N_1 - N_1'}{r_2} + \frac{N_1}{l_1} \cdot \frac{l_1'}{l_1' - d_1'}$$

or on division throughout by N_1

$$\frac{1}{l_2'} = \left(\frac{N_1'}{N_1} - 1\right)\left(\frac{1}{r_1} \cdot \frac{l_1'}{l_1' - d_1'} - \frac{1}{r_2}\right) + \frac{1}{l_1'} \cdot \frac{l_1'}{l_1' - d_1'}.$$

Here N_1'/N_1 is the relative refractive index of the material of the lens referred to the surrounding medium and using the simple symbol N for this index, we find the solution

$$\text{(11p)} \qquad \frac{1}{l_2'} = (N - 1)\left(\frac{1}{r_1} \cdot \frac{l_1'}{l_1' - d_1'} - \frac{1}{r_2}\right) + \frac{1}{l_1} \cdot \frac{l'_1}{l_1' - d_1'}$$

in which

$$\frac{N}{l_1'} = \frac{N - 1}{r_1} + \frac{1}{l_1'}.$$

The second equation following on dividing the original equation of the first surface by N_1 and introducing $N = N_1'/N_1$.

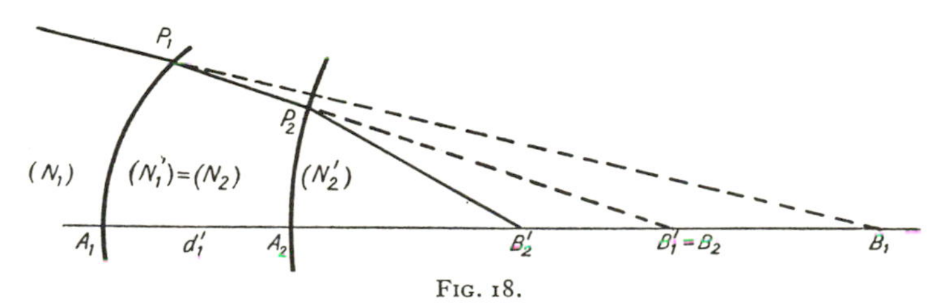

FIG. 18.

The equation will be almost exclusively applied to lenses used in air, and N will then be the ordinary refractive index of the glass. But it should be remembered that the equation can be applied to a lens which, together with its object and image, is immersed in a medium other than air, provided that N is then taken as the relative index of the glass to the medium.

(11p) is an exact paraxial equation : it will be modified on a subsequent occasion so as to render it a useful solution of the problem of a thick lens. For the present we merely use it as a means to deduce the simplest possible equations for use in rough and rapid preliminary studies of lenses and lens systems by restricting the equations to very thin lenses in which d_1' may be treated as negligible in comparison to l_1'. The inconvenient factor $l_1'/(l_1' - d_1')$ in (11p) then becomes $= 1$, but as this would be strictly correct only for an infinitely thin lens, we must note once for all that the convenient equations obtained on this assumption are (with certain rare exceptions) only a more or less rough approximation when applied to actual lenses, and too much dependence must not be placed on the results obtained

[23] by them. As a constant reminder of the inaccuracy of these equations we will distinguish them by putting TL (= thin lens) in front of the numbers assigned to them. The fundamental equation is found by omitting the factor $l_1'/(l_1'-d_1')$ from (11p) as

$$\text{TL (1)} \qquad \frac{1}{l_2'} = (N-1)\left(\frac{1}{r_1} - \frac{1}{r_2}\right) + \frac{1}{l_1}.$$

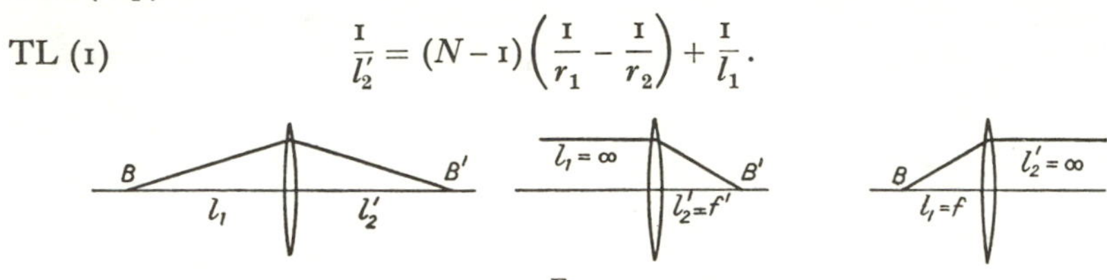

FIG. 19.

This formula is valid for all possible values of the data combined in it, and therefore covers the paraxial properties of all thin lenses, when they are used in air as well as when they are immersed in a denser medium also containing the object and image, provided that we use the proper value of the relative index N in each case. It is easily calculated, especially with the aid of a table of reciprocals, and as its accuracy is limited by the neglecting of the thickness, the slide-rule is usually sufficient for the rest. The equation is rendered still more convenient for most applications by introducing the focal lengths. If we put $l_1 = \infty$ we obtain as the value of l_2' the focal length in the space to the right of the lens as

$$\frac{1}{f'} = (N-1)\left(\frac{1}{r_1} - \frac{1}{r_2}\right),$$

and by putting $l_2' = \infty$ and solving for l_1 we find the focal length in the space to the left of the lens by

$$0 = (N-1)\left(\frac{1}{r_1} - \frac{1}{r_2}\right) + \frac{1}{f} \quad \text{as} \quad \frac{1}{f} = -(N-1)\left(\frac{1}{r_1} - \frac{1}{r_2}\right) = -\frac{1}{f'}.$$

For a thin lens the two focal lengths are therefore numerically equal but of opposite sign.

f', the usually employed focal length of lenses and systems in the space to the right, comes out positive for a convex and negative for a concave lens, in accordance with the universal custom.

The reciprocal of a radius is usually called the curvature of the corresponding surface and it is convenient to have a simple symbol for this. We choose small c with suitable *number*-suffix for this reciprocal of a radius and can then write

$$\frac{1}{f'} = (N-1)(c_1 - c_2) = -\frac{1}{f}.$$

It is obvious from the formula that the focal length of a lens depends solely on the *difference* of the two curvatures, and this difference is therefore the really decisive value. It is convenient to take advantage of this in order still further to simplify the formulae for thin lenses and we adopt plain small c for it, calling this the total or net curvature of a lens. We now further stipulate that in order to distinguish the total curvature of a lens from the individual curvatures of its surfaces, [23] which latter have already been assigned a number-suffix, we shall use letter-suffixes for the data of complete lenses when several are used in combination, the letters being used in alphabetical order from left to right (just like the number-suffixes of surfaces). The same letter-suffixes will be used for the refractive indices and focal lengths of lenses, and as with these thin lenses the intermediate intersection values are of no interest, we shall drop the numerical suffixes of the l, using plain l and l' for a single lens and the proper letter-suffix to l and l' for a combination of thin lenses. The general formulae already deduced will then be

$$\text{TL (1)} \qquad \frac{1}{l'} = (N-1)\left(\frac{1}{r_1} - \frac{1}{r_2}\right) + \frac{1}{l} = (N-1)(c_1 - c_2) + \frac{1}{l} = (N-1)\,c + \frac{1}{l};$$

$$\text{TL (2)} \qquad \frac{1}{f'} = -\frac{1}{f} = (N-1)\left(\frac{1}{r_1} - \frac{1}{r_2}\right) = (N-1)(c_1 - c_2) = (N-1)\,c;$$

and using these values to introduce the focal lengths into TL (1) we find its most useful form as

$$\text{TL (3)} \qquad \frac{1}{l'} = \frac{1}{f'} + \frac{1}{l} = -\frac{1}{f} + \frac{1}{l}.$$

For two separate lenses used in combination we shall write for the lens at the left of the combination

$$\frac{1}{f'_a} = -\frac{1}{f_a} = (N_a - 1)\left(\frac{1}{r_1} - \frac{1}{r_2}\right) = (N_a - 1)(c_1 - c_2) = (N_a - 1)\,c_a$$

$$\frac{1}{l'_a} = \frac{1}{f'_a} + \frac{1}{l_a} = -\frac{1}{f_a} + \frac{1}{l_a};$$

and for the lens at the right of the combination

$$\frac{1}{f'_b} = -\frac{1}{f_b} = (N_b - 1)\left(\frac{1}{r_3} - \frac{1}{r_4}\right) = (N_b - 1)(c_3 - c_4) = (N_b - 1)\,c_b$$

$$\frac{1}{l'_b} = \frac{1}{f'_b} + \frac{1}{l_b} = -\frac{1}{f_b} + \frac{1}{l_b}.$$

We shall hardly ever have more than three or four individual lenses to be discussed in any one lens problem, and there will therefore never be any doubt as to the proper co-ordination of the letter suffixes of the lens-data and the number-suffixes of radii and curvatures.

The fact that the focal properties of a thin lens depend directly on the total curvature c and only indirectly on the individual curvatures of the surfaces is of immense importance in optical designing. As $c = c_1 - c_2$, it follows that when the total curvature c has been determined so as to secure a prescribed focal length with a given kind of glass, we have still perfect freedom to choose the curvature of one of the lens-surfaces; for $c = c_1 - c_2$ gives by transposition $c_1 = c + c_2$ or $c_2 = c_1 - c$, so that for any selected curvature of one surface we can find that curvature of the other surface which will secure the required total curvature. The

[23] nature of this process becomes clear if we assume that we had found c_1 and c_2 as two particular curvatures which satisfied the equation $c = c_1 - c_2$ for a given thin lens. Evidently, if by x we denote any positive or negative number, (c_1+x) and (c_2+x) would equally fulfil the equation, for $(c_1+x)-(c_2+x) = c_1-c_2$. All the possible lenses of a prescribed total curvature can therefore be derived from any one particular form by increasing or diminishing both individual curvatures (algebraically) by the same amount. This would produce the same effect as if we had rendered the original lens plastic and then bent it all around without transferring any material from centre towards edge or vice versa.

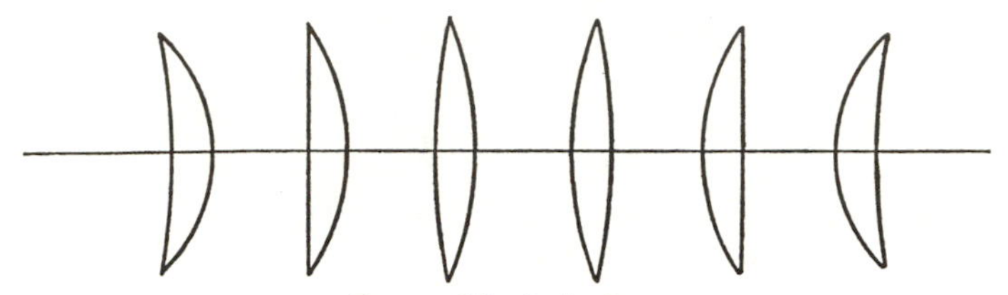

FIG. 20. *'Bending' a lens.*

For that reason the accepted terminus technicus for this choosing of various shapes for a lens of prescribed total curvature is : 'Bending the lens'. It represents the most powerful method of altering the *aberrations* of pencils of finite aperture and comes into use almost exclusively in the later stages of a design. All the preliminary work only requires a knowledge of focal lengths or of total curvatures and refractive indices.

The total curvature c required to produce a lens of given focal length f' is obtained by transposing the first and last terms of TL (2) as

$$c = 1/f'(N-1)$$

and any number of 'bendings' of this lens may then be determined by assuming a succession of values of c_1 and determining the corresponding $c_2 = c_1 - c$, or by assuming c_2 and finding $c_1 = c + c_2$.

In other cases the shape of the lens may be prescribed as plano-convex, equiconvex, or as having the two radii in any given ratio, say $r_2 = k \, . \, r_1$, which covers any conceivable shape by giving to k a suitable value, positive for meniscus-lenses or negative for biconvex or biconcave lenses. TL (2) in its first form then gives

$$1/f' = (N-1)(1/r_1 - 1/k.r_1),$$

and solved for r_1 gives $r_1 = (N-1)f'(k-1)/k$;

and with this $\qquad r_2 = k \, . \, r_1 = (N-1)f'(k-1)$.

For a plano-curved lens having the curve on the first surface k will be infinite and $(k-1)/k = 1$, hence by the equation for r_1

$$r_1 = (N-1)f'.$$

If the curve is on the second surface k will be zero and by the equation for r_2

$$r_2 = -(N-1)f'.$$

For an equiconvex or equiconcave lens k is $= -1$, hence [23]

$$r_1 = -r_2 = 2(N-1)f'.$$

These useful special solutions will be added to TL (2).

When a number of simple thin lenses at finite distances from one another but on the same optical axis are combined to form a lens system we can trace the light from any given object-point through the system step-by-step exactly as for the successive surfaces in calculations by the standard exact equations.

For two lenses (a) and (b) at separation d'_a and an object-point at distance l_a from the first lens we should compute first

$$1/l'_a = 1/f'_a + 1/l_a;$$

then $$l_b = l'_a - d'_a;$$

and finally $$1/l'_b = 1/f'_b + 1/l_b.$$

Obviously this process can be carried on indefinitely for any number of successive thin lenses. Convenient direct solutions for important special cases will be deduced subsequently.

Frequently, as for instance in ordinary achromatic objectives, several lenses are placed close together. If the thicknesses and axial separations of the lenses are treated as negligible, specially simple formulae then result, for, on putting d'_a in the last equations $= 0$, l_b becomes $= l'_a$, and consequently, on putting the value of $1/l'_a$ by the first equation in place of $1/l_b$ in the third, we obtain

$$1/l'_b = 1/f'_a + 1/f'_b + 1/l_a,$$

and this process also admits of indefinite repetition. For any number of thin lenses in contact we therefore have, omitting the unnecessary suffixes,

$$1/l' = 1/f'_a + 1/f'_b + \text{etc.} + 1/l$$

which may be written

$$1/l' = 1/f' + 1/l,$$

with the definition $$1/f' = 1/f'_a + 1/f'_b + \text{etc.}$$

Equation TL (3) may therefore be applied to thin lens *systems* consisting of any number of thin lenses in contact, and a note to this effect will be added to TL (3) in the collected equations.

The reciprocal of a focal length is called the 'power' of the corresponding lens and using this term, the result of the preceding deductions is that the power of a system of closely packed thin lenses is equal to the (algebraical) sum of the powers of the components and that such a system produces the same optical effect as a single lens of the same power; for the simplified formula at which we arrived is identical with TL (3). Moreover, and this should be carefully noted as a valuable means of saving time, the power of such a thin system is utterly independent of the sequence of the components, simply because any addition sum always gives the same result, no matter in what order the items are taken.

We will take the biconvex lens of sections [9] and [20] as a short numerical example, necessarily treating its thickness as negligible. We had

$$l = -24, r_1 = 10, r_2 = -5, N = 1{\cdot}518.$$

[23] Equation TL (1) applied directly gives

$$\frac{1}{l'} = (N-1)\left(\frac{1}{r_1} - \frac{1}{r_2}\right) + \frac{1}{l} = 0{\cdot}518\,(0{\cdot}1 - (-0{\cdot}2)) - \frac{1}{24} = 0{\cdot}518 \times 0{\cdot}3 - 0{\cdot}0417 =$$

$0{\cdot}1554 - 0{\cdot}0417 = 0{\cdot}1137$, or by Barlow's table of reciprocals $l' = 8{\cdot}795$, instead of the exact paraxial result 8·7895 ; the approximation is very good in spite of the considerable thickness (0·6) of the lens, but this is due to the fact that the convergence within the lens is very slight (we found $l'_1 = 149{\cdot}811$) and gives nearly unit-value for the omitted correcting factor $l'_1/(l'_1 - d'_1)$ of the exact formulae (11p). We must not take it for granted that the approximation will usually be so close.

By TL (2) we can calculate the focal length. The net curvature is found as

$$c = \frac{1}{r_1} - \frac{1}{r_2} = 0{\cdot}1 - (-0{\cdot}2) = 0{\cdot}3\,;$$

then

$$\frac{1}{f'} = 0{\cdot}518 \times 0{\cdot}3 = 0{\cdot}1554,$$

or by Barlow $f' = 6{\cdot}435$. We can then find the l' conjugate to the given $l = -24$ by TL (3)

$$\frac{1}{l'} = \frac{1}{f'} + \frac{1}{l} = 0{\cdot}1554 - 0{\cdot}0417 = 0{\cdot}1137 \text{ as before.}$$

In illustration of the 'bending' process we have for our *given* lens $c_1 = 0{\cdot}1$ and $c_2 = -0{\cdot}2$. We can find any number of bendings of our lens by adding the same amount x algebraically to c_1 and to c_2. We can thus form a little table of bendings :

If	$x =$	$-0{\cdot}2$	$-0{\cdot}1$	0	$+0{\cdot}1$	$+0{\cdot}2$	$+0{\cdot}3$
then	$c_1 =$	$-0{\cdot}1$	0	0·1	0·2	0·3	0·4
and	$c_2 =$	$-0{\cdot}4$	$-0{\cdot}3$	$-0{\cdot}2$	$-0{\cdot}1$	0	$+0{\cdot}1$;

and taking reciprocals we find the following prescriptions for lenses, all of which will give the previously calculated focal length and conjugate distances if the lenses are treated as being infinitely thin :

If	$x =$	$-0{\cdot}2$	$-0{\cdot}1$	0	$+1$	$+0{\cdot}2$	$+0{\cdot}3$
then	$r_1 =$	-10	∞	10	5	3·333	2·5
and	$r_2 =$	$-2{\cdot}5$	$-3{\cdot}333$	-5	-10	∞	10.

It will easily be verified that $x = 0{\cdot}05$ would give us an equiconvex lens with $r_1 = -r_2 = 6{\cdot}667$.

We will next discuss the linear magnification produced by a thin lens or a thin system of lenses of power $1/f'$. By the extended theorem of Lagrange we have for object and image in the same medium, as in the present case,

$$h'_k/h_1 = m' = u_1/u'_k\,.$$

Now in a thin (strictly speaking an infinitely thin) lens or system any ray entering at a distance y from the optical axis must emerge at the same distance, simply

[23] because it cannot either rise or fall by a measurable amount in traversing the negligible thickness of the lens or system, and as the whole of the present discussion is limited to the paraxial region, we therefore have

$$u = y/l,\ u' = y/l',\ \text{or}\ u/u' = l'/l.$$

Therefore, omitting the unnecessary suffixes in the general Lagrange equation for the present simple case,

TL (4) $$h'/h = m' = l'/l$$

or $$h/h' = m = l/l' \text{ for right-to-left work.}$$

On reference to Fig. 21 it is seen by similar triangles that this result agrees with the usual graphical determination of the size of an image by drawing straight lines from the ends of the object through the centre of the thin lens.

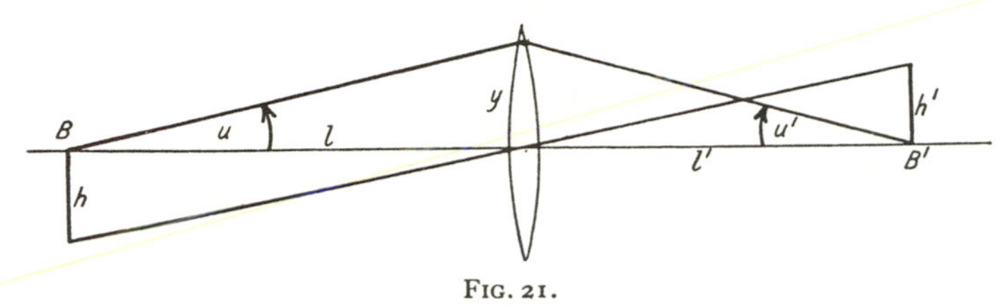

FIG. 21.

This construction shows, in agreement with the equations TL (4), that with a thin lens or system the image is erect if object and image lie on the same side of the lens (i. e. if l and l' have the same sign), and inverted if object and image lie on opposite sides of the lens.

Very convenient formulae for the direct solution of thin lens problems result from a combination of TL (4) with TL (3). TL (4) assumes that l and l' are known ; more usually f' is the known quantity and m' is prescribed. We can then solve the two equations for l and l'. From TL (4) we obtain $l' = l \cdot m'$, and putting this into TL (3) arrive at

$$1/l.m' = 1/f' + 1/l,\ \text{or}\ l = f'\ (1 - m')/m'\ ;$$

and from this, as $l' = l \cdot m'$, at $l' = f'\ (1 - m')$.

The solutions can be transposed to find the focal length which will lead to a prescribed magnification when one of the conjugate distances is given.

The distance BB' from object to image, positive if the image lies to the right of the object, is frequently the important quantity. As in Fig. 21 the distance l is negative, we read off $BB' = l' - l$ as the algebraically correct value, and putting in the above equivalents of l and l' find

$$BB' = -f'\ (1 - m')^2/m'.$$

The linear magnification of the image becomes an inconveniently small number when the object is at a great distance ; moreover in many cases (practically in all star-observations) neither the distance nor the size of the object is known, and the term linear magnification then becomes obviously meaningless. In such cases it

[23] is convenient or even necessary to gauge the object by its angular subtense as seen from the position of the lens, and we must therefore transform the equation of Lagrange so as to substitute this angle of subtense for the actual size h of the object and then to find the actual size h' of the image. For the thin lens or system we found $m' = h'/h = l'/l$ or $h' = h \,.\, l'/l$. If we now refer to Fig. 22 and call

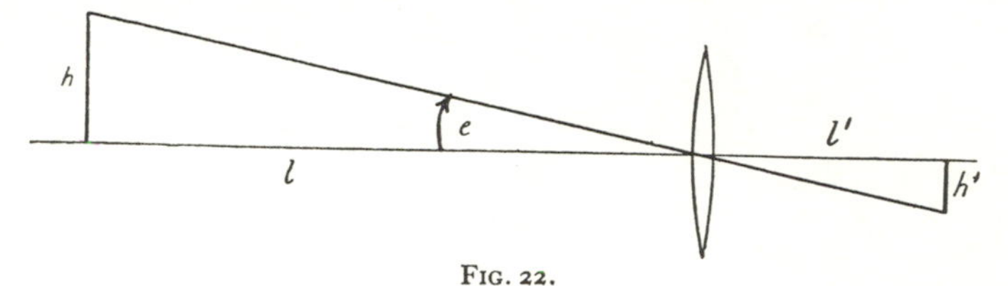

FIG. 22.

the small angle of subtense of h as seen from the lens-centre 'e' (a positive angle e is shown in Fig. 22) we see that within paraxial accuracy and regardless of sign we should have $h/l = e$; but as h is positive, l negative, whilst e must be brought out positive, we must write algebraically $h/l = -e$. (We meet here the first case of the kind referred to in establishing our sign conventions, when the contradiction between the usual optical and the analytical positive sense of angles with the axis necessitates an adjustment of the sign in an equation.) Putting this value of h/l into the previous equation for h', we find

$$h' = -e \,.\, l'.$$

This equation is valid for objects at any distance provided the subtense of the object is taken from the centre of the thin lens. But we use the equation only for objects at a great distance, in which case the image is formed at the distance $l' = f'$. Hence we shall only record the equation in the form

$$h' = -e \,.\, f'.$$

It is easily shown that for right-to-left work from a distant object to the right of the lens and subtending an angle e' we have also

$$h = -e'f = e'f', \text{ because } f = -f'.$$

We can apply the same modification to a centred lens system of any number of surfaces at any separations, with the only restriction that object and image shall both lie in air. For this case Lagrange's equation is

$$m' = h'_k/h_1 = u_1/u'_k \text{ or } h'_k = h_1 u_1/u'_k.$$

If we now draw a diagram, Fig. 23, similar to Fig. 22, but showing the outside surfaces of a thick lens system instead of a thin lens, we have as before $h_1/l_1 = -e$ or $h_1 = -e \,.\, l_1$, and for a ray at the small angle u_1 meeting the first surface at the small distance y_1 above the optical axis we shall have with paraxial accuracy $u_1 = y_1/l_1$.

Therefore $h_1 u_1 = -e \,.\, y_1$ by multiplication of the two equations; and putting this into $h'_k = h_1 u_1/u'_k$ we find for the size of the image

$$h'_k = -e \,.\, y_1/u'_k.$$

This equation, whilst valid for objects at moderate distances, is only used in practice for very distant objects. If for such very distant objects we compare it with the equation $h' = -e . f'$ deduced above for a thin lens or system, we see that with a distant object of given subtense e a lens system of any thickness pro- [23]

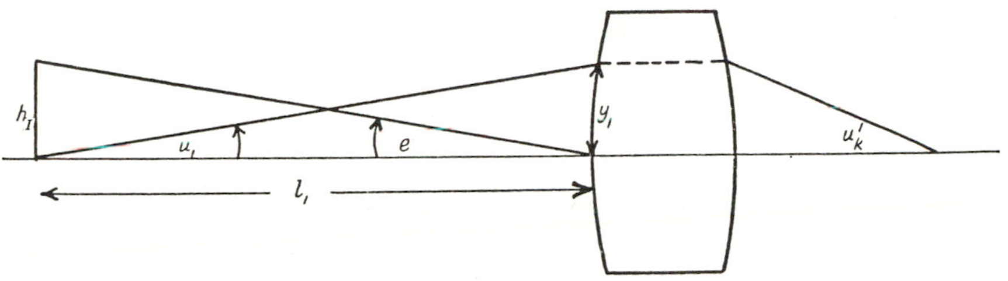

FIG. 23.

duces an image of exactly the same size as that by a thin lens of focal length $f' = y_1/u'_k$, for on putting this value into the equation found above for a thin lens, $h' = -e . f'$, the right side becomes identical with the right side of the equation for h'_k. For very distant objects the lens system therefore produces the same effect or is *equivalent* to a simple thin lens of the stated focal length. For that reason y_1/u'_k is called the equivalent focal length of the lens system, and giving it the usual symbol f' we can write the equations for a lens system applied to very distant objects in the form

$$f' = y_1/u'_k; \; h'_k = -e . f',$$

and add these to (10p)** together with their easily deduced right-to-left equivalents for parallel rays from the right

$$f = y_k/u_1 ; \; h_1 = -e' . f.$$

The equivalent focal length of a lens system has a very simple geometrical signifi-

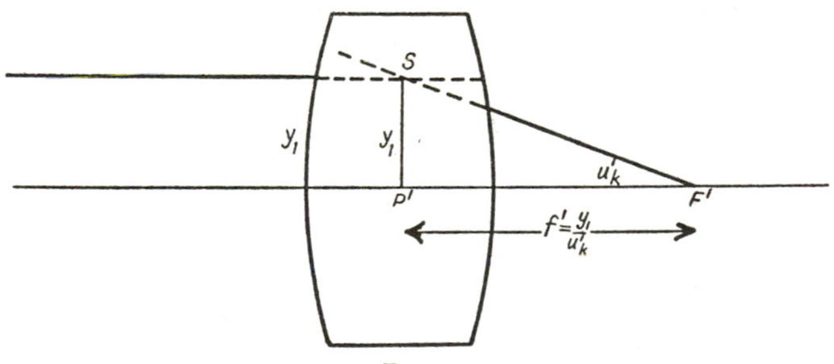

FIG. 24.

cance, for if we re-draw the last diagram with an entering ray from a very distant axial object-point and therefore parallel to the optical axis, we see on producing

[23] the entering and emerging rays to their intersection at S and on dropping the perpendicular SP' upon the axis, that, with paraxial accuracy

$$y_1/P'F' = u'_k,$$

or by transposition $$P'F' = y_1/u'_k = f'.$$

$P'F'$ therefore represents the equivalent focal length of the system.

On account of its general utility we will at once work out a convenient formula for the equivalent focal length of a combination of two thin lenses 'a' and 'b', of individual focal length f'_a and f'_b respectively, and placed at a separation d'_a from each other. Lens 'a' will focus the paraxial rays from a distant object-point at $l'_a = f'_a$. On account of the separation lens 'b' will receive these rays with an intersection length $l_b = f'_a - d'_a$. We can then apply TL (3) to the second lens and find

$$\frac{1}{l'_b} = \frac{1}{f'_b} + \frac{1}{l_b} = \frac{1}{f'_b} + \frac{1}{f'_a - d'_a},$$

which determines the so-called back focus of the combination. The equivalent focal length is defined as $f' = y_a/u'_b$, if y_a is the semi-aperture of a paraxial pencil at the two surfaces of the thin first lens and if u'_b is the angle of convergence at the final focus of that pencil. After passage through the first lens the pencil will have the convergence angle $u'_a = y_a/f'_a$, because y_a is its semi-aperture and f'_a its intersectional-length. On account of the separation the pencil will reach the second lens with a changed semi-aperture $y_b = l_b \cdot u_b$, and as $u_b \equiv u'_a$, this gives

$$y_b = l_b \cdot u'_a = l_b \cdot y_a/f'_a = y_a (f'_a - d'_a)/f'_a.$$

After refraction by the second lens the pencil will have the final convergence angle $u'_b = y_b/l'_b$, and if we introduce the determined values of y_b and $1/l'_b$ this gives

$$u'_b = y_a \frac{f'_a - d'_a}{f'_a}\left(\frac{1}{f'_b} + \frac{1}{f'_a - d'_a}\right)$$

or $$\frac{1}{f'} = \frac{u'_b}{y_a} = \frac{f'_a - d'_a}{f'_a f'_b} + \frac{1}{f'_a} = \frac{1}{f'_a} + \frac{1}{f'_b} - \frac{d'_a}{f'_a \cdot f'_b}.$$

Bringing the last expression to a common denominator gives an alternative equation for the power $1/f'$ of the combination, and inversion of this formula gives a direct formulae for f'. We therefore have the solutions of the problem :

$$\frac{1}{f'} = \frac{1}{f'_a} + \frac{1}{f'_b} - \frac{d'_a}{f'_a \cdot f'_b} = \frac{f'_a + f'_b - d'_a}{f'_a \cdot f'_b}$$

$$f' = \frac{f'_a \cdot f'_b}{f'_a + f'_b - d'_a},$$

which we shall find very useful in subsequent work.

All the equations relating to the magnification produced by a thin lens or system should be collected under TL (4).

We will conclude this chapter on the fundamental computing formulae with an extremely important deduction applicable to all of them, and in fact to practically all subsequent work. In working out any of the equations we introduce the lengths entering into them (radii, thicknesses, intersection-lengths, &c.) as simple numbers, signifying the number of units of length (millimetres, centimetres, inches, or any [23] arbitrary unit whatever) contained in each linear datum. Consequently the calculation would deal with exactly the same numbers and would give exactly the same final numerical result if we subsequently changed our mind as to the unit of length to be employed. There is therefore no need to repeat a calculation if merely a change of linear scale is involved. This means that if we have a satisfactory design say for a photographic objective of 10 inches focal length and are asked for a similar lens of say 6 inches focal length, then no new calculation will be called for; all we have to do is to reduce all linear dimensions of the original design to six-tenths and the resulting prescription will be absolutely identical with the one which we should have found by making a complete new calculation on the lines of the original one. The only reservation to be made with regard to this statement is that the residuals of aberration will also change in linear scale in the same proportion as the regular data of the lens or systems; hence if the residual aberrations are serious in the original design they may become too large if the design is carried out on a larger scale than that originally contemplated, and the working aperture or angle of field of the enlarged system may then call for a suitable diminution.

Special Memoranda from Chapter I

Theoretical. Spherical refracting surfaces can be produced with high precision and render ray-tracing very simple, as the incidence normal of any point is simply a radius of the sphere.

Centred systems of spherical surfaces have the additional advantage that there is perfect symmetry with reference to the optical axis.

A small object at right angles with, and close to, the optical axis of a centred system and at any distance is depicted by paraxial rays in any one colour at a perfectly definite distance from the system and with the linear magnification determined by the theorem of Lagrange. The location and magnification of this paraxial image form the natural standard with which the effects produced by rays at finite angles or by rays of different colour are compared.

The law of axial or longitudinal magnification will be extensively employed in the theory of aberrations.

The fact that 'bending' of a thin lens does not change its power finds numberless applications in the design of lens systems; the meaning of the term should therefore be thoroughly understood.

The results obtained by tracing rays through an optical system remain correct, no matter to what linear scale the system may be executed.

Practical. The computing formulae deduced in this chapter enter into nearly all the routine work of a designer; they should therefore be used in working out examples until the process becomes almost automatic.

The general discussions of numerical calculations will be found helpful in obtaining accurate results in a minimum of time and in tracking down and eradicating errors.

CHAPTER II

SPHERICAL ABERRATION

[24] IN sections [9] and [20] a complete example is worked out for five rays refracted by a simple biconvex lens. We will extract the final results and tabulate them under the respective initial angles U_1. As the calculation for the paraxial rays is mathematically correct only for extremely small angles, we put them under the heading $U_1 = 0$.

Original U_1:	0° (Paraxial)	−1°	−2°	−3°	−4°
L' after first refraction	149·81	148·54	145·15	139·65	132·49
L' after second refraction	8·7895	8·7088	8·4701	8·0545	7·4377

We see at once that at each surface the intersection-lengths steadily shorten as the angle U_1 increases. This becomes still clearer if we form the differences between the paraxial intersection-length and those found for the rays at finite angles. These differences are called the longitudinal spherical aberration, with the symbol LA (with dash for an image-point in the medium to the right of the surface) and have the definition

$$LA' = l' - L' \text{ (i.e. paraxial – marginal)}$$
$$LA = l - L.$$

Introducing these, our table becomes

Original U_1:	−1°	−2°	−3°	−4°
LA' after first refraction:	1·27	4·66	10·16	17·32
LA' after second refraction:	0·0807	0·3194	0·7350	1·3518

The most striking feature of this table is that it shows that the longitudinal aberration after the first refraction is on the average about 14 times as large as it is after the complete passage of the rays through the lens. This *suggests* that the second surface must have acted in opposition to the first and must have largely corrected the aberration caused by the first surface. One of the most important conclusions at which we shall arrive is that this interpretation is quite wrong, and that by far the larger part of the aberration found at the second surface is caused by this surface itself; the explanation being that the longitudinal way of measuring aberration is a very misleading one. In the present case we shall find that, put in a homely way, the aberration, as far as its real importance is concerned, is counted in pennies at the first surface and in pounds sterling at the second surface.

If we examine next the successive figures for each surface we easily detect that they appear to follow a very simple law: the LA' are evidently very nearly proportional to the squares of the corresponding values of the original U_1, for on dividing the LA' by U_1^2, that is, from left to right by 1, 4, 9, and 16 respectively, [24] we find

Original U_1	$-1°$	$-2°$	$-3°$	$-4°$
LA'/U_1^2 first surface	1·27	1·16	1·13	1·08
LA'/U_1^2 second surface	0·081	0·080	0·082	0·084

The quotients are nearly constant at each surface, but show a tendency to diminish at the first and to increase at the second surface with increasing obliquity of the ray.

As spherical aberration is one of the chief defects of lenses and of optical systems, we must try to discover its laws. For this purpose we will deduce an exact equation which contains all the necessary information :

Standard equation (1) may be transposed into the form

$$\frac{r}{L-r} = \frac{\sin U}{\sin I}.$$

To this proportion we can apply a universal algebraical rule, namely that when we are given $\frac{a}{b} = \frac{c}{d}$, then we have also

$$\frac{a}{b+a} = \frac{c}{d+c} \text{ and } \frac{b}{b+a} = \frac{d}{d+c},$$

and it will be known to students that this rule holds for even more complicated corresponding operations on the two sides of the original equation.

By this rule we find from the transposed (1)

$$\text{(a)} \quad \frac{r}{L} = \frac{\sin U}{\sin I + \sin U} \text{ and (b) } \frac{L-r}{L} = \frac{\sin I}{\sin I + \sin U}.$$

We can identically transform (a) by adding $\sin I - \sin I = 0$ in the numerator :

$$\frac{r}{L} = \frac{\sin I + \sin U - \sin I}{\sin I + \sin U} = 1 - \frac{\sin I}{\sin I + \sin U};$$

and if we then multiply throughout by N/r, we obtain

$$\text{(c)} \quad \frac{N}{L} = \frac{N}{r} - \frac{N}{r} \cdot \frac{\sin I}{\sin I + \sin U}.$$

Standard equation (4) exactly corresponds to (1) and can therefore be treated in the same way, with the result that we obtain an equation exactly like (c), but having 'dashed' quantities :

$$\text{(d)} \quad \frac{N'}{L'} = \frac{N'}{r} - \frac{N'}{r} \cdot \frac{\sin I'}{\sin I' + \sin U'}.$$

The two equations (c) and (d) as they stand merely express a general property of any triangle like PCB or PCB' in the standard diagram. In order that the equations

[24] may be restricted to the case of the same ray before and after refraction, we must introduce the conditions

(1) that r shall have the same value in the two equations;
(2) that $N' \sin I'$ shall be equal to $N \sin I$ according to standard equation (2);
(3) that $U'+I'$ shall be equal to $U+I$, for otherwise the point of incidence for equation (c) would be different from the point of emergence for equation (d).

The first two of these conditions will be embodied in the resulting equation if we deduct (c) from (d), treat r as identical in them, and in the long final terms extract $N . \sin I$ in (c) as a common factor because $N \sin I$ must be equal to $N' \sin I'$. This gives us the single equation

$$\text{(e)} \qquad \frac{N'}{L'} - \frac{N}{L} = \frac{N'-N}{r} + \frac{N \sin I}{r}\left(\frac{1}{\sin I + \sin U} - \frac{1}{\sin I' + \sin U'}\right).$$

The final term of this we can now transform further. Taking $1/(\sin I + \sin U)$ outside the bracket, we find:

$$\text{final term} \qquad = \frac{N \sin I}{r(\sin I + \sin U)}\left(1 - \frac{\sin I + \sin U}{\sin I' + \sin U'}\right).$$

By (b) we have $\dfrac{\sin I}{\sin I + \sin U} = \dfrac{L-r}{L}$, which we substitute in the first part. In the bracket we use the trigonometrical formula

$$\sin p + \sin q = 2 \sin \frac{p+q}{2} \cos \frac{p-q}{2}$$

and obtain

$$\text{final term} \qquad = \frac{N(L-r)}{r . L}\left(1 - \frac{2 \sin \dfrac{U+I}{2} \cos \dfrac{I-U}{2}}{2 \sin \dfrac{U'+I'}{2} \cos \dfrac{I'-U'}{2}}\right).$$

This equation gives us the opportunity of introducing the third and final of the conditions enumerated above by cancelling $2 \sin \frac{1}{2}(U+I)$ in the numerator against $2 \sin \frac{1}{2}(U'+I')$ in the denominator. If after this we bring the final bracket to a common denominator, we obtain:

$$\text{final term} \qquad = \frac{N(L-r)}{r . L} \frac{\cos \dfrac{I'-U'}{2} - \cos \dfrac{I-U}{2}}{\cos \dfrac{I'-U'}{2}}.$$

We now apply the trigonometrical formula

$$\cos p - \cos q = 2 \sin \frac{q+p}{2} \sin \frac{q-p}{2},$$

and obtain : [24]

$$\text{final term} = 2\,\frac{N(L-r)}{r\,.\,L}\,\frac{\sin\dfrac{I-U+I'-U'}{4}\,\sin\dfrac{I-U-I'+U'}{4}}{\cos\dfrac{I'-U'}{2}}$$

In the last numerator we introduce, by equation (3), $U' = U+I-I'$ and obtain the final equation :

$$\text{final term} = 2\,\frac{N(L-r)}{r\,.\,L}\,\frac{\sin\dfrac{I'-U}{2}\,\sin\dfrac{I-I'}{2}}{\cos\dfrac{I'-U'}{2}};$$

and putting this into equation (e) in place of the original form of the final term, we arrive at the important equation :

$$(10) \qquad \frac{N'}{L'} = \frac{N}{L} + \frac{N'-N}{r} + 2\,\frac{N(L-r)}{r\,.\,L}\,\frac{\sin\dfrac{I'-U}{2}\,\sin\dfrac{I-I'}{2}}{\cos\dfrac{I'-U'}{2}}.$$

This is a rigorously exact equation, unconditionally true no matter how large the angles may be : it will form a firm foundation for the discussion of spherical aberration.

DISCUSSION OF EQUATION (10)

The equation (10) bears a close resemblance to our general paraxial equation for refraction at a single surface [25]

$$(6p)^{**} \qquad N'/l' = (N'-N)/r + N/l;$$

and we see at once that if we have an object-point at any given distance $l = L$ then the first two terms on the right of (10) become identical with the right of (6p)**. The difference usually found between l' and L' must therefore arise from the long final term in (10), and this term is thus definitely recognized as the one which expresses the spherical aberration and which must contain all its secrets, because (10) is a trigonometrically exact equation.

We will begin the discussion of (10) by seeking the answer to a highly important question : 'Can the aberration arising at a spherical surface become zero for rays at finite angles with the optical axis ? '

There are three cases in which the answer is *Yes.*

(1) The first case is self-evident. If the *object-point B coincides with the pole A* of the refracting surface, all the incident rays necessarily aim at this point and all the refracted rays start out from it : hence there can be no longitudinal aberration. L and L' are both zero no matter how large the angles U and U' may be. In our equation (10) this case makes the first and third terms on the right infinitely large on account of $L = 0$ in the denominators ; hence the left side is also infinitely large, which can only be if L' also $= 0$. See Fig. 25.

[25] 2. The second case of freedom from aberration is also self-evident. If the object-point coincides with the centre of curvature, i.e. if $L = r$, then all rays, regardless of the angle of convergence U, proceed radially through the surface ; hence I and I'

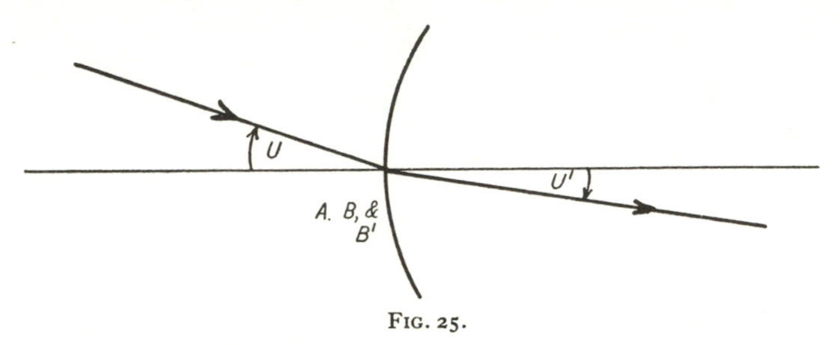

FIG. 25.

are zero, and there is no change in direction of the rays on passing through the surface. Consequently B' also coincides with C for rays at any angle. See Fig. 26.

In our equation (10) this case renders the final aberration term equal to zero because both $(L-r)$ and $\sin \frac{1}{2}(I-I')$ in the numerator are = zero.

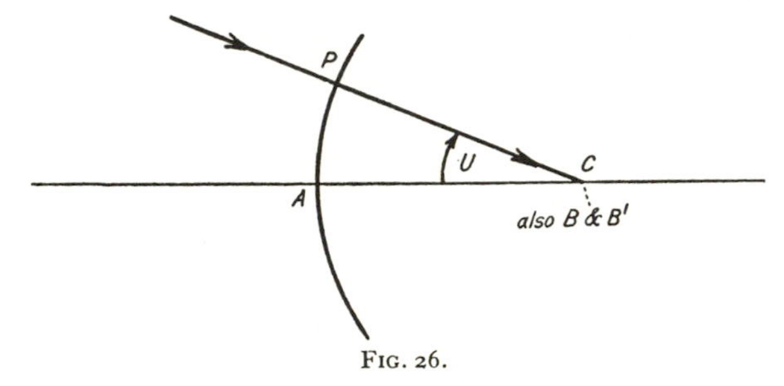

FIG. 26.

3. In all other cases L, r, I, and I' will have finite values and L will be different from r, consequently $(L-r)$ in the aberration-term will also be finite. If we exclude the case when $N = N'$, in which case the surface is optically non-existent as rays at any angle would pass straight through it, there will then also be a difference between I and I'. Consequently all the items in the aberration-term will have finite values, with the possible exception of $\sin \frac{1}{2}(I'-U)$. If this term became zero it evidently would make the whole aberration-term zero and would disclose another case of complete freedom from aberration, and the only case not obvious to mere commonsense. We must therefore carefully investigate this case. $\sin \frac{1}{2}(I'-U)$ can only be zero if the angle itself is zero, and therefore if $I' = U$, or, with equal necessity, if $\sin I' = \sin U$. In section [13] we deduced by combining equations (1) and (2)

$$\sin I' = \frac{N}{N'} \cdot \sin U \cdot \frac{L-r}{r}.$$

If we put in this $\sin I' = \sin U$, we find, as the condition to be fulfilled by our case of freedom from aberration : [25]

$$\frac{L-r}{r}\frac{N}{N'} = 1, \text{ or } \frac{L-r}{r} = \frac{N'}{N}.$$

Applying the algebraical rule :

$$\text{if } \frac{a}{b} = \frac{c}{d}, \text{ then } \frac{a+b}{b} = \frac{c+d}{d} : \text{ this leads to}$$

$$\frac{L}{r} = \frac{N'+N}{N} \text{ or } L = r\frac{N'+N}{N}$$

as the value of L, which realizes the third case of freedom from spherical aberration.

As refractive indices are always positive, the formula shows that the case can only be realized if L and r have the same sign, that is, if B and C lie on the same side of the refracting surface. Taking a case of refraction from air into glass, we shall have $N = 1$ (for air) and N' = the glass-index, say, on an average, 1·55. Hence in this case there will be no aberration if $L = 2{\cdot}55\ r$, just about the proportions of our standard diagram (Fig. 2 *a*), which thus approximately represents our case (3) of no aberration.

This is a property of spherical surfaces which is of the utmost importance, especially in the case of microscope objectives of high power. These are only possible because they can be so constructed that most of the work of bringing the rays round from the object to the image is done at surfaces which more or less nearly come under our third case, combined with others which approach to cases (1) and (2).

It is therefore well worth while to examine case (3) a little further. If we transpose standard equation (3) : $U' = U + I - I'$, into the form $I' - U = I - U'$ we see that the condition of case (3), i.e. $I' = U$, also implies $I = U'$, and therefore also $\sin I = \sin U'$. We thus have the two conditions

$$\sin I' = \sin U$$

$$\text{and} \quad \sin I = \sin U'$$

and dividing the second by the first, obtain

$$\frac{\sin U'}{\sin U} = \frac{\sin I}{\sin I'} = \frac{N'}{N} \text{ by the law of refraction.}$$

This is another important property of the third case of aplanatic refraction : the convergence is changed according to the law

$$\frac{\sin U'}{\sin U} = \frac{N'}{N},$$

therefore the sines of the angles of convergence before and after refraction are in a constant ratio : as we shall learn later, this means that the 'sine-condition' is fulfilled and that case (3) thus leads to a well-defined image not only for the point on the optical axis, but also for the surrounding field.

[25] A simple relation between L and L' in the third case of aplanatic refraction is most easily deduced by equating the two values of PA found in section [12], and transposing into the form

$$\frac{L'}{L} = \frac{\sin U}{\sin U'} \cdot \frac{\cos \frac{U'-I'}{2}}{\cos \frac{U-I}{2}}.$$

As was proved in the preceding paragraph, we have in our case $\sin U/\sin U' = N/N'$. In the cosine-terms, because in this special case we have $U' = I$ and $U = I'$, we can replace the U' by I and the U by I', and the formula becomes

$$\frac{L'}{L} = \frac{N}{N'} \cdot \frac{\cos \frac{I-I'}{2}}{\cos \frac{I'-I}{2}} = \frac{N}{N'} \cdot \frac{\cos \frac{I-I'}{2}}{\cos\left(-\frac{I-I'}{2}\right)} = \frac{N}{N'},$$

the last because a negative and a positive angle of the same magnitude have precisely the same value of the cosine. Hence the intersection-lengths in the third case are in the inverse ratio of the refractive indices. By combining the last equation with the one previously found for L we find a direct expression for L', namely

$$L' = r\,\frac{N'+N}{N'}.$$

Finally there is also a simple relation between L, L', and r. The reciprocals of the values which we determined for L and L' are

$$\frac{1}{L} = \frac{1}{r}\,\frac{N}{N'+N} \text{ and } \frac{1}{L'} = \frac{1}{r}\,\frac{N'}{N'+N}.$$

The simple addition of these two equations gives

$$\frac{1}{L} + \frac{1}{L'} = \frac{1}{r}$$

as yet another property of the third case.

An interesting limiting value is attached to case (3). Its primary conditions are $I = U'$ and $I' = U$. Now I (or I') in the lighter medium may reach 90° as its obvious limit. As U' (or U) in the denser medium must be of the same value, it follows that in the case of aplanatic refraction we can always deal with rays up to but not exceeding U or $U' = 90°$ in the denser medium, that is, with rays filling a complete hemisphere, or as it is usually called, with cones of rays up to an angle of 180° at the apex of the cone.

It is of considerable importance to know what happens when the L and L' do not exactly satisfy the condition which we have deduced for the case of aplanatic refraction. If we assume that Fig. 2 (*a*) represents the exact case when $I' = U$ and $I = U'$ and when the spherical aberration is zero, it is easily seen that for larger values of L and L' rays from the same incidence point P would have larger values of I and I' and smaller values of U and U', and that therefore $(I' - U)$ in the aberration term would be positive, whilst smaller values of L and L' would make $(I' - U)$ negative. As all the *other* terms in the final item of equation (10) evidently do not change their signs for a decidedly large range of L-values—namely from $L = r$ to $L = \infty$—it is clear that the final item, which measures the aberration, changes its sign when we pass from shorter L values, for which $(I' - U)$ is negative, through the aplanatic point to longer L-values for which $(I' - U)$ is positive. The spherical aberration therefore is of opposite sign for object-points to the right and to the left of the aplanatic point. This also is a point of great interest in the design of microscope objectives.

On account of the great importance of the results at which we have arrived, they should be added to the list of formulae, following equation (10), in this form:

(10) *Special Cases*: Refraction at spherical surfaces is quite free from spherical aberration:

(1) When object- and image-point coincide with the pole or apex of the refracting surface.

(2) When object- and image-point coincide with the centre of curvature of the refracting spherical surface.

(3) When object- and image-point do not coincide, provided both lie on the concave side of the refracting surface and fulfil the condition

$$L = r\,\frac{N' + N}{N}.$$

A surface fulfilling this condition also satisfies the equations

$$L' = r\,\frac{N' + N}{N'};\quad \frac{1}{L} + \frac{1}{L'} = \frac{1}{r};\quad \frac{L}{L'} = \frac{N'}{N};$$

$$U = I';\quad U' = I;\quad \frac{\sin U'}{\sin U} = \frac{N'}{N}.$$

Case (3) on account of its very great importance should be verified by a trigonometrical example. For the sake of simplicity we will take $r = 1$ and the refractive indices $N = 1$ (air) to the left and $N' = 1{\cdot}5$ (light crown-glass) to the right of the surface. By the relations worked out above, this surface should be quite free from spherical aberration for $L = r(N' + N)/N = 2{\cdot}5$ in this case. On tracing through a marginal ray at $U = 20°$, the associated paraxial ray at nominal $u = \sin U$ and an intermediate zonal ray at $U = 14°$, precisely the same intersection-length will be found for all three, and in accordance with the theory this will be $= r(N' + N)/N' = 1{\cdot}66667$. Verify the case to the limit of grazing incidence (see Fig. 9) by putting marginal $U = \sin^{-1}(r/(L - r)) = \sin^{-1}(\frac{2}{3})$. I will be found $= 90°$ and U' also $= 90°$, and L' will still remain $= 1{\cdot}66667$. The practical study should be completed by finding out what happens when L is moderately different from the value required for aplanatic refraction by repeating the calculations with $L = 2$ and with $L = 3$ for finite rays at $U = 20°$ and $14°$ and for paraxial rays.

AN EXACT EXPRESSION FOR LA'

[26] Having exhausted the discussion of the three special cases of complete absence of spherical aberration, we will now return to our general equation (10) and render it suitable for the study of the aberration which accompanies refraction at spherical surfaces in all other cases. As a rule the paraxial and marginal rays will arrive at a surface already affected by spherical aberration acquired at previous surfaces: that is to say l will be different from L. But whatever values l and L may have, the refraction of the ray at finite angles must obey our rigorous equation (10), and the refraction of the paraxial rays must be in accordance with the exact paraxial equation (6p)**. We therefore have the two equations

$$(10) \qquad \frac{N'}{L'} = \frac{N}{L} + \frac{N'-N}{r} + 2\,\frac{N(L-r)}{r\,.\,L}\,\frac{\sin\frac{1}{2}(I'-U)\,.\,\sin\frac{1}{2}(I-I')}{\cos\frac{1}{2}(I'-U')};$$

$$(6p)^{**} \qquad \frac{N'}{l'} = \frac{N}{l} + \frac{N'-N}{r};$$

and if we subtract the second from the first and bring the left side of the resulting difference and also the first terms of the right side to common denominators, we obtain

$$N'\,\frac{l'-L'}{l'\,.\,L'} = N\,\frac{l-L}{l\,.\,L} + 2\,\frac{N(L-r)}{r\,.\,L}\cdot\frac{\sin\frac{1}{2}(I'-U)\sin\frac{1}{2}(I-I')}{\cos\frac{1}{2}(I'-U')};$$

and if we now multiply throughout by $l'.\,L'/N'$, the equation becomes a direct expression for $l'-L' = LA'$:

$$(10)^* \quad LA' = LA\cdot\frac{N}{N'}\cdot\frac{l'.\,L'}{l\,.\,L} + 2\,\frac{N}{N'}(L-r)\,\frac{l'.\,L'}{r\,.\,L}\cdot\frac{\sin\frac{1}{2}(I'-U)\sin\frac{1}{2}(I-I')}{\cos\frac{1}{2}(I'-U')}.$$

On the right $l-L$ has been replaced by the simpler symbol LA, as it obviously represents the longitudinal spherical aberration of the arriving pencil of rays.

(10)* consists of two parts of very distinct significance. The first term depends on LA, and tells us how the aberration (if any) of the arriving pencil combines with the new aberration which accompanies the refraction at the new surface and which is given by the second term. The first term of (10)* therefore embodies the *Addition-Theorem* of longitudinal aberration and the second term determines the *new aberration*.

(10)* is not an independent solution of the problem of determining the spherical aberration, for it requires all the data of the trigonometrical ray-tracing to be known. But it does give a much closer approximation to the true value of LA' because it determines it directly instead of finding it as the difference of the separately computed l' and L' by the ray-tracing equations. It will hardly ever be necessary to use (10)* in this way in regular routine-work, and the equation is therefore chiefly of theoretical value, but we will use it here to demonstrate its precision.

We will take the biconvex lens of Example 1 and the ray starting from B at $U = -1°$. At the first surface we have $l = L$, therefore $LA = 0$ and (10)* becomes reduced to its second or new aberration term. To calculate this term we extract from sections [9] and [20]:

[26]

$\log N/N' =$	$-0{\cdot}18127$	$\frac{1}{2}U =$	$-0-30-0$
$\log (L-r) =$	$1{\cdot}53148n$	$\frac{1}{2}I =$	$1-42-3{\cdot}5$
$\log r =$	$1{\cdot}00000$	$\frac{1}{2}I' =$	$1-7-12{\cdot}5$
$L =$	$-24{\cdot}000$	$\frac{1}{2}U' =$	$0-4-51$
$L' =$	$148{\cdot}538$	$\frac{1}{2}(I'-U) =$	$1-37-12{\cdot}5$
$l' =$	$149{\cdot}811$	$\frac{1}{2}(I-I') =$	$0-34-51$
		$\frac{1}{2}(I'-U') =$	$1-2-21{\cdot}5$

and can then calculate

$$LA' = 2\frac{N}{N'}(L-r)\frac{l'\,.\,L'}{r\,.\,L}\frac{\sin\frac{1}{2}(I'-U)\sin\frac{1}{2}(I-I')}{\cos\frac{1}{2}(I'-U')}.$$

$\log 2$	$= 0{\cdot}30103$
$+\log N/N'$	$= +9{\cdot}81873$
$+\log (L-r)$	$= +1{\cdot}53148n$
$+\log l'$	$= +2{\cdot}17554$
$+\log L'$	$= +2{\cdot}17184$
$+\text{colog}\ r$	$= +9{\cdot}00000$
$+\text{colog}\ L$	$= +8{\cdot}61979n$
$+\log\sin\frac{1}{2}(I'-U)$	$= +8{\cdot}45137$
$+\log\sin\frac{1}{2}(I-I')$	$= +8{\cdot}00592$
$+\text{colog}\cos\frac{1}{2}(I'-U')$	$= +0{\cdot}00007$
$\log LA'$	$= 0{\cdot}07577$

Result by exact formula :
$LA' = 1{\cdot}191.$

Inexact result by $l'-L'$ of ray-tracing :
$LA' = 1{\cdot}27.$

The apparently large discrepancy of 0·08 inch is explained by the fact that in the trigonometrical work $(L'-r)$ was found from $U' = 0-9-42$, to the sine of which it is inversely proportional. Owing to the rounded-off seconds of our usual angles the final U' might, with closer interpolation of angles, easily have come out larger or smaller by as much as a whole second, which on an angle of only 582 seconds would amount to one part in 582. $(L'-r)$ is uncertain to the same extent or (as it was found $= 138{\cdot}538$) uncertain to possibly more than 0·2 inch. The 0·08 actually found is therefore perfectly plausible. On looking closer into the table of the LA'/U_1^2 of this lens in the beginning of this chapter it will be noticed that the true value 1·19 for $-1°$ does in fact fit in better with the values for larger angles. Except for theoretical purposes, the discrepancy found need not worry us as it arises from an uncertainty of a mere fraction of a second in the angle U' of our ray. We shall soon learn that there is far more latitude than this in the really necessary accuracy of our calculations when the convergence angle U' is small.

We will now calculate (10)* for the same ray at the second surface of our lens. We extract from sections [9] and [20] :

l	$= 149{\cdot}211$	$\frac{1}{2}U$	$0-4-51$
l'	$= 8{\cdot}7895$	$\frac{1}{2}I$	$-2-28-32$
L	$= 147{\cdot}938$	$\frac{1}{2}I'$	$-3-45-51$
L'	$= 8{\cdot}7088$	$\frac{1}{2}U'$	$1-22-10$
$\log N/N'$	$= 0{\cdot}18127$	$\frac{1}{2}(I'-U)$	$-3-50-42$
$\log (L-r)$	$= 2{\cdot}18452$	$\frac{1}{2}(I-I')$	$1-17-19$
$\log r$	$= 0{\cdot}69897n$	$\frac{1}{2}(I'-U')$	$-5-8-1$

[26] As the $LA' = 1{\cdot}191$ found for the pencil emerging from the first surface is also, obviously, the LA of the same pencil when arriving at our new, second, surface, we have to employ $LA = 1{\cdot}191$ or $\log LA$, as found at the first surface $= 0{\cdot}07577$ in the first part of (10)*, and find:

$$LA \cdot \frac{N}{N'} \cdot \frac{l' . L'}{l . L} \qquad 2\frac{N}{N'}(L-r)\frac{l' . L'}{r . L} \cdot \frac{\sin\frac{1}{2}(I'-U)\sin\frac{1}{2}(I-I')}{\cos\frac{1}{2}(I'-U')}$$

$\log LA$	$= 0{\cdot}07577$	$\log 2$	$= 0{\cdot}30103$
$+\log N/N'$	$= 0{\cdot}18127$	$+\log N/N'$	$= 0{\cdot}18127$
$+\log l'$	$= 0{\cdot}94396$	$+\log (L-r)$	$= 2{\cdot}18452$
$+\log L'$	$= 0{\cdot}93996$	$+\log l'$	$= 0{\cdot}94396$
$+\text{colog}\ l$	$= 7{\cdot}82620$	$+\log L'$	$= 0{\cdot}93996$
$+\text{colog}\ L$	$= 7{\cdot}82992$	$+\text{colog}\ r$	$= 9{\cdot}30103n$
$\log$ (transferred LA)	$= 7{\cdot}79708$	$+\text{colog}\ L$	$= 7{\cdot}82992$
		$+\log\sin\frac{1}{2}(I'-U)$	$= 8{\cdot}82645n$
Transferred LA	$= 0{\cdot}00627$	$+\log\sin\frac{1}{2}(I-I')$	$= 8{\cdot}35196$
New Aberration	$= 0{\cdot}07275$	$+\text{colog}\cos\frac{1}{2}(I'-U')$	$= 0{\cdot}00175$
Final LA'	$= 0{\cdot}07902$	$\log$ (new aberration)	$= 8{\cdot}86185$

By the ray-tracing we found $LA' = l' - L' = 0{\cdot}0807$, nearly $0{\cdot}002$ too large. The discrepancy is again easily accounted for by the uncertainty of the still small final U'. An error of less than 2″ in this angle would explain it, and would doubtlessly be found if the rays were traced with the full accuracy of six-figure logs.

Again, it should be noted that the exact value $LA' = 0{\cdot}079$ fits better into the table of $LA'/U_1'^2$ in the early part of this chapter, and removes an actual irregularity in the successive figures. The ray at $-1°$ was in fact chosen for the demonstration of the occasional practical utility of (10)* as a highly exact equation, because these discrepancies in the table of $LA'/U_1'^2$ called for elucidation in the present theoretical discussion. The other rays would give far smaller percentage errors in their LA' because simultaneously the angles become larger and less affected by the dropping of fractions of a second and the LA' rapidly grow.

The most instructive part of the above calculation for the second surface is contained in the last three numbers showing how the final total aberration $LA' = 0{\cdot}079$ is made up. By far the greater part is new aberration caused by the second surface itself and only $0{\cdot}006$ is contributed by the first surface, although at its own focus that surface showed an $LA' = 1{\cdot}19$, roughly 200 times the amount which it contributes to the final aberration. That is what was referred to in the opening paragraphs when it was suggested that the aberration in this case was counted in pennies at the first, and in pounds sterling at the second surface.

Equation (10)* contains the explanation in its first term. The LA of the incident pencil is by no means simply added to the new aberration: it has to be multiplied by the factor $N . l' . L'/N' . l . L$. This factor will be large if the intersection-lengths l' and L' after refraction are a multiple of those before refraction, and the LA would then make a large contribution to LA'. On the other hand the factor will be small if the dashed and plain l and L are in the inverse relation, and then a large LA of the incident pencil may become quite a small contribution to the LA' after refraction, as happens at the second surface of our biconvex lens.

[26] On comparing the operation by which we obtained (10)* from (10) and (6p)** with that carried out in section [19] determining the axial magnification produced by a single refracting surface, it will be seen that the operations are identical. This shows that the longitudinal aberration is transferred according to the law of axial magnification, just as if it were an actual object of length LA lying in the optical axis. This is important as it is a generally applicable rule for such transfers of differences of focus.

Our example 1, although it was selected in order to bring the misleading changes of magnitude of longitudinal aberration into prominence, does not represent anything approaching an extreme case. Evidently the factors l' or L' in both terms of (10)* will reach infinity if either the paraxial or the marginal ray emerges exactly parallel to the optical axis. In such a case LA' will become infinitely large.

The first surface of our biconvex lens with the object-point at -24 causes all the refracted rays to converge towards the optical axis. If the object-point is brought gradually closer to the lens, the convergence of the refracted rays will diminish, or the l' and L' will become larger, and when (as a simple calculation by (6p)** shows) the object-point is at $-19{\cdot}305$ the emerging paraxial rays will become a thin pencil of parallel rays, or l' will be infinite and therefore also LA'. If the object-point approaches the lens a very little more, l' will become negative whilst L' remains positive, and we shall find large negative values of LA' until the object-point reaches, for a ray starting at $-4°$, the value $L = -18{\cdot}908$. For that value L' becomes infinitive and LA' again reaches infinity. For the least further approach of the object-point L' will become negative, but numerically larger than l', and LA' again returns to positive values. It will well repay the trouble to calculate a paraxial and a marginal ray at $U = -4°$ through the lens for $L = -19{\cdot}4$,

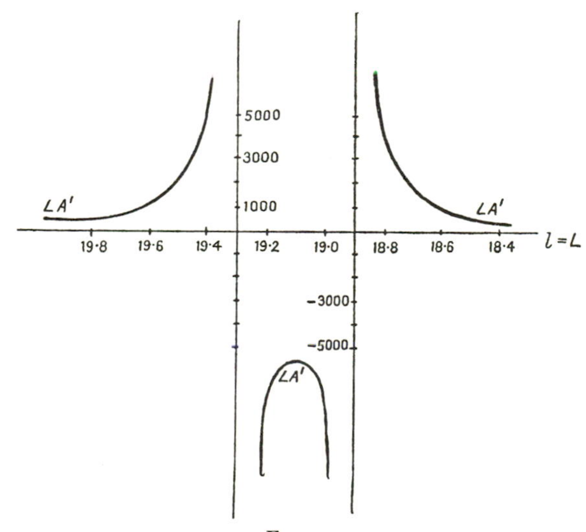

FIG. 27.

[26] $-19{\cdot}2$, $-19{\cdot}0$, and $-18{\cdot}8$, and to plot the resulting aberrations against L. For the aberration after the *first* refraction a curve like Fig. 27 with four infinity-points will result, displaying apparently most alarming discontinuities ; but if the aberration for the complete lens is similarly plotted, it will be found to yield a perfectly smooth curve, utterly unaffected by the wild acrobatics of the LA' at the first surface.

The chief point to be remembered by the computer is therefore that huge LA-values may be met with in pencils of very slight convergence or divergence, and that they need cause no serious alarm. The real defect indicated by longitudinal aberration is then a minute fraction of the numerical value of LA', as transfer to a space with greater convergence will demonstrate. The following chapter will contain safe guidance for the estimation of the real importance of any given LA-value as found by ray-tracing.

One special case, namely that of a plane surface, makes an adaption of (10)* desirable in order to avoid the clumsy expedient of substituting a surface of very long radius for the plane. If in the second part of (10)* we carry out the division of $(L-r)$ by $r \, . \, L$ we obtain

$$LA' = LA \cdot \frac{N}{N'} \cdot \frac{l'}{l} \cdot \frac{L'}{L} + 2\frac{N}{N'}\left(\frac{1}{r} - \frac{1}{L}\right) \cdot l' \, . \, L' \, \frac{\sin\frac{1}{2}(I'-U) \, . \, \sin\frac{1}{2}(I-I')}{\cos\frac{1}{2}(I'-U')} .$$

We can then let r grow beyond all limits, which causes $1/r$ to sink to zero and the spherical surface to become a plane. Introducing also by the Pl. equations $l'/l = N'/N$, $N \, . \, l'/N' = l$, also $I' = -U'$, and $I = -U$, we easily find

$$(10)^* \quad \textit{Plane} : LA' = LA \cdot \frac{L'}{L} + 2\frac{L'}{L} \cdot l \cdot \frac{\sin\frac{1}{2}(U'+U) \, . \, \sin\frac{1}{2}(U'-U)}{\cos U'} .$$

PRIMARY SPHERICAL ABERRATION

[27] Our equations (10)* are perfectly exact for rays at any angle which allows them to pass through a surface. The second term shows that the new aberration diminishes rapidly for rays at smaller and smaller angles on account of the sines in the numerator, and as this holds for all surfaces of a system, the transferred aberration of the first term must also rapidly diminish as the paraxial region is approached. Therefore, with the sole and rather unimportant exception of the case of nearly parallel emergence of the rays from a particular surface to which Fig. 27 refers, the differences between the corresponding marginal and paraxial intersection-lengths which occur on the right of equation (10)* will be small for rays at small angles and will be negligible for rays at indefinitely small angles. In the outer or extended paraxial region we may therefore replace L and L' by the strictly paraxial l and l'. Simultaneously the difference between a sine and the corresponding angle in radians will also become negligible, and the cosine in the denominator of the aberration term will differ by a negligible fraction from its limiting value 'one'. Consequently we can adapt (10)* to the extended paraxial region by replacing L by l, L' by l', by replacing the sines by the angles and by omitting the cosine in the denominator of the aberration term.

Using small letters for the angles, which will be of the paraxial order, and adding small p to the LA and LA' to remind us of the restriction to the extended paraxial region, we obtain from the general equation (10)* [27]

$$LAp' = LAp \cdot \frac{N}{N'}\left(\frac{l'}{l}\right)^2 + \tfrac{1}{2}\frac{N}{N'}\frac{l-r}{r} \cdot \frac{l'^2}{l}(i'-u)(i-i').$$

For convenience in numerical calculations we introduce in the first part by (5p)** $l'/l = u/u'$, and in the second part by (6p)* $N(l-r)/l = N'(l'-r)/l'$, and obtain

$$(10)^{**} \qquad LAp' = LAp \cdot \frac{N}{N'}\left(\frac{u}{u'}\right)^2 + \tfrac{1}{2}\, l' \frac{l'-r}{r}(i'-u)(i-i').$$

The special form of (10)* for plane surfaces can be transformed directly by using in the first term from Pl (3p) $l'/l = N'/N$, and gives

$$(10)^{**} \qquad \textit{Plane}: \; LAp' = LAp \frac{N'}{N} + \tfrac{1}{2}\, l'(u'+u)(u'-u).$$

From the method by which we have arrived at these equations it is rendered evident that they represent the beginnings or first display of spherical aberration in the extended paraxial region, and we shall therefore adopt the name 'Primary Longitudinal Spherical Aberration' for the LAp' calculated by the equations (10)**. The second term which determines the new aberration arising at a surface, consists of the constant factor $\frac{1}{2}\, l'(l'-r)/r$ and of the product $(i'-u)(i-i')$ which depends on the angles. We proved in Chapter I that all the paraxial angles are proportional to the original u with which a calculation is begun, and that, if such a calculation were repeated with $k \cdot u$ as the starting angle, all the other paraxial angles would also come out at k-times the previously found value. Consequently, if we carried out a calculation of (10)** first with a given u and then with $k \cdot u$, the factor $(i'-u)(i-i')$ in the first calculation would in the second calculation become $(k \cdot i' - k \cdot u)(ki - ki') = k^2(i'-u)(i-i')$. It is easily seen that the same argument also applies to the first term of (10)**, for this term has only the constant factors N/N' and $(u/u')^2$ (for $(ku/ku')^2 = (u/u')^2$); but the LAp received from the first surface will, when $k \cdot u$ is used, be k^2 times as large as that found for simply u, and the whole aberration at the second surface will therefore have the factor k^2 and so on for any number of surfaces. We therefore conclude that :

(1) The primary aberration grows in exact proportion to the square of the angle u with which the paraxial calculation is begun.
(2) As by the first of (5p)** the semi-aperture y is proportional to u, the primary aberration also grows as the square of the aperture.
(3) On account of this simple law of the primary aberration, we may calculate (10)** with any fictitious or nominal value of the initial u which pleases us ; we shall of course usually choose the value $u = \sin U$ employed in the standard calculations.

Again this latter choice of the original u has important advantages. Evidently it will give that value of the longitudinal aberration which would be the true one *if* the simple law of primary aberration as being proportional to the square of u or y held up to the full aperture ; in other words, (10)** calculated with our nominal

[27] paraxial angles will yield a more or less close approximation to the actual aberration at full aperture.

As this is a very important conclusion, seeing how simple it is to calculate (10)** when the paraxial data are known, we will carry out the calculation for our biconvex lens of example 1.

For the first surface there is no transferred aberration because we are dealing with a fixed object-point, hence we have only to compute the second part :

$$LAp' = \tfrac{1}{2}\, l' . \frac{l'-r}{r}\,(i'-u)\,(i-i').$$

We extract from section [20] :

$$\begin{array}{lll} l' = 149{\cdot}811 & u = -0{\cdot}069756 & i'-u = 0{\cdot}225995 \\ \log(l'-r) = 2{\cdot}14554 & i = 0{\cdot}237170 & i-i' = 0{\cdot}080931 \\ \log r = 1{\cdot}00000 & i' = 0{\cdot}156239 & \end{array}$$

and can then calculate LAp' thus :

$$\begin{array}{ll} \log \tfrac{1}{2} & = 9{\cdot}69897 \\ +\log l' & = 2{\cdot}17554 \\ +\log(l'-r) & = 2{\cdot}14554 \\ +\text{colog } r & = 9{\cdot}00000 \\ +\log(i'-u) & = 9{\cdot}35410 \\ +\log(i-i') & = 8{\cdot}90812 \\ \log LAp' & = 1{\cdot}28227 \end{array} \qquad LAp' = 19{\cdot}154$$

The result found is larger than that obtained by the trigonometrically traced marginal ray, but only by about 10 per cent. But we can go a step further. We can, by the simple proportion law of the paraxial angles, determine very easily what value of LAp' we should find if we calculated a paraxial ray associated with each of the other three trigonometrical rays at $-1°$, $-2°$, and $-3°$ respectively, and then used the angles found by these calculations in working out equation (10)**. The logs of the starting u_1 would be $8{\cdot}24186n$, $8{\cdot}54282n$, and $8{\cdot}71880n$ respectively (see section [9]), instead of $8{\cdot}84358n$ for the paraxial ray associated with the ray at $-4°$. The *ratios* k of new u_1 to original u_1 would therefore have the logs

	$-1°$	$-2°$	$-3°$
	$8{\cdot}24186n$	$8{\cdot}54282n$	$8{\cdot}71880n$
	$-8{\cdot}84358n$	$-8{\cdot}84358n$	$-8{\cdot}84358n$
$\log k$	$9{\cdot}39828$	$9{\cdot}69924$	$9{\cdot}87522$

and by the square law of primary aberration the latter would be changed in accordance with the square of the ratio. Therefore the results would be

Original U_1	$-1°$	$-2°$	$-3°$
$2.\log k$	$8{\cdot}79656$	$9{\cdot}39848$	$9{\cdot}75044$
$+\log(LAp'$ for $-4°)$	$1{\cdot}28227$	$1{\cdot}28227$	$1{\cdot}28227$
$\log LAp'$	$0{\cdot}07883$	$0{\cdot}68075$	$1{\cdot}03271$
LAp'	$1{\cdot}199$	$4{\cdot}795$	$10{\cdot}782$

[27] By the trigonometrical calculation, supplemented by the exact calculation of the aberration at $U_1 = -1^\circ$ in section [26], we found

Original U_1	-1°	-2°	-3°	-4°
Exact LA'	1·191	4·66	10·16	17·32
LAp'	1·199	4·795	10·782	19·154
Exact – Primary	–0·008	–0·135	–0·622	–1·834

We see that for smaller apertures the approximation improves very rapidly. We will prove, in anticipation of the following section on higher aberrations, that the differences between exact and approximate values are very nearly proportional to the fourth power of the ratios already used above, by applying these fourth powers to the difference $-1·834$ at -4°.

Original U_1	-1°	-2°	-3°	-4°
4. log k	7·59312	8·79696	9·50088	
+ log (–1·834)	0·26340n	0·26340n	0·26340n	
log	7·85652n	9·06036n	9·76428n	
Number	–0·007	–0·115	–0·581	–1·834

It will be seen that these numbers agree with the above for 'Exact – Primary' as nearly as the 'Exact' values can be considered to be correct—for it will be remembered that the U' were very small and only taken to the nearest second.

We will now calculate the primary aberration for the second surface of the biconvex lens by the full equation 10** :

$$LAp' = LAp \cdot \frac{N}{N'}\left(\frac{u}{u'}\right)^2 + \tfrac{1}{2}\, l' \,\frac{l'-r}{r}\,(i'-u)\,(i-i').$$

We extract from section [20] :

log N/N'	= 0·18127	u =	0·011175
log u/u'	= 8·77016	i =	–0·344668
l'	= 8·7895	i' =	–0·523203
log $(l'-r)$	= 1·13955	$i'-u$ =	–0·534378
log r	= 0·69897n	$i-i'$ =	0·178535

and with $LAp = LAp'$ of first surface $= 19·154$ find

First term		Second term	
log LAp	= 1·28227	log $\frac{1}{2}$	= 9·69897
+ log N/N'	= 0·18127	+ log l'	= 0·94396
+ 2. log u/u'	= 7·54032	+ log $(l'-r)/r$	= 0·44058n
log (first term)	= 9·00386	+ log $(i'-u)$	= 9·72785n
first term	= 0·1009	+ log $(i-i')$	= 9·25172
second term	= 1·1563	log (second term)	= 0·06308

LAp' = 1·2572 for final focus and for original $u_1 = \sin(-4^\circ)$.

[27] We can again calculate the corresponding values for the primary aberration for $u_1 = \sin(-1°)$, $-2°$ and $-3°$ by the ratios used at the first surface :

Original U_1	$-1°$	$-2°$	$-3°$
2. log k	8·79656	9·39848	9·75044
+log (LAp' for $-4°$)	0·09940	0·09940	0·09940
log LAp'	8·89596	9·49788	9·84984
LAp'	0·0787	0·3147	0·7077

and comparing with the true aberration by the trigonometrical ray-tracing—corrected by section [26] for $-1°$—obtain the table :

Original U_1	$-1°$	$-2°$	$-3°$	$-4°$
Exact LA'	0·0790	0·3194	0·7350	1·3518
Primary LAp'	0·0787	0·3147	0·7077	1·2572
Exact – Primary	0·0003	0·0047	0·0273	0·0946

We again obtain very close agreement for the finite rays at small angles, and easily see that the residuals are nearly proportional to the fourth power of the original U_1, as was shown in detail at the first surface.

We have now proved both theoretically and by numerical example that the longitudinal spherical aberration grows with the square of the aperture as long as the latter is small and leads to small angles, and that this simple law gives a good approximation even for decidedly large apertures involving big angles (up to 32° in our biconvex lens). We will now show that the simple 'square of the aperture' law gives a very much closer approximation to the true magnitude of the spherical aberration if we follow our regular computing practice by tracing a paraxial and extreme marginal ray trigonometrically and then inferring the probable aberration for intermediate rays by the law. We found the true spherical aberration of the ray starting from the object-point at $-4°$ as 17·32 inches after the first and 1·3518 after the second refraction. If we apply our ratios to these values we obtain for the first refraction :

Original U_1	$-1°$	$-2°$	$-3°$	$-4°$
2 log k	8·79656	9·39848	9·75044	
+log 17·32	1·23855	1·23855	1·23855	
log (Inferred LA')	0·03511	0·63703	0·98899	
Inferred LA'	1·084	4·335	9·750	17·32
True LA'	1·191	4·66	10·16	17·32
True-Inferred	0·107	0·325	0·410	0·00

The maximum error is a little more than 0·4 inch against an error of 1·834 by the primary aberration, obviously a very great improvement. For the result at the final focus we obtain :

Original U_1	$-1°$	$-2°$	$-3°$	$-4°$
2. log k	8·79656	9·39848	9·75044	
+log 1·3518	0·13091	0·13091	0·13091	
log (Inferred LA')	8·92747	9·52939	9·88135	

[27]

Original U_1	-1°	-2°	-3°	-4°
Inferred LA'	0·0846	0·3384	0·7609	1·3518
True LA'	0·0790	0·3194	0·7350	1·3518
True-Inferred	−0·0056	−0·0190	−0·0259	0·0000

Again we have a maximum error of less than 0·03 by the trigonometrical method as compared with one over 0·09 by the primary aberration. We therefore conclude that the latter should only be employed for rough estimates, or when all the angles are known to be small (say under 15°), and that in general the tracing of the actual marginal ray should be preferred in spite of a small amount of extra work required for the trigonometrical ray-tracing.

[28] In opening section [23], on simple lenses, it was pointed out that the line of least resistance in evolving new optical designs is nearly always represented by a succession of approximations, and that a rough scheme, worked out by the simple lens formulae, and merely determining the separations and the proper powers of a combination of thin lenses which would produce the desired result for a sufficiently restricted clear aperture and diameter of field, is the most generally useful first approximation. In simple cases it is practicable to proceed directly from this first scheme to strict trogonometrical calculation ; but in most cases time and trouble will be saved by a second approximation by primary aberration formulae, either with or without inclusion of the requisite thickness of the component simple lenses. Equation (10)** then becomes one of our chief tools because it enables us to secure a close estimate of the spherical aberration on the basis of the tracing of a single paraxial ray through the proposed lens system, for only data of the paraxial ray enter into the equation.

We can greatly increase the utility of the equation (10)** for this important second approximation work by looking more closely into the way in which the equation builds up the total aberration at the final image-point. The equation, valid for any refracting surface, is

$$(10)^{**} \quad \begin{cases} LAp' = LAp\,\dfrac{N\,.\,u^2}{N'u'^2} + \tfrac{1}{2}\,l'\,\dfrac{l'-r}{r}\,(i'-u)\,(i-i'), \text{ or for a plane} \\[2ex] LAp' = LAp\,\dfrac{N'}{N} \quad + \tfrac{1}{2}\,l'\,(u'+u)\,(u'-u), \end{cases}$$

and it was pointed out, when its exact prototype (10)* was deduced, that the first term on the right embodies the addition theorem of longitudinal spherical aberration, by the law of longitudinal magnification, whilst the second term represents the *new aberration* arising at the surface. As an abbreviation we will introduce the simple symbol Sph' for this *new aberration*, so that the equation becomes

$$LAp' = LAp\,\frac{N\,.\,u^2}{N'u'^2} + Sph', \text{ with the definition}$$

$$Sph' = \tfrac{1}{2}\,l'\,\frac{l'-r}{r}\,(i'-u)\,(i-i') \text{ or for a plane } = \tfrac{1}{2}\,l'\,(u'+u)\,(u'-u).$$

It is not necessary to maintain the distinction in the first term, for the factor N'/N

[28] in the formula for the transfer through a plane surface, whilst convenient in the direct application of the original equations, is identical, even for a plane, with the general factor $Nu^2/N'u'^2$ (because for a plane $u/u' = N'/N$) and the latter can therefore always be used.

We now assume that we are given a centred system of k refracting surfaces—k being any whole number—named as usual from left to right 1, 2, &c., to k, and we further assume that a paraxial ray has been traced right through so that all the angles and intersection-lengths are known. No restriction of any kind is made as to thicknesses and separations of the constituent lenses. We can then calculate the 'new aberration' Sph' for every surface of the system, and we can treat all the Sph' as known. In order to secure the utmost generality, we will not assume a perfect object-point. We will assume that our system receives its paraxial ray of the initial u_1 and l_1 used in the ray-tracing from some other optical system, so that there is a primary longitudinal aberration LAp_1 already existing at the object-point to which our system of k surfaces is applied.

At the first surface of our system we shall then have LAp_1 to be transferred and a new aberration $= Sph'_1$; hence we have by the abridged equation (10)**

$$LAp'_1 = LAp_1 \frac{N_1 u_1{}^2}{N'_1 u'^2_1} + Sph'_1.$$

This LAp'_1 is, necessarily, identical with LAp_2 because the second surface receives all the rays exactly as they are delivered to it by the first surface. Therefore our equation (10)**, when applied to the second surface, gives

$$LAp'_2 = \left(LAp_1 \frac{N_1 u_1{}^2}{N'_1 u'^2_1} + Sph'_1\right) \frac{N_2 u_2{}^2}{N'_2 u'^2_2} + Sph'_2\,;$$

and if we multiply this out and remember that by our nomenclature we have $N_2 \equiv N'_1$, and $u_2 \equiv u'_1$, this gives

$$LAp'_2 = LAp_1 \frac{N_1 u_1{}^2}{N'_2 u'^2_2} + Sph'_1 \frac{N'_1 u'^2_1}{N'_2 u'^2_2} + Sph'_2.$$

Proceeding to the third surface, we shall have $LAp'_2 = LAp_3$ or

$$LAp'_3 = \left\{LAp_1 \frac{N_1 u_1{}^2}{N'_2 u'^2_2} + Sph'_1 \frac{N'_2 u'^2_1}{N'_2 u'^2_2} + Sph'_2\right\} \frac{N_3 u_3{}^2}{N'_3 u'^2_3} + Sph'_3\,;$$

and if this is multiplied out and if we again use $N_3 \equiv N'_2$ and $u_3 \equiv u'_2$, we find

$$LAp'_3 = LAp_1 \frac{N_1 u_1{}^2}{N'_3 u'^2_3} + Sph'_1 \frac{N'_1 u'^2_1}{N'_3 u'^2_3} + Sph'_2 \frac{N'_2 u'^2_2}{N'_3 u'^2_3} + Sph'_3.$$

The remarkable law of transfer of the Sph' is already unmistakable, especially when we note that we may give to Sph'_3 the factor $N'_3 u'^2_3 / N'_3 u'^2_3$ without changing its value. We can therefore write down with complete confidence the final result of adding up all the Sph' in the form:

$$LAp'_k = LAp_1 \frac{N_1 u_1{}^2}{N'_k u'^2_k} + Sph'_1 \frac{N'_1 u'^2_1}{N'_k u'^2_k} + Sph'_2 \frac{N'_2 u'^2_2}{N'_k u'^2_k} + \ldots + Sph'_k \frac{N'_k u'^2_k}{N'_k u'^2_k},$$

or more compactly as a sum : [28]

(10)** *for a complete centred system*

$$LAp'_k = LAp_1 \frac{N_1 u_1^2}{N'_k u'^2_k} + \sum_{j=1}^{j=k} Sph'_j \frac{N'_j u'^2_j}{N'_k u'^2_k}; \quad Sph'_j = \tfrac{1}{2} l'_j \frac{l'_j - r_j}{r_j} (i'_j - u_j)(i_j - i'_j);$$

in which j is used as a general symbol for every whole number from 1 to k. The factor of Sph'_k is of course only included for the sake of complete uniformity; being exactly 'one' it is naturally not used in numerical calculations.

As far as numerical calculations are concerned, the transfer of each Sph' directly to the final image instead of by the step-by-step method implied in the original equation (10)** merely means a small saving in time and trouble. But there is profound significance in our sum-formula from the point of view of the designer of new optical systems. Firstly, the direct transfer to the final image removes the difficulty of estimating the relative importance of the aberration arising at each surface, for whilst the Sph' calculated for each surface will be subject to the confusing variation in the actual importance of each unit of LA' which we discussed in [26], the transfer to the same final image puts all the contributions on the same rate of exchange and makes them fairly comparable, so that we can confidently decide which surface is the worst offender. Secondly, and this is of fundamental importance, the sum-formula shows that the amount contributed by each surface to the final aberration depends solely on the Sph', N' and u' of that surface, and on the final N'_k and u'_k, but is utterly independent of all the intervening surfaces; for not a single datum of any of these enters into the transfer. That means that we may make changes in any intervening lens as to curvature or even glass without invalidating the calculation for unchanged surfaces, provided *only* that the changes are such as to leave the final N'_k and u'_k unaltered. Many remarkable solutions will be subsequently based on this property of the primary spherical aberration, especially as we shall prove that other primary aberrations also have the same valuable property.

As a computing example, we will apply the sum-formula to the telescope objective in section [21]. We extract from the paraxial part of the original calculation :

	First Surface.	*Second Surface.*	*Third Surface.*
$\log l'$ (Check !)	1·01793	1·13496	0·88161
$\log (l' - r)$	0·83706	1·23540	1·83003
$-\log r$	−0·55023	−0·55023n	−1·77815n
$\log (l' - r)/r$	0·28683	0·68517n	0·05188n
u	0·000000	0·095956	0·071179
i	0·281689	−0·369539	−0·087128
i'	0·185733	−0·344762	−0·141635
$i' - u$	0·185733	−0·440718	−0·212814
$i - i'$	0·095956	−0·024777	0·054507

For the transfer factors we prepare $\log N'_j u'^2_j$

$\log N'$	0·18087	0·21101	0·00000 (air)
$+2 \log u'$	7·96414	7·70470	8·19858
$\log N'u'^2$	8·14501	7·91571	8·19858

[28] and we then calculate the three sum-items thus :

	First Surface.	*Second Surface.*	*Third Surface.*
$\log \frac{1}{2}$	9·6990	9·6990	9·6990
$+\log l'$	1·0179	1·1350	0·8816
$+\log (l'-r)/r$	0·2868	0·6852*n*	0·0519*n*
$+\log (i'-u)$	9·2689	9·6442*n*	9·3280*n*
$+\log (i-i')$	8·9821	8·3940*n*	8·7365
$+\log N'_j \,.\, u'^2_j$	8·1450	7·9157	0·0000
$+\text{colog } N'_k u'^2_k$	1·8014	1·8014	
log (contribution to final LAp')	9·2011	9·2745*n*	8·6970

Taking the corresponding natural numbers, we find :

$$\text{Final } LAp' = 0{\cdot}15889 - 0{\cdot}18815 + 0{\cdot}04977 = +0{\cdot}02051.$$

We shall use this result subsequently. It may be pointed out that four-figure logs always are amply sufficient for primary aberration work.

Primary Spherical Aberration of a Thin Lens

[29] The method of the preceding section is *exact* as far as the true primary aberration is concerned, and can therefore be applied to any centred system no matter how thick or how widely separated the constituent lenses may be. In many cases the lenses of a system are sufficiently thin to admit of a fairly close estimate of the aberrations by treating the thicknesses as negligible. The strict equation (10)** can then be transformed into a direct *TL* solution for the spherical aberration which does not even call for the tracing of a paraxial ray, but can be applied directly to the preliminary thin lens scheme. We shall show in a subsequent chapter that this *TL* solution is a very powerful weapon when applied with proper judgement and discrimination, but its employment must always be accompanied by a watchful realization of the fact that on account of the total neglecting of the thickness of lenses the solution can never be even a correct second approximation, and that serious disagreements with the inevitable trigonometrical test must be reckoned with. It may therefore often happen, especially to beginners, that the *TL* solution proves inadequate and that the original solution by (10)** must be substituted. It is not possible to lay down universally valid rules in this respect ; the beginner must buy his experience in cases which go beyond the examples which will be given.

In order to deduce this valuable equation, we write down the general equation (10)** for a centred system of two surfaces, so that $k = 2$, and find

$$LAp'_2 = LAp_1 \frac{N_1 u_1^2}{N'_2 u'^2_2} + \tfrac{1}{2}\, l'_1 \frac{l'_1 - r_1}{r_1} (i'_1 - u_1)(i_1 - i'_1) \frac{N'_1 u'^2_1}{N'_2 u'^2_2} + \tfrac{1}{2}\, l'_2 \frac{l'_2 - r_2}{r_2} (i'_2 - u_2)(i_2 - i'_2).$$

As the second surface is also the last one, the transfer factor is exactly 'one' and is therefore omitted. We now specialize this equation for a simple lens (but not yet a thin one) used in air. That gives $N_1 = N_2' = 1$ for air, and as $N_1' \equiv N_2$ [29] is the index of the material of the lens, and the only one remaining in the equation, we introduce simply N as the symbol for it.

For the first surface we have by the law of refraction $i_1 = N \,.\, i_1'$, and by the universal $i+u = i'+u'$ we turn $i_1'-u_1$ into i_1-u_1' and then into $i_1'-u_1 = N \,.\, i_1' - u_1'$. The factor i_1-i_1' we transform into $i_1-i_1' = Ni_1'-i_1' = i_1'(N-1)$.

For the second surface the law of refraction gives $i_2' = N \,.\, i_2$, and with this

$$i_2'-u_2 = N \,.\, i_2-u_2$$
$$i_2-i_2' = i_2-N \,.\, i_2 = -i_2(N-1).$$

On introducing these transformations we obtain

$$LAp_2' = LAp_1 \frac{u_1^2}{u_2'^2} + \tfrac{1}{2}(N-1)\left\{l_1' \frac{l_1'-r_1}{r_1}(Ni_1'-u_1')\,N \,.\, i_1' \frac{u_1'^2}{u_2'^2} - l_2' \frac{l_2'-r_2}{r_2}(N \,.\, i_2-u_2) \,.\, i_2\right\}.$$

In the second term of the big bracket we now have by

(6p)* $\quad N'(l'-r)/l' = N(l-r)/l$, as $N' = 1$, $l_2'-r_2 = N \,.\, l_2'(l_2-r_2)/l_2$,

and as this gives us a factor N in both terms of the bracket, we take N outside. Extending also by $l_2'^4 \,.\, u_2'^2/l_2'^4 \,.\, u_2'^2$, we arrive at the equation

$$LAp_2' = LAp_1 \frac{u_1^2}{u_2'^2} + \tfrac{1}{2}\, l_2'^4 u_2'^2 \,.\, N(N-1)\left\{\frac{l_1'-r_1}{r_1}(Ni_1'-u_1')i_1' \frac{l_1' u_1'^2}{l_2'^4 u_2'^4} - \frac{l_2-r_2}{r_2 l_2}(Ni_2-u_2)\frac{l_2'^2 \,.\, i_2}{l_2'^4 \,.\, u_2'^2}\right\}.$$

We will now transform the final bracket alone. In its first term we have, because at any surface $l \,.\, u = l' \,.\, u' = y$, $l_2'^4 u_2'^4 = l_2^4 \,.\, u_2^4$; and because always $u_2 \equiv u_1'$, we have further

$$l_2^4 u_2^4 = l_2^4 \,.\, u_1'^4 = \left(\frac{l_2}{l_1'}\right)^4 \,.\, l_1'^4 \,.\, u_1'^4.$$

In the second term of the bracket we introduce correspondingly

$$l_2'^4 u_2'^2 = l_2'^2 \,.\, l_2'^2 u_2'^2 = l_2'^2 \,.\, l_2^2 u_2^2,$$

and with slight rearrangement the bracket becomes

$$\{\;\} = \left(\frac{l_1'}{l_2}\right)^4 \frac{l_1'-r_1}{r_1 l_1'}(Ni_1'-u_1')\frac{i_1'}{l_1'^2 u_1'^2} - \frac{l_2-r_2}{r_2 l_2}(Ni_2-u_2)\frac{i_2}{l_2^2 u_2^2}.$$

For a lens of finite thickness l_1' is different from l_2 by the thickness, and it is easily seen that as a consequence we cannot hope to obtain a very simple result for such a lens. But if we now treat the lens as infinitely thin, or at least as so thin that $(l_1'/l_2)^4$ differs only very little from 'one', then we can omit this factor and partially replace l_2 in the second term by l_1', giving

$$\{\;\} = \left(\frac{1}{r_1}-\frac{1}{l_1'}\right)(Ni_1'-u_1')\frac{i_1'}{l_1'^2 u_1'^2} - \left(\frac{1}{r_2}-\frac{1}{l_1'}\right)(Ni_2-u_2)\frac{i_2}{l_2^2 u_2^2},$$

[29] or slightly rearranged

$$\{\} = \left(\frac{1}{r_1} - \frac{1}{l'_1}\right)\left(N\frac{i'_1}{l'_1 u'_1} - \frac{1}{l'_1}\right)\frac{i'_1}{l'_1 u'_1} - \left(\frac{1}{r_2} - \frac{1}{l'_1}\right)\left(N\frac{i_2}{l_2 u_2} - \frac{1}{l_2}\right)\frac{i_2}{l_2 u_2}.$$

By paraxial equation (4p) we now have

$$i' = u'\frac{l'-r}{r}$$, or, on adding suffix 1 and slightly modifying,

$$\frac{i'_1}{l'_1 . u'_1} = \frac{l'_1 - r_1}{r_1 l'_1} = \frac{1}{r_1} - \frac{1}{l'_1},$$

and i_2/l_2u_2 gives in the same way :

$$\frac{i_2}{l_2 u_2} = \frac{1}{r_2} - \frac{1}{l_2} = \frac{1}{r_2} - \frac{1}{l'_1}.$$

The value of the bracket is therefore now

$$\{\} = \left(\frac{1}{r_1} - \frac{1}{l'_1}\right)^2\left(\frac{N}{r_1} - \frac{N}{l'_1} - \frac{1}{l'_1}\right) - \left(\frac{1}{r_2} - \frac{1}{l'_1}\right)^2\left(\frac{N}{r_2} - \frac{N}{l'_1} - \frac{1}{l'_1}\right),$$

or with an obvious further contraction

$$\{\} = \left(\frac{1}{r_1} - \frac{1}{l'_1}\right)^2\left(\frac{N}{r_1} - \frac{N+1}{l'_1}\right) - \left(\frac{1}{r_2} - \frac{1}{l'_1}\right)^2\left(\frac{N}{r_2} - \frac{N+1}{l'_1}\right).$$

We now get rid of all the reciprocals by introducing and slightly extending the simple lens nomenclature of section [23], namely

$$\frac{1}{r_1} = c_1\,;\ \frac{1}{r_2} = c_2\,;\ \frac{1}{r_1} - \frac{1}{r_2} = c \text{ or } \frac{1}{r_2} = c_1 - c,$$

thus introducing the total curvature c of the lens and the curvature c_1 of its *first* surface. As an extension of that nomenclature we introduce into

TL (1) $$\frac{1}{l'} = (N-1)\,c + \frac{1}{l}$$

a corresponding symbol for the reciprocals of the intersection-lengths l and l', and as these are obviously a measure of the *convergence* or *divergence* of the thin pencil, we select small v as an appropriate and easily remembered letter for these reciprocals, giving it a suffix and where necessary a 'dash' agreeing with that of the l. *TL* (1) then takes the form

TL (1) $$v' = (N-1)\,c + v.$$

The same symbol will also be used in the equation (6p)** for the refraction effect of a single surface. For our present first surface with $N_1 = 1$ and $N'_1 = N$ this equation is

$$\frac{N}{l'_1} = \frac{N-1}{r_1} + \frac{1}{l_1}, \text{ or } N\,.\,v'_1 = (N-1)\,c_1 + v_1,$$

and gives [29]

$$\frac{1}{l_1'} = v_1' = \frac{N-1}{N} \cdot c_1 + \frac{v_1}{N}.$$

With this modified nomenclature our last equation for the bracket becomes

$$\{\} = \begin{cases} \left(c_1 - \frac{N-1}{N} \cdot c_1 - \frac{v_1}{N}\right)^2 \left(N \cdot c_1 - \frac{N^2-1}{N} c_1 - \frac{N+1}{N} v_1\right) \\ -\left(c_1 - c - \frac{N-1}{N} c_1 - \frac{v_1}{N}\right)^2 \left(N \cdot c_1 - N \cdot c - \frac{N^2-1}{N} c_1 - \frac{N+1}{N} v_1\right) \end{cases}$$

or, by dividing out $(N-1)/N$ and $(N^2-1)/N$ and contracting,

$$\{\} = \left(\frac{c_1 - v_1}{N}\right)^2 \left(\frac{c_1}{N} - \frac{N+1}{N} v_1\right) - \left(\frac{c_1 - v_1}{N} - c\right)^2 \left(\frac{c_1}{N} - \frac{N+1}{N} v_1 - N \cdot c\right).$$

This has to be multiplied out and ordered. It is at once obvious that the second term produces one partial product which exactly cancels the complete first term, and the only parts to be worked out are therefore

$$\{\} = 2c \frac{c_1 - v_1}{N} \left(\frac{c_1}{N} - v_1 \frac{N+1}{N} - N \cdot c\right) - c^2 \left(\frac{c_1}{N} - v_1 \frac{N+1}{N} - N \cdot c\right) + \left(\frac{c_1 - v_1}{N}\right)^2 \cdot N.c.$$

If this is done, the terms ordered according to descending powers of c and the original outside factor $\frac{1}{2}N(N-1)$ of our bracket included, the result will be

$$LAp_2' = LAp_1 \left(\frac{u_1}{u_2'}\right)^2$$
$$+ l_2'^4 \cdot u_2'^2 \{G_1 \cdot c^3 - G_2 c^2 c_1 + G_3 c^2 v_1 + G_4 c c_1^2 - G_5 c c_1 v_1 + G_6 c v_1^2\}$$

in which the G are pure functions of the refractive index N, defined by

$$G_1 = \frac{N^2(N-1)}{2} \qquad G_2 = \frac{(2N+1)(N-1)}{2} \qquad G_3 = \frac{(3N+1)(N-1)}{2}$$

$$G_4 = \frac{(N+2)(N-1)}{2N} \qquad G_5 = 2\frac{N^2-1}{N} \qquad G_6 = \frac{(3N+2)(N-1)}{2N}.$$

These have been calculated for the whole possible range of indices from 1·43 (fluorite) to 1·76 (densest flint) at intervals of 0·01, and a table of the numerical values and of their logs is given in the appendix. The calculation of the spherical aberration of any thin lens is thus rendered extremely convenient and rapid.

As our equation is intended for use in conjunction with a preliminary thin lens scheme, in which we use only intersection-lengths and no paraxial angles, it is desirable to replace the latter by the former. For any thin lens we have $u_1/u_2' = l_2'/l_1$, which serves to transform the first term. In the second term we have $l_2'^4 \cdot u_2'^2 = l_2'^2 \cdot l_2'^2 \cdot u_2'^2 = l_2'^2 \cdot y^2$, a suffix to the latter being unnecessary for one *thin* lens as there can be no contraction or expansion of the cone of rays in the infinitely small distance between the two surfaces. As we are now dealing

[29] exclusively with thin *lenses*, we can again drop suffixes altogether and write our equation in its final form

TL (10)** *for one thin lens :*

$$LAp' = LAp\left(\frac{l'}{l}\right)^2 + y^2 \,.\, l'^2\,[G\text{-sum}]$$

$$[G\text{-sum}] = G_1 \,.\, c^3 - G_2c^2 \,.\, c_1 + G_3c^2 \,.\, v_1 + G_4c \,.\, c_1^2 - G_5c \,.\, c_1 \,.\, v_1 + G_6c \,.\, v_1^2,$$

c being the total curvature, c_1 the curvature of the first surface, and v_1 the reciprocal of the left intersection-length of the lens.

It is now a simple matter to add up the spherical aberration of a number of thin lenses at any finite separations. Taking the original form of the solution for one thin lens (i. e. before the u were replaced by l) we have for the first lens 'a' of a system

$$LAp'_a = LAp_a\left(\frac{u_a}{u'_a}\right)^2 + l'^4_a u'^2_a\,[G\text{-sum}]_a,$$

and for any following lens we shall have the same equation with merely a change in the letter-suffix to 'b', 'c', &c. The LAp of any following lens will be the LAp' delivered to it by the immediately preceding lens, or $LAp_b = LAp'_a$, $LAp_c = LAp'_b$ &c. Consequently we have for two lenses combined

$$LAp'_b = \left[LAp_a\left(\frac{u_a}{u'_a}\right)^2 + l'^4_a u'^2_a\,[G\text{-sum}]_a\right].\left(\frac{u_b}{u'_b}\right)^2 + l'^4_b u'^2_b\,[G\text{-sum}]_b,$$

and as by the fundamental computing formulae we have again $u'_a \equiv u_b$, the solution of the first term gives

$$LAp'_b = LAp_a\left(\frac{u_a}{u'_b}\right)^2 + l'^4_a\frac{u'^4_a}{u'^2_b}\,[G\text{-sum}]_a + l'^4_b u'^2_b\,[G\text{-sum}]_b.$$

This will be the LAp_c delivered to a third lens and by the general formula will have the transfer factor $(u_c/u'_c)^2 = (u'_b/u'_c)^2$; therefore for a system of three thin lenses at any finite separation

$$LAp'_c = LAp_a\left(\frac{u_a}{u'_c}\right)^2 + l'^4_a\frac{u'^4_a}{u'^2_c}\,[G\text{-sum}]_a + l'^4_b\frac{u'^4_b}{u'^2_c}\,[G\text{-sum}]_b + l'^4_c u'^2_c\,[G\text{-sum}]_c.$$

The general law is already perfectly obvious, especially if we again remember that the final term may be extended by $(u'_c/u'_c)^2$ without changing its value. For a system of 'k' thin lenses at any finite separations the first term will be LAp_a $(u_a/u'_k)^2$, and the general jth term will be $(l'^4_j u'^4_j/u'^2_k)$ $[G\text{-sum}]_j$; but as our equations for thin lenses are intended for use with a preliminary thin lens scheme, we must again modify the solution so as to make it most convenient for actual numerical work. This we do by introducing the directly calculated intersection-lengths and the semi-aperture y of the lenses.

As at any refracting surface we have $l \,.\, u = l' \,.\, u' = y$, we can immediately replace $l'^4_j \,.\, u'^4_j$ by y_j^4. For the u'^2_k which appears everywhere in the denominator we have by the same relation, namely $u'_k \,.\, l'_k = y_k$, $1/u'^2_k = l'^2_k/y_k^2$, and similarly in the first term $u_a^2 = y_a^2/l_a^2$. But as the y do not enter into the original thin lens calculation, we must calculate them, and as we proved that for paraxial rays all the [29] y of a system are in a fixed proportion, we can assume a value for one, conveniently either y_a or y_k and determine the others. We will assume that a definite value has been assigned to y_a, the semi-aperture of the first thin lens. Then we have, because $u'_a \equiv u_b$, $y_a/l'_a = y_b/l_b$, or $y_b = y_a\,(l_b/l'_a)$. In the same way we find $y_c = y_b(l_c/l'_b)$ or on introducing the preceding value of y_b : $y_c = y_a\,(l_b/l'_a)\,(l_c/l'_b)$. Continued to the end, this evidently leads to $y_k = y_a\,(l_b/l'_a)\,(l_c/l'_b)\,\ldots\ldots(l_k/l'_{k-1})$ and all the y can be conveniently calculated in this way. Hence we arrive at the computing formula for the primary aberration of thin lens combinations

TL (10)** *for k thin lenses at any separations :*

$$LAp'_k = LAp_a\left(\frac{y_a}{y_k}\right)^2\left(\frac{l'_k}{l_a}\right)^2 + \frac{y_a^{\,4}}{y_k^{\,2}}\,l'^2_k\,[G\text{-sum}]_a + \frac{y_b^{\,4}}{y_k^{\,2}}\,.\,l'^2_k\,[G\text{-sum}]_b + \;\ldots\ldots + \frac{y_k^{\,4}}{y_k^{\,2}}\,l'^2_k\,[G\text{-sum}]_k$$

in which y_a may be selected ; then $y_b = y_a\,.\,l_b/l'_a$; $y_c = y_a\,(l_b/l'_a)\,(l_c/l'_b)$, &c. In the G-sums for the separate lenses the v_1 is found as $1/l_j$ for the jth lens, the c_1 is the reciprocal of its first (left-hand) radius and the plain c is its total curvature, found by $c = 1/f'_j(N_j - 1)$.

Our general solution becomes simplified for a system of closely packed lenses in which the total thickness may be regarded as still negligibly small, for then all the y_j will be equal and may all be called simply y, the semi-aperture of the *thin system* and the formula becomes

TL (10)** *for k thin lenses in contact :*

$$LAp'_k = LAp_a\left(\frac{l'_k}{l_a}\right)^2 + y^2 l'^2_k\,[G\text{-sum}]_a + y^2 l'^2_k\,[G\text{-sum}]_b + \;\ldots\; + y^2 l'^2_k\,[G\text{-sum}]_k.$$

The two sum-formulae just deduced will find numerous applications in subsequent chapters, but as genuine examples almost invariably include some form of chromatic correction, we will defer numerical work by their aid until later.

Change of Spherical Aberration by Bending a Lens

In the equation TL (10)** for one thin lens the first term in LAp is zero when we [30] apply the lens to a real object-point, so that in this most usual case we have

$$LAp' = y^2 l'^2\,[G_1c^3 - G_2c^2c_1 + G_3c^2v_1 + G_4cc_1^{\,2} - G_5cc_1v_1 + G_6c\,v_1^{\,2}].$$

For a lens of given index the G have perfectly definite values obtainable from our tables ; when we decide upon a definite focal length, then the total or net curvature 'c' of the lens is also fixed by $c = 1/f'(N - 1)$, and when we now apply this lens to an object-point at a fixed distance l, then by definition $v_1 = 1/l$ and this also becomes a fixed quantity. Simultaneously we obtain by TL (3) a fixed value of l', and by putting in all these values of the fixed data, the equation simplifies to

$$LAp' = y^2\,(a\,.\,c_1^{\,2} + b\,.\,c_1 + n),$$

in which a, b, and n have definite numerical values. As c_1 is the curvature $1/r_1$ of

[30] the first surface of our lens, it is evident that the spherical aberration of the latter will vary when we change c_1 by bending the lens. If we calculated LAp' for a succession of values of c_1, that is for a regular series of bendings like those worked out in [23] for our biconvex lens, assuming some definite value of y, and plotted on squared paper the LAp' found as ordinates against the values of c_1 as abscissae, then the graph would assume the form of an ordinary parabola with vertical axis, for it is shown in analytical geometry that any equation of the form

$$y = p \,.\, x^2 + q \,.\, x + r$$

represents a parabola with vertical axis. Moreover the discussion of this general equation shows that the parabola will have its pole below and stretch upwards into positive infinity if the coefficient of the quadratic term is positive, and that the pole will be the highest point and the arms of the curve will reach negative infinity if the quadratic term is negative.

In the case of our lens it is easily seen that in the simplified equation of LAp' $a = G_4 \,.\, c \,.\, l'^2$, and as G_4 and l'^2 are both positive, it is clear that 'a' has the sign of the total curvature c of the lens, and is therefore positive for a convex and negative for a concave thin lens. Hence the spherical aberration of a convex lens has a certain minimum value (at the pole of the parabola) for a particular bending, and can reach any greater value up to positive infinity for sufficiently large changes of the bending, whilst for a concave lens there is a maximum of spherical aberration at the pole of the parabola and extreme bendings give huge negative values of the spherical aberration. It is of course highly important always to remember in mathematical discussions of maximum or minimum values that these do *not* mean the highest or lowest possible *numerical* values, but that the *sign* must be taken into consideration. A mathematical maximum means either the highest possible positive value or the lowest possible negative value, and conversely a minimum means either the lowest possible positive, or highest possible negative value. Moreover, curves of more complicated shape than that of a simple vertical parabola may have a number of alternate maxima and minima at different points of their course, and the above definitions then apply only to the immediate neighbourhood of any one maximum or minimum. For this reason a mathematically determined maximum or minimum may occasionally strike the practical mind as mere hair-splitting or possibly as downright nonsense.

The fact that the primary spherical aberration of a thin lens of given power and working at given conjugate distances may be adjusted by bending to any value between the maximum or minimum value and negative or positive infinity respectively, is of the utmost importance. The correction of spherical aberration in optical instruments is always based on this fact. We must, however, note at once that the simple parabolic law of the change of LAp' by bending only holds for the *primary* aberration of an *infinitely thin* lens ; strictly speaking this is a mere mathematical abstraction and we must estimate the effect of giving a finite thickness to the lens. In section [29] we allowed for thickness up to a certain stage and found that the principal effect was a factor $(l'_1/l_2)^4$ to the contribution of the first surface ; but in addition to this, the contribution of the second surface also contained l_2 instead of l'_1. We obtained the final simple result by neglecting the difference between l_1' and l_2. If we consider only the major inaccuracy of omitting [30] the factor $(l_1'/l_2)^4$ in the first term and restore this factor in the equation arrived at when the c_1 and v_1 were first introduced, we see that then the first term would not be cancelled by one partial product of the second term. Consequently terms in c_1^3, in $c_1^2 v_1$, in $c_1 v_1^2$, and in v_1^3 would be added to the G-sum, and the latter would become a cubic (instead of a quadratic) equation in c_1 and v_1. It is easily seen that the neglecting in the second term of the difference between l_2 and l_1' eliminated further complications, and we must therefore conclude that the parabolic law of the effect of bending is only a first approximation to the truth and is really modified—for lenses of finite thickness—by terms in higher powers of c_1 and chiefly by a term in c_1^3. Whilst we shall almost exclusively adopt the parabolic law in our solutions of lens problems, we must always remember its limited accuracy and must not trust it too much, particularly in extrapolations.

We will calculate the primary spherical aberration of a thin lens of the total curvature and refractive index of our biconvex lens of sections [9], [20], and [23]. We have $N = 1{\cdot}518$ and the total curvature, as determined in [23], $c = 0{\cdot}3000$. As the object-point is given by $l = L = -24$, we have $v_1 = \dfrac{1}{l} = -0{\cdot}0417$, and we found in section [23] $l_2' = l'$ of our present formula $= 8{\cdot}795$. In order to calculate the spherical aberration for *any* bending we introduce the fixed data into the formula at the head of the present section in order to obtain the simplified equation $LAp' = y^2\,(a \,.\, c_1^2 + b \,.\, c_1 + n)$.

Interpolating from the log G table for $N = 1{\cdot}518$ we calculate thus in six parallel columns:

$LAp' = l'^2y^2.\ \{$	$G_1 \,.\, c^3$	$-G_2c^2 \,.\, c_1$	$+G_3c^2\, v_1$	$+G_4c \,.\, c_1^2$	$-G_5c\, v_1 \,.\, c_1$	$+G_6c\, v_1^2\}$
log G	9·7759	0·0193n	0·1579	9·7783	0·2351n	0·0485
+log l'^2	+1·8885	+1·8885	+1·8885	+1·8885	+1·8885	+1·8885
+log c^n	+8·4313	+8·9542	+8·9542	+9·4771	+9·4771	+9·4771
+log v_1^n			+8·6201n		+8·6201n	+7·2402
	0·0957	0·8620n	9·6207n	1·1439	0·2208	8·6543
$LA'p = y^2\ .\ \{$	1·247	$-7{\cdot}278\, c_1$	$-0{\cdot}418$	$+13{\cdot}928 c_1^2$	$+1{\cdot}663\, c_1$	$+0{\cdot}045\}$
Collection of terms	$-0{\cdot}418$	$+1{\cdot}663\, c_1$				
	$+0{\cdot}045$					
$LAp' = y^2\ .\ \{$	0·874	$-5{\cdot}615\, c_1$		$+13{\cdot}928 c_1^2\}$		

It should be noted that in the log G line the minus sign of two of the terms is at once taken into account by adding the 'n' to the log. The outside factor l'^2 of the general equation being taken into the bracket, causes 2 . log l' to appear in each of the six columns. The description log c^n in the third line merely denotes that c appears in the first, second, and third power in different columns; log c as found in the logarithmic table goes directly into the last three columns, doubled into the second and third column, and trebled into the first column. When c is negative, as it is nearly always in the flint components of achromatic lenses, then the first and third powers will be negative and the log c^n in the first and in the last three columns will require the 'n' adding. As v_1 also appears in two different powers, the same remarks apply to log v_1^n. The results give the complete numerical value

[30] of the first, third, and last terms, but only factors of c_1 in the second and fifth column, and the factor of c_1^2 in the fourth column.

Collection of terms then gives the formula for the spherical aberration of any bending of our lens, when the object is at distance $l = -24$:

$$LAp' = (13{\cdot}928\, c_1^2 - 5{\cdot}615\, c_1 + 0{\cdot}874) \, . \, y^2.$$

The bracket alone gives the aberration for $y = 1$, or for a lens of 2″ clear aperture. The last, absolute, term 0·874 obviously is the LAp' of our lens for this aperture when $c_1 = 0$, that is for a planoconvex form with the convexity on the second surface. By putting in $c_1 =$ respectively $-0{\cdot}1$, 0, $+0{\cdot}1$, $+0{\cdot}2$, $+0{\cdot}3$, $+0{\cdot}4$ we shall obtain the primary aberration of each of the six bendings selected in section [23] as

for $c_1 =$	$-0{\cdot}1$	0	$+0{\cdot}1$	$+0{\cdot}2$	$+0{\cdot}3$	$+0{\cdot}4$
$LAp' = y^2 \times$	1·575	0·874	0·452	0·308	0·444	0·858.

For a paraxial ray associated with a marginal ray at $U_1 = -2°$ the initial $y = l \, . \, u$ would be $= (-24) \sin(-2°) = 0{\cdot}8375$, and with this value, or $y^2 = 0{\cdot}7014$, our last results become

$LA'p$ for $U_1 = -2°$:	1·105	0·613	0·317	0·216	0·312	0·601.

As all previous calculations were made for the bending having $c_1 = 1/r_1 = 0{\cdot}1$, we can check the result for this bending; it will be found that in section [27] we found the true primary aberration for $U_1 = -2°$ as 0·3147, and in section [24] the trigonometrically correct value was found as 0·3194, both for the actual lens with a thickness of 0·600. Our thin lens result agrees remarkably well, but this is again largely due to the fact that with this particular bending the rays have only a very slight convergence within the lens. Beginners will be wise if they buy a first instalment of experience by testing trigonometrically some of the other bendings. For an initial $U_1 = -2°$ two thicknesses may be tried; first the approximate minimum possible thickness for that aperture, namely $d_1' = 0{\cdot}12$, and then the full original thickness of 0·600.

The figures in the last numerical table show that the bending with $c_1 = 0{\cdot}2$ gives the smallest spherical aberration. We can find the exact minimum point by differentiating the equation for LAp' with reference to c_1 and equating the differential coefficient to zero. This gives

$$27{\cdot}856\, c_1 - 5{\cdot}615 = 0 \qquad \text{or } c_1 = 0{\cdot}2016.$$

SECONDARY AND HIGHER SPHERICAL ABERRATION

[31] We have proved that the longitudinal spherical aberration in the extended paraxial region grows with the square of the aperture: we have also shown empirically by the numerical results for our biconvex lens that for larger apertures discrepancies from that simple law make their appearance and grow very nearly as the fourth power of the aperture. Moreover, our strict equation (10)* shows that (excluding the discontinuity when l' and L' are very large, which is discussed in section [26]), there will be with increasing aperture a steady growth of the new spherical aberration arising at any surface. Hence if the aberration is plotted against the distance

Y from the optical axis at which rays pass through a surface, or against the angles U' (or their sine or tangent) at which they meet the axis, it will yield a continuous and smooth curve, and the same will be the case for the aberration of the rays after passing through a whole system of centred lens surfaces. [31]

As follows from Taylor's and Maclaurin's theorem, any such smooth and continuous curve can be represented by an infinite series in rising powers of Y (or of U', sin U', tan U', &c.) ; we can therefore claim that our LA' must admit of being represented by the series

$$LA' = a_0 + a_1 Y + a_2 Y^2 + a_3 Y^3 + a_4 Y^4 + \&c.,$$

in which a_0, a_1, &c., are constants for the surface or the system of lenses. We will divide this series into two parts, containing respectively the odd and the even powers of Y, thus

$$LA' = \begin{cases} a_0 + a_2 Y^2 + a_4 Y^4 + \&c. \\ \quad + a_1 Y \ + a_3 Y^3 + \&c. \end{cases}$$

We can then draw an important conclusion from the symmetry of lens surfaces with reference to the optical axis, already referred to in section [6]. LA' by reason of this symmetry is bound to be exactly the same for a ray passing at distance Y above the axis and for a corresponding ray at distance $-Y$ below the axis. Hence our series *also* must give exactly the same result for LA' if $+Y$ is replaced by $-Y$, whatever value of Y between zero and full aperture may be chosen. The terms in the first horizontal line of our last equation satisfy this condition for $(-Y)^2 = (+Y)^2$, $(-Y)^4 = (+Y)^4$, &c. But the terms in the second line contradict our condition derived from the symmetry of centred systems. The odd powers of Y have the sign of Y itself and would therefore cause LA' to come out differently for rays passing at equal distances above and below the optical axis. Therefore the terms in the second line must give zero-value for any value of Y, and this can only be if the coefficient a_1, a_3, &c., are each zero. We thus prove, merely by the symmetry, that the only possible series by which LA' can be represented is one containing only even powers of Y or of any other reasonable measure of the aperture such as PA, U', sin U', tan U', and others. Moreover, as we have agreed to measure LA' from the paraxial focus, the series must give $LA' = 0$ for $Y = 0$; hence the constant first term a_0 must also be zero. The final result is therefore that the longitudinal spherical aberration of any centred optical system must be accurately represented by a sufficient number of terms of the series

$$(11) \qquad LA' = a_2 Y^2 + a_4 Y^4 + a_6 Y^6, \ \&c.$$

in which Y does not necessarily mean the ordinate of the point of penetration of a ray, but may be replaced by other measures of the aperture such as have been enumerated. But such a change in the measure of the aperture will of course call for a new determination of the *numerical value* of the coefficients a_2, a_4, &c.

The reasoning by which we have arrived at (11) should be thoroughly mastered, as it very frequently affords a short cut to important deductions and thus saves long and laborious analytical investigations. We can, for instance, conclude at once that any U, sin U, or tan U, or in fact any plain or 'dashed' angle in a lens system, or sin or tan of such angles or of fractions or multiples of them, must depend on Y or on PA in the form of a series of *odd* powers of Y or PA, simply

[31] because all such angles and their stated functions change sign with Y or PA for corresponding rays above and below the optical axis, and therefore cannot contain terms depending on even powers.

We have already shown that the aberration of our biconvex lens can be very closely represented by the first two terms of (11) when using the sine of the original angle U as the variable denoted by Y in the equation. It is in fact found in computing practice that practically all ordinary optical systems can have their longitudinal spherical aberration represented with ample accuracy over the whole aperture by the first two terms of (11), provided that the trigonometrically determined LA' of the extreme marginal ray is one of the data from which the coefficients a_2 and a_4 are determined. The only, but a very decided and important, exception to this statement is provided by microscope objectives of high power: these call for the use of at least three terms of (11), but special methods will be given in Part II for dealing with these interesting systems. Two terms of (11) will also be insufficient for ordinary systems if, instead of determining the coefficients a_2 and a_4 so as to fit the trigonometrically traced marginal ray, the coefficients are calculated analytically by formulae derived from a second approximation for what we might call the doubly extended paraxial region. Such formulae are not only extremely complicated and therefore difficult to work with, but they leave still higher aberrations entirely out of account, and so are apt to fail conspicuously to give the true aberration for the marginal ray.

Evidently the first term in (11) represents the primary spherical aberration dealt with in the preceding sections. We shall call the aberration represented by the second term 'secondary aberration', and the aberration represented by the third term (if ever wanted) 'tertiary aberration'. In Part I we shall only deal with primary and secondary aberration, and shall therefore always assume that we have nearly enough (provided the true marginal ray is included)

$$LA' = a_2 Y^2 + a_4 Y^4.$$

As this equation involves two constants a_2 and a_4, it is evident that two values of LA' corresponding to different values of Y are sufficient to determine these constants. We have specified that the marginal aberration for the full semi-aperture Y_m which we will henceforth call LA'_m shall always be one of the two. The other one, if trigonometrically determined, will be that of some ray intermediate between axis and margin, or a zonal ray. We will call the value of Y for this ray Y_z (z for zonal), and the corresponding aberration LZA' (longitudinal zonal aberration). We shall then have the two equations

$$LA'_m = a_2 Y_m^2 + a_4 Y_m^4;$$
$$LZA' = a_2 Y_z^2 + a_4 Y_z^4;$$

which are linear with regard to the unknown quantities a_2 and a_4, and give the general solution for these

$$(11)^* \quad \begin{cases} a_2 = (LZA' Y_m^4 - LA'_m . Y_z^4)/[Y_z^2 . Y_m^2 (Y_m^2 - Y_z^2)] \\ a_4 = (LA'_m Y_z^2 - LZA' . Y_m^2)/[Y_z^2 . Y_m^2 (Y_m^2 - Y_z^2)]. \end{cases}$$

We will try this solution for our biconvex lens at its final focus on the assumption that we had only traced the paraxial rays and the rays at original $U_z = -2^\circ$ and

$U_m = -4^\circ$, for which we found in section [24] $LZA' = 0{\cdot}3194$ and $LA'_m = 1{\cdot}3518$, and we will take the original U_1 as the measure of the aperture, so that in the general solution $Y_m = -4^\circ$ and $Y_z = -2^\circ$. This gives [31]

$$a_2 = (0{\cdot}3194 \times 256 - 1{\cdot}3518 \times 16)/[4 \times 16\,(16-4)\,]$$
$$a_4 = (1{\cdot}3518 \times 4 - 0{\cdot}3194 \times 16)/[64 \times 12]$$
$$\text{or } \log a_2 = 8{\cdot}89379 \qquad \log a_4 = 6{\cdot}58740$$
$$a_2 = 0{\cdot}078305 \qquad a_4 = 0{\cdot}000387.$$

The aberrations of the biconvex lens should therefore be closely represented, for any value of original U_1, by the equation

$$LA' = 0{\cdot}078305 \times U_1^2 + 0{\cdot}000387 \times U_1^4.$$

We will try this for all the values originally calculated for, namely -1°, -2°, -3°, and -4°:

Original U_1		-1	-2	-3	-4
$0{\cdot}078305\ U_1^2$	=	0·078305	0·31322	0·70474	1·2529
$0{\cdot}000387\ U_1^4$	=	0·000387	0·00619	0·03132	0·0990
Predicted LA'	=	0·07869	0·31941	0·73606	1·3519
Trigonometrical LA'	=	0·0790	0·3194	0·7350	1·3518
Trig.-predicted	=	+0·0003	–	−0·0011	–

The agreement for 2° and 4° merely proves the accuracy of our solution by (11)*. But the close agreement for 1° and 3° may be taken as evidence of the closeness to the truth of our assumption that the first two terms of the aberration-series, when evaluated from the marginal and one other finite ray, are sufficient to give a practically exact account of the spherical aberration. It should also be noted that the primary aberration terms in U_1^2 for the several angles agree very closely with those found a little earlier by the LAp'-equation.

We can solve the problem of determining a_2 and a_4 with still less trouble by tracing only a marginal and the associated paraxial ray and then calculating LAp' by (10)**. We then have the two equations

$$LA'_m = a_2 Y_m^2 + a_4 Y_m^4$$
$$LA'_p = a_2 Y_m^2.$$

Therefore $\quad a_4 = (LA'_m - LA'p)/Y_m^4$

and $\quad a_2 = LA'p/Y_m^2.$

For our biconvex lens we found $LA'_m = 1{\cdot}3518$ and $LA'p = 1{\cdot}2572$. Using again the original U_1 as the measure of the aperture, we find (the first by slide-rule) $a_4 = 0{\cdot}0946/256 = 0{\cdot}000369$; $a_2 = 1{\cdot}2572/16 = 0{\cdot}078575$, and if we calculate $LA' = a_2 Y^2 + a_4 Y^4$ for $Y = 1, 2, 3, 4$, we obtain:

Original U_1	=	-1°	-2°	-3°	-4°
$a_2 U_1^2$		0·078575	0·31430	0·70716	1·25719
$a_4 U_1^4$		0·000369	0·00590	0·02988	0·09446
Predicted LA'	=	0·07894	0·3202	0·73704	1·35165
Trig. LA'	=	0·0790	0·3194	0·7350	1·3518
Trig.-predicted	=	0·00006	−0·0008	−0·00204	–

[31] We see that the residual errors are again very small fractions of the whole aberration, but not quite so small as in the solution by LA'_m and LZA'.

As a rule the secondary aberration calls for study only in those lens systems which admit of correction of the marginal spherical aberration. As will be proved more conclusively in Part II, the best correction is attained when the paraxial and extreme marginal rays are brought to the same focus, that is, when the system is given such a form that the longitudinal spherical aberration becomes zero for the extreme marginal ray, or $LA'_m = 0$. If we again limit ourselves to the two-term formula for LA' we shall therefore have

$$LA'_m = a_2 Y_m^2 + a_4 Y_m^4 = 0$$

$$\text{or} \quad a_2 = -a_4 Y_m^2 ;$$

and putting this into the general equation for LA' we obtain for this case of trigonometrically corrected systems

$$(a) \quad LA' = -a_4 Y_m^2 Y^2 + a_4 Y^4$$

in which Y_m measures the fixed marginal semi-aperture, whilst Y may have any value between zero for the strictly paraxial rays and Y_m for the extreme margin. We can then ask: for which value of Y will LA' reach its highest value? According to the theory of maxima and minima the answer is found by forming the differential coefficient of equation (a) with reference to Y and equating this to zero. This gives the condition

$$d(LA')/dY = -2\,a_4 Y_m^2 Y + 4\,a_4 Y^3 = 0,$$

or, dividing throughout by $-2a_4 Y$,

$$Y_m^2 - 2Y^2 = 0,$$

which gives the immediate answer that LA' will have a maximum or minimum for

$$Y = \pm Y_m \sqrt{\tfrac{1}{2}} = \pm 0{\cdot}7071\ Y_m.$$

The highest numerical value of LA' is therefore found for the rays passing through 0·7071 of the full aperture: that is the reason why we use this zone when testing a system trigonometrically for zonal aberration, as we did in the case of example 2 in section [21]. We can decide whether the corresponding LZA' is a maximum or a minimum, that is in this case, when we know that LA' is zero for both the paraxial and the marginal zone, whether LZA' is positive or negative, by forming the second differential coefficient and deciding whether it is negative or positive respectively. Differentiating the first differential coefficient again, we find

$$d^2(LA')/dY^2 = -2a_4 Y_m^2 + 12\,a_4 Y^2$$

$$= 2a_4 Y_m^2 \left(6\left(\frac{Y}{Y_m}\right)^2 - 1\right).$$

Putting in the value of Y for which LZA' is a maximum or minimum, namely $(Y/Y_m)^2 = \frac{1}{2}$, this becomes

$$d^2(LA')/dY^2 = 4\,a_4 Y_m^2 ;$$

and as Y_m^2 is necessarily positive, the second differential coefficient must have the sign of a_4, or of the secondary aberration present in the system. Therefore [31]

> LZA' is a maximum and consequently positive if the secondary aberration is negative.
>
> LZA' is a minimum and negative if the secondary aberration is positive.

In the case of all ordinary achromatic and spherically corrected lens systems with cemented components the secondary aberration is negative for positive or collective systems (that is those which produce refraction effects like those of a thin convex lens) and the secondary aberration is positive for negative or dispersive systems.

We can determine the relation between the maximum LZA' and the full secondary aberration present in the system by putting $Y^2 = \frac{1}{2} Y_m^2$ into the general equation (a), with the result

$$LZA' = -\tfrac{1}{2} a_4 Y_m^4 + \tfrac{1}{4} a_4 Y_m^4 = -\tfrac{1}{4} a_4 Y_m^4.$$

$a_4 Y_m^4$ represents the whole of the secondary aberration at the extreme margin of the system (which is there cancelled by an equal amount of primary aberration of opposite sign). LZA' determined for 0·7071 of the full aperture therefore represents in magnitude exactly $\frac{1}{4}$ of the full amount of secondary aberration present in the margin, but is of the opposite sign.

In example 2 we corrected the marginal LA'_m as nearly as five-figure logs allow (within 0·0003 in nearly 8, or to 1 part in 27,000). The 0·7071 zone gave $LZA' = +0{\cdot}0050$. We can therefore now claim :

(1) That 0·0050 is the highest value of zonal aberration for that telescope-objective.
(2) That the secondary aberration present at the margin is four times the zonal, with reversed sign, or $a_4 Y_m^4 = -0{\cdot}0200$.
(3) That this secondary aberration is balanced at the margin by an equal amount of primary aberration ; therefore $a_2 Y_m^2 = +0{\cdot}0200$.
(4) As Y_m was in this case $= 1$, a_4 must be $= -0{\cdot}020$, hence (a) gives the formula for LA' :

$$LA' = 0{\cdot}02\ Y^2 - 0{\cdot}02\ Y^4.$$

By this formula LA' can be calculated for any zone. It will be a useful exercise and will increase faith in the theory, to test the formula, say for $Y = \frac{1}{2}$, for which the formula gives $LA' = 0{\cdot}005 - 0{\cdot}00125 = 0{\cdot}00375$. The calculation should of course be done accurately with full checks and a difference of about 0·0005 may even then be found owing to the limited accuracy of five-figure logs.

We can estimate the secondary aberration of spherically corrected lens systems with decidedly less labour than is required for the tracing of a zonal ray, by tracing only the usual associated paraxial and marginal rays and by then calculating the true primary aberration with the aid of equation (10)**. As at the full aperture of such a system the primary and secondary aberration exactly cancel each other, it follows that for spherically corrected systems we have $LAp' = -a_4 Y_m^4 = 4\,LZA'$. For the 2 inch telescope objective we calculated LAp' in section [28] as $= 0{\cdot}0205$. We can therefore conclude that the LZA' of this objective would be $= 0{\cdot}0051$ if only secondary aberration were present in addition to the primary LAp'. It will

[31] be noticed that the agreement with the directly calculated LZA' is remarkably close. This LAp'-method of testing systems for zonal aberration may be recommended with confidence, as it gives an accurate result by a very short and simple calculation with four-figure logs. But it is necessary to work by the exact equation (10)** ; the TL form of the equation would in nearly all cases give a totally wrong estimate of the zonal aberration!

As we now know the spherical aberration of our telescope objective for three zones, we can determine the secondary *and* tertiary aberration in addition to the primary aberration. In the general equation

$$LA' = a_2 Y^2 + a_4 Y^4 + a_6 Y^6 \tag{11}$$

the first term is the primary aberration and has been calculated in [28] for $y = Y_m$, and the $LAp' = 0{\cdot}0205$ then found therefore represents $a_2 \,.\, Y_m^2$. For the zonal and marginal rays equation (11) gives

$$LZA' = a_2 Y_z^2 + a_4 Y_z^4 + a_6 Y_z^6$$

and

$$LA'_m = a_2 Y_m^2 + a_4 Y_m^4 + a_6 Y_m^6,$$

and if in the first we use our adopted value $Y_z = 0{\cdot}7071\, Y_m$ we obtain, on introducing LAp' in the first term,

$$LZA' = \tfrac{1}{2} LAp' + \tfrac{1}{4} a_4 Y_m^4 + \tfrac{1}{8} a_6 Y_m^6$$

$$LA'_m = LAp' + a_4 Y_m^4 + a_6 Y_m^6.$$

These are two linear equations for the secondary aberration $a_4 Y_m^4$ and the tertiary $a_6 Y_m^6$ of the marginal ray, and give the solution

$$a_4 Y_m^4 = 8\, LZA' - 3\, LAp' - LA'_m \; ; \; a_6 Y_m^6 = -8\, LZA' + 2\, LAp' + 2\, LA'_m.$$

Introduced into the general equation for the LA' corresponding to any value of Y, these give the formula

$$LA' = LAp' \left(\frac{Y}{Y_m}\right)^2 + a_4 Y_m^4 \left(\frac{Y}{Y_m}\right)^4 + a_6 Y_m^6 \left(\frac{Y}{Y_m}\right)^6$$

and this is the most direct and simplest solution of the problem.

By the five-figure calculation in [21] we found $LA'_m = -0{\cdot}0003$ and $LZA' = 0{\cdot}0050$; if we add $LAp' = 0{\cdot}0205$ from [28] we find by the above solution

$$a_4 Y_m^4 = -0{\cdot}0212 \qquad a_6 Y_m^6 = +0{\cdot}0004$$

or the formula for the LA' of *any* zone,

$$LA' = 0{\cdot}0205 \left(\frac{Y}{Y_m}\right)^2 - 0{\cdot}0212 \left(\frac{Y}{Y_m}\right)^4 + 0{\cdot}0004 \left(\frac{Y}{Y_m}\right)^6.$$

A calculation of the same objective by six-figure logs with interpolation of the angles to 0·1 second, made by Mr. J. S. Watkins, gave, with the above $LA'p = 0{\cdot}0205$

$$LA'_m = 0{\cdot}0000 \qquad LZA' = 0{\cdot}0056 \quad \therefore \quad a_4 Y_m^4 = -0{\cdot}0167 \qquad a_6 Y_m^6 = -0{\cdot}0038$$

or the formula for any zone

$$LA' = 0{\cdot}0205 \left(\frac{Y}{Y_m}\right)^2 - 0{\cdot}0167 \left(\frac{Y}{Y_m}\right)^4 - 0{\cdot}0038 \left(\frac{Y}{Y_m}\right)^6.$$

The disagreement of the two formulae for LA' is at first sight alarming, for the coefficients of the secondary and of the tertiary aberration differ by more than 0·004 ; whilst the five-figure work *suggests* a small positive tertiary aberration, the practically exact six-figure calculation *proves* the existence of a decidedly sensible negative tertiary aberration. [31]

The *reason* for this disagreement is easily recognized in the explicit formulae for the secondary and tertiary aberration in terms of LAp', LZA', and LA'_m, for LZA' enters into both with factor 8 ; therefore the residual error in the calculation of LZA', which amounted to 0·0006 in the five-figure calculation, goes into the results also with factor 8 and changes them by 0·0048, and the same applies in lesser degree to the residual errors in LAp' and LA'_m. The conclusions to be drawn are

(1) That, if for some theoretical discussion we require dependable values of the higher aberrations, then very accurate calculations by the ray-tracing formulae are indispensable.
(2) That it is advisable to work out formulae for this purpose directly in terms of the calculated data in order to know the factor (eight in our case) with which the latter go into the result.

From the practical point of view we arrive at more comforting conclusions ; our only requirement then is, to be able to predict the amount of the spherical aberration for any zone of a system with such accuracy as is sufficient to preclude serious disagreement of the predicted aberration with that found by direct calculation. In order to render the remainder of this discussion still more instructive we will add a third solution, namely the two-term formula from Mr. Watkins's results for LZA' and LA'_m, which in accordance with the earlier deductions for the five-figure results gives

$$LA' = 0{\cdot}0224\left(\frac{Y}{Y_m}\right)^2 - 0{\cdot}0224\left(\frac{Y}{Y_m}\right)^4.$$

Calculating by the three formulae for a suitable succession of values of (Y/Y_m) we find :

$\dfrac{Y}{Y_m} =$	0·2	0·4	0·6	0·8	0·9	1·0
By five-figure results : $LA' =$	0·0008	0·0028	0·0047	0·0045	0·0029	−0·0003
Six-figure work and three-term formula	0·0008	0·0029	0·0050	0·0053	0·0037	0
Six-figure work and two-term formula	0·0009	0·0030	0·0052	0·0052	0·0035	0

It will be seen that the greatest disagreement between lines two and three is only 0·0002, a totally negligible amount which could not be detected by any known observational test. And between the results by five-figure logs and by six-figure logs there is a maximum difference of 0·0008, which at most might be suspected by optical tests. Our decision that five-figure logs are sufficiently precise and that ordinary lens systems may be assumed to have their spherical aberration represented with ample accuracy by two terms of (11) is therefore fully justified by

[31] these tests, but with the original reservation that the exact trigonometrically determined LA'_m must be one of the data on which the solution is based. The necessity of this last caution we can also prove by our results: the three-term formula derived from the six-figure results may be regarded as practically exact. We may therefore take it that, if we went through the tremendous labour of calculating the true secondary aberration by a direct analytical equation, we should find it agreeing with the second term (−0·0167) of the formula. We should therefore claim that the aberration of our objective was given by the formula

$$LA' = 0{\cdot}0205\left(\frac{Y}{Y_m}\right)^2 - 0{\cdot}0167\left(\frac{Y}{Y_m}\right)^4,$$

and as this omits the very sensible tertiary aberration (which would utterly defy analytical determination) we should assign to LA'_m the value $+0{\cdot}0038$ instead of zero; which would be a decidedly serious error!

Whilst analytical determination of the exact secondary aberration leaves the whole tertiary aberration as the error of the resulting equation for LA', it is easily shown by a closer investigation that the maximum error is only 4/27th of the marginal tertiary aberration when we solve by the highly convenient LAp' and LA'_m method, and that the maximum error sinks to $\pm 0{\cdot}0481$ of the marginal tertiary aberration when we solve, with a little more labour, by LZA' and LA'_m. In the majority of cases the greater perfection of the smoothing out of the tertiary aberration by the latter method is lost on account of the residual error of LZA' which is very much greater than that of the primary LAp'; for practical purposes the recommendation of the method of solution by LAp' and LA'_m therefore remains justified.

ALTERNATIVE EQUATIONS

[32] The equations deduced in this chapter admit of being transposed by the usual algebraical methods so as to yield direct solutions for any one of the quantities contained in them. Moreover, they all were obtained by straightforward algebraical deductions from the fundamental computing equations of Chapter I, and like these they can also be modified by exchanging throughout each equation 'plain' and 'dashed' symbols. By these legitimate processes we can deduce a number of additional highly useful equations.

It was shown in [26] that the rigorous equation

(10)* LA'

$$= LA\cdot\frac{N}{N'}\cdot\frac{l'L'}{l\,L} + 2\frac{N}{N'}(L-r)\cdot\frac{l'L'}{r\,L}\cdot\sin\tfrac{1}{2}(I'-U)\cdot\sin\tfrac{1}{2}(I-I')/\cos\tfrac{1}{2}(I'-U')$$

is occasionally useful for obtaining highly accurate values of the spherical aberration. If this equation is multiplied throughout by $N'.l.L/N.l'.L'$ it gives in the first term of the right simply LA and can be transposed to

$$LA = LA'\cdot\frac{N'}{N}\cdot\frac{l\,L}{l'L'} - 2(L-r)\cdot\frac{l}{r}\cdot\sin\tfrac{1}{2}(I'-U)\cdot\sin\tfrac{1}{2}(I-I')/\cos\tfrac{1}{2}(I'-U').$$

The minus sign of the second term can be transferred to the factor $\sin\frac{1}{2}(I-I')$;

we thus obtain an equation for accurate determinations of the spherical aberration [32] in right-to-left calculations in the form

(10)* for right-to-left work :

$$LA = LA' \cdot \frac{N'}{N} \cdot \frac{l \cdot L}{l' \cdot L'} + 2\,(L-r) \cdot \frac{l}{r} \cdot \sin \tfrac{1}{2}\,(I'-U) \cdot \sin \tfrac{1}{2}\,(I'-I)/\cos \tfrac{1}{2}\,(I'-U').$$

It will be noticed that this equation is both simpler and more symmetrical than the original (10)*. We can simplify the latter equation correspondingly by simply writing (10)* for right-to-left work with exchange of plain and dashed symbols, which gives for work in the usual direction

(10)* Best form for left-to-right calculations :

$$LA' = LA \cdot \frac{N}{N'} \cdot \frac{l' \cdot L'}{l \cdot L} + 2\,(L'-r)\,\frac{l'}{r} \cdot \sin \tfrac{1}{2}\,(I-U') \cdot \sin \tfrac{1}{2}\,(I-I')/\cos \tfrac{1}{2}\,(I-U).$$

This equation may be obtained in a less elegant, but possibly more convincing, way by substituting in the original (10)* for L'/L in the second term the value obtained by equating the two equivalents of PA in section [12], and for $(L-r)$ N/N' the value following from equation (6), and then replacing $(I'-U)$ by $(I-U')$ in accordance with equation (3).

The special form of (10)* for plane surfaces is easily modified in the same way as the general equation and gives the simplest forms of the equation :

(10)* for Plane Surfaces, right-to-left work :

$$LA = LA' \cdot \frac{L}{L'} + 2\,l \cdot \sin \tfrac{1}{2}\,(U+U') \cdot \sin \tfrac{1}{2}\,(U-U') \cdot \sec U', \quad \text{then}$$

(10)* for Plane Surfaces, left-to-right work :

$$LA' = LA \cdot \frac{L'}{L} + 2\,l' \cdot \sin \tfrac{1}{2}\,(U'+U) \cdot \sin \tfrac{1}{2}\,(U'-U) \cdot \sec U.$$

The equation (10)** for the primary aberration of any centred lens system can be solved for LAp_1 by multiplying it throughout by $N'_k u'^2_k/N_1 u_1^2$ and then rearranging the terms. As the multiplying factor is constant, it can be taken into the sum-term. We thus obtain the two solutions :

(10)** for left-to-right calculation through a lens system :

$$LAp'_k = LAp_1 \frac{N_1 u_1^2}{N'_k u'^2_k} + \sum_{j=1}^{j=k} Sph'_j \cdot \frac{N'_j u'^2_j}{N'_k u'^2_k}$$

or for right-to-left calculations :

$$LAp_1 = LAp'_k \frac{N'_k u'^2_k}{N_1 u_1^2} - \sum_{j=k}^{j=1} Sph'_j \, \frac{N'_j u'^2_j}{N_1 u_1^2}$$

with the definition for both equations

$$Sph'_j = \tfrac{1}{2}\, l'_j \, \frac{l'_j - r_j}{r_j}\,(i'_j - u_j)\,(i_j - i'_j).$$

[32] The marking of the sum in the second equation, as one to be taken from $j = k$ to $j = 1$, is of course a purely formal indication of the direction of the light ; really any sum may be taken in any order of its terms which may be convenient.

If we now write out the equation for LAp_1 in full, including the explicit value of Sph'_j, and then make the exchange of 'plain' and 'dashed' symbols and the change of sequence of the surfaces, we obtain an alternative equation for the usual left-to-right direction, in the form

$$LAp'_k = LAp_1 \frac{N_1 u_1^2}{N'_k u'^2_k} - \sum_{j=1}^{j=k} \tfrac{1}{2} l_j \frac{l_j - r_j}{r_j} (i_j - u') (i'_j - i_j) \frac{N_j u_j^2}{N'_k u'^2_k}.$$

We will take the minus sign of the sum inside by writing $(i_j - i'_j)$ instead of $(i'_j - i_j)$; by also transposing the new equation into a solution for LAp_1 we obtain the alternative equations :

(10)** for left-to-right calculation through a lens system :

$$LAp'_k = LAp_1 \frac{N_1 u_1^2}{N'_k u'^2_k} + \sum_{j=1}^{j=k} Sph_j \frac{N_j u_j^2}{N'_k u'^2_k}$$

or for right-to-left calculations :

$$LAp_1 = LAp'_k \frac{N'_k u'^2_k}{N_1 u_1^2} - \sum_{j=k}^{j=1} Sph_j \frac{N_j u_j^2}{N_1 u_1^2}$$

with the definition, for both equations :

$$Sph_j = \tfrac{1}{2} l_j \frac{l_j - r_j}{r_j} (i_j - u'_j) (i_j - i'_j).$$

These alternative equations are more convenient than the original ones in cases when l'_j is very large and simultaneously u'_j very small ; they must be used if it should ever happen that an infinite l'_j and zero u'_j occurred at a particular surface. As a comparison of the original solution with the alternative one shows that the sums are identical, it is quite legitimate to use the alternative value only for those items in which the above difficulties arise.

We can draw a very valuable conclusion from the corresponding equations for right-to-left and left-to-right work with reference to the magnitude of the primary spherical aberration at the two conjugate points. As the data by which these formulae are worked out are those of a paraxial ray traced through the system from a given point in the medium to the left of the system to the conjugate point in the medium to the right of the system (or vice versa), these data are absolutely definite and fixed. Having obtained them, we could work out LAp'_k on the assumption of a perfect object-point to the left of the system, or $LAp_1 = 0$, and by the alternative set should find

$$LAp'_k = \sum Sph_j \frac{N_j u_j^2}{N'_k u'^2_k} = \frac{1}{N'_k u'^2_k} \sum Sph_j \, . \, N_j \, . \, u^2 :$$

for as $N'_k u'^2_k$ is a perfectly constant factor, we can take it outside the sum. If now [32] we worked through the same system from right to left, assuming the conjugate point at the right to be a perfect object-point, or $LAp'_k = 0$, we should find, by the second equation of the set

$$LAp_1 = -\sum Sph_j \frac{N_j u_j^2}{N_1 u_1^2} = -\frac{1}{N_1 u_1^2} \sum Sph_j \,.\, N_j \,.\, u_j^2.$$

The ratio of these two equations gives

$$\frac{LAp'_k}{LAp_1} = -\frac{N_1 u_1^2}{N'_k u'^2_k} = -\bar{m}'_k.$$

As the longitudinal magnification $\bar{m}$ is always positive, this result proves:

(1) That for any centred lens system working on given paraxial conjugate points the primary spherical aberration at these two points is invariably of opposite sign, for the ratio is always negative.

(2) That the numerical value of the aberration in the two directions is ruled by the law of longitudinal magnification, and as the latter may assume any value between zero and infinity, LAp_1 may be any multiple or fraction of the corresponding LAp'_k.

These conclusions apply *strictly* only to the pure primary aberration, but they lead to a reasonably good estimate in the great majority of cases even when applied to the trigonometrically determined LA and LA' of any lens system with given paraxial conjugates.

We can test this conclusion by the right-to-left example in section [22], which gave $l_1 = -23{\cdot}994$ and $L_1 = -19{\cdot}539$ or $LA_1 = l_1 - L_1 = -4{\cdot}455$ as the trigonometrically exact spherical aberration at the left conjugate point. The nominal u_1 in that calculation was $= -0{\cdot}05121$, whilst we calculated in section [27] $LAp'_2 = 1{\cdot}2572$ for a ray with $u_1 = \sin(-4°) = -0{\cdot}069756$. As the primary aberration grows as the square of u_1, we therefore can confidently claim that for the case in section [22]

$$LAp'_2 = 1{\cdot}2572\left(\frac{0{\cdot}05121}{0{\cdot}069756}\right)^2 = 0{\cdot}6776.$$

In section [20] we found the linear magnification of this lens with the same conjugate distances as $m'_2 = -0{\cdot}3677$, and as this lens is working in air we have

$$\bar{m}'_2 = m'^2_2 = (-0{\cdot}3677)^2 = 0{\cdot}1352.$$

By the law which we have proved above we can now claim that the primary aberration at the left conjugate will be found by

$$\frac{LAp'_k}{LAp_1} = -\bar{m}'_k \quad \text{as} \quad LAp_1 = \frac{LAp'_k}{-\bar{m}'_k} = \frac{0{\cdot}6776}{-0{\cdot}1352} = -5{\cdot}012,$$

and this conclusion may profitably be tested by a direct calculation by one of our right-to-left equations applied to the paraxial data in section [22]. It will be noticed that our result $LAp_1 = -5{\cdot}012$ agrees fairly well with the trigonometrical value

[32] $-4{\cdot}455$. In accordance with the theory developed in section [31] we can conclude that the difference must represent the secondary and higher aberration of our lens for the calculated aperture. This conclusion also is worthy of being tested by calculating a second trigonometrical ray from right to left, say at half the original semi-aperture, and by then solving for primary and secondary aberration by the first method of [31].

We now apply the same methods of transformation to the thin lens equations of section [29]. For one thin lens we found

$$LAp' = LAp\left(\frac{l'}{l}\right)^2 + y^2 l'^2[G_1c^3 - G_2c^2c_1 + G_3c^2v_1 + G_4cc_1^2 - G_5cc_1v_1 + G_6c\,v_1^2].$$

In order to transform this left-to-right equation into the corresponding equation for right-to-left work we must, in accordance with the principles laid down in sections [4] and [5], exchange 'plain' and 'dash', and as a lens with two surfaces is being dealt with, we must also take these surfaces in the right-to-left order in which the light now meets them. Therefore c_1 must be replaced by c_2, and as the true value of the total curvature in this direction (i.e. curvature first met—curvature next met) is now $c_2 - c_1$, we must, in order to adhere to the standardized definition of c, also replace c by $-c$, and this reverses the sign of all terms of the G-sum which contain c in odd powers. Finally, the intersection-length to the right of the lens, from which the light now starts, is, with reference to the lens surfaces, l_2' (called simply l' in the original formula in accordance with our agreed simplification for thin lenses); hence v_1 must be replaced by v_2'. Consequently the right-to-left equivalent of our last equation is

$$LAp = LAp'\left(\frac{l}{l'}\right)^2 + y^2 l^2[-G_1c^3 - G_2c^2c_2 + G_3c^2v_2' - G_4cc_2^2 + G_5cc_2v_2' - G_6c\,v_2'^2].$$

We will change all the signs in the new G-sum by giving to the whole second term a general minus sign. If we then multiply the whole equation by $(l'/l)^2$ and transpose into a solution for LAp', we obtain the alternative solution

$$LAp' = LAp\left(\frac{l'}{l}\right)^2 + y^2 l'^2[G_1c^3 + G_2c^2c_2 - G_3c^2v_2' + G_4cc_2^2 - G_5cc_2v_2' + G_6c\,v_2'^2].$$

Comparison with the starting equation, which is identical with the new one in every item excepting the G-sums, proves that these two sums must unconditionally give the same value for the same lens and conjugate distances. Hence these two forms of the G-sum, containing respectively left-hand and right-hand data of the thin lens, may be used at pleasure according to convenience, and we shall find that a judicious exercise of this liberty greatly shortens many solutions of optical problems. The identity of the two sums may be proved directly in a less subtle way, but at the cost of some complicated algebra, by introducing into the original sum $c_1 = c + c_2$ (from $c = c_1 - c_2$) and $v_1 = v_2' - (N-1)c$ (from TL(3)), multiplying out, ordering according to the sequence in the alternative G-sum and then introducing the explicit G-values in the complicated factors of the first three terms, when the alternative G-sum will result.

[32] We will collect the thin lens formulae originally deduced in section [29] together with their transpositions into solutions for LAp_1:

TL(10)**

For one thin lens:

$$LAp' = LAp\left(\frac{l'}{l}\right)^2 + y^2 l'^2[G\text{-sum}]\,;\quad LAp = LAp'\left(\frac{l}{l'}\right)^2 - y^2 l^2[G\text{-sum}].$$

For k thin lenses in contact:

$$LAp'_k = LAp_a\left(\frac{l'_k}{l_a}\right)^2 + y^2 l'^2_k[G\text{-sum}]_a + y^2 l'^2_k[G\text{-sum}]_b + \ldots\ldots + y^2 l'^2_k[G\text{-sum}]_k$$

$$LAp_a = LAp'_k\left(\frac{l_a}{l'_k}\right)^2 - y^2 l_a^2[G\text{-sum}]_a - y^2 l_a^2[G\text{-sum}]_b - \ldots\ldots - y^2 l_a^2[G\text{-sum}]_k.$$

For k thin lenses at any separations:

$$LAp'_k = LAp_a\left(\frac{y_a}{y_k}\right)^2\left(\frac{l'_k}{l_a}\right)^2 + \frac{y_a^4}{y_k^2}l'^2_k[G\text{-sum}]_a + \frac{y_b^4}{y_k^2}l'^2_k[G\text{-sum}]_b + \ldots + \frac{y_k^4}{y_k^2}l'^2_k[G\text{-sum}]_k$$

$$LAp_a = LAp'_k\left(\frac{y_k}{y_a}\right)^2\left(\frac{l_a}{l'_k}\right)^2 - \frac{y_a^4}{y_a^2}l_a^2[G\text{-sum}]_a - \frac{y_b^4}{y_a^2}l_a^2[G\text{-sum}]_b - \ldots - \frac{y_k^4}{y_a^2}l_a^2[G\text{-sum}]_k.$$

y_a is selected; then

$$y_b = y_a \cdot l_b/l'_a,\ y_c = y_a(l_b/l'_a)\,(l_c/l'_b),\ y_k = y_a(l_b/l'_a)\ldots\ldots(l_k/l'_{k-1}).$$

In all six equations the G-sum may be calculated as either

$$= [G_1c^3 - G_2c^2c_1 + G_3c^2v_1 + G_4cc_1^2 - G_5cc_1v_1 + G_6c\,v_1^2]$$

or

$$= [G_1c^3 + G_2c^2c_2 - G_3c^2v'_2 + G_4cc_2^2 - G_5cc_2v'_2 + G_6c\,v'^2_2].$$

c is the total curvature $= c_1 - c_2$ of the lens concerned; c_1 is the curvature of the left and c_2 that of the right surface; v_1 is the reciprocal of the left intersection-length and therefore for lens $j = 1/l_j$; v'_2 is the reciprocal of the right-hand intersection-length, or for lens $j = 1/l'_j$. It should always be noted that in the alternative sum the second and third terms bear the reverse sign as compared with the original sum.

GRAPHIC REPRESENTATION OF THE ABERRATION

[33] It is very instructive to plot the spherical aberration of a lens or lens system against the aperture, and this can easily be done as soon as the coefficients of the LA'-formula have been worked out. Thus for the biconvex lens we found in section [31] that the aberration at the final focus can be represented by the equation

$$LA' = 0{\cdot}078305 \times U_1^2 + 0{\cdot}000387 \times U_1^4,$$

U_1 being the angle in degrees at which any ray starts out from the original object-

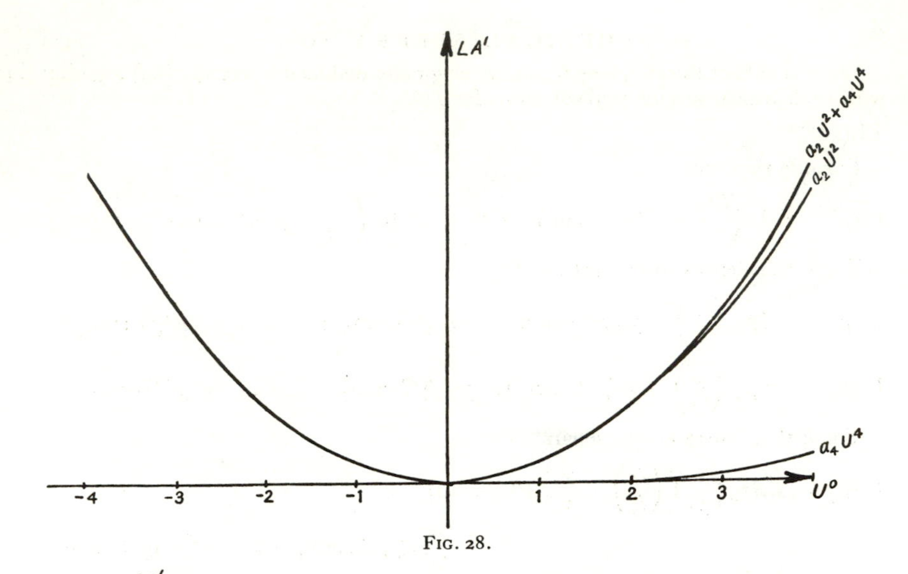

LA'
$a_2U^2+a_4U^4$
a_2U^2
a_4U^4
-4
-3
-2
-1
1
2
3
U^0

FIG. 28.

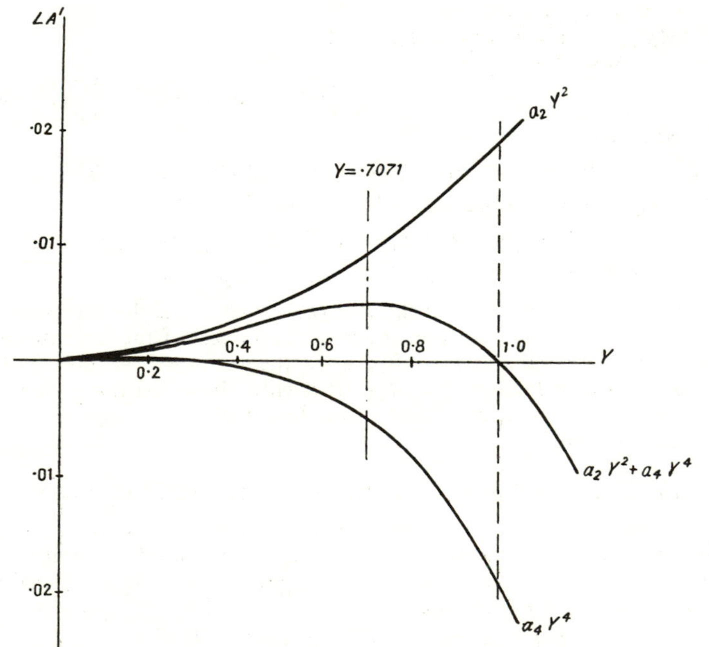

LA'
a_2Y^2
·02
Y=·7071
·01
0·4
0·6
0·8
1·0
Y
0·2
·01
$a_2Y^2+a_4Y^4$
·02
a_4Y^4

FIG. 29

point. If LA' is plotted in the vertical direction against U_1 in the horizontal [33] direction, the first term of the formula will give an ordinary parabola, the second term will give a biquadratic parabola, and the combination of the two will give the true curve of the aberration-graph.

For the achromatic objective we found

$$LA' = 0{\cdot}02\ Y^2 - 0{\cdot}02\ Y^4$$

and this again gives a simple and a biquadratic parabola for the two separate terms and a combined curve which lies above the zero-axis between $Y = 0$ and $Y = 1$ and sinks below it for $Y > 1$.

It is very instructive to draw the separate curves of the two terms in addition to the combined effect, as this shows clearly how the latter is brought about.

OTHER MEASURES OF SPHERICAL ABERRATION

It was shown in the theoretical discussion that the longitudinal spherical aberration [34] may attain any value between $-\infty$ and $+\infty$ without being necessarily of very serious import in its effect on the final image. LA' is in that respect a very misleading and at times inconvenient measure of the aberration. We shall, however, nearly always use it because it is the form in which the aberration naturally comes out in our trigonometrical ray-tracing. We will briefly refer to two other geometrical ways of measuring spherical aberration which have the stated drawbacks in one case to a less extent and in the other not at all. We will deal with these measures only in first approximation, as they are of little interest when accurately expressed for rays at finite angles.

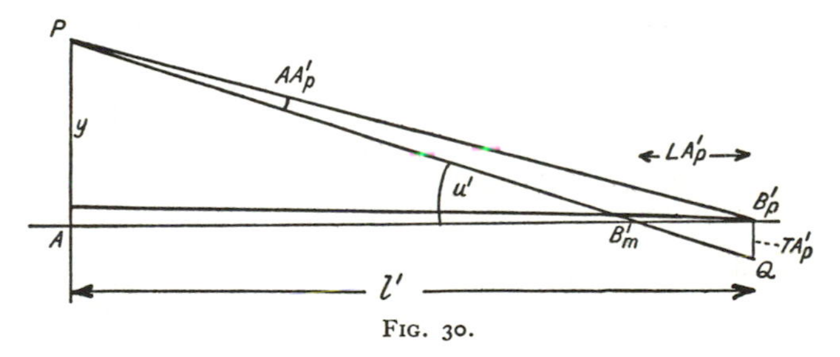

FIG. 30.

Referring to Fig. 30, in which AP represents a refracting surface producing its strictly paraxial focus at Bp' and the intersecting point of rays from a zone of moderate aperture at Bm' with a convergence angle u', we have $Bm'\ Bp' = LAp'$ and the latter is determined by (10)**

$$LAp' = LAp\ .\ \frac{N}{N'}\left(\frac{u}{u'}\right)^2 + \tfrac{1}{2}\,l'\ .\ \frac{l'-r}{r}\,(i'-u)\,(i-i').$$

We obtain one of the alternative measures of the spherical aberration by producing the 'marginal' ray PBm' until it cuts a plane laid through the paraxial focus Bp' and at right angles with the optical axis. The distance QBp', by which the

[34] paraxial focus lies above Q, is called the *transverse* or *lateral aberration* of the ray PBm'; we will give it the symbol TAp' and will give it the positive sign when Bp' falls above Q, the negative sign when Bp' falls below Q. Evidently we now have strictly $TAp' = LAp' . \tan u'$, but as in our primary approximation we must treat u' as a very small angle, we must use $TAp' = u' . LAp'$. Multiplying (10)** throughout by u' we obtain

$$TAp' = LAp' . u' = LAp \cdot \frac{N}{N'} \cdot \frac{u^2}{u'} + \tfrac{1}{2} l'u' \frac{l'-r}{r} (i'-u)(i-i').$$

In the first term on the right $LAp . u$ is by analogy the transverse aberration of the arriving rays, and we therefore substitute TAp for it. The remaining factor $N . u/N' . u'$ is by (10p)* the linear magnification m' produced by the surface, and the whole first term thus becomes $= m'TAp$. In the second term we have by (4p) transposed $u'(l'-r)/r = i'$. Therefore we can write

(10)*** $$TAp' = LAp' . u' = m' . TAp + \tfrac{1}{2} l' . i'(i'-u)(i-i')$$

as a general equation for the primary transverse aberration produced by any one refracting surface. For the special case of a plane surface we have only to introduce $i' = -u'$ and $i = -u$ by Pl (1p) in order to obtain

(10)*** for a plane: $$TAp' = LAp' . u' = m' . TAp + \tfrac{1}{2} l'u'(u'+u)(u'-u).$$

The equations for the transverse aberration are obviously simpler than those for the longitudinal aberration. The second term, which represents the new aberration arising at a surface, depends on the product of three paraxial angles, and as each of these will grow in direct proportion to u or to the semi-aperture AP of a zone, it is clear that the primary transverse aberration grows with the cube of the aperture and therefore at a much faster rate than the longitudinal aberration. According to the first term on the right of (10)*** the transverse aberration is transferred simply by multiplication into the linear magnification m', that is just as if the transverse displacement of the marginal ray, where it cuts a plane laid through the paraxial focus, were a part of an extended object of which the refracting surface forms an image. In the case of the longitudinal aberration we found that it is transferred to a new surface by the law of axial magnification, and as the latter is proportional to the square of the linear magnification it follows that the transfer of transverse aberration will lead to very much smaller changes of magnitude than we found in the case of the longitudinal aberration.

The third geometrical measure of spherical aberration is also represented in Fig. 30 and has a particularly simple definition. If there were no aberration at the surface AP, then all rays would go through the paraxial focus Bp'. We can therefore say that aberration causes rays beyond the strictly paraxial zone to take an undesirable and wrong direction; instead of going to Bp' the ray from P takes a more inclined direction towards Bm'. Evidently the difference between the ideal direction of the ray, towards Bp', and its real direction towards Bm' is another legitimate measure of the spherical aberration. It is represented by the angle $Bp'PBm'$ and is called the *angular aberration*. We will give it the symbol AAp', and we will call it positive when a ruler would have to be turned in the clockwise direction in order to transfer it from the direction of the ideal ray PBp' into that of [34] the real ray PBm'. Fig. 30 shows AAp' as the angle opposite TAp' in the triangle $Bp'PQ$. The angle at Q is evidently $= 90° - u'$ and the trigonometrical sine-relation therefore gives

$$\sin AAp' = \cos u' \,.\, TAp'/PBp'.$$

As primary aberrations must be worked out for indefinitely small angles and apertures, we must replace sin AAp' by the angular aberration AAp' itself; we must also omit the factor cos u', as at the limit it becomes equal to 'one' exactly; finally PBp' differs only very little from $ABp' = l'$ and at the limit becomes equal to l'; therefore, in order to determine the pure primary angular aberration, we not merely may, but must write the last equation as

$$AAp' = TAp'/l' = LAp' \,.\, u'/l',$$

the last by introducing the previously determined value of TAp'. If we introduce TAp' as determined by (10)*** we find

$$AAp' = LAp' \,.\, u'/l' = TAp'/l' = m' \,.\, TAp/l' + \tfrac{1}{2}\, i'(i'-u)\,(i-i').$$

The first term on the right can be transformed further, by introducing by (10p)* $m' = N \,.\, l'/N' \,.\, l$, as follows

$$m' \,.\, TAp/l' = \frac{N \,.\, l' \,.\, TAp}{N' \,.\, l \,.\, l'} = \frac{N}{N'} \cdot \frac{TAp}{l} = \frac{N}{N'} \cdot AAp,$$

the last by analogy with the definition of AAp'. We thus obtain the general equation of the primary angular aberration

$$(10)^{****} \quad AAp' = LAp' \,.\, u'/l' = TAp'/l' = \frac{N}{N'} \cdot AAp + \tfrac{1}{2}\, i'(i'-u)\,(i-i')$$

and use of Pl (1p): $i = -u$, $i' = -u'$ gives the special form

$$(10)^{****} \text{ for a plane:} \quad AAp' = \frac{N}{N'} \cdot AAp + \tfrac{1}{2}\, u'(u'+u)\,(u'-u).$$

The second term on the right again represents the new angular aberration arising at the surface: it contains only paraxial angles and is therefore always a small quantity, utterly unaffected by the magnitude of the intersection-length which latter can cause both the longitudinal and the transverse aberration to become infinitely large. That is a great and characteristic advantage of the angular aberration. The transfer- or addition-theorem expressed by the first term of the right shows that here also we have a very simple rule. The angular aberration of the incident pencil is transferred by the law of refraction, just like an angle of incidence. It therefore never becomes greatly changed and the angular aberration is thus recognized as free from the wild changes in magnitude on being transferred which we found in the case of the longitudinal aberration and which still exist in a less degree in the transverse aberration. Since the new aberration is found as the product of three paraxial angles, the angular aberration resembles the transverse in growing with the cube of the aperture.

Of the three geometrical ways of measuring the spherical aberration, the angular

[34] measure comes nearest to giving a correct idea as to the real seriousness of the defect which it indicates. In all telescopic measuring instruments (theodolites, spectrometers, and telescopes for astronomical measurements) it is in fact the direction of rays which decides the result and the angular aberration, by giving the amount by which a marginal ray deviates from the ideal direction, is then the most direct measure of the possible maximum error.

Whilst the primary angular aberration is the least misleading geometrical measure of spherical aberration and is also expressed by the simplest equation for any one surface, it is subject to one consideration which introduces a slight complication in the transfer from surface to surface when the angular aberration is to be computed for a real lens system with lenses of finite thickness placed at finite separations from each other.

It is obvious by inspection of Fig. 30 that the angular aberration as defined by us represents the angular subtense of the longitudinal aberration as seen from the point P at which a marginal ray pierces the refracting surface under consideration. If we examine Fig. 31 which shows two successive surfaces, numbered 1 and 2, of

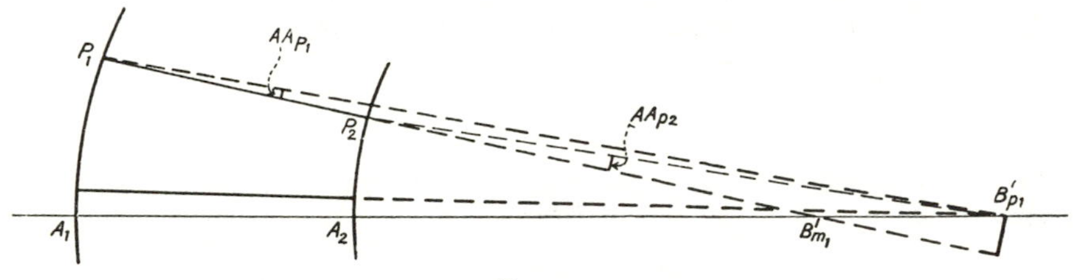

FIG. 31.

a lens-system and assume that the aberration has been calculated for surface 1, then the AAp' for this surface will be the marked angle of subtense, as seen from P_1, of the longitudinal aberration $Bm'Bp'$ existing at the confused focus produced by surface 1. When we want to calculate the total aberration produced by surface 2, the AAp_2 required in working out the first term of (10)**** will be the angle $Bp_1'P_2Bm_1'$ which the longitudinal aberration subtends when seen from point P_2. It is easily shown that in primary approximation we have

$$AAp_2 = AAp_1' \cdot l_1'/l_2 = AAp_1' \cdot y_1/y_2$$

if y_1 and y_2 are the incidence heights at the two surfaces.

All the primary aberration equations in section [32] can be converted into transverse or into angular aberration by using the relations between LAp' and TAp' or AAp' which are stated in the various equations (10)*** and (10)****.

In the study of axial pencils, to which we are as yet limited, we shall not use either the transverse or the angular aberration, but transverse aberrations will play a very important part in the study of oblique pencils of rays.

We will calculate both for our biconvex lens in order to realize the characteristic differences in magnitude and in change from surface to surface. As we have calculated LAp' in section [27], using the original paraxial data of section [20], we can easily derive the transverse and angular aberrations from the fundamental relations

$$TAp' = LAp' \,.\, u' \quad \text{and} \quad AAp' = TAp'/l'.$$

For the first surface we found $LAp' = 19{\cdot}154$, $\log u' = 8{\cdot}04825$, and $l' = 149{\cdot}811$; for the second surface $LAp' = 1{\cdot}2572$, $\log u' = 9{\cdot}27809$, and $l' = 8{\cdot}7895$.

These data give for the transverse aberration :

	First surface.	Second surface.
$\log LAp'$	1·28227	0·09940
$+\log u'$	8·04825	9·27809
$\log TAp'$	9·33052	9·37749

For the angular aberration we obtain :

$\log TAp'$	9·33052	9·37749
$-\log l'$	−2·17554	−0·94396
$\log AAp'$	7·15498	8·43353

We will collect the numerical values of all the aberrations in an instructive little table :

LAp'	19·154	1·2572
TAp'	0·21405	0·23850
AAp'	0·001429	0·027135

In confirmation of our theoretical conclusions we see that whilst the longitudinal aberration for the complete lens is only one-fifteenth of that at the first surface, the transverse aberration comes out nearly equal and the truest geometrical measure, namely the angular aberration, is nineteen times as large for the complete lens as it is for the first surface.

The angular aberrations calculated by our formulae are obtained in radians. They may be turned into degree-measure by the table of circular measure. For our biconvex lens we thus find that the direction of the marginal ray errs by about 5 minutes of arc at the first surface and by 93 minutes after the second refraction.

OVER- AND UNDER-CORRECTION

Any ordinary simple convex lens, when producing a real image of a real object, with the light passing in the usual left-to-right direction, has the paraxial focus situated to the right of the marginal focus, and has therefore positive longitudinal spherical aberration according to our definition. This aberration can only be removed or 'corrected' by compounding the lens of a convex and concave element (as in telescope objectives) or by 'figuring' the simple lens : anyhow by some correcting device. For that reason spherical aberration in the sense of that of a simple convex lens is very commonly referred to as spherical under-correction (i.e. the lens is insufficiently corrected) whilst spherical aberration in our negative sense (the marginal ray of our 2″ telescope-objective displays a minute amount of it) is referred to as spherical over-correction (i.e. the correction has been overdone). The terms are rather expressive but decidedly loose when applied to the great

[34] variety of cases occurring in actual practical designing, and especially when applied to calculations in the right-to-left direction which will play an important part in our methods. It is therefore advisable to shun these popular terms as far as possible and to speak of positive and negative aberration according to the signs resulting from our equations and conventions.

When the transverse or the angular aberration is employed the case becomes even worse, for as these aberrations are proportional to the cube, an odd power, of the aperture, they change sign for rays passing below the optical axis so that even for the same lens and the same conjugates there is no definite sign.

THE DISK OF LEAST CONFUSION †

[35] We now come to the important question of the effect of spherical aberration on the quality of the image of a point. As every zone has a different focal point when spherical aberration is present, we can conclude at once that no perfectly sharp image is then possible ; but there must be a point where the confused bundle of rays has a minimum diameter and this point would geometrically be regarded as the best focus. The disk of light which is formed at this best focus so defined is known as the disk or circle of least confusion.

(1) Disk of least confusion for primary aberration.

Our biconvex lens represents a system with a great predominance of primary aberration. If we draw a diagram of the computed rays, it will look like Fig. 32.

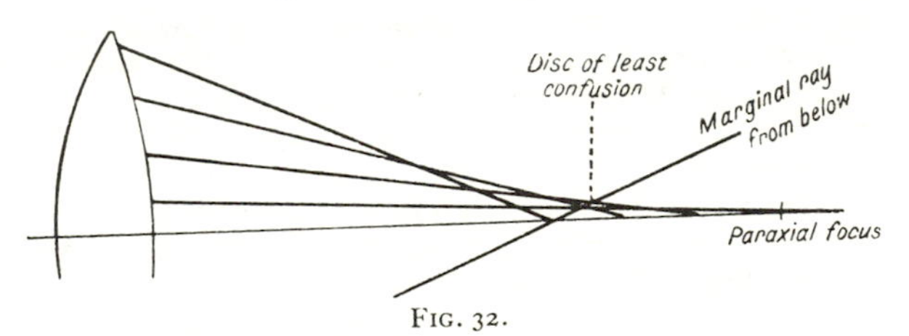

FIG. 32.

The rays from the upper part of the aperture come to successively shorter focal distances. If all the intermediate rays are imagined as added to the diagram they will define by envelopment a continuous curve which is known as a caustic, and is a semi-cubic parabola in the case of pure primary aberration. It is of no interest in practical optics and we therefore merely mention it. In order to find the disk of least confusion we must remember that the rays do not stop at their focus but go straight on. These produced rays cut through the caustic formed by the arriving rays, and it is at once obvious that the produced *marginal* ray from the lower part of the lens will cut the caustic of the arriving rays from the upper part at the point which is farthest from the paraxial focus and therefore marks the position of the smallest diameter of the complete confused bundle of rays. In order to determine this diameter we must decide which one of the arriving rays cuts the produced

† The subject is treated briefly because the disk of confusion gives results which clash hopelessly with deductions from the undulatory theory, as will be shown in the following chapter.

marginal ray from the lower margin of the lens at the greatest distance from the optical axis. As the primary aberration is proportional to the square of any reasonable measure of the aperture, we choose tan U' as the most convenient measure for the present purpose, so that [35]

$$LA' = a_2 \tan^2 U'.$$

We now refer to Fig. 33, in which the horizontal line marks the optical axis, Bm' the marginal, and Bp' the paraxial focus of a lens system. The ray from the

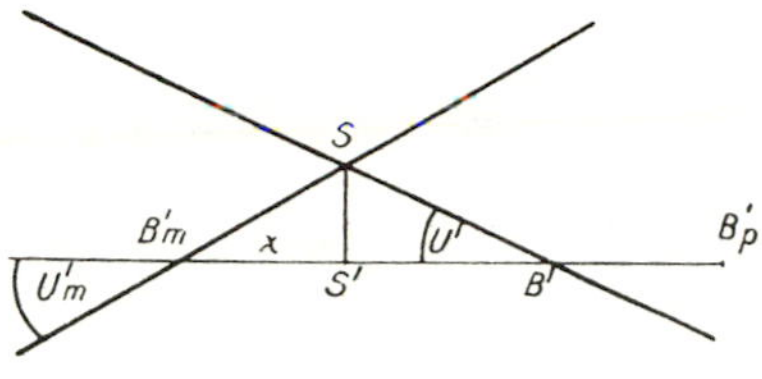

FIG. 33.

lower margin of the system arrives at Bm' under angle Um'. Any intermediate ray from the upper part of the system will arrive at some intermediate focus B' under angle U' and will cut the produced marginal ray from the lower part at S. We have to determine the value of U' or of tan U' for which the distance SS' becomes a maximum.

$Bm'Bp'$ is the longitudinal aberration of the marginal rays, and by our adopted law we have it as

$$LAm' = a_2 \tan^2 Um'.$$

$B'Bp'$ is the longitudinal aberration of an intermediate ray and is defined by

$$LA' = a_2 \tan^2 U'.$$

Calling the distance $Bm'S'$ simply x, we have

$$S'B' = Bm'Bp' - B'Bp' - x$$
$$= a_2 (\tan^2 Um' - \tan^2 U') - x$$

and we can now determine SS' as

$$SS' = x \tan Um' = S'B' \tan U' = [a_2 (\tan^2 Um' - \tan^2 U') - x] \tan U'.$$

From the second and last items of this equation we obtain

$$x (\tan Um' + \tan U') = a_2 (\tan^2 Um' - \tan^2 U') \tan U'$$

or

$$x = a_2 \tan U' (\tan Um' - \tan U'),$$

and putting this into $SS' = x \tan Um'$ we find

$$SS' = a_2 [\tan U' \tan^2 Um' - \tan Um' \tan^2 U'].$$

In this equation tan U' is the only variable quantity and we can find the value of

[35] it which makes SS' a maximum by differentiating the equation with reference to $\tan U'$. This gives

$$d(SS')/d \tan U' = a_2 [\tan^2 Um' - 2 \tan Um' \tan U'].$$

This becomes zero and indicates a maximum of SS' (obviously there can be no minimum) if

$$2 \tan Um' \tan U' = \tan^2 Um'$$

or if

$$\tan U' = \tfrac{1}{2} \tan Um'.$$

Putting this value of $\tan U'$ into the general equation for SS', we find

$$\text{Maximum } SS' = a_2 [\tfrac{1}{2} \tan^3 Um' - \tfrac{1}{4} \tan^3 Um'] = \tfrac{1}{4} a_2 \tan^3 Um'.$$

And as this SS' is the *radius* of the disk of least confusion and $a_2 \tan^2 Um' = LAm'$, we obtain

$$\text{Diameter of Disk of Confusion} = \tfrac{1}{2} LAm' \,.\, \tan Um'$$

as the solution of our problem.

As we found the maximum of SS' for $\tan U' = \frac{1}{2} \tan Um'$, it follows that the disk lies at the point of intersection of the arriving rays from the half-aperture with the produced marginal rays. As by one of our starting equations $x = SS'/\tan Um'$ and as the maximum SS' was found $= \frac{1}{4} a_2 \tan^3 Um'$, we find $x = \frac{1}{4} a_2 \tan^2 Um' = \frac{1}{4} LAm'$. The geometrical disk of least confusion therefore is located at $\frac{1}{4}$ of the distance from the marginal to the paraxial focus.

In the case of the final image of our biconvex lens we found for the ray starting from the object-point under angle $U = -1 : LA' = 0{\cdot}079$ and $U' = 2-44-20$. The diameter of the disk of confusion for this aperture is therefore found as $= \frac{1}{2} \times 0{\cdot}079 \times \tan 2-44-20 = 0{\cdot}00189$ inch.

The disk of confusion due to primary aberration grows with the cube of the aperture. For we found it $= \frac{1}{2} a_2 \tan^3 Um'$, in which $\tan Um'$ grows, in first approximation, proportionately with the aperture. The diameter of the disk is also equal to half the transverse aberration of the marginal ray, for the latter is $= LAm' . \tan Um'$.

(2) Disk of Confusion for Zonal Aberration

In the trigonometrical correction of object-glasses we bring the paraxial and the extreme marginal rays to a common focus. If the aperture is considerable, the intermediate rays then display the zonal aberration which we have already discussed and a disk of least confusion must again result. We learnt that the zonal rays nearly always come to a shorter (under-corrected) focus than the paraxial and marginal rays, hence we can conclude at once that the marginal rays form the most forward and most wide-angled cone and that the disk of least confusion must be defined by the intersection of some produced zonal ray from the opposite side of the optical axis with the marginal ray, say at S in Fig. 34. If we again measure the semi-aperture by $\tan U'$, formula (a) of section [31] assumes the form

$$LA' = B'Bm' = -a_4 \tan^2 Um' \tan^2 U' + a_4 \tan^4 U'$$

and the maximum zonal longitudinal aberration will be

$$LZA' = -\tfrac{1}{4} a_4 \tan^4 Um'.$$

[35] Calling the distance $S'Bm'$ x, as in the case of the primary aberration, we read off Fig. 34

$$SS' = x \tan Um' = (LA' - x) \tan U' = (a_4 (\tan^4 U' - \tan^2 U' \tan^2 Um') - x) \tan U'$$

which gives from the second and fourth equalities

$$x = a_4 \tan^3 U' (\tan U' - \tan Um');$$

and putting this value of x into $SS' = x \tan Um'$ we obtain the general expression for SS'

$$SS' = a_4 \tan^3 U' \tan Um' (\tan U' - \tan Um')$$

in which $\tan U'$ is the only variable quantity. SS' will be a maximum for that value of $\tan U'$ for which $d(SS')/d \tan U' = 0$, which gives

$$a_4 (4 \tan^3 U' \tan Um' - 3 \tan^2 U' \tan^2 Um') = 0$$

$$\text{or } \tan U' = \tfrac{3}{4} \tan Um'.$$

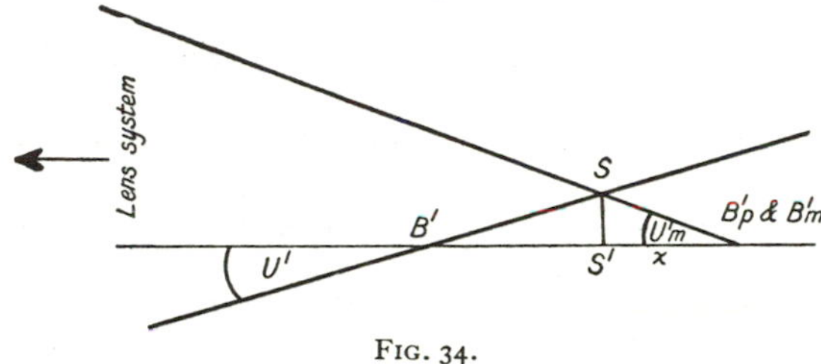

FIG. 34.

The disk of least confusion is therefore defined by the intersection of the produced rays from $\frac{3}{4}$ of the full aperture with the marginal cone.

If we now put $\tan U' = \frac{3}{4} \tan Um'$ into the general equation for SS' we find the radius of the disk of confusion as

$$SS' = a_4 \cdot \frac{27}{64} \tan^4 Um' \left(\tfrac{3}{4} \tan Um' - \tan Um'\right) = -\frac{27}{256} a_4 \tan^5 Um',$$

and if we introduce into this the maximum longitudinal zonal aberration $LZA' = -\frac{1}{4} a_4 \tan^4 Um'$ and double the radius SS' to find the diameter, we obtain

$$\text{Zonal Disk of Confusion} = \frac{27}{32} LZA' \,.\, \tan Um'.$$

For an alternative expression we can introduce the secondary aberration at the margin $= a_4 \tan^4 Um'$ and obtain

$$\text{Zonal Disk of Confusion} = \frac{27}{128} \tan Um'. \text{ (marginal secondary aberration).}$$

In order to discover the zones which have their focus in the centre of the disk we must note that x will be the longitudinal aberration for those zones. Putting $\tan U' = \frac{3}{4} \tan Um'$ into the general equation for x we find

$$x = a_4 \tan^3 U' (\tan U' - \tan Um') = -\frac{27}{256} a_4 \tan^4 Um'.$$

[35] As this is the aberration LAz' of the unknown zone or zones, it must fulfil the general LA' equation : hence

$$-\frac{27}{256} \cdot a_4 \tan^4 Um' = LAz' = -a_4 \tan^2 Um' \tan^2 Uz' + a_4 \tan^4 Uz'$$

which gives

$$(\tan Uz'/\tan Um')^2 = \tfrac{1}{2} \pm \sqrt{\frac{37}{256}}, \text{ nearly } = 0{\cdot}120 \text{ or } 0{\cdot}880.$$

Therefore $\tan Uz' = \tan Um' \times$ either $\sqrt{0{\cdot}12}$ or $\sqrt{0{\cdot}88}$.

The disk lies therefore, nearly enough, at the common focus of the zones whose diameter is respectively 0·346 and 0·938 of the full aperture.

For our cemented telescope objective we found in section [21] $LZA' = 0{\cdot}0050$ and $Um' = 7-17-11$, hence its geometrical disk of confusion is

$$= \frac{27}{32} \times 0{\cdot}0050 \times \tan(7-17-11) = 0{\cdot}00054 \text{ inch.}$$

(3) Disk of Confusion for pure Secondary Aberration

Devotees of the purely algebraical treatment of optical problems very generally adhere to the conviction that aberrations of different orders should not be mixed up : therefore they correct ordinary lens systems, in which the secondary aberration cannot be removed, for primary spherical aberration only, leaving the full amount of the secondary aberration to become manifest in the marginal zone. The equation for the aberration is then, again using $\tan U'$ as the measure of aperture, $LA' = a_4 \tan^4 U'$.

The disk of confusion for this case can be worked out exactly as that for primary aberration by merely changing the law of LA'. The working out gives

$$SS' = a_4 \tan U' \tan Um' (\tan^3 Um' - \tan^2 Um' \,.\, \tan U' + \tan Um' \tan^2 U' - \tan^3 U').$$

The differential coefficient becomes zero and indicates the maximum value of SS' when

$$\tan^4 Um' - 2\tan U' \tan^3 Um' + 3\tan^2 U' \tan^2 Um' - 4\tan^3 U' \tan Um' = 0.$$

The real solution of this cubic equation is very nearly $\tan U' = 0{\cdot}6 \tan Um'$ and, if SS' is worked out and doubled, we find

Disk of Confusion $= 0{\cdot}653\, a_4 \tan^5 Um' = 0{\cdot}653 \tan Um'$. (Secondary Spher. Ab.).

The coefficient of the secondary aberration is more than three times as large as that found above for trigonometrical correction of the marginal ray (i. e. 27/128). Really the case is almost invariably still worse, for the secondary aberration when left outstanding usually comes out decidedly larger than when the objective is modified trigonometrically to produce coincidence of the paraxial and marginal foci : hence the analytical correction, of primary aberration only, leads to a very grave deterioration of the image.

Our telescope-objective becomes free from primary spherical aberration (that is final $LAp' = 0$) if the last surface is changed from $r = -60$ inches to a perfect plane. LAm' is then found $= -0{\cdot}0248$ inch, and $Um' = 6-41-46$. Hence the disk of confusion of the analytically corrected objective will be $= 0{\cdot}653 \times 0{\cdot}0248 \times \tan 6-41-46 = 0{\cdot}00190$ inch. [35]

This compares with a disk of 0·00054 inch for the trigonometrically corrected objective, which latter moreover had a larger convergence angle Um' and would give a greatly reduced disk if its aperture were reduced and the curves modified to give trigonometrical correction for $Um' = 6-41-46$ instead of $7-17-11$.

Special Memoranda from Chapter II

Rays traced from an axial object-point through different zones of the aperture of a centred system will usually give intersection-lengths differing from the corresponding paraxial value. These differences of focus are called longitudinal spherical aberration and are defined as $LA' = l' - L'$; a positive value therefore results when the paraxial focus lies to the right of the focus of rays at finite angles. For any reasonable measure of the aperture Y—see [31]—the spherical aberration obeys the law $LA' = a_2 Y^2 + a_4 Y^4 + a_6 Y^6$, &c., the successive terms being called primary, secondary, tertiary, &c., aberration. Two terms usually give a satisfactory account of the aberration of all zones, provided that the solution is based on the marginal LA' as one of the two values required for the determination of a_2 and a_4. The primary aberration can be calculated by the convenient algebraical formula (10)**. For systems built up of reasonably thin lenses TL (10)** gives extremely useful approximate values. The LA' of a thin lens changes according to a parabolic law when the lens is bent, with a minimum value for a particular convex form, a maximum value for a particular concave form.

The three cases of aplanatic refraction at a spherical surface—[25]—and especially the third, are of high importance.

LA' is not a reliable absolute measure of the seriousness of the spherical aberration ; in pencils of very small convergence a huge value may be almost harmless, whilst in pencils of large angle of convergence a very small fraction of an inch may be very serious. The transverse measure of spherical aberration, TA', is less subject to this objection, and the angular measure AA' is practically free from it.

The law of transfer of LA' is that of longitudinal magnification ; TA' is transferred by the law of linear magnification.

CHAPTER III

PHYSICAL ASPECT OF OPTICAL IMAGES

[36] THE disks of confusion dealt with at the end of the previous chapter would—if correct—enable us to estimate the true capability of any system as regards the sharpness of the image of a point. Unfortunately the geometrically determined size of the image of a point is most seriously wrong in the sense of being too large whenever the residues of aberration are of the moderate magnitude which can be tolerated in respectable optical instruments. It is therefore most desirable that we should become acquainted with a more reliable method of estimating the real effect of small residuals of aberration before the geometrical estimates have become fixed on our minds and have more or less permanently warped our judgement.

We can easily convince ourselves by simple experiments that the geometrical theory of image-formation is quite wrong. Geometrically we found that a lens system when perfectly free from aberration would bring all the rays from an axial object-point to one and the same image-point, and that the latter would therefore be absolutely sharp no matter what the aperture and focal distance might be. In the case of primary spherical aberration we found that a disk of confusion should result, of a diameter proportional to the cube of the aperture. The disk should therefore shrink to 1/8th of its diameter at full aperture if the latter were reduced to one-half, and it should shrink to 1/1,000th of its maximum diameter at 1/10th of the full aperture. In the case of trigonometrically corrected systems with residual zonal aberration the disk of confusion should shrink even more rapidly, i. e. proportionately to the 5th power of the aperture. And in both cases the shrinkage at these tremendous rates should go on indefinitely down to the most minute pinhole aperture of the system.

Experiment proves that all these conclusions are erroneous : no matter what lens system we may try we shall find that when the aperture is gradually reduced the image of a point attains a minimum size at a certain considerable aperture, which will be the full aperture if the spherical aberration is reasonably well corrected. When the aperture of the system is reduced below this critical value the size of the image grows instead of diminishing and becomes very large at small apertures.

We may make the experiment without any apparatus by observing suitable objects such as a page of printed matter at the usual distance of distinct vision through pinholes of smaller and smaller apertures made by piercing dark paper with a fine-pointed needle. The reduction of the amount of light admitted by the small apertures should be compensated by more intense illumination as the apertures become smaller. It will then be seen that as soon as the aperture sinks below about 1 mm. the outlines of objects become softened ; when the aperture is less than about 0·2 mm. ordinary print can no longer be read, and with apertures below 0·1 mm. the page becomes practically uniformly grey, no matter how intense the illumination may be made. If we look at an intense isolated point of light through diminishing apertures, we shall see on an ever-increasing scale the 'Airy spurious disk', namely a central condensation of light with a red margin, surrounded by concentric rings of rapidly diminishing brightness, each ring being blue or green on its inner and red or orange on its outer edge. [36]

More conclusive results will be obtained by producing an image of a real or artificial star by an achromatic telescope objective and by observing this image with a compound microscope of moderate power (×50 or ×100) fitted with an eyepiece micrometer and by then applying a succession of diminishing round diaphragm-openings to the objective. If the spherical aberration of the lens is known, we can then measure the actual size of the brightest central condensation of light obtainable by selecting the focal adjustment and can compare it with that following from the geometrical theory, with the invariable result of proving the latter wrong whilst confirming the deductions from physical optics. With a certain amount of primary spherical aberration we shall find that the actual measured disk of light has only about 1/8 of the diameter following from the disk-of-confusion theory. On the other hand we shall find that with very small apertures, at which the spherical aberration is more or less completely negligible, the real spurious disk as measured is a huge multiple of the minute disk of confusion arrived at geometrically.

As we mean to learn how to design optical instruments so as to secure certain definite results when the instrument is actually tried on its intended kind of work, it is obviously indispensible to elucidate these contradictions between the geometrical theory and the results of direct tests and to secure reliable criteria by which the true value of a new design may be predicted from the calculated data and the residues of aberration.

This can be done by the undulatory theory of light first proposed by Huygens in the seventeenth century and subsequently put into an exact mathematical form suitable for application to all ordinary optical problems by Fresnel in the early years of the nineteenth century.

A systematic mathematical application of this theory to our designing problems is reserved for Part II; but in order to avoid a mere statement of the results without any clue as to their origin we will collect the principal facts in non-mathematical form.

According to the Huygens-Fresnel theory light consists of transverse vibrations of the aether induced by similar vibrations of each material point in a source of light. These vibrations have a considerable resemblance to the undulations set up in a sheet of still water when a pebble is dropped into it. As every one knows, a series of rings start out from the point of impact and spread uniformly in all directions. If there are bits of dry leaves floating on the water, we can see that the outward movement of the rings is not due to a corresponding flow of water in the radial direction, for the floating bits merely bob up and down in a regular rhythm as the rings pass below them without taking any part in their horizontal progression. We easily conclude that the radially spreading rings are an indirect effect of a rhythmical up-and-down movement of all the water particles, and that this latter movement is the real cause of the commotion in the water. What happens when the pebble is dropped into the water, is that it pushes the water down at the point of impact P; this downward movement is communicated to the surrounding water

[36] and a funnel-like depression is the first result: Fig. 35 (*a*). Necessarily a certain small amount of time is required to transmit the downward tendency to the water particles at increasing distances from the point of impact, and the initial depression will therefore spread outward at a certain definite velocity. While this is

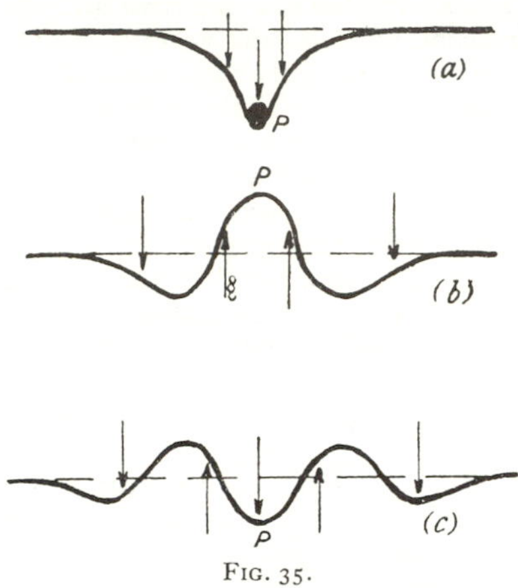

FIG. 35.

going on, the pebble will have produced its maximum effect at *P* and hydrostatic pressure will then come into play and push the water at the point of impact upwards. This upward pressure remains effective until the water has regained the original level, but as the water reaches that level with a certain upward velocity it will overshoot the mark and will now produce a mound of water at *P*. This upward tendency at *P* is again transmitted to the surrounding water and begins to travel outward whilst at a greater distance from *P* the original depression is spreading, thus producing the stage shown in Fig. 35 (*b*). Again, after a little interval, the raised water at *P* will have spent its momentum and will begin to fall and once more will overshoot the original level and produce another depression, Fig. 35 (*c*), now surrounded by a raised ring due to the outward transmission of the upward motion of stage (*b*). It will now be clear how the regular succession of raised rings of ever-increasing diameters is accounted for by a simple slight up-and-down movement of the water-particles transmitted from *P* in all directions at a definite velocity.

The advantage of beginning our study with the wave rings formed on water arises from the fact that the phenomena are on so large a scale and the velocity of propagation is so low that we can see every essential detail with perfect distinctness. The light waves are of the same type, but the distance from crest to crest, or wave-length, is minute, about 0·00075 mm. for extreme red and half of this for extreme violet light, and the velocity of propagation is enormous, namely, about 300,000 kilometres (or 186,000 miles) per second in empty space and sensibly also in air. By dividing the distance travelled in one second by the wave-length, both being of course expressed in the same unit of length, we find the number of vibrations performed by light in one second as 300,000 × 1,000 × 1,000/0·00075 = 400 billions for extreme red light and 800 billions for extreme violet light, the billion being taken as 10^{12}, not as 10^9 as it is usually taken in France and the U.S.A. In denser media such as glass light travels more slowly in inverse proportion to the refractive index, and as the number of vibrations per second is an absolute constant for light of any given colour, the wave-length also is smaller for dense media in inverse proportion to the refractive index. [36]

This change of velocity and wave-length which takes place when light enters a new medium explains the focal effects produced by lenses and lens systems.

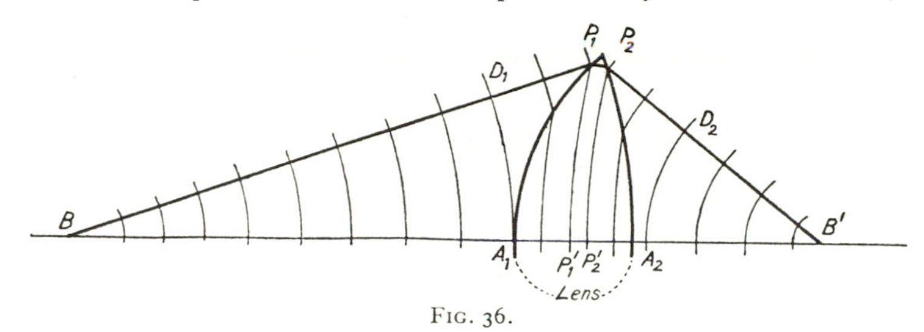

FIG. 36.

A luminous object-point B will send out vibrations or waves towards the lens shown by the thick curves in Fig. 36, and as these vibrations travel at the same speed in all directions as long as there is no change of medium, they will at any given moment be in the same phase—that is at the crest-stage, the trough-stage, or passing through the central position of their total amplitude—on a regular succession of accurately spherical surfaces having B as their common centre. Such surfaces of equal phase are called wave-surfaces or wave-fronts. When these waves, which travel along like the rings on the sheet of water but with the velocity of light, reach the lens, their velocity and wave-length becomes reduced to $1/N$ if N is the relative refractive index of the lens-substance referred to the surrounding medium, and it is easily seen that if the curvature of the first lens surface differs from that of the arriving waves, the curvature of the latter must change owing to the changed velocity. For if we consider the wave A_1D_1 which at a certain moment is in axial contact with the lens surface, we see that whilst its marginal part travels through the first medium from D_1 to the point P_1 on the lens-surface, its axial part travels for the same length of time in the denser medium of the lens and therefore at $1/N$th of the velocity, and so reaches only the point P_1' when the margin of the same wave has reached P_1. As similar considerations apply to all intermediate zones of the wave, it is clear that a new curvature is acquired by the latter. At the second lens surface a further change of curvature occurs, for whilst the marginal zone of the wave on leaving the lens traverses the distance P_2D_2 in the medium surrounding the image B' at the higher velocity belonging to that medium, the axial part is still in the dense medium of the lens and travelling more slowly and the curvature of the emerging waves is increased accordingly. The result is that the waves of negative curvature emanating from B are changed by the lens into waves of positive curvature converging towards B' and produce

[36] a strongly concentrated vibration at and near that point, which thus becomes an image of B.

We have tacitly assumed that the lens turns the spherical entering waves into strictly spherical emerging waves so that the latter again have a common centre at B'. A simple lens would only do this if it were 'figured' on one or both faces to a non-spherical form : a diagram accurately drawn on a large scale with a lens of such bold spherical surfaces as is used in Fig. 36 will in fact reveal that the waves in the lens, and especially those emerging from it, are distorted in the sense that their marginal parts are too far to the right as compared with the more central zones. That is the physical significance of spherical aberration. With a perfect lens all the light starting from B at a certain instant would arrive at B' absolutely simultaneously after performing a certain definite (and in reality a very large) number of vibrations. This is optically expressed by saying that the optical paths leading from an object-point to its perfect image are all equal, the term 'equal optical paths' meaning paths which contain the same number of wave-lengths. Any defect of a lens or system which causes inequalities of the optical paths from an object-point to its desired image-point is called aberration.

To understand the spreading of the image of a point produced by a lens or system of limited aperture which was referred to in the opening paragraphs, we must look a little closer into the mechanism of the propagation of waves. The first tendency of the vibration reaching a point 'a' (Fig. 37) is to transmit that vibration in all directions just as if 'a' were a self-luminous point. This can be easily proved

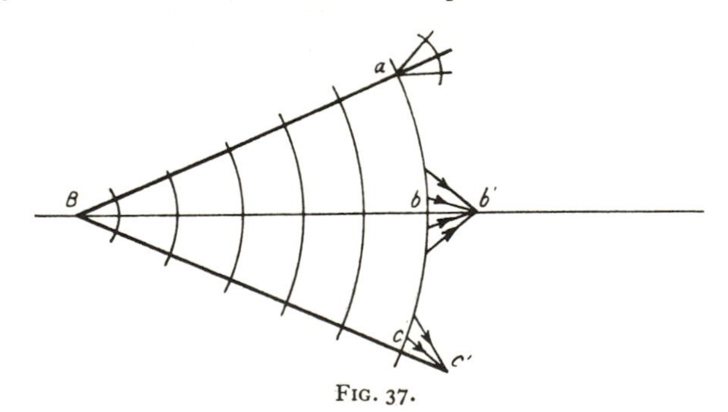

FIG. 37.

experimentally by isolating a very small area of the waves arriving at 'a' by means of a fine pinhole diaphragm ; light is then sent out from 'a' not merely in the direction of the ray Ba, but throughout a considerable angular space surrounding its continuation beyond 'a'. The tendency to break up into new waves of wide extent is therefore present at every point receiving light from a self-luminous point. But if we consider a point like 'b', which is surrounded by an uninterrupted area receiving vibrations from the same source, and ask what effect is produced a little beyond b at point b' on the line $B-b$, we see that the vibration from b will reach b' along the ray Bb. The vibrations from the neighbouring points will reach b' at increasing slants to the direction $b-b'$, but as to each neighbouring point to one side of *b* we can find a corresponding point on the opposite side of *b* which [36] has the opposite slant, it is obvious, on resolving such pairs of vibrations into components, one at right angles to *bb′* and the other in the direction of *bb′*, that the former components will reinforce each other whilst the latter are opposed to each other. Hence only a simple transverse vibration will result at *b′*, just as if it had been transmitted directly from *b* only. The tendency to break up is thus subdued at all interior points of a wave of finite extent. But if only a limited segment of a wave is admitted, as is always the case in optical instruments, then a different state of affairs exists at the extreme margin of the admitted wave-segment. At a point like '*c*' in the lower part of Fig. 37 there will be no vibrations reaching '*c*' from below : hence the deflecting tendency of the vibrations reaching '*c*' from the upper side will not be counteracted and we see that there must be a tendency for the light to spread outside the geometrical cone of rays. This represents the first rough demonstration of the fact that even a perfect optical instrument cannot depict a point as a point, for we have shown that some light is scattered outside the cone of geometrical rays.

The Airy Spurious Disk

[37] For the optical designer interest is limited to the appearance of the image at or near a focus. This can be determined most conveniently by a principle first laid down by Huygens and subsequently greatly extended by Fresnel. It is proved in the mathematical treatment of light-vibrations that whilst the real process of wave propagation is decidedly complicated according to the principles demonstrated by Fig. 37, the same result will be obtained more simply for any point near a geometrical focus by considering a wave-surface on emergence from the system, and therefore at a considerable distance from the focus, and assuming that each point of that surface *really* acts like a luminous point sending out spherical waves in all directions, according to '*a*' in Fig. 37. Let *WW* in Fig. 38 be a spherical light-

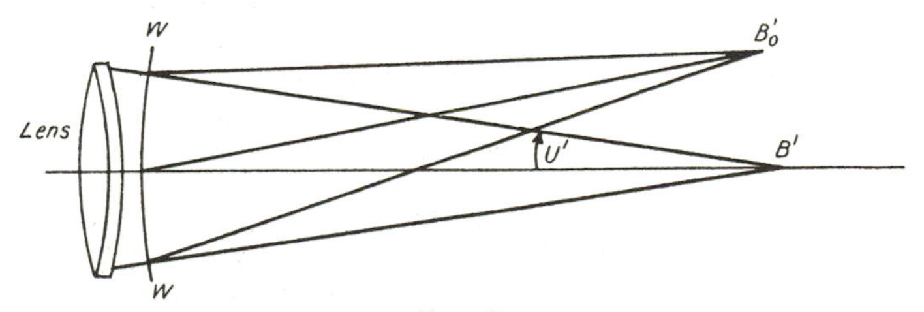

Fig. 38.

wave emerging from a lens system and converging upon the image-point B'. According to the Huygens-Fresnel principle we assume that each point of *WW* sends out light in all directions. Point B' will therefore receive light from all parts of *WW*, and as *WW* is postulated as a spherical wave with B' as its centre, all this light will arrive at B' simultaneously and therefore in the same phase of vibration, and will produce a strong concentration of light. But if we consider a point B'_0 to one side of B', this will also receive light from all parts of *WW*, but along paths

[37] which vary in length, for in its eccentric position B_o' will obviously be nearer to the upper part of WW than it is to its lower part. Hence there will not be at B_o' the perfect co-operation of the light from different parts of WW which we found at the geometrical focus B', and less intensity will result at B_o' than that at B'. It is a subject for an exact integration to determine the change of intensity for the whole neighbourhood of B'. This work was first carried out for a perfectly corrected lens system by the great Astronomer Royal, G. B. Airy, about 1834, and led to the result that the image takes the form of a central strong condensation of light surrounded by a series of rings of light of increasing diameters but of low and rapidly diminishing brightness, as shown roughly in Fig. 39, the lower part giving the actual appearance, the upper a graph of the intensity at different distances from the centre of the disk, all for monochromatic light. The image of a *white* point is represented by the superposition of the disks and rings produced in all the constituent colours of the spectrum, and as each disk or ring has a diameter proportional to the wave-length of the light producing it, the overlapping of the red over the blue disk and rings causes the red margin of the central disk and the blue inner and red outer edges of the rings already described in the opening paragraphs. The diameter of the 'first dark ring', that is of the boundary between the central condensation and the first bright ring, is found to be equal to 1·22 wave-lengths/$N' . \sin U'$, when N' is the index of the medium in which the image is formed, and U' has our usual meaning as the convergence angle under which the extreme marginal ray meets the central ray at the focal point. The diameter of the spurious disk is thus shown to be quite independent of the nature of the instrument, whether microscope, telescope, or photographic lens: it depends only on the wave-length of the light employed and on the product $N' . \sin U'$, but subject to the additional conditions that the lens system is free from aberrations and that the image is observed at the exact focus determined by ray-tracing.

For our telescope-objective (assuming it to be perfect) the formula gives, with the mean wave-length of 0·00002 inch which we shall usually adopt, and with $N' = 1$ and $\sin U' = 0{\cdot}127$ as determined in section [21],

$$\text{Extreme diameter of spurious disk} = 0{\cdot}0000244/0{\cdot}127 = 0{\cdot}00019 \text{ inch.}$$

It is, however, found in practice that the eye does not appreciate the faint light of the outer fringe of the spurious disk, and so records a very much smaller diameter than that found theoretically as the extreme diameter. Numerous careful tests with both microscopes and telescopes have shown that under the most favourable conditions the visual diameter of the central disk of light is closely represented by the formula: Visual Spurious Disk = 0·5 wave-length/$N' \sin U'$, and this we shall usually adopt. It would give, for a perfect object-glass of the convergence-angle of our telescope-objective, a diameter of 0·00008 inch for light of wave-length 0·00002 inch.

Resolving Power of Optical Instruments

The formulae just given for the diameter of the spurious disk contain the full explanation of the experimental fact stated earlier that, when the aperture of a lens or system is cut down, the size of the image of a point of light grows in inverse

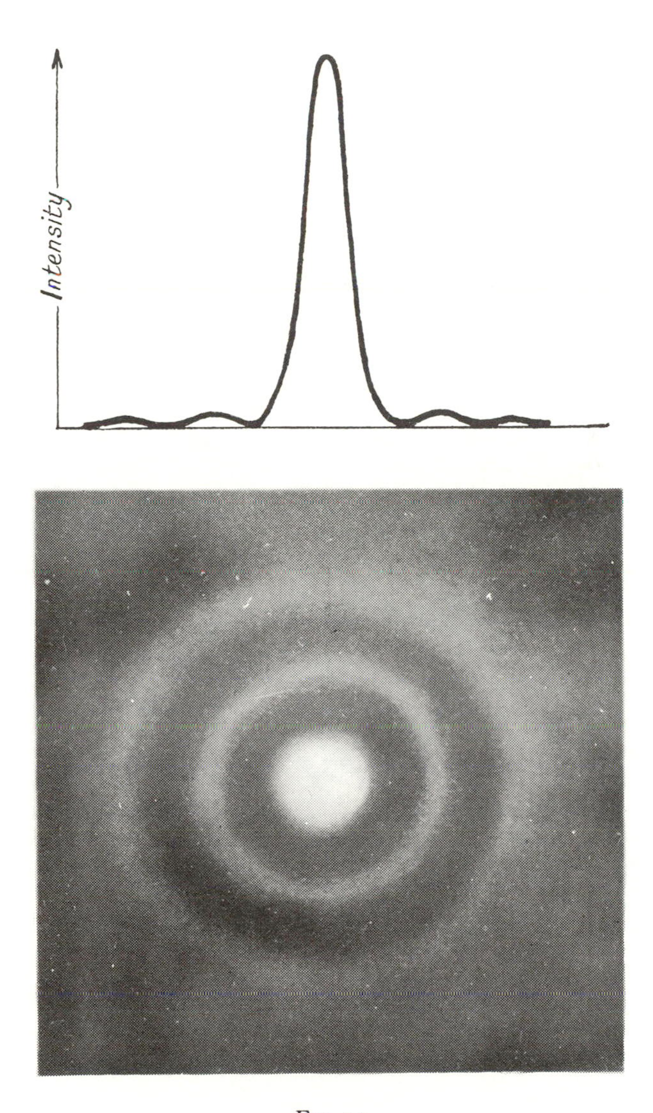

FIG 39

Intensity distribution and visual appearance of the Airy spurious disk

proportion, for we have already made frequent use of the proportionality of any [37] reasonably small U' or sin U' to the aperture, and as by our formulae the diameter of the spurious disk is inversely proportional to sin U' it must also be very nearly inversely proportional to the aperture.

This is a fact of fundamental importance because there is an obvious connexion between the spurious diameter of the image of a point and the least distance at which two close object-points can still be seen separated. Clearly the geometrical images of two such points will be formed at m' times their true separation if m' is the linear magnification produced by the lens system. The finite wave-length of light will cause each image-point to expand into a disk of light of the diameter given by our formulae, and these spurious disks will come into contact or will overlap if their diameter is equal to or greater than their distance apart, and the two points will then cease to be seen clearly separated. We can thus conclude that our telescope objective—if perfect—would fail to show clearly separated points whose images fell within 0·00008 inch of each other, and we should therefore be justified in calling 0·00008 inch the limit of resolving power of this objective referred to the images formed in its focal plane.

It is nearly always of greater interest to know the resolving power referred to the *objects* under examination. We have already stated the general method of solving this problem. The image will be $m' \times$ the object : therefore a separation of two image-points by a certain distance will correspond to a distance of the conjugate object-points of $1/m'$ times. Consequently, as the resolving power in the image-plane is equal to the diameter of the spurious disk, the resolving power for the object-plane is = spurious disk$/m'$ or, using our formula for the visual spurious disk,

$$\text{Least resolvable separation in object-plane} = 0{\cdot}5 \text{ wave-length}/m' \,.\, N' \,.\, \sin U'.$$

In Chapter I we proved by Lagrange's theorem that $m' = N \,.\, u/N' \,.\, u'$. We shall subsequently prove a generalization of this theorem, known as the optical sine condition, according to which we have also, for a perfectly corrected optical instrument, $m' = N \sin U/N' \sin U'$, U being the angle at which a ray leaves the axial object-point and U' the angle at which it reaches the conjugate image-point, and N the index of the medium surrounding the object. Using this value of m' in the last equation, it takes the form

$$\text{Least resolvable separation in object-plane} = 0{\cdot}5 \text{ wave-length}/N \,.\, \sin U.$$

This is a universally useful equation for the resolving power ; it was first given in this form by Abbe for use with the microscope. He introduced a special and now generally adopted name for the term $N \,.\, \sin U$ to which the resolving power is shown to be inversely proportional, calling it the 'Numerical Aperture' of the lens or lens-system with the symbol NA, which should not be interpreted as a product but simply as a compound letter. There is however nothing whatever in our last formula that limits it to microscopical objects or indeed to any particular size or distance of objects, and it may be used whenever the distance of an object is known so that the NA can be calculated. Supposing, for instance, that our telescope-objective of 1″ semi-aperture were applied to objects in air of $N = 1$ at

[37] a distance of 100,000 inches (roughly $1\frac{1}{2}$ miles), we should have $NA = N \,.\, \sin U = 1/100000 = 0{\cdot}00001$, and if we take the wave-length at the rather low value of 0·00002, our formula for the resolving power gives

$$\text{Least resolvable separation} = 0{\cdot}00001''/0{\cdot}00001 = 1 \text{ inch.}$$

We should therefore be justified in expecting our 2″ object-glass (assuming it to be perfectly corrected) to be just capable of showing barely separated, or of 'splitting' as the terminus technicus puts it, two artificial stars one inch apart and placed at 100,000 inches from the object-glass, provided of course that an eyepiece of sufficiently high magnifying power were used and that the conditions of illumination and tranquillity of the atmosphere were the most favourable.

If, on the other hand, we had an oil-immersion microscope objective capable of receiving rays from the object-point up to an inclination $U = 70°$ with a medium of index 1·515 surrounding the object, then we should have $NA = N \sin U = 1{\cdot}515 \times 0{\cdot}9336 = 1{\cdot}41$, and still using the wave-length 0·00002″, should have

$$\text{Least resolvable separation} = 0{\cdot}00001''/1{\cdot}41 = 0{\cdot}000007 \text{ inch.}$$

This objective, with adequate magnifying power and the most favourable conditions as to illumination, should therefore show separated two points 7 one millionths of an inch apart. This does in fact represent about the extreme limit of visual resolving power attainable with the microscope.

For telescopes, and especially for astronomical ones, the resolving power expressed in linear measure is not usually convenient. Telescopic measurements are in the majority of cases expressed in angular measure, and have to be so expressed for that vastly predominating class of astronomical objects whose distance is not known even in rough approximation. We can easily transform our formula to meet this requirement. Supposing the distance of the object to be $= L$ and the full aperture of the telescope objective A, then we shall have with ample approximation $\sin U = A/2L$, and putting this into our equation we find for $N = 1$, which is always realized for telescopic objects,

$$\text{Least resolvable separation} = 0{\cdot}5 \text{ wave-lengths} \times 2\, L/A = L \times 1 \text{ wave-length}/A,$$

the resolving power still being the linear measure of the closest objects which can be separated. If we then throw L upon the left side of the equation we obtain there : least resolvable distance of two objects, divided by their distance from the telescope, and this is obviously the angle of subtense of the closest resolvable objects. Therefore

$$\text{Least angular separation resolvable} = 1 \text{ wave-length}/A,$$

which gives the angle of subtense in radians. To turn this into the more convenient unit of a second of arc we have to multiply by the number of seconds of arc in one radian, or by 206,265 ; and if for inch-measure of the aperture we put the wave-length in inches as 0·000022, which is a nearer value of the wave-length of visually brightest light, we obtain

$$\text{Least angular separation resolvable} = 4{\cdot}5 \text{ seconds of arc/aperture in inches.}$$

This formula agrees closely with one widely adopted by astronomers on the basis of numerous measurements made chiefly by Dawes. For our 2″ object-glass this [37] formula gives $2\frac{1}{4}$ seconds for the closest double star resolvable by it. For the biggest refracting telescope—that of the Yerkes Observatory of 40 inches aperture—the limit comes out at 0·11 second of arc, and for the giant reflector at Mount Wilson, of 100 inches aperture, at 0·045 second.

It should be clearly realized that whilst the formulae for resolving power are very useful guides as to the best results that may be expected, they are not at all certain within even 10 per cent., for they involve firstly an assumption as to what fraction of the theoretical diameter of the spurious disk is appreciated by the human eye, and secondly an assumption as to the mean effective wave-length of ordinary white light. As the latter really contains a long range of wave-lengths, the image of a close double star may be a superposition of a fully resolved violet image upon less perfectly resolved images in the bright central colours of the spectrum with a totally unresolved red image as a background. This accounts for the well-known fact that observers with unusually developed acuity of vision for violet light may see an object fully resolved when other equally good observers without this rather rare gift see hardly a trace of resolution with the same object and instrument.

The formulae are of the utmost value for the optical designer in deciding upon the aperture necessary to attain any prescribed result. If possible we should aim at about twice the theoretical minimum aperture in order to make sure of realizing the required resolution when the conditions of illumination or steadiness of vision are not at their best. Supposing, for instance, we were asked to decide the smallest size of mirror of a sensitive reflecting galvanometer which could still give an image of millimetre divisions at say 2 metres distance from the mirror. A millimetre at 2 metres subtends an angle of 0·0005 radian, and using our first formula for the angular resolving power in radians we have for the least possible aperture

$$0 \cdot 0005 = 1 \text{ wave-length}/\text{aperture}.$$

To find the aperture in millimetres we must put the wave-length also into millimetres or for brightest light as 0·00055 mm., and by transposition obtain

$$\text{Necessary minimum aperture} = 0 \cdot 00055/0 \cdot 0005 = 1 \cdot 1 \text{ mm}.$$

To make sure, we should double this and would confidently say that a mirror 2·2 mm. in diameter would give a clearly recognizable image of a millimetre scale placed at 2 metres from it, provided of course that the observation is made with a telescope of sufficient magnifying power and that the scale is brightly illuminated.

Effect of Imperfections

Our results have been deduced from Airy's determination of the spurious disk produced by an absolutely perfect instrument at its exact focus. In actual practice we cannot hope to produce an absolutely perfect instrument, and in actual observations we cannot locate the exact focus. Hence the extremely important question arises : What is the effect of small residuals of aberration and of observing at an imperfectly adjusted focus ? Until this question is satisfactorily answered our results remain subject to grave uncertainties.

The Rayleigh Limit

[38] It has already been pointed out that at the exact focus of a perfect instrument all the light of any one colour which left the object-point at a given instant arrives absolutely simultaneously after a certain interval of time, which may be only a minute fraction of a second if object and image are both close to the lens system —as in the microscope—or may be an interval of a million years or more if an astronomical telescope receives light from one of the most remote celestial objects.

We also noted that any imperfection of the instrument or of the focal adjustment will cause inequalities of the optical paths of light, and the questions to be answered narrow themselves down to two : Firstly, how large the difference of optical paths leading to a focus may become without serious loss of quality in the image, and secondly, what residual of the various aberrations or what defect of focal adjustment corresponds to the permissible variation of optical paths.

The first of these fundamental questions was answered in 1878 by the late Lord Rayleigh. The intuition of true genius emboldened the great physicist to state as a general deduction from a small number of mathematically discussed test-cases that an optical instrument would not fall seriously short of the performance possible with an absolutely perfect system if the difference between the longest and the shortest optical paths leading to a selected focus did not exceed one quarter of a wave-length. The large amount of additional work done in recent years by way of further searching tests of this remarkable proposition has not shaken its validity ; it remains probably as good a definite single statement as could be desired, although moderate departures in either direction from the adopted quarter-wave are shown to be advisable in certain cases.

Optical Tolerances

When the Rayleigh quarter-wave limit is accepted, its utility depends on the finding of simple formulae which enable the optical designer to convert the defects of lens-systems as they emerge from his calculations into the corresponding difference of optical paths and so to determine whether the Rayleigh limit is exceeded by the residual aberration determined geometrically.

In accordance with a term universally adopted in mechanical engineering, we shall use the name 'Optical *Tolerance*' for the amounts of various lens defects which correspond to the Rayleigh limit, and which therefore do not seriously impair the performance of an instrument.

These optical tolerances will be systematically deduced from the principles of the undulatory theory of light in Part II : we shall here merely record the principal results and show how they are to be applied.

The Focal Range

A perfectly corrected lens system will have a certain range of focal adjustment on either side of the exact geometrical focus which causes differences of optical paths not exceeding the Rayleigh limit. The total range from the nearest to the farthest permissible focus is given by the simple formula

OT (1) $$\text{Focal range} = 1 \text{ wave-length}/N' \sin^2 Um'$$

in which N' stands for the refractive index of the medium in which the image is [38] formed, and is therefore nearly always = 1 for air, and Um' is the angle under which the extreme marginal ray arrives at the focus. To obtain the focal range in inches we must express the wave-length in inches, and may usually take 0·00002 inch as near enough. To obtain the focal range in millimetres the wave-length must also be taken in millimetres and 0·0005 mm. will usually be a suitable estimate. For our telescope objective, assuming it to be perfectly corrected, we should find, with $\sin Um' = 0{\cdot}127$,

$$\text{Focal range in inches} = 0{\cdot}00002/(0{\cdot}127)^2 = 0{\cdot}00124 \text{ inch.}$$

The meaning of the result is that the image would be almost equally good over the whole range of 0·00124 inch, covered by the distance from a near adjustment at 0·00062 inch within the true focus to a far adjustment at 0·00062 inch beyond the true focus.

Astronomical telescopes usually have a focal length equal to 15 times the clear aperture, so that $\sin Um' = 1/30$th. The focal range then rises to 0·018 inch. When a photographic objective is stopped down to $f/32$, we have $\sin Um' = 1/64$th and the focal range (called 'depth of focus' in photography) then rises to 0·082 inch.

The focal range is of very great importance, but as this arises chiefly from the consideration of chromatic aberration and of curvature of the field, we shall refer to these applications at the proper place in later chapters.

Tolerance for Primary Spherical Aberration

This tolerance applies to lens systems not completely corrected for spherical aberration, provided that the latter may be treated as being preponderatingly primary, that is, proportional to the square of the aperture when measured as longitudinal spherical aberration. Spherical aberration causes the waves converging towards a focus to be distorted from the ideal truly spherical form, and no point can then be found at which all the light arrives simultaneously; but there is a point at which the *differences* of optical paths are at their lowest possible value. In the case of primary spherical aberration this point lies exactly midway between the paraxial and marginal foci, and therefore does not coincide with the position of the geometrical disk of least confusion, seeing that we located this at one quarter of the distance from the marginal to the paraxial focus. There is thus a contradiction between the geometrical and the physical theories even as regards the *location* of the best focus.

The tolerance is given by the formula

OT (2) $\quad$ Permissible primary $LA' = 4 \text{ wave-lengths}/N' \sin^2 Um'$,

the meaning of N' and Um' being the same as in OT (1).

We will try this on our biconvex lens for the ray leaving the object-point at $U = -1°$. For the image formed by the first surface we have $N' = 1{\cdot}518$ and found $U' = 0 - 9 - 42$ or $\log \sin U' = \bar{7}{\cdot}4505$. Again taking the wave-length as 0·00002 inch, OT (2) gives

$$\begin{aligned} \log (\text{permissible } LA') &= \log (0{\cdot}00008) - \log 1{\cdot}518 - 2\,(\bar{7}{\cdot}4505) \\ &= \bar{5}{\cdot}903 + \bar{9}{\cdot}819 + 5{\cdot}099 = 0{\cdot}821\,; \\ \therefore \text{permissible } LA' &= 6{\cdot}62 \text{ inches.} \end{aligned}$$

[38] The calculated actual aberration for this ray was 1·27 inches, or only 1/5 of the tolerance.

The apparently huge spherical aberration at the first surface is therefore perfectly innocuous.

For the same ray after passage through the second surface we have $N' = 1$, log sin $U' = 8{\cdot}67929$, and this gives the tolerance for the final focus

$$\begin{aligned} \log(\text{tolerance}) &= \log(0{\cdot}00008) - 2\,(8{\cdot}6793) \\ &= 5{\cdot}903 + 2{\cdot}641 = 8{\cdot}544. \\ \text{permissible } LA' &= 0{\cdot}035 \text{ inch.} \end{aligned}$$

The calculated aberration was 0·081 inch, and therefore more than twice the tolerance. The final image would consequently not be satisfactory if the lens were used with the aperture corresponding to the ray starting from the object-point at $-1°$.

The example is instructive because it shows once more the misleading magnitude of longitudinal aberration. At the first surface of our lens more than 6 inches of longitudinal aberration correspond to the Rayleigh quarter-wave limit, at the second about 1/30th of an inch!

Calculation of the permissible LA' by OT (2) and its comparison with the longitudinal aberration actually found by trigonometrical calculation represents the simplest method of ascertaining the seriousness of the defect.

Tolerance for Zonal Aberration

In lens systems trigonometrically corrected for the spherical aberration of the marginal ray the question to be answered is how large the zonal aberration may become without transgression of the Rayleigh limit.

If the zonal aberration is determined by calculating LZA' for a ray at 0·7071 of the full aperture, then the tolerance on this LZA' is

OT (3) $$\text{Permissible } LZA' = 6 \text{ wave-lengths}/N' \sin^2 U'_m,$$

and is to be calculated with the U'_m of the extreme marginal ray, not with that of the 0·7071 ray. This case applies to our telescope-objective, for which sin $U'_m = 0{\cdot}127$, hence, in inches :

$$\text{Permissible } LZA' = 0{\cdot}00012''/(0{\cdot}127)^2 = 0{\cdot}0075 \text{ inch.}$$

By the 0·7071 ray we found $LZA' = 0{\cdot}0050$, or only 2/3 of the tolerance. The zonal aberration of that objective is therefore well below the Rayleigh limit and will not cause any sensible loss of quality.

In section [31] it was pointed out that a convenient alternative way of searching for zonal aberration consists in calculating the primary aberration of the lens system by (10)**, and that the final LAp' should theoretically come out at four times LZA'. From this we can conclude that the permissible LAp' will also be four times the permissible LZA' and obtain the alternative equation

OT (3)* $$\text{Permissible } LAp' = 24 \text{ wave-lengths}/N' \sin^2 U'_m.$$

For our telescope objective this will of course give four times the above LZA' tolerance or 0·0300 inch. In section [28] the final LAp' by direct calculation comes out at 0·0205 inch, or again at 2/3 the permissible amount.

All the tolerances embody the assumption that the differences of focus are small compared with the intersection-lengths ; they give unreliable results if the light emerges as a nearly parallel beam. [38]

Resolving Power in Presence of Defects

Lord Rayleigh left it an open question what exact effect the slight deterioration of the image corresponding to his limit would produce. The recent work already referred to has answered this question conclusively for the case of spherical aberration.

Up to the Rayleigh limit, and in fact up to nearly twice the Rayleigh limit, the central condensation of the spurious disk does not increase in diameter ; in fact in a few exceptional cases the diameter becomes smaller. Hence there is no loss of resolving power for strongly marked detail even up to the doubled limit. The central condensation however becomes less bright, and the light lost from it appears in the surrounding diffraction rings and—in most cases—in a general luminous halo. This scattered light must obviously diminish the *contrasts* between adjacent bright and dark parts of an image, for the bright parts are diminished in brightness whilst the scattered light appears simultaneously over the dark parts. It is therefore the brilliancy of the image which suffers first : the image becomes foggy. With a perfect instrument 85 per cent. of the total light appears in the central condensation and 15 per cent. is spread out in the surrounding rings. At the Rayleigh limit 20 per cent. (on the average) is lost from the central disk, which thus retains $0 \cdot 8 \times 85$ per cent. or 68 per cent. of the total light, whilst 32 per cent. of the latter is scattered over the surroundings. The loss in contrast is thus decidedly considerable and may render delicate detail invisible, whilst there is no loss of resolving power for strongly marked detail. We must therefore not draw heavily upon the Rayleigh limit unless some overwhelming advantage is to be gained with reference to other defects of the same lens-system. Above all things we should not use the limit at all to cover slipshod work. Whenever it can be done we should completely remove those aberrations which we can control, reserving the Rayleigh limit to cover aberrations which in the particular system are beyond our power, or else to cover technical imperfections such as want of homogeneity of the glass, departures from the prescribed curvatures and thicknesses, and errors of centring.

Defects well below the Rayleigh limit produce a rapidly decreasing loss of light from the central disk, and at quarter of the limit the loss is utterly inappreciable, being only about 1 per cent. Hence we may safely adopt the following rules, which will be found of great value in avoiding needless elaborations and purposeless waste of time on excessively accurate numerical work :

(1) One quarter of the Rayleigh limit as determined by the OT-equations may be used at any time to cover the uncertainties arising from the use of logs of any given number of decimal places. This rule is the justification of our adoption of five-figure logs for practically all optical calculations.

(2) One half of the Rayleigh limit may be drawn upon if a decidedly greater advantage will result with regard to some other aberration.

(3) The whole limit should only be taken advantage of if it will lead to great

[38] improvement in the correction of other aberrations. In extreme cases of this kind we may even feel justified in going up to the doubled limit.

In conclusion we will show how misleading the geometrical disk of confusion is. For our telescope-objective we calculated (section [35]) the diameter of the disk of confusion as 0·00054 inch. As this objective has now been shown to have zonal aberration well below the Rayleigh limit, we are safe in saying that it will give a star-image of the same size as an absolutely perfect objective of its aperture and focal length : this size was calculated above as 0·00008 inch. The true physical image produced by our objective therefore measures only 1/7th of the size concluded by geometrical reasoning! Moreover, the Rayleigh limit would only be reached if LZA' were $1\frac{1}{2}$ times the amount actually found by calculation. The geometrical disk of confusion would then come out also at $1\frac{1}{2}$ times the previously calculated value of 0·00054 inch or = 0·0008 inch, no less than ten times the size which we should find by direct observational test. Yet further, the physically determined spurious disk will not grow in diameter up to the doubled Rayleigh limit. At that point it therefore would measure only 1/20th of the size calculated geometrically by the disk-of-confusion fallacy.

The objection might be, and indeed has been, raised that if the geometrical estimates of the effects of residual imperfections are excessively severe, then surely we shall be absolutely safe in adopting the resulting stringent limits to aperture and field of any given type of optical systems. From the purely theoretical point of view this attitude is perfectly sound and correct. But from the practical and commercial point of view it is fatally wrong. A newly designed instrument has to meet competition with existing ones of the same type, and if, owing to the severity of the geometrical limits, the new instrument has a smaller aperture and smaller field when made to compete in cost, or is more costly on account of a more complicated construction, it has no prospect of success.

Special Memoranda from Chapter III

The interference or diffraction effects which result from the finite wave-length of light have a remarkable levelling-down tendency, by causing the image-points produced by perfect systems to have a finite diameter, and by causing this diameter of the 'spurious disk' to remain almost unchanged by small differences of optical paths up to the Rayleigh limit of one-quarter of a wave-length. As two neighbouring points will cease to be clearly separated when their centre-to-centre distance equals the diameter of their spurious disks, this diameter also determines the limit of resolving power.

Under the most favourable conditions we may expect :

Linear diameter of spurious disk = $\frac{1}{2}$ wave-length/$N' \sin U'_m$
and in the object-space

Least linear separation resolvable = $\frac{1}{2}$ wave-length/$N \sin U_m$.

For distant objects angular measure is preferable :

Least angular separation resolvable = $4\frac{1}{2}$ seconds of arc/aperture in inches.

The chief deductions from the Rayleigh limit may be easily remembered as [38] follows:

Range of focal adjustment within the limit:
Focal range = 1 wave-length/$N' \sin^2 U'_m$.
Simple spherical aberration at the Rayleigh limit:
LA'-tolerance = 4 times focal range.
Pure zonal aberration at $\sqrt{\frac{1}{2}}$ of full aperture:
LZA'-tolerance = 6 times focal range.
Primary LAp' admissible when $LA'_m = 0$:
LAp'-tolerance = 24 times focal range, if $LA'_m = 0$.

The wave-length may usually be taken = 0·00055 mm. = 0·000022 inch.

The Rayleigh limit should only be drawn upon when necessary, and as sparingly as possible, because the brilliance of the image is affected even by fractions of the limit.

CHAPTER IV

CHROMATIC ABERRATION

[39] PRACTICALLY all the equations deduced in the first two chapters contain the refractive index as one of their determining quantities. We may therefore conclude that all the important properties of lenses and lens-systems, such as focal and intersection-lengths, magnification and spherical aberration, will vary with the refractive index, and as the latter varies even for the same substance according to the colour of the light, we may draw the further conclusion that when white light is employed or light containing any sensible range of different colours, then there will in general be a chromatic variation of all the important properties of a lens-system, and the latter will give imperfect and confused images on account of these variations, which are known as 'chromatic aberrations'.

OPTICAL GLASS

For the magnitude of the variation of the refractive index we must refer to glass lists as supplied by the principal makers (Chance Bros., Ltd., in Birmingham; the Parsons Optical Glass Co. in Derby; Mantois in Paris; and Schott & Genossen in Jena), for it is still the universal practice to leave the determination of the optical properties of glass to the manufacturer. A short extract from Chance's list, headed by the properties of the optically priceless mineral fluorite (obtainable from Adam Hilger, Ltd., London) follows:

Number.	Name.	Nd	Nf − Nc	V	Nf − Nd Pfd	Ng′ − Nf Pg′f	Spec. Grav.
	Fluorite	1·4338	0·00454	95·5	0·00321 0·707	0·00256 0·563	3·18
6493	Borosilicate Crown	1·5160	0·00809	63·8	0·00567 0·701	0·00454 0·561	2·54
1203	Hard Crown	1·5155	0·00848	60·8	0·00598 0·705	0·00482 0·568	2·48
3463	Light Barium Crown	1·5407	0·00910	59·4	0·00642 0·705	0·00517 0·568	2·90
4873	Dense Barium Crown	1·6118	0·01037	59·0	0·00733 0·707	0·00590 0·569	3·56
9002	Medium Barium Crown	1·5744	0·00995	57·7	0·00703 0·707	0·00568 0·571	3·23
569	Soft Crown	1·5152	0·00906	56·9	0·00641 0·708	0·00517 0·571	2·55
4277	Telescope Flint	1·5237	0·01003	52·2	0·00708 0·706	0·00575 0·573	2·67
466	Light Barium Flint	1·5833	0·01251	46·6	0·00889 0·711	0·00740 0·592	3·30
407	Light Flint	1·5787	0·01420	40·8	0·01014 0·714	0·00851 0·599	3·26
4626	Dense Barium Flint	1·6236	0·01582	39·4	0·01129 0·714	0·00957 0·605	3·66
360	Dense Flint	1·6225	0·01729	36·0	0·01237 0·715	0·01052 0·608	3·64
337	Very Dense Flint	1·6469	0·01917	33·7	0·01376 0·718	0·01170 0·610	3·87
4141	Do.	1·7167	0·02430	29·5	0·01744 0·718	0·01511 0·622	4·47

The only refractive index given explicitly is the one for the D line, chosen because it lies near the brightest part of the solar spectrum and is easily determined by the light of a soda-flame. The next column gives the difference of the indices for the F and C lines, which are obtained by a vacuum tube containing hydrogen at a suitable low pressure. These lines, bright blue and bright red respectively, roughly correspond to the ends of the visually bright part of the spectrum, and are by long experience very suitable for being brought to a common focus in achromatic lenses for visual observation. [39]

The third column gives under the symbol V a number obtained by dividing $(Nd-1)$ by $(Nf-Nc)$. V therefore grows nearly in inverse proportion with $(Nf-Nc)$, and is largest for the lightest crown glasses and smallest for the dense flint-glasses. It enters into many of our formulae for chromatic aberration, and its definition should be carefully noted.

The fourth and fifth columns give in an upper line for each glass respectively the dispersion $(Nf-Nd)$ between the F and D lines and $(Ng'-Nf)$ between the G' and F lines, G' standing for a deep blue line, near the solar G-line, also obtainable from a hydrogen tube. These additional data enable us to find

$$\begin{aligned} Nf &= Nd+(Nf-Nd) \\ Nc &= Nf-(Nf-Nc) \\ Ng' &= Nf+(Ng'-Nf). \end{aligned}$$

The lower line for each glass in columns 4 and 5 gives the ratio or proportion $(Nf-Nd)/(Nf-Nc)$ and $(Ng'-Nf)/(Nf-Nc)$ respectively. We shall call these numbers Pfd and $Pg'f$ respectively (P for 'proportion' and the small letters for the spectrum lines concerned), when we subsequently use them for estimating the 'secondary spectrum', a residual chromatic aberration of so-called achromatic lenses.

The last column gives the specific gravity of each glass. If the specific gravity is plotted against Nd for all the glasses in the list, it will be found that a very nearly straight-line graph results. Such a graph is useful for forming a close estimate of the refractive index of unknown glass, and so helps to identify it. Before the introduction of the numerous abnormal glasses by the Jena glassworks it was a common practice to order glass by specific gravity instead of its optical properties: even now the copyists rely largely on the specific gravity balance in carrying out their piracies.

Some glass lists give information for additional spectrum lines, especially for the extreme red end represented by the red potassium line A' a little beyond the solar A line. The spectrum beyond the C line is so feeble that this information is of little practical importance.

Longitudinal Chromatic Aberration of a thin Lens

The expressions for the chromatic differences of focus take a reasonably simple form only if we limit ourselves to paraxial rays and moreover treat the lens or lenses as infinitely thin. We therefore use the TL formulae of Chapter I as our starting-point, and must remember that the approximation to the truth may at times be rather unsatisfactory.

[39] The general equation for a thin lens is

TL (1) $$\frac{1}{l'} = (N-1)\left(\frac{1}{r_1} - \frac{1}{r_2}\right) + \frac{1}{l} = (N-1)c + \frac{1}{l},$$

the second form resulting from introducing the simple symbol c for the net curvature $(1/r_1 - 1/r_2)$.

To obtain a perfectly general equation for the chromatic aberration at the focal point we now assume that our lens receives light from an axial object-point of colour r (which may be any colour, but we will call it red without limiting ourselves to that tint) at distance lr from the lens, and also from a similar point of colour v (again any colour, to be called violet for the sake of brevity only) at distance lv from the lens. We therefore assume that the object-point is already afflicted with chromatic aberration. The corresponding image-points shall be formed at $l'r$ and $l'v$ respectively, and the refractive indices for the two colours shall be Nr and Nv respectively. TL (1) then gives

for 'violet' $$1/l'_v = (N_v - 1)\,c + 1/l_v\,;$$

for 'red' $$1/l'_r = (N_r - 1)\,c + 1/l_r.$$

As c depends only on the fixed radii of curvature of the lens, it is of course an absolute constant. The difference of the two equations gives

$$\frac{l'_r - l'_v}{l'_r \,.\, l'_v} = (N_v - N_r)\,c + \frac{l_r - l_v}{l_r \,.\, l_v}$$

which still contains l'_r and l'_v and so is not a complete solution in terms of the given quantities ; to make the solution complete, we introduce the values of $1/l'_v$ and $1/l'_r$ by the original two equations and transpose these to the right, thus obtaining the general equation for longitudinal chromatic aberration :

TL, Chr. (1) $$l'_r - l'_v = \frac{(N_v - N_r)\,c + \quad (l_r - l_v)/l_r l_v}{[(N_v - 1)\,c \; + 1/l_v]\,.\,[(N_r - 1)\,c + 1/l_r]}.$$

We will give to all chromatic equations the letters Chr. to distinguish them from equations applying to light of only one colour.

The general equation can be simplified by replacing our indefinite colours r and v by the usual colours C and F for visual chromatic correction, and by introducing the focal length for the D line :

$$\frac{1}{f'} = (Nd - 1)\,c.$$

Replacing suffix v by f and suffix r by c, the first numerator term of TL, Chr. (1) becomes

$$(N_v - N_r)\,c = (Nf - Nc)\,c = \frac{Nf - Nc}{Nd - 1} \cdot c\,(Nd - 1) = \frac{1}{f'.V},$$

the last, because on extending by $(Nd - 1)$, the first term $(Nf - Nc)/(Nd - 1)$ is the reciprocal of our definition of the glassmakers' V-value and $c\,(Nd - 1)$ is the 'power' $1/f'$ for the D line. The second term in the numerator of TL, Chr. (1)

vanishes if $lr = lv$, and we adopt this restriction which implies that the lens [39] receives light from a single white object-point sending out light of all colours.

In the denominator we note that for all optical glasses the difference between Nc and Nf is a small fraction $(1/V)$ of $(N-1)$, and as the D line is not very far from midway between C and F, we shall introduce only a small inaccuracy if we replace both $(Nc-1)$ and $(Nf-1)$ by $(Nd-1)$ and then write $1/f'$ instead of $(Nd-1)\,c$, and simply $1/l$ in place of $1/lc$ and $1/lf$, which are now all equal. We thus obtain the equation

TL, Chr. (2) $$l'_c - l'_f = \frac{1}{f'V\,(1/_{f'} + 1/_{l})^2}$$

as a fairly close estimate of the longitudinal chromatic aberration of a thin simple lens receiving light from a white object-point at distance l. We will briefly discuss this equation.

For a very distant object-point $1/l$ vanishes and the image will be formed at the principal focus. The equation becomes

$$l'_c - l'_f = \frac{1}{f'(1/_{f'})^2 \,.\, V} = \frac{f'}{V}$$

and clearly shows the significance of V in estimating the chromatic aberration. The latter is inversely proportional to V, and with a given focal length is therefore smallest for the first and largest for the last glass of our list, which for that reason is arranged in order of decreasing V-values. V is sometimes called the 'figure of merit' of a glass because the chromatic aberration becomes smaller and smaller when V grows. In cases when the chromatic aberration cannot be corrected but is nevertheless objectionable (as for instance in ordinary eyepieces) we should therefore give preference to glasses of high V-value.

As V is a positive number, the equation shows that for a positive f', which means for a convex lens, $l'_c - l'_f$ is always positive : the red focus therefore always lies to the right of the blue focus when light passes from left to right through a convex simple lens. This sense of the chromatic aberration is usually referred to as 'chromatic under-correction'.

For object-points at finite distances we can replace $1/f' + 1/l$ in the denominator of TL, Chr. (2) by its equivalent $1/l'$ in accordance with TL (3) of Chapter I. The equation then becomes

$$l'_c - l'_f = \frac{1}{f'V\,(1/l')^2} = \frac{l'^2}{f'\,.\,V} = \frac{f'\,.\,l'^2}{f'^2\,.\,V} = \frac{f'}{V}\left(\frac{l'}{f'}\right)^2.$$

As f'/V measures the longitudinal chromatic aberration at the principal focus, this equation shows that the aberration becomes rapidly larger than that at the principal focus if l' is greater than f', and becomes smaller if l' is less than f'. This is of some importance in estimating the effects obtainable by simple lenses.

Paraxial Achromatism of the Axial Image-point

John Dolland, a London optician, discovered about 1757 that the longitudinal [40] chromatic aberration could be removed by combining a convex lens of crown glass with a weaker concave lens of flint-glass : by this combination he produced the first achromatic telescope objective.

[40] From the equation by which TL, Chr. (1) was obtained, we can deduce the proper proportions of two or more lenses which are to produce achromatism or some other prescribed degree of chromatic correction when they are combined. The equation is :

$$\frac{l'_r - l'_v}{l'_r \, . \, l'_v} = (Nv - Nr) \, . \, c + \frac{l_r - l_v}{lr \, . \, lv} \, .$$

As this is a perfectly general equation we can write it down for any lens or succession of lenses. The images produced by the first lens will act as objects for the second, and so on right through. $(l'r - l'v)$ of any lens will necessarily be identical with $(lr - lv)$ of the following lens, for both expressions apply to the distance between the same 'red' and 'violet' intersection-points. But if there were air-spaces between the successive lenses, then lr of a following lens would be $=$ ($l'r$ of the preceding lens minus the separation) and a complication of the problem would result. We therefore limit ourselves for the present to combinations of thin lenses in axial contact with each other : then any l' of a preceding lens will be strictly equal to the corresponding l of the following lens. If we write down a succession of equations for lenses a, b, c, &c., to k (k being taken as a general symbol for the last lens, however many there may be), they will be

$$\left(\frac{l'_r - l'_v}{l'_r \, . \, l'_v}\right)_a = (N_v - N_r)_a \, . \, c_a + \left(\frac{l_r - l_v}{l_r \, . \, l_v}\right)_a$$

$$\left(\frac{l'_r - l'_v}{l'_r \, . \, l'_v}\right)_b = (N_v - N_r)_b \, . \, c_b + \left(\frac{l_r - l_v}{l_r \, . \, l_v}\right)_b$$

- - - - - - -

$$\left(\frac{l'_r - l'_v}{l'_r \, . \, l'_v}\right)_k = (N_v - N_r)_k \, . \, c_k + \left(\frac{l_r - l_v}{l_r \, . \, l_v}\right)_k ;$$

and as now—owing to the absence of airspaces—l'_r of any preceding lens is equal to l_r of the following lens, and similarly for l'_v and l_v, the left side of the first equation will be equal to the second term on the right of the second equation, the left of the second equal to the final term on the right of the third, and so right through. Hence addition of all the equations leads to extensive cancellations and gives

TL, Chr. (3) $$\left(\frac{l'_r - l'_v}{l'_r \, . \, l'_v}\right)_k = \sum_a^k \left[c \, (N_v - N_r) \right] + \left(\frac{l_r - l_v}{l_r \, . \, l_v}\right)_a$$

as a perfectly general expression for the chromatic aberration of any *thin* lens system.

The most usual problem to be solved in practical optics is to find the proper form of a lens combination which gives a prescribed chromatic correction and a prescribed focal length, the latter being determined by a preliminary calculation with the simple lens equations if a prescribed magnification is to be produced. In such a case two conditions are laid down, and on fundamental mathematical principles two variable quantities—lens-curvatures in our case—will be necessary and sufficient to fulfil these two conditions. TL, Chr. (3) gives for this case of two lenses [40] a and b

$$\left(\frac{l'_r - l'_v}{l'_r \cdot l'_v}\right)_b = c_a (N_v - N_r)_a + c_b (N_v - N_r)_b + \left(\frac{l_r - l_v}{l_r \cdot l_v}\right)_a.$$

We will introduce a simple symbol for the chromatic residuals (which do not come into operation very frequently) by putting

$$R = \left(\frac{l'_r - l'_v}{l'_r \cdot l'_v}\right)_b - \left(\frac{l_r - l_v}{l_r \cdot l_v}\right)_a$$

so that our chromatic equation becomes

$$(1) \qquad c_a (N_v - N_r)_a + c_b (N_v - N_r)_b = R.$$

For the focal length f' of the combination we have by Chapter I: $1/f' = 1/f_a + 1/f_b'$, and if we assume that these focal lengths are to be determined in a colour for which the refractive index is plain N with suffix indicating the lens, we shall have the second equation

$$(2) \qquad c_a (N_a - 1) + c_b (N_b - 1) = \frac{1}{f'},$$

and we now have two linear equations (1) and (2) in c_a and c_b from which these two unknown quantities can be determined.

To determine c_a we eliminate c_b by multiplying (1) by $-(N_b - 1)/(N_v - N_r)_b$ and then adding the resulting equation to (2), with the result

$$c_a \left[(N_a - 1) - \frac{N_b - 1}{(N_v - N_r)_b} (N_v - N_r)_a \right] = \frac{1}{f'} - R \cdot \frac{N_b - 1}{(N_v - N_r)_b}.$$

The first term in the square bracket can be extended into $(N_a - 1)(N_v - N_r)_a / (N_v - N_r)_a$ and the equation then becomes on slight simplification

$$c_a (N_v - N_r)_a \left[\frac{N_a - 1}{(N_v - N_r)_a} - \frac{N_b - 1}{(N_v - N_r)_b} \right] = \frac{1}{f'} - R \cdot \frac{N_b - 1}{(N_v - N_r)_b}.$$

If we now again introduced the three usual spectrum lines C, D, and F, the fractions would become V_a and V_b, and $(N_v - N_r)_a$ outside the square bracket would become $(N_f - N_c)_a$, but as it is sometimes necessary to use different colours for achromatization, we will attach to our formulae the slightly generalized definitions

$$N_v - N_r = \delta N \text{ and } \frac{N-1}{N_v - N_r} = \frac{N-1}{\delta N} = V,$$

and using these with the appropriate lens-suffixes, our solution becomes

$$c_a = \frac{1}{f'} \frac{1}{(V_a - V_b) \cdot \delta N_a} - \frac{R}{\delta N_a} \cdot \frac{V_b}{V_a - V_b}.$$

We can solve in the corresponding way for c_b, but as equations (1) and (2) are

[40] perfectly symmetrical as regards the (*a*) and (*b*) terms, it is legitimate to write down the solution for c_b by merely exchanging (*a*) and (*b*) suffixes wherever they occur. We thus obtain the complete general solution

TL, Chr. (4)
$$\begin{cases} c_a = \dfrac{1}{f'} \cdot \dfrac{1}{(V_a - V_b) \, . \, \delta N_a} - \dfrac{R}{\delta N_a} \cdot \dfrac{V_b}{V_a - V_b}; \\ c_b = \dfrac{1}{f'} \cdot \dfrac{1}{(V_b - V_a) \, . \, \delta N_b} - \dfrac{R}{\delta N_b} \cdot \dfrac{V_a}{V_b - V_a}; \\ \delta N = N_v - N_r \, ; \; V = (N-1)/\delta N \, ; \\ R = \left(\dfrac{l'_r - l'_v}{l'_r \, . \, l'_v}\right)_b - \left(\dfrac{l_r - l_v}{l_r \, . \, l_v}\right)_a ; \end{cases}$$

valid for any three colours, f' being the focal length for plain N.

It should be specially and carefully noted that in the numerator of the final factor on the right of the equations for c_a and c_b, V_b is associated with c_a and vice versa. Such transpositions of suffix are apt to be overlooked or regarded as misprints, with disastrous results!

The above formulae are the most convenient ones for computing purposes, as we then require to know $c = 1/r_1 - 1/r_2$, &c. But occasionally the focal lengths of the components are useful and the formulae are then easily modified. As we have $c_a (N_a - 1) = 1/f_a'$ and $c_b (N_b - 1) = 1/f'_b$, we have only to multiply the first equation by $(N_a - 1)$ and the second by $(N_b - 1)$ to effect the transformation. On the right $(N_a - 1)/\delta N_a$ can then be replaced by V_a and correspondingly for lens (*b*). The solutions then become

TL, Chr. (5)
$$\left.\begin{cases} \dfrac{1}{f'_a} = \dfrac{1}{f'} \cdot \dfrac{V_a}{V_a - V_b} - R \cdot \dfrac{V_a \, . \, V_b}{V_a - V_b}; \\ \dfrac{1}{f'_b} = \dfrac{1}{f'} \cdot \dfrac{V_b}{V_b - V_a} - R \cdot \dfrac{V_b \, . \, V_a}{V_b - V_a}; \\ R = \left(\dfrac{l'_r - l'_v}{l'_r \, . \, l'_v}\right)_b - \left(\dfrac{l_r - l_v}{l_r \, . \, l_v}\right)_a ; \\ V = (N-1)/(N_v - N_r). \end{cases}\right\} \text{All focal lengths are those for plain } N.$$

As has already been stated, the solutions are most frequently required when no chromatic residuals have to be taken into account and when the terms in R disappear. The equations then become still simpler and lead to a further interesting relation when c_a/c_b and f'_a/f'_b are worked out:

TL, Chr. (4)*
$$\begin{cases} \left.\begin{array}{l} c_a = \dfrac{1}{f'} \, \dfrac{1}{(V_a - V_b) \, \delta N_a} \\ c_b = \dfrac{1}{f'} \cdot \dfrac{1}{(V_b - V_a) \, . \, \delta N_b} \end{array}\right\} \dfrac{c_a}{c_b} = - \dfrac{\delta N_b}{\delta N_a} \\ \delta N = N_v - N_r ; \; V = (N-1)/\delta N. \end{cases}$$

[40]

TL, Chr. (5)*
$$\begin{cases} \left.\begin{aligned} f'_a &= f'\,\frac{V_a - V_b}{V_a} \\ f'_b &= f'\,\frac{V_b - V_a}{V_b} \end{aligned}\right\} \dfrac{f'_a}{f'_b} = -\dfrac{V_b}{V_a} \\ V = (N-1)/(N_v - N_r). \\ f'_a \text{ and } f'_b \text{ are those for plain } N. \end{cases}$$

In the starred equations it does not matter in which colour the f' of the complete combination is taken, for as these equations apply to *achromatic* systems, the possible variations of f' are utterly negligible compared to the inaccuracy arising in all TL equations from the neglect of lens thicknesses. By the starting equations it is easily seen that when the difference of focal length in different colours is sensible, the proper colour to be selected is that corresponding to the plain N used in the numerator of the V-value, therefore the D line when the glass-makers' V is used.

The starred solutions, for perfect achromatism, do not contain the l and l' of the two conjugate points in any way. A thin achromatic lens calculated by these solutions therefore is achromatic for objects and images at any conjugate distances. This valuable property of *thin* achromatic lenses is usually expressed by saying that the achromatism of such thin combinations is *stable*. In the general, unstarred, solutions the conjugate distances are brought in by the chromatic residual R. The achromatism is therefore not stable even for thin lenses when chromatic residuals have to be taken into account.

All the equations for c_a and c_b show that the curvatures required to realize a given f' of the complete combination become the greater, the smaller $V_a - V_b$; hence we can at once conclude that in order to keep the curvatures down we must select two glasses with a large difference of their V-values, that is from the opposite ends of the glass list. But we must not go too far in this respect, for the glasses at both ends of the complete list are expensive, and in the case of the very heavy flints they are also wanting in durability and in other desirable physical properties. The hard crown, No. 1,203, and dense flint, No. 360, of our list, mark the neighbourhoods from which most of the ordinary achromatic objectives obtain their components.

Four-figure logarithms or even a well-handled 10-inch slide-rule will do ample justice to the solutions on account of the inherent inaccuracy of all TL equations. The results obtained have almost invariably to be corrected by strict trigonometrical calculation and are only a first approximation, but such a convenient one, that these equations are among the most frequently employed in our whole arsenal. They usually become associated with the simple TL formulae in Chapter I and with the bending principle, and the section referring to these subjects should therefore be carefully revised if not already mastered.

A few examples will show how the formulae are employed:

(1) A perfectly achromatic visual objective is required, having a focal length $f' = 8$ inches and consisting of a crown lens (a) made of No. 1,203 and a flint lens (b) made of No. 360.

[40] As visual achromatism is called for, we employ the glass-maker's $V = 60{\cdot}8$ for crown and $36{\cdot}0$ for flint, and the list-values of $(Nf - Nc)$ for the δN of equations (4) and (4)*. Therefore $\delta N = 0{\cdot}00848$ and $0{\cdot}01729$ respectively. Equations TL, Chr. (4)* then give

$$c_a = \frac{1}{8} \quad \frac{1}{24{\cdot}8 \times 0{\cdot}00848} = 0{\cdot}5944\,;$$

$$c_b = \frac{1}{8} \frac{1}{(-24{\cdot}8) \times 0{\cdot}01729} = -0{\cdot}2915.$$

The calculation is carried out most conveniently in two parallel columns with four-figure co-logs :

co-log 8	$= 9{\cdot}0969$	co-log 8	$= 9{\cdot}0969$
co-log 24·8	$= 8{\cdot}6056$	co-log $(-24{\cdot}8)$	$= 8{\cdot}6056\,n$
co-log 0·00848	$= 2{\cdot}0716$	co-log 0·01729	$= 1{\cdot}7622$
log c_a	$= 9{\cdot}7741$	log c_b	$= 9{\cdot}4647\,n$

which give the above numerical values.

As by definition $c_a = 1/r_1 - 1/r_2$ and $c_b = 1/r_3 - 1/r_4$, we can now select any number of crown and flint lenses which satisfy the condition of achromatism and give a focal length of 8 inches when combined.

A very common choice is to make the crown lens equiconvex, or $r_2 = -r_1$, which gives $1/r_1 = -1/r_2 = \frac{1}{2} c_a = 0{\cdot}2972$, or by taking the reciprocal from Barlow's table, $r_1 = +3{\cdot}365''$, $r_2 = -3{\cdot}365''$. In objectives of this simple type it is very often desired that the flint and crown lens allow of being cemented. That means $r_3 = r_2$ and therefore $1/r_3 = -0{\cdot}2972$. Hence we obtain the last radius from

$$-0{\cdot}2915 = c_b = 1/r_3 - 1/r_4 = -0{\cdot}2972 - 1/r_4,$$

whence $1/r_4 = -0{\cdot}0057$ or $r_4 = -175$ inches. The last surface therefore would be faintly convex.

Another common choice is that the last surface shall be a plane, or $r_4 = \infty$. That gives us

$$c_b = 1/r_3 \qquad \text{therefore } 1/r_3 = -0{\cdot}2915 \qquad \text{or } r_3 = -3{\cdot}431''.$$

If we again wanted a cemented objective, we should have to make $r_2 = r_3$ or $1/r_2 = -0{\cdot}2915$. Hence by the equation defining c_a

$$0{\cdot}5944 = c_a = 1/r_1 + 0{\cdot}2915\,;\; 1/r_1 = 0{\cdot}3029\,;$$

$$r_1 = 3{\cdot}301''.$$

This solution therefore makes the outside (first) surface of the crown lens a little more strongly curved (or 'deeper' as glass-workers call it) than the contact radius.

As we shall soon learn, the choice of the separate radii is usually decided by the demand for correction of the spherical aberration and frequently by the additional demand that the sine condition shall also be fulfilled.

(2) As an example of the use of the more complicated equation TL, Chr. (4) we will assume that an objective of the same focal length and the same glasses as in example (1) is called for, but that a chromatic over-correction $l'_r - l'_v = -0{\cdot}02''$ is

required at its focus. This case arises in practice when a telescope-objective is to [40] be used with an ordinary eyepiece, because the latter is invariably chromatically under-corrected for its focal point. We have then to begin by calculating

$$R = \left(\frac{l'_r - l'_v}{l'_r \cdot l'_v}\right)_b - \left(\frac{l_r - l_v}{l_r \cdot l_v}\right)_a,$$

but as no chromatic aberration $(l_r - l_v)$ at the distant object-point is specified, the formula reduces itself to the first term in which the numerator is given as $= -0{\cdot}02$. l'_r and l'_v in the denominator will be the final intersection-lengths, and as our objective is used at its principal focus, both l' will be very nearly identical with the prescribed f', and it will be quite near enough (seeing that TL-equations always are inaccurate) to calculate

$$R = -0{\cdot}02/f'^2 = -0{\cdot}02/64 = -0{\cdot}00031.$$

When object and image are at finite distances from the lens, it is of course necessary to use the proper intersection-lengths for the evaluation of R, but they may be those for plain N; it is never necessary to use the separate values for 'red' and 'violet'. The remaining data remaining unchanged, we can calculate the total curvature:

$$c_a = \frac{1}{8}\ \frac{1}{24{\cdot}8}\ \frac{1}{0{\cdot}00848} - \frac{-0{\cdot}00031}{0{\cdot}00848} \times \frac{36{\cdot}0}{24{\cdot}8};$$

$$c_b = \frac{1}{8}\ \frac{1}{-24{\cdot}8}\ \frac{1}{0{\cdot}01729} - \frac{-0{\cdot}00031}{0{\cdot}01729} \times \frac{60{\cdot}8}{-24{\cdot}8}.$$

The first term in each is identical with the c_a and c_b of example (1): the second terms being computed by four-figure logs, with careful attention to signs, we obtain

$$c_a = 0{\cdot}5944 + 0{\cdot}0531 = 0{\cdot}6475;$$

$$c_b = -0{\cdot}2915 - 0{\cdot}0450 = -0{\cdot}3365.$$

It will be noticed that the apparently slight over-correction called for increases both curvatures very seriously. If we calculate the data for a cemented objective with plane last surface we shall find

$$1/r_2 = 1/r_3 = c_b = -0{\cdot}3365.$$

Therefore contact-radius $r_2 = -2{\cdot}972$ (against the previous $-3{\cdot}431$), and for the first radius we now have

$$c_a = 0{\cdot}6475 = 1/r_1 + 0{\cdot}3365, \text{ or } 1/r_1 = 0{\cdot}3110;$$

$$r_1 = 3{\cdot}215 \text{ (against the previous } 3{\cdot}301).$$

It is a common practice in such cases, when a telescope-objective is to be used with 'under-corrected' eyepieces (and hardly any eyepiece is perfectly corrected in itself) to treat the eyepiece aberration as 'negligible'. Seven-figure accuracy is then frequently bestowed upon the objective and the glass shop is worried about getting the radii just right to many decimal places; the above quite normal example will show at a glance what an absurd and illusory display of precision this is!

[40] (3) To show the effect of a smaller difference $V_a - V_b$ when the refractive indices remain practically the same, we will solve for an achromatic lens of 8 inches focal length, just as in example (1), but made from the glasses

569 Soft Crown	$Nd = 1{\cdot}5152$	$Nf - Nc = 0{\cdot}00906$	$V = 56{\cdot}9$;
4626 Barium Flint	$1{\cdot}6236$	$0{\cdot}01582$	$39{\cdot}4$.

TL, Chr. (4)* then gives

$$c_a = \frac{1}{8} \; \frac{1}{17{\cdot}5} \; \frac{1}{0{\cdot}00906} = 0{\cdot}7885\,;$$

$$c_b = \frac{1}{8} \; \frac{1}{-17{\cdot}5} \; \frac{1}{0{\cdot}01582} = -0{\cdot}4515\,;$$

and if for easy comparison we again select the externally planoconvex cemented form, we obtain

$$c_3 = c_2 = c_b = -0{\cdot}4515\,; \text{ therefore } r_2 = r_3 = -2{\cdot}215\,;$$

$$c_1 = c_a + c_2 \quad = +0{\cdot}3370\,; \text{ or } \qquad r_1 = 2{\cdot}967.$$

The combination of hard crown and ordinary flint in example (1) gave $r_1 = 3{\cdot}301$ and $r_2 = r_3 = -3{\cdot}431$; the diminished difference of V-values in the present example therefore shortens the contact-radius to about 2/3 of the previous value, and the first radius by 10 per cent.

(4) Another important effect to be noted is that of using glass, more especially crown glass, of higher index. We will choose as reasonably close in V-values to the preceding example:

9002 Medium Barium Crown	$Nd = 1{\cdot}5744$,	$Nf - Nc = 0{\cdot}00995$,	$V = 57{\cdot}7$;
4626 Barium Flint	$1{\cdot}6236$	$0{\cdot}01582$	$39{\cdot}4$;

and for the same focal length of 8 inches obtain

$$c_a = \frac{1}{8} \; \frac{1}{18{\cdot}3} \; \frac{1}{0{\cdot}00995} = 0{\cdot}6866\,;$$

$$c_b = \frac{1}{8} \; \frac{1}{-18{\cdot}3} \; \frac{1}{0{\cdot}01582} = -0{\cdot}4318\,;$$

and these give for the externally planoconvex cemented lens

$$c_3 = c_2 = c_b = -0{\cdot}4318 \qquad \text{or } r_2 = r_3 = -2{\cdot}316\,;$$

$$c_1 = c_a + c_2 \quad = 0{\cdot}2548\,; \qquad r_1 = 3{\cdot}925.$$

The contact radius is slightly longer than in example (3), but as the formulae show, would have come out exactly the same if our barium crown had had the exact V-value of the soft crown. But r_1 for the crown of high index comes out not only longer than in example (3), but actually much longer than in example (1), where the V-difference was larger. This tendency of crown glass of high refractive index to diminish the *external* curvature of *cemented* achromatic lenses is a fact of considerable importance in the design of eyepieces, of microscope-objectives and especially of photographic lenses. We should of course have arrived at a still longer r_1 if we had used flint No. 360 instead of the barium flint. By combining [40] dense barium crown with very light ordinary flint we can actually produce cemented lenses of positive focal length, and therefore producing effects like any ordinary convex simple lens, which are externally concave. This may be verified by the following glasses from Chance's complete list, solving for $f' = 8$ as in our other examples:

8793 Dense barium crown, $Nd = 1{\cdot}6140$, $Nf - Nc = 0{\cdot}01080$, $V = 56{\cdot}9$;

7863 Extra light flint $1{\cdot}5290$ $0{\cdot}01026$ $51{\cdot}6$.

This type of combination is frequently found in photographic anastigmats.

(5) When an achromatic lens is required to produce an image of prescribed magnification, a preliminary calculation by the magnification-formulae for thin lenses is usually necessary. The formulae deduced for this purpose in Chapter I were

TL (4)
$$h'/h = m' = l'/l \text{ or right-to-left } h/h' = m = l/l';$$
$$l = f'(1 - m')/m'; \quad l' = f'(1 - m');$$
$$BB' = -f'(1 - m')^2/m';$$
$$h' = -e \cdot f';$$

in which h and h' stand for the actual size of a small object or image at right angles to the optical axis, l and l' for their respective distances from the lens, m and m' for the magnification, f' for the principal focal length of the lens, BB' for the distance from an object in the medium to the left of the lens to an image in the medium to the right of the lens, and e for the angle of subtense of a very distant object to the left of the lens producing an image of size h' in the medium to the right of the lens. The equations in the second and third lines are the most frequently useful ones: they can of course be transposed to suit any particular purpose. In order to fit in with our usual computing direction it is generally best to assume the object in the medium to the left and the image in the medium to the right of the lens. Numerical examples:

(a) Required: an achromatic lens to give a linear magnification $m' = +10$ times of an object placed at $1{\cdot}1$ inches to the left of the lens; therefore $l = -1{\cdot}1$ inches. $l = f'(1 - m')/m'$ transposed gives the required focal length $f' = l \cdot m'/(1 - m')$, or with our data $f' = -1{\cdot}1 \times 10/(1 - 10) = -11/-9 = +1{\cdot}222$ inches. The TL, Chr. equations can then be worked out on choosing suitable glass. It should be stated in this connexion, in order to avoid subsequent confusion, that this solution is correct only if the magnification is really meant to be the actual *linear* magnification of the *virtual* image. In speaking of an ordinary magnifier as having a magnification of ten times, the usual meaning attached to the statement is that the lens gives an image subtending ten times the angle under which the object itself would appear when placed at 10 inches from the pupil of the observer's eye. The very different conception of 'angular' magnification is thus brought into the problem, and this will be dealt with on a later occasion.

(b) An achromatic lens is required to produce a real and inverted image magni-

[40] fied six times when the distance between object and image is 1·75 metres. In the symbolism of TL (4) the data are in this case : $m' = -6$ (minus because the image is inverted) and $BB' = 1{\cdot}75$ m. We can therefore apply the equation in the third line of TL (4) which gives, when transposed, $f' = -BB' . m'/(1-m')^2$, or with our data

$$f' = (-1{\cdot}75)(-6)/(1-(-6))^2 = 1{\cdot}75 \times 6/7^2 = 10{\cdot}5/49 = 0{\cdot}2146 \text{ m.},$$

or 214·6 mm. by 10-inch slide-rule. The problem is then ready for completion of the solution by TL, Chr. (4)*, or the corresponding unstarred equations if the chromatic aberration of an eyepiece is to be taken into account.

All the examples have called for a positive f' of the required combination. Achromatic lenses with the properties of a simple concave lens are occasionally required. For these the principal focal length f' is negative, and the only change that has to be made in the solution is, therefore, to introduce the required focal length with a negative sign.

Although the calculation of our equations for paraxial achromatism of thin systems is very simple, time may be saved when a number of lenses of the same glasses but of different focal length are required, by noting that in the starred solutions (for perfect achromatism) c_a and c_b are inversely proportional to f' : hence when the solution has been effected for one of the required focal lengths, the c for other focal lengths may be found by this law of inverse proportion. Taking our example (1), we found

for $\quad f' = 8$ inches : $c_a = 0{\cdot}5944 \; c_b = -0{\cdot}2915.$

Supposing we wanted an achromatic lens of the same glasses, but for $f' = 4{\cdot}73$ inches, we should find by a simple slide-rule setting :

$$c_a = (8/4{\cdot}73)(0{\cdot}5944) = 1{\cdot}005 \; ; \; c_b = (8/4{\cdot}73)(-0{\cdot}2915) = -0{\cdot}493.$$

And if we also wanted a negative achromatic of focal length $-6{\cdot}8$ inches we should have

$$c_a = (8/-6{\cdot}8)(0{\cdot}5944) = -0{\cdot}699 \; ; \; c_b = (8/-6{\cdot}8)(-0{\cdot}2915) = +0{\cdot}343.$$

This simplifying rule does not apply to the complete solution by the unstarred equations. In their case it still covers the first and principal item on the right, but the second item which depends on R must be individually worked out.

Beginners will find it of very great assistance in realizing the significance of the examples and of numerous others which they will be wise to work out for themselves, to make a reasonably accurate scale-drawing of the resulting objectives, allowing for the adopted focal length of 8 inches a thickness of the crown lens of 0·15 inch and a thickness of the flint lens of 0·10 inch. If the curves for the crown component are carried to their intersection, it will then be seen at a glance how a low difference of the V-values of crown and flint restricts the aperture and increases the angles of incidence at any given aperture. Beginners usually are very reluctant to adopt this advice, but they make a great mistake in taking that attitude. Practice in neat and accurate drawing, moreover, is essential for the carrying out of the powerful graphical methods of easily solving otherwise difficult problems, which will play a very large and important part in later chapters.

THE SECONDARY SPECTRUM

[41] Our fundamental equations for paraxial achromatism bring the two colours 'r' and 'v' (C and F for visual instruments) to a common focus. The third colour corresponding to plain N was introduced only as that for which the prescribed focal length is to be established ; its presence in our equations must on no account be interpreted to mean that the common focus of colours 'r' and 'v' also coincides with that corresponding to plain N.

The question then arises as to where other colours come to a focus : if their focus differs from that of the two colours selected for exact union, then the achromatism of the system will be imperfect. We now seek the answer to this question.

To simplify the problem, we limit ourselves to the case of perfect union of 'r' and 'v' as established in TL, Chr. (4)*. In practice the departures from the proportions fixed by these equations are never large, and therefore do not materially affect the small residuals which we are seeking.

We can obtain a partial answer to our question from the equation added to the set :

$$c_a/c_b = -\delta N_b/\delta N_a,$$

which is true for corresponding δN of the two glasses in any region of the spectrum. Transposing this condition, we obtain

$$c_b = -c_a \,.\, \delta N_a/\delta N_b,$$

and we see that the same value of c_b can only result for different regions of the spectrum if $\delta N_a/\delta N_b$ has a constant value throughout the spectrum. This then is a condition which two glasses would have to fulfil in order that two given lenses, of curvature c_a and c_b respectively, should bring all colours to the same focus. If we examine the data of the glass list for the two glasses used in example (1), we find the respective dispersions and their ratio as follows :

	$Nf - Nc$	$Nf - Nd$	$Ng' - Nf$
Glass (a)	0·00848	0·00598	0·00482
Glass (b)	0·01729	0·01237	0·01052
Ratio (b)/(a)	2·04	2·07	2·18

Our example, having been worked out for the union of C and F, will have $c_a/c_b = -2{\cdot}04$. Our little table shows that if we had demanded union of D and F (a very small change) the ratio would have to be $-2{\cdot}07$, and for union of G' and F it would rise to $-2{\cdot}18$. The proportion of crown and flint curvatures which suits C and F is, therefore, wrong for D and G', and there must be a sensible departure of the D rays from the common focus established for C and F, and a very decided departure of the G'-rays. Hence we may say with confidence that our objective will have residual chromatic aberration for colours differing from the two for which it was rendered achromatic.

To obtain a general expression for these residual differences of focus for colours

[41] other than the two which are united, we will write down our solutions by TL, Chr. (4)* for the usual case of visual achromatism ; they are

$$c_a = \frac{1}{f'} \frac{1}{(V_a - V_b)(Nf - Nc)_a};$$

$$c_b = \frac{1}{f'} \frac{1}{(V_b - V_a)(Nf - Nc)_b}.$$

These we introduce into the perfectly general equation (1) from which our solution was derived

(1) $$c_a (Nv - Nr)_a + c_b (Nv - Nr)_b = R,$$

and if we limit ourselves to the case when the object-point is free from chromatic aberration, the second term in the equation defining R will be zero and $R = (l'_r - l'_v)/l'_r l'_v$. Introducing this value of R and the values of c_a and c_b into (1) and taking out the common factor $1/f'(Va - Vb)$, we obtain

$$\frac{l'_r - l'_v}{l'_r \cdot l'_v} = \frac{1}{f'(Va - Vb)} \cdot \left[\left(\frac{Nv - Nr}{Nf - Nc}\right)_a - \left(\frac{Nv - Nr}{Nf - Nc}\right)_b\right].$$

In the square bracket we recognize the proportional numbers or relative dispersions to which we assigned the symbol Pvr and which are given in glass lists for certain lines of the spectrum. The numerator on the left is just the quantity which we are seeking, namely, the difference in focus of colours 'v' and 'r' when C and F are brought together, and on introducing the Pvr and slightly transposing we obtain the solution for the secondary spectrum effect :

$$l'_r - l'_v = \frac{l'_r \cdot l'_v}{f'} \cdot \frac{Pvr_a - Pvr_b}{Va - Vb}.$$

A glance at the glass list will show that the effect can never be very large, for the variation of the P is quite small compared to that of the V, hence we can neglect, on the *right*, the small difference between l'_r and l'_v and replace their product by l'^2 of the principal (plain N) colour. And as the secondary spectrum is of most importance in large astronomical telescopes, we will add to the simplified general equation its special form which results when the image is formed at the principal focus and l' becomes equal to f' :

TL, Chr. (6)
Secondary spectrum

$$\begin{cases} l'_r - l'_v = \dfrac{l'^2}{f'} \cdot \dfrac{Pvr_a - Pvr_b}{Va - Vb}; \\ f'_r - f'_v = f' \cdot \dfrac{Pvr_a - Pvr_b}{Va - Vb}. \end{cases}$$

We will work out the ratio on which the magnitude of the secondary spectrum depends for the glasses used in example (1), which are also those usually employed in large telescopes. Extracting the P-values given in our list and adding one for the A' line at the extreme red end of the spectrum, taken from the Jena list for [41] very similar glasses, we have :

	Pda'	*Pfd*	*Pg'f*
Hard crown (*a*)	0·648	0·705	0·568
Dense flint (*b*)	0·604	0·715	0·608
Differences (*a*) − (*b*)	+0·044	−0·010	−0·040

The denominator $V_a - V_b$ being = 24·8 for these glasses, we find for the principal focus by the specialized second equation :

$$f'_{a'} - f'_d = f' \cdot \frac{0\cdot044}{24\cdot8} = 0\cdot00177 f' ;$$

$$f'_d - f'_f = f' \frac{-0\cdot010}{24\cdot8} = -0\cdot00040 f' ;$$

$$f'_f - f'_{g'} = f' \cdot \frac{-0\cdot040}{24\cdot8} = -0\cdot00161 f' .$$

Adding the last two we obtain the difference

$$f'_d - f'_{g'} = -0\cdot00201 f',$$

and if we now refer all the other colours to the focus of the D line as the visually brightest, and remember that the value for the C line must be the same as for the F line because C and F have been brought to a common focus, we obtain the result

$$f'_{a'} = f'_d + 0\cdot00177 f' ;$$
$$f'_c = f'_d + 0\cdot00040 f' ;$$
$$f'_d = f'_d ;$$
$$f'_f = f'_d + 0\cdot00040 f' ;$$
$$f'_{g'} = f'_d + 0\cdot00201 f' .$$

The most striking feature of these results is that all the other colours are shown to come to a focus at a greater distance than the orange D light, so that the extreme red is fairly nearly brought to the same focus as the deep blue light of the G' line, just as C and F are united much closer to the focus of the D line.

As we shall presently show that the magnitude of the secondary spectrum does not vary very much, whichever of the *ordinary* optical glasses may be selected for combination, it is useful to note its magnitude : C and F come to focus at 1/2500th of the focal length beyond the D focus, and the photographically most active rays of the G' region come to focus at 1/500th of the focal length beyond the D focus. Whilst these differences appear small, they become very serious in large astronomical telescopes. We may take 10 m. or 10,000 mm. as an average for the focal length of the large instruments in observatories. The C and F rays will, for such an instrument, be 4 mm. out of focus when D is in focus, and the photographic rays will produce their image 20 mm. beyond the D focus and will be diffused over a considerable diameter at the latter focus. Hence these large instruments must necessarily show an image of a star which is surrounded by a broad halo of scattered

[41] light from both ends of the spectrum, and therefore of the purple tint resulting from the mixing of red, blue, and violet. This bright and extensive halo is the chief drawback of large refractors and sets a fairly definite limit—which has been reached or possibly exceeded in the largest existing instruments—to the size which would yield any advantage over instruments of a smaller aperture. The secondary spectrum is the real cause of the recent turn of the tide in favour of reflectors.

We easily obtain the proof of the statement made above 'that the secondary spectrum varies comparatively little in magnitude for any ordinarily useful combination of glasses' by plotting the *Pvr* numbers of the glass list against the corresponding *V* numbers. With the exception of the numbers for fluorite and to a much less extent of a few special glasses, the *P*-values will be found to fall into a nearly straight line, and as $(Pa-Pb)/(Va-Vb)$ for any two glasses represents—on the

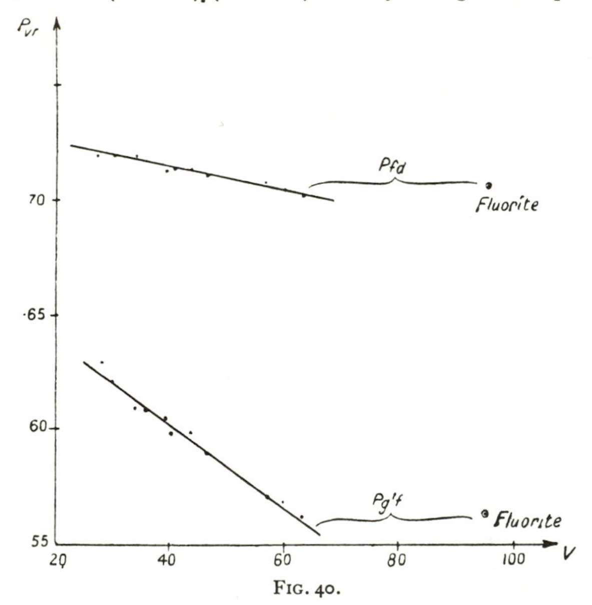

FIG. 40.

graph—the *slope* of the line joining *Pa* and *Pb*, we see that this ratio is practically constant, and as the residual chromatic aberration is strictly proportional to this ratio, the practical constancy of the secondary spectrum is proved.

Useful variations of slope will, however, be noticed for certain pairs of glasses with a comparatively small difference of *V*-values, and as the Telescope Flint, No. 4277, is one which falls most decidedly out of line, we will work out the secondary spectrum given by its combination with Hard Crown 1203. By TL, Chr. (6)—second form—we shall find

$$f'_d - f'_f = f' \cdot \frac{0{\cdot}705 - 0{\cdot}706}{60{\cdot}8 - 52{\cdot}2} = -\frac{0{\cdot}001}{8{\cdot}6} f' = -0{\cdot}00012 f',$$

or 1/3rd of the amount for our previous example; and for the violet end we shall [41] find

$$f'_f - f'_{g'} = f' \cdot \frac{0{\cdot}568 - 0{\cdot}573}{60{\cdot}8 - 52{\cdot}2} = -\frac{0{\cdot}005}{8{\cdot}6} f' = 0{\cdot}0006 f',$$

again about 1/3rd of the amount for ordinary combinations. $(Va - Vb)$ however, is 8·6 for this special pair against 24·8 for Hard Crown and Dense Flint. Hence we shall obtain far deeper curves for objectives of the same aperture and focal length. Moreover we must bear in mind when discussing glass pairs with a small difference in V-value that the P numbers are not absolutely reliable; a more refined spectrometric determination of the refractive indices from which they are deduced might easily alter the P number by $\pm 0{\cdot}001$ or even more. Hence we may easily be deceived. If for instance a more accurate determination of the Pfd used in our last example were to give 0·704 for the crown and 0·707 for the flint, the ratio would become $-0{\cdot}003/8{\cdot}6$ and the whole advantage over ordinary glasses would vanish. It is not likely that the case would turn out quite so badly, but it has been put forward in order to show that the designer should be particularly careful in critically examining the reliability of his data when dealing with cases like the present one.

A far greater and less uncertain advantage results from replacing crown glass by fluorite. If we examine the combination of fluorite with the telescope-flint, No. 4277, we obtain

$$f'_d - f'_f = f' \frac{0{\cdot}707 - 0{\cdot}706}{95{\cdot}5 - 52{\cdot}2} = +\frac{0{\cdot}001}{43{\cdot}3} f' = +0{\cdot}000023 f',$$

or less than 1/17th of the secondary spectrum effect of ordinary glass pairs for the middle of the spectrum; also

$$f'_f - f'_{g'} = f' \frac{0{\cdot}563 - 0{\cdot}573}{95{\cdot}5 - 52{\cdot}2} = -\frac{0{\cdot}010}{43{\cdot}3} f' = -0{\cdot}00023 f',$$

or about 1/9th of the secondary spectrum of ordinary glass pairs for the blue-violet.

On account of the large value of $(Va - Vb)$ for fluorite combinations, the advantage shown is not seriously jeopardized by the small uncertainty of the P-values. Unfortunately fluorite of optical purity is so rare that generally it can only be employed in microscope-objectives; these, however, can be rendered almost absolutely free from chromatic residuals by the judicious use of fluorite. Various likely glasses from the complete lists of the best makers should be examined before finally deciding upon the best selection.

PHOTOVISUAL OBJECTIVES

There is an alternative method of correcting the secondary spectrum which has been known as a theoretical solution of the problem for the best part of a century. On general mathematical principles it is obvious that just as ordinary achromatism (that is, a common focus for *two* colours) can be secured by combining two lenses of different V-value, so it must be possible to bring three colours to a common

[41] focus when *three* suitable glasses are combined to form a triple objective. Attempted solutions with the old kinds of optical glass were rendered futile by the excessive curvature of the resulting lens combination, but a few partial successes were achieved by combining glass lenses with suitable fluids. Fluids are, however, highly objectionable on account of leakage or evaporation, of possible chemical change, and especially because their optical properties change extremely seriously with temperature. The greatly increased range of available optical glass first provided by the Jena glass works included a few varieties which led to greatly diminished curvature, and it became possible to produce triple telescope objectives of the usual ratio of aperture to focal length which were almost absolutely free from secondary spectrum and which could also be highly corrected in other important respects.

Mr. H. Dennis Taylor, best known as the designer of the 'Cooke' photographic objective, was the first to succeed in producing these remarkable triple objectives, which are now widely known as 'Photovisual' lenses because sharp photographs can be obtained at the visual focus without colour screens.

In order to find a thin lens solution of the problem involved, and to discuss the conditions governing favourable proportions of the resulting objective, we will assume that three colours 'r', 'm', and 'v' (red, medium, and violet), which may be of any wave-length, but shall progress from the red towards the violet end of the spectrum, are to be united at a common focus by a thin combination of three lenses 'a', 'b', and 'c', with total curvatures of respective values c_a, c_b, and c_c, and that the prescribed focal length of the combination is f'. To adapt the solution to the usual data of glass lists we will further assume that this focal length and the V-values of the three kinds of glass are to be referred to a fourth colour the indices of which are expressed by plain N without suffix.

As by TL (2) we have for any thin lens $1/f' = (N-1)c$ and by TL (3) for any thin system $1/f' = 1/f'_a + 1/f'_b + \&c.$, the power $1/f'$ of our triple combination in the three colours to be united will be :

$$\frac{1}{f'_r} = c_a\,(N_r-1)_a + c_b\,(N_r-1)_b + c_c\,(N_r-1)_c,$$

$$\frac{1}{f'_m} = c_a\,(N_m-1)_a + c_b\,(N_m-1)_b + c_c\,(N_m-1)_c,$$

$$\frac{1}{f'_v} = c_a\,(N_v-1)_a + c_b\,(N_v-1)_b + c_c\,(N_v-1)_c\,;$$

and as these three powers must be exactly equal in order to secure the desired state of chromatic correction, the subtractions first from second and second from third give us the two conditions of achromatism :

$$0 = c_a\,(N_m-N_r)_a + c_b\,(N_m-N_r)_b + c_c\,(N_m-N_r)_c$$

$$0 = c_a\,(N_v-N_m)_a + c_b\,(N_v-N_m)_b + c_c\,(N_v-N_m)_c\,;$$

and the demand for focal length f' in plain N adds the third equation :

$$\frac{1}{f'} = c_a\,(N-1)_a + c_b\,(N-1)_b + c_c\,(N-1)_c.$$

These are three linear equations for the three unknown quantities c_a, c_b, and c_c, and they therefore solve the problem. But if we took three different kinds of glass at random we should almost certainly obtain impracticable high values, or even infinite ones, for the three lens curvatures. To obtain clear indications as to the proper selection of glass, we make the following modifications :

In the first equation we introduce merely our usual symbol δN for $(N_m - N_r)$.

In the second equation we extend $N_v - N_m$ by writing $N_v - N_m = (N_m - N_r) \cdot \frac{N_v - N_m}{N_m - N_r}$. We recognize the factor as the proportional number given (for certain colours) in glass lists, and will write

$$N_v - N_m = \delta N \,.\, P, \text{ with the definition } P = (N_v - N_m)/(N_m - N_r).$$

In the third equation we write

$$N - 1 = \delta N \frac{N-1}{\delta N}$$

and recognize $(N-1)/\delta N$ as the generalized V-value already introduced by us.

These modifications give us the three equations

$$\begin{aligned} 0 &= c_a\,\delta N_a \quad\;\; + c_b\,\delta N_b \quad\;\; + c_c\,\delta N_c \\ 0 &= c_a\,\delta N_a\,P_a + c_b\,\delta N_b\,P_b + c_c\,\delta N_c\,P_c \\ \frac{1}{f'} &= c_a\,\delta N_a\,V_a + c_b\,\delta N_b\,V_b + c_c\,\delta N_c\,V_c. \end{aligned}$$

The solution of these three equations can be put into the form

$$c_b = 1\Big/\left[f' \,.\, \delta N_b\,(P_c - P_b)\left(\frac{V_b - V_c}{P_c - P_b} - \frac{V_a - V_c}{P_c - P_a}\right)\right]$$

$$\left.\begin{aligned} c_a &= -c_b \cdot \frac{\delta N_b}{\delta N_a} \cdot \frac{P_c - P_b}{P_c - P_a} \\ c_c &= -c_b \;\; \frac{\delta N_b}{\delta N_c} \cdot \frac{P_b - P_a}{P_c - P_a} \end{aligned}\right\} \; c_a = c_c \frac{\delta N_c}{\delta N_a} \cdot \frac{P_c - P_b}{P_b - P_a}.$$

This solution may be verified either by direct repetition or by introducing the stated results into the given equations and ascertaining that they are all satisfied.

To discuss the solution, we will assume that the lens-names 'a', 'b', and 'c' have been chosen so that $P_a < P_b < P_c$, which will practically always imply that $V_a > V_b > V_c$ on account of the tendency which we pointed out of P for the more refrangible rays rising when V falls. As a rule the factors and ratios depending on the P and V values will therefore all be positive when the lens names are selected as advised. This restricts the solution in no respect, for we learnt in section [23] that a combination of thin lenses in contact produces the same effect whatever the *sequence* may be. Hence we may change or reverse the sequence subsequently in any way without having to make a new solution. We can also exclude from the discussion the case when two of the three P-values are equal, for if we could find two kinds of glass with equal P-values and a useful difference of the corresponding

[41] V-values, then these would be suitable for a *binary* combination free from secondary spectrum, and the addition of a lens with a different P-value could only *upset* this highly desirable state of correction.

If we now examine the solution for c_b we see that c_b would become infinite and the solution would be useless if the expression in the square bracket were equal to

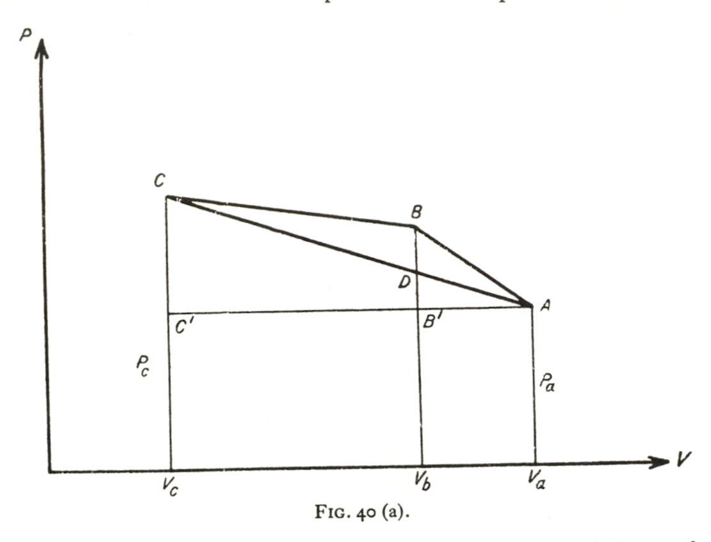

FIG. 40 (a).

zero. This evidently would happen if, and only if, we had $(V_b - V_c)/(P_c - P_b) = (V_a - V_c)/(P_c - P_a)$. In Fig. 40 (a) the P-values of three kinds of glass are plotted at A, B, and C against their V-values, exactly as in the previous Fig. 40, and we see that $(V_b - V_c)/(P_c - P_b)$ is the tangent of the angle $C'CB$ whilst $(V_a - V_c)/(P_c - P_a)$ is the tangent of the angle $C'CA$. These would become equal, and c_b would become infinite and useless, if B fell on the straight line AC. Hence we draw the first and most important conclusion that a photovisual objective with reasonable curvatures is only possible if we can find three kinds of glass for which B falls decidedly to one side of the line joining A and C. We can now prove that the suitability of three glasses is, for practical purposes, directly proportional to the distance DB by which B falls vertically above or below AC:

$$\text{From Fig. 40 (a) we take } V_bB = P_b = V_bB' + B'D + DB.$$

V_bB' is equal to P_a; for $B'D$ we obtain by the similar triangles $CC'A$ and $DB'A$:

$$B'D = CC' \times B'A/C'A = (P_c - P_a)(V_a - V_b)/(V_a - V_c);$$

and if we introduce δP_b as a suitable symbol for DB, we now have

$$P_b = P_a + (P_c - P_a)\frac{V_a - V_b}{V_a - V_c} + \delta P_b.$$

This we introduce into the factor $(P_c - P_b)$ of the equation for c_b and obtain [41]

$$P_c - P_b = P_c - P_a - (P_c - P_a)\frac{V_a - V_b}{V_a - V_c} - \delta P_b = (P_c - P_a)\left(1 - \frac{V_a - V_b}{V_a - V_c}\right) - \delta P_b ;$$

or, on bringing the expression in the second bracket to a common denominator,

$$P_c - P_b = \frac{P_c - P_a}{V_a - V_c}(V_b - V_c) - \delta P_b.$$

The original equation for c_b may be written

$$c_b = 1 / \left[f' . \delta N_b \left((V_b - V_c) - \frac{V_a - V_c}{P_c - P_a}(P_c - P_b) \right) \right],$$

and if we introduce the value of $(P_c - P_b)$ by the previous equation, this becomes

$$c_b = 1 / \left[f' . \delta N_b \left(V_b - V_c - (V_b - V_c) + \delta P_b \frac{V_a - V_c}{P_c - P_a} \right) \right];$$

or, noting an obvious cancellation and slightly rearranging,

$$c_b = \frac{P_c - P_a}{f' . \delta N_b . \delta P_b (V_a - V_c)}.$$

For the two outside glasses 'a' and 'c' the difference $V_a - V_c$ will always be so considerable that, in accordance with the previous discussion $(P_c - P_a)/(V_a - V_c)$ may be regarded as not varying much in value whatever combination may ultimately be selected. Similarly it will be found that suitable 'b' glasses all come from near the middle of a glass list, and therefore δN_b will not vary very greatly. Hence c_b will be primarily dependent on δP_b and is practically inversely proportional to it, and as a small value of c_b is desirable, we should try to find three kinds of glass for which δP_b, the DB of Fig. 40 (a), is as large as possible.

The proper procedure therefore is to plot all promising kinds of glass according to Fig. 40 and then to try all likely lines AC and note the point B which lies farthest above or below. For a first selection it will nearly always be sufficient to aim at union of C, F, and G' light at one common focus, and it will only be necessary to plot the $P_{g'f}$ of the glass list against the corresponding V-values of the list. A scale of 0·2 inch per unit of V and of 0·1 inch per 0·001 of P will be found suitable.

It is, however, not certain that the three glasses which give the absolutely biggest value of δP_b will be the most favourable. As the original solution shows by the minus sign of the equations for c_a and c_c, the two outside lenses tend to be of the opposite curvature as compared with the central lens. If δP_b is positive, as shown in Fig. 40 (a), then the central lens will be convex and both outside lenses will tend to be concave ; if δP_b is negative, that is B below AC, then the central lens will be concave and, as a rule, both outside lenses will be convex. In either case it is favourable for the correction of the spherical aberrations if the two outside lenses do not depart excessively from the relation $c_a = c_c$. Now the added equation of the original solution shows that $c_a = c_c$ demands $\delta N_a(P_b - P_a) = \delta N_c(P_c - P_b)$, and as under our stipulation $(V_a > V_c)$ δN_a will almost certainly be smaller

[41] than δN_c, favourable proportions for the outside lenses call for $(P_b - P_a > (P_c - P_b)$; that is, it is desirable that P_b be greater than the mean of P_a and P_c. It will be found that otherwise desirable glasses persistently refuse to comply with this desideratum. Hence a slightly reduced value of δP_b may be preferable to the absolute maximum.

A few numerical examples will be the best guidance. For the calculations we may use the original solutions or we may use δP_b, either as read off the graph or calculated by our intermediate equation, which gives

$$\delta P_b = (P_c - P_a) \frac{V_b - V_c}{V_a - V_c} - (P_c - P_b).$$

Introduction of our final solution for c_b into the original equations for c_a and c_c then gives the very convenient solution :

Photovisual Objectives :

$$c_b = \frac{1}{f'} \cdot \frac{1}{\delta P_b} \cdot \frac{1}{V_a - V_c} \cdot \frac{1}{\delta N_b} (P_c - P_a);$$

$$c_a = -\frac{1}{f'} \cdot \frac{1}{\delta P_b} \cdot \frac{1}{V_a - V_c} \cdot \frac{1}{\delta N_a} (P_c - P_b);$$

$$c_c = -\frac{1}{f'} \cdot \frac{1}{\delta P_b} \cdot \frac{1}{V_a - V_c} \cdot \frac{1}{\delta N_c} (P_b - P_a).$$

It should be noted that the first three factors are common to all three equations. Owing to the *differences* of *P*-values in all the equations, which are generally uncertain to several per cent., the slide-rule, or rough four-figure log work, is amply sufficient in working out the thin lens scheme.

Plotting all likely glasses of Chance's 1919 list as advised above, the set giving the highest possible value of δP_b was :

	N_d	δN	V	$Pg'f$	Pfd
No. 1203	1·5155	0·00848	60·8	0·568	0·705
No. 3389	1·5376	0·01069	50·3	0·576	0·707
No. 5093	1·6501	0·01936	33·6	0·615	0·719

Worked out by slide-rule for $f' = 8$ and for union of C, F, and G', these give $\delta P_b = -0{\cdot}0102$; $c_b = -1{\cdot}98$; $c_a = +2{\cdot}08$; $c_c = +0{\cdot}187$. It will be seen that lens '*a*' comes out with a greater curvature than lens '*b*', whilst lens '*c*', of the very dense flint, is very weak.

With a view to obtaining a better distribution of curvature, some sacrifice was made in δP_b by selecting :

	N_d	δN	V	$Pg'f$	Pfd
No. 6493	1·5160	0·00809	63·8	0·561	0·701
No. 3389	1·5376	0·01069	50·3	0·576	0·707
No. 466	1·5833	0·01251	46·6	0·592	0·711

which give $\delta P_b = -0{\cdot}00933$; $c_b = -2{\cdot}26$; $c_a = +1{\cdot}54$; $c_c = +0{\cdot}935$. In agreement with our discussion, c_b has gone up, but the great diminution in c_a will lead to a better form of the finished objective. All the glasses in this solution are, [41] according to the list, above reproach, whilst any experienced designer would be very averse to using a very dense flint like No. 5093 in a telescope objective.

As an example of the alternative case when B in Fig. 40 (a) falls above AC, the following is the best selection from the graph :

	N_d	δN	V	$Pg'f$	Pfd
No. 3389	1·5376	0·01069	50·3	0·576	0·707
No. 466	1·5833	0·01251	46·6	0·592	0·711
No. 4743	1·6134	0·01662	36·9	0·606	0·715

These give $\delta P_b = +0{\cdot}0077$; $c_b = +2{\cdot}91$; $c_a = -1{\cdot}59$; $c_c = -1{\cdot}17$. In this case the middle lens is convex and decidedly stronger than the concave middle lens of the two preceding examples.

As in the case of the binary combinations with reduced secondary spectrum, it is essential to remember that the curvatures of the three components are determined by small *differences* of P-values. Any uncertainty in these values therefore throws a considerable doubt upon the value of the solution and the dependability of the dispersion figures will always have to be discussed with critical care. As a rule it will be necessary to determine these data on the glass of the individual disks to at least an accuracy of one unit of the fifth decimal place of the refractive indices, for as the high values of the separate net curvatures show, the final refraction effect is the small difference of the opposite tendencies of very strong lenses.

It is also advisable to test the solution for the so-called 'tertiary spectrum'. When three colours have been brought accurately to a common focal point, other colours will usually display a sensible difference of focus and owing to the great strength of the component lenses this may sometimes approach the magnitude of the ordinary secondary spectrum of binary objectives. We can determine it in close analogy with the method applied to the secondary spectrum. The starting equations for $1/f'$ are absolutely general ; if we write them down for any two colours p and q, form their difference and introduce the proportional numbers, which we will call P', exactly as previously for colours v and m, we obtain

$$\frac{1}{f'_q} - \frac{1}{f'_p} = c_a \,.\, \delta N_a \,.\, P'_a + c_b \,.\, \delta N_b \,.\, P'_b + c_c \,.\, \delta N_c \,.\, P'_c \,.$$

On the left we bring the two fractions to a common denominator ; on the right we introduce the values of c_a, c_b, and c_c, found by the final solution ; separating the common factors, we find

$$\frac{f'_p - f'_q}{f'_p \cdot f'_q} = \frac{1}{f'} \cdot \frac{1}{\delta P_b} \cdot \frac{1}{V_a - V_c}\Big(-P'_a\,(P_c - P_b) + P'_b\,(P_c - P_a) - P'_c\,(P_b - P_a)\Big).$$

As the small quantities on the right can only be found within a comparatively large percentage-accuracy, the $f'_p \,.\, f'_q$ can with perfect safety be replaced by simply f'^2 and thrown on the other side. This gives the perfectly general solution for the

Tertiary Spectrum :

$$f'_p - f'_q = f' \cdot \frac{1}{\delta P_b} \cdot \frac{1}{V_a - V_c}\Big(-P'_a\,(P_c - P_b) + P'_b\,(P_c - P_a) - P'_c\,(P_b - P_a)\Big).$$

[41] In the case of our numerical examples derived form Chance's list P_{fd} is the only available additional P-value given. It enables us to determine the tertiary spectrum effect for the D line. The formula gives

$$\text{1st Example}: f'_d - f'_f = f' \cdot \frac{1}{-0{\cdot}0102} \cdot \frac{1}{27{\cdot}2} (-0{\cdot}705 \times 0{\cdot}039 + 0{\cdot}707 \times 0{\cdot}047 - 0{\cdot}719 \times 0{\cdot}008) = 0{\cdot}000065 f' \; ;$$

$$\text{2nd Example}: f'_d - f'_f = f' \cdot \frac{1}{-0{\cdot}0093} \cdot \frac{1}{17{\cdot}2} (-0{\cdot}701 \times 0{\cdot}016 + 0{\cdot}707 \times 0{\cdot}031 - 0{\cdot}711 \times 0{\cdot}015) = -0{\cdot}000225 f' \; ;$$

$$\text{3rd Example}: f'_d - f'_f = f' \cdot \frac{1}{0{\cdot}0077} \cdot \frac{1}{13{\cdot}4} (-0{\cdot}707 \times 0{\cdot}014 + 0{\cdot}711 \times 0{\cdot}030 - 0{\cdot}715 \times 0{\cdot}016) = -0{\cdot}000077 f' \; .$$

Comparison with the results obtained for the secondary spectrum of ordinary telescope objectives will show that the *tertiary* spectrum in the two first examples is about one quarter and in the third example about one half of the usual *secondary* spectrum for the D line.

THE PRINCIPAL TYPES OF ACHROMATISM

[42] Obviously it is the secondary spectrum which renders it necessary to vary the colours selected for union at a common focus according to the work for which a given instrument is to be used. If all colours automatically came to the same focus on bringing two together, then it would not matter which particular two colours we selected for finding that focus, and every achromatic lens would be equally suitable—as far as freedom from chromatic aberration is concerned—for all possible purposes. We must therefore next study achromatism from this point of view.

(1) *Visual Achromatism.* We have determined the secondary spectrum for the combination of hard crown and dense flint on the assumption that C and F are brought to a common focus, and have also shown that the results would be nearly the same for any other two ordinary glasses with a considerable difference of their V-values. Union of C and F is by long experience a good choice for visual instruments. Taking the focal length of the D-rays as our unit, we found that C and F are focused at a distance from the lens greater by 0·00040, A' at a distance greater by 0·00177, and G' at a distance greater by 0·00201. We will plot these figures against the refractive index of the crown lens for the various colours. The points A', C, D, F, G' can be laid down most conveniently according to their corresponding N-value by noting that the P-values of the glass list give the ratio of the various tabulated differences of refractive indices to that between C and F. Hence, if we place C and F one unit (say 1 inch) apart, then as Pfd = 0·705, D will have to be marked at 0·705 inch from F towards C, similarly A' at 0·648 inch from D towards and beyond C, and G' at 0·568 inch beyond F. Taking the focus for D on the horizontal axis of the diagram, the focal differences will then fall as indicated in Fig. 41, and will suggest a curve very similar to a common parabola as the proper

graph. Additional points of the graph occasionally obtainable from more numerous [42] data fully bear out this conclusion.

We see at once that the focus for D cannot be the absolutely shortest; the position of the five known points irresistibly calls for a slight dip of the graph

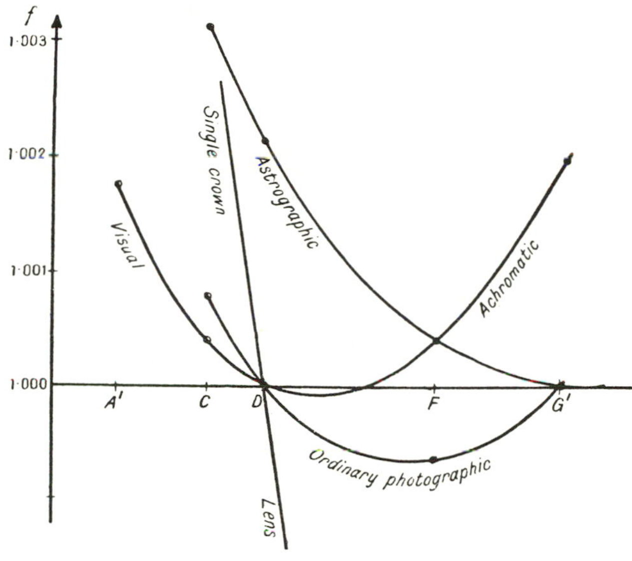

FIG. 41.

below the horizontal axis and the minimum focus is thus located about midway between C and F, but with only a slight difference from the D-focus. We shall locate it more accurately subsequently at wave-length 5555×10^{-7} mm., in the pale yellow-green ('apple-green' is the usual description) region of the spectrum. This minimum focus is the most characteristic feature of achromatic lenses and distinguishes them from simple lenses or wildly over- or under-corrected combinations. By TL, Chr (1) a *simple lens* gives differences of focus which are very nearly proportional to $(Nv - Nr)$ and therefore plot as an inclined straight line; as by the specialized formula for a simple lens we found $f'_c - f'_f = f'/V$ and as our crown glass has $V = 60{\cdot}8$, we can draw this line as one with a drop of $1/V = 0{\cdot}0164$ in the horizontal distance from C to F. We thus see strikingly displayed the immense superiority of an achromatic lens over an unachromatic one as regards chromatic variation of the focus. The graph for the achromatic case also shows that, no matter how great the focal length may be and how badly the more remote regions of the spectrum may be out of focus, an achromatic lens always has very nearly equal focal distances for the colours in the immediate neighbourhood of its minimum focus: that is in fact what saves the giant refractors from being useless, provided that the minimum focus coincides with the brightest region of

[42] the spectrum. We thus recognize the suitable location of the minimum focus as the really important thing in the design of achromatic lenses for ordinary observational purposes.

There is strong and consistent evidence that for the smaller sizes of astronomical telescopes, say from three to ten inches clear aperture, and for the stronger microscope objectives, say over 0·5 *NA*, the union of *C* and *F*, with minimum focus at wave-length 0·5555μ, gives the best definition and the highest resolving power for objects which are white or neutral tinted, like most astronomical objects and also unstained microscopical ones. Achromatization for the *C* to *F* region is therefore the most usual correction for ordinary visual purposes.

In the case of large astronomical telescopes, say of twenty or more inches clear aperture, it is known by direct tests of the best instruments, more especially of Alvan Clark's masterpieces, that a shift of the minimum focus towards the yellow, to about wave-length 0·565μ, gives the best results. This is probably explained by the absorption of blue and violet light in the thick flint-glass lens which naturally would tend to make the less refrangible rays more predominant.

On the other hand, small telescopes like those found in surveying instruments, and low-power microscope objectives not exceeding 0·3 *NA* seem to give the most satisfactory images when the minimum focus is shifted towards the full green, and wave-length 0·535μ might be adopted; this corresponds exactly to the bright green line of thallium. By an interpolation formula to be deduced in Part II the refractive index for this wave-length and the dispersion-value $N_v - N_r = \delta N$ to be used with it are found as

$$N_{.535} = N_d + 0{\cdot}319\,(N_f - N_c)$$
$$\delta N_{.535} = (N_f - N_c) - 0{\cdot}258\,(0{\cdot}607 - P_{g'f})\,(N_f - N_c).$$

The formula for δN has been put into the best form for calculation by slide-rule, as the corrective term only amounts to a moderate number of units of the fifth decimal place for crown glasses and may be neglected for the ordinary kinds of flint glass because these have $P_{g'f}$ between 0·605 and 0·609, so that the term for them does not exceed one unit of the fifth place. When using this altered correction, the V of the glasses must be calculated as $V = (N_d - 1)/\delta N_{.535}$. For example (1) of section [40] we thus find for the crown

$$\delta N_{.535} = 0{\cdot}00839\,; \quad V_{.535} = 0{\cdot}5155/0{\cdot}00839 = 61{\cdot}4.$$

The flint data remain unchanged and the solution for thin lens achromatism gives

$$c_a = 0{\cdot}5866\,; \quad c_b = -0{\cdot}2846.$$

It will be noticed that the changes from the values 0·5944 and $-0{\cdot}2915$ for union of *C* and *F* are quite considerable

The justification of such a shift of the minimum focus in *small* instruments is supplied by the fact that the violet end of the prismatic spectrum beyond *F* is much longer and brighter than the red end beyond *C*, hence the halo of purple light is reduced both in brightness and in diameter, whilst the central colours are not sensibly disturbed on account of the smallness of the differences in focus which could possibly arise.

A still more drastic shift of the minimum focus is becoming prevalent in microscope objectives intended for use on heavily stained histological and medical objects. It is found that a much used blue violet stain defeats objectives corrected for union of C and F. The most successful lenses for this work seem to be corrected for union of C and G', which corresponds to minimum focus for wave-length $0{\cdot}517\mu$. [42]

Systematic experimental work is highly desirable with reference to this important subject.

(2) *Ordinary photographic Achromatism.* The modus operandi in all ordinary photographic work consists in selecting and focusing the subject by visual observation of the image on the ground glass screen and by then substituting the sensitive surface for the screen and making the exposure. The human eye has its maximum sensitivity in the yellow-green region of the spectrum, and therefore locates the focus in or near that tint. Ordinary photographic plates on the other hand have their maximum of sensitivity in the neighbourhood of the G' line. A glance at Fig. 41 shows at once that if we were to use a visually corrected achromatic lens for photography, we should bring into play the whole difference of focus—about 1/500th of the focal length—between the visually brightest rays at the lowest point of the curve and the photographically brightest rays near G', and the image photographed at the visual focus would be correspondingly blurred. This difference of the visual and 'actinic' foci of ordinary achromatic lenses was one of the gravest troubles of the pioneers of photography, for the best approach to a cure was only slowly discovered by the laborious method of trial and error employed by the opticians of their day. If the lens is to be used chiefly on distant objects, a partial cure can be embodied in the camera. It is only necessary to introduce a 'difference of register' between the focusing screen and the sensitive surface of about 1/500th of the focal length so that the plate automatically goes into a position, farther from the lens by that fraction, which represents the best focus for the photographic image. This cure has in fact been frequently applied and is still useful in extemporized apparatus. But our Fig. 41 shows that it is an incomplete remedy, for the graph has a steep slope in the G' neighbourhood and colours to either side of G' itself will still be seriously out of focus in opposite senses. 'The lens is heavily over-corrected for the photographic rays.' Moreover, the first of TL, Chr. (6) shows that for other conjugate distances the focal differences grow as $(l'/f')^2$, so that the required difference of register would be four times as large for an image of unit-magnification as for the image at the principal focus.

Evidently the proper remedy for the 'actinic' focus is, so to compute the photographic lens that the brightest visual and the brightest photographic rays are brought together. According to the preceding paragraph this would call for union of the apple-green and of the G' foci. Present practice, however, points to the union of D and G' as the best compromise, probably because 'warm' tints predominate in the usual subjects and also on account of the ever increasing employment of 'orthochromatic' plates, for both these facts tend to reduce the difference between the visual and actinic foci.

We will calculate a lens of 8 inches focal length for this kind of achromatism. It simply means that we must work out TL, Chr. (4)* completely, with v standing

[42] for G' and r for D. From our glass list we extract for the kinds of glass previously used

Crown: $Ng' - Nd = (Ng' - Nf) + (Nf - Nd) = 0{\cdot}00482 + 0{\cdot}00598 = 0{\cdot}01080$;

Flint: $0{\cdot}01052 + 0{\cdot}01237 = 0{\cdot}02289$;

and still retaining Nd for 'plain' N we find the photographic V-values, distinguished from the visual ones by a 'dash',

$$V'_a = 0{\cdot}5155/0{\cdot}01080 = 47{\cdot}7\,; \quad V'_b = 0{\cdot}6225/0{\cdot}02289 = 27{\cdot}2\,;$$

and these give

$$c_a = \frac{1}{8}\;\frac{1}{20{\cdot}5}\;\frac{1}{0{\cdot}01080} = +0{\cdot}5646\,;$$

$$c_b = \frac{1}{8}\;\frac{1}{-20{\cdot}5}\;\frac{1}{0{\cdot}02289} = -0{\cdot}2664.$$

In example (1) we found for visual achromatism and the same glasses and focal length $c_a = 0{\cdot}5944$, $c_b = -0{\cdot}2915$, so that it is obvious that the change from visual to photographic achromatism alters the proportions of a lens very greatly.

In order to determine the secondary spectrum we must now calculate our own P-values, for they have to give now the ratio of other N-differences to our chosen $Ng' - Nd$. We thus find the photographic P-values, again distinguished by a 'dash',

Crown: $P'_{g'f} = \dfrac{0{\cdot}00482}{0{\cdot}01080} = 0{\cdot}446\,; \quad P'_{fc} = \dfrac{0{\cdot}00848}{0{\cdot}01080} = 0{\cdot}785,$

Flint: $\dfrac{0{\cdot}01052}{0{\cdot}02289} = 0{\cdot}460\,; \quad \dfrac{0{\cdot}01729}{0{\cdot}02289} = 0{\cdot}755,$

and with these from TL, Chr. (6)

$$f'_f - f'_{g'} = f' \cdot \frac{-0{\cdot}014}{20{\cdot}5} = -f'\,0{\cdot}00068,$$

$$f'_c - f'_f = f' \cdot \frac{0{\cdot}030}{20{\cdot}5} = +f'\,0{\cdot}00146,$$

or

$$f'_c - f'_{g'} = +f'\,0{\cdot}00078.$$

Taking the foci of D and G' on the horizontal axis of Fig. 41 the focus of F falls on $-0{\cdot}00068$ and that of C on $+0{\cdot}00078$, and a smooth curve through the four points gives the graph marked 'photographic'.

The residual variations between D and G' are less than one third of those resulting from visual achromatization, so that a much better compromise has been secured for all rays which may make effective contributions to the photographic image even when orthochromatic plates are employed. The slopes of the curve show, however, that the achromatism is imperfect for both the brightest visual rays midway between C and F and for the brightest photographic rays near G'. But the over-correction in the latter region is greatly reduced compared to that in the visually corrected objective, and we therefore have a good compromise in this respect. The graph shows that it would rise steeply if extended to the extreme red end at A' ; that is the reason why ordinary photographic objectives show a red halo around the sharply focused image of a point of light. [42]

(3) *Purely photographic Achromatism.* We have just shown that the ordinary photographic correction represents a compromise and is not perfect for either the visual or the photographic rays. In instruments of large aperture, almost exclusively astronomical ones, this compromise is inadmissible, and perfect correction for the photographic rays must be secured. That means that we must lay the minimum focus upon, or close to, the G' line. The exact desired position is frequently specified by astronomers in ordering such instruments. We will study this case also by working out a specimen lens of the glasses and focal length used in Example (1). It will be proved in Part II that an achromatic combination will have minimum focal length for the G' line if the chromatic equations are worked out with a value of δN found by the formula

$$\delta Ng' = -[9{\cdot}4433]\,(Nf-Nc)+[0{\cdot}3391]\,(Ng'-Nf),$$

the figures in square brackets being the logarithms of the actual numerical factors in order to make the calculation as simple as possible.

We therefore begin by calculating this δN :

	Crown.	*Flint.*
$(Nf-Nc)$	= 0·00848	0·01729
$(Ng'-Nf)$	= 0·00482	0·01052
log $(Nf-Nc)$	7·9284	8·2378
+ log (factor)	9·4433n	9·4433n
log (1st term)	7·3717n	7·6811n
log $(Ng'-Nf)$	7·6830	8·0220
+ log (factor)	0·3391	0·3391
log (2nd term)	8·0221	8·3611
1st term	− 0·002353	− 0·004798
2nd term	+ 0·010522	+ 0·022967
$\delta Ng'$	= 0·00817	0·01817

The separate terms have been taken out to six decimal places merely to make sure of the fifth place in the result. There would be no sense in retaining six places in the latter as the data $(Nf-Nc)$ and $(Ng'-Nf)$ are only known to five places. An observed result cannot be rendered more accurate by playing mathematical tricks with it.

In forming the V-values $= (N-1)/\delta N$ it will be sensible (though not necessary) to use the N for the G' line as that of our most important colour.

From our glass list we obtain :

Nd	= 1·5155	1·6225
$Nf-Nd$	= 0·00598	0·01237
$Ng'-Nf$	= 0·00482	0·01052
Ng'	= 1·5263	1·6454

[42] and with these

$$V''_a = \frac{0{\cdot}5263}{0{\cdot}00817} = 64{\cdot}5\,; \qquad V''_b = \frac{0{\cdot}6454}{0{\cdot}01817} = 35{\cdot}5.$$

The usual equations for c_a and c_b then give

$$c_a = \frac{1}{8}\,\frac{1}{29{\cdot}0}\,\frac{1}{0{\cdot}00817} = 0{\cdot}5276\,; \; c_b = \frac{1}{8}\,\frac{1}{-29{\cdot}0}\,\frac{1}{0{\cdot}01817} = -0{\cdot}2372.$$

Note that c_a and c_b have suffered a further substantial reduction for this purely photographic achromatism.

To complete our example, we have to determine the P-values :

Crown

$$P''_{g'f} = \frac{0{\cdot}00482}{0{\cdot}00817} = 0{\cdot}590\,; \; P''_{fd} = \frac{0{\cdot}00598}{0{\cdot}00817} = 0{\cdot}733\,; \; P''_{fc} = \frac{0{\cdot}00848}{0{\cdot}00817} = 1{\cdot}036\,;$$

Flint

$$\frac{0{\cdot}01052}{0{\cdot}01817} = 0{\cdot}579 \qquad \frac{0{\cdot}01237}{0{\cdot}01817} = 0{\cdot}682 \qquad \frac{0{\cdot}01729}{0{\cdot}01817} = 0{\cdot}952\,;$$

and these give

$$f'_f - f'_{g'} = f'\,\frac{+0{\cdot}011}{29{\cdot}0} = 0{\cdot}00038\,f'$$

$$f'_d - f'_f = f'\,\frac{+0{\cdot}051}{29{\cdot}0} = 0{\cdot}00176\,f'$$

$$f'_c - f'_f = f'\,\frac{+0{\cdot}084}{29{\cdot}0} = 0{\cdot}00290\,f'$$

$$f'_d - f'_{g'} = 0{\cdot}00214\,f'$$

$$f'_c - f'_{g'} = 0{\cdot}00328\,f'$$

We thus have the result that the F-focus lies $0{\cdot}00038\,f'$ beyond the minimum focus, the D-focus $0{\cdot}00214\,f'$ beyond the minimum focus, and the C-focus $0{\cdot}00328\,f'$ beyond the minimum focus. Plotted, these give the third curve, marked 'astrographic', of Fig. 41. Whilst the correction in the vicinity of G' is now ideal, that in the visual region is bad.

The usual method of finding the focus of these purely photographic objectives is to make a series of trial exposures at progressively increased distances. If variation of the focus with temperature has to be reckoned with, then the difference between the best obtainable visual focus and the photographic one must be ascertained once for all by trial exposures : future visual adjustments can then be turned into photographic ones by applying the known difference of focus.

In connexion with astrographic work the problem occasionally presents itself that the owner of a large object-glass for visual observation asks for a correcting lens which shall convert the lens into an astrographic one. If the constructional and optical data of the existing objective are known or can be determined, then the problem is easily solved. Let us assume that the objective of example (1) of the present chapter is the patient. It consists of a crown lens with $c_a = 0{\cdot}5944$ and [42] of a flint lens with $c_b = -0{\cdot}2915$. We have already used the dispersion-values for the G'-region, namely $Ng' = 0{\cdot}00817$ for the crown and $0{\cdot}01817$ for the flint. By the universally valid added equation of TL, Chr. (4)*

$$c_a/c_b = -\delta N_b/\delta N_a,$$

we can determine what value of c_b would produce astrographic achromatism with the original c_a or vice versa. The formula gives

$$c_a = -c_b \,.\, \delta N_b/\delta N_a$$

or

$$c_b = -c_a \,.\, \delta N_a/\delta N_b\,;$$

or with our data

$$\text{for unchanged } c_b: \quad c_a = \frac{0{\cdot}2915 \times 0{\cdot}01817}{0{\cdot}00817} = 0{\cdot}6485\,;$$

$$\text{for unchanged } c_a: \quad c_b = \frac{-0{\cdot}5944 \times 0{\cdot}00817}{0{\cdot}01817} = -0{\cdot}2673.$$

We may therefore either remove the original crown lens and replace it by one of $c_a = 0{\cdot}6485$ or we may leave the original crown lens and replace the flint lens by one of $c_b = -0{\cdot}2673$. In either case the focal length of the modified objective will be decidedly shorter than that of the original one. We can calculate the new focal length by equation (2) of this chapter :

$$\frac{1}{f'} = c_a(N_a - 1) + c_b(N_b - 1),$$

in which we may use the N for the D line as quite near enough. This gives for the unchanged crown

$$\frac{1}{f'} = 0{\cdot}5944 \times 0{\cdot}5155 - 0{\cdot}2673 \times 0{\cdot}6225 = 0{\cdot}1400$$

or

$$f' = 7{\cdot}143.$$

For unchanged flint lens we find

$$\frac{1}{f'} = 0{\cdot}6485 \times 0{\cdot}5155 - 0{\cdot}2915 \times 0{\cdot}6225 = 0{\cdot}1528$$

or

$$f' = 6{\cdot}545.$$

Looked at as inches, the change from $f' = 8$ does not appear very serious. But it must be remembered that the problem is more likely to turn up when the focal length is 8 feet or even 8 metres, and then the reduction in focal length will probably draw the focus into the main tube and there will be trouble and expense in arranging to place a photographic plate into this inconvenient position. A wide-awake designer should not omit to warn the owner of the objective of this eventuality. In any case the figures show that the reduction is least when the change is thrown upon the flint lens and that is therefore the best choice.

The usual method of effecting the correction is, however, to leave the original object-glass untouched so that it may still be used for visual observations and to

[42] add for photographic work a correcting lens which brings either the flint or the crown curvature to the required new value, the correcting lens being mounted close to the original objective and either in front of or behind it. If the added lens is to be of crown-glass its net curvature will have to be $0{\cdot}6485 - 0{\cdot}5944$ or $0{\cdot}0541$, and the focal length of the triple object-glass will be $6{\cdot}545$; if the added lens is flint, then its net curvature will have to be $= -0{\cdot}2673 - (-0{\cdot}2915) = +0{\cdot}0242$ and the combined focal length will be $7{\cdot}143$, the lenses in all cases being treated as thin and in contact. The rough scheme thus arrived at will of course require modification by trigonometrical calculation like all other approximate results derived from TL and paraxial equations.

Effect of Thickness and Separation on the Secondary Spectrum

Seeing that our TL, Chr. equations are admittedly inaccurate, it might at first sight appear that the magnitude of the secondary spectrum derived from them would also be a mere rough approximation of little value. This is not a correct conclusion: on the contrary the estimate of the secondary spectrum obtained from the simple equation TL, Chr. (6) is invariably closely borne out by the final calculation of an optical system after proper corrections have been determined to secure the desired type of achromatism with lenses of finite thickness and separation. The reason is that the various coloured rays never become widely separated in passing through lenses of even more than the usual thickness, simply because the differences of the refractive indices for different colours are so small compared to $(N-1)$ which determines the total refraction effect. As a consequence the little changes in certain radii, which correct the defective achromatism of the crude design, automatically produce an almost precisely corresponding effect on all other colours, and the arrangement of the foci of these is thus restored to practically the same state which the TL, Chr. formulae indicated for the infinitely thin lenses. It is therefore quite safe to rely on TL, Chr. (6) for the estimation of the secondary spectrum in all but rare and exceptional cases.

Hyperchromatic Lens Combinations

Occasionally it becomes apparent in working out a new design that a desirable result could be achieved if a combination of lenses were available which gave a *higher* amount of chromatic aberration than any possible simple lens. This represents a problem of exactly the opposite character to the achromatic one. We require to aggravate the chromatic aberration instead of removing or diminishing it. Our equations TL, Chr. (4) enable us to solve the problem. In example (2) we used this set of equations for the design of an over-corrected lens by introducing an appropriate negative value of R. Obviously a combination under-corrected to any desired extent will result by determining the corresponding positive chromatic aberration and calculating R with this.

Supposing for instance that we desired a combination of 8 inches focal length to give a chromatic aberration between C and F of $0{\cdot}8$ inch at its principal focus. No simple lens could fulfil this condition, for in accordance with the equation for the chromatic aberration of a simple thin lens

$$l'_c - l'_f = f'/V,$$

which, transposed, gives [42]

$$V = f'/(l'_c - l'_f),$$

we should require a glass of

$$V = 8/0{\cdot}8 = 10.$$

Such glass does not exist ; in fact the most dispersive organic compounds like cinnamic ether hardly reach that figure. But R, worked out as in example (2) gives

$$R = 0{\cdot}8/f'^2 = 0{\cdot}8/64 = +0{\cdot}0125,$$

and if we introduce this value into TL, Chr. (4) we find

$$c_a = \frac{1}{8} \cdot \frac{1}{24{\cdot}8} \cdot \frac{1}{0{\cdot}00848} - \frac{0{\cdot}0125}{0{\cdot}00848} \cdot \frac{36{\cdot}0}{24{\cdot}8} = 0{\cdot}5944 - 2{\cdot}1400\,;$$

$$c_b = \frac{1}{8} \cdot \frac{1}{-24{\cdot}8} \cdot \frac{1}{0{\cdot}01729} - \frac{0{\cdot}0125}{0{\cdot}01729} \cdot \frac{60{\cdot}8}{-24{\cdot}8} = -0{\cdot}2915 + 1{\cdot}7726\,;$$

$$\therefore \quad c_a = -1{\cdot}5456\,; \qquad c_b = +1{\cdot}4811.$$

The desired result can therefore be realized by a combination of a strong concave crown lens of $c_a = -1{\cdot}5456$ and a convex flint lens of $c_b = 1{\cdot}4811$. If cemented the combination will have the external appearance of a concave lens, for with $r_4 = \infty$ and $r_2 = r_3$, r_1 will be found by $1/r_1 = -1{\cdot}5456 + 1{\cdot}4811 = -0{\cdot}0645$ or $r_1 = -15{\cdot}5$ inches. The positive power comes out of the higher index of the convex flint-component.

Such hyperchromatic combinations have been used by Rudolph in the Zeiss 'Planar' photographic lens and by the author in the field lens of compensating eye-pieces.

CHROMATIC ABERRATION OF SEPARATED THIN LENSES

[43] In many achromatic systems the constituent simple lenses are separated by considerable air-spaces. The Cooke photographic objective, consisting of a concave lens placed between two convex lenses, may be cited as an example. In such cases it would be futile to apply our solutions for thin lenses in contact, as an utterly fallacious result would be obtained. But if we allow for the air-spaces, the individual lenses may usually still be treated as thin without seriously diminishing the utility of a first rough solution. Even with this simplification the problem leads to a clumsy solution, unsuitable for a rapid first approximation, if we retain the demand that two decidedly different colours like C and F are to be brought to a common focus. But we learnt in the two preceding sections that the most characteristic and important property of achromatic lenses is their minimum focal distance for some particular wave-length, and that they really deserve their name only for a small range of colour in the immediate neighbourhood of that wave-length. Hence it is not merely permissible, but is actually the more philosophical procedure, to work out the condition of achromatism for a small range of colour in the region of the spectrum for which minimum focal distance is to be established.

[43] In section [39] we deduced the condition of achromatism for any range of colours 'r' and 'v' from the fundamental equation for any one thin lens

$$\frac{l'_r - l'_v}{l'_r \cdot l'_v} = (N_v - N_r)c + \frac{l_r - l_v}{l_r \cdot l_v}.$$

When the two colours are taken very close together, then the longitudinal chromatic aberration will become correspondingly small; hence in the denominators l'_r will only differ slightly from l'_v and their product may be replaced by l'^2, especially if the l' is the intersection-length for the minimum focus colour midway between 'r' and 'v'. In practise we need not employ even this refined choice of l', for as thin lens equations are inevitably only approximate, it will make no important difference if we use the l' for the colour used in the preparatory scheme, which will usually mean the D line. The same reasoning applies to $l_r \cdot l_v$ on the right of the equation. We therefore adopt as our general equation the simplified and slightly transposed form, for any one thin lens,

$$\text{(a)} \qquad l'_r - l'_v = l'^2 \cdot (N_v - N_r) \cdot c + l'^2 \frac{l_r - l_v}{l^2}.$$

If we now consider a system of separated thin lenses, named from left to right a, b, c, &c., and assume that a paraxial pencil in the plain l colour has been traced through so that all the l and l' are known, then we have for the first lens 'a', using δN as a short substitute for $(N_v - N_r)$,

$$(l'_r - l'_v)_a = l'^2_a \cdot \delta N_a \cdot c_a + l'^2_a \frac{(l_r - l_v)_a}{l_a^2}.$$

For lens 'b' the equation will be

$$(l'_r - l'_v)_b = l'^2_b \cdot \delta N_b \cdot c_b + l'^2_b \frac{(l_r - l_v)_b}{l_b^2},$$

and here the $(l_r - l_v)_b$ will be the chromatic aberration $(l'_r - l'_v)_a$ delivered by the first lens to the second one; hence we can substitute the right side of the first equation for the $(l_r - l_v)_b$ in the second equation and obtain for the combined effect of the two lenses, on altering the sequence of terms according to the natural order of the suffix-letters

$$\text{(b)} \quad (l'_r - l'_v)_b = l'^2_b \left(\frac{l'_a}{l_b}\right)^2 \cdot \frac{(l_r - l_v)_a}{l_a^2} + l'^2_b \left(\frac{l'_a}{l_b}\right)^2 \cdot \delta N_a \cdot c_a + l'^2_b \cdot \delta N_b \cdot c_b.$$

Repeating the same process for the addition of a third lens, we easily find

$$\text{(c)} \quad (l'_r - l'_v)_c = l'^2_c \left(\frac{l'_b}{l_c}\right)^2 \left(\frac{l'_a}{l_b}\right)^2 \cdot \frac{(l_r - l_v)_a}{l_a^2}$$
$$+ l'^2_c \left(\frac{l'_b}{l_c}\right)^2 \left(\frac{l'_a}{l_b}\right)^2 \cdot \delta N_a \cdot c_a + l'^2_c \left(\frac{l'_b}{l_c}\right)^2 \cdot \delta N_b \cdot c_b + l'^2_c \cdot \delta N_c \cdot c_c,$$

and the law of the coefficients is so obvious that it is not necessary to continue the process, especially as it is very unusual to have more than three separated lenses in a system of this type. The formulae are quite convenient for the *determination* of the chromatic aberration of a given system. The usual problem is, however, [43] to find such a sequence of lenses as will produce a *prescribed* chromatic aberration at the final focus. If a finite amount of aberration is prescribed, a solution is reasonably possible only by a succession of trials, which naturally become least laborious if we seek the total curvature of the *last* lens which will lead to the prescribed value. But if, as is almost invariably the case, the system as a whole is to be perfectly achromatic at its final focus, then the final $(l'_r - l'_v)$ will be zero and a direct solution is easy, provided that it takes the form of determining the required c of the *last* lens.

Thus for a system of two lenses, the left of equation (b) will be zero ; therefore we can divide throughout by l'^2_b and transposition gives the solution for c_b

$$\text{(b)*} \qquad c_b = -\left[\left(\frac{l'_a}{l_b}\right)^2 \frac{(l_r - l_v)_a}{l_a^2} + \left(\frac{l'_a}{l_b}\right)^2 \cdot \delta N_a \cdot c_a\right] / \delta N_b.$$

Equation (c) treated in the same way gives the solution for perfect achromatism at the final focus of a system of three thin lenses

$$\text{(c)*} \quad c_c = -\left[\left(\frac{l'_b}{l_c}\right)^2 \left(\frac{l'_a}{l_b}\right)^2 \frac{(l_r - l_v)_a}{l_a^2} + \left(\frac{l'_b}{l_c}\right)^2 \left(\frac{l'_a}{l_b}\right)^2 \cdot \delta N_a \cdot c_a + \left(\frac{l'_b}{l_c}\right)^2 \delta N_b \cdot c_b\right] / \delta N_c.$$

It will be realized that the difficulty of a direct solution for a prescribed chromatic aberration arises from the fact that then the final l' remains in the equation, and as this changes with the final c it has to be expressed in terms of the final l and final c, and leads to a clumsy quadratic equation which would take more time to solve than would be required for quite a number of trials.

The preceding solutions are not as complete as those for thin lenses in contact, for they do not include the condition that the system shall have a prescribed equivalent focal length ; hence we may have to make several successive attempts in order to attain this additional desideratum. A complete and yet simple solution is possible for achromatic systems of two separated lenses. Systems of this kind were originally introduced for telescopes nearly a century ago under the name of 'dialyte objectives' in order to evade the then predominant difficulty of obtaining large disks of flint glass ; for if the crown lens is placed in front, there is a considerable contraction of the cone of rays before the flint lens is reached, and a smaller diameter of the latter is therefore sufficient. This reason for using separated lenses no longer exists ; but the type has been revived in recent times for the largest astronomical telescopes, firstly, because both sides of each lens can be exposed to the air and the objective thus follows changes of temperature more quickly and especially more uniformly and as a consequence gives sharper images and greater fixity of focus, secondly, because it is possible, by choosing the proper separation between the two components, to correct the spherical aberration for two different colours and thus to satisfy the 'Gauss Condition', so called because the great German mathematician first called attention to the variation of spherical correction in different colours and found one method—although a bad one—of curing it.

If we specialize the solution (b)* for the case when there is no chromatic aberra-

[43] tion at the object-point or when $(l_r - l_v)_a = 0$, but without any restriction as to the distance of the object-point, the solution becomes

$$c_b = -c_a \cdot \frac{\delta N_a}{\delta N_b}\left(\frac{l'_a}{l_b}\right)^2.$$

If we give to the separation our usual name d'_a, then we have $l_b = l'_a - d'_a$ and the equation can be written in the form

$$c_b = -c_a \frac{\delta N_a}{\delta N_b}\left(\frac{1}{1 - \frac{d'_a}{l'_a}}\right)^2.$$

This equation supplies the highly important information that the achromatism of a dialyte objective is not stable ; for if the object-distance l_a changes, then there will be, by TL (3), a corresponding change in l'_a, and that will change the value of the final factor in the equation for c_b. Hence, with a fixed air space d'_a, every change in the distance of the object calls for a different value of c_b, the total curvature of the second lens required to produce achromatism with the given first lens ; therefore any given second lens at a fixed distance d'_a from a given first lens, will give achromatism at only one particular distance of the object-point, and will display chromatic aberration with objects at any other distance. It is a matter more of curiosity than of importance to point out that with two given lenses the achromatism could be maintained if by a delicate automatic movement the air space d'_a were changed, when altering the focus for objects at different distances, so as to keep d'_a/l'_a constant. When the air space is at all considerable, then the instability of the achromatism of dialyte objectives is decidedly serious and restricts their utility to a very small range in l'_a. It may be stated at once that the achromatism is more or less unstable for all systems built up with separated unachromatic lenses or even with *thick* lenses in contact. In some cases, as for instance in the Cooke lens, it is possible to attain achromatism at two *different* distances (but not at the intermediate ones) or for a *short range* of distances, but never for all distances. A close approach to real stability of achromatism is only attainable with *thin* achromatic lenses or with systems composed of *individually achromatized* thin components at any separation, and even then only for a decidedly small aperture.

Direct Solution of the Dialyte Problem

We obtain a complete and simple thin lens solution of the dialyte problem for objects at any distance and for any prescribed chromatic aberration at both the object-point and the image-point, if, instead of aiming at a definite equivalent focal length of the *dialyte* objective, we state the problem in the following way :

'It has been ascertained by a preliminary thin simple lens scheme that a *thin* lens would produce the desired size of image in the available space when the object-distance is l and the conjugate image-distance l'. It is required to substitute a thin lens dialyte objective of prescribed chromatic correction which shall give an image of the same size when its *first lens* is placed at the same distance l from the object.'

[43] Referring to Fig. 42, the upper diagram shows the thin lens prototype giving the desired size of image, or linear magnification, and the lower diagram shows the dialyte objective which is to produce the same magnification when placed at

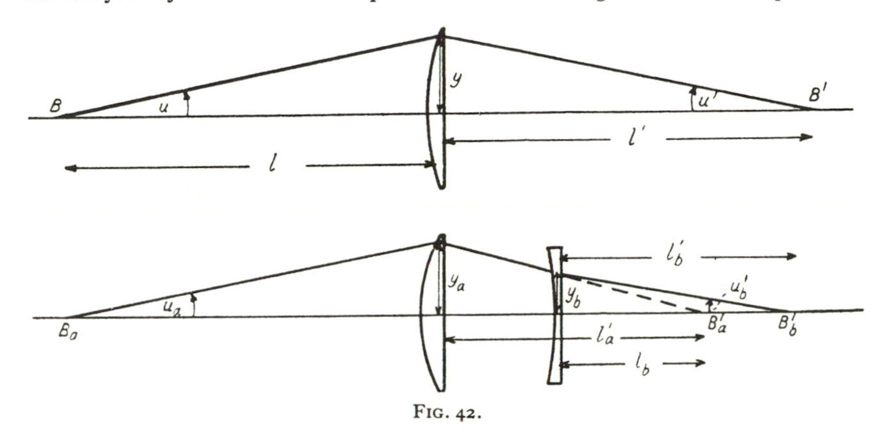

FIG. 42.

the same distance l as the prototype. By Lagrange's theorem the magnification of the simple lens is

$$m'_{TL} = \frac{u}{u'} = \frac{l'}{l}, \text{ as proved in [23].}$$

In the case of the dialyte a paraxial ray starting from the axial object-point under angle u_a will be refracted by the first lens towards some definite intersection-point B'_a at distance l'_a from the first lens. The second lens will receive the ray with an intersection-length $l_b = l'_a - d'_a$ and will finally refract it towards B'_b at distance l'_b from the second lens and with a definite convergence u'_b. The linear magnification produced by the dialyte will then be

$$m'_D = \frac{u_a}{u'_b},$$

and this is to be made equal to the m'_{TL} of the prototype.

Calling the incidence heights of the paraxial ray at the two components of the dialyte y_a and y_b, we have

$$u_a = \frac{y_a}{l}; \; u'_b = \frac{y_b}{l'_b}; \; \text{therefore } m'_D = \frac{u_a}{u'_b} = \frac{y_a}{y_b} \cdot \frac{l'_b}{l}.$$

Now two similar triangles (similar under the usual paraxial and thin lens restrictions) having l'_a and l_b as horizontal and y_a and y_b as vertical sides, give

$$\frac{y_a}{y_b} = \frac{l'_a}{l_b},$$

[43] and introduction of this into the equation for m'_D leads to

$$m'_D = \frac{l'_a}{l_b} \cdot \frac{l'_b}{l}.$$

If we now equate the values of m'_{TL} and m'_D we obtain

$$\text{(d)} \qquad \frac{l'}{l} = \frac{l'_a}{l_b} \cdot \frac{l'_b}{l}, \text{ or } l'_b = l' \cdot \frac{l_b}{l'_a}, \text{ or reciprocally } \frac{1}{l'} = \frac{1}{l'_b} \cdot \frac{l_b}{l'_a}.$$

We have thus determined the value of l'_b which we must aim at in our solution. It should be noted that $y_b/y_a = l_b/l'_a$ is a direct measure of the contraction of the cone of rays in passing from the first to the second lens ; we retain this ratio, or its reciprocal, in our solution as an indirect, but highly convenient, measure of the air space.

We now use the earlier general solution (b) for the condition of achromatism, but replace l_a by the present simple l ; the equation then is

$$(l'_r - l'_v)_b = l'^2_b \cdot \left(\frac{l'_a}{l_b}\right)^2 \cdot \frac{(l_r - l_v)_a}{l^2} + l'^2_b \left(\frac{l'_a}{l_b}\right)^2 \cdot \delta N_a \cdot c_a + l'^2_b \, . \, \delta N_b \cdot c_b.$$

On introducing the value of l'_b by (d) this becomes

$$(l'_r - l'_v)_b = l'^2 \frac{(l_r - l_v)_a}{l^2} + l'^2 \cdot \delta N_a \cdot c_a + l'^2 \left(\frac{l_b}{l'_a}\right)^2 \cdot \delta N_b \, . \, c_b,$$

and division throughout by l'^2 and a transposition gives

$$\frac{(l'_r - l'_v)_b}{l'^2} - \frac{(l_r - l_v)_a}{l^2} = \delta N_a \, . \, c_a + \left(\frac{l_b}{l'_a}\right)^2 \cdot \delta N_b \, . \, c_b.$$

For the 'chromatic residuals' on the left we again use R as a simple symbol. If we then solve for c_b, we obtain

$$c_b = \frac{R}{\delta N_b} \left(\frac{l'_a}{l_b}\right)^2 - c_a \frac{\delta N_a}{\delta N_b} \left(\frac{l'_a}{l_b}\right)^2$$

as *one* equation to be satisfied by c_a and c_b in order to secure the prescribed state of chromatic correction.

In order to obtain the necessary second equation for a complete solution we now work out the final intersection-length l'_b of our dialyte. Equation TL(1) of section [23] gives

$$\text{First Lens : } \frac{1}{l'_a} = (N_a - 1)\, c_a + \frac{1}{l} \text{ ; Second Lens : } \frac{1}{l'_b} = (N_b - 1) c_b + \frac{1}{l_b}.$$

Multiplication of the second equation throughout by l_b/l'_a makes its left side $= 1/l'$ by (d) and the second term of the right equal to $1/l'_a$, so that the right side of the first equation can be substituted, giving

$$\frac{1}{l'} = \frac{1}{l} + (N_a - 1)\, c_a + \frac{l_b}{l'_a} (N_b - 1)\, c_b$$

as the second condition to be satisfied by c_a and c_b. Substitution of the value of c_b

found by the chromatic condition produces an equation containing c_a as the only [43] unknown quantity :

$$\frac{1}{l'} - \frac{1}{l} = (N_a - 1)c_a + R\frac{N_b - 1}{\delta N_b} \cdot \frac{l'_a}{l_b} - c_a \, . \, \delta N_a \frac{N_b - 1}{\delta N_b} \cdot \frac{l'_a}{l_b} \, .$$

This equation invites us by its form to introduce the generalized V-value $(N-1)/\delta N$ for the two kinds of glass and thus leads to

$$\frac{1}{l'} - \frac{1}{l} - R \, . \, V_b \cdot \frac{l'_a}{l_b} = c_a \, . \, \delta N_a \left(V_a - V_b \cdot \frac{l'_a}{l_b} \right) .$$

On the left we can replace $1/l' - 1/l$ by $1/f'$, the power of the simple lens prototype, in accordance with TL(3), and transposition then gives the final solution for c_a

$$c_a = \left(\frac{1}{f'} - R \, . \, V_b \cdot \frac{l'_a}{l_b} \right) \cdot \frac{1}{\delta N_a} \cdot \frac{1}{V_a - \frac{l'_a}{l_b} \, . \, V_b} .$$

Substitution of this value of c_a into the equation for c_b then gives the solution for c_b on slight simplification and we obtain the complete and absolutely general set of equations for dialyte objectives :

$$\text{TL, Chr. (7)} \quad \begin{cases} c_a = \left(\frac{1}{f'} - R \, . \, V_b \cdot \frac{l'_a}{l_b} \right) \cdot \frac{1}{\delta N_a} \cdot \frac{1}{V_a - \frac{l'_a}{l_b} \, . \, V_b} \\ c_b = -\left(\frac{l'_a}{l_b} \right)^2 \left(\frac{1}{f'} - R \, . \, V_a \right) \cdot \frac{1}{\delta N_b} \cdot \frac{1}{V_a - \frac{l'_a}{l_b} \, . \, V_b} \\ R = \frac{(l'_r - l'_v)_b}{l'^2} - \frac{(l_r - l_v)_a}{l^2} . \end{cases}$$

In these equations l, l', and f' are the data of the simple lens prototype ; the remaining quantities, with letter-suffix, are data of the components of the dialyte objective. The resulting air space $d'_a = l'_a - l'_b$ is implied in the factor l'_a/l_b. If a definite air space were required it would be necessary to solve by trials ; but in most applications of dialyte objectives it is the contraction or expansion of the cone of rays between the two components which is primarily important, and then we obtain a direct solution. When the crown lens is in front there will be a contraction of the cone and l'_a/l_b will be greater than one. The equations are equally applicable to the case when the flint lens is in front ; in that case diverging rays will issue from the first lens and l'_a/l_b will be less than one. The mathematically conceivable exceptions which would arise if a very close real object-point were presented to a crown-in-front dialyte, or a very close virtual object-point (i.e. a small positive l) to a flint-in-front combination, are not likely to arise in practice. Similarly it is thinkable that with converging rays issuing from the front lens l'_a might be less than d'_a; in that case l'_a/l_b would become negative, but such an arrangement would be an absolute freak.

As for lenses 'a' and 'b' in contact l'_a will be equal to l_b, it is easily seen that TL, Chr. (7) for this case becomes absolutely identical with the original TL,

[43] Chr. (4), so that the latter solution is merely a special case of the present all-embracing equations.

An interesting limit is attached to dialyte objectives, for if l'_a/l_b approaches the value V_a/V_b the last factor in the equations for c_a and c_b approaches infinity and leads to extremely strong component lenses. Hence we are limited to a comparatively small range in the amount of the contraction or expansion of the cone of rays.

As an example we will calculate a thin lens formula for a dialyte telescope objective of the focal length 8 adopted throughout section [40]. The glasses shall be those of example (1) of that section. l'_a/l_b shall be $= 10/9$, so that the diameter of the cone of rays at the flint lens will be 9/10th of that at the crown lens. With the previously employed $V_a = 60{\cdot}8$ and $V_b = 36{\cdot}0$, $\delta N_a = 00848$ and $\delta N_b = 0{\cdot}01729$ we find, as $R = 0$,

$$c_a = \frac{1}{8} \cdot \frac{1}{0{\cdot}00848} \cdot \frac{1}{60{\cdot}8 - \frac{10}{9} \times 36{\cdot}0} = \frac{1}{8} \cdot \frac{1}{0{\cdot}00848} \cdot \frac{1}{60{\cdot}8 - 40} = 0{\cdot}709\,;$$

$$c_b = -\left(\frac{10}{9}\right)^2 \cdot \frac{1}{8} \cdot \frac{1}{0{\cdot}01729} \cdot \frac{1}{20{\cdot}8} = -0{\cdot}430\,;$$

both by slide-rule. By comparison with example (1) of [40] it will be seen that even this small contraction of the cone has greatly increased the values of c_a and c_b from the previous $0{\cdot}5944$ and $-0{\cdot}2915$. If we try $l'_a/l_b = 1{\cdot}5$, or the effective diameter of the flint lens as 2/3 of that of the crown lens, we find for the same focal length

$$c_a = \frac{1}{8} \cdot \frac{1}{0{\cdot}00848} \cdot \frac{1}{60{\cdot}8 - 54} = \frac{1}{8} \cdot \frac{1}{0{\cdot}00848} \cdot \frac{1}{6{\cdot}8} = 2{\cdot}17\,;$$

$$c_b = -2\tfrac{1}{4} \cdot \frac{1}{8} \cdot \frac{1}{0{\cdot}01729} \cdot \frac{1}{6{\cdot}8} = -2{\cdot}39\,;$$

so that now the flint lens has a higher curvature value than the crown lens. Combinations of this type are of great value in photographic objectives as they admit of the attainment of a flat and well-defined field, as will be proved at a later stage.

It was at one time in the past hoped that the secondary spectrum might be curable on the dialyte principle. It is easily seen that we may explain the secondary spectrum by saying that the flint lens does not exert a sufficient correcting effect on the red end of the spectrum, thus leaving this at too long a focal distance, and that the flint lens acts too strongly on the violet end, thus changing it from too short to too long a focal distance. Now the dispersion of the white light into its constituent colours by the crown lens of a dialyte causes the red to meet the flint lens at a larger diameter than does the violet, and the correcting effect for red is thus increased whilst that for violet is diminished, which is entirely favourable and would make complete correction possible by a sufficient air space. But as we have shown, increased air space calls for increased strength of the components, and actual calculations show that the benefit of the air space is thus lost, owing to the greater curvature of the lenses which are required to produce the same equivalent focal length.

[43] Although the formulae deduced in this section are really valid only for a differential range of colour, it is easily seen from their linear character that the $(l'_r - l'_v)$ will grow in exact proportion with the δN used; therefore we may work out results with any convenient δN, and a value of the order of $N_f - N_c$ will usually be most suitable.

As the formulae are obtained by neglecting the individual thicknesses of the constituent lenses, their use must be restricted to preliminary rough determinations of the chromatic aberration, in conjunction with TL (10)** for the corresponding spherical aberration. In this kind of work we shall find the equations of very great value in the evolution of new designs.

PRIMARY CHROMATIC ABERRATION

[44] As was duly pointed out, the thin lens solutions for chromatic aberration are necessarily inexact when applied to actual lenses, because a certain thickness of these is unavoidable. For telescope objectives and for many ordinary forms of photographic objectives and of eye-pieces the resulting error is not serious; but there are numerous instances, especially among modern photographic objectives, when the effect of the thickness of individual lenses becomes so pronounced that the thin lens solutions would give a very poor approximation or even an absolutely misleading result. In such cases the thickness must be taken into account, and the most generally useful method then consists in calculating the true primary chromatic aberration for the paraxial region surface by surface, in close analogy, and usually in conjunction, with the corresponding method for the true primary spherical aberration by (10)**.

A convenient computing equation is easily deduced from the general equation (6p)**. If we again use 'r' and 'v' for the two colours, this equation gives for any one refracting surface:

$$\text{for 'red'} \quad \frac{N'_r}{l'_r} = \frac{N'_r - N_r}{r} + \frac{N_r}{l_r}; \quad \text{for 'violet'} \quad \frac{N'_v}{l'_v} = \frac{N'_v - N_v}{r} + \frac{N_v}{l_v}.$$

If we introduce δN by putting $N_v = N_r + \delta N$ and $N'_v = N'_r + \delta N'$ the second equation becomes:

$$\text{for 'violet'} \quad \frac{N'_r}{l'_v} + \frac{\delta N'}{l'_v} = \frac{N'_r + \delta N' - N_r - \delta N}{r} + \frac{N_r}{l_v} + \frac{\delta N}{l_v},$$

and if we now form the difference 'violet' – 'red', we obtain

$$N'_r \frac{l'_r - l'_v}{l'_r \cdot l'_v} + \frac{\delta N'}{l'_v} = \frac{\delta N' - \delta N}{r} + N_r \frac{l_r - l_v}{l_r \cdot l_v} + \frac{\delta N}{l_v}.$$

This is a strict paraxial equation, valid for any value of the δN and of the chromatic aberrations $(l_r - l_v)$. But it would become very complicated if we tried to turn it into a complete solution by eliminating l'_r and l'_v in the denominators on the left by use of the starting equations. To avoid this difficulty we again restrict ourselves, as in the previous section, to a very small (strictly speaking a differential) range of colour, in the immediate neighbourhood of the brightest or most important

[44] colour for which minimum focal distance is to be established; then l_r and l_v will differ very little from each other and from the plain l for brightest light, and for the same reason l'_r and l'_v may be replaced by the l' for brightest light. Similarly N_r and N'_r may be replaced by N and N' for brightest light, and the last equation becomes

$$\text{(a)} \qquad N'\frac{l'_r - l'_v}{l'^2} + \frac{\delta N'}{l'} = \frac{\delta N' - \delta N}{r} + N\frac{l_r - l_v}{l^2} + \frac{\delta N}{l},$$

and by solving for $(l'_r - l'_v)$ we obtain

$$l'_r - l'_v = \frac{l'^2}{N'}\left[\delta N'\left(\frac{1}{r} - \frac{1}{l'}\right) - \delta N\left(\frac{1}{r} - \frac{1}{l}\right) + N\frac{l_r - l_v}{l^2}\right].$$

A simple transposition of equation (6p)** gives

$$\frac{1}{r} - \frac{1}{l} = \frac{N'}{N}\left(\frac{1}{r} - \frac{1}{l'}\right),$$

and if this is introduced into the solution for $l'_r - l'_v$ and if the multiplication of the square bracket by its factor is carried out, we find

$$l'_r - l'_v = \frac{N}{N'}\left(\frac{l'}{l}\right)^2\left(l_r - l_v\right) + l'^2\left(\frac{1}{r} - \frac{1}{l'}\right)\left(\frac{\delta N'}{N'} - \frac{\delta N}{N}\right).$$

To render this still more convenient for calculation, we introduce in the first term, by $y = lu = l'u'$, $l'/l = u/u'$; in the second term we make the transformation

$$l'^2\left(\frac{1}{r} - \frac{1}{l'}\right) = l'^2 \cdot \frac{l' - r}{r \cdot l'} = l' \cdot \frac{l' - r}{r}$$

and obtain the final equation for the surface-by-surface calculation of the true primary chromatic aberration of any centred lens system:

$$\text{Chr. (8)} \qquad l'_r - l'_v = \frac{N}{N'}\left(\frac{u}{u'}\right)^2\left(l_r - l_v\right) + l' \cdot \frac{l' - r}{r} \cdot \left(\frac{\delta N'}{N'} - \frac{\delta N}{N}\right).$$

Like the equation for the primary spherical aberration, the present one calls for the tracing of a paraxial ray of brightest light through the system, and this preliminary ray-tracing represents the chief additional labour which is required when the chromatic aberration is to be determined with greater accuracy than that afforded by the highly convenient TL solutions. Nearly always progress towards the desired solution will be quickest if we begin with a rough determination of the proper c-values of the constituent lenses by the TL equations; we then accept all the radii of individual surfaces so arrived at, with the exception of the last, for the tracing of the paraxial pencil and for the calculation of the primary chromatic aberration. Having thus determined the data of the ray arriving at the last surface, and the chromatic aberration delivered to the latter, we complete the work by solving for that value of the curvature of the last surface which will yield the desired state of chromatic correction—nearly always zero-value—at the final focus. We can always arrange the calculation so that the final refraction is one into air, for which N' is equal to one and $\delta N'$ equal to zero when the usual relative refractive indices are employed. Simplifying equation (a) of this section accordingly, a simple transposition gives the solution [44]

Chr. (8)* $$\frac{1}{r} = \frac{1}{l} + \frac{N}{\delta N}\frac{l_r - l_v}{l^2} - \frac{l'_r - l'_v}{\delta N \, . \, l'^2}.$$

In nearly all cases $(l'_r - l'_v)$ will be zero as perfect achromatism is the usual aim, and the solution is then direct and perfectly exact. In the rare cases when a prescribed over- or under-correction has to be realized, the l' in the last term on the right will be unknown, but as the TL preparatory scheme will have aimed at a desired value of l', the latter can be used for a first approximation. The tracing of the paraxial ray through the surface thus defined will give a second and usually sufficiently close value of l' and a corrected final r can be determined. This will be found more convenient and rapid and less subject to numerical errors than the substitution of the value of $1/l'$ by equation (6p)** and solution of the resulting quadratic equation in $1/r$. At first sight it must cause surprise to find that the direct solution for a prescribed residual longitudinal aberration leads to a quadratic equation and therefore to two different values of the final radius. One of these solutions is identical with that found by the recommended approximation method, and gives a radius and a final intersection-length such as we should have been expecting. The second solution gives a very short radius and intersection-length which common sense would immediately reject although it formally satisfies the equation ; the reason is again supplied by the misleading changes in magnitude of all longitudinal aberrations, which were emphasized in Chapter II ; when the aberrations are measured physically as differences of optical path, then there is always only one solution—the one which in geometrical work appeals to common sense.

With reference to numerical calculations by Chr. (8) and (8)*, which will nearly always be accompanied by corresponding work with the aid of (10)** for the primary spherical aberration, the extremely close relationship between Chr. (8) and (10)** should be noted, for in the former $(\delta N'/N' - \delta N/N)$ simply takes the place of $\frac{1}{2}(i' - u)(i - i')$ in the latter. The same logarithms therefore serve for the greater part of the working out. It is preferable, however, to calculate and transfer the chromatic aberration surface by surface and a sum-formula has therefore not been given. In calculating the transfer-factor $Nu^2/N'u'^2$ it should be noted that this may be interpreted as $N'u'^2$ of preceding surface divided by $N'u'^2$ of new surface. The same preparation of the transfer factors is thus rendered available for both the chromatic and the spherical primary aberrations. The special measure of the dispersion $\delta N/N$ should be prepared for each *glass* ; the proper combination of 'following minus preceding' values for each surface is then easily and safely extracted.

We will apply the chromatic formulae to the paraxial ray of the telescope objective in section [21], taking over certain quantities from the example in section [28]. By the differences of the indices for C and F given in [21] we obtain δN of the crown = 0·00817, of the flint = 0·01720, and for brightest light, log N of the crown = 0·18087, of the flint = 0·21101.

[44] These give :

Crown : $\log \delta N$	=	7·9122	*Flint* : $\log \delta N$	=	8·2355
$-\log N$	=	−0·1809	$-\log N$	=	−0·2110
$\log \delta N/N$	=	7·7313	$\log \delta N/N$	=	8·0245
$\delta N/N$	=	0·00539	$\delta N/N$	=	0·01058

Taking over $N'u'^2$ and $l'(l'-r)/r$ from section [28] we then calculate :

	First Surface.	*Second Surface.*	*Third Surface.*
$\delta N'/N'$	0·00539	0·01058	0·00000
$-\delta N/N$	−0·00000	−0·00539	−0·01058
$\delta N'/N' - \delta N/N$	0·00539	0·00519	−0·01058
$\log (l_r - l_v)$	None	9·0361	9·2011*n*
$+\log N'u'^2$ preceding column		8·1450	7·9157
$+\text{colog } N'u'^2$ own column		2·0843	1·8014
log (transferred aberration)	None	9·2654	8·9182*n*
$\log l'(l'-r)/r$	1·3048	1·8202*n*	0·9335*n*
$+\log (\delta N'/N' - \delta N/N)$	7·7313	7·7152	8·0245*n*
log (new aberration)	9·0361	9·5354*n*	8·9580
Transferred aberration	None	+0·1842	−0·0828
+New aberration	0·1087	−0·3431	+0·0908
Total $(l'_r - l'_v)$ for transfer	0·1087	−0·1589	+0·0080

The result is thus, that this objective in its paraxial region is chromatically under-corrected, having a final $(l'_r - l'_v)$ equal to 0·0080″.

If we were beginning an actual design of this object-glass, we should, however, stop the preceding calculation on finding the total $(l'_r - l'_v) = -0{\cdot}1589$ of the last but one, in this case the second, surface. This would be the $(l_r - l_v)$ received by the last surface, and we should then calculate the value of r_3, which would render the system achromatic for its final focus, by using Chr. (8)*. The formula being

$$\frac{1}{r} = \frac{1}{l} + \frac{N}{\delta N}\,\frac{l_r - l_v}{l^2},$$

we must take the l of the last surface from the paraxial ray-tracing in [21] where it is found = 13·4446, or its log = 1·12855. $N/\delta N$ being that of the flint glass, we calculate the second part of the equation thus :

$\text{colog } \delta N/N$	= 1·9755			
$+\log (l_r - l_v)$	= 9·2011*n*	Second part	=	−0·08308
$+\text{colog } l$	= 8·8714	First part	=	+0·07438
$+\text{colog } l$	= 8·8715	$1/r$	=	−0·00870
log (second part)	= 8·9195*n*	$r_3 = -115''$ by Barlow's table.		

To render this object-glass achromatic for its paraxial region would therefore call for a last radius of −115″ in place of the radius of −60″ which, as we shall see, gives the best compromise as regards achromatism for the intended full aperture of 2 inches. [44]

With reference to the values of δN to be used in working with Chr. (8), the observation at the end of the previous section again holds. As the equation proves a linear relation between the δN and the $(l_r - l_v)$, the latter will always be in strict proportion to the former ; therefore δN-values of the usual order may be employed. They will, however, not give results which *accurately* agree with a direct paraxial ray tracing in the two separate colours, for the latter will bring out the secondary spectrum effect which our differential formula excludes.

CHROMATIC ABERRATION AT FINITE APERTURE

[45] It was proved in Chapter I that a paraxial pencil traced through any centred lens system always comes to a perfectly definite final focus. If several paraxial pencils, in different colours, are traced, then each will give such a definite focus, but on account of the change in refractive indices corresponding to a change of colours, the foci obtained for the different colours will as a rule not coincide ; they will be separated by finite, though usually small, intervals which we have called longitudinal chromatic aberration.

For any given lens system, colours, and distance of the object-point the longitudinal chromatic aberration is therefore a constant quantity within the paraxial region.

If we trace a white ray at a *finite* angle U from an object-point through a centred lens system, the different colours contained in it will again be assigned separate tracks by the law of refraction on account of the corresponding variations of the refractive indices, and as the final result we must again expect a difference of focus for different colours. But this difference will now be affected by longitudinal spherical aberration, and as all our equations for spherical aberration showed that it depends on N and therefore varies for different colours, we must expect to find that the same two colours, traced from the same object-point through the same system at paraxial and at finite angles respectively, will give different values for the paraxial final $l'_r - l'_v$ and the 'marginal' final $(L'_r - L'_v)$. By way of example we will take the calculation of a blue ray (suffix v) through the biconvex lens of Example 1 and compare it with the results of the tracing of a corresponding yellow ray (suffix r), given in detail in sections [9] and [20] :

Original U_1	= 0 (Paraxial)	$-1°$	$-2°$	$-3°$	$-4°$
L'_r	= 8·790	8·709	8·470	8·054	7·438
L'_v	= 8·585	8·506	8·269	7·857	7·246
$L'_r - L'_v$	= 0·205	0·203	0·201	0·197	0·192
Paraxial – 'Marginal'	=	0·002	0·004	0·008	0·013

The longitudinal chromatic aberration $L'_r - L'_v$ shows a steady diminution with increasing aperture, which is most clearly disclosed by taking the difference between the result for the paraxial rays and that for the successive rays at finite angles.

[45] The calculations for C and F light through the telescope objective, Example 2 a of section [21], give another illustration :

Semi-aperture :	0	0·7071	1·000
Final L' for C	7·621	7·613	7·614
Final L' for F	7·614	7·613	7·622
$L'_c - L'_f =$	+0·007	0·000	−0·008
Paraxial − 'Marginal' =		+0·007	+0·015

The longitudinal chromatic aberration is positive ('under-corrected') for the paraxial rays, zero for the zonal rays, and negative for the extreme marginal rays, which looks like a result very different from that for the biconvex lens ; but if we again study the *change* of the longitudinal chromatic aberration by comparing the last two with the paraxial value, we obtain differences of 0·007 for the zonal and 0·015 for the marginal ray, or again a steady growth.

We must discover the law which governs this change of the longitudinal chromatic aberrations with increasing aperture. We can do so without recourse to complicated algebra by the reasoning based on the symmetry of a centred lens system with reference to the optical axis. On account of its extraordinary potency and simplicity this method of reasoning was given very fully in Chapter II section [31].

In the present case the argument runs in an exactly corresponding manner. By reason of the symmetry around the optical axis coloured rays traced at $+Y$ and at $-Y$, or $+U$ and $-U$, must give the same longitudinal chromatic aberration, hence the formula for $L'_r - L'_v$ can only have the form

$$L'_r - L'_v = b_0 + b_2 Y^2 + b_4 Y^4 + \&c.,$$

in which b_0, b_2, &c., are constants and Y is any reasonable measure of the aperture. We must retain the constant first term of the series in this case because it obviously represents the value of $L'_r - L'_v$ for very small apertures, or the paraxial longitudinal chromatic aberration, which we have shown to be usually a finite quantity.

We will test the formula by our two examples. In the first we use U_1 as the aperture measure, and employing only the first two terms, which are ample in all but the exceptional case of high-power microscope objectives and certain freak-systems, we have, with b_0 = the paraxial $l'_r - l'_v = 0{\cdot}205$:

$$L'_r - L'_v = 0{\cdot}205 + b_2 \,.\, U_1^2.$$

We will use the ray at $-4°$ to determine b_2 ; it gives

$$0{\cdot}192 = 0{\cdot}205 + b_2\,(-4)^2 = 0{\cdot}205 + 16\,b_2$$

or
$$b_2 = -0{\cdot}013/16 = -0{\cdot}00081,$$

and we then can calculate :

$$\text{for } U_1 = -1° : L'_r - L'_v = 0{\cdot}205 - 0{\cdot}00081\,(-1)^2 = 0{\cdot}204\,;$$

$$\text{for } U_1 = -2° : L'_r - L'_v = 0{\cdot}205 - 0{\cdot}00081\,(-2)^2 = 0{\cdot}202\,;$$

$$\text{for } U_1 = -3° : L'_r - L'_v = 0{\cdot}205 - 0{\cdot}00081\,(-3)^2 = 0{\cdot}198.$$

We see that the values obtained by the formula agree in every case within 0·001″

(which is not more than the uncertainty of our data) with the direct trigonometrical calculation. [45]

For the telescope objective we use the actual Y employed in the trigonometrical calculations as the aperture measure and with $b_0 = l'_c - l'_f = 0{\cdot}007$ find from the $L'_c - L'_f = -0{\cdot}008$ of the marginal ray

$$-0{\cdot}008 = 0{\cdot}007 + b_2\,(1)^2$$

or

$$b_2 = -0{\cdot}015,$$

and this gives the computed $L'_c - L'_f$ for the zonal rays as $(L'_c - L'_f) = 0{\cdot}007 - 0{\cdot}015\,(0{\cdot}7071)^2 = 0{\cdot}007 - 0{\cdot}0075 = -0{\cdot}0005''$, which agrees with the zero-value found trigonometrically well within the dependability of the five-figure work.

The variation of the longitudinal chromatic aberration for different zones of the aperture obviously renders it impossible to establish perfect achromatism for the aperture as a whole. We must therefore be satisfied with a compromise, but naturally desire to choose the best possible compromise. The determination of the latter is our next task. By our formula

$$L'_r - L'_v = b_0 + b_2 Y^2$$

the variation is shown to be proportional to the square of the aperture ; it therefore follows the law of primary spherical aberration, and we may say that the variation is due to a difference of the magnitude of the primary spherical aberration for different colours or that there is relative primary spherical aberration between different colours. Now it was pointed out in Chapter III, and will be fully proved in Part II, that in the presence of primary spherical aberration the best compromise focus (namely the one which gives the least residual differences of optical paths) lies midway between the paraxial and the marginal focus, and therefore coincides with the focus of the rays from $\sqrt{\tfrac{1}{2}} = 0{\cdot}7071$ of the full aperture. Therefore, in order to obtain the best possible union of two colours afflicted with a difference of primary spherical aberration, we must bring their best compromise foci together, which will be effected if we unite the rays which both colours send through the 0·7071 zone of the full aperture.

We have thus arrived at the highly important

RULE : 'A prescribed state of chromatic correction will be most perfectly realized for the whole aperture if it is established trigonometrically for the 0·7071 zone.'

On referring back to the numerical data for our telescope objective it will be seen that the latter satisfies our rule and therefore represents the best possible compromise with regard to achromatism. We may add that the calculations of this object-glass also bear out our conclusions as to the secondary spectrum, for in the 0·7071 zone the C and F rays come to focus at $L' = 7{\cdot}613$, whilst for the brightest rays midway between C and F the result for the zone in question (see section [21]) was $L' = 7{\cdot}609$, or $0{\cdot}004''$ shorter. As the focal length is practically 8 inches, the secondary spectrum amounts to $0{\cdot}004/8 = 1/2000$th of the focal length. Our estimate for the D line was $1/2500$th, but as we found by plotting that in objectives

[45] designed for the union of C and F rays the brightest rays must have a slightly shorter focus than the D rays, the agreement obviously is as close as could be desired.

TRIGONOMETRICAL METHODS OF CHROMATIC CORRECTION

It was pointed out when the TL, Chr. equations were deduced that their extreme simplicity is obtained by neglecting the thickness of the lenses and that they are not even paraxially correct as a consequence. The data obtained from these equations therefore require correction by more exact methods in practically every case. These methods must now be described.

We will first show by our telescope objective that the variations to be expected are not negligible. The crown lens of the objective had $r_1 = -r_2 = 3{\cdot}55$. Hence the net curvature is $c_a = (1/3{\cdot}55) - 1/(-3{\cdot}55) = 2/3{\cdot}55 = 0{\cdot}5634$. As by the data we have $\delta N_a = 0{\cdot}00817$ and $\delta N_b = 0{\cdot}01720$, it follows by the added equation of TL, Chr. (4)* that the flint lens should have

$$c_b = -c_a \frac{\delta N_a}{\delta N_b} = -0{\cdot}5634 \frac{0{\cdot}00817}{0{\cdot}01720} = -0{\cdot}2676.$$

As $c_b = 1/r_3 - 1/r_4$, this gives $1/r_4 = 1/r_3 - c_b$, and as the lens is cemented we have $r_3 = r_2 = -3{\cdot}55$, and the last radius is found by

$$1/r_4 = -0{\cdot}2817 - (-0{\cdot}2676) = -0{\cdot}0141 \text{ or } r_4 = -70{\cdot}9 \text{ inches.}$$

We found the objective perfectly corrected with the last radius equal to -60 inches; a difference of nearly 11 inches.

As a more complete test we will carry out the whole solution by the TL, Chr. (4)* equations. The equivalent focal length of our actual objective was found $= 7{\cdot}956$ inches. The δN have already been given as $0{\cdot}00817$ and $0{\cdot}01720$. The refractive indices were $1{\cdot}5166$ and $1{\cdot}6256$, consequently

$$V_a = \frac{0{\cdot}5166}{0{\cdot}00817} = 63{\cdot}2 \qquad V_b = \frac{0{\cdot}6256}{0{\cdot}01720} = 36{\cdot}4.$$

Therefore

$$\begin{cases} c_a = \dfrac{1}{7{\cdot}956} \cdot \dfrac{1}{26{\cdot}8} \cdot \dfrac{1}{0{\cdot}00817} = 0{\cdot}5741 \\[2ex] c_b = \dfrac{1}{7{\cdot}956} \cdot \dfrac{1}{-26{\cdot}8} \cdot \dfrac{1}{0{\cdot}01720} = -0{\cdot}2727. \end{cases}$$

For an equiconvex crown lens c_a would give

$$1/r_1 = -1/r_2 = 0{\cdot}28705\,;\; r_1 = 3{\cdot}484\,;\; r_2 = -3{\cdot}484\,;$$

and for a cemented flint lens we should find

$$1/r_4 = -0{\cdot}28705 - (-0{\cdot}2727) = -0{\cdot}01435\,;\; r_4 = -69{\cdot}7.$$

It will be seen that r_1 and r_2 have come out shorter than those which really give the desired focal length with perfect achromatism, and the last radius comes out decidedly too long.

Inaccuracies of this order must be expected in working with the TL, Chr. solutions, and it will now be granted that the results of the latter can rarely be accepted as final. [45]

According to our rule the trigonometrical correction of a TL solution for chromatic aberration must aim at bringing the two selected colours accurately to the same focus (or to the required difference of focus) for the 0·7071 zone of the aperture. No absolutely exact trigonometrical solution of this problem is at present known, but the following approximate solutions usually come very close to the desired result in all cases which are suitable for purely trigonometrical treatment. A really rigorous direct solution based on the optical path method will be given in Part II. All our solutions of the problem aim at the determination of that value of the radius of curvature of the last surface which will secure the desired state of chromatic correction ; those now to be given assume that the two rays of different colour for which correction is to be established have been trigonometrically traced through all except the last surface, and that the data L' and U' of both rays on emergence from the last-but-one surface are therefore known. As in the preceding sections, we will distinguish the two colours by suffix (r) and (v) respectively. It is assumed also that the rays emerge from the last surface into air of index 'one'.

First method, for strict union of the two coloured rays. By equation (10), Chapter II, we have for any ray

$$(10) \qquad \frac{N'}{L'} = \frac{N}{L} + \frac{N'-N}{r} + \text{spherical aberration term.}$$

This equation gives a very close approximation to a solution when we assume that the final spherical aberration term of the complete equation is of the same value for the two colours. This is not strictly correct, and a second approximation may therefore be occasionally required, but in most cases the last surface of an achromatic lens is comparatively shallow, with reasonably small angles of incidence, and the first solution is then practically exact.

By applying the thickness of the last lens, the L' of the last-but-one surface give us L_r and L_v for the last surface. Remembering that N' is taken as $= 1$ for air, equation (10) gives

$$\frac{1}{L'_r} = \frac{N_r}{L_r} + \frac{1-N_r}{r} + \text{spherical aberration term,}$$

$$\frac{1}{L'_v} = \frac{N_v}{L_v} + \frac{1-N_v}{r} + \text{ditto, assumed identical,}$$

and the difference of the two equations gives zero on the left because $L'_r = L'$ for the perfect achromatism aimed at. Hence

$$0 = -\frac{N_v - N_r}{r} + \frac{N_v}{L_v} - \frac{N_r}{L_r}$$

or

$$0 = -\frac{N_v - N_r}{r} + \frac{N_v \,.\, L_r - N_r \,.\, L_v}{L_v \,.\, L_r};$$

[45] or putting, in the second term, $N_v = N_r + (N_v - N_r)$,

$$0 = -\frac{N_v - N_r}{r} + \frac{N_r(L_r - L_v) + (N_v - N_r)\, L_r}{L_v \,.\, L_r}.$$

Solved for $\frac{1}{r}$, the required curvature of the last surface, this gives

Chr. (1) $$\frac{1}{r} = \frac{(L_r - L_v)\quad N_r}{L_v \,.\, L_r\,(N_v - N_r)} + \frac{1}{L_v},$$

which is our solution by the first method.

As a numerical example we will take example (1) of section [40], assuming that the lens is to be used as a telescope objective, and is to be of 1·6 inches clear aperture. Hence the two coloured rays C and F have to be calculated for 0·7071 of that aperture or with $Y = 0{\cdot}7071 \times 0{\cdot}8 = 0{\cdot}5657''$.

The crown must be given a thickness of 0·2″ to allow of the stated clear aperture, and giving to the flint lens 0·15″, the formula is

$$r_1 = 3{\cdot}365 \qquad d_1' = 0{\cdot}200$$
$$r_2 = -3{\cdot}365 \qquad d_2' = 0{\cdot}150$$
$$(r_3 = -175).$$

The refractive indices for the C and F lines will be found from the glass list as

$$N_r = 1{\cdot}51300 \text{ and } 1{\cdot}61758\,;$$
$$N_v = 1{\cdot}52148 \text{ and } 1{\cdot}63487.$$

Tracing the two rays through the first two surfaces the results will be

$$L_r' = 13{\cdot}0141,\; L_v' = 13{\cdot}1634,\; U_r' = 2-27-13,\; U_v' = 2-25-32.$$

Allowing 0·15 for the thickness of the flint lens, we have for the rays incident upon the last surface $L_r = 12{\cdot}8641$,

$$L_v = 13{\cdot}0134, \text{ hence } L_r - L_v = -0{\cdot}1493.$$

From the data of the flint lens we also obtain

$$N_v - N_r = 0{\cdot}01729,$$

and Chr. (1) therefore gives for the last curvature

$$\frac{1}{r} = \frac{-0{\cdot}1493 \times 1{\cdot}61758}{0{\cdot}01729 \times 12{\cdot}8641 \times 13{\cdot}0134} + \frac{1}{13{\cdot}0134},$$

or, largely by cologs :

log (−0·1493)	= 9·17406n		
+log 1·61758	= 0·20887		
+colog 0·01729	= 1·76220		
+colog 12·8641	= 8·89062	First Term	= −0·083437
+colog 13·0134	= 8·88561	+Second term	= +0·076844
log first term	= 8·92136n	$\frac{1}{r}$	= −0·006593

By Barlow $r = -151{\cdot}68$

Really ample justice would be done to the data by using four-figure logs, for [45] $L_r - L_v$ and $N_v - N_r$ are only known to four figures.

Tracing the two rays through the last surface with the radius as determined $= -151{\cdot}68$, the final results will be found (by the check formula):

$$L'_r = 7{\cdot}6895,\ L'_v = 7{\cdot}6897,\ U'_r = U'_v = 4-6-0.$$

By the calculation there appears to be a chromatic over-correction of $-0{\cdot}0002''$, but as the final L must be considered doubtful—with five-figure logs and angles to the nearest second—by several ten thousandths of an inch, the result may be accepted as perfectly satisfactory.

It will be highly instructive to try the last radius $= -175$ found by the TL, Chr. solution, which is apparently wildly wrong. It will be found, by the check-formula which must always be used for long radii like these, that this radius gives $L'_r = 7{\cdot}7222$ and $L'_v = 7{\cdot}7231$, or an over-correction of $-0{\cdot}0009''$, which in fact could hardly be *guaranteed* as real with the number of decimal places employed in the calculations. L'_r might conceivably be too small by 0·0004 and L'_v simultaneously too large by 0·0005! The fact always to be borne in mind is that the condition of achromatism does not put a very close limit on the last radius, especially not when the latter comes out very long. As all optical formulae show, it is really the curvature, or reciprocal of the radius, which counts, and the reciprocals of 175 and of 151·68 only differ by 0·00088, which is very small compared with the total curvature 0·2915 of our flint lens.

When the highest accuracy is to be secured with five-figure logs, the angles should be taken to fractions of a second: telescope objectives, as the only ones for which the utmost accuracy is desirable, nearly always have such small angles of incidence that half seconds or even 1/5th of a second can be taken out with some real gain in precision, especially when the five-figure logs are taken from a six-figure table.

Second method, also for strict union of two coloured rays.

This method also implies one assumption which prevents it from being a rigorous solution and which will in unfavourable circumstances cause it to give a worse approximation than the first method.

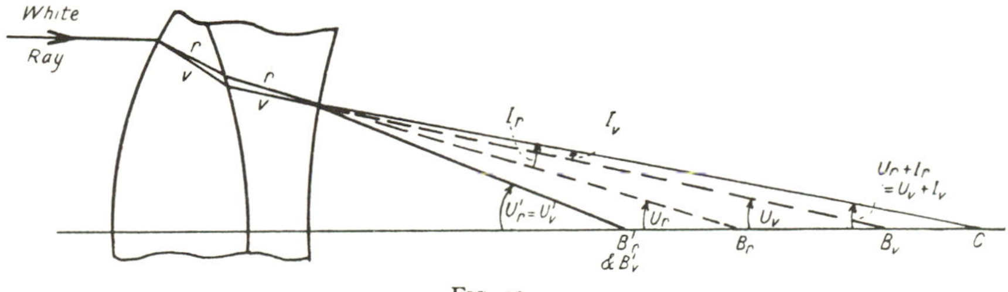

FIG. 43.

A white ray of light incident at the first surface of a lens system is broken up into its constituent colours, because 'red' has a lower index than 'violet'. The red and violet rays consequently become separated. In any ordinary cemented object-

[45] glass the second surface produces dispersion in the opposite sense, and in general more than corrects the dispersion of the first surface, so that usually the separated red and violet rays leave the dispersive second surface with a converging tendency. In more complicated systems the same antagonistic tendencies may come into operation several times. The *assumption* which underlies our second method is that the thicknesses of the lenses are so adjusted that the red and violet rays intersect each other again in the last surface, and so have a common point of incidence P (Fig. 43) at that surface. Although this is a highly desirable feature of a lens system because, as will be shown subsequently, it leads to achromatism of magnification in addition to achromatism of the image-point, it is rarely realized with mathematical accuracy. In ordinary objectives the separation of red and violet at the last surface cannot easily reach more than a very few ten-thousandths of an inch, and the error which is introduced by the assumption of a common point of incidence P at the last surface is therefore, as a rule, extremely small.

The two rays having been traced through all but the last surface, the final data for the last-but-one surface will, by allowance for the thickness of the last lens, give us L_r, U_r, L_v, and U_v of the rays incident at that surface. Whatever radius $r = PC$ may eventually result from the solution, we shall then have, on our stated assumption,

$$U_r + I_r = U_v + I_v = \text{angle } ACP, \quad \text{hence}$$

(a) $$I_r - I_v = -(U_r - U_v),$$

and the difference between the angles of incidence for the two colours is therefore known.

As the point P is assumed to be common to the two rays, the two refracted rays obviously will coincide and produce an achromatic focus if we choose r so as to cause I'_r to become equal to I'_v. The first part of our solution therefore consists in determining what value of I_r or I_v will lead to that result, and the second part of the solution determines the value of r which will yield the angle of incidence found to be required in the first part. The refraction at the last surface will, by the law of refraction, produce the results

$$\sin I'_v = N_v \,.\, \sin I_v$$

and $$\sin I'_r = N_r \,.\, \sin I_r,$$

and as our first condition $I'_v = I'_r$ implies also equality of the sines, the two equations give us one:

(b) $$N_v \sin I_v = N_r \,.\, \sin I_r.$$

To solve this equation for I_v we introduce on the left the identity $N_v \equiv N_r + (N_v - N_r)$ and multiply out. On the right we introduce $I_r \equiv I_v + (I_r - I_v)$ and resolve the sine by the sin $(a+b)$ formula. These operations give

$$N_r \,.\, \sin I_v + (N_v - N_r) \sin I_v = N_r \sin I_v \,.\, \cos (I_r - I_v) + N_r \cos I_v \,.\, \sin (I_r - I_v).$$

The dispersion-angle $(I_r - I_v)$ is always small in lens systems, rarely reaching even 10 minutes for which the difference between cos and Unity is still insensible in five-figure logs. Hence we can replace cos $(I_r - I_v)$ in the first term on the right by 'one', and this term then cancels out against the first term on the left. We thus [45] obtain the simplified equation

$$(N_v - N_r) \sin I_v = N_r \cos I_v \sin (I_r - I_v)$$

and on dividing throughout by $(N_v - N_r) \cos I_v$ and replacing $(I_r - I_v)$ by $(U_v - U_r)$ according to (a), we obtain the first part of our solution

$$\text{(c)} \qquad \tan I_v = \frac{N_r \sin (U_v - U_r)}{N_v - N_r}$$

by which the required angle of incidence of the 'violet' ray can be calculated. To find the radius of curvature which will give this angle of incidence we use a transposition of standard computing equation (1)

$$\frac{r}{L_v - r} = \frac{\sin U_v}{\sin I_v}$$

from which we derive by the well-known algebraical rule already frequently employed:

$$\frac{r}{L_v} = \frac{\sin U_v}{\sin I_v + \sin U_v} = \frac{\sin U_v}{2 \sin \frac{1}{2}(U_v + I_v) \cos \frac{1}{2}(U_v - I_v)}$$

which gives a solution for r by transposing L_v to the right. Combining this with (c) we obtain the complete solution by the second method:

$$\text{Chr. (2)} \qquad \begin{cases} \tan I_v = \dfrac{N_r \,.\, \sin (U_v - U_r)}{N_v - N_r} \\[2ex] r \;\; = \dfrac{L_v \sin U_v}{2 \sin \frac{1}{2}(U_v + I_v) \cos \frac{1}{2}(U_v - I_v)}. \end{cases}$$

As a numerical example we will apply this solution to the lens already used in connexion with the first method, for which we had

U_v	$= 2-25-32$	also L_v	$= 13{\cdot}0134$
U_r	$= 2-27-13$	$N_v - N_r$	$= 0{\cdot}01729$
$U_v - U_r$	$= -0-\ 1-41$		

With these data we calculate the first part of Chr. (2) by four-figure logs:

$\log N_r$	$= 0{\cdot}2089$	I_v	$= -2-37-23$
$+\log \sin (U_v - U_r)$	$= 6{\cdot}6899n$	U_v	$= \ \ 2-25-32$
$+\text{colog} (N_v - N_r)$	$= 1{\cdot}7622$	$\frac{1}{2}(U_v + I_v)$	$= -0-\ 5-56$
$\log \tan I_v$	$= 8{\cdot}6610n$	$\frac{1}{2}(U_v - I_v)$	$= \ \ 2-31-28$

We then obtain the radius by the second part of Chr. (2):

$\log L_v$	$1{\cdot}1144$	
$+\log \sin U_v$	$8{\cdot}6266$	
$+\log \frac{1}{2}$	$9{\cdot}6990$	
$+\text{colog} \sin \frac{1}{2}(U_v + I_v)$	$2{\cdot}7630n$	
$+\text{colog} \cos \frac{1}{2}(U_v - I_v)$	$0{\cdot}0004$	
$\log r$	$= 2{\cdot}2034n$	$r = -159{\cdot}7$

[45] Calculation through the last surface with this radius gives $L'_r = 7{\cdot}7018$, $L'_v = 7{\cdot}7019$, $U'_r = U'_v = 4-5-37$, or a chromatic over-correction of 0·0001 inch, utterly negligible.

It is instructive to notice that although the second method has given a diminished convexity of the last surface as compared with the first method, and corresponding longer intersection-lengths, it has given—*on paper*—less over-correction: −0·0001 against −0·0002. This cannot possibly be correct, and a six-figure calculation for both radii would undoubtedly give more over-correction for the second than for the first solution. It illustrates the absurdity of carrying interpolations farther than the accuracy of the logarithmic work warrants; the discrepancy is entirely due to the fortuitous effects of the rounding off of angles to the nearest second and to the inaccuracy of the last decimal place of the log sin of small angles when the latter are only known to the nearest second. The last section of this chapter, on chromatic tolerances, will give the highly desirable guidance as to what departures from absolute equality of L'_r and L'_v are to be regarded as harmless.

Both methods, but particularly the second, call for care in watching the signs of all the quantities which enter into the equations. It is worthy of notice that both methods call for no preliminary determination of the radius of the last surface. It is therefore unnecessary to calculate the total curvature (c_b in the usual two-lens achromatics) of the last lens when one of these two methods is to be used. In the rare cases when the methods give an insufficiently perfect achromatism, the radius is most conveniently corrected by the equation Chr. (3) given for the third method, simply by putting the required $(L'_r - L'_v) = 0$. The first method will in general be found preferable to the second.

Third method, for chromatically over- or under-corrected lens systems. It is not difficult to adapt the equations for the first or second method so as to meet this case, but the equations then become so much more complicated that a different method is almost sure to give the desired final result both more quickly and with fewer worries about signs. In using the third method we determine all the radii of the system by the equation TL, Chr. (4), and having assigned appropriate thicknesses to the constituent lenses, we trace the two coloured rays completely through and determine the final $(L'_r - L'_v)_{TL}$ which corresponds to the TL solution. If this should happen to agree with the required over- or under-correction, within the accuracy to be expected from the calculation, the work would be finished. But it will usually prove wrong by rather more than can be overlooked, and the last radius will then call for correction. We can look upon this alteration of the last radius as equivalent to the adding to the last lens of a very thin and feeble lens of the same material and having the original radius of the TL solution on its contact

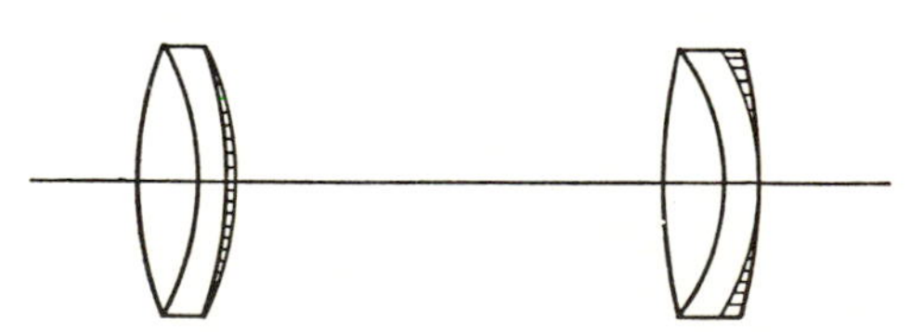

FIG. 43 (a).

face and the correct radius on its external face, as indicated with great exaggeration [45] by the shaded parts in the sketches Fig. 43 a. So frail a lens will satisfy the TL equations practically exactly : it will also have very little spherical aberration, and especially no appreciable variation of spherical aberration for two different colours. We may therefore apply TL equations to this added lens—nearly always a meniscus —for the L-values of rays at finite angles without sensible loss of accuracy.

The first equation in section [40] on paraxial achromatism will then be for this lens :

$$\frac{L'_r - L'_v}{L'_r \cdot L'_v} = (N_v - N_r) \cdot c + \frac{L_r - L_v}{L_r \cdot L_v}.$$

As the added lens is to be in contact with the last TL-surface of the system, its L will be identical with the L'_{TL} found trigonometrically in tracing through the last TL-surface, and as the added lens is a very feeble one but receives rays of comparatively strong convergence, the difference between its corresponding L and L' values will be so small that it will usually be safe to neglect it by putting—in the denominators only—the L' on the left equal to the L on the right and therefore both equal to the TL-results. Distinguishing all the latter results by the TL suffix and using plain r for the second radius of the added lens, the previous equation, transposed, becomes

$$(N_v - N_r)\left(\frac{1}{r_{TL}} - \frac{1}{r}\right) = \frac{(L'_r - L'_v) \text{ prescribed } - (L'_r - L'_v)_{TL}}{(L'_r \cdot L'_v)_{TL}}$$

and this can be immediately transposed into the required solution for the last radius

$$\text{Chr. (3)} \qquad \frac{1}{r} = \frac{1}{r_{TL}} + \frac{(L'_r - L'_v)_{TL} - (L'_r - L'_v) \text{ prescribed}}{(N_v - N_r)(L'_r \cdot L'_v)_{TL}}.$$

Excepting rare cases when the total curvature of the added lens represented by the long second term is relatively large, the radius of the last surface obtained from Chr. (3) will be found to give the prescribed chromatic correction. In the rare cases referred to we should treat the results obtained by trigonometrical calculation from the r of the first solution as TL results for a second approximation by Chr. (3), using the r of the first solution as the r_{TL} for the second solution.

As a numerical example we will take the second example of the section on paraxial achromatism, for a prescribed chromatic over-correction of $-0{\cdot}02$ inch. We found for the externally planoconvex form of a cemented lens having $f' = 8''$:

$$r_1 = 3{\cdot}215 \qquad d'_1 = 0{\cdot}25$$
$$r_2 = -2{\cdot}972 \qquad d'_2 = 0{\cdot}15$$
$$(r_3 = \infty)$$

Refractive indices as in example (1). The thickness of the crown component has been put at 0·25 to allow of 1·6 inches clear aperture with the deeper curves of this over-corrected lens.

[45] A complete trigonometrical tracing of C and F rays with initial $Y = 0{\cdot}7071 \times 0{\cdot}8 = 0{\cdot}5657''$ will be found to give the final intersection-lengths

$$L'_c = 7{\cdot}6934\ L'_f = 7{\cdot}7151\ ; \quad \text{therefore, in the symbols of Chr. (3) :}$$

$$(L'_r - L'_v)_{TL} = -0{\cdot}0217 \text{ instead of the prescribed } -0{\cdot}0200.$$

The difference 0·0017 is too large to be attributed to inaccuracy of the five-figure calculation and also too large to be left in a respectable design. Therefore we use Chr. (3) to obtain a closer approximation. As $r_3 = r_{TL}$ of the equation $= \infty$ the equation gives

$$\frac{1}{r} = 0 + \frac{-0{\cdot}0217-(-0{\cdot}02)}{0{\cdot}01729 \times 7{\cdot}69 \times 7{\cdot}71} = \frac{-0{\cdot}0017}{0{\cdot}01729 \times 7{\cdot}69 \times 7{\cdot}71}.$$

The rounded-off values in the denominator are amply justified because the numerator is, at best, dependable to one part in 17! Slide-rule or rough four-figure logarithmic calculation gives $1/r = -0{\cdot}00166$ or, nearly enough, $r = -600''$. Tracing the trigonometrical rays from the second surface through the third surface with this radius, the result, by the check-formula, will be

$$L'_c = 7{\cdot}6320,\ L'_f = 7{\cdot}6519 \text{ or } L'_c - L'_f = -0{\cdot}0199\ ;$$

closer to the prescribed value than we could usually expect with five-figure logarithms.

When the desired state of chromatic correction has been established by one of the three methods, the spherical aberration has next to be determined, and the system will have to be modified by 'bending' so that the desired state of spherical correction is also secured. This part of the art of designing simple lens systems will be the subject of the next chapter.

CHROMATIC TOLERANCES

[46] The tolerances based upon the Rayleigh limit which were given in Chapter III, find very important applications in judging the seriousness of the chromatic aberration of lens systems.

We have learnt that the chromatic aberration of the focal point consists of a difference of focus for different colours which is nearly constant for the whole aperture. If follows that if, in the presence of chromatic aberration, we focus the image-point in one particular colour, all or most other colours will be out of focus, and would from the geometrical point of view cause the formation of a circle of confusion. It was shown in Chapter III that physical optics leads to a different conclusion ; whatever the angle U'_m of a pencil of rays may be, there is a certain 'focal range' along the optical axis within which the image remains practically equally good. Obviously comparison with this focal range is the proper method of gauging the seriousness of any given residue of chromatic aberration, for as long as the two extreme colours which are considered as important contributors to the image form their geometrical foci within the focal range distance of each other, the sharpness and brightness of the image at a point midway between those foci will not be seriously diminished.

The last sentence implies a very considerable uncertainty in estimates of chromatic residuals. These estimates depend on a preliminary decision as to the two [46] points towards the red and the violet end of the spectrum at which the luminosity of the latter is considered still sufficiently great to be capable of making a useful contribution to the defined image. As the fall in intensity is usually a perfectly gradual one to both sides of the brightest region, our decision in this respect must necessarily be a more or less arbitrary one and must ultimately be checked by direct visual or photographic tests. We will limit our present estimates to visual instruments, and will assume that the C and F rays mark the limits beyond which the brightness of the spectrum is too low to make much difference in the quality of the image.

The equation for the focal range given in Chapter III was

OT (1) $$\text{Focal range} = 1 \text{ wave-length}/N' \sin^2 U'_m,$$

N' being the index of the medium in which the image is formed and therefore nearly always $= 1$ for air, and U'_m being the angle under which the *extreme marginal ray* meets the axial ray at the image-point.

We will apply it first to the case which arises occasionally, that for some good and sufficient reason we want to leave a system chromatically under- or over-corrected to the greatest extent compatible with good definition. As the focal range defines the distance between the nearest and the farthest focal adjustment at which the image remains good, it follows that the focal ranges of our selected colours C and F will just meet at the midway focus if the distance between the C and F foci is equal to the focal range. Hence

$$\text{Permissible longitudinal chromatic aberration} = 1 \text{ wave-length}/N' \sin^2 U'_m.$$

For the 2″ objective with 8″ focal length we calculated the right of this equation in section [38] as = 0·00124 inch. Hence 1/800th of an inch would be the maximum chromatic aberration which might be left uncorrected in an objective of these proportions. The equation shows, however, that the tolerance is inversely proportional to the square of sin U'_m, and as for a given aperture sin U'_m is nearly enough inversely proportional to the focal length, it follows that the tolerance for a given aperture is sensibly proportional to the square of the focal length. An astronomical 2″ object-glass rarely has less than 24″ focal length, or three times that of our very bold numerical example. By the law of proportionality to the square of the focal length a 2″ object-glass of 24″ focal length would therefore have a chromatic tolerance of 9/800th, or say 1/90th of an inch.

As was pointed out at the end of Chapter III, one quarter of any of the standard tolerances leads to an absolutely insensible deterioration of the image, hence we may make the most important deduction from the above that even for object-glasses having a focal length of only four times the clear aperture 0·0003 inch or 0·008 mm. of residual chromatic aberration may be ignored as utterly unimportant. For the usual sizes of object-glasses of such bold proportions this just about represents the uncertainty of a careful calculation by five-figure logs, and the use of the latter is thus proved as sufficient even for the calculation of such object-glasses.

A most important application of the chromatic tolerance is that to the fixing of the minimum focal length which may be given to a lens or lens system of given

[46] clear aperture. In order to find approximately the true limit, and one which agrees fairly well with ordinary observational experience, we will draw upon the statement at the end of Chapter III, that in accordance with numerous direct integrations the image remains well defined up to the doubled standard tolerances, although it is then surrounded by a considerable halo of scattered light. In fixing our limits we will therefore allow the doubled chromatic tolerance for the C and F rays.

(A) Minimum Focal Length for Simple Lenses

Simple lenses suffer from the primary chromatic aberration determined in section [39] as equal to

$$f'_c - f'_f = \frac{f'}{V}.$$

The doubled focal range is given by

$$\text{Doubled focal range} = 2 \text{ wave-lengths}/N' \sin^2 U'_m.$$

As the simple lenses can only be used with cones of rays of small convergence because all the aberrations would become too large for wide-angled cones, we may put

$$\sin U'_m = Y/f' = \tfrac{1}{2} A/f',$$

if A is introduced as a convenient symbol for the clear aperture of the lens. Introducing this value of $\sin U'_m$ and noting that if our lens is to give a satisfactory image its chromatic aberration must not exceed the doubled focal range, f'/V takes the place of the latter in the second equation and leads to

$$\frac{f'}{V} = 2 \text{ wave-lengths}\Big/\left(N' \,.\, \tfrac{1}{4}\left(\frac{A}{f'}\right)^2\right).$$

Putting $N' = 1$ for air and simplifying, we next find

$$\frac{f'}{V} = 8 \text{ wave-lengths} \left(\frac{f'}{A}\right)^2$$

which by transposition gives the minimum focal length as

$$\text{Minimum } f' = A^2 \frac{1}{8 \text{ wave-lengths} \times V}.$$

It will be amply near enough to put the wave-length = 0·00002″ and V for light crown glass = 62·5 ; our last equation for f' and A in inches then becomes

$$\text{Minimum } f' = 100 \,.\, A^2 \text{ (for inch-measure and light crown glass).}$$

This formula gives the full explanation of the fantastic length of the astronomical telescopes of pre-achromatic days. A telescope of only 1″ aperture—just about enough to show distinctly the phases of Venus, the belts and moons of Jupiter, and the ring of Saturn—requires to be 8 feet long to reduce the chromatic aberration to comparatively harmless proportions ; a 2″ objective ought to have 33 feet focal length ; a 3″ 75 feet ; and a 4″ 133 feet focal length. There is a seventeenth-century record of the making of a simple lens object-glass of 300 feet focal length which, by our formula, would have admitted of an aperture of 6 inches, but it does not appear to have ever been erected and used. [46]

The formula is still useful to estimate the minimum focal length compatible with really good definition in simple magnifying glasses. The effective aperture of these is restricted to that of the observer's pupil, and may be taken as 1/6 inch in moderate and 1/10 inch in very bright light, corresponding to minimum focal lengths by our formula of 2·8″ and 1″ respectively. One inch may therefore be taken as the least focal length at which a simple magnifier can be expected to give good images: this agrees remarkably well with practical experience. Ordinary Huygenian and Ramsden eye-pieces also come more or less under this rule, but their clear working aperture is usually restricted to less than that of the observer's pupil by the limiting effect of the object-glass to which they are applied. Eye-pieces are therefore best discussed individually by the general focal range equation.

B. Minimum Focal Length for ordinary Achromatic Lenses

The most serious outstanding chromatic aberration of all ordinary achromatic lenses is represented by the secondary spectrum. We have shown that the magnitude of the latter is nearly constant for a given focal length, and causes the combined C and F foci of a visually corrected system to fall about 1/2000th part of the focal length beyond the focus of the brightest yellow green light. $f'/2000$ therefore takes the place of f'/V in the preceding case of a simple lens. We must, however, look carefully into the question of the allowable focal range. For the feeble light of the C and F rays we can still allow the doubled focal range, and the selected compromise focus may therefore be moved to the extent of the single focal range away from the *combined* C and F foci. But the other extremity of the focal difference to be covered by the tolerance is now represented by the brightest light of the whole spectrum. Evidently we must not allow this to be substantially weakened by differences of optical paths: we will therefore only draw upon half the normal focal range for the brightest light, which means that the compromise focus may lie at $\frac{1}{4}$ the normal focal range from the focus of the brightest light.

We thus arrive at the conclusion that the whole distance from the combined C and F foci to the brightest focus may be $1\frac{1}{4}$ times the normal focal range, and by the process employed for the simple lens obtain the equation

$$\frac{f'}{2000} = 1\tfrac{1}{4}\ \text{wave-lengths}/N' \sin^2 U'_m,$$

and on introducing, as before, $N' = 1$ for air and $\sin U'_m = \frac{1}{2} A/f'$, we obtain next

$$\frac{f'}{2000} = 5\ \text{wave-lengths}\left(\frac{f'}{A}\right)^2$$

or

$$f' = \frac{A^2}{10{,}000\ \text{wave-lengths}}.$$

For f' and A in inches we have to put the wave-length as, nearly enough, 0·00002 inch and obtain the final equation

$$\text{Minimum } f' = 5\,A^2 \text{ (for inch-measure).}$$

[46] Comparing this with the corresponding equation for the minimum focal length of a simple crown glass lens, we see exactly what Dolland accomplished by his discovery of achromatism. He reduced the necessary focal length to 1/20th of that required for a simple lens. A 1-inch achromatic lens may be made of only 5 inches focal length, a 2″ may have 20″ focal length, a 3″ calls for 45″, a 4″ for 80″, and so on. We see also that the bold 2″ object-glass of section [21] has a focal length far too short to keep the secondary spectrum within proper bounds : it ought to have at least 20″ focal length instead of the 8″ which we gave it. With reference to astronomical object-glasses we see that we reach the usual ratio of A to f' as 1 : 15 with a 3″ aperture. Hence all astronomical object-glasses of this focal ratio transgress our tolerance when they have more than a modest 3 inches of clear aperture. The Lick telescope of 36″ aperture ought to have 5×36^2 inches, or 540 feet focal length ; its real focal length is about 54 feet, or one-tenth of that which would make the secondary spectrum comparatively harmless. From the fact which we proved fairly conclusively in the discussion of the nature of the secondary spectrum, that the latter plots as a parabolic curve and therefore grows very nearly as the square of the difference of refractive index between the colours brought to a common focus, we can easily deduce that in the Lick telescope only a range of the middle of the spectrum equal to about one-third of the C to F range contributes to the sharply defined image. All the rest of the light is scattered into the surrounding halo. The magnificent work accomplished with the Lick telescope, in spite of the extreme seriousness of its secondary spectrum, teaches us, however, that whilst we should always try to keep the secondary spectrum within the bounds assigned by the Rayleigh limit, we need not despair when we have to transgress that limit. There will be severe drawbacks, but they will not by any means nullify the increased resolving power and light gathering power secured by the increased aperture of the resulting lens system.

The tolerances for the spherical variation of the chromatic aberration are chiefly of importance in the higher powers of microscope-objectives. These will be dealt with in Part II.

Special Memoranda from Chapter IV

THEORETICAL. As refracting media have different indices for different colours, lens systems usually have differences of focus for different colours or suffer from longitudinal chromatic aberration, which obeys the law $L'_r - L'_v = b_0 + b_2 Y^2 + \&c.$, the constant first term being called 'primary chromatic aberration', the second term 'spherical variation of the chromatic aberration'. The higher terms are usually unimportant. This aberration can be corrected for one zone of the aperture by combining glasses of different V-value, the zone at $\sqrt{\frac{1}{2}}$ of the full aperture being the most favourable. With all ordinary glass-combinations only two colours can be brought to a common focus, because the denser glasses disperse the blue end of the spectrum relatively too much. The 'secondary spectrum' which results is characterized by a minimum focal distance for one particular colour or wavelength and by union in pairs of other colours at increasing distances. The choice of the minimum focus colour distinguishes the different types of achromatism. [46] For visual purposes yellow-green is chosen ; C and F are then focused (for distant objects) at 1/2000th of the focal length beyond the minimum, G' and A' at about 1/500th of the focal length beyond the minimum. The secondary spectrum can be corrected very perfectly by using fluorite, it can be reduced to about 1/3 of the normal amount by special glass-pairs with a small V-difference, and it can also be corrected by combining three carefully chosen kinds of glass on the 'photo-visual' principle. The two latter methods are only suitable for telescope objectives, but fluorite renders 'apochromatic' microscope objectives possible.

PRACTICAL. The numerous computing formulae in this chapter are extremely useful and should be thoroughly mastered. The thin lens solutions TL, Chr. (4) and (4)*, and the trigonometrical correcting formulae Chr. (1) and (3) are the most important ones. Section [46] on tolerances is worthy of most careful attention.

CHAPTER V

DESIGN OF ACHROMATIC OBJECT-GLASSES

[47] THE general ray-tracing methods of Chapter I, together with the theory of the spherical and chromatic aberration at the axial image-point as given in Chapters II, III, and IV, enable us to solve in a simple and systematic manner a problem in optical designing which is of extremely frequent occurrence, namely, to find that form of a proposed lens system which has a desired focal length, a prescribed amount of chromatic aberration, and also a prescribed spherical correction. This does not mean that we are fully equipped to tackle any optical designing problem, for in many cases the three stated conditions are not even the most important ones; in photographic objectives and in eye-pieces the aberrations of oblique pencils are by far the most serious and must therefore take first place. But in ordinary telescopes and even in the lowest powers for microscopes the field of view to be covered is so small that there is not much scope for oblique aberrations, and the methods now to be given are then usually sufficient to attain a perfectly satisfactory state of correction. We shall therefore limit the examples given in the present chapter to simple systems of these types, although the method itself covers a much wider field.

The proper sequence of operations is determined by our theoretical studies. We learnt that for a prescribed focal length the chromatic condition prescribes definite net curvatures for the components, and that bending of these components does not alter the chromatic correction at all if the lenses are treated as infinitely thin, and alters it only to a very moderate extent in the case of any ordinary thickness. Therefore we must begin by establishing the required state of achromatism, which gives us a definite strength of each component lens, and we must follow this up by trying bendings until the desired spherical correction is also reached. Obviously a reversal of this sequence would merely waste time. The finding of the correct bending is greatly simplified by the first approximation law proved in section [30], according to which a simple vertical parabola results when the primary spherical aberration of a thin lens submitted to the bending process is plotted as ordinate against the curvature of a selected surface of the lens as abscissa. We shall frequently employ the graphical method in our solutions as it is both the simplest and the most instructive. Three known points completely determine a parabola with vertical axis; therefore we must calculate the spherical aberration for three bendings in order to be able to draw the graph. When the thickness of a lens is considerable, or when considerable amounts of higher aberration have to be reckoned with, the true graph may depart sensibly from a simple parabolic form; we saw that a cubic term is likely to be the chief disturber, and this can be included by basing the graph on four bendings. Mere common sense would suggest the selection of bendings at equal intervals of the abscissae; but there is a further good reason for this choice, because the geometrically exact drawing of the curve can then be put into the following extremely simple and convenient form:

The equation of a parabola including a cubic term is, in the usual rectangular coordinates,

$$y = a + bx + cx^2 + dx^3$$

and the four constants a, b, c, and d completely define the curve. If there is no cubic term ($d = 0$) then only three constants enter into the problem. With reference to Fig. 44 our method of drawing the curves depends upon a remarkable property of the distance PQ from an intermediate point P of the parabola to the [47]

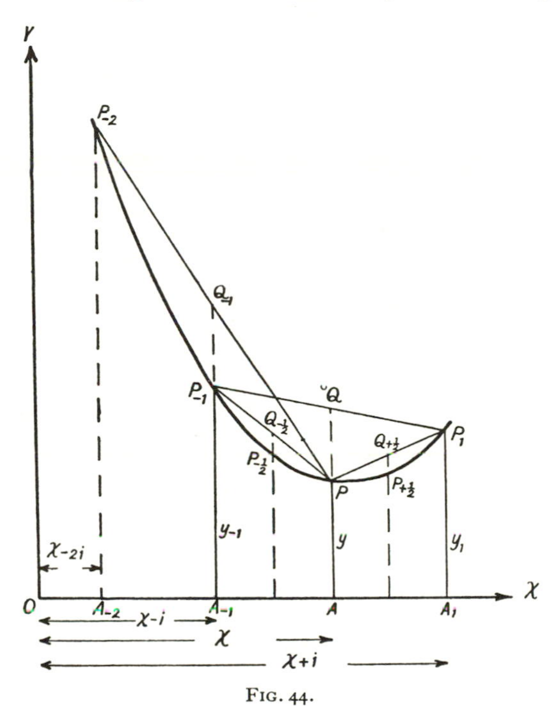

FIG. 44.

chord $P_{-1} P_1$ joining two neighbouring points of the curve. If we call the abscissa of P 'x' and the uniform interval of successive ordinates 'i', then P_{-1} will have the abscissa $(x-i)$ and P_1 will have $(x+i)$ and the ordinates will be

$$y_1 = a+b(x+i)+c(x+i)^2+d(x+i)^3$$
$$y = a+bx \qquad +cx^2 \qquad +dx^3$$
$$y_{-1} = a+b(x-i)+c(x-i)^2+d(x-i)^3.$$

Now we have by inspection $PQ = AQ - AP = \frac{1}{2}(y_1+y_{-1})-y$, and if we work this out by the preceding three equations we find

$$PQ = \tfrac{1}{2}(y_1+y_{-1})-y = 0+0 \qquad +c \,.\, i^2 \quad +3\,d \,.\, x \,.\, i^2 = i^2(c+3\,d \,.\, x).$$

We shall call PQ the 'dip' of the curve and the last expression for it leads to

[47] a simple and geometrically exact construction for any desired number of points of the curve when the necessary minimum number of points is given :

(1) *Case of the common parabola* $y = a+bx+cx^2$. As $d = 0$ we have

$$\text{Dip} = c \cdot i^2.$$

As this is independent of x, it follows that for *any* three ordinates at the original interval i of the abscissae the dip is absolutely constant. This enables us to continue the parabola in either direction, or to extrapolate from the three given points. Let Fig. 44 apply to a case when we know, or feel justified in assuming, that the curve of which P_{-1}, P, and P_1 are given points at equal intervals of the abscissae is a common parabola with vertical axis. By joining $P_{-1}P_1$ and producing (if necessary) the middle ordinate we find the dip PQ. To find the parabola point P_{-2} on the next ordinate to the left, or for abscissa $(x-2i)$, we mark on the ordinate of P_{-1} the distance $P_{-1}Q_{-1} = PQ$ and in the same direction. The straight line PQ_{-1} when produced to intersection with the ordinate at $(x-2i)$ defines P_{-2}, for P_1P_{-1} and P_{-2} will have the fixed value of the dip which is characteristic of a common parabola.

The process obviously admits of indefinite continuation, but the error due to inaccuracy of the drawing will rapidly grow even when it is known that the true curve is an exact parabola. When a parabolic law is only a first approximation to the truth or is merely assumed, then extrapolation is always a highly risky kind of prophecy, and may prove to be disastrously wrong. We should therefore endeavour to choose the three directly determined points of the curve so as to *include* the interesting region in order to reduce the problem to the much safer one of interpolation. The equation $\text{Dip} = c \cdot i^2$ then tells us that for any other interval i of the abscissae the dip changes as the square of the interval. Hence the dip of the curve at any ordinate midway between two of the original ordinates will be $\frac{1}{4}PQ$ with reference to the neighbouring original points. All that is required in order to locate these intermediate points is therefore to draw the chords $P_{-1}P$ and PP_1 and the midway ordinates and to apply one quarter of the original dip from the points $Q_{-\frac{1}{2}}$ and $Q_{+\frac{1}{2}}$ in the same sense as at the original point Q. Points $P_{-\frac{1}{2}}$ and $P_{+\frac{1}{2}}$ of the parabola are thus located with geometrical accuracy. If still closer intervals between the known parabola points appear desirable we have only to repeat this bisecting process, but as the intervals will now be one quarter of the original one, the dip to be applied will be one sixteenth of PQ.

(2) *Case of the cubic parabola* $y = a+bx+cx^2+dx^3$. The equation for the dip is now

$$\text{Dip} = i^2(c+3\,d \cdot x)\,;$$

it is therefore still strictly proportional to the square of the interval, but it is a linear function of x and varies in magnitude for different values of x, according to a law which is graphically expressed by an inclined straight line.

With reference to Fig. 44 a we must now calculate four ordinates of points P_{-1}, P, P_1, and P_2, because there are four constants in the equation of the curve. The four known points then enable us to find two values of the dip, namely PQ at abscissa x and P_1Q_1 at abscissa $(x+i)$. As the dip follows a straight line law of change, we begin by drawing an auxiliary graph above or below the principal one, [47] with a horizontal zero axis for the dip. We then plot PQ on its proper ordinate over this axis at $P'Q'$ and similarly P_1Q_1 at $P_1'Q_1'$. Owing to the straight line

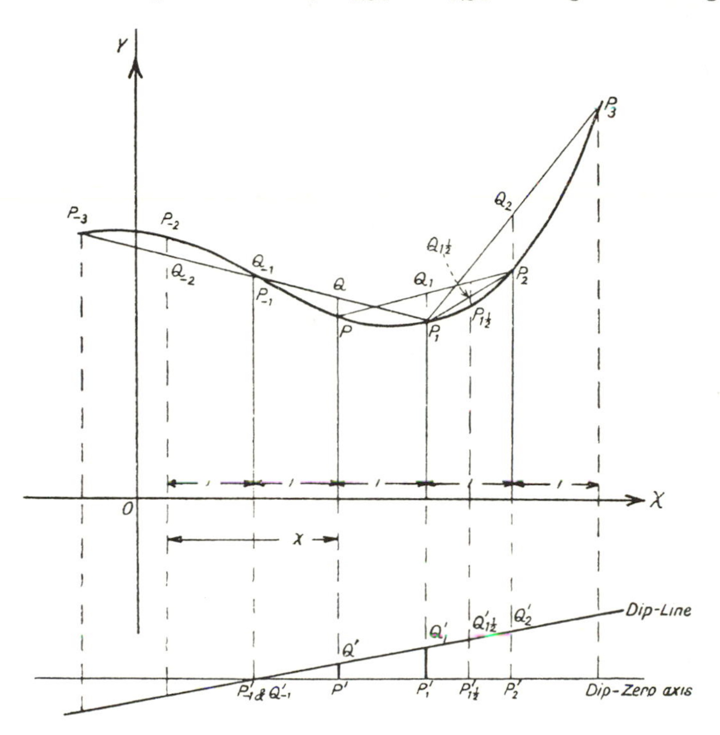

FIG. 44 (a).

law, the dip for any value of the abscissa can then be read off the inclined straight line drawn through points Q' and Q_1'. We can therefore continue the curve to the next ordinate to the right by transferring $P_2'Q_2'$ of the auxiliary graph to P_2Q_2, drawing the chord P_1Q_2 and producing this till it cuts the next ordinate at P_3, which must lie on the true curve because it has the dip following from the equation of the curve. At the left end of the directly determined range of values it happens that the dip line cuts its zero axis at the ordinate of P_{-1}; therefore P_{-2} falls on the line PP_{-1}. It will be obvious to any one familiar with analytical geometry that the point where the dip line cuts its zero axis indicates the abscissa of the 'point of inflexion' at which the curve changes from concave to convex curvature, and at which three consecutive points of it lie in a straight line. If ever an actual graph for an optical problem should include this point of inflexion—which is highly improbable—care must be taken in applying the dip in the correct direction.

[47] The extrapolation process with which we have again begun comes under the ban of being a venturesome experiment. For interpolation by continued bisection the process is exactly the same as for the common parabola, with the important exception that the 'original dip' must be taken from the auxiliary graph on the line of each bisecting ordinate on which an interpolated point is to be determined. Thus for the point $P_{1\frac{1}{2}}$ midway between the ordinates of P_1 and P_2 we must draw the chord P_1P_2, bisected at $Q_{1\frac{1}{2}}$; the auxiliary graph then gives $P'_{1\frac{1}{2}}\ Q'_{1\frac{1}{2}}$ as the value of the 'original dip' at that point and we apply one quarter of this at $Q_{1\frac{1}{2}}$ in the sense indicated by the dip line.

On squared paper the processes described will be found to give a beautifully smooth and true curve with surprisingly little trouble. The methods may also be used to great advantage in finding the real roots of quadratic and cubic equations, more especially the latter. The results obtained by this graphical process are in every case exactly the same as those which would result numerically by the ordinary process of interpolation by first and second differences when three values of the function at equal intervals of the argument are known, or by first, second, and third differences when four values of the function are known. What we have called the 'dip' of the curve is, in fact, easily seen to represent half the second difference which would be used in numerical interpolation. It is not difficult to deduce an exact method for a quartic or quintic; but when five or six values of the function are known, a curve of ample approximation to the truth will be obtained by merely plotting the three or four values of the 'dip', laying a common or cubic parabola through them by the methods already given and then using the dip read off this auxiliary graph for either extra- or interpolation exactly as described above.

In accordance with the principles which we have now established we must begin by determining the total curvature of the components of the proposed system which will give the prescribed focal length and chromatic correction. For the simple objectives contemplated in the present chapter the solutions by the extremely simple and convenient equations TL, Chr. (4) or (4)* are nearly always the best, because they give in the majority of cases a closer approximation to the correct proportion of crown to flint curvature than the theoretically exact paraxial solution by the true primary chromatic aberration according to equations Chr. (8) and (8)*. The reason is that the exact paraxial solution almost invariably yields a combination which proves chromatically over-corrected at full aperture, whilst the neglecting of the thickness in the TL solutions tends to compensate this tendency towards over-correction at full aperture to a useful extent. In the bolder and more complicated systems which we shall study at a later stage this curious superiority of the TL solutions breaks down and the stricter solution frequently will have to be substituted for it.

As a numerical example we will take a cemented telescope objective of 10″ focal length, to be made of the glasses of our list:

3463	$Nd = 1{\cdot}5407$	$Nf - Nc = 0{\cdot}00910$	$V = 59{\cdot}4$
360	$1{\cdot}6225$	$0{\cdot}01729$	$36{\cdot}0$

For perfect achromatism TL, Chr. (4)* gives [47]

$$\text{Crown}: c_a = \frac{1}{10} \cdot \frac{1}{23{\cdot}4} \cdot \frac{1}{0{\cdot}00910} = 0{\cdot}470\,;$$

$$\text{Flint}: \quad c_b = \frac{1}{10} \cdot \frac{1}{-23{\cdot}4} \cdot \frac{1}{0{\cdot}01729} = -0{\cdot}247.$$

The components thus determined have now to be bent into such a form as to yield the desired perfect achromatism and the prescribed spherical correction at full aperture. This can be done either entirely by trigonometrical calculation or we can make a rapid TL solution for spherical correction by TL (10)**, using the *G*-sums, and correct the rough approximation thus obtained trigonometrically. For the present simple problem there is not much to choose between the two methods ; but for the more complicated problems to be attacked later the second method will prove by far the best.

A. THE PURELY TRIGONOMETRICAL METHOD

[48] From experience it may be taken that for all suitable ordinary glasses the spherically corrected forms of a cemented telescope objective with crown lens on the side of the distant objects are about equidistant in curvature from that form in which the crown lens has 1/3 of its net curvature on the first surface and that a bending by about 1/5 of the net curvature in either direction will give suitable extreme forms. For convenience in the plotting it is desirable to have round numbers for the selected curvatures $c_1 = 1/r_1$ of the first surface ; as 1/3 of 0·470 is nearly 0·160, we choose this as the middle value of c_1, and as 1/5 of 0·470 is nearly 0·10 we choose $0{\cdot}160 \pm 0{\cdot}10$ for the extreme values. As by $c_a = c_1 - c_2$ we have $c_2 = c_1 - c_a$, we obtain the data for our three bendings :

$c_1 =$	0·060	0·160	0·260
$c_2 =$	−0·410	−0·310	−0·210

and taking reciprocals from Barlow's table we find :

$r_1 =$	16·67	6·25	3·846
$r_2 =$	−2·439	−3·226	−4·762

As we do not require a first approximation for r_3 in the methods Chr. (1) and Chr. (2) for perfect achromatism, we need not carry the TL preparation any further.

Equally good, and sometimes preferable achromatic object-glasses may be designed with the flint lens on the side of the distant objects, or in 'front', as it is usually called. That means, in the strict working of our TL, Chr. equations, that we call the flint lens (a) and the crown lens (b) : but it is easily seen that the working out will give the same values for flint and crown as before, therefore for flint in front, or the 'Steinheil' form of objective of the same focal length of 10″ :

$$c_a = -0{\cdot}247.$$

Objectives of this type tend to have a stronger curvature of the first surface than

[48] those with crown in front.† We will therefore choose for c_1 somewhat larger values, 0·10, 0·20, and 0·30, and with c_2 again $= c_1 - c_a = c_1 + 0{\cdot}247$ find the three prescriptions for the selected bendings :

$c_1 =$	0·10	0·20	0·30
$c_2 =$	0·347	0·447	0·547
or $r_1 =$	10·000	5·000	3·333
$r_2 =$	2·882	2·237	1·828

There is no need to worry unduly about the selection of the bendings ; we can easily add a fourth bending at the adopted interval if the graph from the originally selected three does not reach far enough at one end or the other.

The most accurate trigonometrical way of establishing perfect chromatic correction is always that prescribed in the preceding chapter, namely, to trace the two coloured rays which are to be brought to a common focus through the 0·7071 zone of the intended full aperture and then to calculate the spherical aberration by tracing the 'brightest light' through the paraxial and extreme marginal zones. That, however, requires the tracing of four rays through each selected bending. As the graphical process to be adopted is not absolutely accurate and as, moreover, the secondary aberrations which dictate the choice of rays are usually small in the types of lenses to which the method will generally be applied, it is permissible and saves twenty-five per cent. of computing time to adopt a compromise-method which calls for the tracing of only three rays, and which gives such good results that it may safely be adopted even for the *final* calculation of nearly all objectives for ordinary terrestrial telescopes, for surveying instruments and for small 'finders' of astronomical telescopes. It consists firstly in establishing spherical and chromatic correction for the same zone instead of doing this for the full and the 0·7071 aperture respectively ; by way of compromise we shall usually adopt 0·8 of the full aperture as the selected outer zone. Secondly, we avoid the difficulty caused by the secondary spectrum by using 'brightest light' for the paraxial and for one of the 0·8 aperture rays and by adopting as the index of refraction of the coloured ray, also traced through the 0·8 aperture, one which would be correct if the two glasses had proportional dispersion throughout the spectrum ; for visual instruments this is conveniently represented by $Ny + (Nf - Nc)$ and for ordinary photographic lenses by $Nf + (Ng' - Nd)$. We thus arrive at the rule for this method :

'*Three-ray method of trigonometrical correction* : For visual instruments calculate $Ny = Nd + 0{\cdot}188\,(Nf - Nc)$ and $Nv = Ny + (Nf - Nc)$. Trace a paraxial and a 0·8 aperture ray in Ny and a 0·8 aperture ray only in Nv. For photographic correction use Nf and $Nv = Nf + (Ng' - Nd)$ in the same way.

In either case the three rays should be brought to a common focus or to the required state of over- or under-correction, the chromatic aberration being obtained as the difference of the intersection-lengths of the two 0·8 rays and the spherical aberration as the difference of the intersection-lengths of the paraxial and 0·8 rays in brightest light.'

† The best value of the middle bending for flint-in-front telescope objectives of any usual types of optical glass is found by the formula $c_1 = c_a + c_b$. This for the above example gives $c_1 = -0{\cdot}247 + 0{\cdot}470 = +0{\cdot}223$, which for convenience in plotting would be rounded off to $c_1 = 0{\cdot}220$.

The formula for the refractive index of brightest light, $N_y = N_d + 0{\cdot}188(N_f - N_c)$, [48] depends on the accurate interpolation method already referred to in section [42]. N_y is the index for wave-length $0{\cdot}5555\mu$.

In the case of our numerical examples the application of this method calls for the determination of N_y and N_v:

	Crown.	*Flint.*
N_d from list	1·5407	1·6225
$+0{\cdot}188(N_f - N_c)$ by slide rule	0·00171	0·00325
N_y	1·54241	1·62575
$+(N_f - N_c)$	0·00910	0·01729
N_v	1·55151	1·64304
$\log N_y$	0·18820	0·21105
$\log N_v$	0·19075	0·21565

It then only remains to decide upon the full clear aperture of our objectives and upon a corresponding thickness of the lenses. As a fair average of modern practice for small telescopes we will fix the clear aperture at 1·6 inches, or the full *semi*-aperture at 0·8 inch. A suitable thickness for the concave flint lens will then be 0·2 inch, but for the crown component we will choose 0·25 inch in order to secure a reasonable thickness at the extreme edge. Lenses with a sharp edge are detested in the workshop!

As by the three-ray method we calculate the trigonometrical rays at 0·8 of the full aperture, the initial Y will be $0{\cdot}8 \times 0{\cdot}8 = 0{\cdot}64$ inch, and in accordance with our universal computing practice we use the same value $y = 0{\cdot}64$ as the 'nominal' or 'fictitious' value for the paraxial ray. The calculations for each bending can then be carried out to the pattern in section [21] in three parallel columns for 0·8 ray in N_y, paraxial ray in N_y and 0·8 ray in N_v. When the three rays have been taken through the first two surfaces of each bending we apply the thickness of the second lens in order to find the l and L for the last surface, and can then determine the last radius by whichever of the Chr.-solutions we like. Chr. (1) is probably the most suitable for strictly achromatic objectives.

The principal results for the three bendings of the crown-in-front objective are:

$c_1 =$	0·06	0·16	0·26
$L_{3y} =$	− 582·50	26·2380	12·5881
$l_{3y} =$	−1119·07	25·9383	12·6176
$L_{3v} =$	− 286·313	27·0263	12·6545
By solution $r_3 =$	− 5·904	−15·024	+25·10
$L'_{3y} =$	9·5414	9·6086	9·5908
$l'_{3y} =$	9·5664	9·5853	9·6232
$L'_{3v} =$	9·5407	9·6097	9·5908
By (first y)/final u' : $f'_y =$	9·6152	9·7974	10·0413
$\log \sin U'_{3y} =$	8·82280	8·81409	8·80896
$\log u'_{3y} =$	8·82322	8·81507	8·80600
Sph. Aberr. $(l'_{3y} - L'_{3y}) =$	+ 0·0250	− 0·0233	+ 0·0324

[48] As has already been pointed out, the chromatic solution does not fix r_3 within very close limits, and students repeating the calculations need not be perturbed if they find slightly different values; the test of the calculation is the agreement of L'_{3y} and L'_{3v} which should not be allowed to differ by more than about 0·001 inch even in these semi-final calculations. If the first solution leads to a greater disagreement (this happened for $c_1 = 0·06$ when Chr. (1) was used) the result should be corrected by using Chr. (3) as advised in the previous chapter. For our present purpose the important results are those in the last line. They show that the two extreme bendings have spherical under-correction while the middle one has over-correction, and they suggest the existence of two spherically corrected intermediate forms.

The three selected bendings of the flint-in-front or Steinheil type give the results:

$c_1 =$	0·10	0·20	0·30
$L_{3y} =$	45·5145	16·9849	10·2524
$l_{3y} =$	44·9193	16·9149	10·2718
$L_{3v} =$	47·4219	17·1261	10·2564
By Chr. (1) : $r_3 =$	−7·769	−41·88	10·98
$L'_{3y} =$	9·5523	9·6179	9·8966
$l'_{3y} =$	9·6010	9·6026	9·9246
$L'_{3v} =$	9·5515	9·6179	9·8966
By (first y)/final u' : $f' =$	9·7295	9·8969	10·4064
$\log \sin U'_{3y} =$	8·81829	8·80958	8·79118
$\log u'_{3y} =$	8·81809	8·81068	8·78888
Sph. Aberr. $(l'_{3y} - L'_{3y}) =$	+0·0487	−0·0153	+0·0280

In working by the specimen calculation, or by any others as yet mentioned in these pages, beginners should carefully note that the reversed sequence of the lenses introduces unfamiliar modifications; the second radius is always positive; the flint-indices must be used at the first and the crown-indices at the last surface, and at the contact surfaces the refraction is from a denser into a lighter medium, and I' will therefore be larger than I and both nearly always positive. Moreover, in the present case the first, or flint lens, will have 0·2 thickness and the crown lens 0·25 thickness.

The last line shows the same kind of variation in the spherical aberration as that found for the crown-in-front type: again, we may expect two intermediate forms for which the spherical aberration will be zero.

To find the four spherically corrected cemented lenses which can be made from our two selected glasses we plot the spherical aberration of the three bendings of each type as ordinates against the corresponding values of the curvature of the first surface c_1 as abscissae. In accordance with the discussion in section [30] the graph will closely resemble a common parabola with vertical axis, and can therefore be quickly drawn by the method deduced in the preceding section. As the theory given in section [30] shows that the true graph only *approximates* to a common parabola, it should be realized that some importance is attached to the selection of [48] the bendings; they should not cover too long a range, as intermediate points might be located with a sensible error if the interval of the known ordinates were unduly large.

Solutions of Numerical Examples

The results arrived at were:

For the crown-in-front type:

If $c_1 =$	0·06	0·16	0·26
then $l_3' - L_3' =$	+0·0250	−0·0233	+0·0324

For the flint-in-front type:

If $c_1 =$	0·10	0·20	0·30
then $l_3' - L_3' =$	+0·0487	−0·0153	+0·0280

The plotting paper should be decimally divided and lines at 0·1 inch or else at 2 mm. intervals are the most convenient; we will assume that the former is used. It is a mistake to draw graphs on a huge scale. Owing to the inaccuracy of the assumption that our graph will be an exact ordinary parabola, coupled with the uncertainty of five-figure work, we can hardly hope to secure solutions which, when tested trigonometrically, will have less than 0·001 inch of spherical aberration; hence it will be sufficient if we adopt a scale in which 0·001 inch of actual aberration will be clearly discernible. That may be taken for granted if we plot the aberrations at twenty times the calculated figures or 0·050 of actual aberration = 1·00 inch (or double cm.) on the diagrams. We shall secure a nice shape of the parabolas (neither too flat nor too much like a hairpin) if we lay down the curvatures c_1 at ten times the tabulated values, that is, at intervals of one inch for 0·1 of c_1.

When both types have been submitted to calculation, they should be laid down side by side. The complete graph will then take the form of Fig. 45.

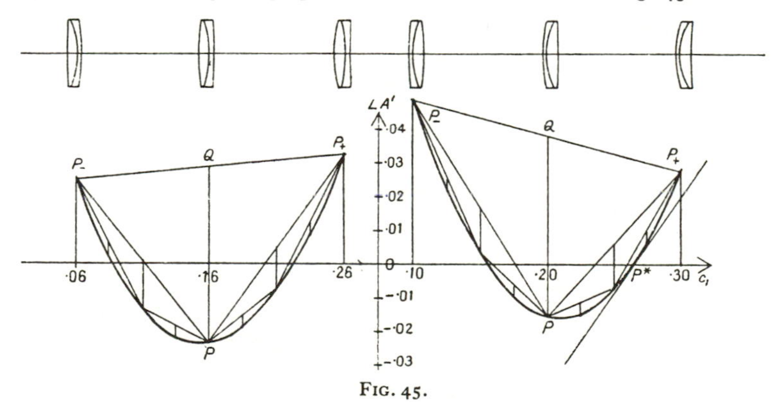

Fig. 45.

[48] We lay down the three computed points of each set. We then lay a ruler across the first and last, mark the point Q where the middle ordinate is cut and read off the paper-divisions the dip PQ. We then similarly mark the bisecting points of the chords $P_{-1}P$ and PP_1 and apply one quarter of the dip downwards, thus fixing two intermediate parabola points. We again draw short chords between the five points now known and apply 1/16th of the original dip at their bisecting points. The nine points fixed will be so close together that we can draw the curves without difficulty. It is highly advisable to add a reasonably accurate drawing of the calculated lens forms over the original points of the graph so as to visualize clearly the gradual change of lens form throughout the extent of the graph. On a carefully drawn graph it will be found that the left parabola—for crown-in-front—cuts the horizontal axis and indicates zero spherical aberration for $c_1 = 0{\cdot}0895$ and for $c_1 = 0{\cdot}224$. With $c_a = 0{\cdot}470$ these give the values of $c_2 = c_1 - c_a$ as $c_2 = -0{\cdot}3805$ and $c_2 = -0{\cdot}246$ respectively. Taking reciprocals, we obtain the two prescriptions:

$r_1 =$	11·17		$r_1 =$	4·464	
		0·25			0·25
$r_2 =$	−2·628		$r_2 =$	−4·065	
		0·20			0·20
$r_3 =$	?		$r_3 =$	?	

The third radius must be determined so as to give perfect chromatic correction.

The right-hand parabola for the flint-in-front type gives the two readings for zero spherical aberration $c_1 = 0{\cdot}154$ and $0{\cdot}265$, and from these by $c_2 = c_1 + 0{\cdot}247$: $c_2 = 0{\cdot}401$ and $0{\cdot}512$. Taking reciprocals, we obtain the two spherically corrected Steinheil-objectives:

$r_1 =$	6·494		$r_1 =$	3·774	
		0·20			0·20
$r_2 =$	2·494		$r_2 =$	1·953	
		0·25			0·25
$r_3 =$	?		$r_3 =$	?	

The last radius must be determined by the achromatic condition. In many cases of ordinary routine work the solving of the Chr. equation for the determination of the last radius may be avoided by adding to the parabolic graph of the spherical aberration a nearly straight line determined by laying down the $c_3 = 1/r_3$ found for the three bendings and reading off the curvature of the third surface of the selected solution from this added line of the graph. The same scale of one inch equal to 0·1 of curvature should be used; the graph will then give a line at nearly 45° with the horizontal axis; the slight departure from straightness must be allowed for on the dip-principle if the readings from this auxiliary graph are to be reliable.

If we required objectives with a prescribed amount of spherical aberration we could of course pick these out—if they are possible—at the points where the parabola reaches the proper value of its ordinate. The correct graph will show at once that for our selected glasses, focal length and aperture the highest attainable spherical over-correction at 0·8 of the intended aperture would be that at the pole

of the parabolas or $-0{\cdot}0235''$ for the first and $-0{\cdot}0165$ for the Steinheil-type. [48]
There never is any limit to the amount of under-correction as the parabolas extend upwards to infinity.

Final Correction of the Solutions

As has already been stated, we must be prepared to find the prescriptions read off the graphs slightly defective in not realizing exactly the desired state of spherical correction. As a rule the residual will amount to only a moderate fraction of the spherical tolerance, and it may then be ignored, especially if only a few lenses are to be made from the design. For mass-production it is always wise to go to the small trouble of determining the best possible solution. A small additional bending of the objective will secure this result, and we can determine the proper amount with ample accuracy from the graph by the slope of the parabola at the point from which the solution was taken. For inasmuch as the graph represents LA' as a continuous function of c_1, the slope of the curve, indicated by the direction of the tangent at the point from which a solution has been obtained, represents the rate of change of LA' or the differential coefficient dLA'/dc_1. By Taylor's theorem we then have in first and sufficient approximation the change in LA' resulting from a change of c_1 by δc_1

$$\delta LA' = \delta c_1 \cdot \frac{dLA'}{dc_1}.$$

If therefore the trigonometrical test of the direct graph solution has disclosed a remnant of spherical aberration which we will call LA'_{graph}, whilst the desired objective is required to have an aberration $LA'_{required}$, the necessary bending is determined by the equation

$$LA'_{required} = LA'_{graph} + \delta c_1 \frac{dLA'}{dc_1}$$

and this gives by transposition

$$\delta c_1 = (LA'_{required} - LA'_{graph}) \Big/ \frac{dLA'}{dc_1}.$$

This is a first example of a general method for the correction of approximate solutions which will be very extensively employed in subsequent work. As a numerical example we will take the fourth of the solutions obtained from Fig. 45 at the point P^*, namely a flint-in-front or 'Steinheil' objective having $r_1 = 3{\cdot}774$ and $r_2 = 1{\cdot}953$. From the correctly drawn original graph having the added line for the determination of c_3 the curvature of the last surface was read as $0{\cdot}050$, giving $r_3 = 20{\cdot}00$. In the case of the trigonometrically determined graph it pays to add the c_3 line as the value obtained will usually prove right within the admissible limits. In the rare cases when the residual chromatic aberration is found to exceed the probable precision of a five-figure calculation, r_3 must be corrected by Chr. (3). A calculation through the complete objective by the three-ray method, with angles taken out to ½ second, gave

$$L'_{3y} = 9{\cdot}7941\,; \qquad l'_{3y} = 9{\cdot}7925\,; \qquad L'_{3v} = 9{\cdot}7946\,;$$

[48] or a chromatic over-correction $L'_{3y} - L'_{3v} = -0{\cdot}0005$ which is within the reliability of a five-figure calculation and may therefore be ignored. For the spherical aberration we find, however, an over-correction $l'_{3y} - L'_{3y} = -0{\cdot}0016$ which, though far within the tolerance, is too large to be attributed to uncertainty of the calculation and which therefore ought to be removed. Drawing the tangent to the parabola at P^* we find that it rises 1·24 inches between two successive vertical lines at 1 inch interval. As the scale of the ordinates of the graph is 1″ = 0·05″ of longitudinal spherical aberration and the scale of the abscissae 1″ = 0·1 of c_1, we have the rate of change

$$\frac{dLA'}{dc_1} = \frac{1{\cdot}24 \times 0{\cdot}050}{1{\cdot}00 \times 0{\cdot}100} = \frac{0{\cdot}062}{0{\cdot}100} = 0{\cdot}62.$$

As we require complete correction of the spherical aberration, or $LA'_{required} = 0$, the formula for δc_1 gives

$$\delta c_1 = (0 - (-0{\cdot}0016))/0{\cdot}62 = 0{\cdot}0016/0{\cdot}62 = 0{\cdot}0026.$$

We apply this bending to all three surfaces of the objective :

Curvatures from graph	$c_1 =$	0·265	$c_2 =$	0·512	$c_3 =$	0·050
$+\delta c_1$		+0·0026		+0·0026		+0·0026
Corrected curvatures	$c_1 =$	0·2676	$c_2 =$	0·5146	$c_3 =$	0·0526
Or by Barlow's table	$r_1 =$	3·737	$r_2 =$	1·943	$r_3 =$	19·01

A trigonometrical test of this corrected prescription by the three-ray method gives the results :

$$L'_{3y} = 9{\cdot}7783\,; \qquad l'_{3y} = 9{\cdot}7781\,; \qquad L'_{3v} = 9{\cdot}7783\,;$$

or Chromatic Aberration 'Nil'

Spherical Aberration $= -0{\cdot}0002$.

The almost absolute removal of the residual aberrations of the graph solution is of course partly to be attributed to a friendly conspiracy of the rounding-off errors in the final calculation, more especially with regard to the chromatic residual, for we have done nothing to bring about the zero result in this respect. But the effectiveness of the differential correcting method is clearly proved in the case of the spherical aberration and it will be realized that the trigonometrical method of finding solutions is absolutely systematic, and bound to give the right result even to the inexperienced beginner. The worst that could happen to the latter would be a bad choice of the original bendings ; but a plotting of these would immediately show in which direction the curve required continuing by an added bending.

If the prescribed focal length is to be exactly realized, we must make a final adjustment based on the concluding paragraph of Chapter I ; we must alter all the radii in the proportion of focal length required to actually found focal length of the direct solution. Our Steinheil objective in its final form has an equivalent focal length of 10·196″ instead of the prescribed 10·000″. To realize the latter, we merely multiply all the radii by 10·000/10·196. Strictly, the thicknesses of the components should be similarly reduced, but this will rarely be necessary because ordinary telescope objectives are not highly sensitive in this respect. The correction may be applied to the direct graph solution by plotting yet another line on the graph Fig. 45, namely, a parabola laid through points corresponding to the focal lengths of the original bendings. In our case this line reads directly 10·20″ as the focal length of the graph solution and the corresponding correction can be applied before beginning the trigonometrical test ; this is probably the best method. [48]

If the best possible correction of an objective of deep curvatures or of unusually large diameter is to be secured, then the *solution* derived from the graph should be tested and adjusted, not by the convenient and time-saving three-ray method, but by the strict four-ray method, by establishing chromatic correction for the 0·7071 zone and then determining the spherical correction by a paraxial and extreme marginal ray in brightest light. In determining the final slight bending to secure exact spherical correction we must then allow for the difference in aperture at which the spherical correction has been determined—namely, at full aperture for the final solution and at 0·8 aperture for the three bendings of the graph. As primary spherical aberration grows with the square of the aperture, we must therefore multiply the spherical residual found in the first test of the selected solution by $0{\cdot}8^2 = 0{\cdot}64$ before we return to the graph for the determination of the necessary alteration δc_1 of the curvatures.

Selection of the Best Solution

We have learnt that with two given kinds of glass there are normally four solutions for a cemented objective which either completely correct the spherical and chromatic aberration or give a prescribed moderate amount of both. The question therefore arises which we should select.

When the highest quality of the product is the primary consideration, our choice must be determined by the optical merit of the four solutions with reference to defects not included in the present solution. We shall soon learn that the solutions differ characteristically with reference to the quality of the images of extra-axial object-points which we shall learn to judge by the optical sine condition. Again, the solutions differ very sensibly with regard to the magnitude of the secondary spherical and chromatic aberrations, and these afford another criterion which will usually decide in favour of the Steinheil-type. These cases we will leave for subsequent occasions.

For ordinary commercial work the primary considerations are ease and cheapness of manufacture and high immunity of the lenses from damage by exposure to air and dampness or by injudicious cleaning As a rule the choice then falls upon the second (right hand in the graph) form of the crown-in-front type which with telescope objectives of any usual glasses has a nearly equiconvex crown lens and a nearly plano-concave flint lens, both of which are easily and cheaply produced. Moreover, a crown face is exposed to the outside, and as ordinary crown is harder and less tarnishable than flint, this type has a decided advantage in this respect also ; it is therefore most frequently met with in ordinary instruments. With regard to higher aberrations, it is usually the worst of the four solutions, but that objection is an almost purely theoretical one when applied to the usual types of commercial telescopes.

[48]

Simplifications in Special Cases

When a new batch of glass has to be used for a regularly manufactured type of object-glass, then the necessary changes will usually be quite small. The least troublesome way of determining them will be to compute through the objective with the original r_1, r_2, and thicknesses but with the new glass data, and to solve for r_3 so as to secure the required chromatic correction. If focal length and spherical correction prove satisfactory the work is finished. Wrong focal length can be immediately adjusted by a proportional change of all the radii. If there is also an objectionable residue of spherical aberration a small bending will be required. If the above method of solution was used in the original design, the graph will be available to determine δc_1 as has been described. Otherwise very little experience will tell us whether the particular type of the objective came from the rising or the falling arm of the parabola. We then make an experimental bending by say 1/50th of the net crown curvature in the proper direction and, as the parabola may be regarded as a straight line over so short a length, find the correct bending by simple proportion. For *new* designs of the usual crown-in-front type we can quickly solve by taking advantage of the fact that these tend to be externally plano-convex. We therefore solve first by TL, Chr. (4) or (4)*, assume $c_1 = c_a + c_b$ and test this trigonometrically. If it turns out spherically under-corrected we diminish c_1 by about 1/20th of c_a and try again. If the first attempt proves over-corrected we increase c_1 by a similar amount. Linear interpolation and, if necessary, adjustment of the focal length then complete the work.

Effect of Change in Distance of Objects

Our numerical examples have been computed for objects at a very great distance, that being the usual condition under which telescope objectives are used. When an objective is used on near real objects or on virtual objects, the L at the first surface becomes a finite negative or positive quantity, and all the angles of incidence become different from those resulting from an incident pencil of parallel rays. Remembering that the spherical aberration which arises at a refracting surface depends chiefly upon the various angles of the rays, we easily conclude that an objective spherically corrected for very distant objects will as a rule prove imperfect for objects at any other distance, and that a lens system will therefore have to be specially designed for the distance of the objects to which it is to be applied. But as the *angles* of the rays are the deciding quantities, we must not judge the disturbance of the correction by the value of L but by its reciprocal, and the latter will be very small for all distances from infinity to about 50 or 100 times the focal length of the objective. We may therefore compute the objectives of an opera-glass or of a surveyor's level for objects at infinity without any fear of imperfect correction, although the objects may be only 50 or 100 feet away. When the distance of the objects is reduced to a moderate multiple of the focal length, then $1/L$ becomes comparable with the curvature of the surfaces and the change in the angles of incidence must be taken into account. This makes no serious difference in the calculation of the three bendings on which our solution depends, for we merely begin with the required value of the initial L instead of using the opening for parallel rays. But because the spherical aberration will come out different the aberration-parabola will shift its position, and if we selected the values of c_1 by the rules given earlier, which are applicable to distant objects, there would be a risk that both cuts of the parabola with the c_1-axis would not be included within the calculated range. No exact formula can be given here to fix the proper change in the c_1-values of the three bendings, but an approximate rule will assure inclusion of the interesting range in practically all cases. This rule is: [48]

Select the three values of c_1 by the rule given for distant objects and then increase all three *algebraically* by $1 \cdot 5/L$ if L is the distance of the object-point for which the objective is to be corrected. Thus supposing that our 10-inch objectives had been required for real objects at 50 inches from the lens, we should have $L = -50$, therefore $1 \cdot 5/L = -0 \cdot 03$, and as for crown-in-front the curvatures c_1 for distant objects were 0·06, 0·16, and 0·26 respectively, we should use for the close objects $0 \cdot 06 + (-0 \cdot 03) = 0 \cdot 03$ for the first bending, 0·13 for the second, and 0·23 for the third bending.

Effect of a Change of Glass

In telescope objectives it is nearly always desirable to use glasses with a large difference of V-value in order that the curvature of the surfaces may be kept down. Ordinary dense flint is therefore used almost exclusively, but there is a considerable choice of crown glasses even in the limited range from $V = 64$ to $V = 56$ to which it is generally desirable to restrict oneself. We chose a light barium crown with $Nd = 1 \cdot 54$ for our example. The glass most often employed is hard crown of $Nd = 1 \cdot 516$ or thereabouts. The only important difference which a change to that glass makes is that the aberration-parabolas fall decidedly lower on our graph. Their pole therefore gives lenses of considerably larger spherical over-correction, and the two solutions for zero spherical aberration fall correspondingly further apart. As we shall learn in a subsequent chapter, this is undesirable for telescope objectives because it means a greater offence against fulfilment of the sine-condition and it also leads to higher secondary aberrations. The best cemented telescope objectives are in fact obtained by changing the refractive index of the crown in the upward direction beyond the 1·54 of our example. A trial with the medium barium crown 9002 combined with our original flint will show that for this combination the aberration parabolas dip only very slightly below the horizontal zero-axis and that the two solutions fall much closer to the pole. On the whole this is highly desirable, and this combination is therefore worthy of being borne in mind. If we raise the index of the crown still more, as for instance, by using the dense barium crown 4873 of our abridged glass list, we shall find that the whole aberration parabola falls above the zero-axis and that no solution for a spherically corrected cemented objective can be found. This therefore is a too drastic change.

For photographic purposes cemented objectives of pronounced meniscus form are nearly always required in order to obtain good correction for the outer part of the large field to be covered. In photographic lenses we therefore usually find glass-combinations with a rather small V-difference but a considerable difference of re-

[48] fractive indices, because such combinations cause the aberration-parabolas to sink to a very low position, and all the spherically corrected cemented combinations then acquire deep external meniscus-forms. These forms are utterly unsuitable for telescopes.

B. SOLUTIONS BY ALGEBRAICAL APPROXIMATION

[49] The alternative method of finding solutions of optical problems by algebraical thin lens or primary aberration formulae is usually called the 'analytical method', but as the trigonometrical method of section [48] is also a truly analytical one, the name at the head of the present section appears more appropriate ; it implies the essential distinction between algebraical and trigonometrical formulae and points out the danger inherent in the approximate character of the process. The deep suspicion with which the great majority of manufacturing opticians still look upon optical systems arrived at entirely by calculation has in fact been created by the bitter experience that systems, offered to them by believers in the purely analytical method as perfectly corrected, proved defective or even hopelessly bad when turned into glass and brass. With the possible exception of ordinary thin telescope objectives of very considerable focal length, say ten or more times the clear aperture and therefore of very gentle curvature, no optical system calculated by the analytical method should ever find its way into the workshop until it has been corrected trigonometrically. Thus used, the algebraical method is of the highest value, especially in the design of complicated systems, because it greatly shortens the time required by quickly finding an approximate solution capable of differential correction by the trigonometrical method.

The approximate solution can be worked out either by the strict primary aberrations or by the thin lens equations for chromatic and spherical aberration. For the present we will limit ourselves to the latter method as by far the most rapid one ; moreover the TL method very usually gives a closer first approximation in the case of the ordinary comparatively thin object-glasses to which we are restricting ourselves in the present chapter, owing to the tendency of the neglected thicknesses to work in partial compensation of the higher aberrations. It should, however, be realized at once that the relative merit of the two methods is likely to become reversed in the case of thick systems of bold curvatures, and that the more laborious strict primary aberrations will then have to be used.

The first step again consists in working out the net curvature of the components which will give the prescribed focal length and the prescribed chromatic correction ; this is done by applying the equations TL, Chr. (4) or (4)*. For our numerical examples we will take the glasses specified in section [47] :

	Nd	$N_f - N_c$	V
3463 :	$= 1\cdot5407$	$= 0\cdot00910$	$= 59\cdot4$
360 :	$1\cdot6225$	$0\cdot01729$	$36\cdot0$

and will again solve for perfectly achromatic telescope objectives of 10" focal length, with crown in front, so as to be able to compare the results yielded by the two methods of solution. We therefore have $c_a = 0\cdot470$ and $c_b = -0\cdot247$.

In order to determine the spherical aberration of any conceivable bending of each of these components we now apply the equation TL (10)** for thin lenses in contact. As our present system has only two component lenses, the equation is [49]

$$LAp'_b = y^2 l_b'^2 \,[G\text{-sum}]_a + y^2 \,.\, l_b'^2 \,[G\text{-sum}]_b,$$

and the two terms of this equation give the contributions of the two component lenses to the total spherical aberration at the final focus. For the contribution of the first lens the effect of the transfer across the second lens is therefore included, and for that reason the value of the first term does not agree at all with the spherical aberration found trigonometrically at the own focus of the first lens only; if we wanted to determine an algebraical approximation to the latter, it would be given by $y^2 \,.\, l_a'^2 \,[G\text{-sum}]_a$.

It was shown in section [32] that the G-sum of a lens may be calculated in two ways which always give precisely the same result, namely, either in terms of the left-hand data c_1 and v_1, or in terms of the right-hand data c_2 and v'_2. As special interest is attached to the nature of the contact between two adjacent components, and as the possibility of cementing them together depends on equality of the right-hand curvature of the first with the left-hand curvature of the second component, it is nearly always advisable to use the alternative G-sum, in terms of right-hand data, for the first component and the original G-sum for the second component. With the closely packed lenses with which we are dealing we have the additional advantage that $v'_2 = 1/l'_2$ will be identical with $v_3 = 1/l_3$, as the separation is treated as negligible. In accordance with TL (1) written in the v-notation we then have in general $v'_2 = (N_a - 1)\, c_a + v_1$, and as for telescope objectives we treat the object-distance l_1 as infinite, we have $v_1 = 0$, and therefore, *for this case only*,

$$v'_2 = v_3 = (N_a - 1)\, c_a.$$

For our numerical example we have $N_a = 1{\cdot}5407$ and $c_a = 0{\cdot}470$, and four-figure logs give $\log v'_2 = \log v_3 = 9{\cdot}4051$.

We have now to calculate

$$\begin{aligned} LAp'_b = y^2 l_b'^2 \,[&G_1^a \,.\, c_a^3 + G_2^a \,.\, c_a^2 \,.\, c_2 - G_3^a \,.\, c_a^2 \,.\, v'_2 + G_4^a \,.\, c_a \,.\, c_2^2 \\ &- G_5^a \,.\, c_a \,.\, c_2 \,.\, v'_2 + G_6^a \,.\, c_a \,.\, v_2'^2)] \\ + y^2 l_b'^2 \,[&G_1^b\, c_b^3 - G_2^b \,.\, c_b^2 \,.\, c_3 + G_3^b \,.\, c_b^2 \,.\, v_3 + G_4^b \,.\, c_b \,.\, c_3^2 - G_5^b \,.\, c_b \,.\, c_3 \,.\, v_3 \\ &+ G_6^b \,.\, c_b \,.\, v_3^2], \end{aligned}$$

the first line with the G values found with $N_a = 1{\cdot}5407$ as argument, the second line with the G values for $N_b = 1{\cdot}6225$. As l'_b is the final intersection-length, it is equal to f', or in our case 10, for telescope objectives. Remember, however, that for thin systems applied to objects at a finite distance the proper thin lens conjugate distance must be used, and on no account the focal length. For the semi-aperture y we should use the value which is to be adopted in the trigonometrical correction, hence the full semi-aperture if the strict four-ray method is to be applied, or 0·8 of the full semi-aperture if the convenient—if slightly less perfect—three-ray method is contemplated. In order to keep all results strictly comparable with those in section [48] we will use the three-ray method with $y = 0{\cdot}64$. This gives $y^2 \,.\, l_b'^2 = 40{\cdot}96$ as the outside factor of both G-sums.

[49] Keeping the contributions of the two components strictly separated, we now calculate them as quadratic equations in c_2 and c_3 respectively, exactly as in the specimen calculation in section [30], with four-figure logs and numbers which may conveniently be taken from the less bulky five-figure table, provided that the latter gives directly the log of every four-figure number without interpolation. As the differences in the table of the logs of the G-values average about 100, the interpolation is most expeditiously done by a slide-rule ; the numbers to be multiplied are too large to be dealt with by the average computer's mental arithmetic.

Crown lens. $N = 1{\cdot}5407$; $c_a = 0{\cdot}470$; $\log v_2' = 9{\cdot}4051$; $y^2 l_b'^2 = 40{\cdot}96$

Formula : $y^2 l_b'^2 \times$	$G_1 c_a^3$	$+G_2 c_a^2 . c_2$	$-G_3 c_a^2 . v'_2$	$+G_4 c_a . c_2^2$	$-G_5 c_a v'_2 . c_2$	$+G_6 c_a v'^2_2$
$\log G$	9·8074	0·0427	0·1818$_n$	9·7932	0·2512$_n$	0·0652
$\log y^2 l_b'^2$	1·6124	1·6124	1·6124	1·6124	1·6124	1·6124
$\log c_a^n$	9·0163	9·3442	9·3442	9·6721	9·6721	9·6721
$\log v_2'^n$			9·4051		9·4051	8·8102
log-sum	0·4361	0·9993 $\times c_2$	0·5435$_n$	1·0777 $\times c_2^2$	0·9408$_n$ $\times c_2$	0·1599
Collection of Terms	2·730					
	+1·445	9·984 c_2				
	4·175	−8·726 c_2				
	−3·495					
a-contribution :	0·680	+1·258 c_2		+11·96 c_2^2		

Flint lens. $N = 1{\cdot}6225$; $c_b = -0{\cdot}247$; $\log v_3 = 9{\cdot}4051$; $y^2 l_b'^2 = 40{\cdot}96$

Formula : $y^2 l_b'^2 \times$	$G_1 c_b^3$	$-G_2 c_b^2 . c_3$	$+G_3 c_b^2 v_3$	$+G_4 c_b . c_3^2$	$-G_5 c_b v_3 . c_3$	$+G_6 c_b v_3^2$
$\log G$	9·9135	0·1209$_n$	0·2616	9·8420	0·3037$_n$	0·1197
$\log y^2 l_b'^2$	1·6124	1·6124	1·6124	1·6124	1·6124	1·6124
$\log c_b^n$	8·1781$_n$	8·7854	8·7854	9·3927$_n$	9·3927$_n$	9·3927$_n$
$\log v_3^n$			9·4051		9·4051	8·8102
log-sum	9·7040$_n$	0·5187$_n$ $\times c_3$	0·0645	0·8471$_n$ $\times c_3^2$	0·7139 $\times c_3$	9·9350$_n$
Collection of Terms	−0·5058					
	−0·8610	−3·301 c_3				
	−1·3668	+5·175 c_3				
	+1·1601					
b-contribution :	−0·207	+1·874 c_3		−7·032 c_3^2		

The result of these extremely simple calculations is that the spherical aberration of a telescope objective of the given kinds of glass and focal length 10 at semi-aperture 0·64 can be determined for any conceivable independent bendings of the two components by putting the proper values of c_2 and c_3 into the equation

$$\begin{aligned} LAp_b' = {} & 0{\cdot}680 + 1{\cdot}258\, c_2 + 11{\cdot}96\, c_2^2 \\ & -0{\cdot}207 + 1{\cdot}874\, c_3 - 7{\cdot}03\, c_3^2. \end{aligned}$$

The contributions of the components are still kept strictly separated. At present we can explore only a few of the solutions obtainable from the general equation.

(1) *Cemented telescope objectives.* In a cemented objective c_2 is equal to c_3 ; we can therefore simply add together the corresponding terms in the two lines of our equation. Calling the contact curvature c_2, we thus obtain

$$LAp_b' = 0{\cdot}473 + 3{\cdot}132\, c_2 + 4{\cdot}93\, c_2^2.$$

This is a quadratic equation in c_2 which can be solved for any desired value of the [49] LAp'_b at the final focus. We will solve it for the most usual case of perfect spherical correction by putting $LAp'_b = 0$; the equation then is

$$4{\cdot}93\, c_2^2 + 3{\cdot}132\, c_2 + 0{\cdot}473 = 0,$$

and solution by the usual rule gives

$$c_2 = -\frac{3{\cdot}132}{9{\cdot}86} \pm \sqrt{\left(\frac{3{\cdot}132}{9{\cdot}86}\right)^2 - \frac{0{\cdot}473}{4{\cdot}93}} = -0{\cdot}3180 \pm 0{\cdot}0714$$

when worked out by the slide-rule, which is quite satisfactory for this purpose as high precision would be absolutely wasted on these rough approximations by TL equations. We now have the result that our cemented objectives will be free from spherical aberration if we make $c_2 =$ either $-0{\cdot}3894$ or $-0{\cdot}2466$. It should be noted by comparison with the trigonometrical solution in [48] that the results agree fairly well, for in [48] we found respectively $-0{\cdot}3805$ and $-0{\cdot}246$. We will submit the first solution, $c_2 = -0{\cdot}3894$, to the trigonometrical testing and correcting process. With the fixed $c_a = 0{\cdot}470$ and $c_b = -0{\cdot}247$ we find

$$c_1 = c_a + c_2 = \quad 0{\cdot}470 - 0{\cdot}3894 = +0{\cdot}0806$$

$$c_3 = c_2 - c_b = -0{\cdot}3894 + 0{\cdot}247 = -0{\cdot}1424$$

and on taking reciprocals and adding the previously chosen thicknesses we obtain the complete prescription

$$\begin{array}{ll} r_1 = 12{\cdot}407 & \\ & 0{\cdot}25 \\ r_2 = -2{\cdot}568 & \\ & 0{\cdot}20 \\ r_3 = -7{\cdot}022 & \end{array}$$

The three radii of the cemented objective have been numbered consecutively, as is usual in optical calculations.

Beginners are strongly advised to submit the *complete* TL prescription to the trigonometrical test in order to acquire personal experience as to the degree of approximation. In the present example a five-figure test with angles to half seconds, by the three-ray method carried out according to the examples in section [48], gives the final results

$$L'_{3y} = 9{\cdot}9370\,;\; l'_{3y} = 9{\cdot}9296\,;\; L'_{3v} = 9{\cdot}9475\,;$$

or

$$\text{spherical over-correction } l'_{3y} - L'_{3y} = -0{\cdot}0074\,;$$

$$\text{chromatic over-correction } L'_{3y} - L'_{3v} = -0{\cdot}0105.$$

Both residuals, but more especially the chromatic one, exceed the amount which a self-respecting designer could possibly countenance; trigonometrical correction must therefore follow. As we have a complete ray-tracing, this is carried out most conveniently by Chr. (3), which gives corrected $r_3 = -6{\cdot}731$. The rays issuing from the unchanged second surface are now traced through the last surface with this corrected radius. It is advisable to add the PA-check, for as the directly found final $(L' - r)$ and $(l' - r)$ are nearly twice the resulting L' and l' on account of the

[49] negative radius, the precision of the result will be approximately doubled by the check. The results found with $r_3 = -6{\cdot}731$ then are :

$$L'_{3y} = 9{\cdot}5571 \; ; \; l'_{3y} = 9{\cdot}5638 \; ; \; L'_{3v} = 9{\cdot}5562 :$$

or spherical under-correction $l'_{3y} - L'_{3y} = +0{\cdot}0067$;

chromatic under-correction $L'_{3y} - L'_{3v} = +0{\cdot}0009$.

We see that the removal of the huge chromatic error of the TL solution has been very nearly successful, but is overdone to a somewhat undesirable extent, as the 0·0009 is not likely to be due entirely to rounding-off errors. Chr. (3) was therefore applied again, giving $r_3 = -6{\cdot}757$, and with this the trigonometrical results, again by *PA*-check,

$$L'_{3y} = 9{\cdot}5912 \; ; \; l'_{3y} = 9{\cdot}5967 \; ; \; L'_{3v} = 9{\cdot}5916 :$$

or spherical under-correction $l'_{3y} - L'_{3y} = +0{\cdot}0055$;

chromatic over-correction $L'_{3y} - L'_{3v} = -0{\cdot}0004$.

The chromatic correction now is perfectly satisfactory ; but there is a heavy residual of spherical aberration which can only be removed by bending the objective as a whole. In order to avoid a series of more or less haphazard trials we use the differential method described in the previous section by calculating the proper δc by the equation

$$\delta c_2 = (LA'_{required} - LA'_{found}) / \frac{dLA'}{dc_2},$$

but as we now have the nearly accurate TL equation

$$LAp'_b = 0{\cdot}473 + 3{\cdot}132\, c_2 + 4{\cdot}93\, c_2^2$$

on which our solution was based, we can determine the required differential coefficient directly from this equation as

$$\frac{dLAp'_b}{dc_2} = 3{\cdot}132 + 9{\cdot}86\, c_2,$$

and putting in the c_2 of our objective $= -0{\cdot}3894$, find

$$\frac{dLAp'_b}{dc_2} = -0{\cdot}704.$$

As $LA'_{required}$ is zero and $LA'_{found} = 0{\cdot}0055$, we obtain

$$\delta c_2 = -0{\cdot}0055/(-0{\cdot}704) = +0{\cdot}0078.$$

As the objective is to be bent as a whole, this δc must be applied to all three surfaces, and as we had $c_1 = 0{\cdot}0806$, $c_2 = -0{\cdot}3894$, while our final $r_3 = -6{\cdot}757$ gives $c_3 = -0{\cdot}1480$, we obtain the corrected curvatures

$$\textit{corrected}\; c_1 = 0{\cdot}0884 \; ; \; c_2 = -0{\cdot}3816 \; ; \; c_3 = -0{\cdot}1402 :$$

or by reciprocals the radii which should give spherical correction

$$r_1 = 11{\cdot}312 \; ; \; r_2 = -2{\cdot}6205 \; ; \; r_3 = -7{\cdot}133.$$

[49] A moderate bending of a complete objective is not likely to alter the chromatic correction to any serious extent, and it is therefore advisable to determine all three radii in the manner shown and to test the complete formula trigonometrically. In the present case the results are :

$$L'_{3y} = 9{\cdot}5893\ ;\ l'_{3y} = 9{\cdot}58915\ ;\ L'_{3v} = 9{\cdot}5892:$$

spherical over-correction $l'_{3y} - L'_{3y} = -0{\cdot}00015$;

chromatic under-correction $L'_{3y} - L'_{3v} = +0{\cdot}0001$.

The capricious rounding-off errors have once more entered into a friendly conspiracy and have given us a result surpassing reasonable expectations. It may be pointed out that the final solution at which we have arrived agrees very well with that read directly off the trigonometrical graph in section [48], namely, $r_1 = 11{\cdot}17$ and $r_2 = -2{\cdot}628$; this shows that the trigonometrical method gives an incomparably closer direct result than the TL solution, which was $r_1 = 12{\cdot}407$, $r_2 = -2{\cdot}568$, with an error in r_2 of 0·0537 against one of only 0·0075 from the trigonometrical graph.

(2) *Telescope objectives with a difference of curvature at the contact surfaces.* The general equation

$$\begin{aligned} LAp'_b = \; & 0{\cdot}680 + 1{\cdot}258\, c_2 + 11{\cdot}96\, c_2^2 \\ & -0{\cdot}207 + 1{\cdot}874\, c_3 - 7{\cdot}03\, c_3^2 \end{aligned}$$

now applies to the crown-in-front objectives which we have selected for our numerical examples, and as c_2 and c_3 are independent variables, it is at once evident that the number of possible solutions for any prescribed spherical aberration at the final focus is enormously increased. We shall learn later that the proper use to be made of this increased liberty is to satisfy two aberrational conditions simultaneously, most usually to secure the desired spherical correction *and* fulfilment of the optical sine condition. At the present stage it will be highly instructive to obtain a comprehensive knowledge of the possibilities with reference to spherical correction only ; for that purpose a graphical representation of our general solution is extremely useful. As the contributions of the two components to the final aberration are both quadratic equations in c_2 or c_3 it is evident that they will plot as exact parabolas with vertical axis against a suitable horizontal scale of curvature values. For the application of the convenient dip method we have therefore only to calculate three values of each line of our general equation for equal increments of curvature in order to enable us to draw the graph. In order to cover a wide range without excessive extrapolation we choose

c_2 or c_3 =	−0·3	+0·1	+0·5

and find by slide-rule, which is quite sufficiently accurate, the values

a-contribution	1·379	0·926	4·299
b-contribution	−1·402	−0·090	−1·028

As the contributions reach huge values, natural size on the usual inch plotting paper will be suitable for them, with 0·1 of curvature per horizontal inch in the x-direction. If we plotted both contributions with their true sign we should obtain a graph

[49] like the reduced-scale one inset in Fig. 45a, namely a crown parabola entirely above the horizontal zero axis and a flint parabola below the axis, and we should find combinations with zero aberration at the final focus by seeking points on the two curves which had numerically equal ordinates of opposite sign. That would be a tedious process and would not give us a very clear general conception of the possibilities. We therefore plot one of the contributions, usually that with predominating negative values, with reversed sign, so that points of the two curves at the same distance from the axis indicate *oppositely* equal contributions and therefore spherical correction of the corresponding lens combination. The main part of Fig. 45a shows the resulting graph when the contributions of the flint lens are

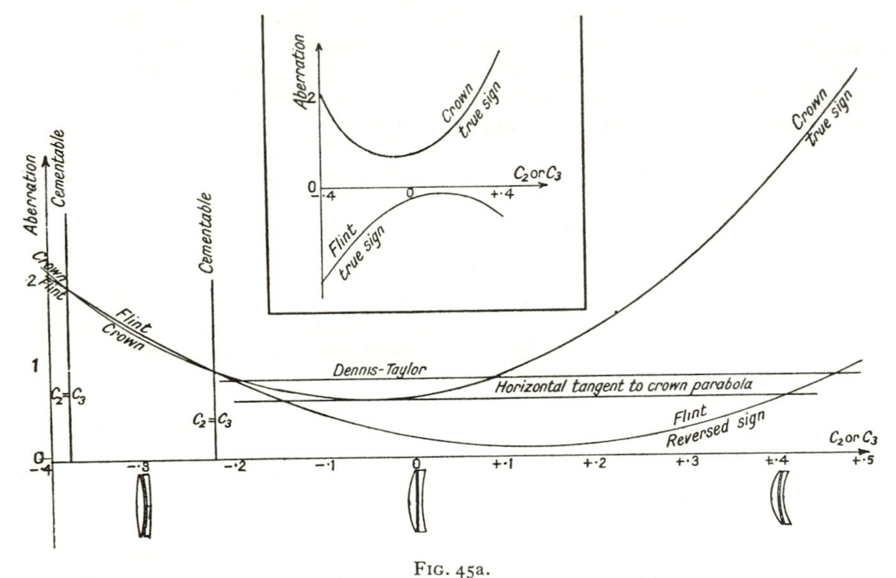

FIG. 45a.

plotted with reversed sign. We obtain two parabolas which form a very shallow intersection at the left side of the diagram. As the abscissae represent the contact curvatures c_2 and c_3, it is evident that an ordinate erected at any point corresponds to a crown lens and flint lens having $c_2 = c_3$ and equal to the value read off the horizontal scale. In general such an ordinate will cut the two parabolas at different heights, and as the latter measure the contributions of the two lenses to the total spherical aberration, but with reversed sign for the flint lens, it is clear that if at the point chosen the crown parabola lies higher than the flint parabola, the resulting cementable lens will be spherically under-corrected, whilst we shall obtain over-corrected cementable lenses from the short region between the two intersection-points of the parabolas, because in that region the flint parabola lies above the crown parabola, and the negative contribution of the flint lens to the final aberration is therefore in excess of the positive contribution of the crown lens.

Manifestly the two intersection-points of the parabolas themselves mark the only [49] two cementable lenses which are spherically corrected, and it will be noticed that they correspond to readings of the horizontal scale which agree closely with the values of the contact curvature which we obtained by the algebraical solution on which subsection (1) was based.

With reference to lens combinations with a difference between c_2 and c_3 it is easily seen that a horizontal line drawn across the two parabolas marks four lenses which have the same numerical value of their contribution to the final aberration, but with opposite sign for crown and flint respectively. Hence either of the two crown lenses placed in front of either of the flint lenses will give a spherically corrected telescope objective, therefore four different possibilities. It may be stated at once that the great majority of these forms are undesirable on account of the seriousness of other aberrations, and that combinations of a crown lens from the descending branch of the crown parabola with a flint lens from the ascending branch of the flint parabola, or vice versa, are almost unconditionally bad. It is possible to find for any conceivable crown lens two different flint lenses which will give spherical correction. But the converse is not unconditionally true, for as the flint parabola lies considerably lower than the crown parabola, it is clear that the negative aberration of all flint lenses from the polar region of the flint parabola is too low to be matched by any possible crown lens ; the points where the horizontal tangent of the crown parabola cuts the flint parabola obviously define the excluded region and show that in the case of our numerical example all flint lenses having c_3 between $-0{\cdot}145$ and $+0{\cdot}410$ give insufficient negative aberration to produce a spherically corrected object-glass even with the crown lens of least possible positive aberration.

The violent meniscus form which results for all possible flint lenses from the right-hand part of our graph robs this part of nearly all interest from the practical point of view as far as telescope objectives are concerned ; the latter all come from the short region where the two parabolas lie very close to each other. As in the case of Fig. 45, it will be found very helpful to add a few reasonably accurate drawings of the actual lens forms at intervals along the horizontal axis.

The general character of a graph like Fig. 45a does not change very much for any of the glass pairs usually employed in telescope objectives. The optically most significant change is that of the depth of the intersection between the two parabolas ; the intersection becomes deeper for pairs with approximately the same value of $(V_a - V_b)$ when the difference of the refractive indices is increased, and it becomes still shallower than in our example if that difference is diminished. When the index of the crown-glass is only about two units of the second decimal place less than that of the flint-glass (the V-difference remaining substantially unaltered) the two parabolas merely touch or become very slightly separated, and in the latter case spherically corrected *cemented* lenses become impossible.

Changes of object-distance also do not change the character of the graph very much as long as the object-distance is a multiple of the focal length. The conclusions drawn from Fig. 45a therefore apply to all ordinary telescope objectives and simple microscope objectives ; but it is of course necessary to calculate the equation of the parabolas for the actual object-distance and the actual refractive

[49] indices if the graph is to be used for the finding of actual *numerical* data for proposed lens systems.

(3) As an actual example of a telescope objective with a difference between c_2 and c_3, or, as it is usually called, with a 'broken contact', we can at present only take a case in which the form of one of the components is definitely fixed. We might, for instance, have a valuable large crown or flint lens and desire to turn it into a complete object-glass by matching it with a new second component, the optical properties of the glass of the existing lens being of course assumed to be known. We will, however, select a more interesting case, namely, the type of ordinary astronomical object-glass favoured by Mr. Dennis Taylor, in which, according to an example given in his *System of Applied Optics*, the crown lens has about 0·6 of its total curvature on the first, and therefore −0·4 on its second surface. The reasons for this choice are mainly practical ones: firstly, that the crown lens makes a fairly close approach to the minimum of spherical aberration, and is therefore less likely to be prejudicially affected by flexure from its own weight when the telescope is pointed upwards; secondly, that the minute air-space between the two components assumes the form of a convex meniscus, so that the two lenses can be allowed to come into actual and airtight contact at their edges, thus excluding dust permanently and also assuring absolute relative centring; thirdly, that the flint lens assumes biconcave form, so that the perfection of both its surfaces can be most sensitively tested by examining the image by reflection of a point of light placed close to the centre of curvature. From the point of view of other optical aberrations the type is unfavourable, but this is unimportant in its use for astronomical telescopes because the high ratio of focal length to aperture which is usual—about 15 to 1—renders all other residual aberrations comparatively harmless in relation to the inevitable secondary spectrum.

By the fixed data of our numerical examples we have c_a of the crown lens $= 0{\cdot}470$, therefore

$$c_2 = -0{\cdot}4\, c_a = -0{\cdot}188.$$

This enables us to locate the crown lens on the graph, Fig. 45a, and a horizontal line drawn through the proper point of the crown parabola cuts the flint parabola at $c_3 = -0{\cdot}198$ and at $c_3 = +0{\cdot}469$, thus giving the two possible solutions for a Dennis Taylor objective. Although these solutions would prove amply near enough to the truth for rapid correction by the trigonometrical method, they do not do justice to the TL method of solution on account of the scale of our graph, which can hardly guarantee $\pm 0{\cdot}01$ of aberration. In order to obtain another fair test of the degree of approximation we therefore solve algebraically for the true TL values of c_3. The general equation

$$\begin{aligned} LAp'_b = {} & 0{\cdot}680 + 1{\cdot}258\, c_2 + 11{\cdot}96\, c_2^2 \\ & -0{\cdot}207 + 1{\cdot}874\, c_3 - 7{\cdot}03\, c_3^2 \end{aligned}$$

again applies, and as we have fixed c_2 at the value $-0{\cdot}188$ we can determine the first line numerically as $= 0{\cdot}866$, in good agreement with the graph, and combining this with the absolute term $-0{\cdot}207$ of the second line we obtain the simple quadratic equation in c_3

$$LAp'_b = 0{\cdot}659 + 1{\cdot}874\, c_3 - 7{\cdot}03\, c_3^2.$$

Solving this in the usual way for perfect spherical correction, i.e. $LAp'_b = 0$, [49] we obtain

$$c_3 = 0{\cdot}1333 \pm 0{\cdot}3339 \,; \quad \therefore c_3 = \text{either } -0{\cdot}2006 \text{ or } +0{\cdot}4672.$$

The second solution is of the useless type coming from opposite slopes of the two parabolas, and moreover it defeats two of the three advantages aimed at. We therefore select the first solution, which with $c_a = 0{\cdot}470$ and $c_b = -0{\cdot}247$ gives the complete prescription

$$c_1 = 0{\cdot}282\,; \quad c_2 = -0{\cdot}188\,; \quad c_3 = -0{\cdot}2006\,; \quad c_4 = +0{\cdot}0464\,;$$

or by Barlow's table of reciprocals the radii

$$r_1 = 3{\cdot}546\,; \quad r_2 = -5{\cdot}319\,; \quad r_3 = -4{\cdot}985\,; \quad r_4 = +21{\cdot}55.$$

We retain the adopted thicknesses of 0·25 for the crown and 0·20 for the flint; but we must on no account forget to assign a correct value to the thickness of the axial air-space which results from the difference of the contact curvatures when the two lenses are in contact at their extreme edges, for an insufficient value of this small separation would introduce into the *trigonometrical* calculation the particularly treacherous error referred to at the end of section [14]. This air-space is equal to the difference of the X-values of the two contact surfaces for the Y-value of the marginal contact line. In the case of deep curvatures and a big difference it would be advisable to calculate by the exact equations (8), choosing whichever is most easily applied. But in our present case the curvatures are quite moderate and differ only very little; it will therefore be permissible to use the first approximation formula (8p): $x = \frac{1}{2} y^2/r = \frac{1}{2} y^2 \cdot \frac{1}{r}$ which gives for the difference between x_2 and x_3, or the air-space: $x_2 - x_3 = \frac{1}{2} y^2 \left(\frac{1}{r_2} - \frac{1}{r_3}\right) = \frac{1}{2} y^2 (c_2 - c_3)$. For our objective we have $c_2 - c_3 = -0{\cdot}188 - (-0{\cdot}2006) = +0{\cdot}0126$; the intended clear aperture is 1·6 and the gross diameter of the lenses may be estimated as 1·7. This gives the value of y at the extreme edge $= 0{\cdot}85$, and therefore

$$\text{Air-space} = \tfrac{1}{2}(0{\cdot}85)^2 \times 0{\cdot}0126 = 0{\cdot}00455 \text{ by slide-rule.}$$

We may safely round this off *upwards* to 0·005, for a few units of the fourth place will not have any sensible effect on the aberrations of a lens of focal length 10. The prescription to be tested trigonometrically now is:

$r_1 =$	3·546	
		0·25
$r_2 =$	−5·319	
		0·005
$r_3 =$	−4·985	
		0·20
$r_4 =$	21·55	

A calculation right through—merely in order to buy some more experience—by the three-ray method gives

$$L'_{4y} = 9{\cdot}3539\,; \qquad l'_{4y} = 9{\cdot}3598\,; \qquad L'_{4v} = 9{\cdot}3514\,;$$

[49] or spherical under-correction $l'_{4y} - L'_{4y} = +0{\cdot}0059$;
chromatic under-correction $L'_{4y} - L'_{4v} = +0{\cdot}0025$.

Unlike the example in subsection (1) the present TL solution proves to be under-corrected; this is due to the very considerable contraction of the cone of rays brought about by the strong convexity of the first surface; in the ordinary forms of object-glasses the negative or over-correcting tendency of the higher aberrations usually predominates.

To remove the chromatic aberration we reject the calculation through the original r_4 and solve by Chr. (1) for the correct last radius, finding $r_4 = 20{\cdot}68$, which now gives

spherical under-correction $l'_{4y} - L'_{4y} = +0{\cdot}0054$;
chromatic under-correction $L'_{4y} - L'_{4v} = +0{\cdot}0003$.

To get rid of the almost undiminished spherical aberration we must bend the flint lens by the differential correcting method. As the crown lens remains unchanged, we have only to differentiate the contribution of the flint lens, namely

$$-0{\cdot}207 + 1{\cdot}873\, c_3 - 7{\cdot}03\, c_3^2;$$

and find

$$\frac{dLAp'_b}{dc_3} = 1{\cdot}873 - 14{\cdot}06\, c_3,$$

which, with our $c_3 = -0{\cdot}2006$, gives $\dfrac{dLAp'_b}{dc_3} = 4{\cdot}69$.

The universal correcting equation

$$\delta c_3 = (LA'_{required} - LA'_{found}) \Big/ \frac{dLA'}{dc_3}$$

now gives, with $LA'_{required} = 0$ and $LA'_{found} = +0{\cdot}0054$, also $\dfrac{dLA'}{dc_3} = 4{\cdot}69$

$$\delta c_3 = (0 - 0{\cdot}0054)/4{\cdot}69 = -0{\cdot}00115.$$

As we had		$r_3 = -4{\cdot}985$		$r_4 = 20{\cdot}68$
or		$c_3 = -0{\cdot}2006$		$c_4 = 0{\cdot}04835$,
the bending by		$\delta c = -0{\cdot}00115$		$-0{\cdot}00115$
gives:	corrected	$c_3 = -0{\cdot}20175$,	corrected	$c_4 = 0{\cdot}04720$;
or by Barlow:	corrected	$r_3 = -4{\cdot}957$,	corrected	$r_4 = 21{\cdot}19$

Caution is highly necessary at this stage, for as c_3 has increased numerically, the quantity $(c_2 - c_3)$, which determines the necessary air-space, has grown from the original 0·0126 to 0·01375, or by 9 per cent., but as the originally calculated air-space was 0·00455, the increase by 9 per cent., still keeps it just below the 0·005 which we used in our prescription, and we can retain that value. But it should be well noted that an alteration of the air-space might be imperatively called for with a larger change in c_3, lest the treacherous error referred to above should creep into the *final* calculation.

Tracing the unaffected rays issuing from the crown lens through the slightly bent flint lens, the results are [49]

$L'_{4y} = 9{\cdot}4683$; $l'_{4y} = 9{\cdot}4683$; $L'_{4v} = 9{\cdot}4686$;

or spherical aberration Nil ;

chromatic over-correction $L'_{4y} - L'_{4v} = -0{\cdot}0003$;

once more a final result rather better than could normally be expected from a five-figure calculation.

All our examples of the application of the TL solution have referred to crown-in-front telescope objectives. For flint-in-front or 'Steinheil' objectives the concave flint lens will receive the rays from the object-point and becomes component 'a', to be worked out by the alternative form of the G-sum, while v'_2 will usually be negative. For negative or dispersive combinations the total curvature becomes positive for the flint lens and negative for the crown lens, but the *method* calls for no change. What is essential is unrelenting and pedantic watchfulness as to the sign of every quantity that enters into the calculation.

C. HYBRID METHODS OF SOLUTION

[48.] The strong point in favour of trigonometrical ray-tracing is the absolute accuracy and finality of the process ; but it takes decidedly more time than the elegant though inaccurate TL method. It is therefore well worth while to inquire whether a trigonometrically exact final solution might be obtainable in less total time by combining the two methods more freely than we have done in the preceding two sections. The exceptional intermediate number of the present section is justified by this purpose. We will restrict ourselves to cemented objectives of two components, leaving extensions of the methods for later occasions. As we then have $c_2 = c_3$ and $v'_2 = v_3$, the general TL solution for the spherical aberration of such a system of three refracting surfaces can be written in the form :

$$LAp'_b = y^2 \,.\, l'^2_b \left\{ \begin{array}{l} G_1{}^a c_a{}^3 + G_1{}^b c_b{}^3 - G_3{}^a c_a{}^2 v'_2 + G_3{}^b c_b{}^2 v'_2 + G_6{}^a c_a v'^2_2 + G_6{}^b c_b v'^2_2 \\ + c_2(G_2{}^a c_a{}^2 - G_2{}^b c_b{}^2 - G_5{}^a c_a v'_2 - G_5{}^b c_b v'_2) \\ + c_2^2(G_4{}^a c_a + G_4{}^b c_b) \end{array} \right\}$$

by giving to the G-values the proper distinguishing letter a or b and by ordering the equation as a quadratic equation in c_2. We shall derive our simplified methods from this equation.

(1) In section [47] we defined the 'dip' of a vertical parabola as $= \frac{1}{2}(y_{-1} + y_1) - y$, y_{-1}, y, and y_1 being three successive ordinates at equal intervals 'i' of the abscissae, and we proved that it represents $c \,.\, i^2$ if c is the coefficient of the quadratic term in the equation of the parabola. In our general TL solution the coefficient of the quadratic term is $y^2 l'^2_b (G_4{}^a c_a + G_4{}^b c_b)$, hence we have for the graph of any cemented objective

$$(\text{TL})\ \text{Dip} = y^2 l'^2_b (G_4{}^a c_a + G_4{}^b c_b) \,.\, i^2.$$

For our numerical examples we had $y^2 l'^2_b = 40{\cdot}96$, $c_a = 0{\cdot}470$, and $c_b = -0{\cdot}247$; with $\log G_4{}^a = 9{\cdot}7932$ and $\log G_4{}^b = 9{\cdot}8420$ this gives

$$(\text{TL})\ \text{Dip} = 4{\cdot}93\, i^2.$$

48.5] The three bendings for the trigonometrical graph were at intervals $i = 0{\cdot}1$ of curvature ; hence our thin lens formula gives for the dip of that graph :

$$\text{Dip} = 4{\cdot}93 \times 0{\cdot}1^2 = 0{\cdot}0493.$$

As the dip is also defined by $\frac{1}{2}(y_{-1}+y_1)-y$, in which the y represent the plotted spherical aberration of the three bendings, we have for the true trigonometrical dip of the crown-in-front parabola

$$\text{trig. dip} = \tfrac{1}{2}(0{\cdot}0250+0{\cdot}0324)-(-0{\cdot}0233) = +0{\cdot}0520.$$

This agrees with the TL approximate result within 0·0027 or within about $5\frac{1}{2}$ per cent. It follows that if we had calculated only the first and third bendings trigonometrically and had then calculated in a few minutes the TL dip instead of spending about one hour on the trigonometrical calculation of the middle bending, we should have found the spherical aberration of the latter correctly within 0·0027, and the errors in the actually wanted cuts of the parabola with the zero-axis would have been still smaller and easily within the range for which the final differential correction gives a practically exact result. The total time would therefore have been reduced by about 20 per cent. Alternatively, the TL dip may be used as a check on the results of the *three* trigonometrical bendings, sufficiently searching to render calculation without full checks safe even for a moderately erratic computer.

(2) The general TL solution gives the differential coefficient

$$\frac{dLA'}{dc_2} = y^2 l_b'^2(G_2{}^a c_a{}^2 - G_2{}^b c_b{}^2 - G_5{}^a c_a v_2' - G_5{}^b c_b v_2') + 2\, c_2 y^2 l_b'^2 (G_4{}^a c_a + G_4{}^b c_b)$$

which is required for the differential correction. As this equation contains only 6 terms and G-values against 12 in the complete equation for LAp_b', it can be calculated in about half the time, and in a small fraction of the time required for a complete trigonometrical calculation. In the cases referred to in section [48], subsection on 'simplifications in special cases', when the first trigonometrical ray-tracing will practically always give a close approach to the required solution, the calculation of the differential coefficient by the above equation and its employment in the differential correcting formula will usually lead directly to a final solution of ample accuracy.

(3) With the aid of the TL differential coefficient stated in the preceding subsection we can find both solutions of a given problem from a single trigonometrical calculation for any reasonably chosen bending by the following method, which is also of considerable mathematical interest. With reference to the all-positive diagram, Fig. 45b, let P represent the plotted position of the selected and trigonometrically computed bending with contact curvature c_2 and spherical aberration LA'_{found}. Let P^* represent the required solution giving $LA'_{required}$ with an additional bending δc_2 of our cemented lens. P and P^* must both lie on the aberration parabola (as yet quite unknown) of our lens, and we treat this as a common parabola with *vertical* axis, which is its ideal form for thin lenses of small aperture. If the slope of the chord PP^* is defined by the angle O, we have

$$\tan O = (LA'_{required} - LA'_{found})/\delta c_2,$$

and we see that we could determine δc_2 from this equation if we knew tan O. If we now bisect the chord PP^* at Q so that Q has the abscissa $(c_2 + \frac{1}{2}\,\delta c_2)$, then the ordinate of Q will be a 'diameter' of the aberration parabola, because it is parallel

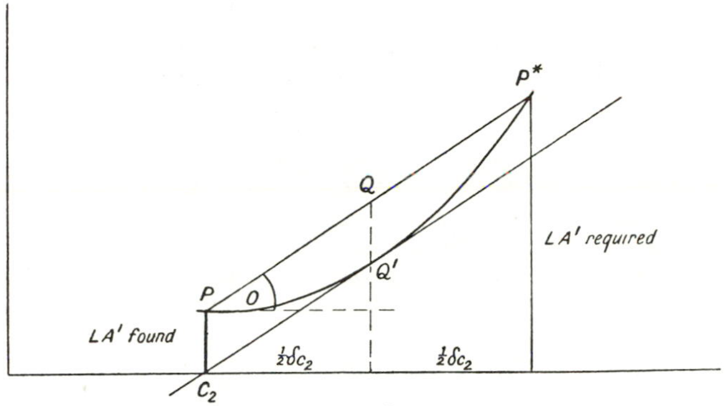

FIG. 45b.

to its vertical axis; and as it bisects the chord PP^* it will, by a well-known property of the common parabola, also bisect all other chords of the parabola which are *parallel* to PP^*, including the infinitely short chord, or tangent of the parabola, at the point Q' where the ordinate of Q cuts the aberration parabola. Hence the chord PP^* is parallel to the tangent of the parabola at Q', and we can determine tan O by the direction of this tangent or by the differential coefficient of our curve at Q', that is for the value $(c_2 + \frac{1}{2}\,\delta c_2)$ of the abscissa. Equating the value of tan O derived directly from Fig. 45b and the value of the differential coefficient by sub-section (2), the latter for the abscissa $(c_2 + \frac{1}{2}\,\delta c_2)$, we obtain the equation

$$\frac{LA'_{required} - LA'_{found}}{\delta c_2} = y^2 l_b'^2 (G_2{}^a c_a{}^2 - G_2{}^b c_b{}^2 - G_5{}^a c_a v_2' - G_5{}^b c_b v_2') + 2\,(c_2 + \tfrac{1}{2}\,\delta c_2) y^2 l_b'^2 (G_4{}^a c_a + G_4{}^b c_b).$$

In this equation δc_2 is the only unknown quantity, and it can therefore be determined by it. To shorten the expressions we will call the first bracket containing G-values P, the final bracket containing G-values Q, and will introduce also $LA'_{required} - LA'_{found} = \delta LA'$. If we slightly rearrange the terms, we then obtain

$$\frac{\delta LA'}{y^2 l_b'^2} = \delta c_2 (P + 2\,c_2 \,.\, Q) + (\delta c_2)^2 \,.\, Q,$$

a quadratic equation in δc_2 which gives the two possible solutions of the problem as

$$\delta c_2 = -\left(\tfrac{1}{2}\frac{P}{Q} + c_2\right) \pm \sqrt{\left(\tfrac{1}{2}\frac{P}{Q} + c_2\right)^2 + \frac{\delta LA'}{y^2 \,.\, l_b'^2 \,.\, Q}}.$$

Provided that the c_2 of the trigonometrical raytracing was reasonably chosen, say by the rules for 'middle-bendings' in section [48], the two solutions will be

[48.5] within easy range of the differential correcting method. As the contact curvature of the solutions is $c_2+\delta c_2$, with care to use the correct δc_2 for each solution, the differential coefficient for the final correction will be

$$= y^2 l_b'^2 \,.\, P + 2\, y^2 l_b'^2 (c_2+\delta c_2) \,.\, Q$$

and is therefore obtainable by a few logs or slide-rule settings.

We will collect the formulae and instructions for this extremely convenient method in concise form :

(a) Go through the usual TL, Chr. preparatory work, finding c_a and c_b ; also select thicknesses suitable for the proposed aperture and prepare the necessary N-values.

(b) Select a value of c_2 for the trigonometrical raytracing, trying for a 'middle-bending' if both solutions are to be explored, or shifting the position suitably towards the desired form if only one solution is required. If the LA' comes out very close to the required value, proceed at once to differential correction by sub-section (2) above ; otherwise :

(c) *From the preparatory TL data* calculate

$$v_2' = \frac{1}{l_1} + (N_a - 1)c_a \,;$$

then with the aid of the table of G-values

$$P = G_2{}^a c_a{}^2 - G_2{}^b c_b{}^2 - G_5{}^a c_a v_2' - G_5{}^b c_b v_2' \,; \qquad Q = G_4{}^a c_a + G_4{}^b c_b \,;$$

and from the trigonometrical result $\delta LA' = LA'_{required} - LA'_{trig.}$. Note that it is *not* advisable to try to improve the preparatory TL data by substituting those found in the trigonometrical work.

(d) Calculate $\delta c_2 = -\left(\frac{1}{2}\frac{P}{Q} + c_2\right) \pm \sqrt{\left(\frac{1}{2}\frac{P}{Q} + c_2\right)^2 + \frac{\delta LA'}{y^2 l_b'^2 \,.\, Q}}$.

c_2 + either value of δc_2 then gives the contact curvature of the solutions ; then c_1 = corrected $c_2 + c_a$.

(e) Test the selected solution trigonometrically ; if it is not considered close enough to the prescribed value of LA', apply the correcting method with

$$\frac{dLA'}{dc} = y^2 l_b'^2 \,.\, P + 2\, y^2 l_b'^2 (c_2+\delta c_2) \,.\, Q.$$

In the trigonometrical work the last radius must in each case be adjusted so as to give the required chromatic correction before proceeding to the next step.

We will test this method on the results for flint-in-front telescope objectives in section [48].

(a) We take over $c_a = -0{\cdot}247$, $c_b = 0{\cdot}470$, also the adopted thicknesses, aperture, and N-values.

(b) We assume that only the middle bending has been calculated trigonometrically. It had $c_2 = 0{\cdot}447$ and gave $LA'_{trig.} = -0{\cdot}0153$.

(c) As $l_1 = \infty$ we have $v_2' = 0{\cdot}6225 \,.\, (-0{\cdot}247)$ giving $\log v_2' = 9{\cdot}1868_n$. With $LA'_{required} = 0$ we have $\delta LA' = +0{\cdot}0153$; the sign-reversal should be well noted.

We now calculate P and Q with $c_a = -0{\cdot}247$, $c_b = 0{\cdot}470$, $\log v_2' = 9{\cdot}1868_n$ [48.5] thus, remembering that 'a' is of index 1·6225, 'b' of index 1·5407 :

$P =$	$G_2^a c_a^2$	$-G_2^b c_b^2$	$-G_5^a c_a v_2'$	$-G_5^b c_b v_2'$	$Q =$	$G_4^a c_a$	$+G_4^b c_b$
$\log G$	0·1209	$0{\cdot}0427_n$	$0{\cdot}3037_n$	$0{\cdot}2512_n$		9·8420	9·7932
$\log c^n$	8·7854	9·3442	$9{\cdot}3927_n$	9·6721		$9{\cdot}3927_n$	9·6721
$\log v'_2$			$9{\cdot}1868_n$	$9{\cdot}1868_n$			
log-sums	8·9063	$9{\cdot}3869_n$	$8{\cdot}8832_n$	9·1101		$9{\cdot}2347_n$	9·4653
Collection of terms	0·0806	−0·2437					+0·2919
	0·1289	−0·0764					−0·1717
	0·2095	−0·3201	$\therefore P = -0{\cdot}1106$			$Q =$	0·1202

(d) We now calculate δc_2 :

$$\log \tfrac{1}{2} = 9{\cdot}6990 \qquad \therefore \tfrac{1}{2}\frac{P}{Q} = -0{\cdot}4602 \qquad \log \delta LA' = 8{\cdot}1847$$

$$+\log P = 9{\cdot}0438_n \qquad +c_2 = 0{\cdot}447 \qquad +\text{colog}\, y^2 l_b'^2 = 8{\cdot}3876$$

$$+\text{colog}\, Q = 0{\cdot}9201 \qquad \tfrac{1}{2}\frac{P}{Q} + c_2 = -0{\cdot}0132 \qquad +\text{colog}\, Q = 0{\cdot}9201$$

$$\log \tfrac{1}{2} \cdot \frac{P}{Q} = 9{\cdot}6629_n \qquad \frac{\delta LA'}{y^2 \cdot l_b'^2 \cdot Q} = 0{\cdot}003107 \qquad \log \frac{\delta LA'}{y^2 l_b'^2 Q} = 7{\cdot}4924$$

Taking the square of $\frac{1}{2}\frac{P}{Q} + c_2$ from Barlow's table we now have

$$\delta c_2 = +0{\cdot}0132 \pm \sqrt{0{\cdot}000174 + 0{\cdot}003107} = +0{\cdot}0132 \pm 0{\cdot}0573,$$

$$\therefore \quad \delta c_2 = \text{either} -0{\cdot}0441 \text{ or } +0{\cdot}0705.$$

With the $c_2 = 0{\cdot}447$ of the trigonometrical result these give the solutions

$$c_2 + \delta c_2 \text{ either } 0{\cdot}4029 \text{ or } 0{\cdot}5175\,;$$

and by applying $c_a = -0{\cdot}247$ we find the corresponding

$$c_1 = \text{either } 0{\cdot}1559 \text{ or } 0{\cdot}2705.$$

The values of c_1 read from the trigonometrical graph in section [48] were c_1 either 0·154 or 0·265, and for the latter the trigonometrical verification in [48] gave 0·2676 as the true value. It will be seen that the approximation is remarkably good. When it is realized that the above very short and simple calculation of the two values of $c_2 + \delta c_2$ takes the place of two complete trigonometrical raytracings through the 'flanking bendings' plus the drawing and reading of a graph, and that for ordinary telescope objectives it gives practically as good a result, the great advantage of this hybrid method will be properly appreciated.

(4) The preceding method is a perfectly general one, for we can base it on any bending which is judged to lie near the desired solution. In photographic and in eyepiece lenses that bending may be located high up one of the branches of the aberration parabola. For spherically corrected simple cemented microscope and telescope objectives, and especially for the best type of these in which the pole of the aberration parabola lies closely below the zero-axis owing to the use of crown-

[48.5] glass of high index, we can render the general method yet a little more elegant and simple by specializing it in the following way :

Having made the usual TL, Chr. preparations, we begin the serious part of the work by locating the *pole* of the TL aberration parabola. As at that point $LA'p$ is a minimum, we have

$$\frac{dLA'}{dc} = y^2 l_b'^2 \, . \, P + 2\, y^2 l_b'^2 \, . \, c_2 \, . \, Q = 0$$

and this gives us the solution for c_2 :

$$c_2 = -\tfrac{1}{2} \cdot \frac{P}{Q}.$$

We calculate this, and from it $c_1 = c_a + c_2$, and use the resulting 'minimum aberration lens' as the trigonometrical basis of our solution by tracing the appropriate rays through it and finding the last radius which will give the prescribed chromatic correction. If the lens then comes out with spherical under-correction we can at once stop further work, for it will be necessary to lower the entire aberration parabola by substituting a crown of somewhat lower refractive index with the same V-value or one with the same index but a lower V-value. On the other hand heavy over-correction would render it advisable to make the opposite change in the crown-glass. Here we have one of the great advantages of this method. When the negative LA' is judged small enough to promise a good solution, but not small enough to consider the work finished, we proceed to the solution for δc_2. Here we have another simplification, for as the trigonometrically computed lens has $c_2 = -\tfrac{1}{2}\dfrac{P}{Q}$, the term $\tfrac{1}{2}\dfrac{P}{Q} + c_2$ in the general solution for δc_2 is zero and the equation becomes simply

$$\delta c_2 = \pm \sqrt{\frac{\delta LA'}{y^2 \, . \, l_b'^2 \, . \, Q}} \, .$$

When trigonometrically corrected for chromatic aberration, the solution so obtained will usually be far within tolerances ; but if a final differential correction should be considered desirable we have yet another simplification, for in the general equation for the required differential coefficient as given in subsection (3)

$$\frac{dLA'}{dc} = y^2 l_b'^2 \, . \, P + 2\, y^2 l_b'^2 (c_2 + \delta c_2) \, . \, Q,$$

the special value of $c_2 = -P/2\,Q$ again cancels nearly everything and gives, of course for this minimum aberration method only, the differential coefficient

$$\frac{dLA'}{dc} = 2\, y^2 l_b'^2 \, . \, \delta c_2 \, . \, Q.$$

The concise instructions now are :

(a) The usual TL, Chr. preparation is immediately followed by the determination of v_2', P and Q by the formulae in the preceding subsection.

(b) Calculate $c_2 = -\frac{1}{2}P/Q$ and $c_1 = c_a + c_2$, and submit this 'minimum aberration lens' to the trigonometrical test with adjustment of the last radius for chromatic correction. With the LA'_{found} calculate $\delta LA' = LA'_{required} - LA'_{found}$.

(c) With the TL values of y and l'_b calculate

$$\delta c_2 = \pm \sqrt{\frac{\delta LA'}{y^2 l'^2_b \cdot Q}},$$

and submit the selected solution to trigonometrical calculation with adjustment of the last radius for chromatic correction.

(d) In the final correction by the differential method, if it is considered to be necessary, use $dLA'/dc = 2\,y^2 l'^2_b \cdot \delta c_2 \cdot Q$.

This method is *not* suitable for spherically corrected lenses when the glasses to be employed are such as to yield an aberration parabola which dips deeply below the zero-axis, nor for solutions which would lie far from the pole of the aberration parabola by reason of a large value of '$LA'_{required}$'. For such solutions the preceding method, subsection (3), will answer if the key-bending for the trigonometrical foundation is skilfully selected. In difficult cases the purely trigonometrical method of section [48] will be the line of least resistance.

As a final numerical example illustrating method (4) we will select the following :

'Required, a simple cemented microscope objective of 1 inch focal length and giving a linear magnification of − 5 times.'

The stated magnification is of course that produced by the objective alone ; it does not include the further magnification of the primary image by the eyepiece. We shall not include eyepiece aberrations, as in *microscopes* of every kind they are nearly always really negligible from the Rayleigh limit point of view on account of the very slender cones of rays which the objective delivers to the eyepiece. The case is very different (as was indicated in section [40]) in the modern small telescope in which the objective delivers cones of rays with a marginal U' of from 4° up to 7° and even 8°!

At first sight it would appear that the only justifiable mode of procedure in computing any lens system must be to trace the rays as they really will go in the finished instrument, that is, from the object towards the image. But in the case of systems free from spherical and chromatic aberration all rays from the axial object-point meet again in the image-point, and it is self-evident that if we traced these rays backwards from the image-point towards the object-point, they would all be found to meet again in the latter, and would prove the system to be a spherically and chromatically corrected one in the reverse as well as in the intended direction of the rays. We can therefore establish full correction equally well by tracing the rays in whichever direction we please. Now there are very important advantages in working in the direction from the longer towards the shorter conjugate distance, or in the reverse of the actual direction in the case of microscope objectives. The principal advantages are :

(a) In proceeding from the original TL scheme towards the final fully corrected system, we have to introduce finite thicknesses and various bendings which, as we have already seen, change the focal length by a very perceptible percentage. If we

[48.5] worked towards the long conjugate, the latter would vary by a vastly increased percentage, and might indeed be altered from real to virtual, or vice versa. As in microscopes the long conjugate roughly corresponds to the tube-length, a very important datum, we could not adhere to the prescribed value without irksome complications in the computations. Working in the reverse direction we have the long conjugate distance as our fixed starting value and the little variations of focal length will only have a small effect on either the short conjugate or on the magnification, and these changes are quite unimportant and can be compensated by a final small adjustment.

(b) At the long conjugate distance the longitudinal aberrations would be subject to the misleading magnitude and wild acrobatic changes discussed in Chapter II, whilst at the short conjugate they are rendered small, and steady in change, by the strong convergence of the pencils of rays.

(c) Small final adjustments of the system are generally, and always in microscope systems, more easily effected by alteration of the lens—in microscopes the 'front-lens'—next to the short conjugate.

For these reasons we reverse our microscope objective so that the long conjugate distance is on the left. In the usual left-to-right computing direction the magnification will then be the reciprocal of that in the actual working direction, or $m' = -0{\cdot}200$. As $f' = 1{\cdot}000$ is prescribed, we can calculate the conjugate distances by TL (4) and find $l = f'(1-m')/m' = 1{\cdot}2/-0{\cdot}2 = -6$; $l' = f'(1-m') = 1{\cdot}2$; the latter will be the l'_b in the later stages of our work, while $l = -6$ is the initial l-value for all the calculations and by its reciprocal determines $v_1 = -\frac{1}{6} = -0{\cdot}167$.

We must now choose the glass for the two components of our achromatic lens, and in order to secure the advantages to be derived from crown-glass of high index we select from the abridged list in Chapter IV:

(a)	No. 9002	$Nd = 1{\cdot}5744$	$Nf - Nc = 0{\cdot}00995$	$V = 57{\cdot}7$
(b)	No. 360	$1{\cdot}6225$	$0{\cdot}01729$	$36{\cdot}0.$

For our computing direction we choose the crown-in-front type, which means that on the microscope the crown lens will be next the eyepiece and the flint lens next the object. TL, Chr. (4)* then gives, with $f' = 1$: $c_a = 4{\cdot}631$, $c_b = -2{\cdot}665$. We now calculate $v'_2 = v_1 + (N_a - 1)c_a = -0{\cdot}167 + 2{\cdot}660 = +2{\cdot}493$ and obtain, exactly as in subsection (3),

$$P = 7{\cdot}856; \qquad Q = 1{\cdot}167,$$

which give $c_2 = -\frac{1}{2}P/Q = -3{\cdot}928/1{\cdot}167 = -3{\cdot}366$;

and by adding $c_a = 4{\cdot}631$

$$c_1 = 1{\cdot}265.$$

Barlow's table then gives $r_1 = 0{\cdot}7905$, $r_2 = -0{\cdot}2971$ as the data of the 'minimum aberration lens' which will form the basis of our solution. We must now decide on a safe aperture in order to be able to fix an appropriate thickness of the components. An aperture of 1/4 of the focal length is usually quite safe for small lenses of this type and is therefore suitable for the first attempt. This gives $2Y = 0{\cdot}25$ or $Y = 0{\cdot}125$. The corresponding gross or 'edged' diameter would be about 0·30 in order to provide a shoulder for the lens to rest upon in its cell. Making a drawing [48.5] of the lens, a scale of 100 mm. to the inch being suitable, we find that a thickness of 0·055 would give a sharp edge, and as this is detested by the workshop, we decide on $d_1' = 0{\cdot}07$ as a practically suitable thickness. For the flint lens $d_2' = 0{\cdot}04$ will be appropriate.

We thus arrive at the prescription for the minimum aberration lens :

$$\begin{array}{lll} r_1 = & 0{\cdot}7905 & \\ & & 0{\cdot}07 \\ r_2 = & -0{\cdot}2971 & \\ & & 0{\cdot}04 \\ r_3 = & ? & \end{array}$$

and this has to be computed trigonometrically and the last radius has to be determined so as to produce chromatic correction.

On account of the low ratio of focal length to aperture, namely, four to one, it was decided to employ the strict four-ray method, which calls for the tracing of a paraxial and an extreme marginal ray in brightest light with index $Ny = Nd + 0{\cdot}188\,(Nf - Nc)$, and of a red and a blue ray, usually C and F, through 0·7071 of the full aperture. As the initial $l = -6$ of our lens is finite we require corresponding values of the initial U. It is quite unnecessary to make a highly precise calculation for this purpose ; we may safely use the approximate expression $\sin U = Y/l$, which for the extreme marginal ray gives $\sin U = 0{\cdot}125/-6 = -0{\cdot}02083$ and corresponds nearly enough to $U = -1\text{–}11\text{–}30$, which is the adopted value. The associated paraxial ray will as usual have $u = \sin(-1\text{–}11\text{–}30)$. For the two coloured rays we adopt the nearest angle, rounded off to 10 seconds, to 0·7071 of the marginal U, or $U_r = U_v = -0\text{–}50\text{–}50$. A five-figure ray-tracing in four parallel columns through the two first surfaces gives

$$L_{3r} = 3{\cdot}9238\,; \qquad L_{3v} = 4{\cdot}0949\,; \qquad L_{3y} = 4{\cdot}0927\,; \qquad l_{3y} = 3{\cdot}9230\,;$$

and from the first two by Chr. (1) $r_3 = -1{\cdot}3296$. Tracing the four rays through the third surface of this curvature we obtain the final results :

$$L_{3r}' = 1{\cdot}13128\,; \qquad L_{3v}' = 1{\cdot}13128\,; \qquad L_{3y}' = 1{\cdot}13360\,; \qquad l_{3y}' = 1{\cdot}12986\,;$$

$$L_{3r}' - L_{3v}' = 0\,; \qquad\qquad LA'_{trig.} = -0{\cdot}00374.$$

The chromatic correction is perfect and the spherical over-correction is so small that it is worth while to calculate the tolerance by OT (2) of section [38]. The marginal ray having given log sin $U_3' = 9{\cdot}02621$, we have with the usual wave-length in inches $= 0{\cdot}00002$:

$$\text{Permissible } LA' = 4 \text{ wave-lengths}/\sin^2 U_m' = \pm 0{\cdot}0071.$$

This is nearly twice the amount found for our minimum aberration lens, and we should be perfectly justified in accepting the result as satisfactory. The correcting process was applied, however, and with $\delta LA' = +0{\cdot}00374$, $y = 0{\cdot}125$, $l_b' = 1{\cdot}2$ (the TL value, *not* the above true value 1·13) the formula $\delta c_2 = \pm\sqrt{\dfrac{\delta LA'}{y^2 l_b'^2 \cdot Q}}$ gave $\delta c_2 = \pm 0{\cdot}377$.

[48.5] The high value of the change of curvature required to remove the small residual aberration is of course explained by the fact that our lens is located at the pole of the aberration parabola where the rate of change is very small. Using the positive value of δc_2, the corrected prescription is :

$$r_1 = 0{\cdot}6090$$
$$r_2 = -0{\cdot}3346$$
$$r_3 = -2{\cdot}8525,$$

the last radius being that found by Chr. (1) in the course of the trigonometrical calculation through the corrected lens.

The final results now obtained were :

$$L'_{3r} = 1{\cdot}12298\,; \qquad L'_{3v} = 1{\cdot}12289\,; \qquad L'_{3y} = 1{\cdot}12298\,; \qquad l'_{3y} = 1{\cdot}12363\,;$$
$$L'_{3r} - L'_{3v} = 0{\cdot}00009\,; \qquad LA' = +0{\cdot}00065.$$

The spherical aberration is now less than one tenth of the tolerance, and as the chromatic tolerance is always one quarter of the spherical one, or in our case $\pm 0{\cdot}00178$, the chromatic residual amounts to only one-twentieth of its tolerance. On the basis of our present knowledge the corrected lens would appear to be greatly superior to the one obtained from the direct solution for the minimum aberration. We shall discuss this case more closely a little later, when we shall find that the 'corrected' and apparently highly perfect lens would be nearly useless on account of its grave offence against the inexorable optical sine condition, whilst the direct solution is practically perfect in this respect.

As a first exercise in the application of the method (4), the solution for the flint-in-front form of the same microscope objective is recommended. It will probably give a smaller and therefore more favourable spherical over-correction of the minimum aberration lens, but this tendency of the flint-in-front form may prove so pronounced as to lead to a small under-correction.

USE OF TOLERANCES IN LENS-DESIGN

[50] It has already been stated that in designs for regular wholesale production we should seek the best solution possible at the price obtainable for the article, in order to have the widest margin to cover technical imperfections of the actual lenses and the effect of variation in the glass. But when only a few specimens are to be made at a reasonable cost, then the number of new tools and gauges must be kept down to the utmost, and this will only be possible if theoretical perfection is sacrificed and a clever compromise based on the proper application of our tolerances is substituted. A designer with a good stock of aberration-parabolas for the usual glasses will thus manage with very little trouble to produce almost any ordinary objective that may be asked for, entirely from stock-curves or at most with only one set of new tools and gauges. A time-limit for the completion of a design may similarly force the designer to depend upon stock-curves, for the making of tools and gauges is expensive because it takes a long time. In working out the tolerances from the figures obtained in the computations we must allow for the fact that part or all of the calculations are not made for the full aperture, but for 0·8 or 0·7071 of it.

For the chromatic aberration we have the focal range tolerance [50]

OT (1) Focal range $= 1$ wave-length$/N' \sin^2 U_m$;

but it must be worked out with the *marginal* U'_m, and as it diminishes inversely as the square of the aperture, we must multiply its value found for the U' of the 0·8 ray by 0·64 and its value for the 0·7071 ray by 0·5.

Using the sin U' of the actually traced coloured rays, we therefore have :

For the 0·8 aperture 3-ray method

chromatic tolerance $= 0{\cdot}64$ wave-length$/\sin^2 U'_{0\cdot8}$;

chromatic tolerance in inches $= 0{\cdot}000013''/\sin^2 U'_{0\cdot8}$.

For the strict 0·7071 aperture calculation

chromatic tolerance $= 0{\cdot}5$ wave-length$/\sin^2 U'_{0\cdot7071}$;

chromatic tolerance in inches $= 0{\cdot}00001''/\sin^2 U'_{0\cdot7071}$.

In our numerical examples by the 3-ray method log sin U' averaged about 8·81, therefore

log chromatic tolerance in inches $= \log(0{\cdot}000013) - 7{\cdot}62$

$= 5{\cdot}11 - 7{\cdot}62 = 7{\cdot}49$,

or chromatic tolerance $= \pm 0{\cdot}0031$ inch.

L'_{3y} might therefore differ from L'_{3v} up to this amount without making any *serious* difference in performance.

For the spherical tolerance we have

OT (2) spherical tolerance $= 4$ wave-lengths$/N' \sin^2 U'_m$.

For calculations at full aperture this is immediately available. In the 3-ray method we calculate for 0·8 of the full aperture, and the $\sin^2 U'$ for this will be only 0·64 of the marginal $\sin^2 U'_m$. At the same time the spherical aberration at 0·8 of the aperture is only 0·64 of the marginal. Combining the two corrections, we obtain $0{\cdot}64^2 = 0{\cdot}41$. Therefore the tolerance worked out with the sin U' of the 0·8 ray must be put at

spherical tolerance $= 1{\cdot}64$ wave-lengths$/\sin^2 U'_{0\cdot8}$;

spherical tolerance in inches $= 0{\cdot}000033''/\sin^2 U'_{0\cdot8}$.

For our numerical examples with $\log \sin U'_{0\cdot8} = 8{\cdot}81$ this gives

spherical tolerance $= 0{\cdot}0079''$ at 0·8 aperture.

On our plotting scale of $\times 20$ this means that any solutions would be available which lie on points of the parabolas within 0·16 inch of the horizontal axis, and it will at once be obvious that this greatly extends the range of available radii even if we draw upon only one half or one quarter of the full tolerance.

Once more we have a striking illustration of the possibilities of the apparently ridiculously small Rayleigh limit of $\frac{1}{4}$ wave-length or 1/200,000th of an inch for the permissible variation of the optical paths from object to image.

[50] The tolerances worked out for our telescope objectives show also that the five-figure calculations, which enable us to guarantee the chromatic and spherical aberration within 0·001 inch, are amply sufficient to exclude any risk of loss of quality through inaccuracy of the results.

It was forcibly pointed out in Chapter III that the tolerances should never be used to cover slipshod computing work. But there are the cases mentioned in the opening paragraph of this section, and the more numerous and important cases of which the microscope objective in section [48·5], subsection (4) is a first example, namely, when there are more conditions to be fulfilled than can be systematically and fully satisfied with the given type. In these cases we must seek the best possible compromise, and that will nearly always lead to drawing most heavily upon the *chromatic* tolerance, for the reason that achromatism is a decidedly indefinite thing when the ordinary optical glasses are employed. As was shown in section [42], quite considerable changes in the ratio of c_a to c_b merely shift the colour or wave-length for which the combination has minimum focal length, whilst the other colours are brought to a common focus in pairs from opposite sides of the minimum focus colour, thus continuing to satisfy the usual definition of achromatism. Moreover, we learnt that the normal visual correction of union of C and F, or minimum focus for wave-length $0{\cdot}5555\mu$, can only be claimed to be *satisfactory* by long experience ; there is no proof that it is the only admissible choice, nor even that it is the best possible one. On the contrary, the modern tendency in systems of ordinary size is emphatically towards a change of the minimum focus colour towards the full green or even blue green, or of the minimum focus wave-length down to $0{\cdot}515\mu$. Now a change of this order from the usual C and F correction amounts to several times our chromatic tolerance in telescope objectives like those used as numerical examples in this chapter, and we can safely conclude that we may use the full chromatic tolerance, more especially in the direction of under-correction, without any fear of adverse criticism with regard to the resulting type of achromatism. The procedure to be recommended in the cases referred to is therefore to try the effect of a change of the chromatic correction by the full tolerance, and this is easily done by Chr. (3) by putting the calculated tolerance into the second numerator, making the formula

$$\frac{1}{new\ r} = \frac{1}{original\ r} \pm \frac{\text{tolerance}}{(N_v - N_r)\,(L'_r\ .\ L'_v)_{\,original}}.$$

For our microscope objective this formula, when applied to the direct solution, changes r_3 from $-1{\cdot}3296$ to $-1{\cdot}2011$, and alters the final LA' from $-0{\cdot}00374$ to $+0{\cdot}00079$, thus showing that less than the single chromatic tolerance would lead to perfect spherical correction. The best compromise would probably be to use about one half of the chromatic tolerance, which would bring the spherical residual down to less than one quarter of its more important and definite tolerance. This problem will be again referred to in Chapter VII on the optical sine condition, because it is usually this highly important condition which renders it necessary to seek a compromise of the kind under discussion.

As the present chapter is the first one which deals with the solution of actual optical problems purely by calculation, it may appropriately be concluded by a few condensed remarks on certain technical aspects with which a designer should be [50] acquainted, as they affect the degree of precision to which solutions may profitably be carried and the extent to which we may expect agreement between the state of correction of the calculated system and its ultimate realization in glass and brass.

The first of these technical aspects concerns the radii of curvature and separations of the refracting surfaces. The accuracy to which these can be fixed by *calculation* obviously is only limited by the number of decimal places in the logarithms employed ; but as actual work can only be done with the glass and brass instrument produced in the workshops, the real question evidently is, how nearly a prescribed radius or separation can be executed in the *workshop*. It may be accepted that in a properly equipped and efficiently supervised modern 'glass shop', radii up to 10 mm. can be produced correctly within 0·001 mm., radii from 10 to perhaps 200 mm., with an accuracy of 1 part in 10,000, still longer radii with a gradually diminishing percentage accuracy which may be roughly estimated at perhaps an average of 1 part in 3,000, up to radii of several metres, and which then sinks lower and lower as the plane is approached. These data will show that there is no sense in prescribing radii more closely than to 1 part in 10,000 at the uttermost ; that is the reason why radii in our numerical examples are usually stated with four significant figures. In the vast majority of cases the precision stated as attainable in the workshop is amply sufficient to render the resulting departures from the computed data insensible, but there are exceptions. The front lens of an oil immersion microscope objective has, for instance, a radius of about 0·9 mm., which can therefore only be depended upon to about 1 part in 900 ; when a change of that order is tried by calculation without any other alteration, it is found to affect the correction seriously ; but fortunately both calculation and direct experiment prove that sufficient compensation can be secured either by a small change of the thickness of the front lens or by a small change of its separation from the lens behind it, or merely by using the complete objective with a moderately changed tube length. Small departures from the calculated tube length are in fact usually due to these inevitable errors in radius, thickness, and separation of the most deeply curved lenses of microscope objectives.

Small thicknesses and separations can be executed within 0·001″ or 0·025 mm. without greatly increasing the cost of production, and for high-power microscope objectives really competent workmen with proper tools will even work within less than 0·01 mm. when necessary. But in ordinary telescope work the demand for such precision would mean a prohibitive increase of costs, and a tolerance of ±0·01 inch is more nearly required and usually admissible. Computers should study designs, especially those intended for wholesale production at competitive prices, with reference to their sensitiveness to departures from the exact values of radii and separations, in order to be able to fix safe and yet not unnecessarily severe tolerances for these data.

With reference to the geometrical precision of the actual polished surfaces on lenses and prisms, this is of a high order in properly organized modern 'glass shops' because the surfaces are polished so as to produce broad rings or bands of interference colour on the principle of 'Newton's rings' when brought into close contact with a carefully prepared 'test-plate' of polished glass. An extremely close

[50] and amply sufficient approach to the ideal spherical or plane surface is thus secured. But there is a small but just sensible exception to be made with reference to the marginal part of any one polished surface. By reason of the fluid nature of the pitch now universally used for polishing optical surfaces, there is a slight excess of polishing action at these margins—the edge is polished or 'rounded' down—and although this effect is barely visible on any one surface even with the test-plate, it is cumulative when a pencil of rays passes successively through a number of glass-air surfaces at their full aperture, and produces the well-known effect that large lenses or systems have a strong tendency to prove spherically under-corrected when on paper they were perfectly corrected. A designer who has chiefly to deal with large work will be wise if he studies this effect, and he will probably find that the need of final 'figuring'—that is deliberate polishing down of selected zones—will be greatly reduced if these systems of large aperture are designed so as to be spherically over-corrected, *on paper*, to the extent of a considerable fraction, or even a low multiple, of the tolerance given by our formulae.

In large work another extremely serious disturbing factor comes in owing to actual variation of the refractive index in different parts of the same plate, due partly—but usually to a very small extent—to imperfect annealing and partly to imperfect mixing of the molten glass. These variations of the index may easily reach one unit in the fourth decimal place in any one plate, and a number of such units in different parts of the same melting. They are one of the strongest justifications of calculations even for the largest objectives with not more than six-figure logarithms. The greater and the most difficult part of the figuring invariably required by large objectives is due to this variation of the refractive index.

A few practical points affecting the cost and difficulty of executing designs may finally be briefly stated :

(1) Deeply curved surfaces are more expensive than those of moderate curvature, and should therefore be avoided or reduced in number as far as may be practicable.

(2) Very thin lenses should be avoided because they are easily distorted from their natural unstrained form. It is desirable that the axial thickness should be not much less than 1/10th of the diameter.

(3) Lenses with a very low value of $c_1 - c_2$, especially if they are meniscus-shaped, are an abomination from the glassworker's point of view ; hence they should only be used when there is absolutely no way of avoiding them.

(4) Certain types of optical glass, though of desirable optical properties, lead to greatly increased cost of production owing to excessive hardness or softness.

(5) Triple and multiple cemented combinations are very troublesome if they require to be centred with high precision ; preference should therefore be given to binary cemented combinations.

(6) Lenses with a very close *approach* to equiconvexity or equiconcavity are a terrible nuisance, as it is so difficult to distinguish quickly which is the deeper or shallower face.

To become really expert and efficient, an optical designer should seek opportunities of seeing, and if possible of actually doing himself, genuine glasswork, and he should try to keep in close and friendly touch with the workshops.

Special Memoranda from Chapter V

Strictly systematic methods are developed for the calculation of two-lens, usually cemented, object-glasses of prescribed focal length, chromatic and spherical correction, which are extensively used in all kinds of telescopes and low-power microscopes. A simple solution by TL and TL, Chr. formulae fixes the total curvatures of the crown and flint components which will give, with fair approximation, the desired focal length and chromatic correction. The prescribed spherical correction is then secured by bending either the cemented combination as a whole or one component in the case of a 'broken contact'. For a cemented lens there are usually two solutions with 'crown-in-front' and two more with 'flint-in-front', or a total of four solutions. With a broken contact the number of possible solutions becomes infinite.

By the method of [48] three bendings of a cemented objective at equal intervals of c_1 are chosen by simple rules and trigonometrically tested with correction of the last radius by Chr. (1) or (3). A parabolic graph of LA' plotted by the dip-method against c_1 as abscissa then gives directly a very close solution of the problem, which can be rendered exact by differential correction.

The algebraical solution by TL (10)**, section [49], yields approximate results with remarkably little trouble. The selected solution must, however, be tested trigonometrically and corrected by the differential method, possibly twice in succession ; the saving in time is thus rendered problematical.

The hybrid methods of section [48.5] will very frequently prove the most efficient.

The 'three-ray method' of trigonometrically determining chromatic and spherical aberration is sufficiently exact for the majority of ordinary problems, and then saves time.

The general discussion [49] (2) and the tolerances in [50] will repay careful study.

CHAPTER VI

EXTRA-AXIAL IMAGE-POINTS

[51] WE have hitherto limited our study of the aberrations to object- and image-points on the optical axis of centred lens systems. For object-points which do not lie on the optical axis we have had to rely on Lagrange's theorem which determines the linear magnification of the image, but only for the paraxial region of both the lens system and of object and image. We must be prepared to find that oblique pencils of finite aperture, especially when they come from object-points at a considerable distance from the optical axis, will not conform very closely to the paraxial magnification theorem of Lagrange, but will be affected by aberrations which alter the location as well as the sharpness of the corresponding image-points. The latter may therefore prove to be quite unsatisfactory even when excellent correction has been secured for the axial image-point by the methods which have already been given or by others to be dealt with subsequently. Hence we must study the properties of oblique pencils of rays before we can undertake the designing of optical systems which shall give satisfactory images throughout a field of view of finite extent.

As we can only give a certain limited aperture (at the utmost twice the shortest radius of curvature) to our lens systems, and as in many cases we have to introduce 'diaphragms' or 'stops' which still further restrict the light actually allowed to reach the image from any one object-point, we must begin our study of oblique pencils by noting how this inevitable limitation of the effective aperture affects the various pencils of rays.

Limitation of Pencils of Rays

A clear fundamental notion of the effects of a diaphragm will be most readily acquired by a careful study of the following diagrams, most of which refer to the same thin lens applied to objects at a fixed distance from that lens.

In Fig. 46 (a) the thin lens of given clear aperture is shown unobstructed by any diaphragm, and pencils corresponding to the full aperture are therefore transmitted both to the axial and to the extra-axial image-point.

Fig. 46 (b) shows the aperture of the lens restricted by a diaphragm placed close to it ; the effect is substantially the same as if the lens itself were edged down to the smaller diameter, and all the admitted pencils still pass practically through the same central part of the lens.

If we now proceed to Fig. 46 (c) we see the reduction of the aperture of the effective pencils effected by a diaphragm at a considerable distance from the first surface of the lens. It is evidently possible to give such a bore to this diaphragm as to admit an axial pencil of the same angular extent as in case (b), and the diaphragm of case (c) is shown dotted in Fig. 46 (b) to render this clear. If we now examine Fig. 46 (c) we see that whilst the axial pencil is exactly the same as in case (b), the effective oblique pencil now passes through the upper part of the lens and its rays meet the surfaces at angles of incidence very different from those in case (b) ; we may therefore safely anticipate that the aberrations at the extra-axial

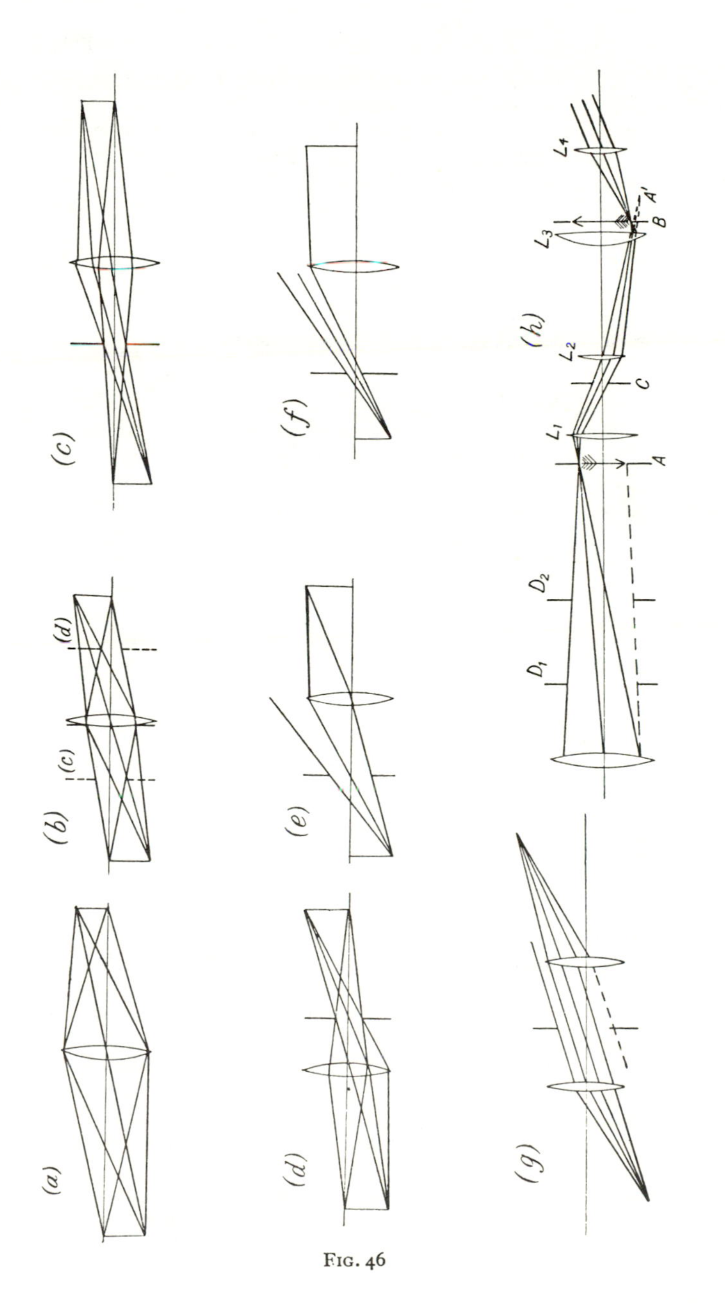

FIG. 46

[51] image-point will be likely to be very different from those in case (b), whilst the axial aberrations are necessarily identical in both cases.

A similar dislocation of the oblique pencil, but in the opposite sense, occurs when we place the diaphragm of appropriate diameter (see dotted diaphragm '*d*' in Fig. 46 (b)) behind the lens according to Fig. 46 (d). The oblique pencil now passes eccentrically through the lower part of the lens, and its aberrations are likely to prove different from those of either case (b) or case (c).

These simple diagrams show clearly that the position of the diaphragm must be taken into account when selecting the rays to be traced from an extra-axial object-point, and also that shifts of the diaphragm may prove a highly effective means of altering the state of correction of the oblique pencils.

Limitation of the Field by the Diaphragm

Another effect of the aperture-diaphragm should also be carefully noted, namely, the *limitation of the field of view* to which its presence leads whenever there is an appreciable axial separation between the diaphragm and the refracting surfaces of the lens or lens system. In Fig. 46 (c) and (d) the proportions have been so chosen that the complete oblique pencil which the diaphragm allows to pass just clears the extreme edge of the lens. It is at once obvious that if we traced the rays from an object-point farther from the optical axis than the one selected in cases (c) and (d), some or all of the rays admitted by the diaphragm would miss the lens. Taking Fig. 46 (c) as the starting-point, Fig. 46 (e) shows that for an object-point a little lower only about one half of the pencil admitted by the diaphragm falls upon the lens so as to be transmitted to the image, and Fig. 46 (f) represents the final limit of the available field of view, for the pencil admitted by the diaphragm from the new object-point, still lower than that in Fig. 46 (e), is seen just to miss the lens entirely, so that the latter cannot form any image of this point. Evidently the object-point shown in Fig. 46 (f) marks the extreme limit of the field which our lens with the chosen diaphragm can depict.

Unequal Illumination of the Field

An inevitable and important consequence of the limitation of the field of view by the aperture-diaphragm is that for all object-points below that shown in Fig. 46 (c) there will be a steady diminution of the light which reaches the image-point, simply because only a part of the rays admitted by the diaphragm is received by the lens. Consequently the image will be of approximately uniform brightness only up to the point of the field chosen in Fig. 46 (c) ; at the image-point represented in Fig. 46 (e) there will be only about half that brightness, and the latter dies down to zero at the limiting case, Fig. 46 (f). This falling off in brightness towards the margin of the illuminated field, which is due to the mutilation of the most oblique pencils by the combined effects of diaphragm and lens aperture, is a well-known source of trouble in photographic objectives when they are used at large aperture or close to the limit of their full illuminated field. It is usually referred to as the 'vignetting effect' on account of its obvious similarity to the gradual fading away of the outer parts (body and background) of photographic portraits which is produced by a suitably cut mask over the printing frame and which bears the same name.

In ordinary telescopes we usually obtain unobstructed light from the full clear [51] aperture of the comparatively thin object-glass in all parts of the field of view actually utilized. The clear aperture is then the common base of all the cones of rays which are admitted into the instrument ; the simple case of Fig. 46 (a) or (b), of 'central passage' of all the effective cones of rays, is therefore very nearly realized and there is no sensible mutilation of the oblique pencils and no appreciable vignetting effect.

Photographic 'landscape lenses' or 'meniscus lenses', with which the vast majority of inexpensive hand cameras are fitted, are usually fairly thin simple, or cemented achromatic, lenses with a diaphragm at some distance on the side of the distant objects ; they then realize case (c). Occasionally landscape lenses are designed for use with a diaphragm behind the lens, and case (d) is then realized. In either case the vignetting effect becomes very obvious and troublesome in the outer part of the field if a large diaphragm opening is used.

Modern photographic lenses of more ambitious design practically always consist of two more or less widely separated 'components' distinguished by the names 'front component' and 'back component' respectively, and the aperture-diaphragm is located between these components. The front component then comes under case (d) and the back component under case (c), and the vignetting effect in the outer parts of the illuminated field becomes more complicated because the two components usually mutilate the actually admitted pencils exceeding a certain critical obliquity from opposite sides, as is indicated in Fig. 46 (g).

The practical computer should carefully note that the eccentric passage of the oblique pencils will call for a much larger lens *diameter* than that which would suffice for the corresponding axial pencil. The consequence is that the convex elements of systems covering a large field call for a greatly increased thickness, which, as in all other cases, is most easily and safely ascertained by a reasonably accurate scale-drawing of the projected system.

Abbe's Treatment of the Diaphragm Problem

Having realized the fundamental importance of the position of the diaphragm in any centred lens system, we must next decide upon a convenient method of tracing rays with due regard to the location and aperture of the diaphragm. Assuming our usual direction of ray-tracing from left to right, it is easily seen that there would be no special difficulty in cases (b) and (c) of Fig. 46, for object-point and diaphragm being located in the same medium, there would be no geometrical difficulty in determining the direct course of the ray from the object-point (either axial or extra-axial) to any desired point in the diaphragm-aperture. But if we look at Fig. 46 (d) or (g) we find the lens system or one of its components intervening between the object-point and the diaphragm so that every ray will suffer a succession of bendings by refraction at lens surfaces before it reaches the diaphragm, and it looks as if nothing but a wearisome series of trials could enable us to find *that* ray from a given object-point which would eventually pass through some particular point in the diaphragm-aperture. This would in fact be the state of affairs if the selecting of the proper ray called for the full trigonometrical precision which we aim at in studying the residual aberrations at an image-point. But fortunately this

[51] high precision is quite unnecessary, for as a lens system, in order to be useful, must have only small residuals of aberration in its final image, *all* the rays forming that image must necessarily pass within a very short distance of the ideal location of the image, and, as a consequence, there can be only very little difference between the points of arrival of a ray from some *particular* point in the diaphragm and of rays from reasonably close *neighbouring* points of the diaphragm. For that reason an approximate method is nearly always admissible, and a very simple solution of the apparently formidable problem becomes available, which Abbe put into a convenient form accompanied by a most apt nomenclature.

Iris and Pupils of Optical Systems

The human eye is a complete optical instrument, closely resembling a photographic camera in its properties. It possesses an adjustable diaphragm in the centre of a disk of complicated structure which lies just in front of its lens. This disk, which also determines the colour of the eye by the predominating tint of its outside surface, is called the iris, and Abbe therefore introduced this name 'iris' for the *actual*, *material*, diaphragm which in any given lens system determines the diameter of the axial pencil which can pass through the system. When we look at a person's eye, we cannot see the actual iris directly because the cornea and the liquid which fills the space between cornea and lens are interposed ; what we see is a *virtual image* of the actual iris, slightly magnified by the lens-effect of the intervening media, and the central hole of the iris as it appears in this virtual image represents what we always call the 'pupil' of the eye.

If we looked from the side of the object towards the lenses in Fig. 46 (d) and (g), or from the side of the image in Fig. 46 (c) and (g), we should again obtain no direct view of the actual diaphragm, but should see the image of it (nearly always an erect virtual one) produced by the intervening lens, and on account of the complete analogy with the eye Abbe introduced the name 'Pupil' for these images of the actual diaphragm, calling the pupil seen from the object-space the 'Entrance pupil', and that seen from the image-space the 'Exit pupil', with symbolic abbreviations *EP* and *AP* respectively, the latter derived from the German name 'Austrittspupille'. We shall slightly depart from these symbols by adhering to the fundamental principle of our nomenclature concerning the use of the 'dash', and shall use *EP* as the symbol for the pupil in the medium to the *left* of any lens system or surface, and *EP′* for that on the *right*. We shall use these symbols even when the real diaphragm or iris itself is located in one of the media, for as the iris then obviously acts as pupil, no harm is done by calling it *EP* or *EP′* in such cases. The pupils solve the problem of defining the ray which will pass through a given point in the actual iris opening simply *because* they are images of the latter ; for by the very definition of an optical image all rays from a given object-point must pass through the conjugate image-point. The process of defining the entering ray therefore consists in first finding the point of the entrance pupil which is conjugate to the point of the diaphragm-aperture through which the ray is required to pass after refraction, and then determining the initial data of the ray from the object-point so that the ray aims directly at the proper point in the entrance pupil. The ray so determined will *not* pass with *absolute precision* through the desired point of the diaphragm, but it will do so within small quantities of the order of the transverse aberrations, and that is practically always sufficiently near, for the reasons already given.

Methods of determining the Pupils

As the problem now is simply that of determining the location and the magnification of the image of a given object, namely the diaphragm or iris, and as only a moderate degree of precision is required, we can usually proceed as follows :

(1) By paraxial ray-tracing. Knowing the distance of the diaphragm from the nearest lens surface, we trace a paraxial ray with that distance as initial l (or l' if right-to-left), and with any initial nominal u (or u') we please ($u = 0{\cdot}1$ being convenient as we need not turn up its logarithm in the book) from the centre of the diaphragm through all intervening refracting surfaces to the image- or object-space, finding a final l' or l which locates the centre of the pupil at that distance from the last refracting surface, and we then calculate the magnification of the pupil by Lagrange's theorem as $m' = u/u'$ (or $m = u'/u$). Any point at distance Y from the centre of the actual diaphragm, and in its plane would then be conjugate to a point at distance $m'Y$ (or mY) from the centre and in the plane of the pupil, and the ray from the object-point would have to be laid through the point so determined. Examples of this most generally useful method will be given subsequently.

(2) Occasionally, when the lens intervening between the diaphragm and the space from which oblique pencils are to be traced is a thin one and of known focal length, we can locate the centre of the pupil sufficiently closely by the thin lens formula ($1/l' = 1/f' + 1/l$) with a few figures from Barlow's table or from a slide-rule, and $m' = l'/l$ or $m = l/l'$, also by slide-rule.

(3) When the main object is the tracing of an oblique pencil at a finite angle of inclination, and when it is known (or can be estimated sufficiently nearly) at what angle of inclination the central ray of this pencil will cross the actual diaphragm, we can save time and obtain somewhat higher precision than by (1) if we trace this central ray (which will have to be done in any case) by the usual trigonometrical method. The magnification is then obtained by $M' = N \sin U/N' \sin U'$ in accordance with the optical sine condition which will be proved in the following chapter.

Other Types of Diaphragms

A few additional remarks on the important subject of diaphragms will be helpful in removing any remaining doubts and difficulties. In the first place it should be realized that the *aperture*-diaphragm exclusively discussed in the preceding sections must be placed at a considerable distance from either the object- or the image-plane ; a few additional diagrams which the student should draw himself will show that such a diaphragm, if placed near these planes, would require an extremely small aperture, and with any practicable lens-diameter would restrict the field severely and would correspondingly aggravate the vignetting effect and the inequality of illumination. Such a position of the 'iris' or pupils must therefore be avoided at almost any cost.

All types of diaphragms usually met with in optical instruments are represented in the diagram of a terrestrial telescope, Fig. 46 (h). As in practically all telescopes,

[51] the clear diameter of the object-glass acts as 'iris' or effective aperture-diaphragm, and the rest of the instrument should (with rare and undesirable exceptions) be so designed as to allow to pass, to every point in the actually utilized field of view, the whole of the light from the conjugate object-point which the object-glass admits. The characteristic three rays by which a point at the extreme margin of the intended field is depicted are traced right through the instrument. The object-glass forms the primary inverted image at A. This image acts as object for the two lenses L_1 and L_2, which together form the 'Erector' and act exactly like a microscope objective. By itself the Erector would produce at A' a secondary erect image of the external objects. This image, however, is intercepted by the field lens L_3 of the Huygenian eyepiece L_3L_4, and a reduced real secondary image is formed at B and viewed by the eye lens L_4.

If there were no additional diaphragms in this telescope, then the limitation of the field would be produced on the vignetting principle by the edge of one or more of the lenses L_1 to L_4, and on looking through the instrument we should see a field gradually fading away towards its extreme limit, and in all probability there would also be very defective definition and unpleasant colour effects near the margin of the field. To remove these defects, a diaphragm is always placed between the lenses L_3 and L_4, so that its edge is seen sharply defined through the eye lens L_4, thus putting the desirable definite limit to the field of view. Such a diaphragm may appropriately be called a 'field stop'; its clear aperture should be adjusted so as to cut off the vignetted and possibly ill-defined margin of the full field afforded by the undiaphragmed eyepiece. A field stop may be placed into any plane where a reasonably sharp image of the external objects is formed, and it is frequently advisable to place one into every image-plane, especially in long narrow instruments like submarine periscopes. But any duplicate field stops should have a little excess diameter compared with the principal field stop, which latter is usually the one nearest to the eye-end of the instrument; for if the field stops were all calculated to embrace exactly the same extent of view, then any slight defect of centring or a slight bending of the instrument would cause mutual overlappings and an unsightly non-circular final field of view. In the terrestrial telescope there is an opportunity for the placing of a secondary field stop at A, but it is hardly ever used.

The aperture-diaphragm may also be duplicated with beneficial results in instruments in which real images of the actually effective iris are formed. The terrestrial telescope offers an opportunity of that kind which is invariably taken advantage of. At C where all the oblique pencils cross the optical axis, a real image of the clear aperture of the object-glass is formed by the first lens L_1 of the Erector. A diaphragm placed into this position, with an aperture slightly in excess of the diameter of the image of the object-glass, will not obstruct the really useful light in any way, but it will stop all false light reflected by the inside of the telescope tube from going any farther; this 'erector stop' thus greatly increases the brilliancy of the image. Unfortunately, this stop is frequently misused by cheapjack opticians for the purpose of surreptitiously cutting down the effective aperture of a bad object-glass, simply by making the hole *smaller* than the full diameter of the image of the object-glass, thus improving the definition of the final image at the expense of its brightness.

A third type of diaphragm is chiefly found in astronomical and in very long terrestrial telescopes. It is shown (although not usually employed in the type) at D_1 and D_2 in Fig. 46 (h). The purpose of these diaphragms is to stop light from the sky, which is reflected by the always more or less shiny internal surface of the telescope-tube from reaching the eye of the observer. These 'glare stops' should be so placed and dimensioned that they do not cut off any useful light; they will satisfy this important condition if they just clear a truncated cone constructed with the clear aperture of the object-glass as base and the required clear diameter of the primary image as top. [51]

THE PRIMARY ABERRATIONS OF OBLIQUE PENCILS

[52] We can now secure a sound and clear knowledge of the nature and of the principal laws of the oblique aberrations, based upon the foundations laid in previous chapters, if we restrict our investigation to the *primary* aberrations. It was shown in Chapter II that the laws of spherical aberration for axial object-points assume their simplest form if we restrict the aperture to what we called the extended paraxial region, and this simplification becomes even more pronounced in the case of the aberrations of oblique pencils; for whilst for the axial pencil in any one colour there is only one primary and one secondary spherical aberration, a corresponding oblique pencil suffers from five different primary aberrations (usually called the Seidel aberrations), to which the next approximation adds no fewer than nine different secondary aberrations, which latter have in fact never yet been rendered amenable to convenient numerical determination.

In order to realize clearly the problem confronting us, we will begin by examining Fig. 47 (a), in which a spherical refracting surface is shown with centre at C

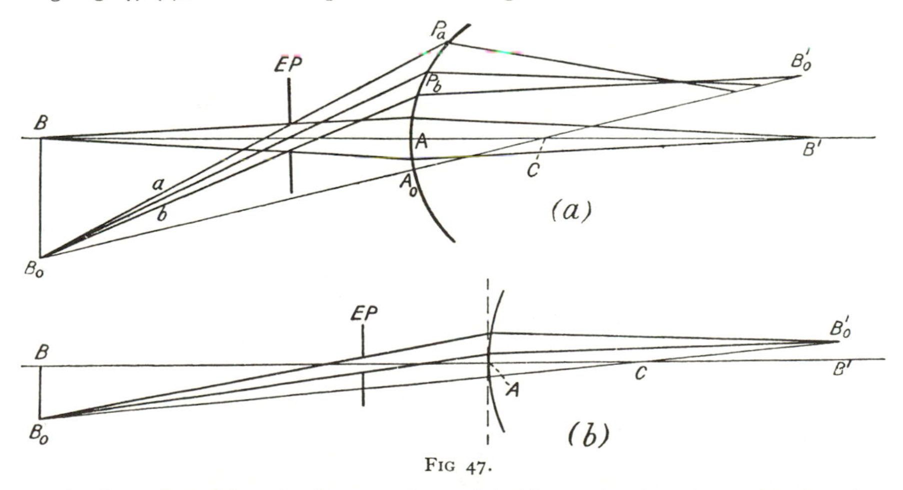

FIG 47.

and pole at A, and forming images of an axial object-point B at the conjugate point B', and of an extra-axial point B_o at the conjugate B'_o, the pencils admitted to the

[52] surface being regulated and limited by the diaphragm or entrance pupil *EP*. The axial pencil will as a rule be subject to spherical aberration following the law deduced in Chapter II :

$$LA' = a_2 Y^2 + a_4 Y^4, \text{ \&c.,}$$

and its marginal rays will display a longitudinal spherical aberration found by putting the distance *PA* of *P* from the optical axis *AC* into this equation as the value of *Y*. If we now turn our attention to the extra-axial object-point B_o, we can (as in the proof of the Lagrange theorem in Chapter I) draw the auxiliary optical axis B_oA_oC. With reference to this axis the rays from B_o will be subject to laws exactly corresponding to those applying to *B* when that point is referred to the principal optical axis *BAC*. It is therefore perfectly legitimate to trace a paraxial pencil from B_o, and thus to determine the conjugate point at which rays very close to the auxiliary axis would come to a focus. Rays leaving B_o under considerable angles with the auxiliary axis, say the extreme rays B_oP_a and B_oP_b admitted by our entrance pupil, will then be subject to the ordinary law of longitudinal spherical aberration according to the distance of the points of penetration like P_a and P_b from the auxiliary axis, and the point of intersection of these rays with the auxiliary axis, after refraction, can be determined by that law when the coefficients of the latter are known. For a surface of given radius, which separates media of given refractive indices, the coefficients of the law of spherical aberration depend solely upon the distance of the object-point from the surface, and as even in the case of our present diagram with its decidedly bold obliquities the distance B_oA_o evidently differs very little from the *BA* of the axial pencil, we may safely assume that the coefficients of the *LA'* equation for the oblique pencil will not differ very seriously from those for the axial pencil. Ignoring whatever moderate difference there may be, we then arrive at once at one of those simple fundamental facts which every optical designer should always have clearly in his mind, namely, that oblique pencils are far more seriously affected by aberration than the corresponding axial pencil ; for in our diagram, which is not a highly abnormal one, the ray *b* of the oblique pencil passes the surface at about three times, and the ray *a* at about five times the distance from the auxiliary axis as compared with the corresponding distance *PA* of the extreme rays of the axial pencil. Therefore by the law $LA' = a_2Y^2 + a_4Y^4$, ray *b* will have a primary *LA'* 9 times as large and a secondary *LA'* 81 times as large as the marginal rays of the axial pencil, and for the still more remote ray '*a*' the corresponding multiples will be 25 and 625 for primary and secondary aberration respectively.

There is one solitary exception from this extremely serious aggravation of the aberration in the oblique pencils, namely, the case when the entrance pupil is located at the centre of curvature of the refracting surface, in which case the conditions for axial and oblique pencils are obviously identical. This case is rarely realized completely ; but the equally obvious conclusion that a position of the pupil reasonably near the centre of curvature keeps the aggravation of the aberrations comparatively low, supplies the explanation why photographic objectives, and especially those for a field of unusually wide angle, nearly always take the form of two meniscus-shaped components which turn their concavities towards the diaphragm placed more or less midway between them. The usually very serious [52] aggravation of secondary aberrations in the oblique pencils must be well remembered, for it will be found over and over again that systems with barely discoverable traces of secondary aberration in the axial pencil display huge amounts of it in pencils of great obliquity: 'Freedom from sensible secondary aberration in the axial pencil admitted by the diaphragm by no means implies similar freedom in the corresponding oblique pencils.'

For the purpose of our investigation we shall employ the method illustrated by Fig. 47 (a) of treating the oblique aberrations as a special form of the ordinary spherical aberration. But as we have seen that the secondary aberrations are far more serious in the oblique pencils than they are in the corresponding axial pencil, it will be realized that we shall have to restrict both aperture and field very drastically in order to limit our investigation to the pure primary or 'Seidel' aberrations with which we seek acquaintance. We shall in fact have to treat mathematically the limiting case when both aperture and field are indefinitely small, and that will justify a number of simplifying assumptions. Taking Fig. 47 (b) as representing more nearly (although still with gross exaggeration) the case to be discussed, we see in the first place that the difference between the distances BC and B_0C becomes very small; therefore the *same* value of the coefficient of the primary LA' formula may be used for both the axial and the oblique pencil. In the second place the small segment of the refracting surface which now comes into action will depart so slightly from its tangent plane at the pole A that we shall make a negligible error by measuring the diameter of the axial and of the oblique pencil on that tangent plane instead of on the curved surface, and in also measuring the distance from the auxiliary axis of points of incidence like P_a and P_b in Fig. 47 (a) in the same tangent plane instead of along the true perpendicular to the auxiliary axis. In the third place all angles of rays with either axis or with each other, and the angle between the principal and the auxiliary axis, will be of paraxial magnitude, and their sines, tangents, or radian-value may be treated as interchangeable and their cosines or secants as equal to 'one'. These perfectly justified, and in fact necessary, simplifications resulting from the restriction of aperture and field should be well remembered throughout the following deductions, because we shall now have to return to hugely exaggerated diagrams in order to render visible the small aberrations which we seek to determine. In the more rigorous treatment of the oblique aberrations which is reserved for Part II, it will be shown that the simplifying assumptions which have been laid down have no effect on the value of the *primary* aberrations, but that due consideration of the disturbing effects (which these assumptions exclude in the present investigation) leads to the vast complication of the problem of determining the true *secondary* aberrations which up to the present has rendered the complete evaluation of the latter a task of almost hopeless complexity. Our present aim is not to allow the vital fundamental facts to become drowned in a vast mass of complicated algebra, but to gain a clear, simple, and yet perfectly reliable knowledge of the primary aberrations, and to prove the effectiveness of the usual practical computing methods which are employed in the design of optical instruments.

It should be clearly understood that whilst the primary oblique aberrations repre-

[52] sent the *exact truth* only for a very small aperture and a very small field of view of any ordinary centred lens system, they share with the primary spherical aberration of axial pencils the valuable additional property of giving a good *first approximation* to the real state of correction of pencils of quite considerable aperture and obliquity ; they will therefore be available in the case of the oblique pencils for the purpose of quickly finding the approximate neighbourhood in which the solution of any given problem is to be looked for, so that the final search by exact trigonometrical calculations can be greatly shortened by differential correcting methods like those used in the preceding chapter. This is an extremely valuable *practical* application of the Seidel aberrations. On the other hand the Seidel aberrations as true primary aberrations of the oblique pencils enable us to prove the presence and to determine the magnitude of higher aberrations ; for any disagreement between the defects indicated by the Seidel aberrations and those found by rigorous trigonometrical calculations can only be due to secondary and still higher aberrations. For these reasons the primary oblique aberrations are of fundamental importance in the design of all instruments which are required to cover a considerable field of view ; they must on no account be looked upon as mathematical abstractions of such severely limited validity as to render them worthless for practical purposes. The following pages should therefore be studied and mastered with the utmost care and attention.

GENERAL SOLUTION OF THE PROBLEM

[53] We restrict the present solution to the case of light of one definite colour, thus leaving the question of achromatism for separate treatment. With regard to the lens system we only assume that it is centred, and therefore perfectly symmetrical with reference to its optical axis, and that all its refracting surfaces are strictly spherical, but we make no restrictive assumptions as to the thickness of the constituent simple lenses or as to their number or separations. As it is the object of our investigation to determine the *defects* of the images produced by the system, the complete image corresponding to any one point in the original object will as a rule be a more or less confused patch of light without any very definite location ; therefore we must adopt definite ideal locations of the images to which the confused bundles of rays can be referred. As the location of the image depends on that of the original object we must first define the latter. It would be manifestly impossible to adopt an ordinary view such as usually confronts a photographic camera, because the great variety of distances in such a view would complicate the problem hopelessly by involving questions of depth of focus. We therefore take as our object a flat drawing or test-diagram placed at some definite distance from the lens system and at right angles to the optical axis. The ideal rendering of such an object is obviously represented by a flat image which may be either larger or smaller than the object, but which should be sharp and true to scale throughout its whole extent ; in other words, the image should be geometrically similar to the object. On account of the symmetry of a centred lens system with reference to its optical axis we can at once conclude that to the already adopted object-plane at right angles with the optical axis there can only correspond an ideal image-plane also at right angles with the optical axis. Although in the actual use of the system

these strict definitions are only important with reference to the original object and the final image, it will be convenient to use corresponding ideal image-planes for all the intermediate images produced after refraction through the successive surfaces of the system.

Having decided that our objects and the ideal images shall all be located on planes at right angles with the optical axis, we can make only one reasonable choice as to the points at which the successive ideal image-planes shall cut the optical axis, for we found in Chapter I that the one perfectly definite image-point that can be easily calculated is the paraxial one. Therefore we place the ideal image-planes at the successive intersection-points of a paraxial ray traced from the original axial object-point right through the entire system. In all that follows we assume that this ray has been thus traced and that all the l, l', u and u' of it are known.

It only remains to fix a convenient value of the magnification of the images in each of our ideal image-planes. Again the fundamental equations of Chapter I indicate the proper selection, for we proved in section [18] that in the immediate neighbourhood of the optical axis the linear magnification is definitely determined by Lagrange's theorem ; therefore the only possible way of obtaining identical magnification throughout the whole of any one of our adopted paraxial image-planes is to take the value given by Lagrange's theorem, simply because no other value is possible in the central part of the field. This inevitable selection of the value of m' fits in perfectly with our adopted method of tracing the oblique pencils by reference to a thin radial pencil along the auxiliary optical axis. For, with reference to Fig. 48, in which A is the pole and C the centre of curvature of a refracting

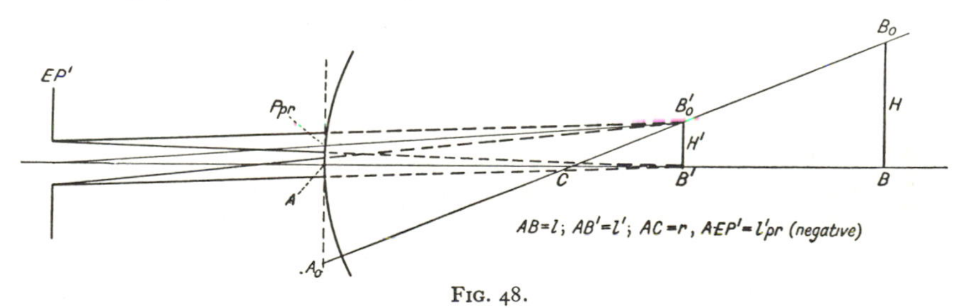

FIG. 48.

surface of radius r, B the axial object-point, and B' its image by paraxial rays, we see that if we draw the auxiliary optical axis from C to any extra-axial object-point B_o at distance H from the principal optical axis, then this auxiliary axis cuts the paraxial image-plane at a point B'_o at a distance H', such that by similar triangles we have

$$\frac{H'}{H} = \frac{CB'}{CB} = \frac{l'-r}{l-r}.$$

This ratio of image- to object-height agrees exactly with one of the alternative values of m' derived from Lagrange's theorem in equation (10p)* of section [18], namely, $m' = \dfrac{h'}{h} = \dfrac{l'-r}{l-r}$, and the cut of the auxiliary axis with the paraxial image-

[53] plane therefore marks directly and automatically our adopted ideal location of the image and its ideal magnification.

In accordance with the general principles laid down in section [52] we must now determine the constant of the longitudinal primary spherical aberration for each surface of our centred lens system. Putting for the present investigation

$$LA'p = a \, . \, SA^2,$$

'a' taking the place of the previous a_2 and SA ('semi-aperture') that of the previous y or Y, we have to determine 'a'. Having traced a paraxial ray through the system, we have all the data for the calculation of the new primary aberration arising at each surface by the equation of section [28]

$$Sph'_j = \tfrac{1}{2} \, l'_j \, \frac{l'_j - r_j}{r_j} \, (i'_j - u_j) \, (i_j - i'_j),$$

and this represents $LA'p_j = a_j \, . \, SA_j^2$ for the jth surface if in accordance with the fundamental paraxial equations we put $SA_j = l'_j \, . \, u'_j$. We therefore have

$$a_j \, . \, l'^2_j \, u'^2_j = \tfrac{1}{2} \, l'_j \, \frac{l'_j - r_j}{r_j} \, (i'_j - u_j) \, (i_j - i'_j),$$

and this gives by a simple transposition

(a) $$a_j = \tfrac{1}{2} \, \frac{l'_j - r_j}{r_j \, . \, l'_j \, . \, u'^2_j} \, (i'_j - u_j) \, (i_j - i'_j),$$

by which the constant of the primary longitudinal spherical aberration can be calculated for every surface of the system.

Turning our attention to the oblique pencils, we have seen in section [51] that their course through the system depends upon the position of the aperture-limiting diaphragm and of its images, the entrance and exit pupils, as formed by all the surfaces of the system. This part of the work must therefore begin with the tracing of a paraxial ray from the centre of the actual diaphragm through the entire system. The intersection-points of this ray for the media to right and left of each refracting surface locate the exit and entrance pupils EP' and EP of that surface. We shall call this ray the 'principal ray' and shall distinguish its usual data by adding the suffix *pr* (for principal) to the symbols, so that u_{pr} signifies the convergence angle, i_{pr} the angle of incidence, and l_{pr} the intersection-length of the principal ray in the medium to the left of any surface, with the usual dash for the corresponding data in the medium to the right of the surface. The principal ray will have to be traced left-to-right if the diaphragm is entirely to the left of the system, right-to-left if the diaphragm is to the right of the complete system, and partly to left and partly to right if the actual diaphragm is located within the complete system, as is most usual.

For the purpose of locating the pupils the original nominal convergence angle of the principal ray at the actual diaphragm may be assumed at any convenient value, such as 0·1000, for it was proved in section [17] that the same paraxial intersection-lengths are found, no matter which initial u-value may have been adopted.

Returning to Fig. 48, EP' shall be the position of the exit pupil for the refracting [53] surface at A, determined by the value of l'_{pr} for this surface as found by the above tracing of the principal ray. In accordance with the definition of the pupils we then have an axial cone of rays converging upon B', and an oblique cone converging upon B'_o, both cones having the same base defined by the diameter of the pupil. Under our simplifying assumptions of a very small aperture and a very small value of H' the refracting surface will depart only very slightly from its tangent plane at A, and the sections of the two cones with the tangent-plane and with the refracting surface will differ only by insensible amounts. Selecting the intersection of the cones of rays with the tangent-plane as the one most easily determined, we see that this tangent-plane is parallel to the plane of the common base of the two cones at EP' and also parallel to the paraxial image-plane at B' in which both cones come to a point at B' and at B'_o respectively. By well-known theorems of solid geometry it follows at once that the tangent-plane cuts both cones in figures which are equal to each other, and similar to the aperture of the exit pupil in the fixed proportion of AB' to $EP' - B'$. As this conclusion applies to every surface of the system with reference to its particular exit pupil, and as all the exit pupils are images of the actual diaphragm at the fixed linear magnification following from the theorem of Lagrange, we have now proved that within the extended paraxial region, in which the primary aberrations may be regarded as representing the exact truth, both the axial and all admissible oblique pencils cut any one refracting surface in geometrically equal figures which are similar to the aperture of the actual diaphragm, and which therefore grow or diminish in size in strict proportion to every change in the actual diaphragm. This is the most valuable of the simplifications which result from the restriction of our investigation to small apertures and a small field. When the aperture becomes considerable, and especially when the obliquity for extra-axial object-points increases simultaneously, the difference between the intersections of a bundle of rays with the tangent-plane and with the refracting surface itself becomes very sensible and leads to a first batch of awkward correcting terms. Simultaneously aberrations come in with considerable magnitude, affecting on the one hand the location and configuration of the exit pupils in a degree which varies for different points in the field of view, and on the other hand the individual bundles of rays cease to be cones with a practically sharp point and cause yet further complications. It is easily seen that all these complications will become particularly grave when the exit pupils are located at a considerable distance from the respective surfaces and when the separations between successive surfaces or complete components of a system are considerable. The restrictive effect of all these complications can in fact become so severe as to preclude the primary aberrations from giving even a useful first approximation in extreme cases, and trigonometrical methods may have to be employed from the very beginning. Such highly unfavourable cases are rare exceptions, however, and are not likely to lead to a favourable final result by any method, because the bad approximation by the primary aberrations can only be due to the presence of exceptionally heavy higher aberrations.

We will now use Fig. 48 to determine the most fundamental datum for the calculation of the oblique aberrations. The oblique cone of rays with base at EP'

[53] and apex at B'_o has as its axis the ray from the centre of the exit pupil to B'_o. By reason of its passing through the centre of the exit pupil this ray is a principal ray, and as it also passes through the image-point B'_o we shall call it the principal ray of the oblique pencil under consideration. In accordance with the general explanations in section [52] the spherical aberration imparted to this ray by refraction at the surface will be proportional to the constant 'a' of spherical aberration as determined by equation (a), and proportional to the square of the distance of its point of incidence P_{pr} from the auxiliary axis $A_oCB'_o$. We stipulated at the end of section [52] that we shall measure the incidence height of all rays in the tangent-plane at A, and we have therefore to find a general algebraical expression for the distance $A_oP_{pr} = A_oA + AP_{pr}$. For A_oA the similar triangles A_oAC and $B'_oB'C$ give the proportion

$$A_oA/r = H'/(l'-r)$$

or

$$A_oA = H'.r/(l'-r).$$

For the part AP_{pr} of the incidence height the similar triangles $A-P_{pr}-EP'$ and $B'-B'_o-EP'$ give, on bearing in mind that the l'_{pr} in the diagram is negative according to our sign-conventions, whilst all the other data are positive,

$$AP_{pr}/(-l'_{pr}) = H'/(l'-l'_{pr})$$

or

$$AP_{pr} = -H'.l'_{pr}/(l'-l'_{pr}),$$

and on combining A_oA and AP_{pr} we obtain the incidence height of the principal ray of any oblique pencil as

$$\text{(b)} \quad \begin{cases} A_oP_{pr} = H'\left(\dfrac{r}{l'-r} - \dfrac{l'_{pr}}{l'-l'_{r}}\right) = q'H', \text{ if we put} \\[2ex] q' = \dfrac{r}{l'-r} - \dfrac{l'_{pr}}{l'-l'_{pr}}. \end{cases}$$

We shall use $q'H'$ in our work to shorten the expressions, but must remember that q' has to be calculated by equation (b) just as the aberration constant 'a' has to be calculated by equation (a). The importance of equation (b) arises from the fact that it shows how the image height H' finds its way into the formulae for the oblique aberrations.

We are now ready to deduce our fundamental equations for one refracting surface. In Fig. 49 let C be the centre of curvature of the refracting surface at A, B the paraxial or ideal object-point, B' the corresponding image-point. We lay our ideal object- and image-planes through these points. To find the ideal image-point corresponding to the extra-axial object-point B_o at object height H we draw the auxiliary axis A_oCB_o; its cut at B'_o with the ideal image-plane is the ideal image-point to which we have to refer our traced rays. If EP is the entrance pupil of our surface, then $EP - B_o$ will be the ideal principal ray of the arriving oblique pencil; it will cut the refracting surface, and under our restrictions sensibly also its tangent plane, at a definite point P_{pr} in the plane of our diagram, and we have determined its incidence height A_oP_{pr} with reference to the auxiliary axis as $= q'H'$ by equation (b). Any outer ray of the oblique pencil will cut the refracting surface at some

[53] definite distance from P_{pr}. Referring to the end-view (as seen from any axial point to the *right* of the surface) of the refracting surface, or strictly of its tangent-plane, as shown in the supplementary diagram to the left of the principal one, we select

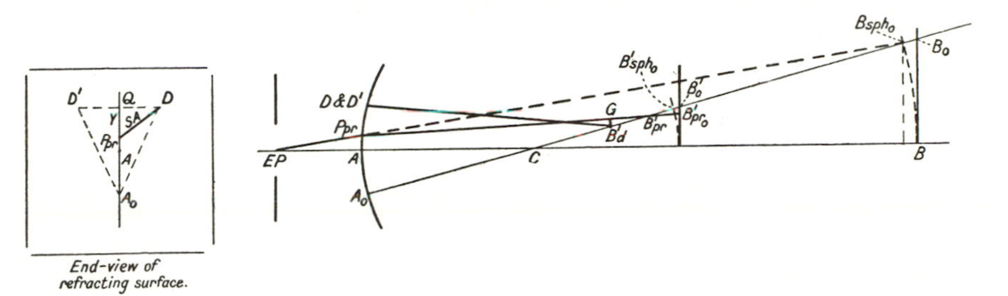

FIG. 49.

as our 'general ray' one passing through the point D at distance SA from P_{pr}, but as this 'semi-aperture' alone would apply to all points of a circle with P_{pr} as centre and SA as radius, we must add a second coordinate, for which we choose the distance $P_{pr}Q = Y$ of the projection of D upon the plane of the main diagram from P_{pr}. SA shall have the sign of the corresponding $y = lu = l'u'$ of the traced paraxial ray, and Y shall have the positive sign if it falls above P_{pr} and the negative sign if it falls below P_{pr}. As the incidence height of any ray of an oblique pencil must be measured from the auxiliary axis, we now have A_oD as the incidence height of our general ray and the spherical aberration of the latter with reference to a thin radial pencil traced along the auxiliary axis will be proportional to $(A_oD)^2$. Evidently this same incidence height applies also to the symmetrically placed incidence point D' which is found by producing DQ by its own length. Therefore the rays from D and D' will unconditionally (for the assumed *centred* system) have the same longitudinal aberration and must cut each other accurately in one point, not only after one, but after any number of successive refractions, and this intersection-point or focus of any 'D-pair', as we shall call it, must lie in the plane of our principal diagram; moreover, it follows from symmetry that the optical paths of the two rays of such a pair must always be exactly equal so that the focus formed by them is not merely a geometrical intersection but also a true physical focus resulting in a strong concentration of light.

For the determination of $(A_oD)^2$ the right-angled triangles A_oQD and $P_{pr}QD$ give

$$\text{(c)}\quad (A_oD)^2 = (A_oQ)^2 + (QD)^2 = (q'H' + Y)^2 + (SA^2 - Y^2)$$
$$= SA^2 + 2\,q'H'.\,Y + (q'H')^2.$$

We now return to the principal diagram of Fig. 49, and examine for the present, not the actual ideal object-point B_o, but the point $Bsph_o$ at which the auxiliary optical axis for point B_o cuts a sphere struck with C as centre and $CB = l - r$ as radius. A thin radial pencil along the auxiliary axis with $Bsph_o$ as focus will then have the same intersection-length as the corresponding paraxial pencil aiming at B, and the intersection-lengths resulting from refraction at our surface AA_c must also be equal. Therefore the thin radial pencil aiming before refraction at $Bsph_o$

[53] must after refraction find its focus at $B'sph_o$, on a sphere struck with C as centre and $CB' = l'-r$ as radius. As the axial and the oblique thin pencils have the same intersection-lengths, the constant of spherical aberration must also be the same and equal to the 'a' determined for our surface by its paraxial data introduced into equation (a). Hence the spherical aberration of our D-pair of rays will be

$$a\,.\,(A_oD)^2 = a\,.\,SA^2+2\,a\,.\,q'H'\,.\,Y+a(q'H')^2,$$

which means that with incident rays aiming at $Bsph_o$, the refracted rays from D and D' would cut each other on the auxiliary axis at a point to the left of $B'sph_o$ by the stated amount if the spherical aberration be positive.

We now ask, what happens when the object-point is shifted from the assumed position at $Bsph_o$ to its true position at B_o, that is by the *small* distance $Bsph_o\,B_o$. Evidently this small longitudinal shift is subject to the law of longitudinal magnification determined by equation (10p)*** of section [19] as

$$\bar{m}' = \frac{N\,.\,u^2}{N'.\,u'^2}.$$

The distance $Bsph_oB_o$ is, under our restrictions, sensibly equal to its projection upon the principal optical axis and is thus recognized as the x-value of the spherical surface $BBsph_o$ of radius $l-r$. The y-value is sensibly equal to $BB_o = H$, therefore equation (8p) gives $Bsph_o\,B_o = \frac{1}{2}\frac{H^2}{l-r}$, and applying the law of longitudinal magnification we find

$$\text{transferred value of } Bsph_o\,B_o = \tfrac{1}{2}\frac{H^2}{l-r}\cdot\frac{Nu^2}{N'u'^2};$$

and as this must be applied in the same direction as the original value, or to the right, it is subtractive with reference to the longitudinal aberration determined for the original object-point $Bsph_o$ and produces, as the distance from the D-focus to $B'sph_o$,

$$B'_dB'sph_o = a\,.\,SA^2+2\,a\,.\,q'H'\,.\,Y+a(q'H')^2-\tfrac{1}{2}\frac{H^2}{l-r}\cdot\frac{Nu^2}{N'u'^2},$$

where we have introduced B'_d as the adopted symbol for the true focus of the refracted D-rays when the incident rays aim at B_o. It now only remains to add the distance $B'sph_oB'_o$ in order to refer the aberration to our adopted ideal image-plane. By analogy with the proof given above for the corresponding distance $Bsph_oB_o$ we can write $\frac{1}{2}\frac{H'^2}{l'-r}$ for this, and adopting LA'_d as a suitable symbol for the longitudinal spherical aberration $B'_dB'_o$ of our D-pair of rays, we have

$$LA'_d = aSA^2+2\,a\,.\,q'H'\,.\,Y+a(q'H')^2+\tfrac{1}{2}\frac{H'^2}{l'-r}-\tfrac{1}{2}\frac{H^2}{l-r}\cdot\frac{Nu^2}{N'u'^2}.$$

The two last terms are capable of great and highly significant simplification. The Lagrange theorem $HNu = H'N'u'$ gives on being squared and transposed

[53]

$H^2 = H'^2 \frac{N'^2u'^2}{N^2u^2}$. Introducing this into the last term and adding the last but one, we find

$$\text{Two last terms} = \tfrac{1}{2} H'^2 \left(\frac{1}{l'-r} - \frac{N'}{N} \cdot \frac{1}{l-r}\right).$$

In the bracketed expressions we now eliminate l' by (6p)*,

$$N'(l'-r)/l' = N(l-r)/l, \text{ or transposed } \frac{l'}{l'-r} = \frac{N'.l}{N(l-r)},$$

which by the well-known rule gives

$$\frac{r}{l'-r} = \frac{N'l - N(l-r)}{N(l-r)} \text{ or } \frac{1}{l'-r} = \frac{(N'-N)l + N.r}{N.r.(l-r)}.$$

Using this in the first term of the bracketed expression and extending the second term by r in order to secure a common denominator, we obtain

$$\frac{1}{l'-r} - \frac{N'}{N} \cdot \frac{1}{l-r} = \frac{(N'-N)l - (N'-N)r}{N.r(l-r)} = \frac{N'-N}{N.r},$$

showing that the expression, although originally it contained both l and l', is really quite independent of both and depends only on the radius of curvature of the refracting surface, and on the indices of the media separated by it. Multiplying the remarkable expression by $\frac{1}{2} H'^2$ and substituting it for the original last two terms in the equation for LA'_d, the latter assumes its final form

$$\text{(d)} \qquad LA'_d = a.SA^2 + 2a.q'H'.Y + a(q'H')^2 + \tfrac{1}{2} H'^2 . \frac{N'-N}{N.r}.$$

This equation determines the distance from the D-focus B'_d to the ideal image-plane at B', strictly speaking when measured along the auxiliary optical axis ; but as we must treat the angle between the principal axis and the auxiliary axis as very small, we may measure it also by its projection upon the principal axis or upon any other line forming only a small angle with the latter, and this consequence of our simplifying assumptions we must carefully note as it will prove highly useful.

Equation (d) would be perfectly satisfactory for the determination of the quality of the image produced by a single refracting surface, but we intend to apply it to any centred system, which will imply a change to a new auxiliary axis at every surface. To avoid the easily visualized complications attached to this change, we will refer the point B'_d to the principal ray of the oblique pencil itself, and we must therefore locate the principal ray. By regarding the latter as produced by the fusion of an infinitely close D-pair of rays passing, in the end-view, immediately to right and left of the ultimate principal ray, we see that equation (d) will give us its intersection with the auxiliary axis by putting $SA = 0$ and also $Y = 0$, and if we call the intersection-point B'_{pr} and the corresponding longitudinal aberration LA'_{pr}, we thus find

$$\text{(e)} \qquad LA'_{pr} = a(q'H')^2 + \tfrac{1}{2} H'^2 \cdot \frac{N'-N}{N.r}.$$

LA'_{pr} differs from LA'_d by the first two terms of the latter ; hence B'_{pr} and B'_d

[53] will in general not coincide, and as the principal ray meets the auxiliary axis under a definite angle it follows that B'_d will as a rule lie above or below the principal ray and not on it. We can express this by saying that our D-pair of rays has a *transverse aberration* with reference to the principal ray in addition to its longitudinal aberration with reference to the ideal image-plane. We will define the transverse aberration by $B'_d G$, which shall be taken at right angles with the principal optical axis, and in analogy with the definition in section [34] we will give to it the *positive* sign if the principal ray passes *above* B'_d, and we will call it TA'_d. We can now determine TA'_d by noting that under our convention of measuring the incidence heights in the tangent-plane of the refracting surface we have similar triangles $P_{pr}A_oB'_{pr}$ and $GB'_dB'_{pr}$ which give

$$GB'_d/B'_dB'_{pr} = A_oP_{pr}/A_oB'_{pr} \quad \text{or} \quad TA'_d = B'_dB'_{pr} \times A_oP_{pr}/A_oB'_{pr}.$$

We have by inspection

$$B'_dB'_{pr} = LA'_d - LA'_{pr} = a\,.\,SA^2 + 2\,a\,.\,q'H'\,.\,Y \text{ by (d) and (e),}$$

and by equation (b) $A_oP_{pr} = q'H'$.

With reference to $A_oB'_{pr}$, this differs from $AB' = l'$ only by the small quantity $B'_{pr}\,B'sph_o$, which in our primary approximation we must treat as negligible compared with the full distance l'; therefore we not only may, but must, use l' as the true ultimate value of $A_oB'_{pr}$. Introducing these equivalents, the equation for TA'_d becomes

$$\text{(f)} \qquad TA'_d = \frac{a}{l'} \cdot q'H'\,.\,SA^2 + 2\,\frac{a}{l'}\,(q'H')^2\,.\,Y.$$

The location of the focus of any D-pair of rays with reference to the principal ray and to the ideal image-plane is now completely defined by equations (d) and (f), and it only remains to determine the point where the principal ray cuts the ideal image-plane; for if this point is found not to coincide with the ideal image-point B'_o, it will evidently indicate yet another defect of the extra-axial image, which will persist even when the diaphragm is closed to the tiniest pinhole aperture so as to reduce the admitted light to the principal ray and its immediate neighbouring rays, for which TA'_d would be zero and the longitudinal aberration would be covered by depth of focus. Fig. 49 at once shows that because the principal ray crosses the auxiliary axis at B'_{pr}, its intersection with the ideal image-plane at the point $B'pr_o$ cannot as a rule coincide with the intersection-point B'_o of the auxiliary axis, which represents our ideal and perspectively correct image-point. We may regard the undesirable displacement $B'pr_o\,B'_o$ as a transverse aberration of the principal ray with reference to the ideal image; we will give to it the symbol *Dist'* because we shall recognize it as distortion, the fifth of the Seidel aberrations. We can determine this aberration in closest analogy with TA'_d from the similar triangles $P_{pr}\,A_o\,B'_{pr}$ and $B'pr_o\,B'_o\,B'_{pr}$ which give the proportion: $B'pr_o\,B'_o/B'_{pr}\,B'_o = P_{pr}\,A_o/A_oB'_{pr}$. By introducing $B'pr_o\,B'_o = Dist'$, $B'_{pr}\,B'_o = LA'_{pr}$ as determined in (e), $P_{pr}A_o = q'H'$ and $A_oB'_{pr}$ as before with its limiting value l', we obtain

$$Dist' = LA'_{pr} \times q'H'/l',$$

or on introduction of the value of LA'_{pr} [53]

(g) $$Dist' = \frac{a}{l'}(q'H')^3 + \frac{1}{2}\frac{q'}{l'}H'^3 \cdot \frac{N'-N}{N \cdot r},$$

which will be seen to grow as the cube of H'.

The problem of tracking an oblique pencil through one refracting surface is now completely solved by equations (d), (f), and (g) used in conjunction with the data obtained from the postulated preparatory work. For referring once more to Fig. 49, but imagining the Exit pupil EP' added to it at the calculated distance l'_{pr} from A, we can proceed as follows with reference to any point B'_o of the *ideal* image. A straight line drawn from EP' to the selected B'_o will be the *ideal* principal ray (i.e. not affected by aberration) and its intersection with the refracting surface will fix its incidence point P_{pr}. Next the value of $Dist'$ calculated by (g) applied at B'_o—downwards if $Dist'$ is positive, upwards if $Dist'$ is negative—will fix the point $B'pr_o$. Then the straight line joining P_{pr} to $B'pr_o$ is the *true* principal ray to which the D-foci are referred by equation (f). To locate any one D-focus we determine the distance of the selected incidence point D from P_{pr} and its projection $P_{pr}\,Q$ upon the plane of the optical axis; using these as SA and Y respectively we can calculate LA'_d and TA'_d. A point on the true principal ray at distance LA'_d from the ideal image-plane, to the left if LA'_d is positive, then fixes the point G and the true focus B'_d is found by applying TA'_d at G, in a direction at right angles with the optical axis and downwards if TA'_d is positive. The purpose of this detailed explanation is to render it clear to the beginner that our final equations, by reason of being in general algebraical form, cover not one particular pair of rays, but enable us to track, in primary approximation, the course of *every* ray that can be admitted by the clear aperture of the refracting surface. That is why we called the outer ray originally selected a 'general' ray of the oblique pencil.

SUMMATION FOR A COMPLETE CENTRED SYSTEM

[54] In order to sum up the oblique aberrations for a complete system we must discover their addition theorems. For that purpose we repeat the necessary parts of the deductions in the previous section on the assumption that the arriving rays are already affected by the aberrations determined by equations (d), (f), and (g) of [53].

Leaving the distortion for separate treatment, we begin with the relations between the D-focus and the principal ray.

Referring to Fig. 50 and giving suffix 1 to the arriving rays, without necessarily implying that they come from the first surface of the lens system, the ideal object-plane of the new surface will be identical with the ideal image-plane of the preceding surface and shall be located at $B'_1 \equiv B_2$. The arriving D-pair of rays shall have its focus at B'_{d1} and shall have a transverse aberration with reference to the principal ray $B'_{d1}G_1 = TA'_{d1}$, and a longitudinal aberration LA'_{d1} represented by the distance of B'_{d1} from the normal plane at B_2. As A_2 and C_2 are the pole and centre of curvature of the new refracting surface, we lay an auxiliary axis through the actual object-point B'_{d1}, cutting the surface at A_{o2} and the object-plane at Bdo_2; note that the latter point will in general not coincide *accurately* with the ideal object-point as defined by us, but that it will do so nearly enough to allow us to treat

[54] $B_2 Bdo_2$ as practically $= H_2 = H'_1$. We can now state that if the actual object-point were situated at Bdo_2, equation (d) of the previous section would be applicable and would give us the *new* longitudinal aberration produced by our surface; we

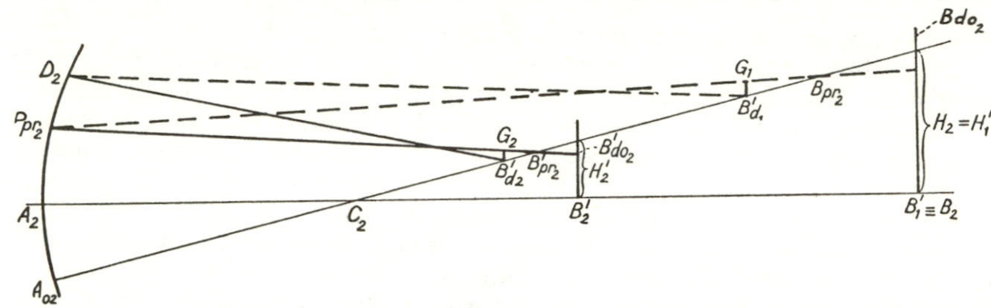

FIG. 50.

will call this 'new $LA'_{d\,2}$'. The actual object-point really lies on the auxiliary optical axis at $B'_{d\,1}$, at distance $LA'_{d\,1}$ from the object-plane and sensibly also from Bdo_2, for we must always remember the inevitable gross exaggeration of our diagrams with reference to the magnitude of all the aberrations and all the angles of rays. $LA'_{d\,1}$ may therefore be regarded as a small longitudinal displacement of the actual object-point and its transfer to the new image will be subject to the law of longitudinal magnification which for our surface with suffix 2 is $\bar{m}'_2 = N_2 u_2^2 / N'_2 u'^2_2$. Hence we have for the true location of $B'_{d\,2}$ with reference to the image-plane at B'_2:

$$\text{Total } LA'_{d\,2} = \text{New } LA'_{d\,2} + LA'_{d\,1} \cdot \frac{N_2 u_2^2}{N'_2 u'^2_2};$$

and as we have, by our nomenclature, $N_2 \equiv N'_1$ and $u_2 \equiv u'_1$, we can write this also

$$\text{Total } LA'_{d\,2} = \text{New } LA'_{d\,2} + LA'_{d\,1} \cdot \frac{N'_1 u'^2_1}{N'_2 u'^2_2}.$$

The law of transfer expressed by this equation agrees absolutely with that deduced in section [28] for the ordinary primary spherical aberration of an axial pencil; consequently we can accept the proof, given in detail in that section, that repeated application of the transfer of any given longitudinal aberration through any number of successive surfaces leads to a transfer factor of the general form $N'_j u'^2_j / N'_k u'^2_k$, and that therefore none of the intervening surfaces have any effect on the transferred value. For our summation we now have the general law as to the transferred value of the new longitudinal aberration arising at the jth surface of a system of k surfaces

(a) Transferred value of New $LA'd_j$ = (new $LA'd_j$) . $N'_j u'^2_j / N'_k u'^2_k$.

With a view to the discovery of the law of transfer of $TA'd$ we study next the course of the true principal ray after refraction at the new, nominally second, surface. If the principal ray delivered to our surface aimed at the point Bdo_2 in which the auxiliary axis $Ao_2 C_2 B'd_1$ cuts the object-plane at B_2, then equation (e) of the previous section would suffice to determine the $LA'pr_2$ of the refracted [54] principal ray and thus to locate it with reference to the image-plane at B_2'. Really, the true principal ray delivered to our surface cuts our chosen auxiliary axis at a point marked in Fig. 50 as Bpr_2, which is *not* identical with the point $B'pr_1$, because the latter point lay on the intersection of the true principal ray with the *previous* auxiliary axis joining C_1 to $B'd_1$. It would therefore be a grave error to use $LA'pr_1$ as the value of the distance Bpr_2Bdo_2 which is evidently required for our present purpose. But we can determine this distance correctly by using $G_1B'd_1 = TA'd_1$ in the first instance to determine $B'd_1Bpr_2$, and then $Bpr_2Bdo_2 = LA'd_1 - B'd_1Bpr_2$. Under our simplifying assumptions similar triangles $B'd_1G_1\ Bpr_2$ and $Ao_2Ppr_2Bpr_2$ give

$$B'd_1Bpr_2 = TA'd_1 \times Ao_2Bpr_2/Ao_2Ppr_2.$$

As before, Ao_2Bpr_2 in the limit becomes equal to $A_2B_2 = l_2$; Ao_2Ppr_2 is the incidence height of the principal ray at our new surface and therefore $= q_2'H_2'$ in accordance with equation (b) of the previous section; hence we have

$$B'd_1Bpr_2 = TA'd_1 \,.\, l_2/q_2'H_2',$$

and then $$Bpr_2Bdo_2 = LA'd_1 - TA'd_1 \,.\, l_2/q_2'H_2'.$$

Bpr_2Bdo_2 again represents a small longitudinal displacement of the actual intersection-point Bpr_2 with reference to the ideal point Bdo_2, and is subject to the law of longitudinal magnification $\bar{m}_2' = N_2u_2^2/N_2'u_2'^2$ for its transfer. Therefore the intersection-point $B'pr_2$ of the refracted true principal ray with the adopted auxiliary axis is determined by:

$$\text{Total } LA'pr_2 = B'pr_2B'do_2 = \text{new } LA'pr_2 + (LA'd_1 - TA'd_1 \,.\, l_2/q_2'H_2') \cdot \frac{N_2u_2^2}{N_2'u_2'^2}.$$

Having accurately located the points $B'd_2$ and $B'pr_2$ by their total longitudinal aberrations we can determine $B'd_2G_2$, the true $TA'd_2$, in exact accordance with the method used at that stage in section [53]. The similar triangles $B'd_2G_2B'pr_2$ and $Ao_2Ppr_2B'pr_2$ give

$$B'd_2G_2 = B'd_2B'pr_2 \times Ao_2Ppr_2/Ao_2B'pr_2,$$

and as $B'd_2G_2$ is the total $TA'd_2$, $Ao_2Ppr_2 = q_2'H_2'$, and the limit value of $Ao_2B'pr_2 = l_2'$, whilst $B'd_2B'pr_2$ represents total $LA'd_2$ − total $LA'pr_2$, we obtain next

$$\text{Total } TA'd_2 = (\text{total } LA'd_2 - \text{total } LA'pr_2) \,.\, q_2'H_2'/l_2'.$$

We determined earlier

$$\text{Total } LA'd_2 = \text{New } LA'd_2 + LA'd_1 \cdot \frac{N_2u_2^2}{N_2'u_2'^2}$$

and $$\text{Total } LA'pr_2 = \text{New } LA'pr_2 + (LA'd_1 - TA'd_1 \,.\, l_2/q_2'H_2') \cdot \frac{N_2u_2^2}{N_2'u_2'^2}.$$

By simple subtraction these give

$$\text{Total } LA'd_2 - \text{Total } LA'pr_2$$
$$= (\text{New } LA'd_2 - \text{New } LA'pr_2) + TA'd_1 \cdot \frac{l_2}{q_2'H_2'} \cdot \frac{N_2u_2^2}{N_2'u_2'^2},$$

[54] and if we introduce this value into the equation for total $TA'd_2$ we obtain

$$\text{Total } TA'd_2 = (\text{New } LA'd_2 - \text{New } LA'pr_2) \cdot \frac{q_2'H_2'}{l_2'} + TA'd_1 \cdot \frac{l_2}{l_2'} \cdot \frac{N_2 u_2{}^2}{N_2' u_2'^2}.$$

The first part on the right obviously represents the transverse aberration which we should really have if $TA'd_1$ were zero, that is, in the absence of aberration in the arriving pencil ; we are therefore justified in replacing the first part by 'New $TA'd_2$'. In the second part on the right we have by the fundamental paraxial equations $l_2 u_2 = l_2' u_2'$, which leads to a considerable cancellation ; thus the last equation becomes

$$\text{Total } TA'd_2 = \text{New } TA'd_2 + TA'd_1 \cdot \frac{N_2 u_2}{N_2' u_2'}$$

and we see that the transverse aberration of the incident oblique pencil is transferred by the law of linear magnification exactly like the simple transverse aberration of an axial pencil in section [34].

It is now quite easy to prove that the transverse aberration shares the valuable property of the longitudinal aberration of being transferable directly from one surface to any following surface without any reference to the intervening surfaces. By replacing N_2 by N_1' and u_2 by u_1' and reversing the sequence on the right, the last equation becomes

$$\text{Total } TA'd_2 = TA'd_1 \frac{N_1' u_1'}{N_2' u_2'} + \text{New } TA'd_2,$$

and we stipulated that suffix 1 does not necessarily restrict us to the absolutely first surface of a system ; the equation holds for transfer from any one surface to the next following surface. If now we apply it successively to the surfaces of a complete system we shall have the first surface applied to the original objects which are of course free from aberrations. Therefore

$$\text{Total } TA'd_1 = \text{New } TA'd_1$$

and this will have to be combined with the new aberration of the second surface by our law of transfer from surface to surface. Therefore

$$\text{Total } TA'd_2 = \text{New } TA'd_1 \cdot \frac{N_1' u_1'}{N_2' u_2'} + \text{New } TA'd_2$$

and this has to be transferred to the third surface, giving

$$\text{Total } TA'd_3 = \text{Total } TA'd_2 \cdot \frac{N_2' u_2'}{N_3' u_3'} + \text{New } TA'd_3,$$

or on introducing the explicit value of total $TA'd_2$

$$\text{Total } TA'd_3 = \text{New } TA'd_1 \frac{N_1' u_1'}{N_3' u_3'} + \text{New } TA'd_2 \frac{N_2' u_2'}{N_3' u_3'} + \text{New } TA'd_3.$$

The proof is already sufficiently complete, especially if we use the previously employed trick of giving to the final term the factor $N_3' u_3' / N_3' u_3' = 1$. We can therefore state the general law of transfer, in analogy with the equation (a) for the [54] longitudinal aberration, in the form

(b) Transferred value of New $TA'd_j$ = (New $TA'd_j$) . $N'_j u'_j / N'_k u'_k$.

We now consider the case of distortion. In section [53] we recognized the distortion as the transverse aberration of the true principal ray measured in the ideal image-plane, and as the proof of the addition theorem of transverse aberration in the preceding paragraphs was obtained without any restrictive assumption as to the origin or explicit algebraical form of the TA' to be transferred, we can immediately claim it as applicable to the transfer of the distortion. We therefore have the law of transfer for distortion

(c) Transferred value of New $Dist'_j$ = (New $Dist'_j$) . $N'_j u'_j / N'_k u'_k$.

By using the three transfer laws (a), (b), and (c) we can now sum up for a complete system. In accordance with equations (d), (f), and (g) of [53] the new aberrations arising at the jth surface are :

$$\text{New } LA'd_j = a_j SA_j^2 + 2\, a_j q'_j H'_j \,.\, Y_j + a_j q'^2_j H'^2_j + \tfrac{1}{2} H'^2_j \cdot \frac{N'_j - N_j}{N_j \,.\, r_j};$$

$$\text{New } TA'd_j = \frac{a_j}{l'_j} \,.\, q'_j H'_j \,.\, SA_j^2 + 2 \frac{a_j}{l'_j} q'^2_j H'^2_j \,.\, Y_j;$$

$$\text{New } Dist'_j = \frac{a_j}{l'_j} \cdot q'^3_j \,.\, H'^3_j + \tfrac{1}{2} \frac{q'_j}{l'_j} \cdot H'^3_j \cdot \frac{N'_j - N_j}{N_j \,.\, r_j}:$$

and by working these out for $j = 1, 2, 3 \ldots k$, multiplying each by its proper transfer factor and then adding up, we shall obtain the three aberrations at the final focus of a system of k centred surfaces. But there remains a great inconvenience owing to the variation of the values of SA, Y, and H' from surface to surface, and it is obviously desirable to express these in terms of a suitably selected fixed value. We will select the values SA_k, Y_k, and H'_k of the three variables at the last surface as those to which the values at all other surfaces shall be referred.

With regard to the H' we have no difficulty, for Lagrange's theorem gives directly $H'_j N'_j u'_j = H'_k N'_k u'_k$ or $H'_j = H'_k N'_k u'_k / N'_j u'_j$, by which all the H' can be converted into H'_k.

With reference to SA it was shown in section [53] by reference to Fig. 48 that the diameters of the circles, in which the axial *and oblique* cones of rays admitted by a given diaphragm cut any one surface, are exactly equal and in a fixed proportion to the aperture of the actual diaphragm. Now the paraxial ray-tracing through the system, on which our deductions are based, gives for the chosen nominal initial u_1 the semi-aperture at every surface as $y_j = l_j u_j = l'_j u'_j$ for $j = 1, 2, 3 \ldots k$; therefore by the fixed proportion just recalled, we have

$$\frac{SA_j}{SA_k} = \frac{y_j}{y_k} = \frac{l'_j u'_j}{l'_k u'_k}$$

which gives the necessary equation for the replacement of SA_j by the fixed SA_k as

$$SA_j = SA_k \,.\, l'_j u'_j / l'_k u'_k .$$

[54] Finally we have with regard to the Y the similarity of all the cross-sections of a given cone of rays at all the pupils and surfaces, which gives

$$\frac{Y_j}{Y_k} = \frac{SA_j}{SA_k} = \frac{l'_j u'_j}{l'_k u'_k} \quad \text{or} \quad Y_j = Y_k \,.\, l'_j u'_j / l'_k u'_k.$$

For our final form of the solution of the problem we have now to make the three substitutions

$$H'_j = H'_k \frac{N'_k u'_k}{N'_j u'_j}; \qquad SA_j = SA_k \frac{l'_j u'_j}{l'_k u'_k}; \qquad Y_j = Y_k \,.\, \frac{l'_j u'_j}{l'_k u'_k}:$$

and we have to turn the 'new aberrations' into their values when transferred to the final image by the transfer factors for LA' : $\frac{N'_j u'^2_j}{N'_k u'^2_k}$; for TA' and for $Dist'$: $\frac{N'_j u'_j}{N'_k u'_k}$.

All the transforming terms have been collected in close juxtaposition in order to render it easy to verify the result of the transformations :

Contribution of *j*th surface to the final $LA'd =$

$$a\, SA_k^2 \cdot \frac{N'_j l'^2_j u'^4_j}{N'_k l'^2_k u'^4_k} + 2\; Y_k H'_k \,.\, l'_k \,.\, a_j \,.\, q'_j \frac{l'_j u'^2_j}{l'^2_k \,.\, u'^2_k} + H'^2_k \,.\, a_j \,.\, q'^2_j \,.\, \frac{N'_k}{N'_j}$$
$$+ \tfrac{1}{2} H'^2_k \,.\, N'_k \,.\, \frac{N'_j - N_j}{N_j \,.\, N'_j \,.\, r_j};$$

Contribution of *j*th surface to the final $TA'd =$

$$SA_k^2 H'_k \,.\, a_j q'_j \frac{l'_j u'^2_j}{l'^2_k u'^2_k} + 2\; Y_k H'^2_k \cdot \frac{1}{l'_k} \cdot a_j \,.\, q'^2_j \cdot \frac{N'_k}{N'_j};$$

Contribution of *j*th surface to the final $Dist' =$

$$H'^3_k \left[a_j \,.\, q'^3_j \cdot \frac{1}{l'_j} \cdot \frac{N'^2_k u'^2_k}{N'^2_j u'^2_j} + \tfrac{1}{2} q_j \cdot \frac{1}{l'_j} \cdot \frac{N'_j - N_j}{N_j r_j} \cdot \frac{N'^2_k u'^2_k}{N'^2_j u'^2_j} \right].$$

By taking a factor l'_k, which is constant, out of the coefficient of the second term of the LA' contribution it has been rendered clear that the remainder of that coefficient is identical with the coefficient of the first term in the TA' contribution ; and in the second term of the latter $1/l'_k$ has been similarly separated out in order to show that the remaining coefficient of this term is identical with the coefficient of the third term in the LA' contribution. There are therefore only five really different coefficients contained in the three equations. Each of the k surfaces will yield three contributions of the type of these equations, and as the factors depending on aperture and image height are common to them all, they can be taken outside in summing up and give the final result :

Seidel Aberrations I

$$\text{Final } LA'd = g_1 \,.\, SA_k^2 + 2\, g_2 \,.\, l'_k \,.\, Y_k \,.\, H'_k + g_3 \,.\, H'^2_k + \tfrac{1}{2} g_4 \,.\, H'^2_k;$$

$$\text{Final } TA'd = \qquad g_2 \,.\, SA_k^2 \,.\, H'_k \quad + 2\, g_3 \cdot \frac{1}{l'_k} \cdot Y_k \,.\, H'^2_k;$$

Final $Dist' = g_5 \,.\, H_k'^3$: [54]

with the definition of the five coefficients $g_1 \ldots g_5$:

$$g_1 = \sum_{j=1}^{j=k}\left[a_j \frac{N'_j l'^2_j u'^4_j}{N'_k l'^2_k u'^4_k}\right]; \quad g_2 = \sum_{j=1}^{j=k}\left[a_j q'_j \frac{l'_j u'^2_j}{l'^2_k u'^2_k}\right]; \quad g_3 = \sum_{j=1}^{=k}\left[a_j q'^2_j \frac{N'_k}{N'_j}\right];$$

$$g_4 = \sum_{j=1}^{j=k}\left[N'_k \frac{N'_j - N_j}{N_j N'_j r_j}\right]; \quad g_5 = \sum_{j=1}^{=k}\left[a_j q'^3_j \cdot \frac{1}{l'_j} \cdot \frac{N'^2_k u'^2_k}{N'^2_j u'^2_j} + \tfrac{1}{2} q'_j \frac{1}{l'_j} \frac{N'_j - N_j}{N_j r_j} \frac{N'^2_k u'^2_k}{N'^2_j u'^2_j}\right].$$

The five coefficients are closely related to the famous five sums by which *von Seidel* first clearly defined the primary aberrations of oblique pencils about 1857. Taken singly, the coefficients are a direct measure :

g_1 of the ordinary *longitudinal spherical aberration.*
g_2 of the unsymmetrical deformation of the image known as *Coma.*
g_3 of the *Astigmatism.*
g_4 of the *Petzval curvature* of the field of view.
g_5 of the *Distortion* or false perspective of the image.

We shall begin our discussion by establishing the exact meaning of the three equations for LA'_d, TA'_d, and *Dist'*. On comparing these equations with the equations (d), (f), and (g) of section [53] it will be seen that there is the closest possible correspondence, and as *LA'd*, *TA'd*, and *Dist'* in our final equations have precisely the same significance as the corresponding quantities in the equations for a single refracting surface, we can locate the final focus of a *D*-pair of rays by precisely the same steps which were specified in the last paragraph of section [53], that is, a line drawn from the centre of the final exit pupil of the system to the ideal final image-point B'_o determined by Lagrange's theorem fixes the point of emergence *Ppr* of the principal ray of the oblique pencil from the last surface of the lens system. Applying the final *Dist'* at B'_o, downwards if *Dist'* is positive, we find the point $B'pr_o$ in which the true principal ray of the oblique pencil penetrates the final ideal image-plane. Then LA'_d locates the *D*-focus B'_d longitudinally—to the left of the image-plane if LA'_d is positive, and TA'_d locates it with reference to the true principal ray—below the latter if TA'_d is positive. We have now to determine the precise character of the imperfect image which results when this determination of any one *D*-focus is extended to all the rays of an oblique pencil of given aperture.

For this discussion we need not trouble about the exact *numerical* values of the coefficients g_1 to g_5 ; it is sufficient to realize that for a given system applied to objects at a fixed distance and having a diaphragm in a fixed position the coefficients have perfectly definite values as they only depend on the fixed radii and indices of refraction, on the intersection-lengths of the traced paraxial and principal rays, and

[54] on *ratios* of certain paraxial angles ; the 'g' are therefore true *constants* of the system. For actual numerical work our discussion will lead to still simpler methods of determining these constants ; for that reason no detailed computing instructions have been added to our fundamental equations.

DISCUSSION OF THE OBLIQUE ABERRATIONS

[55] As we have expressed the oblique aberrations in the form of longitudinal and transverse departures from the ideal positions of the final D-foci at the ideal image-point, we must begin by excluding the case when the final ideal image-plane falls at an infinite distance, or even at a very great distance, from the last surface, for in that case the aberrations would assume the repeatedly urged misleading magnitude and disconcerting rapidity of change. We will in fact stipulate again, as in the case of the microscope objective in [48.5], (4), that systems with very unequal conjugate distances should be studied and designed in the direction from the long towards the short conjugate regardless of the actual direction of the light in the practical use of the system.

We must next note that, like all other methods, our method of deducing the oblique aberrations becomes invalid in certain highly specialized cases. For our method of proof there are two of these exceptional cases, namely :

(1) When at some *intermediate* surface l' becomes infinite and consequently u' zero, the formulae for that surface become meaningless ; if it ever arose, the case could be met in practice by a slight alteration of the radius of the surface without altering the transferred value of the aberrations to any sensible extent.

(2) When an image-plane falls close to the centre of curvature of a surface, the auxiliary optical axis would form a large angle with the principal axis even for extra-axial image-points quite close to the centre of the field, and it would not be permissible to treat this angle as small. This case is fairly closely approached in certain types of eye-pieces.

Both cases can be formally met by a modified geometrical proof, but as we shall subsequently, in Part II, arrive at identical equations by a method for which the stated cases are not in any way exceptional ones, the student is asked for the present to accept an assurance that although the *proofs* given in this chapter break down, the *results* obtained by our computing formulae are exact and to be depended upon in every possible case, provided only that we avoid an infinite, or even a very long, *final* conjugate distance.

The equations which we have to discuss are

$$\text{Final } LA'_d = g_1 SA_k^2 \,.\, + 2\, g_2 l'_k Y_k H'_k + g_3 H'^2_k + \tfrac{1}{2}\, g_4 H'^2_k\,;$$

$$\text{Final } TA'_d = \qquad g_2 SA_k^2 H'_k + 2\, g_3 \frac{1}{l'_k} Y_k H'^2_k\,;$$

$$\text{Final } Dist' = g_5 H'^3_k.$$

TA_d measures the distance from any D-focus to the true principal ray ; the equation for TA'_d shows that this aberration can be made as small as we please by diminishing SA_k and Y_k more and more, that is, by closing the diaphragm, eventually to a mere pinhole. Therefore the rays of a sufficiently narrow pencil [55] will cluster very closely around the true principal ray and their departures from it will be quite inappreciable in the neighbourhood of the final image-plane. Hence the intersection of the principal ray with the final image-plane may, at small apertures, be regarded as the true location of the image, and we will discuss it first.

A. Distortion

With reference to Fig. 51, which is drawn on the principle of an engineer's drawing in three orthogonal projections, the 'side-view', which represents our usual

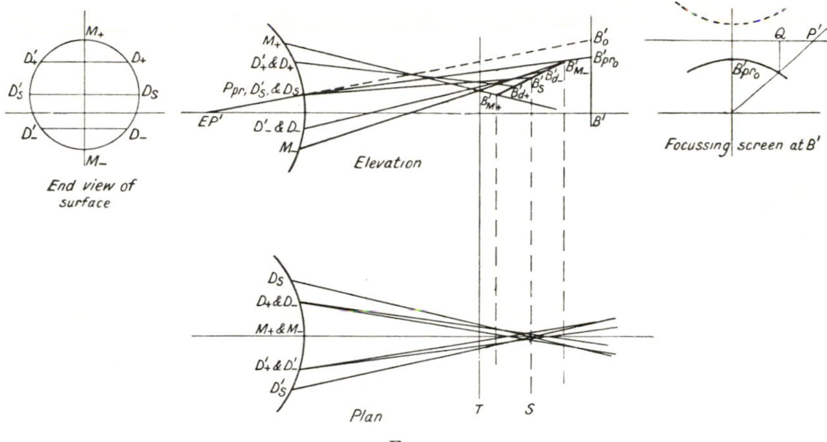

Fig. 51.

diagram, shows the ideal image-point of the oblique pencil selected at B'_o, whilst the point at which the true principal ray cuts the image-plane is marked $B'pr_o$. Then $B'pr_oB'_o$ is the transverse aberration of the principal ray in the image-plane and represents our $Dist' = g_5H'^3_k$. We imagine a focusing screen placed into the image-plane and the right-hand end-view is to represent the appearance of the image as seen from a point to the right of B′ in the side-view. Instead of seeing the particular extra-axial object-point depicted at B'_o, where it should be according to Lagrange's theorem, we should see it at $B'pr_o$. Now let us assume that our selected object-point marks the centre of a horizontal line in the postulated test diagram ; its image should also be a horizontal straight line on the focusing screen, with its centre at B'_o. We have already located the real position of this centre at $B'pr_o$. If we now consider another point P' on the ideal image of our straight line, we can, by reason of the symmetry of a centred lens system with reference to its optical axis, lay a section-plane of our system through its optical axis and the point P' without changing any of the constants of the system. For the ideal image-point B'_o we had $Dist' = g_5H'^3_k$, H'_k being the distance $B'B'_o$. For the new ideal image-point P' we shall have by the same formula a distortion $= g_5 . (B'P')^3$, and as $B'P'$ is greater than $B'B'_o$, this distortion will be very much larger than that at B'_o,

[55] giving, say, $P'pr$ as the location of the image-point. In order to locate $P'pr$ accurately we will call the distance $B'_oP' = z$. Then we have, by Pythagoras, $B'P' = (H'^2_k + z^2)^{\frac{1}{2}}$, and therefore

$$P'P'pr = g_5(H'^2_k + z^2)^{\frac{3}{2}}$$

by the cubic law of the distortion. For the discussion it will be more convenient to use the direct distance $P'prQ$ of $P'pr$ from the ideal straight line. Similar triangles $P'P'prQ$ and $P'B'B'_o$ give $P'prQ = P'P'pr \times B'B'_o/B'P'$, or on introducing the values already determined,

$$P'prQ = g_5 \, . \, H'_k(H'^2_k + z^2) = g_5 H'^3_k + g_5 \, . \, H'_k \, . \, z^2.$$

As $g_5 H'^3_k$ is equal to $B'pr_oB'_o$, the last equation shows that $P'pr$ is farther than Bpr_o from the ideal straight line by $g_5 \, . \, H'_k \, . \, z^2$, and as this increases as the *square* of the horizontal distance $B'_oP' = z$, it proves that the real image-line $B'pr_oP'pr$ is curved ; in first approximation, which is all that we can decide by our primary aberrations, the curve is a parabola with pole at $B'pr_o$. As the curvature of the real image-line is proportional to $g_5 H'_k$, in which H'_k is the least distance of the ideal line from the centre of the field at B', it follows that the only straight lines which a system with a finite value of g_5 can depict as straight lines are those for which H'_k is zero, that is lines passing through the centre of the field. All other straight lines are rendered as parabolic curves, and as this is obviously a false representation, the defect measured by g_5 is called the *distortion* of the lens system. Our diagram is drawn on the assumption of a positive value of g_5 ; it is evident that a negative value would give a real image-line of the opposite curvature and lying beyond the ideal image-line, as is indicated by the dotted parabola at the top of the diagram. If the straight ideal image-line of the diagram were one side of a square placed symmetrically with reference to the centre of the field at B', the image of the complete square really seen on the focusing screen would be bounded by four parabolas turning their concave sides towards the centre of the field in the case of a positive g_5. The resulting figure would have some resemblance to the perspective view of a barrel, and positive distortion is for that reason usually called barrel distortion. In the case of a negative g_5 the parabolic sides of the real image would turn their convexity inwards ; the figure would have the outline of a well-filled pin-cushion, and negative distortion is therefore called pin-cushion distortion. It should be remembered that a barrel-distorted image lies entirely within the ideal square which it misrepresents, whilst a pin-cushion distorted image lies entirely outside the ideal square. The former is on too small a scale, the latter is on too large a scale as compared with the magnification in the centre of the field.

We have now shown that, for light of one definite colour to which the present discussion is restricted, distortion is the only defect which is recognizable at very small apertures of the diaphragm. We can at once add that this will also be true at any admissible or possible aperture if the coefficients g_1, g_2, g_3, and g_4 are all zero, for then both LA'_d and TA'_d will be unconditionally zero for any finite values of SA_k, Y_k, and H'_k, all possible D-foci will be united at the point $B'pr_o$, and we shall have a perfectly sharp focus at that point of the ideal image-plane.

This shows that we may have a perfectly sharp and flat image of our postulated test diagram which is imperfect only with regard to its configuration. [55]

B. Tangential and Sagittal Rays

Our next task is to determine the imperfections of the image of any one object-point which result from finite values of the coefficients g_1, g_2, g_3, and g_4. Referring to the end-view of the last refracting surface of our system in the upper left corner of Fig. 51, we shall have the principal ray of the oblique pencil emerging from a definite point *Ppr*, and at any fixed aperture of the diaphragm the other rays of the pencil will completely fill a circle with *Ppr* as centre and the resulting maximum value of SA_k as radius. In accordance with our principle of maintaining throughout the closest possible contact of theory with actual practice, we shall pay special attention to the rays which are usually submitted to the ray-tracing process. In addition to the true principal ray which naturally takes first place, the traced rays are almost exclusively selected from the two principal diameters of the complete pencil, namely, from the vertical diameter M_+M_- and the horizontal diameter $D_sD'_s$ respectively. As the vertical diameter lies in the plane containing the ideal object- and image-point and also the principal optical axis, or the plane of all our usual diagrams, the rays passing through it can be traced trigonometrically by the straightforward process to which we have become accustomed. These rays, which are called *meridional* or *tangential* rays of the oblique pencil, are therefore first favourites. Two such rays taken at equal distances above and below *Ppr*, as for instance the rays through M_+ and M_-, are called a *tangential pair of rays*, and the totality of all the tangential rays which cover continuously the whole extent of the vertical diameter of a given aperture is called a *tangential fan of rays*. One of the most important practical conclusions from the present discussion will be that we can derive a complete knowledge of the *primary* oblique aberrations by relying exclusively on the convenient tangential rays, and we shall see later that suitably chosen tangential rays will also give valuable and usually sufficient information concerning the *higher* oblique aberrations.

Rays from the horizontal diameter of the aperture are called *sagittal* rays, and the names 'sagittal pair of rays' and 'sagittal fan of rays' have a meaning corresponding precisely to that specified for tangential rays. As sagittal rays do not proceed in the plane laid through the optical axis and the original object-point, they cannot be traced trigonometrically by the formulae which we have hitherto used ; they call for special 'skew-ray' formulae which will be given subsequently. Sagittal rays, however, are the more convenient ones for treatment by our present primary aberration method, and it will be clear that our *D*-pairs of rays on which the method is based may be looked upon as a slight generalization of the definition of a sagittal pair, inasmuch as they come from a horizontal *chord* of the aperture.

C. Focal Effect of a Circular Zone

In order to gain a clear understanding of the ray distribution in an oblique pencil, it is best to consider the complete area of the circle through which the rays emerge from the last refracting surface, subdivided into narrow annular zones by a large number of concentric circles having *Ppr* as their centre, and to study the course of

[55] the rays from *one* such zone, say from the circle $M_+D_sM_-D'_s$ in the end-view of the refracting surface in Fig. 51. By taking the rays issuing from this zone in pairs like $D_+D'_+$ we can apply to them our fundamental equations :

$$LA'_d = g_1SA_k^2 + 2\,g_2l'_k\,Y_kH'_k + g_3H'^2_k + \tfrac{1}{2}\,g_4H'^2_k\,;$$

$$TA'_d = \qquad g_2SA_k^2H'_k + 2\,g_3\frac{1}{l'_k}\,Y_kH'^2_k.$$

These equations are valid for any value of H'_k, including zero. Now $H'_k = 0$ takes us to the axial point B', and the simplified form of our equations resulting from $H'_k = 0$, namely

$$LA'_d = g_1SA_k^2 \quad \text{and} \quad TA'_d = 0,$$

therefore gives the aberrations of the axial pencil. As TA'_d is zero and LA'_d does not contain Y_k, by which the various possible D-pairs are differentiated, the equations tell us that *all* the rays of the *axial* pencil for the zone of semi-aperture SA_k come to one common sharp focus at the distance $LA' = g_1SA_k^2$ from B'. This is in accordance with the theory of the axial spherical aberration in Chapter II; we have extracted the information from our general equations, firstly because this clearly establishes the fact that the first term in the equation of LA'_d represents simply the ordinary spherical aberration of the axial pencil, that it therefore can be calculated with the greatest ease, and that it appears in exactly the same magnitude in all the pencils, both axial and oblique, which are admitted by any one zone of the clear aperture. Secondly, the knowledge that in the axial pencil all the rays of our zone form a single sharp focus will serve as a standard of comparison in judging the state of affairs in the oblique pencils.

We now apply the general equations to the selected zone of an *oblique* pencil and begin with its sagittal pair of rays from D_s and D'_s in Fig. 51. For this pair Y_k is zero and gives the results

$$LA'_S = g_1SA_k^2 + g_3H'^2_k + \tfrac{1}{2}\,g_4H'^2_k\,; \quad TA'_S = g_2SA_k^2\,.\,H'_k\,;$$

where the suffix S has been introduced for the data of the sagittal pair. The focus resulting from the values of LA'_S and TA'_S is marked B'_S in the side-view of Fig. 51. For any other D-pair at distance Y_k from *Ppr* the complete general equations will be applicable and will fix the focus at a point like $B'd_+$ in Fig. 51. We immediately obtain a result of the greatest significance on determining the *changes* in LA' and TA' by taking the differences

$$LA'_d - LA'_S = 2\,g_2l'_k\,Y_kH'_k \quad \text{and} \quad TA'_d - TA'_S = 2\,g_3\frac{1}{l'_k}\,Y_kH'^2_k.$$

for these differences will be of some definite magnitude if g_2 and g_3 are not zero. There are no such differences in the axial pencil, for we showed that in its case all rays of any one zone have a sharp focus. Hence we can safely conclude that the extra-axial image is bound to be less perfect than the axial image if g_2 and/or g_3 have finite values, and that the extra-axial image produced by any one zone will be equal in sharpness to the axial image if, and only if, both g_2 and g_3 are zero. The constants g_2 and g_3 are thus recognized as those which are responsible for the usual falling-off in sharpness of definition of individual image-points in the extra-axial [55] part of the field of lens systems.

As under our restrictions to small angles of all rays with the optical axis we may measure the TA' as perpendiculars on the principal ray, it is clear that the ratio of $(TA'_d - TA'_S)$ to $(LA'_d - LA'_S)$ defines the tangent of the angle between the principal ray and a straight line joining the particular D-focus to the S-focus. If we call that angle O, we find

$$\tan O = -\frac{TA'_d - TA'_S}{LA'_d - LA'_S} = -\frac{g_3}{g_2} \cdot \frac{H'_k}{l'^2_k},$$

a negative sign having been added because in the all-positive case it will be a counter-clockwise angle when referred to the principal ray. In the equation for $\tan O$, Y_k has cancelled out, and as the value of Y_k is the quantity that distinguishes the various D-pairs, we conclude that the angle O has the same value for all possible D-pairs, and that implies that *all the D-foci of any one zone lie on a straight line.* As we have reached this result without having made any restrictive assumptions as to the value of the aberration constants g_1 to g_5, it is of absolutely universal validity as far as primary aberrations are concerned ; we will state the result in this form :

'In a centred lens system the rays of a zone concentric with the principal ray of any oblique pencil meet in pairs on a straight line in the plane of the optical axis ; we will call this focal line the *characteristic focal line* of the zone.'

This focal line takes the place of the simple point focus of the corresponding rays of the axial pencil. It is very easy to form a mental picture of this line, with rays crossing it at every point of its length, and it is for that reason of peculiar value in helping one to realize the nature of the confusion of rays in an uncorrected oblique pencil. In order to complete the study of the characteristic focal line we note first that in the equations from which we deduced it,

$$LA'_d - LA'_S = 2\,g_2 l'_k\, Y_k H'_k \quad \text{and} \quad TA'_d - TA'_S = 2\,g_3 \frac{1}{l'_k}\, Y_k H'^2_k,$$

the first defines the longitudinal displacement and the second the transverse displacement of any D-focus with reference to the sagittal focus of the zone, both displacements being referred to the principal ray of the particular oblique pencil. As both are proportional to Y_k and as the highest possible value of Y_k for our circular zone is $\pm SA_k$, it follows that the two ends of the characteristic focal line are defined by the ultimate infinitely close D-pairs from M_+ and M_- in the end-view of Fig. 51 ; they are found by replacing Y_k by $\pm SA_k$ in the last equations, giving the extreme values

$$LA'_{\pm M} - LA'_S = \pm 2\,g_2 l'_k SA_k H'_k \quad \text{and} \quad TA'_{\pm M} - TA'_S = \pm 2\,g_3 \frac{1}{l'_k}\, SA_k H'^2_k,$$

or by forming the difference (upper—lower sign)

$$LA'_{+M} - LA'_{-M} = 4\,g_2 l'_k SA_k H'_k \quad \text{and} \quad TA'_{+M} - TA'_{-M} = 4\,g_3 \cdot \frac{1}{l'_k}\, SA_k H'^2_k;$$

[55] and these equations determine the extreme longitudinal and transverse extent of the complete characteristic focal line.

As the displacements for $+Y_k$ and for $-Y_k$ are always equal but of opposite sign, it follows that the characteristic focal line is bisected by the sagittal focus B'_S and that the latter represents a kind of centre of gravity of all the D-foci of the zone. The sagittal focus was located earlier by the data

$$LA'_S = g_1 SA_k^2 + g_3 H_k'^2 + \tfrac{1}{2} g_4 H_k'^2 \quad \text{and} \quad TA'_S = g_2 SA_k^2 \,.\, H'_k.$$

There is an interesting relation between TA'_S and the extreme longitudinal range $LA'_{+M} - LA'_{-M}$, for the ratio of these gives

$$\frac{TA'_S}{LA'_{+M} - LA'_{-M}} = \tfrac{1}{4}\,\frac{SA_k}{l'_k};$$

and as by reference to the corresponding axial zone we have $SA_k/l_k = u'_k$, the convergence angle of the marginal rays of the axial pencil, we find

$$TA'_S = \tfrac{1}{4}\,(LA'_{+M} - LA'_{-M})u'_k;$$

there is thus a perfectly definite relation between these two, and as the angle u'_k is always decidedly small and is further multiplied by the fraction $\frac{1}{4}$, we see that the longitudinal range of the characteristic focal line is always a large multiple of the distance of its centre from the principal ray. It is therefore not legitimate to lay down an assumed position of the sagittal focus B'_S on a diagram like Fig. 51 and then to claim that an arbitrary line through B'_S at any slope, but bisected by it, may be treated as the characteristic focal line of a selected zone. But for mere general study even considerable departures from the proper ratio expressed by our last equation do not change the character of the result very seriously, and the proportions of Fig. 51 itself are in fact wrong; the focal line of length $B'_{+M}B'_{-M}$ should, with the chosen convergence of rays, lie much closer to the principal ray, but as that would have led to a very confusing tangle of lines, strict accuracy was sacrificed.

The characteristic focal line is not by any means merely a convenient mathematical conception; on the contrary, it is an easily observed real optical phenomenon. The remarkable fact that it has never before been recognized and described as the most characteristic property of the oblique pencils of any centred optical system is probably explained by the circumstance that in the vast majority of cases it lies nearly parallel to the principal ray and cannot be shown, as the bright and extremely sharp line of light which it really is, on a focusing screen in the usual position nearly at right angles with the principal optical axis. But if any imperfectly corrected centred lens or system is fitted with a ring-diaphragm so as to transmit a reasonable approximation to our postulated annular zone of rays and is then placed *obliquely* across the rays from a strong point source of light, no difficulty will be found in finding the focal line. The most suitable focusing screen consists of a sheet of tissue paper stretched across a U-shaped frame, for as the line usually lies almost in the direction of the rays which form it, rays may arrive at both faces of the receiving screen and the complete phenomenon could not then be shown on the usual ground glass. This experiment is well worth making, and may profitably be accompanied by a study of the light-distribution when the screen is used in the orthodox position nearly at right angles with the rays; the remarkable looped [55] figures to be referred to presently can thus be conveniently verified by direct observation.

D. Graphic Determination of the Ray Distribution

We have proved that all possible D-pairs of rays of a given zone come to a focus on the characteristic focal line of the zone, that the sagittal focus of the zone bisects the characteristic focal line, and that the distance of any other D-focus from this bisecting point is exactly proportional to the Y_k of the corresponding D-points on the last refracting surface. Consequently the D-foci can be accurately laid down on a calculated or assumed characteristic focal line by this law of proportional distribution. This has been done in the side-view of Fig. 51 and the rays from the selected points of emergence have been drawn; note that in this strict side elevation the rays of any one D-pair become accurately superposed. In the plan shown below the side elevation the D-foci can be transferred in the usual drawing-board way. The points of emergence of the rays will be distributed according to their horizontal distance from the vertical centre-line of the left end-view, and for the symmetrically distributed points selected D_+ and D_- will become superposed, and similarly D'_+ and D'_-; but as $B'd_+$ and $B'd_-$ are separated points of the characteristic focal line, there will be two different rays to be drawn from each of the superposed D-points in the plan. A most significant property of the rays of the zone immediately leaps to the eyes.

In the plan there is a constriction of the bundle of rays at the point B'_S, the sagittal focus, whilst the elevation shows at the same position a very considerable spreading out of the bundle. On a focusing screen we should therefore see an oblong vertical patch pointing like an arrow (*sagitta*) towards the centre of the field. That is the origin of the name 'sagittal focus' for the point B'_S.

The side-view or elevation shows a similar constriction of the bundle of rays at the point marked B'_T at which the extreme rays from the highest and lowest points M_+ and M_- of the zone cut each other; this constriction therefore coincides with the tangential focus of the zone as defined previously. The plan shows a considerable spreading out of the rays in the same position, hence a focusing screen placed at B'_T would show an elongated horizontal patch which, with reference to the optical axis, or the centre of the field of view, would point in a tangential direction, whence the name of this approximation to a focus.

It should be noted that Fig. 51 refers to the absolutely general case when no restrictions of any kind are made as to the numerical values of the five constants g_1 to g_5 which determine the state of correction of a centred lens system. The distinctive properties of the sagittal and tangential foci of any zone which we have just discovered are therefore characteristic of all centred systems and can only vary *in degree* according to the relative values of the five constants.

Although Fig. 51 clearly brings out the remarkable properties of an oblique pencil affected by primary aberrations, it is not a really accurate representation, on account of the enormous exaggeration of the longitudinal and transverse displacements of the D-foci in comparison with the dimensions of the system. As we insisted from the beginning, it is necessary to treat the displacements as extremely

[55] small compared with the dimensions of the system in order to limit the investigation to the *pure* primary aberrations ; as it would be impossible to do this in a complete drawing like Fig. 51, we must modify the method so as to show the correct relations in the interesting focal region on the necessary large scale by drawing the rays at the slopes which they would have when coming from a very large aperture at a very great distance to the left of the focal region. We can find these slopes from an auxiliary small-scale drawing of the aperture and focal distance by treating the aberrations as negligibly small to this reduced scale, that is, by assuming a sharp focus. Every ray in the large-scale drawing of the focal region can then be correctly laid down as a line parallel to its prototype in the small-scale drawing.

Convenient proportions will result by assuming a semi-aperture of 'one' (say 1 inch) and a focal distance of 'five' for the small-scale drawing, and by selecting the rays to be drawn at equal intervals around the circumference of the zone. For actual detailed study sixteen rays at $22\frac{1}{2}°$ intervals are recommended, but in the specimen diagrams, Fig. 52, only half that number is used.

We begin by drawing two centre lines for elevation and plan respectively, and representing the principal ray in the diagrams of the ray distribution. At the extreme left of each centre line we draw a circle with the selected semi-aperture as radius, mark the chosen D-points on each and project them upon the vertical diameter of the circles. From the centre of each circle we mark the chosen focal distance towards the right, and join the focus so fixed to the projected D-points on the vertical diameters ; these lines give the slopes of the rays for elevation and plan respectively. We must now choose the cases to be studied. In the specimen three are taken, namely, at the left the general case when both g_2 and g_3 are finite and positive, yielding an inclined characteristic focal line. To obtain a correct result we must observe the relation between TA'_S and $(LA'_{+M} - LA'_{-M})$ which is the longitudinal extent of the line or its projection upon the principal ray. The relation being $TA'_S = \frac{1}{4} u'_k(LA'_{+M} - LA'_{-M})$ and our selected proportions giving $u'_k =$ semi-aperture/focal length $= \frac{1}{5}$, we have to make the projection of the characteristic focal line equal to twenty times the chosen value of TA'_S. $TA'_S = 0{\cdot}1$ was chosen, hence the line will extend to one unit (say inch) to either side of the sagittal focus.

For the central case, $g_3 = 0$ was selected, hence the characteristic focal line will be parallel to the principal ray because $\tan O = -g_3 H'_k / g_2 l'^2_k = 0$. TA'_S was chosen $= 0{\cdot}15$, which gives the total length of the focal line $= 20 \times 0{\cdot}15 = 3$.

The right-hand case represents $g_2 = 0$. The focal line assumes a position at right angles with the principal ray because $\tan O = \infty$. We may in this case choose any length of it which we like, but 0·6 was selected. As $g_2 = 0$ causes the sagittal focus to fall upon the principal ray (because $TA'_S = g_2 SA_k^2 H'_k = 0$), the perpendicular focal line will be bisected by the principal ray.

The selected focal lines must now have the chosen D-foci clearly marked on them by the law of simple proportion to the Y-values in the end-view of the surface, and it will be highly advisable to put appropriate corresponding letters to the points selected. All the rays can then be drawn at the slopes given by the small-scale drawing at the extreme left, taking care to associate the points correctly. We thus obtain the correct projected position of every ray in plan and elevation, and can build up the cross-section or end-view of the bundle of rays at any point along the principal ray in the usual draughtsman's way, the elevation giving the vertical [55] coordinate with reference to the principal ray, and the plan giving the horizontal coordinate with reference to the vertical centre line of each end-view. The two sections of greatest interest are those at the sagittal and at the tangential focus, followed next by that at the position midway; but additional sections are well worth adding. A smooth curve drawn through the determined points will in most cases give looped lines of trochoidal type, but the special cases depicted in the central and right-hand diagrams provide exceptions. In the case depicted in the centre, when $g_3 = 0$, the tangential and sagittal foci fall vertically above each other and there is only one constriction of the bundle of rays, at which the rays from *opposite points* of the zone meet *in pairs* around the circumference of a circle with TA'_S as radius and with centre at $2\,TA'_S$ below the principal ray. At equal distances to either side of the constriction the cross-section of the bundle of rays assumes the shape of equal double loops. This is the case of *pure Coma*. The right-hand diagram shows the case of *pure Astigmatism* and gives a sharp horizontal focal line at the tangential focus, a sharp vertical focal line of exactly the same length at the sagittal focus, and distribution around a circle at the midway point. At all other positions the rays will be found distributed on ellipses. In this case of pure astigmatism it should be specially noted that all the figures covered by the rays are absolutely symmetrical with reference to the principal ray, whilst in the case of pure coma the distribution of the rays is totally unsymmetrical with reference to the principal ray. The general case of mixed coma and astigmatism will be seen to partake of the characteristics of both the pure aberrations, for the existence of two well separated and strongly elongated constrictions is obviously astigmatic, whilst the want of symmetry with reference to the principal ray is as clearly of the coma type.

The want of symmetry in the coma image is its worst feature; it causes a point of light to be rendered as a comet-like smudge with a strong concentration of light near the principal ray and a gradual fading away towards the broad tail of the figure. Such a caricature of what should be a small round dot is not only extremely unsightly in ordinary visual observations or in photographs, but also renders exact measurements quite impossible, as it is difficult to decide which part of such an unsymmetrical patch should be regarded as the location of the mean image-point. For that reason coma has been looked upon as the worst of all the aberrations from the earliest days of practical optics, and even in the days of pure empiricism its removal to the utmost practicable extent was regarded as imperative. The beginner will be wise if he accepts this result of long experience and declares relentless war against any easily visible residue of coma. This attitude finds additional justification in the fact that coma is more easily corrected—or at least rendered almost invisible—than any other aberration, and that in the great majority of cases its removal has a strong tendency to reduce other aberrations. For these reasons we shall assume in the remainder of this discussion that the coma is of small magnitude, not sufficient to cause an objectionable want of symmetry in the individual image-points or to affect their location and character to a notable extent.

On that understanding astigmatism will be the only serious defect with which we have to reckon. Astigmatism is also a severe drawback when present, as any

[55] one point in the image will be rendered either as a line or as an ellipse, or at the midway point as a circular patch ; but all these false renderings are at any rate symmetrical, so that they can be bisected by a micrometer wire with considerable accuracy and certainty ; and in visual observation or in photographic pictures the symmetrical deformation is unquestionably less intolerable than the smudged appearance of a coma image. This is fortunate, for we shall presently discover that in whole classes of important optical instruments we have to admit a considerable remnant of astigmatism in order to minimize an even more disturbing curvature of the field. The simultaneous removal of astigmatism and of curvature of the field in fact remained an unattainable ideal until 1883, and is even at the present time only approximately realized by the so-called anastigmats.

E. Mathematical Determination of the Tangential Focus

We defined the tangential focus in an earlier subsection as the intersection-point of the rays from the highest and the lowest point of the selected aperture, but as a careful inspection of the side-views in the instructive Fig. 52 shows that the pairs of rays from $D_+D'_+$ appear to cross those from $D_-D'_-$ at the same axial position, we will solve the little problem in the more general form of seeking this crossing point of two D-pairs from $+Y_k$ and $-Y_k$ respectively.

In Fig. 53, which is drawn on the principles of Fig. 52, B'_{+d} shall be the focus of the D-pair from $+Y_k$, and B'_{-d} the focus of the D-pair from $-Y_k$. These

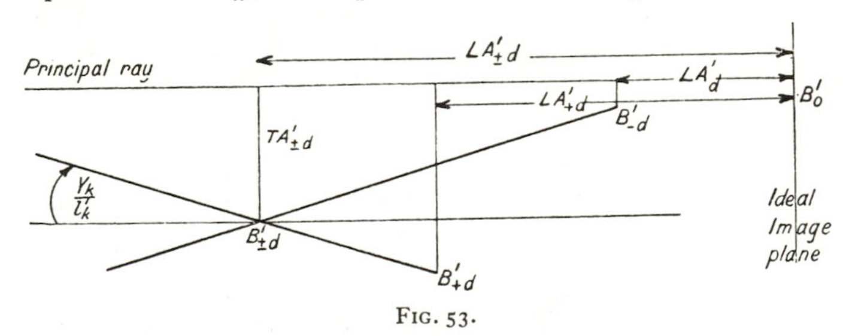

Fig. 53.

pairs of rays as projected upon the plane of the side-view will have oppositely equal slopes with reference to the principal ray, defined by Y_k/l'_k and $-Y_k/l'_k$ respectively, and their projections will cross each other at some definite point, which we will call $B'_{\pm d}$ and which we will locate on our usual system by its transverse aberration called $TA'_{\pm d}$ and its longitudinal aberration with reference to the ideal image-plane at B'_o, with the symbol $LA'_{\pm d}$. B'_{+d} and B'_{-d} will have corresponding coordinates which we can write down by our fundamental equations for any D-pair. If we draw a parallel to the principal ray through $B'_{\pm d}$ we obtain a small right-angled triangle with $B'_{\pm d}B'_{+d}$ as hypotenuse and with the angle at $B'_{\pm d} = Y_k/l_k$, and this triangle gives

$$TA'_{+d} - TA'_{\pm d} = (LA'_{\pm d} - LA'_{+d}) \cdot Y_k/l'_k.$$

The corresponding triangle with $B'_{\pm d} - B'_{-d}$ as hypotenuse gives

$$TA'_{-d} - TA'_{\pm d} = -(LA'_{\pm d} - LA'_{-d}) \cdot Y_k/l'_k.$$

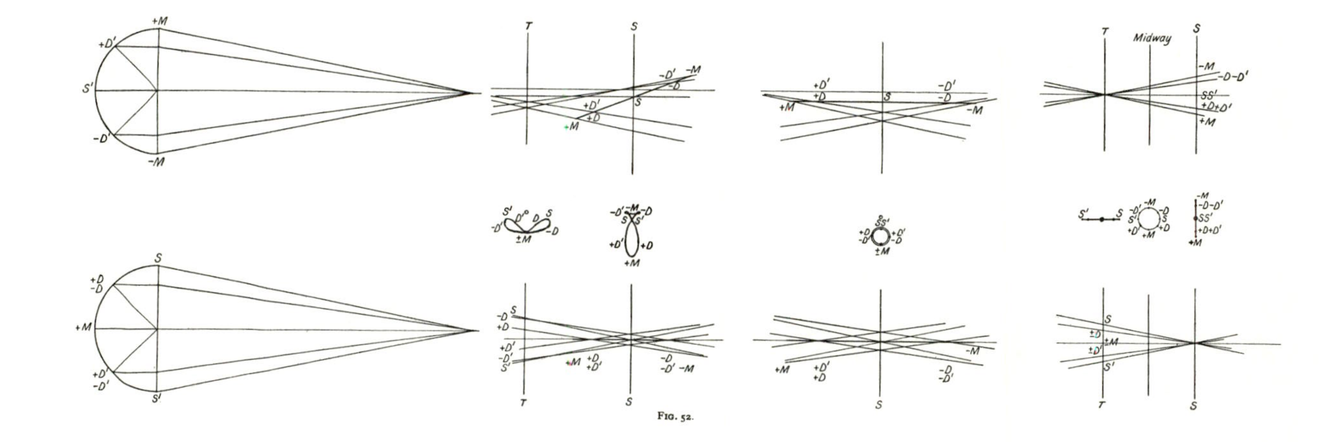

Fig. 52.

These are two equations containing the two unknown quantities $TA'_{\pm d}$ and [55] $LA'_{\pm d}$ which therefore can be determined from them.

Subtraction of the second equation from the first eliminates $TA'_{\pm d}$ and gives for the determination of $LA'_{\pm d}$:

$$TA'_{+d} - TA'_{-d} = 2\, LA'_{\pm d} \frac{Y_k}{l'_k} - (LA'_{+d} + LA'_{-d}) \cdot \frac{Y_k}{l'_k}$$

or $$LA'_{\pm d} = \tfrac{1}{2}(TA'_{+d} - TA'_{-d}) \frac{l'_k}{Y_k} + \tfrac{1}{2}(LA'_{+d} + LA'_{-d}).$$

Now we have by Seidel aberrations I the values of LA'_d for $+Y_k$ and $-Y_k$ respectively:

$$LA'_{+d} = g_1 SA_k^2 + 2\, g_2 l'_k Y_k H'_k + g_3 H'^2_k + \tfrac{1}{2} g_4 H'^2_k$$
$$LA'_{-d} = g_1 SA_k^2 - 2\, g_2 l'_k Y_k H'_k + g_3 H'^2_k + \tfrac{1}{2} g_4 H'^2_k\,;$$

or the mean required in the equation for $LA'_{\pm d}$

$$\tfrac{1}{2}(LA'_{+d} + LA'_{-d}) = g_1 SA_k^2 \qquad + g_3 H'^2_k + \tfrac{1}{2} g_4 H'^2_k.$$

For TA' the fundamental equations give correspondingly:

$$TA'_{+d} = g_2 SA_k^2 H'_k + 2\, g_3 \frac{1}{l'_k} Y_k H'^2_k$$

$$TA'_{-d} = g_2 SA_k^2 H'_k - 2\, g_3 \frac{1}{l'_k} Y_k H'^2_k\,;$$

or the half difference required in the equation for $LA'_{\pm d}$

$$\tfrac{1}{2}(TA'_{+d} - TA'_{-d}) = \qquad 2\, g_3 \frac{1}{l'_k} Y_k H'^2_k\,;$$

and if these explicit values are introduced we find

(a) $$LA'_{\pm d} = g_1 SA_k^2 + 3\, g_3 H'^2_k + \tfrac{1}{2} g_4 H'^2_k.$$

As this equation does not contain the Y_k characterizing the particular D-pairs for which we worked out the equations, it already proves that the projections of all D-pairs taken at equal distances above and below the horizontal or sagittal diameter of the aperture cross each other at the same distance from the ideal image-plane.

We now determine $TA'_{\pm d}$ by transposing the second of our starting equations into the form

$$TA'_{\pm d} = TA'_{-d} + (LA'_{\pm d} - LA'_{-d}) \frac{Y_k}{l'_k},$$

and taking from the equations already used

$$TA'_{-d} = g_2 SA_k^2 H'_k - 2\, g_3 \frac{1}{l'_k} Y_k H'^2_k$$

$$LA'_{\pm d} - LA'_{-d} = 2\, g_2 l'_k Y_k H'_k + 2\, g_3 H'^2_k, \quad \text{we easily find}$$

(b) $$TA'_{\pm d} = g_2 H'_k (SA_k^2 + 2\, Y_k^2).$$

[55] As this equation contains the Y_k which differentiates the various possible D-pairs, it shows that whilst the projections upon the plane of the optical axis of symmetrically placed D-pairs cross always at the same distance from the ideal image-plane in accordance with equation (a), they do so at varying distances $TA'_{\pm d}$. For the true sagittal pair of rays we have $Y_k = 0$, and therefore for the sagittal pair: $TA'_{\pm d} = g_2 H'_k SA_k^2$.

For other pairs the distance grows as Y_k^2, and evidently attains its maximum for the infinitely close D-pairs from the highest and lowest points of the zone, for which $Y_k = \pm SA_k$. As these extreme pairs coincide with the two tangential rays of the zone, we have for this pair $TA'_{\pm M} = 3\,g_2 H'_k SA_k^2$. We have thus proved algebraically that at the tangential focus all the rays of the zone pass within the limiting distances $g_2 H'_k SA_k^2$ and $3\,g_2 H'_k SA_k^2$ from the principal ray, and as these expressions do not involve g_3, we see also that these limits apply no matter what the value of g_3 may be.

It is easily proved mathematically, by applying the above methods to the plan-view of Fig. 52 and by solving a simple maximum problem, that the constriction of the bundle of rays at the sagittal focus is of precisely the same width ($2\,g_2 H'_k SA_k^2$) as the constriction at the tangential focus. But as the graphic method of Fig. 52 brings out these relations in the clearest possible manner, we will omit the formal proof.

F. Coma and Astigmatism

We have learnt that the rays which are almost exclusively used in the exact calculations, namely the sagittal pair and the tangential pair of a selected zone, are not merely convenient but that they accurately locate the two most characteristic points in the course of the rays from such a zone, namely the two constrictions which mark the nearest approach to a focal effect in presence of the specific oblique aberrations. We have already introduced suffix S for the data of the sagittal pair and shall henceforth use suffix T for those of the tangential pair, and we will in future also use the simple letters S and T to mark these foci on our diagrams.

We then have these foci determined by the coordinates:

for the sagittal focus: $$LA'_S = g_1 SA_k^2 + g_3 H_k'^2 + \tfrac{1}{2}\,g_4 H_k'^2$$
$$TA'_S = g_2 H'_k SA_k^2\,;$$

for the tangential focus: $$LA'_T = g_1 SA_k^2 + 3\,g_3 H_k'^2 + \tfrac{1}{2}\,g_4 H_k'^2$$
$$TA'_T = 3\,g_2 H'_k SA_k^2.$$

We recognized the transverse displacements as indicative of the highly objectionable one-sided and unsymmetrical confusion of the rays which is called Coma, and we shall usually refer to TA'_S as the sagittal Coma and to TA'_T as the tangential Coma, and as both are determined by the constant g_2, we shall call g_2 the Coma-Constant of the system. The equations for TA'_S and TA'_T immediately show a remarkable and highly important one-to-three ratio of sagittal and tangential Coma, and this ratio should be firmly committed to memory, because our computing methods determine sometimes one and sometimes the other, and any failure to remember the ratio would lead to false conclusions. Another important property of Coma to be well remembered is the law of change expressed by the [55] coefficient $H'_k \,.\, SA_k^2$, which means that with a given aperture Coma grows directly as the distance H'_k of an image-point from the optical axis, whilst when the aperture is changed, the Coma at any one point in the field changes as the *square* of the aperture. Unless these laws are borne in mind, we may be misled in judging the correction of Coma in the earlier stages of a design when we may be calculating for an aperture and a diameter of the field very different from the contemplated final values.

With reference to the longitudinal displacements of the sagittal and tangential foci, we must leave their *total* values for discussion in the following subsection on curvature of the field, thus limiting ourselves for the present to the *difference* $LA'_T - LA'_S$, which represents the distance separating the two characteristic constrictions of the cone of rays from any one zone. We recognized this difference as the effect of Astigmatism ; our equations give

$$LA'_T - LA'_S = 2\, g_3 H_k'^2$$

and show that the astigmatic difference of focus of an oblique pencil depends solely on the constant g_3, which we shall therefore call the astigmatic constant of the system. We see that this astigmatic difference of focus grows as the square of the distance H'_k of an image-point from the optical axis ; it is for that reason almost insignificant in the close vicinity of the centre of the field, but becomes rapidly increased in pencils of considerable obliquity and can reach huge values in so-called wide-angle systems. Although the astigmatic difference of focus itself does not depend in any way upon the aperture, we must not draw the conclusion that its effect upon the appearance and sharpness of the extra-axial image-points is independent of the aperture, for the image which we see through an eyepiece or upon a focusing screen is, geometrically considered, a *cross-section* of the bundle of rays. If, in accordance with the earlier remarks on the intolerable nature of any considerable coma effect, we limit ourselves to the case of practically pure astigmatism, reference to the right-hand diagram of Fig. 52 will make it clear that what we should see would be the focal lines, the circle of rays at the midway point, and the various elliptical cross sections of the bundle of rays, and also that all these appearances would grow or diminish in their linear dimensions directly as the aperture of the zone. Hence we have a means of reducing the effect of a given value of g_3 in reduction of the aperture ; but as there is only simple proportionality, the aperture has to be greatly reduced in order to obtain reasonable definition in the outer part of the field of a system suffering from heavy astigmatism. That was the great drawback of the earlier photographic objectives.

We can determine the length of the focal lines by reference to Fig. 52, for the right-hand diagram shows that half of each focal line is equal to the distance between the two astigmatic foci S and T multiplied by the convergence angle or slope of the marginal ray, and as we have determined the distance from T to S as $= 2\, g_3 H_k'^2$, whilst the angle of convergence is $= SA_k/l'_k$, we have

$$\text{Length of the focal lines} = 4\, g_3 \,.\, H_k'^2 \,.\, SA_k/l'_k \,.$$

If we can adjust a system so that the astigmatic circle of rays midway between S

[55] and T may be regarded as the actually utilized position, then we have the diameter of this circle obviously equal to half the length of the focal lines, hence

$$\text{Diameter of astigmatic circle} = 2\, g_3 \,.\, H_k'^2 \,.\, SA_k / l_k'.$$

By these equations we can estimate the deterioration of the image in the presence of astigmatism.

Whilst all the equations which have been given are of universal validity as far as the *primary* aberrations of oblique pencils are concerned, the diagrams all represent as nearly as possible the all-positive case which we usually select in order to render the fixing of the correct sign as easy as possible. In actual *calculations* the correct result will invariably be obtained if our usual sign conventions are observed ; but it appears highly desirable that the beginner should have a *clear mental picture* of the distribution of the rays and the relative position of the astigmatic foci in all cases that can occur in practice. We will therefore add a diagram, Fig. 54, which represents the four possible combinations of sign of coma and astigmatism for our usual orientation of the diagrams, with the lens system well to the left and for an image-point above the optical axis.

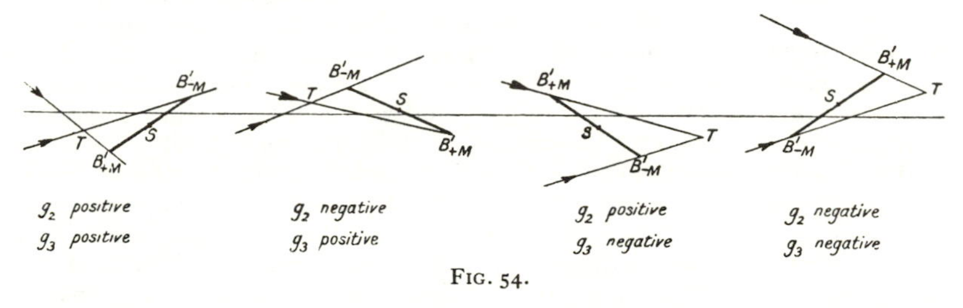

FIG. 54.

In each of the four cases the characteristic focal line of the zone is shown as a heavy line bisected at the sagittal focus S, and its two ends are marked respectively B'_{+M} for the point at which the tangential ray from the highest point of the zone aims, and B'_{-M} for the point of arrival of the tangential ray from the lowest point of the zone. The intersection-point of these extreme rays then marks the tangential focus. The location of S with reference to the principal ray is given by $TA'_S = g_2 H'_k SA_k^2$ and therefore changes sign with g_2. The slope of the characteristic focal line with reference to the principal ray is determined by $\tan O = -g_3 H'_k / g_2 l_k'^2$, and is therefore counterclockwise when g_2 and g_3 have the same sign, and clockwise when g_2 and g_3 have opposite signs. The question whether B'_{+M} occupies the left or the right extremity of the characteristic focal line is answered by the equation

$$LA'_{+M} - LA'_S = 2\, g_2 l'_k SA_k H'_k,$$

which shows that for the usual positive values of l'_k, SA_k, and H'_k, B'_{+M} is to the left for a positive g_2 and to the right for a negative g_2. These data enable us to draw the two tangential rays and thus to locate the tangential focus T. It should be noted that the relative position of S and T thus found agrees in every case with the independent determination of the same relation by the equation for the astigmatic difference of focus : [55]

$$LA'_T - LA'_S = 2\,g_3 H'^2_k,$$

according to which S is to the right of T for a positive g_3 and to the left of T for a negative g_3.

G. Curvature of the Field

We have hitherto studied the effect of the aberration constants on the quality of the individual image-points. We must now examine the location of the extra-axial image-points with reference to the corresponding axial image-point and to the ideal image-plane, and as we have learnt that the nearest approaches to a focal effect in the oblique pencils are found at the sagittal focus S and at the tangential focus T, we shall be chiefly interested in the relations of these two points with the ideal image-plane.

Referring to Fig. 55, which inevitably shows the aberrational displacements with grossly exaggerated magnitude, the pole of the last surface of the lens system is

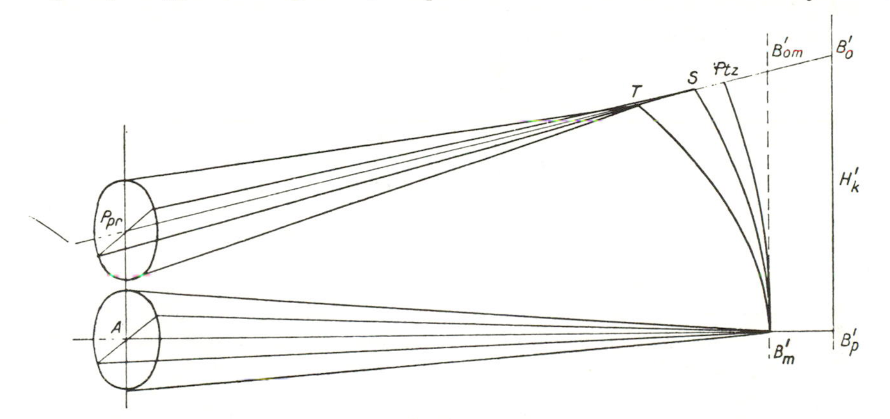

Fig. 55.

indicated at A and the point of emergence of the principal ray of an oblique pencil at P_{pr}. The circles in which the rays of a selected zone pass through the last surface are shown in perspective with their horizontal and vertical diameters in order to make it possible to draw the four cardinal rays of the axial and of the oblique pencil. The paraxial focus shall be at B'_p and our ideal image-plane will cut the optical axis at that point. $B'_p B'_o$ will then be the H'_k of our equations for the selected oblique pencil, and for a zone of semi-aperture SA_k the longitudinal displacements of the sagittal and tangential foci will be determined by the equations

$$LA'_S = g_1 SA_k^2 + g_3 H'^2_k + \tfrac{1}{2}\,g_4 H'^2_k \quad \text{and} \quad LA'_T = g_1 SA_k^2 + 3\,g_3 H'^2_k + \tfrac{1}{2}\,g_4 H'^2_k.$$

These equations are valid for any point in the field of view including the axial point; and as for the latter H'_k is zero, both equations give $LA' = g_1 SA_k^2$ for

[55] the axial pencil, which, as we pointed out earlier, will have a perfectly sharp focus of our selected zone at this distance from the paraxial focus, say at the point marked B'_m. As our primary aberrations are strictly correct only for pencils of very small obliquity, we may—and in future always shall—measure the longitudinal displacements of points on the principal rays of the oblique pencils by their projection upon the principal optical axis; hence we can get rid of the term $g_1SA_k^2$ in LA'_T and LA'_S by erecting a new image-plane at the true axial focus B'_m of our selected zone, and we thus see that $g_3H_k'^2 + \frac{1}{2}g_4H_k'^2$ measures the net amount by which the sagittal focus is removed from the plane laid through the axial focus, and that $3g_3H_k'^2 + \frac{1}{2}g_4H_k'^2$ represents the corresponding measure for the tangential focus. If we imagine any number of oblique pencils with steadily diminishing values of H'_k added to our diagram, filling the space between the axial pencil and the one oblique pencil actually shown, they will have displacements of the S and T foci with reference to the image-plane at B'_m which vary as the successive values of $H_k'^2$. Therefore these displacements follow a parabolic law, and we are justified in claiming that two parabolas with the principal optical axis as axis and B'_m as their common pole, and passing respectively through S and T of our selected oblique pencil, will also pass through the S- and T-foci of any other oblique pencil whose principal ray falls into the plane of our diagram. Owing to the symmetry of a centred system with reference to its optical axis, we can extend this conclusion to the field as a whole by stating that all the S-foci will be found on a paraboloid of rotation having its axis coinciding with the optical axis and its pole at B'_m, and that all the T-foci will lie on a similar paraboloid differing only in the value of its parameter or axial radius of curvature. This curvature of the focal surfaces on which the sagittal and tangential rays come to a focus is another serious defect of an imperfectly corrected lens system, and we must study it more closely.

The longitudinal displacements $g_3H_k'^2 + \frac{1}{2}g_4H_k'^2$ for sagittal and $3g_3H_k'^2 + \frac{1}{2}g_4H_k'^2$ for tangential rays which lead to the curvature of the field depend on the two constants g_3 and g_4, but although these two act in a closely similar manner in producing curvature of the field, they differ fundamentally in other respects, for whilst we had to discuss g_3 extensively in previous subsections as the cause of astigmatism, g_4 invariably cancelled out of the equations and therefore has no effect on the sharpness or definition of individual image-points, but is purely a source of curvature of the field. The distinctive character of g_4 also reveals itself in the mathematical expression by which it is defined, for whilst the other four g-coefficients all involve the constant of spherical aberration 'a' at each surface, and with the exception of g_1 also the quantity q' which depends on the position of the diaphragm, the coefficient g_4 is independent not only of 'a' and q' but also of the conjugate distances of object and image and of all intermediate intersection-lengths; for it came out in the final summation of the oblique aberrations with the value

$$g_4 = \sum N'_k \frac{N'_j - N_j}{N_j N'_j \cdot r}$$

which depends only on refractive indices and on radii of curvature. This explicit equation for g_4 immediately proves yet another remarkable property, for although we put no restriction of any kind on the thicknesses of the individual lenses in the system nor on their separations, the sum by which g_4 is to be calculated contains these thicknesses and separations in no form either expressed or implied; hence g_4 is also utterly independent of thicknesses and separations, and we may change these to any possible extent in trying to impart desirable properties to a system in course of evolution without having to recalculate g_4. Finally, it is easily proved that g_4 is also not affected in its value by *bending* the constituent lenses of a system working in air; this is of extreme importance, as we have already learnt that bending is the most powerful means of altering the state of correction of a given system. To furnish the proof we have only to remember that every lens system is built up of a number of simple lenses, and that even when some of these lenses are united by balsam we may imagine the thin balsam film removed and replaced by an optically equally ineffective thin layer of air. Now for any one simple lens of refractive index N and bounded by surfaces with r_1 and r_2 as radii, the surrounding air will have index 'one' and the equation for g_4 gives

$$g_4 = \frac{N-1}{N \cdot r_1} + \frac{1-N}{N \cdot r_2} = \frac{N-1}{N}\left(\frac{1}{r_1} - \frac{1}{r_2}\right) = \frac{N-1}{N} \cdot c,$$

when we use our usual symbol c for the total or net curvature of the lens, and as the latter is not changed by bending, the proof required is already complete for one lens. If there are several lenses, the same argument holds for each one, and as g_4 is in the form of a simple sum, we can write the alternative form

$$g_4 = \sum \left(\frac{N_j - 1}{N_j} \cdot c_j\right)$$

which is usually the most convenient one for actual calculations. In this equation j stands for the successive *letter*-suffixes of the constituent *lenses* of the system.

Yet another form of the equation of g_4 is obtained from the last one by noting that in accordance with TL (2) of section [23] we have $(N-1)c = 1/f'$, in which f' is the focal length of a *thin* lens having the same radii as the actual thick lens. As it is of the highest importance to remember that the f' which thus enters into the g_4 equation *must* be that calculated by the TL equation and *not* the true equivalent focal length of the actual thick lens, we will attach suffix TL to it and write this third form of the equation

$$g_4 = \Sigma(1/N_j \cdot f'_{TLj}),$$

in which j again stands for the letter-suffix of the successive lenses.

The aberration constant g_4, which we have shown to be independent of the conjugate distances of object and image, of thicknesses and separations, and even of the shape into which the constituent lenses of a system may be bent, and which for these reasons indicates a singularly fixed and unyielding property of any given lens system, is known as the Petzval sum. It is so named in honour of Joseph Petzval (1807–91), a Hungarian and professor of mathematics at Vienna, who investigated the aberrations of oblique pencils about 1840, and apparently arrived at a complete theory not only of the primary, but also of the secondary oblique

[55] aberrations ; but as he never published his methods in any complete form, he lost the priority which undoubtedly would have been his. It is, however, perfectly clear from his occasional brief publications that he had a more accurate knowledge of the profound significance of the Petzval theorem than any of his successors in the investigation of the oblique aberrations for some eighty years after his original discovery. Petzval is best known as the designer of the famous Petzval portrait lens which he computed by his methods about 1840, and which is still the favourite lens for studio work with the majority of professional photographers.

We now return to the discussion of the curvature of the field. We showed that the distance $TB'o_m = 3\,g_3H_k'^2 + \frac{1}{2}\,g_4H_k'^2$ and that the corresponding distance $SB'o_m = g_3H_k'^2 + \frac{1}{2}\,g_4H_k'^2$. We see that the singularly fixed and unchangeable constant g_4 enters into the dislocation of both the sagittal and the tangential focus by the same amount $\frac{1}{2}\,g_4H_k'^2$, which at once suggests the separation of this term in the same way as we originally separated the spherical aberration $g_1SA_k^2$ by shifting our image-plane from the paraxial focus to that of our selected zone of the aperture. But as the Petzval term $\frac{1}{2}\,g_4H_k'^2$ follows a parabolic law with reference to the image height H_k', we must now lay a parabolic curve through B_m' and the point Ptz of our selected principal ray, which is found by marking off the distance $Ptz\,B'o_m = \frac{1}{2}\,g_4H_k'^2$. The curve so determined will be the trace of a curved image surface corresponding to the value of the Petzval sum of our system ; we will call this curved image surface the Petzval surface of the system. It is at once evident that we should obtain sharp extra-axial image-points on this surface if (in the absence of coma assumed throughout the present discussion) g_3 were zero, for the additional displacements of S and T by $g_3H_k'^2$ and $3\,g_3H_k'^2$ would then vanish. Hence we have a first result of profound importance, namely, that when a lens system is free from astigmatism, then the sharp images will lie on the curved Petzval surface instead of on the ideal and highly desirable image-plane. It follows that sharp images on a flat surface are only obtainable by making g_4 also equal to zero, and as we learnt that the Petzval sum cannot be altered (as all the other aberration constants can) by changing thicknesses or separations or by bending the constituent lenses, this zero value of g_4 cannot be achieved by the usual tricks of the optical designer ; like achromatism, a low value of the Petzval sum must be provided for in the first rough scheme for a proposed system ; to try to do so at a later stage is absolute waste of time. For the present it may merely be stated that the highly desirable low value of g_4 is only attainable in photographic objectives of the anastigmatic type. In telescope and microscope objectives, and in all ordinary forms of eyepieces and of inexpensive photographic lenses, a decidedly serious value of g_4 must be accepted ; we must therefore complete the discussion of curvature of the field by seeking the best way of minimizing the effects of the Petzval curvature.

For any given system the Petzval curvature will be practically a fixture, but the astigmatism measured by the constant g_3 is usually easily controlled, and our problem is to decide upon the most favourable value of g_3. We have seen that if we make g_3 zero, we shall secure sharp extra-axial image-points for our selected zone, which in itself would be a most desirable state of correction ; but the field will be curved, and if we focus the axial image-point, extra-axial points will be out

[55] of focus by the full depth of the Petzval surface $= \frac{1}{2} g_4 H_k'^2$, and this will in many cases reduce the sufficiently sharp part of the field below the desirable extent. If now we admit a finite value of g_3, the extra-axial image-points will become defective owing to the astigmatism which will cause the characteristic constrictions at S and T, but up to a certain value of g_3 this loss of sharpness will not be very serious. We will accept for the present an assurance that both by practical experience and on the strength of the undulatory theory of light there is not much to choose, for reasonably *low* amounts of astigmatism, between any selected adjustments of the focus throughout the range from S to T, but that outside this range the image rapidly deteriorates; also that the loss of quality in the image is approximately proportional to the distance from S to T and therefore to the value of g_3, and that this loss of quality is of the same magnitude as that of a sharp image-point which is out of focus by the same S to T distance.

We must next note that with reference to the Petzval surface, point S is at a distance $= g_3 H_k'^2$ and point T at a distance $= 3\, g_3 H_k'^2$, that is, S and T always lie on the same side of the Petzval surface and at distances from it which are always in the fixed ratio of one to three. If we now look at Fig. 55, which represents the case of a positive value of g_3, we see at once that this is an unconditionally bad state of correction, for even the point S is farther from the image-plane laid through the axial point B'_m than the point *Ptz*, so that we have an aggravated curvature of the field and in addition a loss of sharpness of the extra-axial images. Hence positive astigmatism in presence of a positive Petzval curvature must be avoided whenever possible. We shall however see subsequently that this is the case invariably present in all ordinary telescope objectives, and that it is the positive astigmatism of telescope objectives which renders them quite incapable of giving satisfactory images for a field exceeding a few degrees in angular extent.

In order to discover to what extent we can improve on the unsatisfactory state of correction represented in Fig. 55, we will study the four diagrams, Fig. 56, all of which show the same positive Petzval sum, but in association with successive increments of negative astigmatism. In all the diagrams the lower horizontal line

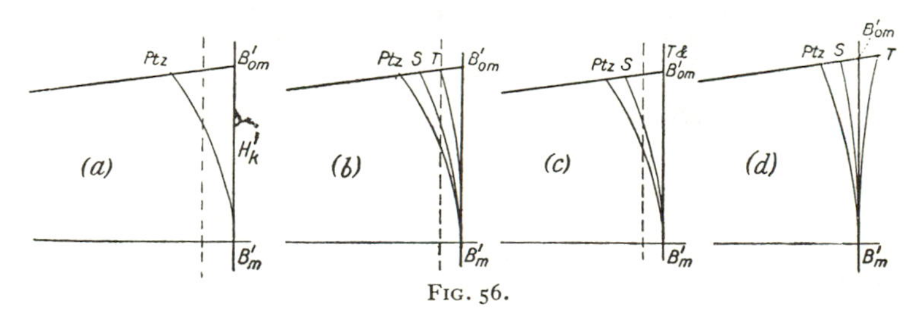

FIG. 56.

represents the optical axis of the system, the upper inclined line represents the principal ray of an extra-axial pencil, and the arc B'_m *Ptz* shall be the trace of the curved image surface resulting from the value of the Petzval sum in the absence of astigmatism. The distance *Ptz* $B'o_m$ therefore is $= \frac{1}{2} g_4 H_k'^2$.

In Fig. 56 (a) the system is assumed to be quite free from astigmatism, so that

[55] our selected zone of the aperture will produce sharp images on the Petzval surface itself. If the image were received on a plane laid through the axial point B'_m we should obtain sharp image-points close to the axis, but this sharpness would rapidly deteriorate in the outer part of the field, for the extra-axial images would be out of focus by $\frac{1}{2}g_4H_k'^2$ according to the value of H'_k for the individual points. In order to reduce the *maximum* want of sharpness to a minimum for a given field, say up to the drawn principal ray, we would evidently have to shift our focusing screen or photographic plate to the dotted position through half the marginal value of $\frac{1}{2}g_4H_k'^2$, for then the extreme marginal as well as the axial point would be out of focus by only $\frac{1}{4}g_4H_k'^2$ calculated with the marginal value of H'_k; this therefore represents the best adjustment possible under the circumstances. By the parabolic law there would then be a zone of sharp definition at 0·7071 of the marginal value of H'_k.

Fig. 56 (b) represents the best possible adjustment of the negative astigmatism for cases like that of astronomical photography, when only *very slight* imperfection of the image can be tolerated within the field of view actually utilized. In such cases the focal line tendency of astigmatism is largely neutralized by the general spreading out of image-points by diffraction in accordance with the theory of the Airy spurious disk, and the rule already given applies that there is nothing to choose between the images obtained anywhere within the S to T range of a slightly astigmatic pencil and that the small loss in sharpness due to a given distance from S to T is equal to the loss in sharpness of a *sharp* image-point when the latter is out of focus by the same distance. Consequently we shall obtain the least maximum value of imperfection within a given field if the axial point is out of focus to the extent of the S to T distance at the margin of the field, and by the one-to-three ratio of $PtzS$ to $PtzT$ it is easily seen that this optimum will be attained when $PtzS = 0{\cdot}2$ of $PtzB'o_m$ and $PtzT = 0{\cdot}6\,PtzB'o_m$, leaving for the distance $TB'o_m$ the value $0{\cdot}4\ PtzB°o_m = 0{\cdot}2\,g_4H_k'^2$, and when the photographic plate is placed into the dotted position so as to receive the marginal point T. The maximum imperfection will then correspond to the out-of-focus effect of $0{\cdot}2\,g_4H_k'^2$ against one of $0{\cdot}25\,g_4H_k'^2$ in the case of no astigmatism represented in Fig. 56 (a), and there is a decided though not very large advantage. The necessary value of g_3 is easily found from the fact that

$$PtzS = -g_3H_k'^2 = 0{\cdot}2\ PtzBo_m = 0{\cdot}1\,g_4H_k'^2,$$

which shows that this state of correction calls for

$$g_3 = -0{\cdot}1\,g_4.$$

The reasoning which leads to the choice represented in Fig. 56 (b) is not legitimately applicable to ordinary photographic objectives or to eyepieces, for in these the field demanded is of very considerable angular extent, and the negative astigmatism which must be admitted for the best compromises is of such serious magnitude that very obvious focal lines are produced at the points S and T in the outer part of the field; hence a focus near the midway point of S and T will be preferable for that region because it leads to approximately *round* diffused images, which are less disturbing and misleading than strongly elongated misrepresentations of the individual image-points. Long and varied experience has led practical [55] designers to the two principal compromises represented in Fig. 56 (c) and (d).

The most favourable compromise in the great majority of cases, and also that which fits in most conveniently with our usual computing practice, is represented in Fig. 56 (c). The lens system is so adjusted that the tangential foci T fall into the plane of the axial focus B'_m. On a focusing screen placed into that position we should then obtain a sharp axial image, but tangential focal lines of rapidly growing length towards the margin of the field, which would be a very undesirable result. But if we change the focus to the dotted position, namely so that at the margin of the actually utilized field the round image midway between S and T falls upon the screen, then the definition throughout the field will become approximately uniform, for, geometrically considered, the round images at the margin will be equal in diameter to the out-of-focus images in the central part of the field, and at all intermediate positions in the field the diffused images will have the same diameter in the sagittal direction simply because the focusing screen is parallel to the tangential focal surface. There will be, however, an elliptical deformation of the intermediate images which will reach a maximum in the form of sharp sagittal focal lines at 0·7071 of the full field where the focusing screen cuts through the sagittal focal surface. Fairly uniform though 'soft' definition throughout the field is nearly always preferable—at any rate in photographic objectives—to 'needle-sharp' images in the centre which deteriorate to severe diffusion at the edge of the field, and that is the reason why the state of correction shown in Fig. 56 (c) is most usually chosen by experienced designers when anastigmatic correction cannot be attained. It is important to know the ratio of g_3 to g_4 which leads to this adjustment. As $PtzS$ is always equal to one-third of $PtzT$, and as in this case $PtzT = PtzB'o_m = \frac{1}{2} g_4 H_k'^2$, we have

$$PtzS = -g_3 H_k'^2 = \frac{1}{3} PtzB'o_m = \frac{1}{6} g_4 H_k'^2;$$

therefore the desired flat tangential field calls for

$$g_3 = -\frac{1}{6} g_4.$$

The compromise between astigmatism and Petzval curvature which is shown in Fig. 56 (d) is one which experienced designers have only used in photographic wide-angle objectives of the older types. As these objectives are expected to cover an angular field of from 70° to over 90°, the curvature effects become extremely serious, and tolerable results can only be hoped for at very small apertures of 1/16th of the focal length or less. Moreover these lenses are chiefly used for architectural studies of the interior of buildings, and owing to the combined effect of low aperture and feeble light it is frequently impossible to see the outer parts of the image on the focusing screen. The correct focus of the outer part has therefore to be inferred from the more easily visible central part of the image, and this implies that the least objectionable extra-axial images must lie in the plane of the sharpest axial image. We have seen that in the presence of heavy astigmatism the round images midway between S and T are to be preferred, hence these should be brought into the plane of the axial image. On account of the fixed one-to-three ratio of the distances of S and T from the Petzval surface this adjustment of a lens system

[55] requires point S to lie midway between Ptz and $B'o_m$, and point T at an equal distance beyond $B'o_m$. As we shall then have $PtzS = \frac{1}{2} PtzB'o_m$, the ratio of g_3 to g_4 will be found from

$$PtzS = -g_3 H'^2_k = \tfrac{1}{2} PtzB'o_m = \tfrac{1}{4} g_4 H'^2_k,$$

which gives for this case

$$g_3 = -\tfrac{1}{4} g_4.$$

As the previous (and usually preferable) case of a flat tangential field required a value of $g_3 = -\frac{1}{6} g_4$, it will be seen that the flattening of the field for the midway points of S and T calls for one-and-a-half times the astigmatism, and therefore gives diffusion patches at the margin of the field which are also one-and-a-half as large as in the case of a flat tangential field. This represents a heavy penalty attached to the case of Fig. 56 (d), which is aggravated by the fact that the serious diffusion of the marginal image-points will be rendered more painfully obvious by the simultaneous sharpness of the axial region of the field.

It will now be evident that nothing but disadvantages could be expected from a still further increase of the negative astigmatism, and we have thus reached the highly important conclusion that whilst a moderate amount of negative astigmatism is beneficial in lens systems afflicted with a considerable positive value of the Petzval sum, this amount should never exceed the limit $g_3 = -\frac{1}{4} g_4$, and that on the whole $g_3 = -\frac{1}{6} g_4$ represents the best compromise for ordinary photographic objectives and also for the majority of eyepieces. On the other hand, we have learnt that astigmatism of the sign of the Petzval sum is unconditionally bad and to be avoided whenever possible.

The significance of the preceding discussion will be most readily appreciated if we calculate g_4 for a typical case and apply our conclusions to the result. We shall see a little later that the 10-inch achromatic objectives studied in Chapter V can be turned into satisfactory photographic objectives of the 'landscape lens' type by bending them into meniscus form: we will therefore choose these objectives as our example. They had $c_a = 0{\cdot}470$, $N_a = 1{\cdot}5407$, $c_b = -0{\cdot}247$, and $N_b = 1{\cdot}6225$, therefore the Petzval sum for any conceivable bending of these objectives will be

$$g_4 = \frac{N_a - 1}{N_a} \cdot c_a + \frac{N_b - 1}{N_b} \cdot c_b = \frac{0{\cdot}5407}{1{\cdot}5407} \times 0{\cdot}470 - \frac{0{\cdot}6225}{1{\cdot}6225} \times 0{\cdot}247$$

or

$$g_4 = 0{\cdot}1650 - 0{\cdot}0948 = 0{\cdot}0702.$$

In accordance with the discussion we shall have the distance from the sagittal focus to the image-plane laid through the axial focus of a given zone determined by

$$LA'_S - LA' = g_3 H'^2_k + \tfrac{1}{2} g_4 H'^2_k,$$

and for the tangential focus the corresponding distance will be

$$LA'_T - LA' = 3\, g_3 H'^2_k + \tfrac{1}{2} g_4 H'^2_k.$$

A photographic landscape lens of 10 inches focal length would usually be expected to cover the 'half plate' measuring $6\frac{1}{2} \times 4\frac{3}{4}$ inches. Allowing a little for the unused edges of the plate, the value of H'_k at the corners of the plate would be about 4 inches; at the middle of the short sides H'_k would be about 3 inches, and at the middle of the long sides it would be about 2 inches. We will therefore [55] calculate the differences of focus for these three values of H'_k and for the three cases discussed theoretically, namely, when $g_3 = 0$, when $g_3 = -\frac{1}{6}g_4$ and when $g_3 = -\frac{1}{4}g_4$. These give the respective formulae :

$$\begin{aligned} LA'_S - LA' &= \tfrac{1}{2} g_4 H'^2_k \\ &= (-\tfrac{1}{6} g_4 + \tfrac{1}{2} g_4) H'^2_k = \tfrac{1}{3} g_4 H'^2_k \\ &= (-\tfrac{1}{4} g_4 + \tfrac{1}{2} g_4) H'^2_k = \tfrac{1}{4} g_4 H'^2_k \end{aligned}$$

$$\begin{aligned} LA'_T - LA' &= \tfrac{1}{2} g_4 H'^2_k \\ &= (-\tfrac{3}{6} g_4 + \tfrac{1}{2} g_4) H'^2_k = 0 \\ &= (-\tfrac{3}{4} g_4 + \tfrac{1}{2} g_4) H'^2_k = -\tfrac{1}{4} g_4 H'^2_k \end{aligned}$$

and calculation with $g_4 = 0{\cdot}0702$ gives by slide-rule :

		for H'_k =	2″	3″	4″
First case :	$LA'_S - LA' = LA'_T - LA'$	=	0·14″	0·32″	0·56″
Second case	$LA'_S - LA' = \frac{1}{3} g_4 H'^2_k$	=	0·09″	0·21″	0·375″
	$LA'_T - LA' = 0$		0	0	0
Third case	$LA'_S - LA' = \frac{1}{4} g_4 H'^2_k$	=	0·07″	0·16″	0·285″
	$LA'_T - LA' = -\frac{1}{4} g_4 H'^2_k$	=	−0·07″	−0·16″	−0·285″

Even when we note that in the first two cases the maximum differences of focus will be halved by choosing the compromises which were described, it will be clear that a lens will have to be reduced to a decidedly small aperture in order to render these out-of-focus effects comparatively harmless. It may however be accepted that all ordinary photographic objectives—exclusive of anastigmats—conform fairly closely to our little table when the figures are altered to scale for objectives whose focal length differs from the assumed 10 inches. In well-designed anastigmats the residual effects of astigmatism and curvature of field are about one-eighth to one-twentieth of those in our table and are of a zonal character, that is, they display their maximum amount at about seven-tenths of the full diameter of the field for which the lens has been designed.

H. Extension to all Zones of a given Aperture

The preceding discussion of the effects produced by the Seidel aberrations has been restricted to one circular zone of the clear aperture of the centred system of spherical surfaces. If we treat the zone hitherto dealt with as the extreme marginal zone for the given system, then we have to estimate the effects resulting from the addition of all the inner zones. For the most general case when g_1, g_2, g_3, and g_4 are all of considerable magnitude, which is really of very little practical interest as a system so grievously afflicted would be almost useless, we can reach sufficiently comprehensive conclusions most easily and quickly by studying the characteristic focal lines of all the zones. For a given point in the field of view determined by its distance H'_k from the optical axis, we found for the characteristic focal line

[55] produced by a zone of semi-aperture SA_k, with reference to the principal ray and to the paraxial image-plane :

Location of centre of the characteristic focal line at

$$TA'_S = g_2 SA_k^2 H'_k \quad \text{and} \quad LA'_S = g_1 SA_k^2 + g_3 H'^2_k + \tfrac{1}{2} g_4 H'^2_k ;$$

Inclination of the line with reference to the principal ray

$$\tan O = -\frac{g_3}{g_2} \cdot \frac{H_k}{l'^2_k} ;$$

Longitudinal extent in the direction of the principal ray

$$LA'_{+M} - LA'_{-M} = 4 g_2 l'_k SA_k H'_k ;$$

Transverse extent at right angles with principal ray

$$TA'_{+M} - TA_{-M} = 4 g_3 \frac{1}{l'_k} SA_k H'^2_k .$$

As H'_k is constant for a given point in the field, we note first of all that tan O is constant for all zones ; therefore all the characteristic focal lines of a given oblique pencil are parallel to each other. If we write down the coordinates of the centres of two of these parallel focal lines for any two different apertures SA_{k1} and SA_{k2} and form their differences

$$\begin{aligned} TA'_{S1} &= g_2 SA_{k1}^2 H'_k \\ TA'_{S2} &= g_2 SA_{k2}^2 H'_k \\ \hline TA'_{S1} - TA'_{S2} &= g_2 H'_k (SA_{k1}^2 - SA_{k2}^2) ; \\ LA'_{S1} &= g_1 SA_{k1}^2 + g_3 H'^2_k + \tfrac{1}{2} g_4 H'^2_k \\ LA'_{S2} &= g_1 SA_{k2}^2 + g_3 H'^2_k + \tfrac{1}{2} g_4 H'^2_k \\ \hline LA'_{S1} - LA'_{S2} &= g_1 (SA_{k1}^2 - SA_{k2}^2) \end{aligned}$$

we can determine the slope, with reference to the principal ray, of the line joining the two centres exactly as in the case of the characteristic focal line. Calling the angle O_S, with clock-sense as in the case of angle O, we find

$$\tan O_S = -(TA'_{S1} - TA'_{S2}) / (LA'_{S1} - LA'_{S2}) = -H'_k \cdot g_2 / g_1 ;$$

and as tan O_S is found to depend only on H'_k, g_2, and g_1 we conclude that the centres of *all* the characteristic focal lines, and therefore also all the sagittal foci of a given aperture, lie on another straight line at a fixed inclination O_S with the principal ray. This 'line of the sagittal foci' for a given full aperture begins at the sagittal focus of the indefinitely small zone immediately surrounding the principal ray for which $SA_k = 0$ and therefore

$$TA'_{So} = 0 \qquad LA'_{So} = g_3 H'^2_k + \tfrac{1}{2} g_4 H'^2_k ;$$

and it ends at the sagittal focus of the marginal zone, for which we have

$$TA'_S = g_2 SA_k^2 H'_k ; \; LA'_S = g_1 SA_k^2 + g_3 H'^2_k + \tfrac{1}{2} g_4 H'^2_k .$$

The longitudinal extent of this line is therefore $= g_1 SA_k^2$, and the sagittal foci as well as the centres of the successive characteristic focal lines are distributed on it [55] in proportion to the square of the aperture of the zones.

With regard to the length of the successive focal lines, the equation for $LA'_{+M} - LA'_{-M}$ immediately shows that it is proportional to the aperture.

The resulting relations are shown on a large scale in Fig. 57. S represents the sagittal focus of the marginal zone, S_0 that of the indefinitely small zone closely

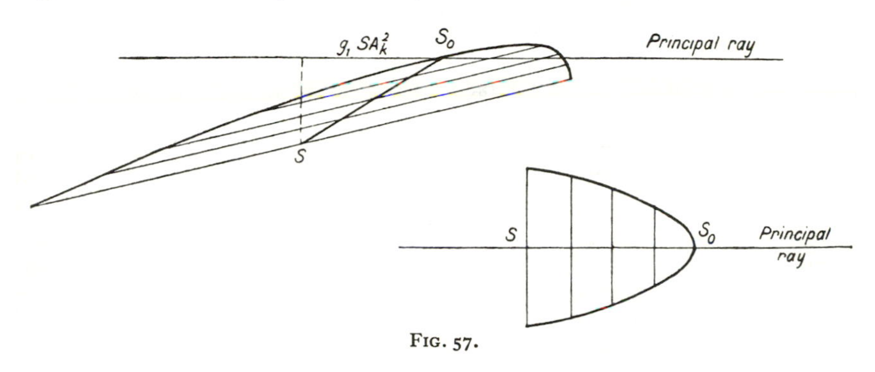

FIG. 57.

surrounding the principal ray. The length and inclination of the characteristic focal line of the marginal zone are arbitrarily assumed. The corresponding data for smaller zones then follow from the laws which we have proved. Focal lines are drawn for zones with semi-diameters equal to the marginal SA_k times 1, $\sqrt{\frac{3}{4}}$, $\sqrt{\frac{1}{2}}$, $\sqrt{\frac{1}{4}}$, and practically zero respectively, and as the distribution of their centres on the line SS_0 is proportional to the square of the aperture, the crossing points will be at S, at $\frac{3}{4}$, $\frac{1}{2}$, $\frac{1}{4}$ of SS_0 and at S_0 itself, whilst the lengths of the lines will be equal to that of the marginal one times $\sqrt{\frac{3}{4}}$, $\sqrt{\frac{1}{2}}$, $\sqrt{\frac{1}{4}}$, and practically zero. It is easily seen from the theory of the conic sections, and follows in fact from the proof given in section [47] for the 'dip' method of drawing parabolas, that the ends of all the characteristic focal lines lie on an arc of a common parabola of which SS_0 is a diameter; hence the axis of this parabola is parallel to SS_0. All possible D-pairs of the entire aperture have their foci within the area bounded by the parabolic arc and the characteristic focal line of the marginal zone, and the strongest concentration of light will therefore be found at some mean position of this area. More detailed information will be most easily obtained by drawing a considerable number of rays, say for the zones selected in Fig. 57, by the method of projection used in Fig. 52. The complete ray-distribution and accurate cross sections of the bundle of rays can thus be worked out, and the results obtained by the student's own effort will be more readily assimilated by the mind and more easily remembered than many pages of complicated algebra.

In the absence of coma the characteristic focal lines become perpendicular to the principal ray and are bisected by it. The bounding parabola then becomes symmetrical, with the principal ray as its axis, as shown in the lower right-hand corner of Fig. 57. A complete drawing in elevation and plan by the method of

[55] Fig. 52 will show that a precisely equal parabolic distribution results at the tangential focus when coma is absent.

I. Curvature of the Field at Full Aperture

On account of the extremely objectionable appearance of an image affected by an easily visible amount of coma, the only case met with in respectable optical instruments, and hence the only one of real practical importance, is that of fairly complete correction of coma. We will therefore discuss the curvature of the field at full aperture on the assumption that the small residuals of coma may safely be considered as negligible in the presence of the considerable Petzval curvature and usually also of astigmatism, which we have shown to be inevitable in all ordinary optical systems which are expected to cover a field of considerable angular extent.

Under the stated restriction we have then to discuss the longitudinal differences of focus

$$LA'_S = g_1 SA_k^2 + g_3 H_k'^2 \quad + \tfrac{1}{2} g_4 H_k'^2$$

and

$$LA'_T = g_1 SA_k^2 + 3\, g_3 H_k'^2 + \tfrac{1}{2} g_4 H_k'^2$$

for all values of SA_k from zero to the full semi-aperture. As only the first term in each equation depends on SA_k it is at once clear that nothing remains to be added to the discussion for a single zone of the aperture in the greatly predominating cases when the system is a spherically corrected one, for g_1 is then zero and LA'_S and LA'_T will both have a perfectly constant value for all zones ; hence the curvature of the field at full aperture will be the same as that determined for a single zone.

There are, however, many optical systems which give perfectly satisfactory results in spite of considerable residues of ordinary spherical aberration ; nearly all eyepieces and many simple types of photographic objectives are of this class, which we must therefore discuss.

In the highly exaggerated Fig. 58, drawn in elevation and plan, the axial pencil is supposed to have its paraxial focus at B'_p and its marginal focus at B'_m. In the oblique pencil S_0 is assumed as the focus of the ultimate very close pair of sagittal rays and T_0 is the corresponding tangential ultimate focus. As the $g_1 SA_k^2$ in the equations for LA_S and LA'_T is identical with the spherical aberration of an axial pencil of the same aperture, the marginal S and T foci must satisfy the condition $TT_0 = SS_0 = B'_m B'_p$, and the same relation holds for all intermediate zones. It follows—with the ever necessary allowance for the exaggeration in our diagrams, which really should show very small aberrations and inclinations of rays—that the curvatures of the sagittal and tangential fields are exactly the same for all zones of the full aperture. Moreover, it is evident that the confusion of rays caused by the spherical aberration is of precisely the same type in the region $B'_m B'_p$ of the axial pencil, in the region TT_0 of the elevation of the oblique pencil, and in the region SS_0 of the plan of the oblique pencil, hence we can conclude that the best compromise focus must lie at the same fraction of the distance from the marginal to the ultimate focus in all three cases, and that the diameter of the diffused axial image at its best focus must be equal to the diffused thickness of the sagittal and tangential focal lines at their best focus. It follows that the best compromise foci throughout the field must lie on curves parallel to those for the marginal and paraxial zones, say on the curves dotted in the elevation of Fig. 58. The final [55] result therefore is that in the absence of considerable amounts of coma, the curvatures of the sagittal and tangential fields are the same as those worked out earlier for a single zone of the aperture, and that the favourable compromises between astigmatism and Petzval sum also apply for the full aperture.

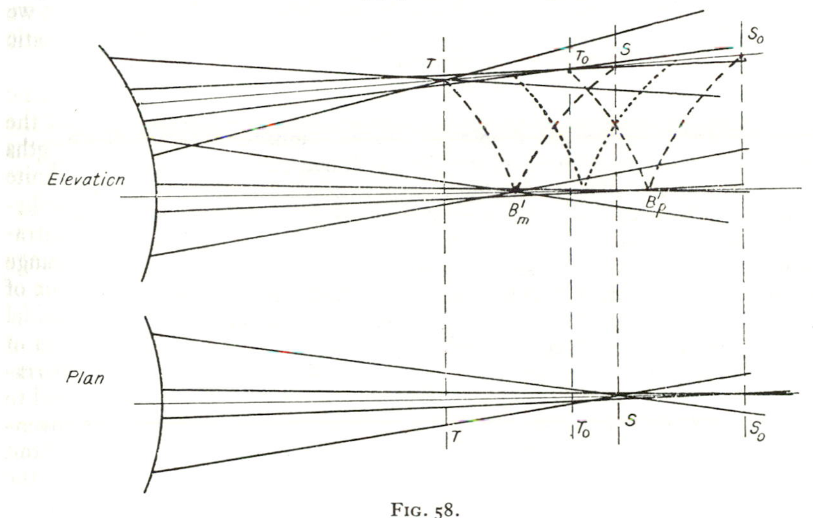

FIG. 58.

It will be easily seen by reference to the diagrams representing the general case when considerable coma is also present that the same *tendency* must exist even in that case ; and as the extremely ill-defined and unsymmetrical images which would result under those conditions cannot be accurately located, we may safely conclude that any residual departures from the above simple rules would probably be incapable of verification by direct observation and would therefore be quite unimportant.

CHROMATIC ABERRATIONS OF OBLIQUE PENCILS

[56] As the course of a ray through a lens system is determined by the law of refraction, and as the index of refraction varies with the colour, it is evident that we must be prepared to find a chromatic variation of all the important optical properties of a given system. With reference to the final axial image-point we shall have the longitudinal chromatic aberration which we studied in Chapter IV, and as we lay our ideal image-planes for the study of the oblique aberrations through the paraxial focus, it follows that in general different colours will have different image-planes and that the longitudinal chromatic aberration must affect the extra-axial image-points to the same extent as the axial image-point. But for the extra-axial image-points we must also consider their distance H'_k from the optical axis, which determines the linear magnification of the system. It is clear that this also will usually

[56] vary in different colours and that it will then constitute another and entirely distinct form of chromatic aberration, usually known as the chromatic variation—or difference—of magnification. We thus realize that in systems expected to cover a field of considerable extent, achromatism implies the simultaneous fulfilment of *two* conditions : there must be *achromatism of the axial image-point* so as to cause the ideal image-planes in all colours to coincide, and there must also be *achromatism of magnification* in order that rays of all colours sent out by a given extra-axial object-point shall be reunited in a single sharp and white image-point. Hence we shall have to deal with two more or less independent forms of primary chromatic aberration.

From the strictly mathematical point of view we ought to include the chromatic variation of the five Seidel aberrations, for as all five depend on the value of the refractive indices, and all excepting the Petzval sum also on the intersection-lengths and angles of the rays, their values must vary for different colours. For a finite difference of the refractive indices in two colours these variations would, mathematically considered, be of the same 'order of magnitude' as the full Seidel aberrations for any one colour. But as the change in refractive index for the usual range of colours from C to F or from D to G' is only about 1 per cent. of the value of N_d, the variations are numerically only a very small fraction of the full Seidel aberrations, and for that reason it has been the custom from the earliest days of analytical optics to treat the chromatic variations of the primary spherical aberrations as comparable in magnitude with the secondary spherical aberrations and to neglect them when only primary aberrations are taken into account. This convention has always worked well in practice, and we shall adopt it. We therefore limit the present investigation to the two primary chromatic aberrations named in the preceding paragraph.

The secondary spectrum effect which has to be reckoned with in nearly all cases justifies a further simplification ; as was shown in sections [43] and [44] we not merely may, but we ought to restrict the investigation to a very small range of colours in the immediate neighbourhood of the colour for which the system is to have minimum focal length or intersection-length. The adoption of this restriction gives us Chr. (8) of section [44] as the formula for the longitudinal chromatic aberration at any one refracting surface :

$$\text{Chr. (8)} \qquad l'_r - l'_v = \frac{Nu^2}{N'u'^2}(l_r - l_v) + l'\,\frac{l'-r}{r}\left(\frac{\delta N'}{N'} - \frac{\delta N}{N}\right).$$

The second term on the right represents the new aberration arising at the surface, whilst the first term shows that transfer is ruled by the law of longitudinal magnification. In accordance with the general proof in section [54] we can therefore effect the summation for a complete system by transferring the new aberration arising at each surface directly to the final image-plane, and obtain the equation :

$$\text{(a)} \quad \text{Final } (l'_r - l'_v) = \sum_{j=1}^{j=k}\left[l'_j\,\frac{l'_j - r_j}{r_j}\left(\frac{\delta N'_j}{N'_j} - \frac{\delta N_j}{N_j}\right) N'_j u'^2_j / N'_k u'^2_k\right] = \text{Final } L'ch.$$

This equation contains neither SA_k nor H'_k, hence the primary longitudinal chromatic aberration is independent of both ; it is a fixed difference of focus between different colours. [56]

With reference to the 'chromatic difference of magnification' we can acquire a sound fundamental notion of the nature of this defect by studying Fig. 59. The centred system indicated at the left will in general be afflicted with chromatic

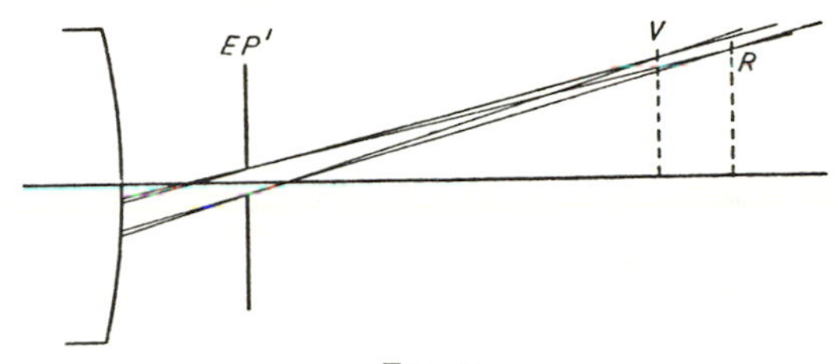

Fig. 59.

aberration, and will then give images in 'violet' at V and in 'red' at some other point R. The longitudinal difference of focus is of course the Lch' which we have already determined. To find out the relative position of the 'red' and 'violet' images as received on a focusing screen or photographic plate, or as seen in the field of an eyepiece, we must draw the respective pencils of rays with the exit pupil EP' as base, for the images received or seen will represent a cross-section of all the pencils. It is immediately obvious that under the conditions shown in our diagram we should obtain a violet image patch farther from the optical axis—or centre of the field—than the corresponding red patch ; hence violet would be more highly magnified than red. The most significant point to note is that this is the result obtained, although the *sharp* focus at V has been assumed closer to the optical axis than the *sharp* focus at R ; this demonstrates that we should commit a grave error if, in the presence of longitudinal chromatic aberration, we tried to determine the chromatic difference of magnification by comparing the image heights of the different colours measured at the respective sharp foci. The evident further conclusion that the true result depends on the location of the exit pupil will be followed up in a subsequent section.

To determine the primary term of the chromatic difference of magnification, that is, its value at indefinitely small aperture, we now refer to Fig. 59 (a).

The refracting surface at A shall have its centre of curvature at C, and the entrance pupil for this surface shall be at EP. B shall be the axial and B_o a corresponding extra-axial object-point, both for the present assumed to be free from chromatic aberrations. By drawing the principal ray $EP - B_o$ and the auxiliary optical axis CB_o we determine the point of incidence Ppr of the principal ray and the auxiliary pole A_o, and the distance A_oPpr will be the incidence height $q'H'$ of the principal ray as introduced in section [53]. Longitudinal chromatic aberration will lead to the formation of separate foci of the refracted rays in different colours, and as in our primary approximation we treat this as constant throughout the field, we shall have

$$B'_vB'_r = B'pr_vB'pr_r = l'\frac{l'-r}{r}\left(\frac{\delta N'}{N'} - \frac{\delta N}{N}\right) \text{ by Chr. (8).}$$

[56] We see that the incident principal ray is split up into its constituent colours, and as at small aperture the remaining rays of the complete pencil in any one colour will cluster very closely around its principal ray throughout the focal region in

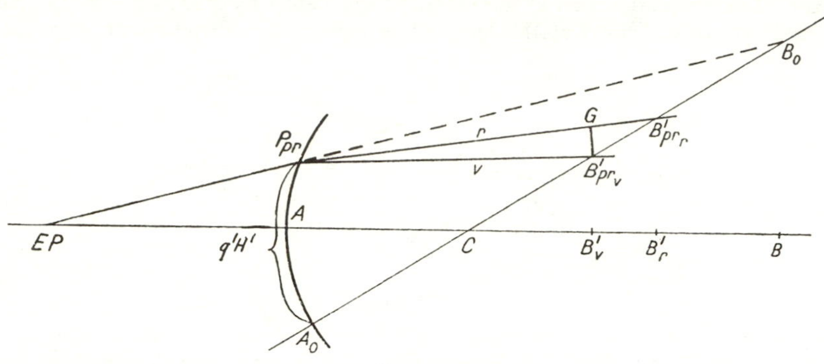

FIG. 59 a.

accordance with the discussion in section [55], the 'red' and 'violet' image-points will fall apart on a focusing screen by a distance which is sufficiently closely measured by $B'pr_vG$. By analogy with the coma and distortion we thus recognize the chromatic difference of image height as a transverse aberration, and can determine it by the similar (at the limit) triangles $PprA_oB'pr_r$ and $GB'pr_vB'pr_r$ as

$$GB'pr_v = B'pr_vB'pr_r \times A_oPpr/A_oB'pr_r.$$

$B'pr_vB'pr_r$ is the new longitudinal chromatic aberration produced by our surface; using $L'ch$ as a suitable and easily remembered symbol for it, we have

$$L'ch = B'pr_vB'pr_r = l'.\frac{l'-r}{r}\cdot\left(\frac{\delta N'}{N'} - \frac{\delta N}{N}\right).$$

A_oPpr we have already identified as $= q'H'$ and $A_oB'pr_r$ will at the limit become $= l'$. If we also introduce $T'ch$ as a convenient symbol for the transverse chromatic aberration, or chromatic difference of image height, we thus find

$$T'ch = L'ch\,.\,q'H'/l' = q'H'.\frac{l'-r}{r}\left(\frac{\delta N'}{N'} - \frac{\delta N}{N}\right),$$

the last on introducing the explicit value of $L'ch$.

As a transverse aberration $T'ch$ can be summed up for a whole system in accordance with the general proof in section [54] by transferring the new aberration arising at each surface directly to the final image-plane, for the law of linear magnification will be applicable. The new aberration arising at the jth surface will thus contribute

$$q'_j\,.\,H'_j\,.\,\frac{l'_j - r_j}{r_j}\left(\frac{\delta N'_j}{N'_j} - \frac{\delta N_j}{N_j}\right) N'_ju'_j/N'_k\,.\,u'_k$$

to the final aberration, and if we put $H'_j = H'_k . N'_k . u'_k / N'_j u'_j$ by the theorem [56] of Lagrange, we obtain the final sum

(b) $$\text{Final } T'ch = H'_k \sum_{j=1}^{j=k} \left[q'_j \frac{l'_j - r_j}{r_j} \left(\frac{\delta N'_j}{N'_j} - \frac{\delta N_j}{N_j} \right) \right].$$

In analogy with the other transverse aberrations (coma and distortion) the final $T'ch$ will be positive when the principal image (in our case 'red') falls above the secondary coloured image, in our case 'violet'. As $T'ch$ is proportional to H'_k, the *ratio* of corresponding image heights in 'red' and ' violet' will be constant throughout the field ; that is the justification of the usual name of this aberration : 'Chromatic difference of magnification'. When this aberration is the only one present in the system, the evident effect is that the image of an extra-axial object-point is drawn out into a linear spectrum represented by $B'pr_vG$ in Fig. 59 between the colours 'v' and 'r'. This defect hardly ever attains visible magnitude in the case of ordinary reasonably thin telescope objectives, but it becomes easily visible in dialyte telescope objectives and in the majority of microscope objectives, and it calls for close attention in eyepieces and in photographic objectives, especially when the latter are intended for use in the three-colour process of photo-engraving.

SUMMARY OF THE DISCUSSION

[57] Before we enter upon the practical applications of the theory of the oblique aberrations it appears advisable to collect in concise form the principal results of the long discussion.

(1) In any one colour of light, for an object-plane at a fixed distance, and for the aperture diaphragm in a given position, the properties of a centred optical system are completely defined by five constants g_1 to g_5 which measure the five Seidel aberrations called 'Spherical Aberration', 'Coma', 'Astigmatism', 'Petzval Curvature', and 'Distortion '.

(2) *Spherical aberration* is a longitudinal difference of focus between rays from successive zones of the clear aperture. It is of the same magnitude for all image-points throughout the field, and is therefore most conveniently determined for the axial pencil by the methods detailed in Chapter II. It is positive when the paraxial focus lies to the right of the marginal focus, and it grows as the square of the aperture.

The other four Seidel aberrations affect only the extra-axial image-points.

(3) *Coma* causes an unsymmetrical deformation of the individual extra-axial image-points to a comet-like appearance. As the most objectionable of all aberrations, it must be reduced below easily visible magnitude at almost any cost. It is positive when the principal ray passes above the dislocated outer rays. It grows as the distance of an image-point from the optical axis and as the square of the aperture. In numerical calculations it is usually determined either as the distance from the focus of the marginal sagittal pair of rays to the principal ray $= g_2 SA_k^2 H'_k$ or by the corresponding tangential distance $= 3\, g_2 SA_k^2 H'_k$. The fixed one-to-three ratio of these two measures must be borne in mind.

[57] (4) *Astigmatism* is characterized by two longitudinally separated constrictions of the bundle of rays from any one zone of an oblique pencil. At the tangential focus the cross-section of the bundle is elongated in a direction at right angles with the plane of the optical axis containing the object-point, whilst at the sagittal focus the elongation is in the direction towards the optical axis. In the absence of coma the two constrictions become sharp focal lines. If spherical aberration is also absent, the focal lines remain sharp even at full aperture. The astigmatism is positive when the sagittal focus lies to the right of the tangential focus. The astigmatic difference of focus is independent of the aperture, but grows as the square of the distance of an image-point from the optical axis. The length of the focal lines grows directly as the aperture and also as the square of the distance from the optical axis.

(5) *Petzval curvature* of the field. The singularly obstinate defect measured by the Petzval sum has no effect of any kind upon the sharpness or definition of individual image-points ; in the absence of astigmatism it simply causes the image of a plane object to be formed upon a curved surface instead of the usually desirable flat surface. When astigmatism is also present, then the sagittal and tangential foci of any one oblique pencil are formed always on the same side of the Petzval surface and at distances which are in a one-to-three ratio—just as for sagittal and tangential coma. If g_3 and g_4 are of the same sign, the curvature of the field is aggravated ; but if the astigmatism is under control, then a judicious amount of g_3 with the opposite sign of g_4 affords the only means of minimizing the Petzval curvature in that great majority of optical systems in which the Petzval sum must be accepted at whatever value it assumes. Correction for flatness of the tangential field is usually the best compromise, and calls for $g_3 = -\frac{1}{6}g_4$.

(6) *Distortion* is the departure from strict geometrical similarity of any configuration in the object-plane with its image in the paraxially conjugate image-plane. The magnification in the central part of the field being fixed by the theorem of Lagrange, we measure the distortion by the distance from the actual extra-axial image-point to its ideal position as dictated by Lagrange's theorem, and we give the positive sign to the distortion when the ideal image-point falls above the actual one. Thus defined, the distortion grows as the cube of the distance of the image-point from the optical axis. Positive distortion causes the outer parts of the image to be on too small a scale, and is called barrel distortion ; in the case of negative distortion the outer parts of the image are on too large a scale, and we describe it as pin-cushion distortion. Distortion has no effect upon the sharpness of the individual image-points ; it only dislocates them towards or away from the optical axis.

(7) *The characteristic focal line*. Whilst all the rays from any one concentric zone of the aperture form a single sharp focus in the case of the axial image-point, they only intersect in pairs in the case of an oblique pencil, and lead to the formation of a straight focal line in the plane of the optical axis as the nearest approach to a focal effect. With reference to the principal ray of the oblique pencil, the longitudinal projection of the characteristic focal line is proportional in length to the value of g_2, whilst the transverse projection is proportional to g_3. The focal line can therefore shrink to a point only when both g_2 and g_3 (coma *and* astigmatism)

are zero, and these two aberrations are thus solely responsible for the usual [57] deterioration of the definition of the individual image-points in the outer part of the field. The characteristic focal line is of great value :

(*a*) because it enables one clearly to visualize the ray distribution in an oblique pencil, no matter in what magnitude or combination the five aberrations may be present ;

(*b*) because it supplies a longitudinal measure of the coma which we shall find useful in shortening the proof of certain propositions ;

(*c*) because it is particularly convenient in extracting the greatest possible amount of information from the results of the Hartmann test as applied to oblique pencils.

In the usual case when the object sends out light containing a whole range of different colours, two primary chromatic aberrations become added to the five Seidel aberrations, namely,

(8) *Longitudinal chromatic aberration*, which is uniform throughout the field and therefore identical with the chromatic aberration studied in Chapter IV.

(9) *Transverse chromatic aberration* or chromatic variation of the image height, which affects only the extra-axial part of the field, and which causes white object-points to be depicted in the form of linear spectra disposed radially with reference to the centre of the field and proportional in length to the distance of the individual object-points from the optical axis.

From the point of view of the designer of optical systems, the most significant conclusion to be drawn from our study of the primary aberrations is that no less than seven conditions have to be satisfied in order to render a system perfect even at moderate aperture and for a field of moderate extent, for we have discovered seven aberrations which must be reduced to practically zero value. Even then there are left out of account nine secondary aberrations of the Seidel type, the five chromatic variations of the ordinary Seidel aberrations, and the secondary spectrum, and some of these have to be seriously considered in the design of highly corrected systems. As hitherto we have only learnt how to correct two aberrations simultaneously, namely the ordinary spherical and chromatic aberration of the axial pencil, we see that we shall have to develop our methods very extensively before we shall be ready to attack the more ambitious problems in an effective and systematic manner.

The problem of the primary aberrations has been solved in quite a number of different ways ever since Petzval's time ; it may, however, be pointed out that all the correct solutions are necessarily identical in the mathematical sense ; they differ only in the method adopted for the measurement of the departure of any ray from the desirable direction and in sign-conventions. The various methods must therefore be judged by the degree of convenience with which both theoretical conclusions and numerical results can be derived from them. The treatment adopted in this chapter represents the final development of the method of looking upon the Seidel aberrations as mere modified manifestations of the ordinary spherical aberration which the author gave in his earlier lectures, including the initial widely circulated Summer Courses of 1917.

CALCULATION OF THE SEIDEL ABERRATIONS

[58] In accordance with the theory we base our calculations upon the tracing of a paraxial pencil from the axial object-point through the given system and upon a paraxial principal ray through the centre of the aperture diaphragm. As was proved in Chapter I, a paraxial ray-tracing gives the same intersection-lengths whatever initial value of the angle u or of the incidence height may be adopted, whilst all the other angles are in a fixed proportion to the adopted initial angle or incidence height. The calculations will nearly always be considerably shortened and simplified, especially at the later stage when the primary results are trigonometrically corrected by the differential method, if we take advantage of this property of paraxial rays by choosing the nominal starting angle of the axial pencil so as to correspond to the intended full aperture of the system, and by choosing the nominal obliquity of the principal ray so as to correspond to the point in the field at which trigonometrical correction is to be established.

Another general principle, the adoption of which will nearly always reduce the total computing time when working by the strict primary aberration method, consists in beginning the work by establishing true achromatism of the axial focal point at full aperture of the system or of each bending of it; for we saw in section [44] that correction of the primary or paraxial chromatic aberration gives a decidedly different value of the last radius (in the particular example $r_3 = -115$ instead of $r_3 = -60$) from that which is best at full aperture, with the necessary consequence that if the calculation of the Seidel aberrations had been based on the primary chromatic correction, the first thing that would happen in the trigonometrical correction would be a change in the last radius which would alter all the other calculated aberrations, thus calling for an irritating batch of corrections.

When the paraxial pencil and the principal ray have been traced, we could calculate the five g's directly by the formulae collected in Seidel Aberrations I of section [54], to which equations (a) and (b) of section [53] would have to be added for the determination of the spherical aberration constant and of q' at each surface. This method would probably be at least as simple as any other hitherto known. The calculations can be greatly simplified, however, by taking advantage of the fact that the g's are constants of the system which can be derived from judiciously selected terms of the formulae for LA'_d, TA'_d, and $Dist'$ calculated for known values of SA_k and H'_k. We make the following selection:

For the determination of g_1 we calculate $g_1 SA_k^2$, which is simply the spherical aberration of the axial pencil $= LA'p$ of Chapter II.

For the determination of g_2 we calculate $g_2 SA_k^2 H'_k$ equal to the sagittal coma with symbol $Coma'_S$.

For g_3 we select $g_3 H'^2_k$, the distance from S to Ptz in Fig. 56, which we call the sagittal astigmatism with symbol Ast'_S.

If the distortion is *not* to be calculated, g_4 is found directly by $g_4 = \Sigma c (N-1)/N$. But if the distortion is also to be calculated, we calculate the Petzval term $\frac{1}{2} g_4 H'^2_k$ surface by surface by the sum in Seidel Aberrations I, and then g_5 by the two separate terms of the sum formula, with inclusion of the factor H'^3_k.

For SA_k we take its value $l'_k u'_k$ for the traced paraxial pencil, and for H'_k we accept [58] the value corresponding to the traced principal ray. A remarkable relation between the values of SA and of $q'H'$ at any one surface then becomes available. For the axial pencil SA is the incidence height, called y in Chapter I, and $= l \,.\, u = l' \,.\, u'$. By paraxial equation (4p) transposed we have $u' = i' \,.\, r/(l'-r)$, therefore $SA = i' \,.\, l' \,.\, r/(l'-r)$. For the oblique pencil $q'H'$ is the incidence height of the principal ray with reference to the auxiliary axis, and the same relation must therefore be applicable. Moreover our primary aberration method depends upon the use of the limiting values of the intersection-lengths of the rays of oblique pencils; therefore we not only may, but must, treat the l and l' of the oblique pencil as equal to those of the axial pencil, for they become equal at the limit of indefinitely small obliquity. Hence we have

$$SA = i' \,.\, l' \,.\, r/(l'-r) \quad \text{and} \quad q'H' = i'_{pr} \,.\, l' \,.\, r/(l'-r)$$

which give $q'H'/SA = i'_{pr}/i'$ or $q'H' = SA \,.\, i'_{pr}/i'$.

The angles of refraction i'_{pr} for the principal and i' for the paraxial ray are known by the preliminary ray tracings, hence we can use this relation to replace $q'H'$ by SA at any one surface. This leads to a surprising simplification by bringing the other aberrations into a most simple relation with the ordinary spherical aberration.

Introducing the explicit g-values, the aberrations to be directly calculated are:

$$g_1 SA_k^2 = SA_k^2 \sum \left[a_j \frac{N'_j l'^2_j u'^4_j}{N'_k l'^2_k u'^4_k} \right]; \quad g_2 SA_k^2 H'_k = SA_k^2 H'_k \sum \left[a_j q'_j \frac{l'_j u'^2_j}{l'^2_k u'^2_k} \right];$$

$$g_3 H'^2_k = H'^2_k \sum \left[a_j q'^2_j \frac{N'_k}{N'_j} \right] \quad ; \tfrac{1}{2} g_4 H'^2_k = \tfrac{1}{2} H'^2_k \sum \left[N'_k \frac{N'_j - N_j}{N_j N'_j r_j} \right]$$

$$g_5 H'^3_k = H'^3_k \sum \left[a_j q'^3_j \frac{1}{l'_j} \cdot \frac{N'^2_k u'^2_k}{N'^2_j u'^2_j} \right] + \tfrac{1}{2} H'^3_k \sum \left[q'_j \frac{1}{l'_j} \frac{N'_j - N_j}{N_j r_j} \cdot \frac{N'^2_k u'^2_k}{N'^2_j u'^2_j} \right];$$

in which $SA_k = l'_k u'_k$ and $a_j = \frac{1}{2} \dfrac{l'_j - r_j}{r_j l'_j u'^2_j} (i' - u_j)(i_j - i'_j)$:

and as these are all surface-by-surface sums we have only to work out the contribution of any one surface to each sum. We will assign to these contributions the easily remembered symbols, in the usual order of the aberrations:

SC' = spherical contribution; CC' = coma contribution;
AC' = astigmatic contribution; PC' = Petzval contribution;
DCI' = first distortion contribution; $DCII'$ = second distortion contribution.

Omitting the unnecessary suffix j, we then have

$$SC' = SA_k^2 \,.\, a \,.\, \frac{N' l'^2 u'^4}{N'_k l'^2_k u'^2_k} = l'^2_k u'^2_k \,.\, a \,.\, \frac{N' l'^2 u'^4}{N'_k l'^2_k u'^4_k} = a \,.\, \frac{N' l'^2 u'^4}{N'_k u'^2_k} \,.$$

The coma contribution is

$$CC' = SA_k^2 H'_k \, a \, q' \frac{l' u'^2}{l'^2_k u'^2_k} = l'^2_k u'^2_k H'_k \, a \, q' \frac{l' u'^2}{l'^2_k u'^2_k} = a \,.\, q' H'_k l' u'^2.$$

[58] In the $q'H'_k$ of this equation we replace H'_k by the H' of the particular surface, for Lagrange's theorem gives $H'_k N'_k u'_k = H'N'u'$ or $H'_k = H'N'u'/N'_k u'_k$, or

$$CC' = a \,.\, q'H' \,.\, N'l'u'^3/N'_k u'_k.$$

For $q'H'$ we then have the relation worked out above :

$$q'H' = SA \,.\, i'_{pr}/i', \text{ in which } SA = l'u' \therefore q'H' = l'u' \,.\, i'_{pr}/i'.$$

If we introduce this into the last equation for CC' and extend also by u'_k/u'_k, we obtain

$$CC' = a\, N'l'^2u'^4 \,.\, u'_k \,.\, i'_{pr}/N'_k u'^2_k i' = a\, \frac{N'l'^2u'^4}{N'_k u'^2_k} \,.\, u'_k\, \frac{i'_{pr}}{i'} = SC' \,.\, u'_k \,.\, \frac{i'_{pr}}{i'},$$

and the last is our computing equation for CC'.

The astigmatic contribution $AC' = H'^2_k \,.\, a \,.\, q'^2 N'_k/N'$ becomes, by introduction of $H'_k = H'N'u'/N'_k u'_k$,

$$AC' = a\, (q'H')^2 N'u'^2/N'_k u'^2_k,$$

and if we then introduce $q'H' = l'u'i'_{pr}/i'$ we obtain

$$AC' = a\, N'l'^2u'^4 i'^2_{pr}/N'_k u'^2_k i'^2 = a\, \frac{N'l'^2u'^4}{N'_k u'^2_k} \,.\, \left(\frac{i'_{pr}}{i'}\right)^2 = SC' \left(\frac{i'_{pr}}{i'}\right)^2,$$

which is the computing equation for AC'.

If the distortion is to be included among the actually calculated primary aberrations, we calculate the Petzval term surface-by-surface as

$$PC' = \tfrac{1}{2}\, H'^2_k N'_k (N' - N)/NN'r$$

and the two parts of the distortion can then also be put into a very simple form. For DCI' we have

$$DCI' = H'^3_k \,.\, a \,.\, q'^3 \,.\, N'^2_k u'^2_k / l'N'^2 u'^2.$$

Introducing $H'_k = H'N'u'/N'_k u'_k$, we obtain from this

$$DCI' = a\, (q'H')^3\, N'u'/l'N'_k u'_k,$$

and using next $q'H' = l'u'i'_{pr}/i'$ and extending by u'_k/u'_k, we find

$$DCI' = a\, N'l'^2u'^4 u'_k i'^3_{pr}/N'_k u'^2_k i'^3 = a\, \frac{N'l'^2u'^4}{N'_k u'^2_k} \,.\, u'_k\, \frac{i'_{pr}}{i'} \left(\frac{i'_{pr}}{i'}\right)^2 = CC' \left(\frac{i'_{pr}}{i'}\right)^2,$$

a very conveniently calculated expression.

DCII$'$ we can write in the form

$$DC\text{II}' = \tfrac{1}{2}\, H'^2_k N'_k\, \frac{N' - N}{N \,.\, N' \,.\, r} \,.\, q'H'_k\, \frac{N'_k u'^2_k}{l'N'u'^2} = PC' \,.\, q'H'_k\, \frac{N'_k u'^2_k}{l'N'u'^2} = PC' \,.\, q'H'\, \frac{u'_k}{l'u'},$$

and if we then use $q'H' = l'u'i'_{pr}/i'$, we find

$$DC\text{II}' = PC'u'_k\, \frac{i'_{pr}}{i'}.$$

To complete the set of computing equations we have only to put the formula for

SC' into its final form by introduction of the explicit value of the constant 'a'. [58]
This gives

$$SC' = \tfrac{1}{2}\,\frac{l'-r}{rl'u'^2}\,(i'-u)\,(i-i')\,N'l'^2u'^4/N'_k u_k'^2.$$

Here we introduce by (4p) $(l'-r)/r = i'/u'$, which leads to considerable cancellations and by a rearrangement of sequence produces

$$SC' = \tfrac{1}{2}\,l'u'N'i'\,(i'-u)\,(i-i')/N'_k u_k'^2.$$

This equation will give an inaccurate result when there is only a small difference between N and N', and therefore also between i and i', at a cemented surface. This case is met by using the law of refraction $Ni = N'i'$ to put $i - i'$ into the form

$$i - i' = i - iN/N' = i\,(N'-N)/N',$$

which gives the alternative form

$$SC' = \tfrac{1}{2}\,l'u'\,.\,i\,.\,i'\,(i'-u)\,(N'-N)/N'_k u_k'^2.$$

For a plane surface we have, by Pl (1p), $i = -u$ and $i' = -u'$, hence *for Plane:*
$SC' = \tfrac{1}{2}\,l'u'^2\,(u'+u)\,(u'-u)\,N'/N'_k u_k'^2 = \tfrac{1}{2}\,l'u\,u'^2\,(u'+u)\,(N-N')/N'_k u_k'^2.$

The computing equations which we have deduced bring out certain important facts in the clearest possible manner. In the first place they show that the five Seidel aberrations are by no means independent of each other, for the coma contribution and the astigmatic contribution of any one surface are obtained by applying simple factors to the spherical contribution, and the distortion in its two terms is similarly related to both the spherical and the Petzval terms. For the designer of optical systems it is particularly worth noting that the astigmatic contribution is derived from the spherical one by the necessarily positive factor $(i'_{pr}/i')^2$, which means that these contributions always have the same sign at any one surface. We saw that a moderate amount of negative astigmatism is very usually desirable in order to establish the best compromise with regard to flatness of field. We have now learnt that this negative astigmatism can only be produced by a surface which also produces negative or 'over-corrected' spherical aberration, hence we can lay it down as a fundamental principle that a system which is to have a flat field must have in it at least one surface with negative spherical aberration. A simple biconvex or planoconvex lens, for instance, produces positive spherical aberration at both surfaces whenever it is used to give real images of real objects; by the principle just deduced such a lens cannot possibly be made so as to yield a flat image. But if the lens is bent into meniscus form, then the concave side will usually yield negative spherical aberration, and flattening of the field becomes possible. The prevalence of meniscus forms in photographic objectives is due to this important relation.

The equations given above are the most convenient ones for calculation when all the aberrations are to be included. We will, however, now work them out explicitly in terms of the original data, as the resulting formulae are more suitable when only one or two aberrations are to be determined, and chiefly because the explicit equations lead to some further interesting deductions. We will adopt the alternative form of the SC' equation, with replacement of i by $i'N'/N$:

$$SC' = \tfrac{1}{2}\,(N'-N)\,(N'/N)\,l'u'i'^2\,(i'-u)/N'_k u_k'^2.$$

[58] Introducing this into $CC' = SC' . u'_k\ (i'_{pr}/i')$ we obtain

$$CC' = \tfrac{1}{2}(N' - N)(N'/N)\, l'u'i'i'_{pr}\ (i' - u)/N'_k u'_k,$$

and $AC' = SC'\ (i'_{pr}/i')^2$ gives

$$AC' = \tfrac{1}{2}(N' - N)(N'/N)\, l'u'i'^2_{pr}\ (i' - u)/N'_k u'^2_k.$$

These three equations display the close relationship of the first three Seidel aberrations in a particularly striking manner. They also show directly that, provided we act on the advice at the beginning of [55] by assuring that u'_k is not zero or nearly zero, the contributions to the first three aberrations are always finite; for as l', which might be very large or even infinite at a particular surface, occurs only multiplied by u', the resulting $l'u' = y$ is simply the incidence height of the axial pencil and is therefore always of moderate magnitude and not subject to violent changes. Another interesting deduction refers to the three cases when in accordance with [25] refraction at a surface is not accompanied by spherical aberration. In the first case, when the axial object-point coincides with the pole of the surface, we have $l' = 0$, and SC', CC', and AC' are all zero. In the second case, when the object-point coincides with the centre of curvature, we have radial passage of the rays of the axial pencil, and therefore $i = i' = 0$, whilst all the other data are finite. In this case SC' and CC' are zero, but AC' is finite, for it does not contain i' alone as a factor. In the second case we shall therefore obtain astigmatism of the oblique pencils, and this is important, for the curved face of the field-lens of ordinary eyepieces frequently approximates closely to this case. In the third case of truly aplanatic refraction we have $(i' - u) = 0$, and all three contributions are zero. With regard to the second case it may be recalled that the proof of the Seidel aberrations given in this chapter breaks down; we must therefore accept for the present the statement already made that although the present *proof* breaks down, the *formulae* are correct even for this case, as will be shown in Part II.

The Petzval contribution in itself is of such simple form as not to lend itself to any further useful deductions. But we must examine the distortion. Working out $DCI' = CC'\ (i'_{pr}/i')^2$ we obtain the explicit expression

$$DCI' = \tfrac{1}{2}(N' - N)(N'/N)\, l'u'\ (i'^3_{pr}/i')\ (i' - u)/N'_k u'_k;$$

or on carrying out the multiplication by $(i' - u) = (i - u')$ by (3p) and noting that in the first resulting term we have $i/i' = N'/N$,

$$DCI' = \tfrac{1}{2}(N' - N)\,(N'/N)^2\, l'u'i'^3_{pr}/N'_k u'_k - \tfrac{1}{2}\,(N' - N)\,(N'/N)\, l'u'^2(i'^3_{pr}/i')/N'_k u'_k.$$

It will be noticed that the second term would become infinitely large for the second case of aplanatic refraction, because i' would then become zero.

To determine $DC\text{II}'$ we first transform PC' by extending the computing formula by $N'_k u'^2_k$ and by then using the theorem of Lagrange to replace the resulting $H'_k N'_k u'_k$ by $H'N'u'$, thus finding

$$PC' = \tfrac{1}{2}\, H'^2 N'u'^2\ (N' - N)/NN'_k u'^2_k r;$$

and if we now form $DC\text{II}' = PC'u'_k(i'_{pr}/i')$, we obtain

$$DC\text{II}' = \tfrac{1}{2}(N' - N)(N'/N)\, H'^2 u'^2\ (i'_{pr}/i')/N'_k u'_k r,$$

which also becomes infinite for $i' = 0$.

The complete contribution of the surface to the distortion will be $DC' = DCI'$ [58] $+DCII'$ and comes out as

$$DC' = \begin{cases} \tfrac{1}{2}(N'-N)(N'/N)^2\, l'u'i'^3_{pr}/N'_k u'_k \\ +[\tfrac{1}{2}(N'-N)(N'/N)\, l'u'^2\,(i'_{pr}/i')/N'_k u'_k]\,(H'^2/l'r - i'^2_{pr}). \end{cases}$$

We modify this further by expressing H' in terms of the data of the traced rays.

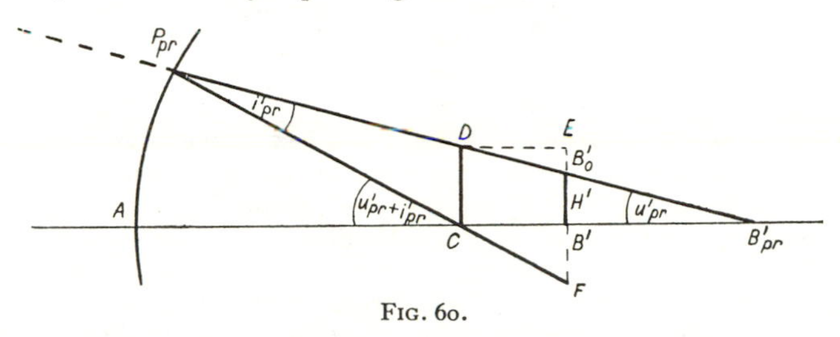

FIG. 60.

Referring to Fig. 60 and remembering that we must treat H' and the angles as indefinitely small, we have two alternative values of H', namely,

$$H' = CD - B'_oE$$

or

$$H' = FB'_o - FB'.$$

With due allowance for the highly exaggerated slopes in the diagram, we have $CD = r\,.\,i'_{pr}$; $B'_oE = (l'-r)\,u'_{pr}$; $FB'_o = l'\,.\,i'_{pr}$; and $FB' = (l'-r)\,(u'_{pr}+i'_{pr})$. Hence :

$$H' = r\,.\,i'_{pr} - (l'-r)\,u'_{pr} \text{ and also } H' = l'\,.\,i'_{pr} - (l'-r)\,(u'_{pr}+i'_{pr})$$

or

$$H'/r = i'_{pr} - (l'-r)\,u'_{pr}/r \quad\text{and}\quad H'/l' = i'_{pr} - (l'-r)\,(u'_{pr}+i'_{pr})/l'.$$

Multiplication of the last two equations gives

$$H'^2/rl' = i'^2_{pr} + \frac{l'-r}{r}\left[-i'_{pr}u'_{pr} - \frac{r}{l'}\,i'_{pr}(u'_{pr}+i'_{pr}) + \frac{l'-r}{l'}\,(u'_{pr}+i'_{pr})\,u'_{pr}\right],$$

and by putting $(l'-r)/l' = 1 - r/l'$ and then simplifying, this gives

$$H'^2/rl' = i'^2_{pr} + \frac{l'-r}{r}\left[u'^2_{pr} - \frac{r}{l'}(u'_{pr}+i'_{pr})^2\right].$$

The final round bracket in the equation for DC' represents this expression less i'^2_{pr} and the latter is thus cancelled. In the surviving term we have $(l'-r)/r = i'/u'$ by (4p), and the equation for DC' becomes

$$DC' = \begin{cases} \tfrac{1}{2}(N'-N)(N'/N)^2\, l'u'i'^3_{pr}/N'_k u'_k \\ +\left[\tfrac{1}{2}(N'-N)(N'/N)\ l'u'i'_{pr}/N'_k u'_k\right].\left[u'^2_{pr} - \frac{r}{l'}(u'_{pr}+i'_{pr})^2\right], \end{cases}$$

or by a final rearrangement of terms

$$DC' = \left[\tfrac{1}{2}(N'-N)(N'/N)\,l'u'i'_{pr}/N'_k u'_k\right].\left[(N'/N)\ i'^2_{pr} + u'^2_{pr} - \frac{r}{l'}(u'_{pr}+i'_{pr})^2\right].$$

[58] The significance of this equation lies in the fact that it no longer contains i or i' in the denominator; it therefore proves that DC' is always finite although the two separate parts, in which it will usually be calculated by the general solution first given, can both become infinite. The equation is therefore suitable for the determination of the distortion at surfaces at which DCI' and $DCII'$ separately would become undesirably large.

In order to render easy and convenient the comparison of the results by the present primary aberration method with those by strict trigonometrical calculations we must add a few formulae which give the curvature of the field directly in the terms which we shall introduce subsequently for the trigonometrical work. We found in section [55] that with reference to a focal plane laid through the sharp focus of an axial pencil of the same aperture, the distance from the sagittal focus S to this plane is $= g_3 H_k'^2 + \frac{1}{2} g_4 H_k'^2$, and the distance from the tangential focus T to this plane is $= 3\, g_3 H_k'^2 + \frac{1}{2} g_4 H_k'^2$, both distances being subject to the sign convention of longitudinal aberrations, that is positive when the focal plane lies to the right of the astigmatic foci. In the trigonometrical work we shall calculate these same distances accurately (that is, including the effects of all the higher aberrations), but shall deal with them as abscissae referred to the axial focus of the zone as origin; they will therefore be then subject to the sign convention of analytical geometry, or positive when S and T lie to the right of the adopted focal plane, which means reversed signs. The symbols adopted for these differences of focus will be X'_S and X'_T respectively; hence the formulae to be added to our set are

$$\text{primary } X'_S = -g_3 H_k'^2 \quad -\tfrac{1}{2} g_4 H_k'^2 \quad \text{and}$$
$$\text{primary } X'_T = -3\, g_3 H_k'^2 - \tfrac{1}{2} g_4 H_k'^2,$$

and as these are built up of the contributions AC' and PC' which we have already defined, no further instructions are required.

We now only have to put the expressions for the two chromatic aberrations into the most convenient form for numerical evaluation. Calling the contribution of each surface to the final longitudinal chromatic aberration $LchC'$ in analogy with the nomenclature adopted for the other aberrations, equation (a) of section [56] gives directly, omitting the unnecessary suffix j,

$$LchC' = l' \, \frac{l'-r}{r} \left(\frac{\delta N'}{N'} - \frac{\delta N}{N} \right) N' u'^2 / N'_k u_k'^2;$$

and if we again replace $(l'-r)/r$ by i'/u', this gives the most convenient computing formula

$$LchC' = l'u' . N'i' \left(\frac{\delta N'}{N'} - \frac{\delta N}{N} \right) / N'_k u_k'^2.$$

The equation preceding (b) of section [56] gives the contribution of any one surface to the final transverse chromatic aberration as

$$TchC' = q'H' \, \frac{l'-r}{r} \left(\frac{\delta N'}{N'} - \frac{\delta N}{N} \right) N'u' / N'_k u'_k;$$

and if we introduce, as in the previous transformations, $(l' - r)/r = i'/u'$ and [58] $q'H' = l'u'i'_{pr}/i'$, this gives the computing formula

$$TchC' = l'u'N'i'_{pr}\left(\frac{\delta N'}{N'} - \frac{\delta N}{N}\right)/N'_k u'_k.$$

The extremely close similarity of this to the equation for $LchC'$ should be noted, but also that in the transverse aberration the denominator contains u'_k in the first power whilst $LchC'$ has u'^2_k in the denominator.

As has already been pointed out, it is very usually preferable to establish exact achromatism for the axial image-point before the calculation of the Seidel aberrations is begun ; the inclusion of Lch' in the latter work is then more or less superfluous. Hence the explicit equation for Tch' will usually be the most convenient one ; but if Lch' is to be included in the work, then Tch' is obtained more readily and with a greatly reduced number of logarithms by expressing Tch' in terms of Lch'. A mere glance at the respective equations gives the formula

$$TchC' = LchC'.\,u'_k.\,i'_{pr}/i',$$

which also shows that the relation between $LchC'$ and $TchC'$ is precisely the same as that between the contributions of each surface to the final spherical aberration and the final coma.

We will now collect the computing formulae, adding concise instructions which appear desirable.

Seidel Aberrations II

(1) Four-figure logarithms are always amply sufficient provided that the special formulae are used in the exceptional cases referred to below.

(2) A paraxial ray must be traced from the axial object-point to the final axial image-point. As far as the Seidel aberrations are concerned, four-figure accuracy will again be ample, but as in most cases the same ray will also be required to determine the exact LA' by addition of a trigonometrically traced marginal ray, it may save time to trace the ray with five-figure logs. It is convenient, but not necessary, to use an initial u or y corresponding to the intended full working aperture.

(3) A principal ray must be traced with four-figure accuracy through the centre of the aperture diaphragm or of one of the corresponding pupils, either left-to-right, right-to-left, or partly in one and partly in the other direction according to the location of the selected starting point. If the position of the diaphragm has not yet been decided upon, a few figures will be saved by assuming either an entrance pupil in contact with the first refracting surface or an exit pupil in contact with the last refracting surface, as the initial l_{pr} or l'_{pr} will then be zero. Again it is convenient, but not necessary, to assume a fictitious initial u_{pr} or u'_{pr} corresponding to the intended actual field of the lens system. To reduce the chances of mistakes it is advisable to arrange the successive columns of ray-tracing in the usual left-to-right order of the surfaces. In any right-to-left part of the ray-tracing the surface-by-surface calculations should therefore be continued contrary to one's natural instinct, namely towards the left. As only the angles i'_{pr} enter into the equations

[58] of the Seidel aberrations, the logs of these angles should be conspicuously marked as the tracing of the principal ray proceeds.

(4) The two preceding paragraphs supply all the data to calculate, in all ordinary cases, for each refracting surface :

(a) The contribution to the final primary spherical aberration

$$SC' = \tfrac{1}{2}\, l'u'N'i'\,(i'-u)\,(i-i')/N'_k u_k'^2, \text{ or for a plane}$$

$$(\text{Plane } only) = \tfrac{1}{2}\, l'u'^2\,(u'+u)\,(u'-u)\,.\,N'/N'_k u_k'^2.$$

(b) The contribution to the final primary sagittal coma

$$CC' = SC'\,.\,u'_k\,.\,i'_{pr}/i'.$$

(c) The contribution to the final primary sagittal astigmatism

$$AC' = SC'\,(i'_{pr}/i')^2.$$

(d) The contribution to the final Petzval Curvature

$$PC' = \tfrac{1}{2}\, H_k'^2 N'_k\,(N'-N)/NN'r,$$

in which $H'_k = -(l'_k - l'pr_k)\,u'pr_k.$

(e) The contribution to the final primary distortion

$$DC' = CC'\,(i'_{pr}/i')^2 + PC'\,.\,u'_k\,.\,i'_{pr}/i'.$$

(f) The contribution to the final longitudinal chromatic aberration

$$LchC' = l'u'N'i'\left(\frac{\delta N'}{N'} - \frac{\delta N}{N}\right)/N'_k u_k'^2.$$

(g) The contribution to the final transverse chromatic aberration

$$TchC' = LchC'\,.\,u'_k\,.\,i'_{pr}/i',$$

or if $LchC'$ is not calculated because exact chromatic correction has already been established by the methods of Chapter IV, by the explicit formula

$$TchC' = l'u'N'i'_{pr}\left(\frac{\delta N'}{N'} - \frac{\delta N}{N}\right)/N'_k u'_k\,.$$

The aberrations in the final images are then found as simple sums (with due regard to the individual signs of the contributions), namely :

$$\text{Spherical aberration} = g_1 SA_k^2 = \Sigma SC'\,;$$
$$\text{Sagittal Coma} = g_2 SA_k^2\,.\,H'_k = \Sigma CC'\,;$$
$$\text{Sagittal Astigmatism} = g_3 H_k'^2 = \Sigma AC'\,;$$
$$\text{Petzval Curvature} = \tfrac{1}{2}\, g_4 H_k'^2 = \Sigma PC'\,;$$
$$\text{Distortion} = g_5 H_k'^3 = \Sigma DC'\,:$$

and the five aberration constants, g_1 to g_5 can be found, if required, by dividing the respective sum-values by the proper powers of $SA_k = l'_k u'_k$ and/or of $H'_k = -(l'_k - l'pr_k)u'pr_k$. With regard to astigmatism and curvature of field the most useful final results usually are the differences of focus of the sagittal and of the tangential rays with reference to the axial focus, subject to the sign convention of analytical geometry, that is positive when S or T lies to the right of the axial focus: [58]

$$\text{primary } X'_S = -g_3 H_k'^2 \quad -\tfrac{1}{2} g_4 H_k'^2\,;$$
$$\text{primary } X'_T = -3\, g_3 H_k'^2 - \tfrac{1}{2} g_4 H_k'^2.$$

A positive value of these is usually described as a 'hollow' field, a negative value as a 'round' field.

With reference to the two chromatic aberrations it will be convenient to introduce symbols g_6 for the constant of the longitudinal and g_7 for the constant of the transverse chromatic aberration, and as the former measures a fixed difference of focus which varies neither with the aperture nor with the position in the field of view, whilst the transverse chromatic aberration, as the fundamental equations show, grows simply as H'_k, we have:

$$\text{Longitudinal chromatic aberration} = g_6 \qquad = \Sigma LchC'\,;$$
$$\text{Transverse chromatic aberration} \quad = g_7 \,.\, H'_k = \Sigma TchC'.$$

To find any one of the seven primary aberrations for an aperture and point in the field differing from those corresponding to the two traced rays, we have only to apply the laws expressed by the coefficients of the several g's in the above equations. Thus the Coma for a doubled value of SA_k and doubled value of H'_k would be $2^2 = 4$ times on account of $SA_k{}^2$, and would again be doubled on account of H'_k; it would therefore be eight times as large as for the single values of SA_k and of H'_k.

(5) *Special cases.*

(a) The standard equation for SC' in (4) will give an ill-determined value if at deep contact surfaces i and i' are decidedly large but nearly equal. A closer value can then be found by calculating it by

$$SC' = \tfrac{1}{2}\, l'u' \,.\, i \,.\, i'\, (i' - u)\, (N' - N)/N'_k u_k'^2,$$

because $(N' - N)$ is known with the full accuracy of the available glass-data.

(b) If only one (or a few) of the seven aberrations is to be calculated, time will be saved by using the explicit formulae for the second to the fifth aberrations:

$$CC' = \tfrac{1}{2}\, (N' - N)\, (N'/N)\, l'u'i'i'_{pr}\, (i' - u)/N'_k u'_k\,,$$
$$AC' = \tfrac{1}{2}\, (N' - N)\, (N'/N)\, l'u'i'^2_{pr}\, (i' - u)/N'_k u_k'^2,$$
$$g_4 = \Sigma(\, (N - 1) \,.\, c/N) \text{ calculated lens-by-lens,}$$
$$DC' = [\tfrac{1}{2}\, (N' - N)\, (N'/N)\, l'u'i'_{pr}/N'_k u'_k] \,.\, [(N'/N)\, i'^2_{pr} + u'^2_{pr} - \frac{r}{l'}\, (u'_{pr} + i'_{pr})^2].$$

(c) The last equation, for DC', must also be used when the usual equation (4) e of this summary would give it as a small difference of two *large* numbers owing to i' being very small.

Numerical Example

As a numerical example of the calculation of the primary aberrations by the exact formulae, we will choose an ordinary photographic 'landscape' lens of the

[58] type usually found in inexpensive cameras, made of the glasses used in most of the examples of Chapter V and with the prescription:

$$r_1 = -\ 5{\cdot}556 \qquad \text{Flint in front.}$$
$$d_1' = 0{\cdot}10$$
$$r_2 = +14{\cdot}92 \qquad \text{Clear working aperture } 0{\cdot}600.$$
$$d'_2 = 0{\cdot}30$$
$$r_3 = -\ 2{\cdot}506$$

The last radius may be accepted as giving the usual photographic achromatism at the stated aperture. In accordance with the rules laid down in Chapter IV we must work with the refractive indices for F and with the dispersion between D and G', hence we begin by extracting these data from the glass list:

	Crown.	*Flint.*
$N_d =$	1·5407	1·6225
$+(N_f - N_d)$	0·00642	0·01237
$N_f =$	1·54712	1·63487
$N_f - N_d$	0·00642	0·01237
$N_{g'} - N_f$	0·00517	0·01052
δN	0·01159	0·02289

The object being assumed at infinity, we begin by tracing a paraxial ray parallel to the optical axis at nominal y equal to half the prescribed clear aperture, or 0·300, completely through the objective, finding for:

	First Surface.	*Second Surface.*	*Third Surface.*
l'	− 14·3074	− 12·9618	+9·8367
u	0	− 0·020968	− 0·023306
i	− 0·053996	0·041215	− 0·100030
i'	− 0·033028	0·043553	− 0·154757
u'	− 0·020968	− 0·023306	0·031421

A simultaneously traced marginal ray gives $U'_3 = 1\text{–}49\text{–}31$ and $L'_3 = 9{\cdot}6713$ or a true trigonometrical $LA' = 0{\cdot}1654$, which we shall be able to compare with the primary aberration. We must next trace a principal ray, and as the best position of the diaphragm is not specified but is to be determined, we can make an arbitrary selection; to save a few figures we assume the diaphragm at the pole of the last refracting surface, or $l'pr_3 = 0$, and have then to fix a suitable nominal value for the corresponding angle $u'pr_3$ in order to be able to trace the principal ray in the right-to-left direction. We may take it that such an unpretentious lens would cease to give satisfactory definition beyond 3 or 4 inches from the optical axis, and we will therefore calculate for $H'_k = 3{\cdot}000$ as our nominal value. That gives the paraxial nominal $u'pr_k = -H'_k/(l'_k - l'pr_k)$, or in our case, as $l'pr_3 = 0$, $u'pr_3 = -3{\cdot}000/9{\cdot}8367 = [9{\cdot}4843n]$. The tracing through the third surface is simplified, because $l'_{pr} = 0$ leads to $i'_{pr} = u'_{pr}(0-r)/r = -u'_{pr}$, so that only the law of [58] refraction has to be applied, giving $upr_3 = -0{\cdot}19715$ $lpr_3 = 0$, and the only datum going into the primary aberration work $i'pr_3 = [9{\cdot}4843]$. At the second surface we shall find $lpr_2 = 0{\cdot}318$, $upr_2 = -0{\cdot}18677$, $i'pr_2 = [9{\cdot}2859]$, and at the first surface the entrance pupil is located at $lpr_1 = 0{\cdot}249$, with $upr_1 = -0{\cdot}31427$ and $i'pr_1 = [9{\cdot}3028]$. The two traced rays now supply all the data required for the calculation of the Seidel aberrations of our lens. In accordance with Seidel Aberrations II (4) we have to determine for each refracting surface $SC' = \frac{1}{2} l'u'N'i'(i'-u)(i-i')/N'_k u_k'^2$ as the key-quantity, then $CC' = SC'.u'_k.i'pr/i'$ and $AC' = SC'(i'pr/i')^2$. If all seven aberrations are to be calculated, we have to add $PC' = \frac{1}{2} H_k'^2 N'_k (N'-N)/NN'r$, then $DC' = CC'(i'pr/i')^2 + PC'.u'_k(i'pr/i')$ and the two chromatic aberrations

$$LchC' = l'u'N'i'(\delta N'/N' - \delta N/N)/N'_k u_k'^2\,; \quad TchC' = LchC'.u'_k.i'pr/i'.$$

The sum of the three contributions to each aberration then gives the total amount of the latter at the final focus.

As the calculation of the Seidel aberrations will be a frequently recurring standard operation, we will build up a regular scheme which gives the results with a maximum of safety and a minimum of labour. We begin with a record of two logs which occur a number of times :

$$\log u'_k = 8{\cdot}4972\,; \quad \text{colog}\, u'_k = 1{\cdot}5028.$$

All the rest of the work should then be done in as many vertical columns as there are refracting surfaces—three in the present case—beginning with the preparation of data which go into the formulae :

	First Surface.	*Second Surface.*	*Third Surface.*
$\log i'_{pr}$ =	9·3028	9·2859	9·4843
$-\log i'$ =	$-8{\cdot}5189n$	$-8{\cdot}6390$	$-9{\cdot}1896n$
$\log (i'_{pr}/i')$ =	$0{\cdot}7839n$	0·6469	$0{\cdot}2947n$
Paraxial angles i	$-0{\cdot}053996$	0·041215	$-0{\cdot}100030$
Paraxial angles i'	$-0{\cdot}033028$	0·043553	$-0{\cdot}154757$
Paraxial angles u	0	$-0{\cdot}020968$	$-0{\cdot}023306$
$i-i'$	$-0{\cdot}020968$	$-0{\cdot}002338$	0·054727
$i'-u$	$-0{\cdot}033028$	0·064521	$-0{\cdot}131451$

Now we are ready to compute SC', CC', and AC'. The scheme includes a few simplifying tricks. As $N'i' = Ni$ by the law of refraction, we can use either at pleasure, and as at *glass-air* surfaces the air index is exactly one, we save one log by using simply the air angle, easily recognized as the one with the bigger log. Secondly, in deriving CC' and AC' from SC', we first form $SC'.i'pr/i'$, which in itself has no direct optical significance. But by writing $\log u'_k$ below it, we can form $\log CC'$ by adding $\log SC'.i'pr/i'$ to the log below it and $\log AC'$ by adding $\log SC'.i'pr/i'$ to the log next above it, which makes the calculation as compact as is possible. The calculations thus become :

[58]

	First Surface.	*Second Surface.*	*Third Surface.*
$\log \frac{1}{2}$	9·6990	9·6990	9·6990
$\log l'$	1·1556n	1·1127n	0·9928
$\log u'$	8·3216n	8·3675n	8·4972
$\log N'$	}	0·1895	}
$\log i'$	8·7324n	8·6390	9·1896n
$\log (i'-u)$	8·5189n	8·8097	9·1188n
$\log (i-i')$	8·3216n	7·3688n	8·7382
$2 \text{ colog } u'_k$	3·0056	3·0056	3·0056
$\log SC'$	7·7547n	7·1918n	9·2412
$+\log (i'pr/i')$	0·7839n	0·6469	0·2947n
$\log SC' . i'pr/i'$	8·5386	7·8387n	9·5359n
$\log u'_k$	8·4972	8·4972	8·4972
$\log CC' = \log SC'i'pr/i' + \log u'_k$	7·0358	6·3359n	8·0331n
$\log AC' = \log SC'i'pr/i' + \log (i'pr/i')$	9·3225n	8·4856n	9·8306

We now take out the numbers and sum up the totals of LA', $Coma'_S$, and Ast'_S at the final focus. It is of course convenient to collect the positive and negative items separately so as to reduce the work to a minimum. The surfaces from which the successive items are taken are added in bracketed numbers :

Final	LA'		$Coma'_S$		Ast'_S
(1)	−0·00568	(2)	−0·00022	(1)	−0·21014
(2)	−0·00156	(3)	−0·01079	(2)	−0·03059
sum	−0·00724		−0·01101		−0·24073
(3)	+0·17426	(1)	+0·00109	(3)	+0·67702
	0·16702		−0·00992		+0·43629

It will be noticed that the primary spherical aberration, 0·16702, agrees very well with the more exact trigonometrical amount, 0·1654, showing that the higher spherical aberration is small.

For the calculation of the Petzval contributions we prepare $\frac{1}{2} H_k'^2 N'_k$, which in our case, with $H'_k = 3$ and N'_k (as nearly always) $= 1$, gives $\frac{1}{2} H_k'^2 N'_k = 4·500$. We then calculate in three columns :

	First Surface.	*Second Surface.*	*Third Surface.*
N'	1·63487	1·54712	1·00000
$-N$	−1·00000	−1·63487	−1·54712
$N'-N$	0·63487	−0·08775	−0·54712
$\log \frac{1}{2} H_k'^2 N'_k$	0·6532	0·6532	0·6532
$\log (N'-N)$	9·8027	8·9432n	9·7381n
$\text{colog } N$	0·0000	9·7865	9·8105
$\text{colog } N'$	9·7865	9·8105	0·0000
$\text{colog } r$	9·2552n	8·8262	9·6010n
$\log PC'$	9·4976n	8·0196n	9·8028
PC'	−0·3145	−0·0105	+0·6350

The Petzval curvature is therefore $\frac{1}{2}g_4H_k'^2 = \Sigma PC' = +0{\cdot}3100$. We then obtain [58] the two parts of the distortion :

log CC'	7·0358	6·3359n	8·0331n
+2 log (i'_{pr}/i')	1·5678	1·2938	0·5894
log DCI'	8·6036	7·6297n	8·6225n
DCI'	+0·04014	−0·00426	−0·04193
log PC'	9·4976n	8·0196n	9·8028
+log u'_k	8·4972	8·4972	8·4972
+log (i'_{pr}/i')	0·7839n	0·6469	0·2947n
log $DC\text{II}'$	8·7787	7·1637n	8·5947n
$DC\text{II}'$	+0·06008	−0·00146	−0·03933

and collection of terms gives $\Sigma DCI' = -0{\cdot}00605$; $\Sigma DC\text{II}' = 0{\cdot}01929$ or the distortion at the final focus of our oblique pencil $g_5H_k'^3 = \Sigma DCI' + \Sigma DC\text{II}' = 0{\cdot}01324$.

For the two chromatic aberrations we have to prepare the values of $\delta N/N$ for the two kinds of glass in our objective :

	First Lens.	*Second Lens.*
δN	0·02289	0·01159
N	1·63487	1·54712
log δN	8·3596	8·0641
−log N	−0·2135	−0·1895
log $\delta N/N$	8·1461	7·8746
$\delta N/N$	0·013999	0·007492

and can then again calculate in three columns :

	First Surface.	*Second Surface.*	*Third Surface.*
$\delta N'/N'$	0·013999	0·007492	0
$-\delta N/N$	0	−0·013999	−0·007492
$\delta N'/N' - \delta N/N$	0·013999	−0·006507	−0·007492
log $l'u'$	9·4772	9·4802	9·4900
+log $N'i'$	8·7324n	8·8285	9·1896n
+log $(\delta N'/N' - \delta N/N)$	8·1461	7·8134n	7·8746n
+2 colog u'_k	3·0056	3·0056	3·0056
log $LchC'$	9·3613n	9·1277n	9·5598
+log (i'_{pr}/i')	0·7839n	0·6469	0·2947n
+log u'_k	8·4972	8·4972	8·4972
log $TchC'$	8·6424	8·2718n	8·3517n

The calculation of $LchC'$ is slightly contracted, because $l'u'$ and $N'i'$ were already used in the calculation of SC' and are easily taken over as a single log

[58] for each product. Taking and combining numbers, we now complete the calculation:

	LchC'		*TchC'*
(1)	$-0{\cdot}2298$	(2)	$-0{\cdot}01870$
(2)	$-0{\cdot}1342$	(3)	$-0{\cdot}02248$
	$-0{\cdot}3640$		$-0{\cdot}04118$
(3)	$+0{\cdot}3629$	(1)	$+0{\cdot}04389$
Total *Lch'*	$-0{\cdot}0011$		$+0{\cdot}00271$

The collected results for the aberrations with 0·6 aperture and $H'_k = 3$ are therefore, for diaphragm in contact with the last surface:

$LA' = g_1 SA_k^2 = 0{\cdot}1670$; $Coma'_S = g_2 SA_k^2 H'_k = -0{\cdot}00992$;
$Ast'_S = g_3 H_k'^2 = 0{\cdot}4363$;
Petzval curvature $= \frac{1}{2} g_4 H_k'^2 = 0{\cdot}3100$; Distortion $= g_5 H_k'^3 = 0{\cdot}01324$;
Longitudinal chromatic aberration $= g_6 = -0{\cdot}0011$;
Transverse chromatic aberration $= g_7 H'_k = 0{\cdot}00271$.

The astigmatism is by far the most serious of these aberrations, as becomes very obvious when we calculate the differences of focus between the axial and the oblique pencil:

$$\text{primary } X'_S = -g_3 H_k'^2 - \tfrac{1}{2} g_4 H_k'^2 = -0{\cdot}4363 - 0{\cdot}3100 = -0{\cdot}7463;$$

$$\text{primary } X'_T = -3\, g_3 H_k'^2 - \tfrac{1}{2} g_4 H_k'^2 = -1{\cdot}3089 - 0{\cdot}3100 = -1{\cdot}6189.$$

These figures mean that if we first focused the axial image, we should have to shorten the camera by 0·75″ in order to obtain a sharp sagittal focus at 3″ from the centre of the field, and we should have to shorten by 1·62″ in order to reach the corresponding tangential focus, whereas our ideal correction calls for flatness of the tangential field, or $X'_T = 0$. We shall learn in a following section how we can easily and quickly determine that shift of the diaphragm which will give $X'_T = 0$ or any other possible adjustment of the final aberrations which may appear desirable.

THE SEIDEL ABERRATIONS FOR THIN LENSES

[59] It was shown in section [29] that the spherical aberration of the axial image-point can be very conveniently determined by treating the thickness of the individual lenses of a system as negligible, and we learnt in section [49] how the equations arrived at in [29] lend themselves to very rapid solutions of the problem of designing simple telescope objectives so as to give a prescribed amount (most usually zero) of spherical aberration, chiefly because the formulae include the effect of bending the individual lenses and thus admit of direct algebraical solutions for the correct form of a proposed combination.

It is at once realized that simple equations of corresponding type, that is, *including the effect of bending*, would be even more desirable for the more intricate problems of the other Seidel aberrations. We therefore proceed now to seek the solution of this problem and begin with the simplest case, which will subsequently be shown to be easily extended to cover the whole ground, namely the case of *a simple thin lens somewhere within a complete lens system*, but in such a position that all the oblique pencils pass centrally through it ; the thin lens will therefore be assumed to be coincident with, or to act as, the iris of the system.

The case of spherical aberration is of course fully covered by the formulae in section [29] and by their extension in section [32], and these formulae are therefore available for our present purpose without any further modification. With reference to the other Seidel aberrations, the perfectly general solutions of the preceding section are necessarily applicable to the present specialized case ; hence we have only to transform the general solutions in order to obtain the equations which we are seeking.

With reference to Fig. 61, the spherical surface AP shall be one refracting surface (it does not matter whether the first or the second, but the first is implied

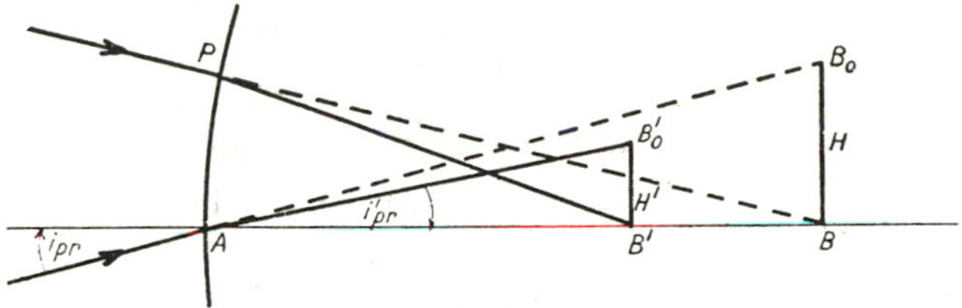

FIG. 61.

by the diagram) of our assumed thin lens. PB shall be the arriving paraxial ray with intersection-length $AB = l$, and PB' shall be the refracted ray with $AB' = l'$. If the principal ray arrives at A under angle i_{pr} and is refracted so as to form angle i'_{pr}, both being positive in the diagram because they are clockwise from ray to incidence normal, then the intersections of this principal ray with the focal planes at B and B' will define the extra-axial object- and image-points B_o and B'_o and the object- and image-heights of the surface $BB_o = H$ and $B'B'_o = H'$. The diagram shows at sight that in the special case of a principal ray passing through the pole A of the refracting surface, we have

$$i'_{pr} = H'/l'.$$

By Lagrange's theorem we have $H'N'u' = H'_k N'_k u'_k$, or transposed $H' = H'_k N'_k u'_k / N'u'$, and using this, the preceding equation becomes

(a) $$i'_{pr} = H'_k N'_k u'_k / l'N'u'.$$

This relation will be our principal means for the simplification of the general solution.

A. Coma of a Thin Lens

The universally valid equation for the coma contribution of any one refracting surface by [58] summary (5)b is

$$CC' = \tfrac{1}{2}(N' - N)(N'/N)l'u'i'i'_{pr}(i' - u)/N'_k u'_k.$$

If we introduce into this the apparently complicated value of i'_{pr} by equation (a), very extensive cancellations result, and give, *for our special case only*,

$$CC' = \tfrac{1}{2} H'_k i'(i' - u)(N' - N)/N.$$

[59] As our lens is to be treated as infinitely thin, this special equation applies to both its surfaces ; distinguishing these by the usual suffixes 1 and 2, we therefore have

$$CC'_1 = \tfrac{1}{2} H'_k i'_1(i'_1 - u_1)\,(N'_1 - N_1)/N_1$$
$$CC'_2 = \tfrac{1}{2} H'_k i'_2(i'_2 - u_2)\,(N'_2 - N_2)/N_2,$$

and the coma contribution of the complete lens will be simply the sum of these two items. The lens being again treated (as in section [29]) as standing in air, we have $N_1 \equiv N'_2 = 1$ for air and $N'_1 \equiv N_2 = N$ of the material of the lens, and we obtain

$$\text{Total } CC' = \tfrac{1}{2} H'_k(N-1)\,[i'_1(i'_1 - u_1) - i'_2(i'_2 - u_2)/N].$$

It now only remains to bring this equation into harmony with that for the spherical contribution of a thin lens by replacing the paraxial angles in the square bracket by the curvature and convergence values. For this purpose we first express all the angles in terms of those *inside the lens*. In the first part of the bracket we replace $i'_1 - u_1$ by $i_1 - u'_1$, which are equal by (3p) of the fundamental computing formulae, and then i_1 by $N \,.\, i'_1$ by the law of refraction, thus arriving at

$$i'_1(i'_1 - u_1) = i'_1(N \,.\, i'_1 - u'_1).$$

In the second part of the square bracket we merely have to replace i'_2 by $N \,.\, i_2$, again by the law of refraction at a glass-air surface, and obtain

$$i'_2(i'_2 - u_2)/N = i_2(N \,.\, i_2 - u_2).$$

The coma contribution of the lens now is

$$\text{Total } CC' = \tfrac{1}{2} H'_k(N-1)\,[i'_1(N \,.\, i'_1 - u'_1) - i_2(N \,.\, i_2 - u_2)].$$

As the next step, we extend the first term in the square bracket by $l'^2_1 \,.\, u'^2_1/l'^2_1 \,.\, u'^2_1 = 1$ and the second term by $l_2^2 u_2^2/l_2^2 u_2^2$ also $= 1$, and as $l'_1 u'_1$ is the incidence height of the paraxial ray at the first surface, $l_2 u_2$ that at the second surface, these being necessarily equal for our assumed infinitely thin lens, we take the numerators of both extending factors outside the bracket under the name of SA^2 in accordance with the nomenclature of the present chapter ; the denominators of the extending factors are equally distributed among the two factors of each term, and we thus obtain

$$\text{Total } CC' = \tfrac{1}{2} H'_k \,.\, SA^2 \,.\, (N-1)\left[\frac{1}{l'_1}\frac{i'_1}{u'_1}\left(N \,.\, \frac{1}{l'_1} \cdot \frac{i'_1}{u'_1} - \frac{1}{l'_1}\right) - \frac{1}{l_2}\frac{i_2}{u_2}\left(N\frac{1}{l_2} \cdot \frac{i_2}{u_2} - \frac{1}{l_2}\right)\right].$$

The paraxial angles now only occur in the general form i/u, and the paraxial ray-tracing formulae give

by (4p) $\quad i'_1/u'_1 = (l'_1 - r_1)/r_1 = l'_1/r_1 - 1$ by dividing out ;

by (1p) $\quad i_2/u_2 = (l_2 - r_2)/r_2 = l_2/r_2 - 1$:

and as for an infinitely thin lens we have $l_2 = l'_1$, we can use the latter symbol throughout, and our equation becomes

Total CC'

$$= \tfrac{1}{2} H'_k SA^2(N-1)\left[\left(\frac{1}{r_1} - \frac{1}{l'_1}\right)\left(\frac{N}{r_1} - \frac{N}{l'_1} - \frac{1}{l'_1}\right) - \left(\frac{1}{r_2} - \frac{1}{l'_1}\right)\left(\frac{N}{r_2} - \frac{N}{l'_1} - \frac{1}{l'_1}\right)\right].$$

We now introduce the simple symbols employed in the corresponding equation [59] for the spherical aberration, namely $1/l'_1 = v'_1$, $1/r_1 = c_1$, $1/r_2 = c_2$, the total curvature $c = c_1 - c_2$, and by this $c_2 = 1/r_2 = c_1 - c$; and as it would be inconvenient to express the object-distance by the convergence within the lens, we use (6p)** for the first surface, giving

$$\frac{N}{l'_1} = \frac{N-1}{r_1} + \frac{1}{l_1},$$

or in the c and v notation,

$$N \cdot v'_1 = (N-1)c_1 + v_1$$

to substitute for $1/l'_1$ its equivalent,

$$1/l'_1 = v'_1 = \frac{N-1}{N}\, c_1 + \frac{v_1}{N}.$$

By using these equivalents in the square bracket of our last equation for the coma contribution of our lens, we find

$$\frac{1}{r_1} - \frac{1}{l'_1} = c_1 - \frac{N-1}{N}\, c_1 - \frac{v_1}{N} = \frac{c_1 - v_1}{N}$$

$$\frac{N}{r_1} - \frac{N+1}{l'_1} = N \cdot c_1 - \frac{N^2-1}{N}\, c_1 - \frac{N+1}{N}\, v_1 = \frac{1}{N}(c_1 - (N+1)v_1)$$

$$\frac{1}{r_2} - \frac{1}{l'_1} = c_1 - c - \frac{N-1}{N}\, c_1 - \frac{v_1}{N} = \frac{c_1 - v_1}{N} - c$$

$$\frac{N}{r_2} - \frac{N+1}{l'_1} = N \cdot c_1 - N \cdot c - \frac{N^2-1}{N}\, c_1 - \frac{N+1}{N}\, v_1 = \frac{1}{N}(c_1 - (N+1)v_1 - N^2 \cdot c)$$

and the equation becomes

$$\text{Total } CC' = \tfrac{1}{2} H'_k \cdot SA^2(N-1)\left[\frac{c_1 - v_1}{N} \cdot \frac{c_1 - (N+1)v_1}{N} - \left(\frac{c_1 - v_1}{N} - c\right) \cdot \frac{c_1 - (N+1)v_1 - N^2 \cdot c}{N}\right].$$

It is clear that the whole of the first term in the square bracket is cancelled by a similar partial product of the subtracted second term. The surviving terms give

$$\text{Total } CC' = \tfrac{1}{2} H'_k \cdot SA^2(N-1)\left[c(c_1 - v_1) + \frac{c}{N}(c_1 - (N+1)v_1 - N^2 \cdot c)\right]$$

and simple final reductions give

$$\text{Total } CC' = H'_k \cdot SA^2\left[\tfrac{1}{2}\frac{N^2-1}{N} \cdot c \cdot c_1 - \frac{(2N+1)(N-1)}{2N} \cdot c \cdot v_1 - \tfrac{1}{2}N(N-1)c^2\right].$$

The coefficients in the square bracket are again pure functions of the refractive index which can be calculated and tabulated once for all to cover the usual range of indices in optically useful materials. This has been done in the table originally

[59] given with the corresponding equations for the spherical aberration of a thin lens, which include

$$G_5 = 2\frac{N^2-1}{N};\quad G_7 = \frac{(2N+1)(N-1)}{2N};\quad G_8 = \frac{N(N-1)}{2}.$$

The final extremely simple solution for the coma contribution of a thin lens passed centrally by the oblique pencils then becomes

(b) $$\text{Total } CC' = H'_k SA^2(\tfrac{1}{4}G_5 \,.\, c \,.\, c_1 - G_7 \,.\, c \,.\, v_1 - G_8 \,.\, c^2).$$

As in the case of the spherical aberration, it will frequently be even more convenient to calculate CC' in terms of c_2 and v'_2. By putting into the explicit formula (i.e. before introduction of the G-notation) $c_1 = c_2 + c$, and by the thin lens formula $1/l'_2 = (N-1)c + 1/l_1 : v_1 = v'_2 - (N-1)c$, and then simplifying, the desired equation is found as

(b)′ $$\text{Total } CC' = H'_k \,.\, SA^2\,[\tfrac{1}{4}G_5 \,.\, c \,.\, c_2 - G_7 \,.\, c \,.\, v'_2 + G_8 \,.\, c^2]\,;$$

and it will be seen to differ only by the sign of the last term, which however will call for careful attention.

B. Astigmatism of a Thin Lens

We apply to the two surfaces of the thin lens the general equation for the astigmatic contribution

$$AC' = \tfrac{1}{2}(N'-N)(N'/N)l'u'i'^2_{pr}(i'-u)/N'_k u'^2_k,$$

again putting $i'_{pr} = H'_k N'_k u'_k / l'N'u'$ by equation (a), when we find

$$AC' = \tfrac{1}{2}H'^2_k \,.\, N'_k(N'-N)(i'-u)/NN'l'u'.$$

Applying this equation with suffix 1 to the first surface of the thin lens, we shall have $N_1 = 1$ for air and $N'_1 = N$ of the glass. In this case the remarkable final result is most readily obtained by using the *outside* data of the lens, and we therefore further replace i'_1 by i_1/N, and $l'_1u'_1$ in the denominator by $l_1 \,.\, u_1$; this is permissible because both express the incidence height of the paraxial ray. We thus find

$$AC'_1 = \tfrac{1}{2}H'^2_k \,.\, N'_k(N-1)(i_1/N - u_1)/Nl_1u_1,$$

or on distributing the denominator,

$$AC'_1 = \tfrac{1}{2}H'^2_k \,.\, N'_k \cdot \frac{N-1}{N}\left(\frac{1}{N\,.\,l_1}\cdot\frac{i_1}{u_1} - \frac{1}{l_1}\right).$$

For the second surface we write the general equation with suffix 2 and have $N_2 = N$ of the glass and $N'_2 = 1$ for air. To obtain outside (or in this case 'dashed') angles throughout, we first replace $i'_2 - u_2$ by $i_2 - u'_2$, and then in the latter i_2 by i'_2/N, thus producing

$$AC'_2 = \tfrac{1}{2}H'^2_k N'_k(1-N)(i'_2/N - u'_2)/Nl'_2u'_2,$$

or by rearrangement

$$AC'_2 = -\tfrac{1}{2}H'^2_k N'_k\,\frac{N-1}{N}\left(\frac{1}{N\,.\,l'_2}\cdot\frac{i'_2}{u'_2} - \frac{1}{l'_2}\right):$$

and the total astigmatic contribution of the lens becomes

$$\text{Total } AC' = \tfrac{1}{2} H_k'^2 N'_k \frac{N-1}{N}\left(\frac{1}{N \cdot l_1}\cdot\frac{i_1}{u_1} - \frac{1}{l_1} - \frac{1}{N \cdot l_2'}\cdot\frac{i_2'}{u_2'} + \frac{1}{l_2'}\right).$$

We now introduce by the fundamental paraxial formulae $i_1/u_1 = (l_1 - r_1)/r_1 = l_1/r_1 - 1$ and $i_2'/u_2' = (l_2' - r_2)/r_2 = l_2'/r_2 - 1$, and find

$$\text{Total } AC' = \tfrac{1}{2} H_k'^2 N'_k \frac{N-1}{N}\left(\frac{1}{N \cdot r_1} - \frac{1}{N \cdot l_1} - \frac{1}{l_1} - \frac{1}{N \cdot r_2} + \frac{1}{N \cdot l_2'} + \frac{1}{l_2'}\right).$$

By obvious collections of terms this becomes

$$\text{Total } AC' = \tfrac{1}{2} H_k'^2 N'_k \frac{N-1}{N}\left[\frac{1}{N}\left(\frac{1}{r_1} - \frac{1}{r_2}\right) + \frac{N+1}{N}\left(\frac{1}{l_2'} - \frac{1}{l'}\right)\right],$$

or by including the outside factor $(N-1)/N$,

$$\text{Total } AC' = \tfrac{1}{2} H_k'^2 N'_k \left[\frac{N-1}{N^2}\left(\frac{1}{r_1} - \frac{1}{r_2}\right) + \frac{N^2-1}{N^2}\left(\frac{1}{l_2'} - \frac{1}{l_1}\right)\right].$$

Now we have by TL(2) of section [23]

$$(N-1)\left(\frac{1}{r_1} - \frac{1}{r_2}\right) = \frac{1}{f'} \text{ or } = \text{the power of the thin lens,}$$

and by simple transposition of TL(3) of the same section, restoring the suffix-numbers of the initial l and the final l',

$$\frac{1}{l_2'} - \frac{1}{l_1} = \frac{1}{f'} \text{ or also } = \text{the power of the thin lens.}$$

Our last equation for total AC' therefore becomes reduced to

$$\text{Total } AC' = \tfrac{1}{2} H_k'^2 N'_k \left[\frac{1}{N^2}\cdot\frac{1}{f'} + \frac{N^2-1}{N^2}\cdot\frac{1}{f'}\right].$$

Combining the two items in the square bracket, we arrive at the extraordinary and extremely important result for the astigmatic contribution of a thin lens passed centrally by the oblique pencils :

(c) $$\text{Total } AC' = H_k'^2 \cdot N'_k \cdot \frac{1}{2f'}.$$

It will be seen that this contribution is thus proved to be independent not only of the conjugate distances l_1 and l_2' at which the lens may be working, but even independent of its refractive index ; it depends solely upon the focal length!

C. The Petzval Curvature for a Thin Lens

If we apply the general solution for the Petzval contribution of any one refracting surface, as given in [58] Summary 4 d, to the two surfaces of our thin lens, we obtain

$$\text{Total } PC' = \tfrac{1}{2} H_k'^2 N'_k \left(\frac{N_1' - N_1}{N_1 \cdot N_1' \cdot r_1} + \frac{N_2' - N_2}{N_2 \cdot N_2' \cdot r_2}\right):$$

[59] and as for the thin lens surrounded by air we have $N_1 = N_2' = 1$ and $N_1' = N_2 =$ the index of the glass, which we denote by plain N, the equation becomes

$$\text{Total } PC' = \tfrac{1}{2} H_k'^2 N_k' \left(\frac{N-1}{N \cdot r_1} + \frac{1-N}{N \cdot r_2}\right) = \tfrac{1}{2} H_k'^2 N_k' \frac{N-1}{N}\left(\frac{1}{r_1} - \frac{1}{r_2}\right).$$

By TL(2) of [23] we can replace $(N-1)(1/r_1 - 1/r_2)$ by $1/f'$ or the power of the thin lens, and we thus arrive at the final equation for the Petzval contribution of one thin lens :

(d) $$\text{Total } PC' = \tfrac{1}{2} H_k'^2 N_k' \cdot 1/Nf'.$$

It should be noted that the expression is very similar to that for the astigmatic contribution, but differs from it by having the index of the glass in its denominator and therefore varying with the index. This is of great importance in the design of anastigmatic photographic objectives.

D. The Distortion of a Thin Lens

It is easily shown that a thin lens passed centrally by the oblique pencils does not make any contribution to the distortion of the lens system of which it is a part. This follows without mathematics from the fact that the axial element of any thin lens is indistinguishable from a thin plano-parallel plate, hence the principal ray leaves the lens under the angle u_{pr} with which it entered, and at the limit of infinite thinness the ray goes straight through just as if it had passed through a pin-hole ; hence there can be no distortion. The formal algebraical proof is easily furnished by working out the explicit equation for DC', section [58] summary (5) b, for the two surfaces of a thin lens, noting that for a centrally passing principal ray we have $(u_{pr}' + i_{pr}')$ necessarily zero and $i'pr_2 = N \cdot i'pr_1$.

E. The Longitudinal Chromatic Aberration of a Thin Lens

By [58] summary (4) f, any one surface makes a contribution to the final longitudinal chromatic aberration :

$$LchC' = l'u'N'i'\left(\frac{\delta N'}{N'} - \frac{\delta N}{N}\right)/N_k' u_k'^2.$$

In order to apply this equation to the two surfaces of our thin lens, we extend it by the factor

$$\frac{l_k'^2 u_k'^2}{l_k'^2 u_k'^2} \cdot \frac{l'u'}{l'u'} = 1,$$

and as $l'u'$ always represents the paraxial semi-aperture of a refracting surface, we replace $l' \cdot u'$ in the numerator by SA and $l_k' u_k'$ in the denominator by SA_k. Replacing the $l'u'$ in the equation for $LchC'$ also by SA, and applying the extending factor, the equation becomes

$$LchC' = l_k'^2 \cdot \frac{SA^2}{N_k' SA_k^2} \cdot \frac{N'i'}{l'u'}\left(\frac{\delta N'}{N'} - \frac{\delta N}{N}\right),$$

and by the law of refraction we are at liberty to replace $N'i'$ by Ni, and always

$l'.u'$ by $l.u$. As the two surfaces of a *thin* lens necessarily have the same value of [59] SA, we can write the total contribution of our lens as

$$\text{Total } LchC' = l_k'^2 \frac{SA^2}{N'_k SA_k^2}\left[\frac{N_1' i_1'}{l_1' u_1'}\left(\frac{\delta N_1'}{N_1'} - \frac{\delta N_1}{N_1}\right) + \frac{N_2 i_2}{l_2 u_2}\left(\frac{\delta N_2'}{N_2'} - \frac{\delta N_2}{N_2}\right)\right].$$

As we always employ refractive indices measured in air, we have $N_1 = N_2' = 1$ and $\delta N_1 = \delta N_2' = 0$, and using again plain N for the index of the glass and plain δN for its dispersion, the last equation simplifies to

$$\text{Total } LchC' = l_k'^2 \frac{SA^2}{N'_k SA_k^2} \cdot \delta N \left(\frac{i_1'}{l_1' u_1'} - \frac{i_2}{l_2 u_2}\right).$$

Here we introduce again $i_1'/u_1' = (l_1' - r_1)/r_1 = l_1'/r_1 - 1$ and correspondingly $i_2/u_2 = l_2/r_2 - 1$, and find

$$\text{Total } LchC' = l_k'^2 \frac{SA^2}{N'_k SA_k^2} \cdot \delta N \left(\frac{1}{r_1} - \frac{1}{l_1'} - \frac{1}{r_2} + \frac{1}{l_2}\right).$$

As for a thin lens we have $l_1' = l_2$, the two l-terms in the bracket cancel each other and the final equation becomes

$$\text{(e)} \qquad \text{Total } LchC' = l_k'^2 \frac{SA^2}{N'_k SA_k^2} \cdot \delta N \left(\frac{1}{r_1} - \frac{1}{r_2}\right) = l_k'^2 \frac{SA^2}{N'_k SA_k^2} \cdot c \cdot \delta N$$

on introducing the total curvature of the thin lens.

F. The Transverse Chromatic Aberration of a Thin Lens

As we have defined this aberration as the chromatic dispersion of the different colours in an entering principal ray of white, or in general mixed, light, it is at once clear that the same argument applies which we used in the case of distortion. For as the axial part of a *thin* lens may be regarded as equivalent to a *thin* plano-parallel plate, principal rays of any colour and at any inclination will go straight through; hence a thin lens passed centrally by the oblique pencils cannot make any contribution to the transverse chromatic aberration of the system to which it belongs. This may easily be proved in a more formal and laborious manner by working out the equation for $TchC'$ for the two surfaces of a thin lens in the manner already repeatedly exemplified.

G. The Spherical Aberration of a Thin Lens

The equations deduced in this section for the aberrations of a thin lens surrounded by air are in one respect more general than those given in section [29] for the spherical aberration of such a lens, for by treating the thin lens as forming part of a complete system we have on the present occasion included the possibility that the system may form its final image in a medium other than air, whilst the earlier proof was only applicable to a final image in air. Although the possibility of a final image formed in a medium other than air is rather remote, it appears worth while to examine the case of spherical aberration again along the line taken in the present section.

In accordance with the most general result of section [29] the spherical con-

[59] tribution of a thin lens somewhere within a system of thin lenses at any separation was, in our present symbols,

$$\text{Total } SC' = l_k'^2 \frac{SA^4}{SA_k^2} \text{ [spherical } G\text{-sum]}.$$

Starting with the perfectly general solution: Summary of [58], 5a, for the contribution of any one refracting surface

$$SC' = \tfrac{1}{2} l'u'ii'(i'-u)(N'-N)/N'_k u_k'^2,$$

we can secure an outside factor very similar to the earlier one by extending the right of the last equation by

$$1 = \frac{l'^3u'^3}{l'^3u'^3} \cdot \frac{l_k'^2u_k'^2}{l_k'^2u_k'^2} = \frac{SA^3}{l'^3u'^3} \cdot \frac{l_k'^2u_k'^2}{SA_k^2},$$

and by also putting the initial $l'u'$ of the equation itself $= SA$. The equation for the spherical contribution thus becomes

$$SC' = l_k'^2 \frac{SA^4}{N'_k SA_k^2} \cdot \tfrac{1}{2} \frac{i}{l'u'} \cdot \frac{i'}{l'u'} \cdot \frac{i'-u}{l'u'} (N'-N).$$

It is then easily shown that when this equation is applied to the two surfaces of a thin lens with the same algebraical tricks used for the other aberrations, the final factors of the last equation beginning with $\frac{1}{2}$ simply become the spherical G-sum, hence the most general solution for the spherical contribution of a thin lens forming part of a complete centred system of any construction is

$$\text{(f)} \qquad \text{Total } SC' = l_k'^2 \frac{SA^4}{N'_k SA_k^2} \text{ [spherical } G\text{-sum]}.$$

It will be seen that the only change from the earlier solution, and the complete generalization sought by us, consists in the appearance of N'_k in the denominator. It is remarkable that this index of the medium in which the final image is formed appears in the denominator in the case of the longitudinal spherical and chromatic aberrations, in the numerator of the astigmatic and of the Petzval contributions, and not at all in the Coma-term.

H. Extension to Thin Systems of Lenses

As we have deduced our equations for the perfectly general case of a thin lens forming part of a centred system which it was not necessary to submit to any special conditions, excepting the one restriction that the oblique pencils should pass centrally through the thin lens, and as the contributions to the total of any one aberration by the successive surfaces are simply additive, it is clear that if instead of one thin lens we have *a closely packed set of thin lenses* in the postulated position, so that the oblique pencils may be regarded as passing centrally through the whole set, then our equations for one thin lens will apply to each lens of the thin set, and the total effect of the latter will be the simple sum of the effects produced by the constituent lenses. Moreover, we measure the semi-aperture of the thin lens by the paraxial pencil, and as the whole set is to be treated as infinitely thin, SA will have precisely the same value for all the constituent thin lenses of the set and can be taken out as a common factor together with the data of the last surface of the complete system which enter into the equations.

We can therefore write down the equations for such a thin set of lenses by simply adding a sum-sign to the lettered equations in the preceding deductions :

Seidel Aberrations III

For a thin combination of thin lenses, forming part of a centred system of k surfaces, and placed in contact with the diaphragm of that system :

(1) Total contribution by the thin combination to the final longitudinal spherical aberration

$$SC' = l_k'^2 \frac{SA^4}{N'_k SA_k^2} \cdot \quad \Sigma \text{(spherical } G\text{-sums of individual lenses).}$$

(2) Total contribution of the thin combination to the final Coma

$$CC' = H'_k \,.\, SA^2 \,. \quad \Sigma \text{(Coma } G\text{-sums of individual lenses).}$$

(3) Total contribution of the thin combination to the final Astigmatism

$$AC' = H_k'^2 N_k' . \quad \Sigma \left(\frac{1}{2f'} \text{ of individual lenses}\right) = H_k'^2 N'_k \cdot \frac{1}{2f'},$$

if in the last form f' denotes the focal length of the complete thin combination.

(4) Total contribution of the thin combination to the final Petzval-curvature

$$PC' = \tfrac{1}{2} H_k'^2 N'_k \,.\, \Sigma \left(\frac{1}{N \,.\, f'} \text{ of the individual lenses}\right).$$

(5) Total contribution of the thin combination to the final distortion

$$DC' = 0.$$

(6) Total contribution of the thin combination to the final longitudinal chromatic aberration

$$LchC' = l_k'^2 \frac{SA^2}{N'_k SA_k^2} \Sigma \, (c \,.\, \delta N \text{ of the individual lenses}).$$

(7) Total contribution of the thin combination to the final transverse chromatic aberration

$$TchC' = 0.$$

SA_k denotes the semi-aperture of the paraxial pencil at the last surface of the complete system of which the thin combination forms a part.

N'_k is the index of the medium in which the complete system forms its final image ; it is therefore nearly always $= 1$ for air.

l'_k is the final intersection-length of the paraxial pencil on emergence from the last surface of the complete system.

For the individual lenses of the thin combination the total curvature $c = c_1 - c_2$, the refractive index N and the dispersion δN must be given or selected, also the distance l_1 of the axial object-point presented to the first lens of the combination.

[59] Using the v-notation, we then have $v_1 = 1/l_1$ and must calculate $v_2' = v_1 + c_a(N_a - 1)$; $v_4' = v_2' + c_b(N_b - 1)$; $v_6' = v_4' + c_c(N_c - 1)$, &c. Also, if the astigmatism and the Petzval curvature are to be included in the calculation, the power $1/f'$ of the individual lenses and of the complete combination must be calculated as

$$1/f_a' = c_a(N_a - 1)\,; \quad 1/f_b' = c_b(N_b - 1)\,; \quad 1/f_c' = c_c(N_c - 1), \text{ \&c.}$$

and $1/f' = 1/f_a' + 1/f_b' +$ &c. for the complete combination.

The G-sums of the individual thin lenses are then obtained, separately for each lens of total curvature c, curvature of first surface c_1 or of second surface c_2, convergence of arriving rays v_1 or of emerging rays v_2', as

$$\text{Spherical } G\text{-sum} = G_1c^3 - G_2c^2c_1 + G_3c^2v_1 + G_4cc_1^2 - G_5cc_1v_1 + G_6cv_1^2$$

$$or \quad ,, \quad = G_1c^3 + G_2c^2c_2 - G_3c^2v_2' + G_4cc_2^2 - G_5cc_2v_2' + G_6cv_2'^2\,;$$

$$\text{Coma } G\text{-sum} = \tfrac{1}{4}\,G_5cc_1 - G_7cv_1 - G_8c^2$$

$$or \quad ,, \quad = \tfrac{1}{4}\,G_5cc_2 - G_7cv_2' + G_8c^2.$$

Mathematically it does not matter which one of the alternative G-sums is used, as they always give precisely the same result, but very decided advantages in the computing work are secured by a judicious selection, as was indicated for the spherical aberration in section [49]. The G-values, or more usually their logs, are taken directly from the table with the refractive index as argument.

In a great many cases the thin combination will be used by itself and not in conjunction with other parts of a more elaborate thick system; the formulae then become a little more simple, for N_k' will then be $= 1$ for air and SA_k will become identical with the SA of the thin combination; therefore the outside factor of SC' will be $l_k'^2SA^2$, that of $LchC'$ simply $l_k'^2$, k being now taken as the number of surfaces in the thin combination.

It would not be worth while to use these thin lens equations in the study of the properties of a finished lens system of given construction; for that purpose the necessarily limited accuracy of the equations renders them unsuitable, and the primary aberration equations, Seidel Aberrations II, must be given preference, together with strict trigonometrical ray-tracing by the methods to be given in the two following chapters. The proper, and the only intended, use of the approximate thin lens equations is again, as in the case of simple chromatic and spherical aberration, to shorten and to simplify the calculations required in the working out of new designs. The simple algebraical form of the thin lens equations then renders it possible to arrive at direct though approximate solutions of numerous problems which would call for wearisome, more or less blind, trials by the stricter methods. For these purposes the value of the new equations can scarcely be estimated too highly, and their adaptation to the solution of different types of optical problems will form the subject of a considerable part of our subsequent work.

For the present we can only take the simplest case of the ordinary two-lens achromatic combination, most usually a telescope or microscope objective. We learnt in Chapter V how the equation for the spherical aberration can be used to find in a remarkably short time that form of a *cemented* objective which will have a prescribed amount—usually zero—of spherical aberration; but we could find [59] no full use for the two liberties afforded when the objective is not to be a cemented one and when each constituent lens might be bent independently so as to satisfy two conditions of correction simultaneously; we only dealt with the obvious possibility of adopting an arbitrary form for one constituent and then correcting its spherical aberration by determining the proper bending of the other constituent.

Our equations show that Coma is the only other aberration which depends on the bending of the lens, for the coma-equation contains c_1 or c_2 in the first power, and shows that the coma of a thin lens passed centrally by the oblique pencils changes *in direct proportion with the bending* and therefore *plots as an inclined straight line* against either c_1 or c_2. As we have also learnt in the earlier parts of this chapter that coma is a most objectionable defect in the images produced by optical instruments, we realize that correction of the coma will be a highly suitable second condition to be coupled with correction of the spherical aberration. In accordance with a now widely adopted suggestion by Abbe, lens systems simultaneously free from spherical aberration and from coma are called aplanatic systems. We will now seek a general solution for thin two-lens *aplanatic objectives*.

The preparatory work for the determination of the total curvatures c_a and c_b of the two components by the achromatic condition and of the required v-values will be exactly the same as in the simpler cases dealt with in sections [49] and [48·5]. The application of the equation for the spherical aberration also remains exactly as then exemplified, and gives a quadratic equation in c_2 and c_3 for the final value of this aberration. The small amount of added work consists in determining the sagittal coma of the objective by Seidel Aberrations III (2). Applying that equation to the two lenses of the objective, and using the second form of the coma G-sum for the first and the first form for the second lens, in order to express the coma in terms of c_2 and c_3 as was advised for the spherical aberration, we find

$$CC' = H'_k SA^2 \left\{ \begin{array}{l} \quad \tfrac{1}{4}\, G_5^a c_a c_2 - G_7^a c_a v_2' + G_8^a {c_a}^2 \\ + \tfrac{1}{4}\, G_5^b c_b c_3 - G_7^b c_b v_3 - G_8^b {c_b}^2 \end{array} \right\}$$

the first line giving the contribution of the first lens and the second that of the second lens, the G-values in the two lines being of course interpolated separately from the table for the respective refractive indices of the two lenses. By putting in also the actual numerical values of H'_k, SA, the c-values, and the value of $v_2' = v_3$ in a manner exactly similar to that exemplified in section [49], we obtain on collection of the absolute terms into a single one an equation for the total coma contributions of the form

$$CC' = a \,.\, c_2 + b \,.\, c_3 + d.$$

This, combined with the quadratic equation for the spherical aberration, gives the necessary two equations which enable us to solve for c_2 and c_3 and thus to determine the two forms of the objective which give the desired values of SC' and CC', most usually zero for both; but it should be noted that it is just as easy to solve for any desired amount and sign of spherical aberration and/or coma by putting the proper numerical values on the left of the two equations. For SA we shall of course introduce the value of the semi-aperture for which the objective is

[59] to be trigonometrically corrected, and the same applies to H'_k if we know its intended value. If we are not working for a definite value of the final image-height, then we can choose an arbitrary value for H'_k. The most judicious choice will nearly always be $H'_k = 1$, which not only saves the turning up of one logarithm but has another more important advantage which will become revealed in the next chapter on the optical sine condition. As the latter will most usually be used in the trigonometrical correction of the thin lens solution, we will deal with the final adjustment of the objective in Chapter VII.

As a short numerical example we will determine the aplanatic forms of the telescope objectives in section [49]. For these we had the data :

$$\begin{array}{ll} \text{First lens}: N_a = 1{\cdot}5407\,;\ c_a = 0{\cdot}470 & \multirow{2}{*}{\Big\}}\ SA = \text{previous } y = 0{\cdot}64\,; \\ \text{Second lens}: N_b = 1{\cdot}6225\,;\ c_b = -0{\cdot}247 & \ \log v_2' = \log v_3 = 9{\cdot}4051\,; \end{array}$$

and if we adopt $H'_k = 1$ as suggested above, we shall have $H'_k SA^2$ equal to $0{\cdot}64^2 = 0{\cdot}4096$. As there are only three items in the contribution of each constituent lens, we can place the two parts side-by-side and find :

	First Lens.			*Second Lens.*		
Formula : $H'_k SA^2 \times$	$\frac{1}{4} G_5^a c_a c_2$	$-G_7^a c_a v_2'$	$+G_8^a c_a^2$	$\frac{1}{4} G_5^b c_b c_3$	$-G_7^b c_b v_3$	$-G_8^b c_b^2$
log $\frac{1}{4}$	9·3979			9·3979		
log $H'_k SA^2$	9·6124	9·6124	9·6124	9·6124	9·6124	9·6124
log G	0·2512	9·8550n	9·6196	0·3037	9·9108n	9·7033n
log c^n	9·6721	9·6721	9·3442	9·3927n	9·3927n	8·7854
log v		9·4051			9·4051	
Sums	8·9336($\times c_2$)	8·5446n	8·5762	8·7067n($\times c_3$)	8·3210	8·1011n
Numbers	0·08582 c_2	$-$0·03504	0·03769	$-$0·05090 c_3	0·02094	$-$0·01262

Combining the four purely numerical terms according to their signs, we have the result :

$$\text{Total } CC' = 0{\cdot}08582\, c_2 - 0{\cdot}05090\, c_3 + 0{\cdot}01097,$$

a linear equation in c_2 and c_3. In section [49] we found for the spherical aberration, in our present symbols,

$$\text{Total } SC' = 11{\cdot}96\, c_2^2 + 1{\cdot}258\, c_2 - 7{\cdot}03\, c_3^2 + 1{\cdot}874\, c_3 + 0{\cdot}473,$$

and these two equations have to be solved for c_2 and c_3. Assuming that perfect correction of spherical aberration and coma is aimed at, we put the left of the equations as zero and for the solution use the linear equation to express c_3 in terms of c_2, finding

$$0{\cdot}05090\, c_3 = 0{\cdot}08582\, c_2 + 0{\cdot}01097$$

or

$$c_3 = 1{\cdot}686\, c_2 + 0{\cdot}2155.$$

We use this to eliminate c_3 from the quadratic equation and find

$$0 = \text{required } SC' = 8{\cdot}03\, c_2^2 + 0{\cdot}692\, c_2 - 0{\cdot}550.$$

The solution of this simple quadratic equation then gives

$$c_2 = \text{either } -0{\cdot}3084 \text{ or } +0{\cdot}2222,$$

and using this in $c_3 = 1{\cdot}686\, c_2 + 0{\cdot}2155$ we find the associated

$$c_3 = \text{either } -0{\cdot}3045 \text{ or } +0{\cdot}5901.$$

[59] As $c_a = c_1 - c_2$, we then calculate $c_1 = c_a + c_2 = 0{\cdot}470 + c_2$ and find

$$c_1 = \text{either } +0{\cdot}1616 \text{ or } +0{\cdot}6722,$$

and as $c_b = c_3 - c_4$, we have $c_4 = c_3 - c_b = c_3 + 0{\cdot}247$ or

$$c_4 = \text{either } -0{\cdot}0575 \text{ or } +0{\cdot}8371.$$

Collecting the results and adding the radii = the reciprocals of the curvatures, we obtain the two thin lens prescriptions :

First Solution.		*Second Solution.*	
$c_1 = 0{\cdot}1616$	$r_1 = 6{\cdot}188$	$c_1 = 0{\cdot}6722$	$r_1 = 1{\cdot}488$
$c_2 = -0{\cdot}3084$	$r_2 = -3{\cdot}243$	$c_2 = 0{\cdot}2222$	$r_2 = 4{\cdot}500$
$c_3 = -0{\cdot}3045$	$r_3 = -3{\cdot}284$	$c_3 = 0{\cdot}5901$	$r_3 = 1{\cdot}695$
$c_4 = -0{\cdot}0575$	$r_4 = -17{\cdot}39$	$c_4 = 0{\cdot}8371$	$r_4 = 1{\cdot}195$

It will be seen that the first solution gives radii of reasonable length and only a very slight difference of curvature at the broken contact. In the second solution three of the radii are extremely short, and this form could only come into question if cost were of no importance and if at the same time the form had some decided advantage over the alternative one. As a matter of fact, the second solution has no advantage of any kind ; moreover, it very usually proves incapable of trigonometrical correction when sufficient thickness is given to the components. The second solution, as far as telescope objectives are concerned, may therefore be safely ignored and regarded as a freak ; but in photographic objectives of certain types such combinations of pronounced meniscus form are sometimes useful.

As for the good form of an aplanatic objective the difference of curvature at the adjacent second and third surfaces is always very small for all kinds of glass which are suitable for such objectives, the approximate solution required for trigonometrical final correction can be obtained far more quickly and yet with sufficient accuracy by solving for a cemented objective free from coma and ignoring the spherical aberration. That means merely the working out of CC', and in our case would give

$$\text{Total } CC' = 0{\cdot}08582\, c_2 - 0{\cdot}05090\, c_3 + 0{\cdot}01097$$

as found above. Putting $c_3 = c_2$ for a cemented objective and demanding zero value of CC', this gives

$$0{\cdot}08582\, c_2 - 0{\cdot}05090\, c_2 + 0{\cdot}01097 = 0$$

$$0{\cdot}03492\, c_2 + 0{\cdot}01097 = 0$$

$$\therefore c_2 = -0{\cdot}01097/0{\cdot}03492 = -0{\cdot}314,$$

which differs very little from the value $-0{\cdot}3084$ found above by the strict and complete solution, and would probably be as readily corrected by the differential method as the latter.

FUNDAMENTAL LAWS OF OBLIQUE PENCILS

[60] There are three general properties of oblique pencils which are of such importance and utility as to be singled out for ready reference as fundamental laws ; they are firstly the relations between curvature of field of the sagittal and tangential fans

[60] and the Petzval curvature ; secondly a remarkable law which enables us to deduce the change in the oblique aberrations brought about by merely shifting the aperture diaphragm ; and thirdly the surprising fact proved in the preceding section, that the astigmatism of a thin lens passed centrally by the oblique pencils depends solely upon the focal length of the lens. The third law is necessarily restricted to thin lenses ; the first two are absolutely general as far as primary aberrations are concerned.

(1) The First Fundamental Law

We included in Seidel Aberrations II the two important equations

$$\text{Primary } X'_S = -g_3 H'^2_k \quad - \tfrac{1}{2} g_4 H'^2_k$$
$$\text{Primary } X'_T = -3\, g_3 H'^2_k - \tfrac{1}{2} g_4 H'^2_k,$$

which determine the amounts by which a focusing screen would have to be shifted along the optical axis in order to change from the sharp focus of the axial pencil at a given aperture to the sagittal and tangential foci respectively, at the same aperture. The second term on the right of both equations represents that part of this focal shift which is due to the Petzval term, and we may call this X'_{Ptz} in strict analogy with the symbols adopted for the total shifts, thus making the equations

$$\left.\begin{aligned} X'_S &= -g_3 H'^2_k \quad + X'_{Ptz} \\ X'_T &= -3\, g_3 H'^2_k + X'_{Ptz} \end{aligned}\right\} X'_{Ptz} = -\tfrac{1}{2} g_4 H'^2_k.$$

If we now eliminate $g_3 H'^2_k$ from these two equations by multiplying the first by 3 and the second by (-1) and then adding, we obtain

$$3\, X'_S - X'_T = 2\, X'_{Ptz},$$

and this is the relation which we shall refer to as the first fundamental law ; it embodies the remarkable three-to-one ratio of tangential and sagittal astigmatism and the influence of the Petzval term, and enables us by transposition to find any one of the three X' values when the other two are known. As in the practical computing work X'_T and X'_{Ptz} are nearly always the most readily determined data, we will record the relation in the form :

$$\textit{First Fundamental Law}: \begin{cases} 3\, X'_S = 2\, X'_{Ptz} + X'_T \\ X'_{Ptz} = -\tfrac{1}{2} g_4 H'^2_k. \end{cases}$$

(2) The Second Fundamental Law

The second law is by far the most important of the three, and will be extensively employed subsequently in the systematic solution of problems in the design of optical systems.

As the oblique aberrations of an optical system are completely determined—as far as primary aberrations are concerned—by the seven g-values, it follows that these g-values, when once calculated, enable us to determine the course of any ray which can pass through the system from any point in the given object-plane ; we have only to put into the equations the proper values of SA_k and of H'_k. A longitudinal shift of the aperture diaphragm evidently cannot admit any rays which could not also pass through a sufficiently large aperture of the original diaphragm, [60] and which for that reason admit of being tracked by the original g-values. Hence it follows that it must be possible to determine the value of the aberrations with the shifted diaphragm from those for the original diaphragm without repeating the whole calculations required in the original determination ; there must be a shorter way. The second law determines the relations between the original and the new aberrations by simple algebraical equations which are eminently suitable for analytical solutions.

With reference to Fig. 62, the last refracting surface of a centred system shall cut the optical axis at A, and the primary aberrations of the system shall have been

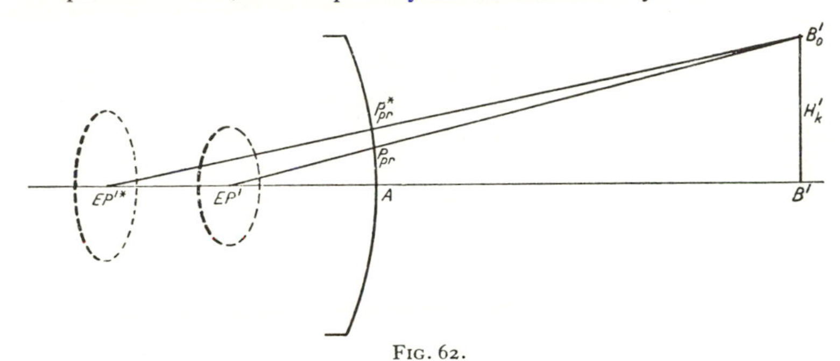

FIG. 62.

determined by the methods already dealt with for a position of the diaphragm yielding the final exit pupil EP'. We seek expressions for the aberrations which will exist when the aperture diaphragm is shifted along the optical axis so as to yield a final exit pupil at EP'^*. We will note at once that all the data for the shifted diaphragm will be distinguished by a 'star'. If B' is the final paraxial image-point of the system and B'_o a corresponding extra-axial ideal image-point, then the original principal ray $EP'B'_o$ will cut the last surface at P_{pr}, whilst the new principal ray $EP'^*B'_o$ will cut the last surface at P^*_{pr}. The distance $P_{pr}\,P^*_{pr}$ is the key-quantity in our transformations. We can at once prove that this distance bears a fixed proportion to $B'B'_o = H'_k$. The distances from A to EP' and to EP'^* evidently are the final intersection-lengths of the respective principal rays and represent l'_{pr} and l'^*_{pr}, both *constant* lengths. The distance from A to B' is the final l' of the system, also a constant. Similar right-angled (in the limit) triangles then give

$$H'_k/(l'-l'_{pr}) = AP_{pr}/-l'_{pr} \qquad \text{or } AP_{pr} = -H'_k\,.\,l'_{pr}/(l'-l'_{pr})$$

and $$H'_k/(l'-l'^*_{pr}) = AP^*_{pr}/-l'^*_{pr} \qquad \text{or } AP^*_{pr} = -H'_k\,.\,l'^*_{pr}/(l'-l'^*_{pr}),$$

which prove that both AP_{pr} and AP^*_{pr} are simply proportional to H'_k; hence their difference $P_{pr}P^*_{pr}$ is also proportional to H'_k, q.e.d. We may therefore put $P_{pr}P^*_{pr} = W'H'_k$, in which W' is a purely numerical constant depending on l', l'_{pr}, and l'^*_{pr}, which can be determined from the equations for AP_{pr} and AP^*_{pr} (but we shall find a way of avoiding even that trouble in all usual cases).

[60] With the aid of $P_{pr}P^*_{pr} = W'H'_k$ we can now deduce the values of the aberrations referred to the new principal ray from the known values for the original principal ray as follows.

(1) Spherical Aberration

As the spherical aberration of all the oblique pencils is identical with that of the axial pencil, and as the latter is not changed in any way by a longitudinal shift of the diaphragm if we adhere to our adopted method of measuring the clear aperture on the last surface, there can be *no change* in the spherical aberration ; and using the star-nomenclature, we must have

$$g_1^* = g_1 \text{ or final } LA' = g_1 SA_k{}^2 = g_1^* SA_k{}^2.$$

(2) Coma

To determine the coma for the shifted position of the diaphragm, we utilize the fact proved in the early part of section [55] that the longitudinal extent of the characteristic focal line depends purely on the coma-coefficient g_2, being determined as

$$LA'_{+M} - LA'_{-M} = 4\, g_2 l'_k \,.\, SA_k \,.\, H'_k.$$

Being part of our absolutely general theory, this equation necessarily applies to the coma coefficient g_2^* to be determined for the new position of the exit pupil, and it is moreover valid for any values of either SA_k or H'_k.

With reference to Fig. 63, which is our usual end-view of the last surface shown in section in Fig. 62, we choose as the semi-aperture SA_k to be used in the present deductions the distance $P_{pr}P^*_{pr} = W'H'_k$; therefore the LA'_{+M} to be used in the above equation will be the longitudinal aberration of an infinitely close D-pair of rays from the extreme top of the resulting zone and LA'_{-M} will be the longitudinal aberration of a similar close D-pair from the lowest point of the same zone. With reference to the original principal ray at P_{pr}, for which the aberrations are known, the upper of these D-pairs is one with $Y_k = SA_k = 2\, W'H'_k$, and the lower is one with $Y_k = 0$ (being simply to right and left of P_{pr}) and with SA_k indefinitely small. If we now take our general equation for any D-pair at the final focus by Seidel Aberrations I, section [54],

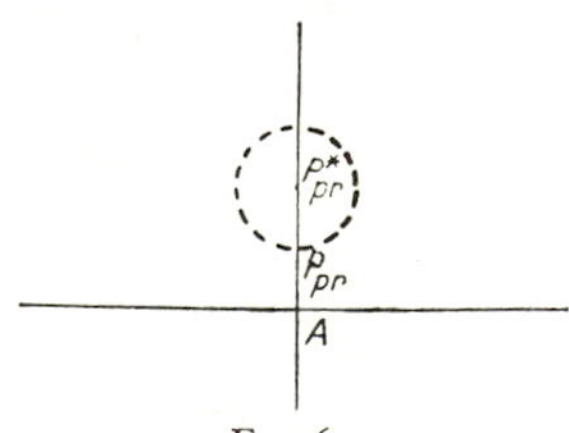

Fig. 63.

$$\text{Final } LA'_d = g_1 SA_k{}^2 + 2\, g_2 l'_k Y_k H'_k + g_3 H'^2_k + \tfrac{1}{2}\, g_4 H'^2_k,$$

and put in the special values of Y_k and SA_k just stated, we find for use in the first equation of this subsection :

$$LA'_{+M} = 4\, g_1 (W'H'_k)^2 + 4\, g_2 l'_k \,.\, W'H'_k \,.\, H'_k + g_3 H'^2_k + \tfrac{1}{2}\, g_4 H'^2_k$$
$$LA'_{-M} = g_3 H'^2_k + \tfrac{1}{2}\, g_4 H'^2_k$$

$$\therefore LA'_{+M} - LA'_{-M} = 4\, g_1 (W'H'_k)^2 + 4\, g_2 l'_k \,.\, W'H'_k \,.\, H'_k$$

The g_2 in our opening equation may be taken as the g_2^* referred to the new principal ray at P^*_{pr}, and we selected for SA_k in the equation the special value $W'H'_k$; [60] hence the opening equation now takes the form

$$LA'_{+M} - LA'_{-M} = 4\,g_2^* \,.\, l'_k \,.\, W'H'_k \,.\, H'_k,$$

and if we equate the two alternative values of $LA'_{+M} - LA'_{-M}$ we obtain

$$4\,g_2^* \,.\, l'_k \,.\, W'H'_k \,.\, H'_k = 4\,g_1\,(W'H'_k)^2 + 4\,g_2 \,.\, l'_k \,.\, W'H'_k \,.\, H'_k,$$

and division throughout by $4\,l'_k \,.\, W'H'_k \,.\, H'_k$ gives the remarkable relation

(a) $$g_2^* = g_2 + W' \,.\, g_1 / l'_k$$

by which the new coma coefficient g_2^* can be calculated from the original and known coefficients g_1 and g_2.

In our practical applications we usually work with the actual sagittal coma and spherical aberration for definite, selected, values of SA_k and H'_k. To find the corresponding relation for these quantities, we have only to multiply throughout by the coma factor $SA_k{}^2 \,.\, H'_k$, obtaining

$$g_2^* SA_k{}^2 H'_k = g_2 SA_k{}^2 H'_k + g_1 SA_k{}^2 \,.\, W'H'_k / l'_k.$$

This becomes still more convenient for numerical calculations by extending the second term on the right by SA_k/SA_k, which by definition of SA_k may be written $l'_k u'_k / SA_k$, and then cancelling out the l'_k. The equation thus becomes

(a)* $$g_2^* SA_k{}^2 H'_k = g_2 SA_k{}^2 H'_k + g_1 SA_k{}^2 \,.\, u'_k \,.\, (W'H'_k / SA_k).$$

We will defer the discussion until we have deduced the corresponding relations for the other primary aberrations.

(3) and (4) Astigmatism and Petzval Term

To find the changes in these primary aberrations we proceed in a similar manner. We consider the close sagittal rays near the new principal ray at P^*_{pr} in Fig. 63. The general LA'_d equation will apply if we use the starred g-coefficients and $Y_k = 0$ with SA_k indefinitely small; hence, for these rays, with reference to the new principal ray

$$LA'_d = g_3^* H'^2_k + \tfrac{1}{2}\, g_4^* H'^2_k,$$

the two first terms in the general equation being zero.

With reference to the original principal ray at P_{pr} this same close D-pair is one at $Y_k = SA_k = P_{pr}P^*_{pr} = W'H'_k$; hence with reference to the original principal ray and the original and known g-values, the LA'_d of these rays comes out as

$$LA'_d = g_1\,(W'H'_k)^2 + 2\,g_2 l'_k \,.\, W'H'_k \,.\, H'_k + g_3 H'^2_k + \tfrac{1}{2}\, g_4 H'^2_k,$$

and if we now equate the two alternative values of the LA'_d we find

$$g_3^* H'^2_k + \tfrac{1}{2}\, g_4^* H'^2_k = g_1\,(W'H'_k)^2 + 2\,g_2 l'_k \,.\, W'H'_k \,.\, H'_k + g_3 H'^2_k + \tfrac{1}{2}\, g_4 H'^2_k.$$

We proved in the discussion of curvature of field in section [55] that the Petzval term g_4 depends solely on the radii of curvature of the surfaces of a system and on the refractive indices of the various lenses, and as all these are obviously unaffected by a shift of the diaphragm, it follows that g_4 and g_4^* must be identical, and that

[60] therefore the last terms on right and left of our last equation cancel each other ; if we then divide throughout by $H_k'^2$, we obtain

(b) $$g_3^* = g_3 + 2\,g_2 \,.\, l'_k \,.\, W' + g_1 \,.\, W'^2$$

as the equation by which the new astigmatic coefficient can be calculated from the original g-values. It is desirable to put the relation into terms of the actual aberrations for adopted values of $SA_k = l'_k u'_k$ and of H'_k with which we shall usually carry out our actual calculations. To do so we restore the factor $H_k'^2$ in equation (b), making it

$$g_3^* H_k'^2 = g_3 H_k'^2 + 2\,g_2 l'_k \,.\, W' \,.\, H_k'^2 + g_1 W'^2 \,.\, H_k'^2,$$

and then extend the second and third terms on the right by SA_k^2/SA_k^2, writing this, in the second term only, $SA_k^2/SA_k l'_k u'_k$. A mere arrangement of the resulting terms, with cancellation of l'_k in the second, then gives the computing equation

(b)* $$g_3^* H_k'^2 = g_3 H_k'^2 + 2\,g_2 SA_k^2 H'_k \,(W'H'_k/SA_k)/u'_k + g_1 SA_k^2\,(W'H'_k/SA_k)^2.$$

Again we defer the discussion till later.

(5) Distortion

We can determine the distortion which results when the diaphragm of a lens system is shifted longitudinally in a manner closely similar to that employed in the preceding cases.

With reference to Fig. 64, in which B'_o represents the ideal extra-axial image-point as determined by the theorem of Lagrange, the true principal ray from the

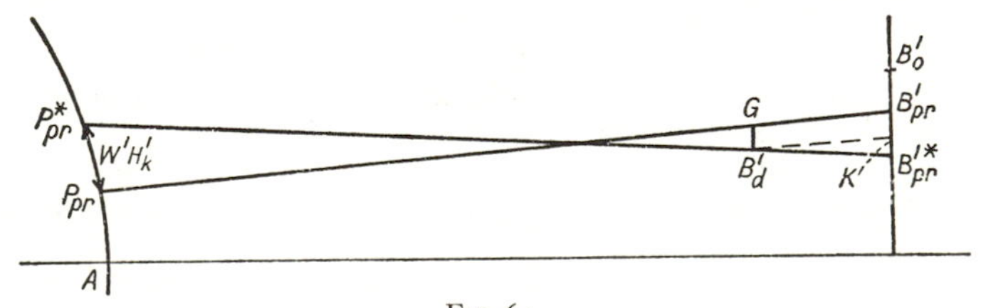

FIG. 64.

original exit pupil will be affected by aberrations and will cut the ideal focal plane at some point B'_{pr}, and $B'_{pr}B'_o$ will be the distortion for the original position of diaphragm and pupils, and represents the $g_5 H_k'^3$ of sections [54] and [55]. The new principal ray emerging at P^*_{pr} from the *shifted* exit pupil will in general cut the ideal focal plane in yet another point B'^*_{pr}, and in accordance with our definition $B'^*_{pr}B'_o$ will be the distortion for the shifted diaphragm and will represent the $g_5^* H_k'^3$ to be determined. We can determine this quantity if we succeed in defining the relative position of the old and new principal rays. Now the new principal ray from P^*_{pr} may be regarded as the fusion of an indefinitely close D-pair of rays, which with reference to the original principal ray at P_{pr} has the co-ordinates $Y_k = SA_k = W'H'_k$, just as in the case of astigmatism. The new principal ray must pass through the focus B'_d of this D-pair, and in accordance with

Seidel Aberrations I B'_d is located with reference to the ideal image-point and to [60] the principal ray at the coordinates LA'_d and TA'_d of the general solution

$$LA'_d = g_1 SA_k^2 + 2g_2 l'_k Y_k H'_k + g_3 H'^2_k + \tfrac{1}{2} g_4 H'^2_k$$

$$TA'_d = \qquad g_2 SA_k^2 H'_k + 2\, g_3\, (1/l'_k)\, Y_k H'^2_k.$$

Putting in the special value $Y_k = SA_k = W'H'_k$, these equations give

$$LA'_d = g_1\, (W'H'_k)^2 + 2g_2 l'_k \,.\, W'H'_k \,.\, H'_k + g_3 H'^2_k + \tfrac{1}{2} g_4 H'^2_k$$

$$TA'_d = \qquad g_2 (W'H'_k)^2 H'_k \quad + 2\, g_3\, (1/l'_k)\, W'H'_k \,.\, H'^2_k$$

and point B'_d is definitely located.

We now draw through B'_d a parallel to the original principal ray, cutting the ideal image-plane at K, so that $B'_{pr}K = GB'_d = TA'_d$ and $B'_d B'_{pr} = LA'_d$. We then see that the new distortion $B'_o B'^*_{pr} = g^*_5 H'^3_k$ is made up of three parts $B'_o B'_{pr} = g_5 H'^3_k$, $B'_{pr}K = TA'_d$, and KB'^*_{pr}. To determine this last part we must once more remember that for our primary aberration work all the finite quantities must be used at their limiting value for very small aperture and field, and all the aberrations must be treated as small even compared with the aperture. We then have by triangles similar in the limit

$$KB'^*_{pr}/KB'_d = W'H'_k / P_{pr}G,$$

and have to put $P_{pr}G$ at its limiting value $= l'_k$. Putting in the explicit values already determined for the remaining terms of the proportion, and transposing, we find

$$KB'^*_{pr} = LA'_d \,.\, W'H'_k / l'_k,$$

and with this introduced into the building up of $g^*_5 H'^3_k$ we obtain

$$g^*_5 H'^3_k = g_5 H'^3_k + TA'_d + LA'_d \,.\, W'H'_k / l'_k :$$

introduction of the explicit values of LA'_d and TA'_d then gives

$$\begin{aligned} g^*_5 H'^3_k = g_5 H'^3_k \qquad & + g_2\, (W'H'_k)^2 H'_k + 2\, g_3\, (1/l'_k)\, W'H'_k \,.\, H'^2_k \\ + g_1\, (W'H'_k)^3 / l'_k + 2\, g_2\, (W'H'_k)^2 \,.\, H'_k & + g_3\, (1/l'_k)\, W'H'_k \,.\, H'^2_k \\ & + \tfrac{1}{2} g_4 (1/l'_k)\, W'H'_k \,.\, H'^2_k \end{aligned}$$

and collection of terms, with arrangement in descending order of the original g-values and division throughout by H'^3_k, gives

$$\text{(c)} \qquad g^*_5 = g_5 + 3\, W'\, (1/l'_k)\, (g_3 + \tfrac{1}{6} g_4) + 3\, W'^2 \,.\, g_2 + W'^3\, (1/l'_k)\, g_1.$$

For use in our numerical calculations it is more convenient to express the relation in terms of the actual aberrations. We then restore the factor H'^3_k throughout and extend the terms in g_3 and g_4 by $SA_k/SA_k = l'_k u'_k / SA_k$, the term in g_2 by SA_k^2 / SA_k^2, and that in g_1 by $SA_k^2 l'_k u'_k / SA_k^3$, thus finding

$$\text{(c)*} \quad g^*_5 H'^3_k = g_5 H'^3_k + 3\, (g_3 + \tfrac{1}{6} g_4)\, u'_k H'^2_k \,.\, \frac{W'H'_k}{SA_k} + 3\, g_2 SA_k^2 H'_k \left(\frac{W'H'_k}{SA_k}\right)^2 + g_1 SA_k^2 \,.\, u'_k \left(\frac{W'H'_k}{SA_k}\right)^3.$$

(6) Longitudinal Chromatic Aberration

For this aberration we have the same argument which we applied in the case of the spherical aberration : The longitudinal chromatic aberration is a fixed difference of focus throughout the entire field and its amount is most readily found for the axial pencil ; as the latter is not affected by the location of the aperture diaphragm, the longitudinal chromatic aberration also must be unaffected by diaphragm shifts.

(7) Transverse Chromatic Aberration

As we have accepted the usual convention of treating the variation of the five true Seidel aberrations with colour as negligible when primary aberrations only are to be considered, we have only to inquire what effect a shift of the diaphragm produces when longitudinal chromatic aberration is present.

Referring to Fig. 65, the coloured principal ray from the original exit pupil will proceed from P_{pr} to some definite point B'_{pr} in the ideal image-plane, and its

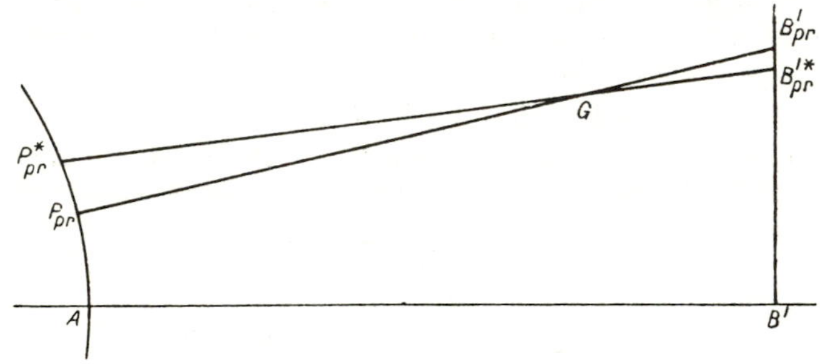

FIG. 65.

transverse chromatic aberration with reference to the corresponding principal ray in brightest light will have been calculated for the original diaphragm position. A shift of the diaphragm will cause the new coloured principal ray to emerge at P^{*}_{pr}, and if the system is not free from longitudinal chromatic aberration, this new principal ray, as one of the rays which originally started out from the given extra-axial object-point, will be affected by this longitudinal chromatic aberration and will cut the original ray at say G, so that GB'_{pr} is equal to the calculated longitudinal chromatic aberration. We see at once that this crossing of the two principal rays will produce an additional amount of transverse chromatic aberration $= B'^{*}_{pr}B'_{pr}$, and the similar triangles with a common corner at G, with the usual limiting values, give

$$B'^{*}_{pr}B'_{pr} = P_{pr}P^{*}_{pr} \times GB'_{pr}/l'_k = Lch' . W'H'_k/l'_k :$$

and if with the original diaphragm position there was already a transverse chromatic aberration Tch', the total new Tch'^{*} will be the sum $Tch' + B'^{*}_{pr}B'_{pr}$, and is found as

$$Tch'^{*} = Tch' + Lch' . W'H'_k/l'_k ,$$

or on extension of the second term by $l'_k u'_k / SA_k$, as

(d)* $$Tch'^* = Tch' + Lch' . u'_k . \frac{W'H'_k}{SA_k} .$$

To put the two chromatic aberrations into terms of g-values and of the constant W', we have only to remember that $Lch' = g_6$ and $Tch' = g_7 . H'_k$; hence we have

$$g_6^* = g_6 \quad \text{or unchanged.}$$

The equation preceding (d*) becomes

$$g_7^* H'_k = g_7 H'_k + g_6 . W' . H'_k / l'_k ,$$

or on dividing throughout by H'_k,

(d) $$g_7^* = g_7 + g_6 (1/l'_k) . W' .$$

Before collecting and discussing the equations, we must first note that they represent a perfectly exact and universally valid solution of the problem of determining the primary aberrations for a shifted position of the diaphragm ; they therefore give exactly the same result as if we had traced the new principal ray right through the system and had then recalculated the aberrations by Seidel Aberrations II. It will be obvious that the latter operation would be far more laborious than the application of the simple equations of the second fundamental law, and the value of the latter begins to reveal itself. But that is only a small part of the advantages afforded by the law. By far its greatest feature is represented by its analytical power. By Seidel Aberrations II we could only calculate for definite positions of the diaphragm, and we should have to do this for at least three positions in order to solve for that particular position which would bring the most objectionable aberration to zero or to a prescribed value ; even then we should still have to trace yet another principal ray through the position found in order to determine the corresponding value of the other aberrations. In the equations of the second fundamental law, on the other hand, the diaphragm position is represented by the single constant W' if we work in terms of g-values, or $(W'H'_k/SA_k)$ if we use the actual aberration-values ; hence these equations are simply linear (for coma and transverse chromatic), quadratic (for astigmatism), or cubic (distortion) equations in W' or $(W'H'_k/SA_k)$, and can be solved directly for that value of any one aberration which is known to be desirable. For this reason we shall find the second law a most powerful tool in the solution of actual designing problems.

The value of W' follows from the opening paragraphs, in which we found

$$P_{pr}P_{pr}^* = AP_{pr}^* - AP_{pr} = H'_k . W' = H'_k [l'_{pr}/(l' - l'_{pr}) - l'^*_{pr}/(l' - l'^*_{pr})],$$

for the last two equations give at once

$$W' = l'_{pr}/(l' - l'_{pr}) - l'^*_{pr}/(l' - l'^*_{pr}),$$

in which l' is the final intersection-length of the system, l'_{pr} the distance of the original exit pupil, and l'^*_{pr} the distance of the shifted exit pupil.

In the expressions for the second law in terms of the actual calculated aberrations, the effect of the diaphragm shift is embodied in the apparently clumsy form of $(W'H'_k/SA_k)$; but we can quickly prove that this is in reality an extremely

[60] convenient and time-saving indirect measure of the diaphragm position in the cases of analytical solutions for which the second law is primarily intended. When we merely require to know the aberrations for a shift of the exit pupil to distance l'^*_{pr}, we must calculate $W'H'_k$ as defined above, but can gain a slight simplification by noting that $H'_k/(l'_k - l'_{pr}) = -u'pr_k$. Hence the most convenient formula is

$$P_{pr}P^*_{pr} = W'H'_k = -l'pr_k u'pr_k - H'_k l'pr^*_k/(l'_k - l'pr^*_k),$$

the divisor SA_k being of course $= l'_k u'_k$. But in the analytical use of the formulae we shall find directly that value of $W'H'_k/SA_k$ which will give some favourable state of correction, and the problem then is to find the corresponding position of the diaphragm and/or pupils.

Referring to Fig. 66, the three characteristic rays of an oblique pencil are drawn thinly through a schematic lens system, showing rather excessive contractions and

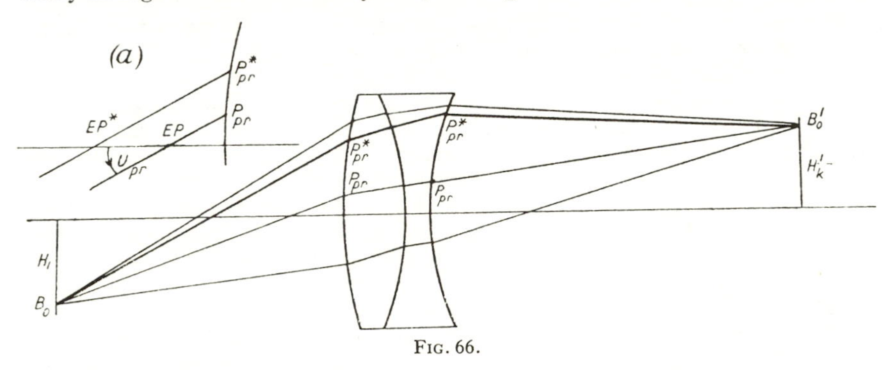

Fig. 66.

expansions of the diameter of the pencil at the successive surfaces. The thickly drawn ray is to be regarded as the new or starred principal ray resulting from a diaphragm shift. As all these rays pass, necessarily, from the object-point B_o to the final image-point B'_o, they are all rays of the oblique pencil and come under the paraxial law, already extensively employed in this chapter, that every such pencil cuts all the surfaces of a system in accurately similar configurations. It follows that the distance $P_{pr}P^*_{pr}$ at every surface is in a perfectly fixed proportion to the semi-aperture of the pencil at the same surface; in other words, the ratio $P_{pr}P^*_{pr}/SA$ is an invariant throughout the system, and as $W'H'_k$ is merely another name for $P_{pr}P^*_{pr}$ at the last surface, the peculiar ratio $W'H'_k/SA_k$ also has the invariable value just demonstrated. We will therefore now introduce the simple symbol Q for this invariant, and we then have

$$P_{pr}P^*_{pr} \text{ at any surface} = Q \, . \, (SA \text{ of the same surface}).$$

Hence we can locate the point of penetration of the new principal ray at *any* surface as soon as $Q = W'H'_k/SA_k$ has been determined for the last surface, and the whole course of the new principal ray is thus determined. This means that the latter does not need to be traced by the usual ray-tracing formulae, and that the apparently clumsy $(W'H'_k/SA_k)$ leads to a very considerable saving of labour. Further, it is clear from the original proof, in connexion with Fig. 62, of the relations between [60] $P_{pr}P^*_{pr}$ and the data of the traced rays, that these relations must hold equally for any surface, and not only for 'dashed' but also for 'plain' data. Therefore we can now generalize the above equation for $P_{pr}P^*_{pr} = W'H'_k, = Q \cdot SA_k$ with our new symbol, and write it, valid for *any* surface,

$$Q \cdot SA = -l'_{pr} \cdot u'_{pr} - H'l'^*_{pr}/(l' - l'^*_{pr}) = -l_{pr} \cdot u_{pr} - H \cdot l^*_{pr}/(l - l^*_{pr}).$$

To find the value of Q corresponding to a *given* shift of any of the pupils we put in the values of the data on the right, which will all be known, and divide the result by the SA at that surface. Q can then be used directly to find the changed aberrations by the second fundamental law.

In the far more important analytical use of the law, Q will be the directly determined quantity, and therefore known, and l^*_{pr} or l'^*_{pr} will be the quantity to be determined. Therefore the above equations must then be solved for these data. Taking plain data, a simple transposition gives

$$\frac{l^*_{pr}}{l - l^*_{pr}} = \frac{Q \cdot SA + l_{pr} \cdot u_{pr}}{-H} \therefore \frac{l^*_{pr}}{l} = \frac{Q \cdot SA + l_{pr} \cdot u_{pr}}{Q \cdot SA + l_{pr} \cdot u_{pr} - H}$$

and a final transposition and the corresponding equation in dashed quantities give the solutions

$$l^*_{pr} = l\,\frac{Q \cdot SA + l_{pr} \cdot u_{pr}}{Q \cdot SA + l_{pr} \cdot u_{pr} - H}\,; \qquad l'^*_{pr} = l'\,\frac{Q \cdot SA + l'_{pr} \cdot u'_{pr}}{Q \cdot SA + l'_{pr} \cdot u'_{pr} - H'}.$$

$l'_{pr} \cdot u'_{pr}$ is of course unconditionally $= l_{pr} \cdot u_{pr}$ and the change of symbols has therefore been merely retained for the sake of symmetry.

One case which occurs nearly always at the first surface of telescope and photographic objectives, and which may conceivably occur at some intermediate surface, calls for a special solution. It is that of a pencil of parallel rays. In such a case l and H or l' and H' would be infinite, and the general formulae would break down. The solution is easily taken from Fig. 66 (a), in which $P_{pr}P^*_{pr} = Q \cdot SA$ is positive, whilst l_{pr}, l^*_{pr}, and the uniform slope u_{pr} of all the rays are negative. Bearing these signs in mind and writing the result in both 'plain' and 'dash', we find

$$\text{For} \parallel \text{rays}: \; l^*_{pr} = l_{pr} + Q \cdot SA/u_{pr}\,; \; l'^*_{pr} = l'_{pr} + Q \cdot SA/u'_{pr}.$$

In collecting the formulae we will replace $(W'H'_k/SA_k)$ by the new symbol Q for this invariant, and we will also use the descriptive symbols LA', $Coma'_S$, &c., instead of the explicit forms in terms of g-values, SA_k and H'_k.

Seidel Aberrations IV. The Second Fundamental Law

A. In terms of the actual oblique aberrations for a definite point in the field:

The seven aberrations LA', $Coma'_S$, Ast'_S, Ptz', $Dist'$, Lch', and Tch' being known for any centred lens system by Seidel Aberrations II, their values for any shift of the diaphragm will be

LA', Ptz', and Lch' remain unchanged;

$Coma'^*_S = Coma'_S + LA' \cdot Q \cdot u'_k$;

$Ast'^*_S \;\; = Ast'_S + 2\,Coma'_S \cdot Q/u'_k + LA' \cdot Q^2$;

[60] $Dist'^* = Dist' + 3\,(Ast'_S + \frac{1}{3}\,Ptz')\,.\,Q\,.\,u'_k + 3\,Coma'_S\,.\,Q^2 + LA'\,.\,Q^3\,.\,u'_k$;

$Tch'^* = Tch' + Lch'\,.\,Q\,.\,u'_k$.

As we showed in section [55] that in most cases the most desirable correction of curvature of the field consists in producing a flat tangential field, and as by Seidel Aberrations II this calls for $X'_T = -3\,g_3 H'^2_k - \frac{1}{2}\,g_4 H'^2_k = -3\,Ast'_S - Ptz' = 0$, we can add a very useful modification of the above equation for the astigmatism by adding $\frac{1}{3}\,Ptz'$ on both sides, producing the equation

$$Ast'^*_S + \tfrac{1}{3}\,Ptz' = (Ast'_S + \tfrac{1}{3}\,Ptz') + 2\,Coma'_S\,.\,Q/u'_k + LA'\,.\,Q^2,$$

for solution of this equation with the left put equal to zero will obviously secure a flat tangential field with the shifted diaphragm.

If the problem to be solved is merely to find the changed aberrations resulting from the shift of any of the pupils from old position at l_{pr} from a given surface to new position at l^*_{pr} from the same surface, or correspondingly in 'dashed' data, then Q is found by

$$Q = [-l'_{pr}\,.\,u'_{pr} - H'\,.\,l'^*_{pr}/(l' - l'^*_{pr})]/SA \text{ or } = [-l_{pr}\,.\,u_{pr} - H\,.\,l^*_{pr}/(l - l^*_{pr})]/SA,$$

or, *only* in the case when the rays are *parallel*,

$$Q = (l'^*_{pr} - l'_{pr})\,.\,u'_{pr}/SA \text{ or } = (l^*_{pr} - l_{pr})\,.\,u_{pr}/SA \text{ for } \| \text{ rays.}$$

$SA = l\,.\,u = l'\,.\,u'$ is the semi-aperture of the surface to which the particular pupils are referred.

In analytical work we solve for Q that equation which is to lead to a desirable value of the starred aberration on its left; in the case of astigmatism there may of course be an imaginary result, indicating that the desired value is an impossible one. Otherwise the resulting value of Q can be put into the remaining equations to determine the new magnitude of the other aberrations, and the new position of any pupil can be found by:

$$l^*_{pr} = l\,\frac{Q\,.\,SA + l_{pr}\,.\,u_{pr}}{Q\,.\,SA + l_{pr}\,.\,u_{pr} - H} \quad \text{or} \quad l'^*_{pr} = l'\,\frac{Q\,.\,SA + l'_{pr}\,.\,u'_{pr}}{Q\,.\,SA + l'_{pr}\,.\,u'_{pr} - H'},$$

or when the bundle of rays is an exactly parallel one, by

$$l^*_{pr} = l_{pr} + Q\,.\,SA/u_{pr} \text{ or } l'^*_{pr} = l'_{pr} + Q\,.\,SA/u'_{pr} \text{ for } \| \text{ rays.}$$

B. In terms of the aberration *constants* g_1 to g_7, also obtainable by Seidel Aberrations II. The changed values are:

g_1, g_4, and g_6 remain unaltered.

$g^*_2 = g_2 + W'\,.\,g_1/l'_k$; $\qquad g^*_3 = g_3 + 2\,g_2\,.\,l'_k\,.\,W' + g_1\,.\,W'^2$;

$g^*_5 = g_5 + 3\,W'\,(1/l'_k)\,(g_3 + \frac{1}{6}\,g_4) + 3\,W'^2\,.\,g_2 + W'^3\,(1/l'_k)\,.\,g_1$;

$g^*_7 = g_7 + W'\,(1/l'_k)\,.\,g_6$.

In these equations the definition of W' is, *in data of the last surface only*,

$$W' = l'_{pr}/(l' - l'_{pr}) - l'^*_{pr}/(l' - l'^*_{pr}). \text{ N.B. Data of last surface.}$$

In analytical work this equation must be solved for l'^*_{pr} to find the final exit pupil corresponding to W'.

NUMERICAL EXAMPLE

It will be easier to appreciate the discussion of the second fundamental law if we apply it first to the results obtained in section [58] for an ordinary photographic landscape lens. We found that with the diaphragm in contact with the last refracting surface, with SA at the first surface $= 0{\cdot}3$, and for an extra-axial final image-point at distance $H'_k = 3{\cdot}000$ from the optical axis, the five Seidel aberrations were, in our adopted symbols,

$$LA' = 0{\cdot}16702\,;\ Coma'_S = -0{\cdot}00992\,;\ Ast'_S = 0{\cdot}4363\,;\ Ptz' = 0{\cdot}3100\,;$$
$$Dist' = 0{\cdot}01324.$$

The two chromatic aberrations were so small that we need not include them in the present example. The angle u'_k under which the traced paraxial ray arrives at the final focus was $= 0{\cdot}03142$. As we recognized the astigmatism and curvature of the field as by far the most serious aberrations of the lens, we will begin by determining the value of the constant Q which will render the tangential field flat, by solving the equation

$$Ast'^*_S + \tfrac{1}{3}\,Ptz' = (Ast'_S + \tfrac{1}{3}\,Ptz') + 2\,Coma'_S\,.\,Q/u'_k + LA'\,.\,Q^2,$$

for Q when the left side $= -\tfrac{1}{3}\,X'^*_T$ is put equal to zero.

The solution is $$Q = -\frac{Coma'_S}{LA'\,.\,u'_k} \pm \sqrt{\left(\frac{Coma'_S}{LA'\,.\,u'_k}\right)^2 - \frac{Ast'_S + \frac{1}{3}\,Ptz'}{LA'}}\,;$$

worked out with the numerical values collected above, it gives $Q = 1{\cdot}8902 \pm 0{\cdot}5848$ $\therefore Q$ either $= 1{\cdot}3054$ or $= 2{\cdot}4750$. By putting these two values of Q, which will secure a flat tangential field, into the equations for $Coma'^*_S$ and $Dist'^*$, we can find the values which these aberrations will then have, and obtain the results :

$$\text{for } Q = 1{\cdot}3054:\ Coma'^*_S = -0{\cdot}00307 \quad Dist'^* = +0{\cdot}04060$$
$$\text{for } Q = 2{\cdot}4750:\ Coma'^*_S = +0{\cdot}00307 \quad Dist'^* = +0{\cdot}03640.$$

The Coma also is greatly diminished in magnitude, but the positive or barrel distortion is increased, especially for the smaller value of Q. From the purely optical point of view the second solution, $Q = 2{\cdot}4750$, would therefore be the better one. The simplest solution is that for the value of Q which secures freedom from Coma, for putting $Coma'^*_S = 0$, we obtain the single solution by

$$0 = Coma'_S + LA'\,.\,Q\,.\,u'_k:\ Q = -Coma'_S/LA'\,.\,u'_k,$$

which with our numerical values gives

$$Q = 0{\cdot}00992/(0{\cdot}16702 \times 0{\cdot}03142) = 1{\cdot}8902,$$

exactly midway between the two values which give a flat tangential field. We can then calculate $-\tfrac{1}{3}$ of the tangential curvature of the field for $Q = 1{\cdot}8902$ as

$$Ast'^*_S + \tfrac{1}{3}\,Ptz' = (Ast'_S + \tfrac{1}{3}\,Ptz') + 2\,Coma'_S\,.\,Q/u'_k + LA'\,.\,Q^2 = -0{\cdot}0570\,;$$

whence $X'^*_T = +0{\cdot}1710$. Hence our lens will have a 'hollow' tangential field if we render it quite free from Coma by the diaphragm position corresponding to $Q = 1{\cdot}8902$. By the $Dist'^*$ equation we find for this value of Q : $Dist'^* = +0{\cdot}03850$.

[60] It is extremely instructive to plot these results so as to obtain a continuous picture of all the possibilities. *Coma'** is a linear function of Q and therefore plots as an inclined straight line, really requiring only two points to determine it. *Ast'** is a quadratic function of Q; hence it plots as a parabola with vertical axis, and requires three known points, just the number which we have determined, and as we have them for three values of Q at equal intervals, we can plot the curve by the convenient 'dip'-method of Chapter V. But *Dist'** is a cubic function of Q, which requires four known points to enable us to draw the true curve. Consequently we must calculate *Dist'** for a fourth value of Q, and in order to be able to use the dip method we will choose another step of the same magnitude $= 0{\cdot}5848$, and calculate for $Q = 1{\cdot}3054 - 0{\cdot}5848 = 0{\cdot}7206$, finding

$$\text{for } Q = 0{\cdot}7206 \qquad Dist'^{*} = 0{\cdot}03640.$$

The astigmatism will be most instructively plotted as X'^{*}_{T} and may be significantly supplemented by calculating also X'^{*}_{S} by the first fundamental law as

$$X'^{*}_{S} = \tfrac{2}{3}\, X'_{Ptz} + \tfrac{1}{3}\, X'^{*}_{T};$$

and as X'_{Ptz} is merely another name for $-Ptz' = -0{\cdot}3100$, we have

$$X'^{*}_{S} = -0{\cdot}2067 + \tfrac{1}{3}\, X'^{*}_{T}.$$

The values to be plotted are therefore:

for Q	=	0·7206	1·3054	1·8902	2·4750
$Coma'^{*}_{S}$	=		−0·00307	0	+0·00307
X'^{*}_{T}	=		0	+0·1710	0
X'^{*}_{S}	=		−0·2067	−0·1497	−0·2067
$Dist'^{*}$	=	0·03640	0·04060	0·03850	0·03640

As the transverse aberrations Coma and Distortion run in very much smaller values than the longitudinal X' values, it is advisable to adopt a much larger scale for the former, say 20 × or 50 ×. The graph, Fig. 67, shows at once that it would be hopeless to try to render our lens free from distortion by the choice of the proper value of Q, for we see that, although the cubic parabola will eventually cut the zero axis far to the left, all the other aberrations would have huge values at that point, so that the image-points would be extremely diffused. From the point of view of astigmatic correction we concluded in section [55] that the only reasonably admissible range was from the stage where the S and T foci coincide (= the intersecting points of the X'^{*}_{T} and X'^{*}_{S} parabolas) to the point where the S and T foci lie at equal distances on opposite sides of the focal plane, which in our graph means the positions where the X'^{*}_{T} curve is as far above the zero axis as the X'^{*}_{S} curve is below it. We thus find that a good solution can only be obtained from the two strips enclosed between pairs of vertical parallel lines. All other values of Q give unsatisfactory astigmatic correction. We shall learn later that the coma shown for these spaces is not really serious. We have pointed out already that the right-hand solutions—for large Q—would be *optically* preferable on account of the lower distortion. But as $Q = P_{pr}P^{*}_{pr}/SA$ or $P_{pr}P^{*}_{pr} = Q\,.\,SA$, and as our original principal ray emerged from the pole of the last surface whilst [60] in our case $SA_k = l'_k \cdot u'_k = 0{\cdot}309''$, the right-hand solutions with Q about $= 2{\cdot}5$ would cause the new principal ray to emerge at $2{\cdot}5 \times 0{\cdot}309 = 0{\cdot}77''$ above the

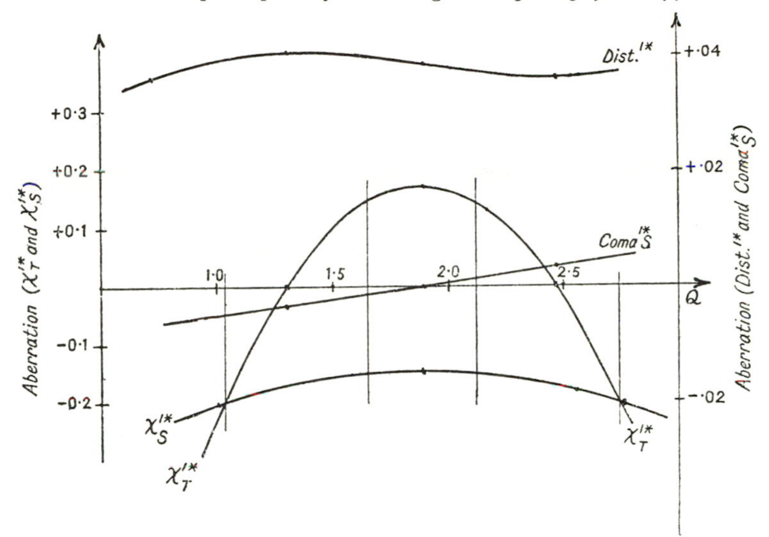

FIG. 67.

optical axis, and another $0{\cdot}309''$ would be required to pass also the marginal rays of the oblique pencil ; hence the lens would require $1{\cdot}08''$ semi-aperture and would be large and correspondingly expensive. For the left-hand solution we have $Q =$ about $1{\cdot}3$, or $P_{pr}P^*_{pr} = 1{\cdot}3 \times 0{\cdot}309'' = 0{\cdot}40''$, and adding $0{\cdot}309''$ to admit the whole pencil, we arrive at a clear semi-aperture of only $0{\cdot}709''$, or a very much smaller and cheaper lens. For that reason the left-hand solution would be selected in practice. We will therefore complete the example by calculating the positions of the pupils for the left-hand case of a flat tangential field, or for $Q = 1{\cdot}3054$. In accordance with Seidel Aberrations IV A we have

Final exit pupil at $l'^*_{pr} = l' \dfrac{Q \cdot SA + l'_{pr}u'_{pr}}{Q \cdot SA + l'_{pr}u'_{pr} - H'}$ with data of last surface ;

Initial entrance pupil, || rays, at $l^*_{pr} = l_{pr} + Q \cdot SA/u_{pr}$ with data of first surface.

Section [58] gives the necessary data, namely $SA_3 = l'_3u'_3 = [9{\cdot}4900)$; SA_1 as assumed $= 0{\cdot}3000$; H'_3 as assumed $= 3{\cdot}000$; $l_{pr1} = +0{\cdot}249$; $u_{pr1} = -0{\cdot}31427$, and $l'_3 = 9{\cdot}8367$. As $l'_{pr\,3}$ was assumed at zero, the term $l'_{pr\,3} \cdot u'_{pr\,3}$ is also zero ; that is another advantage of tracing the original principal ray through the pole of the last surface. We therefore have

$$l'^*_{pr} = 9{\cdot}8367 \frac{1{\cdot}3054 \times [9{\cdot}4900]}{1{\cdot}3054 \times [9{\cdot}4900] - 3{\cdot}000} = -1{\cdot}528,$$

$$l^*_{pr} = 0{\cdot}249 + 1{\cdot}3054 \times 0{\cdot}3/(-0{\cdot}31427) = -0{\cdot}997.$$

[60] The final exit pupil is virtual, but the initial entrance pupil is real; hence the actual diaphragm of 0·6 aperture must be placed at 0·997 to the left (or in front) of the first surface of the lens. It will be an instructive exercise to calculate through two other *bendings* of the lens, say by 0·04 of curvature in either sense, and to study these in the same way as shown above and in section [58].

Discussion of the Second Law

The law enables us to calculate the exact primary aberrations of any centred system of spherical surfaces for *any conceivable* position of the diaphragm when these aberrations are known for *one* position of the diaphragm; by solving its equations for the eccentricity constant Q or W' we can use it analytically to determine directly, without repeated trials, that position of the diaphragm which will bring any one of the aberrations subject to the law to a desirable value, provided of course that this value is capable of being attained by shifting the diaphragm.

The equations expressing the law demonstrate that the extremely close relationship between the five Seidel aberrations, which we previously found existing at any one surface and which gave us the surprisingly simple method of calculating them by Seidel Aberrations II, persists for the summed-up totals at the final focus at which the five aberrations might be described as 'quick-change-artists', seeing that the distortion, to take the most striking instance, becomes, with a shift of the diaphragm, made up of the original distortion plus the original astigmatism and Petzval curvature with a simple factor, plus the original coma with another simple factor, plus a final term arising from the original spherical aberration. The equations also justify the order into which we have always placed the five aberrations, for the changes of any one aberration invariably depend only on the preceding ones, never on a following one. Seidel adopted this order because for the telescope objective, in which he was chiefly interested, it is also the order of importance of the aberrations. From that point of view, the order would become almost completely reversed in the case of systems to be used for a large field of view. The second law thus supplies a more valid reason for the adopted sequence.

The factors or divisors u'_k or l'_k, which appear in many terms of the equations, have a very simple significance: they are necessary because we measure spherical aberration, astigmatism, and the Petzval curvature as longitudinal aberrations, coma and distortion as transverse aberrations. Now we found in section [34] that for spherical aberration $TA'_p = LA'_p \,.\, u' = LA'_p \,.\, y/l'$, and it will be seen that u' or l' always appear in accordance with these relations.

We can at present draw only a few of the simpler conclusions from the second law, reserving many others for later, and largely for Part II. In drawing these conclusions we must bear in mind that $Q = P_{pr}P^*_{pr}/SA$ is necessarily a finite quantity, for as our systems must have a certain definite clear aperture, SA is necessarily finite. Then Q could become zero only if $P_{pr}P^*_{pr}$ were zero, and as that would mean identity of the new principal ray with the original one, there would be nothing to discuss. Q could become ∞ only if $P_{pr}P^*_{pr}$ were ∞, and that is absurd, for as $P_{pr}P^*_{pr}$ is the separation between the new and the original principal rays on the lens surfaces (or more strictly on the normal planes substituted for them in our primary approximation), an infinitely large lens would be implied. Remembering this, we can conclude : [60]

(1) If the system is a perfectly corrected one so that all seven aberrations are zero, then all the terms on the right-hand sides of the equations will be zero for any possible value of Q, hence the new aberrations are also zero. This conclusion may look like an insult to the intelligence of the reader, but its far-reaching consequences render it extremely important, for it means that if we are working out a new design and know that only a perfectly corrected solution will be acceptable, then we may place the diaphragm wherever we like in the primary aberration work, and that liberty can be used with startling effect to shorten the preliminary analytical solution.

(2) *With regard to coma.* If the lens system is spherically corrected ($LA' = 0$), the coma retains the same value no matter where the diaphragm may be placed. Hence the coma of spherically corrected systems is incurable by diaphragm shifts. If there is spherical aberration, then the coma-equation, being linear in Q, gives one and only one real solution for Q which will produce any prescribed value—nearly always zero—of the coma. Hence systems with spherical aberration can be corrected for coma by shifting the diaphragm, which however may assume an inconvenient or even impracticable position. If Q is large, a remarkably small amount of spherical aberration may suffice. This case arises in achromatic eyepieces.

(3) *With regard to astigmatism.* We will base the discussion on the alternative equation

$$Ast'^{*}_{S} + \tfrac{1}{3}\,Ptz' = (Ast'_{S} + \tfrac{1}{3}\,Ptz') + 2\,Coma'_{S}\,.\,Q/u'_{k} + LA'\,.\,Q^{2},$$

because the left of this equation represents $-\frac{1}{3}\,X'^{*}_{T}$, which it is nearly always desirable to reduce to zero. If coma and spherical aberration are both zero, that is, if the system is *aplanatic* in Abbe's sense of the adjective, diaphragm shifts leave the astigmatism entirely unchanged and render this usually powerful means of modification ineffective. If only LA' is zero (spherically corrected system), but coma finite, then the latter will be independent of Q and therefore incurable, but the astigmatic equation will be a linear one in Q and can be solved for that value of Q which will yield zero-value of X'^{*}_{T}. For a system with spherical aberration but free from coma the astigmatic equation becomes a pure quadratic in Q, which however may have imaginary roots ; in any case the diaphragm shift will reintroduce coma by (2). In the most general case, when with the original diaphragm position LA' and $Coma'$ are both finite—as in our numerical example—the resulting quadratic equation in Q will have two numerically different roots, which, if real, yield two different diaphragm positions with which the system gives a flat tangential field.

(4) *With reference to distortion*, it is interesting to note that if the system has spherical aberration, the equation for $Dist'^{*}$ will be cubic, and must have either one (the usual case) or three (unusual) real roots. A system with spherical aberration therefore can always be freed from primary distortion by a mere shift of the diaphragm, but the position is in most cases either inconvenient or such as leads to prohibitive values of other aberrations. There is again certainty of a real solution

[60] in the case of aplanatic systems (LA' and *Coma'* both zero), for *Dist'** is then given by a linear equation in Q, *provided the astigmatic term on the right is finite.* The solution has a strong tendency to be unfavourable in other respects. If only LA' is zero, the *Dist'** equation becomes quadratic in Q and gives either two, or no real, solutions.

We obtain additional results of very great interest and importance if we search for maximum or minimum values. As the special cases when any of the original aberrations are zero are easily singled out, we will discuss this aspect on the assumption that all the aberrations have finite values. As the spherical aberration is unchangeable and the coma is a linear equation in Q, the question of maxima or minima only arises for astigmatism and distortion.

(5) *Maximum or minimum of astigmatism.* Differentiating

$$Ast'^{*}_{S} + \tfrac{1}{3}\,Ptz' = (Ast'_{S} + \tfrac{1}{3}\,Ptz') + 2\,Coma'_{S}\,.\,Q/u'_{k} + LA'\,.\,Q^{2}$$

with reference to Q, we find

$$d\,(Ast'^{*}_{S} + \tfrac{1}{3}\,Ptz')/dQ = 2\,Coma'_{S}/u'_{k} + 2\,Q\,.\,LA'\ldots = 2\,Coma'^{*}_{S}/u'_{k},$$

the last by comparison with the equations Seidel Aberrations IV. As a maximum or minimum corresponds to zero value of the first differential coefficient, we have a first remarkable result that the astigmatism of a given system with given conjugates is a maximum or minimum for that value of Q, or that position of the diaphragm, which renders the new *Coma'** equal to zero. To decide between maximum and minimum we form the second differential coefficient

$$d^{2}\,(Ast'^{*}_{S} + \tfrac{1}{3}\,Ptz')/(dQ)^{2} = 2\,LA',$$

which therefore has the sign of LA'. Hence the astigmatism attains a minimum if LA' is positive and a maximum if LA' is negative. In our numerical example LA' was positive. The graph at first sight appears to contradict the conclusion just drawn that the astigmatism should attain a minimum at the value of Q for which *Coma'** is zero, for our drawn parabola has its *highest* point there ; but we have only to remember that the drawn curve represents $X'^{*}_{T} = -3\,Ast'^{*}_{S} - Ptz'$ to see that this reversal of the sign inverts the parabola and that the graph therefore confirms our deduction. *This conclusion is highly important.* Many simple types of systems, especially the ordinary Huygenian and Ramsden eyepieces, suffer from a considerable amount of positive or under-corrected spherical aberration, and at the same time it is not easy to flatten their field, so that there is little risk of reaching an undesirably hollow field. Our conclusion then means that if we so fix the proportions of systems of these simple types as to secure zero-coma, we have not only got rid of *one* highly objectionable aberration but have simultaneously and *automatically* corrected their curvature of the field of view to the utmost possible extent. It will be seen subsequently that the design of ordinary eyepieces is rendered very simple by this remarkable result of the interdependence of the Seidel aberrations.

(6) *Maximum and minimum of distortion.* The first differential coefficient of *Dist'** is easily seen to be $= 3\,(Ast^{*\prime}_{S} + \tfrac{1}{3}\,Ptz')\,.\,u'_{k} = -X'^{*}_{T}\,.\,u'_{k}$; hence maximum and minimum of distortion coincide with the two values of Q which secure a flat tangential field. The second differential coefficient is found $= 6\,.\,Coma'^{*}_{S}$, and

therefore has the sign of *Coma'**. Hence the maximum of distortion occurs for the [60] value of Q which makes X'^{*}_{T} zero and the *Coma'** negative, the minimum at the value for which the *Coma'** is positive. The curve in the graphical representation of our numerical example confirms these conclusions. It also recalls an earlier reminder in these pages that when a function contains the variable in higher powers than the square, the mathematical conception of maxima and minima clashes with the colloquial meaning of highest and lowest *possible* values. Our cubic parabola for distortion goes to negative infinity on the left and to positive infinity on the right. The two positions of maximum and minimum deserve these names only with reference to the closely neighbouring points of the curve. As the second differential coefficient becomes zero with *Coma'**, it follows that the point of inflexion of the distortion-curve lies there.

III. The Third Fundamental Law

As was stated in the opening paragraph of this section, the third law refers to the remarkable special properties of a *thin* lens or system when the oblique pencils pass *centrally* through it. In such a lens or thin system distortion and transverse chromatic aberration are zero, astigmatism has the *fixed* value $H'^2_k/2f'$, so that coma and spherical aberration, together with the longitudinal chromatic aberration in some cases, are the only aberrations which require calculation by the G-sums, being thus quickly determined for all possible bendings of the lens or system. Hence we secure simultaneously a great reduction in the computing work and a vast extension of the analytical possibilities when we can adopt the thin lens simplification.

We will record these valuable facts in this form :

Third fundamental law :

> For a *thin* lens or system passed centrally by the oblique pencils, distortion and transverse chromatic aberration are zero, the astigmatism has the fixed and *very large* value $H'^2_k/2f'$, spherical aberration and coma can be readily calculated for all possible bendings by the G-sums.

The numerical example in subsection II refers to a system which may be regarded as reasonably thin, and the graph derived from its discussion by the second fundamental law is typical for all thin positive lenses when they are required to have a reasonably flat field and to be also sufficiently free from coma.

The decisive fact in the case of thin systems is their huge astigmatism when the diaphragm is in contact with the system. If the latter is to give tolerable definition over a field exceeding the few degrees of angular extent which suffice for telescopes, this astigmatism must be reduced by a shift of the diaphragm, and if the field is to be rendered fairly flat, the astigmatism must not merely be reduced to zero but must be turned into a moderate amount of astigmatism of the opposite sign so as to bring to zero the left of our equation

$$(Ast'^{*}_{S} + \tfrac{1}{3}\,Ptz') = (Ast'_{S} + \tfrac{1}{3}\,Ptz') + 2\,Coma'_{S}\,.\,Q/u'_{k} + LA'\,.\,Q^2,$$

the right side of which contains the huge initial astigmatism of the system with the

[60] original central passage of the oblique pencils, further increased in practically all cases by the item $\frac{1}{3}$ Ptz', which with rare exceptions has the same sign as Ast'_S. As a diaphragm shift corresponds to a finite value of Q, the last two terms show that the right side of the complete equation can be brought to zero only if either $Coma'_S$ or LA' is finite. That means that it is impossible to secure a flat field with an aplanatic thin lens. This is a first highly important and singularly little-known conclusion.

We obtain additional enlightenment most readily if we remember that coma also is a highly objectionable aberration; but the coma which interests us is that with the *shifted* diaphragm, or

$$Coma'^{*}_S = Coma'_S + LA' . Q . u'_k,$$

and we therefore transform the astigmatic equation by introducing into it

$$Coma'_S = Coma'^{*}_S - LA' . Q . u'_k,$$

when the equation becomes

$$(Ast'^{*}_S + \tfrac{1}{3}\, Ptz) = (Ast'_S + \tfrac{1}{3}\, Ptz') + 2\, Coma'^{*}_S . Q/u'_k - LA' . Q^2.$$

We must also watch the distortion, for although a moderate amount of this aberration is very usually tolerated in simple systems such as we are considering, there are limits to this tolerance. For the thin systems the first term in the distortion equation, $Dist'$, is zero. By adding *and* subtracting suitable multiples of $Coma'_S . Q^2$ and of $LA' . Q^3 . u'_k$ we can put the equation into the form

$$\begin{aligned} Dist'^{*} = 3\,(Ast'_S + \tfrac{1}{3}\, Ptz') . Q . u'_k + 6\, Coma'_S . Q^2 &+ 3\, LA' . Q^3 . u'_k \\ - 3\, Coma'_S . Q^2 &- 3\, LA' . Q^3 . u'_k \\ &+ LA' . Q^3 . u'_k \end{aligned}$$

without altering the value of the right-hand side, as the vertical columns add up to the items of the standard equation. But the first horizontal line now represents $3\,(Ast'^{*}_S + \frac{1}{3}\, Ptz) . Q . u'_k$, which must be zero for a flat tangential field, and the second line represents $-3\, Coma'^{*}_S . Q^2$. Hence the transformed equation is

$$Dist'^{*} = 3\,(Ast'^{*}_S + \tfrac{1}{3}\, Ptz')\, Q . u'_k - 3\, Coma'^{*}_S . Q^2 + LA' . Q^3 . u'_k,$$

which we can discuss in conjunction with the astigmatic one

$$(Ast'^{*}_S + \tfrac{1}{3}\, Ptz') = (Ast'_S + \tfrac{1}{3}\, Ptz') + 2\, Coma'^{*}_S . Q/u'_k - LA' Q^2.$$

In these two equations $(Ast'_S + \frac{1}{3}\, Ptz')$ is known and fixed for a *thin* system as soon as its combined focal length and the total curvature and refractive indices of its components have been fixed; but there remain four quantities, LA', $Coma'^{*}_S$, $(Ast'^{*}_S + \frac{1}{3}\, Ptz')$, and $Dist'^{*}$, which will vary greatly according to the adopted bending of the thin system, because bending changes both spherical aberration and $Coma'_S$ and through them the starred aberrations resulting from diaphragm shifts. We may therefore assume for the present that we can produce any value of any one of the four variable quantities which is likely to prove of interest simply by suitable bendings. But as we have four variables and only two equations, we must assign definite desirable values to two of the variables in order to be able to solve for the other two. We will stipulate as one fixed condition that the solution must yield a flat tangential field, that is $(Ast'^*_S + \frac{1}{3}\,Ptz')$ shall be zero. This does not [60] restrict the discussion to any serious extent, because only small values of residual curvature of field can be tolerated, and these correspond to a very small variation in the value of $(Ast'^*_S + \frac{1}{3}\,Ptz')$. By adding a second demand that one or other of the remaining three variables shall also be zero, we obtain the following results :

(1) $Coma'^*_S$ shall also be brought to zero. As coma is a highly objectionable aberration, this is a good choice. The equations will then be

$$Dist'^* = LA'\,.\,Q^3\,.\,u'_k;\quad 0 = (Ast'_S + \tfrac{1}{3}\,Ptz') - LA'Q^2.$$

The second gives $LA' = (Ast'_S + \frac{1}{3}\,Ptz')/Q^2$, and as for the usual positive system the numerator is positive, we learn that flatness of field *and* freedom from coma are only obtainable if we choose a bending with what usually proves to be a very considerable amount of spherical under-correction. If we now put the required value of LA' into the $Dist'^*$ equation, we find

$$Dist'^* = (Ast'_S + \tfrac{1}{3}\,Ptz')\,Q\,.\,u'_k,$$

and as Q is positive for a system with the usual position of the diaphragm in front, we see that another inevitable result of simultaneous correction for flat field and zero coma is a usually very pronounced barrel distortion. The landscape lens of our numerical example comes fairly close to this type.

(2) In addition to having a flat tangential field, the thin system shall also be spherically corrected. The equations will now be

$$Dist'^* = -3\,Coma'^*_S\,.\,Q^2;\quad 0 = (Ast'_S + \tfrac{1}{3}\,Ptz') + 2\,Coma'^*_S\,.\,Q/u'_k.$$

The second gives $Coma'^*_S = -(Ast'_S + \frac{1}{3}\,Ptz')\,.\,u'_k/2\,Q$, showing that this correction demands negative coma in the finished system. Putting this value of $Coma'^*$ into the first equation gives

$$Dist'^* = 1\tfrac{1}{2}\,(Ast'_S + \tfrac{1}{3}\,Ptz')\,.\,Q\,.\,u_k,$$

again barrel-distortion, which with the same values of Q and u'_k as in the first case would be $1\frac{1}{2}$ times as large. But the zero spherical aberration almost invariably calls for a higher value of Q than in the first case, and whilst this diminishes $Coma'^*_S$ owing to Q being in the denominator of its value, it further increases the distortion. That is the reason why a spherically corrected *thin* landscape lens has never yet been successfully produced.

(3) In addition to having a flat tangential field, the system shall be free from distortion. The equations will be

$$0 = -3\,Coma'^*_S\,.\,Q^2 + LA'\,.\,Q^3\,.\,u'_k;\quad 0 = (Ast'_S + \tfrac{1}{3}\,Ptz') + 2\,Coma'^*_S\,.\,Q/u'_k - LA'\,.\,Q^2;$$

with the solution

$$Coma'^*_S = (Ast'_S + \tfrac{1}{3}\,Ptz')u'_k/Q;\quad LA' = 3\,(Ast'_S + \tfrac{1}{3}\,Ptz')/Q^2.$$

The coma of the finished system requires to be positive and of twice the amount found for case (2). The longitudinal spherical aberration must also be positive and three times as large as in case (1). For landscape lenses this case is a hopeless one. But 'thin' achromatic eyepieces, which are very desirable for astronomical

[60] telescopes, come also under this type; in their case Q is very large, and as the solutions for $Coma'^{*}_{S}$ and for LA' have Q and Q^2 respectively in the denominator, this case becomes entirely practicable and indeed the ideal to be aimed at for eyepieces of the kind referred to. It should be remembered that we found in the discussion of our landscape lens that positive coma diminished the distortion; but we rejected the solution on account of the large value of Q which would have demanded an undesirably large lens. In eyepieces that objection does not arise.

Whilst the algebraical solutions given above are perfectly general for all thin lenses and systems, the discussion has been based on the assumption that the thin system has positive focal length and that the diaphragm finds its place in front of the lens, which means that Q is positive. Negative thin systems are hardly ever used by themselves, and we need not discuss them from the present point of view. But there are interesting cases which lead to a negative value of Q, or to 'diaphragm behind lens' in the case of positive systems. The general result is that the signs of the coma and of the distortion associated with the several types of correction become reversed. As an example of this kind and also of a purely thin lens solution we will take the Dennis Taylor objective of section [49], having a focal length of 10 and a semi-aperture of 0·64, and giving spherical correction—by the TL solution—with $c_2 = -0{\cdot}188$ and $c_3 = -0{\cdot}2006$. In section [59] we determined the coma of all the objectives in section [49] by the equation, for $H'_k = 1$,

$$\text{Total } Coma'_S = 0{\cdot}08582\, c_2 - 0{\cdot}05090\, c_3 + 0{\cdot}01097.$$

Merely introducing the values of c_2 and c_3 for the Dennis Taylor objective, we find its considerable positive coma as

$$Coma'_S = 0{\cdot}00504 \text{ for } H'_k = 1.$$

The presence of this coma makes it possible to find a position of a diaphragm which will yield a flat tangential field (or any other desired astigmatic correction). We have only to solve for Q the equation

$$(Ast'^{*}_{S} + \tfrac{1}{3}\, Ptz') = (Ast'_S + \tfrac{1}{3}\, Ptz') + 2\, Coma'_S \,.\, Q/u'_k + LA' \,.\, Q^2,$$

putting the left = the desired zero value, and as the spherical aberration is corrected, the last term on the right is also zero.

Hence
$$Q = -(Ast'_S + \tfrac{1}{3}\, Ptz') \,.\, u'_k/2\, Coma'_S.$$

As we use paraxial data in all TL work, we have $u'_k = SA/f' = 0{\cdot}64/10 = 0{\cdot}064$. Ast'_S is, by the third law, $= H'^2_k/2f'$, or for our $H'_k = 1$: $Ast'_S = 0{\cdot}05000$. Ptz' represents $\tfrac{1}{2}\, g_4 H'^2_k$; with g_4 as determined in section [55] = 0·0702 and $H'_k = 1$, we find $Ptz' = 0{\cdot}0351$; hence

$$Q = -0{\cdot}0617 \times 0{\cdot}064/2 \times 0{\cdot}00504 = -0{\cdot}3917.$$

We can then calculate the resulting distortion from

$$Dist'^{*} = 3\,(Ast'^{*}_{S} + \tfrac{1}{3}\, Ptz')\, Q \,.\, u'_k - 3\, Coma'^{*}_{S} \,.\, Q^2 + LA' Q^3 \,.\, u'_k,$$

in which the first and last terms are zero, on account of the flat tangential field already secured and because the spherical aberration is corrected, whilst for the latter reason $Coma'^*_S = Coma'_S = 0{\cdot}00504$. Hence $Dist'^* = -3 \times 0{\cdot}00504 \times (0{\cdot}3917)^2 = -0{\cdot}00232$. [60]

The value found appears small compared with the 0·04060 found for our landscape lens. But we must remember that whilst the latter had nearly the same focal length as the Dennis Taylor lens, it was calculated for $H'_k = 3$; as distortion grows with the cube of H'_k, the Dennis Taylor lens at $H'_k = 3$ would have

$$Dist'^* = -0{\cdot}00232 \times 3^3 = -0{\cdot}0627,$$

thus confirming the statement that the distortion in a spherically corrected thin system with a flat tangential field is of very serious magnitude.

To find the position of the diaphragm which will realize the flat tangential field, we now put the value of Q into the equation of Seidel Aberrations IV for l'^*_{pr}, and as the original principal ray passed centrally, l'_{pr} is zero ; hence

$$l'^*_{pr} = l' . Q . SA/(Q . SA - H')$$

$$= 10 \times -0{\cdot}3917 \times 0{\cdot}64/(-0{\cdot}3917 \times 0{\cdot}64 - 1) = +2{\cdot}004.$$

The diaphragm therefore must be fixed at 2·004 distance behind the lens. In old terrestrial telescopes, consisting of an object-glass with positive coma and the usual four-lens terrestrial eyepiece, one occasionally finds a diaphragm so close to the position yielding a flat field as to suggest that it was deliberately placed there for that purpose.

SYSTEMS OF SEPARATED THIN COMPONENTS

The second fundamental law enables us now to build up a system of equations by which we can calculate the combined effect of any number of separated thin lenses or components. [61]

The individual focal lengths and the separations of the thin components having been decided upon, we begin by determining the paraxial l and l' for all of them by using $1/l' = 1/f' + 1/l$ successively from the given object-distance through the entire system. As the method will most usually be applied to systems of thin *compound* lenses, we will reserve the usual small-letter suffixes for the ultimate thin simple lenses, and will introduce capital-letter suffixes for the data of the thin components of our complete system. The initial paraxial calculation therefore begins with l_A, the original object-distance, and gives successively l'_A, then $l_B = l'_A - d'_A$, and so right through to the final l'_K. We must then select a suitable semi-aperture for one of the components, say the first, calling it in that case SA_A ; then we obtain all other semi-apertures by the usual ratios : $SA_B = SA_A . l_B/l'_A$, $SA_C = SA_B . l_C/l'_B$, and so right through. The construction of the individual components of the system has then to be settled ; usually this will merely call for application of the formulae for thin lens achromatism for such of the components as require building up of two or more thin lenses. We then have all the data which are necessary to apply Seidel Aberrations III to the individual components, thus finding the contributions which each of them *would* make to the final aberrations *if* the diaphragm were placed in contact with it. But as the principal ray of an oblique pencil will as a rule cross the optical axis in only one point, Seidel Aberrations III

[61] will give the *correct* contributions only for a component which might happen to be located at that exact point. Clearly we have arrived at the stage of the work when we must call in the powerful aid of the second fundamental law. Referring to Fig. 68, we will first assume that it represents the case for which we deduced the contributions of a thin component somewhere within a complete system, but in

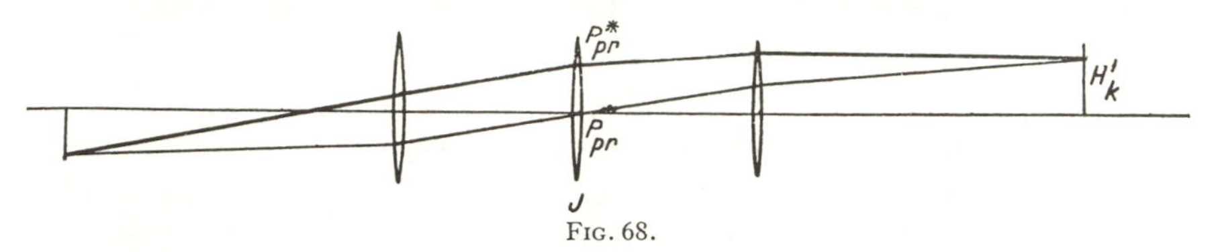

Fig. 68.

contact with its diaphragm. Let suffix J mark this component. The aberrations at the final focus will be the sum of the contributions of all the components, or

$$\text{Final } LA' = SC'_A + SC'_B + \ldots + SC'_J + \ldots \,; \quad \text{Final } Coma'_S = CC'_A + CC'_B + \ldots + CC'_J + \ldots$$

and similarly for the other final aberrations. If we then shift the diaphragm by an amount corresponding to an eccentricity invariant Q, the second law will apply; for the astigmatism, as an example, it will give

$$\text{Final } Ast'^{*}_S = (AC'_A + AC'_B + \ldots + AC' + \ldots) + 2\,(CC'_A + CC'_B + \ldots + CC'_J + \ldots)\, Q/u'_k + (SC'_A + SC'_B + \ldots + SC' + \ldots)\, Q^2.$$

Merely multiplying out and rearranging the terms then leads to:

$$\text{Final } Ast'^{*}_S = \begin{cases} AC'_A + 2\,CC'_A \,.\, Q/u'_K + SC'_A \,.\, Q^2 \\ + AC_B + 2\,CC'_B \,.\, Q/u'_K + SC'_B \,.\, Q^2 \\ \ldots\ldots\ldots\ldots\ldots\ldots\ldots\ldots \\ \ldots\ldots\ldots\ldots\ldots\ldots\ldots\ldots \\ + AC'_J + 2\,CC'_J \,.\, Q/u'_K + SC'_J \,.\, Q^2 \,; \\ \ldots\ldots\ldots\ldots\ldots\ldots\ldots\ldots \\ \ldots\ldots\ldots\ldots\ldots\ldots\ldots\ldots \end{cases}$$

and as each horizontal line in this arrangement shows precisely the same application of the second law as in the previous equation, but now to the contributions of each surface or component, we can conclude that it is perfectly legitimate to apply the second law separately to the contributions of each surface or component and then to add up for the complete system, and this necessarily is true for any possible value of Q. But for a new principal ray traced straight through a system, Q is an invariant; it has the same value at each surface; hence we may use the Q found for each surface or component and apply it to the contributions of that part without changing the final result in the least. As this again holds for any possible value of Q at the individual surface or component, we have the final result that whatever value of Q may result for a surface or component from a shift of the diaphragm, the contributions of that part to the final aberrations must be in strict accordance with the second fundamental law worked out with the value of Q for the part of the system under consideration. [61]

For our system of thin separated components we know the contributions which each component *would* make if the principal ray passed centrally through it. If we now trace a principal ray through the actual diaphragm position, this ray will pass most or all of the components eccentrically and will lead to a definite positive or negative value of Q at each component. On applying the second law to the original contributions of each component with its own value of Q, we shall then find the true contributions, and the problem is solved.

When the final diaphragm position of the system is not given or capable of being estimated, we may make a labour-saving choice by assuming it in contact with one of the thin components. We then apply the thin lens formula in the form $1/l'_{pr} = 1/f' + 1/l_{pr}$ from the selected diaphragm position, if necessary to right and left, thus locating all the pupils by the values of l_{pr} and l'_{pr} for each component. We must then determine the point P^*_{pr} of each component at which the true principal ray cuts it, and as H'_K has been used in working out the original contributions of the components, we must carry this part out in the right-to-left direction. With reference to Fig. 69, the principal ray will reach the final image from the final exit pupil

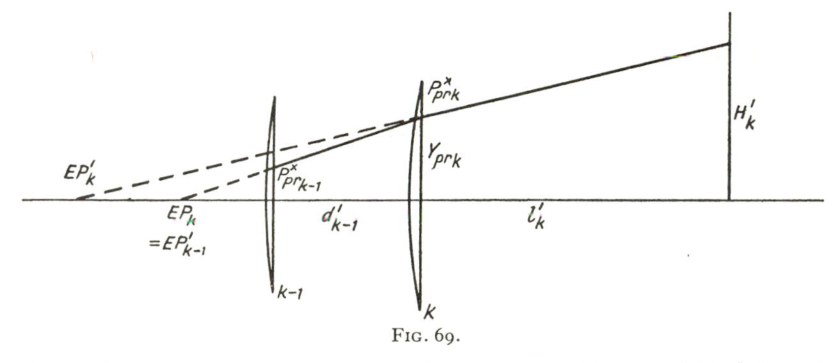

FIG. 69.

EP_K at distance $l'pr_K$ (negative in the diagram). The distance of Ppr^*_K from the optical axis, to which we will apply the symbol Ypr_K, is then given by similar triangles as

$$Ypr_K = H'_K \frac{l'pr_K}{l'pr_K - l'_K},$$

and this must be calculated with careful attention to signs. The principal ray will reach the last component from the exit pupil EP'_{K-1} which is identical with EP_K and at distance $l'pr_{K-1}$ from the last but one component. Similar triangles then give the distance of Ppr^*_{K-1} from the optical axis as

$$Ypr_{K-1} = Ypr_K \frac{l'pr_{K-1}}{l'pr_{K-1} - d'_{K-1}} \text{ or } = Ypr_K \frac{lpr_K + d'_{K-1}}{lpr_K},$$

and this formula with successive lowering of the suffix number will apply for all

[61] the remaining components. Careful watching of the signs is necessary, for the principal ray will most usually cross the optical axis somewhere, and the *Ypr* will then change their sign. A reasonably accurate scale drawing will be found very helpful. As the original position of the point *Ppr* at each component is in our present case on the optical axis, the *Ypr* are identical with the distance *PprPpr** ; hence we can now determine the correct value of *Q* for each component as

$$Q_A = Ypr_A/SA_A \; ; \; Q_B = Ypr_B/SA_B, \text{\&c.}$$

These values must be used in applying the second law to the originally calculated contributions of each component to find the starred contributions ; addition of the latter then gives the final aberrations for the system.

The general formula for the *Ypr* breaks down for use beyond a component standing exactly at the crossing point of the principal ray with the optical axis, for as for such a component *Ypr* would be zero, the general formula would give the next following *Ypr* as infinity × zero, or indeterminate. But as under the thin lens fiction, to which we must adhere when once we have adopted it, the principal ray would go straight through the component which it crosses on the axis, similar triangles having the Y_{pr} of the preceding and the following components as vertical sides will give

$$Ypr_{J-1} = -\,Ypr_{J+1} \cdot \frac{d'_{J-1}}{d'_J}$$

if the principal ray crosses lens or component *J* centrally. If the diaphragm has been assumed coincident with the last component, then we shall have $Ypr_K = 0$ and can find

$$Ypr_{K-1} = -H'_K \, \frac{d'_{K-1}}{l'_K}.$$

We will collect the equations for systems of separated thin components under the reference :

SEIDEL ABERRATIONS V

(1) Having decided upon the focal lengths and the separations of the components, trace a paraxial pencil from the given object-distance l_A right through to the final image by TL (3), thus finding all the l and l'. Select a suitable semi-aperture SA_A of the first component and calculate $SA_B = SA_A \cdot l_B/l'_A$; $SA_C = SA_B \cdot l_C/l'_B$, &c.

(2) Determine the composition of such of the components as are not simple thin lenses—usually by the achromatic condition. Then calculate Seidel Aberrations III for each component separately, using the above SA and l or l' values and a suitably selected value of the final image-height H'_K.

(3) Trace a principal ray from the centre of the selected diaphragm by TL (3) right through the system, thus finding all the l_{pr} and l'_{pr}. Then calculate all the *Ypr* by the equations

$$Ypr_K = H'_K \, \frac{l'pr_K}{l'pr_K - l'_K} \quad \text{and} \quad Ypr_{J-1} = Ypr_J \, \frac{l'pr_{J-1}}{l'pr_{J-1} - d'_{j-1}} = Ypr_J \, \frac{lpr_J + d'_{J-1}}{lpr_J};$$

or if at a preceding surface Ypr is $= 0$, then by

$$Ypr_{K-1} = -H'_K \, . \, d'_{K-1}/l'_K \text{ and } Ypr_{J-1} = -Ypr_{J+1} \, . \, d'_{J-1}/d'_J,$$

and with the above SA values calculate

$$Q_A = Ypr_A/SA_A \, ; \; Q_B = Ypr_B/SA_B \ldots Q_J = Ypr_J/SA_J \ldots$$

(4) Apply the second fundamental law, separately for each component, by calculating from the contributions SC', CC', AC', PC', and $LchC'$ for central passage the true contributions to the final aberration :

$$SC' \, ; \; CC'^* = CC' + SC' \, . \, Q \, . \, u'_K \, ; \; AC'^* = AC' + 2\, CC' \, . \, Q/u'_K + SC' \, . \, Q^2$$

$$DC'^* = 3\,(AC' + \tfrac{1}{3}\, PC')\, Q \, . \, u'_K + 3\, CC' \, . \, Q^2 + SC' \, . \, Q^3 \, . \, u'_K \, ; \; LchC' \text{ unchanged}$$

$$PC' \text{ unchanged} \, ; \; TchC'^* = LchC' \, . \, Q \, . \, u'_K.$$

Its own value of Q must of course be used for each component. The sums of the respective contributions thus determined will be the true final primary aberrations of the complete system for the selected position of the diaphragm.

(5) If the results are such as to suggest that another shift of the diaphragm would improve matters, the second fundamental law may now be applied to the final aberrations of the complete system in the way exemplified in section [60].

Seidel Aberrations V can be adapted to the solution in first approximation of a vast number of optical problems, as will be shown in Part II. For the present we will be satisfied with only two applications of considerable interest and importance.

A. *Systems of two separated thin components* which in their final form are to be free from longitudinal *and* transverse chromatic aberration and also free from spherical aberration, coma, and tangential curvature of field. Such systems conform to conclusion (1) of section [60], that is, we may place the diaphragm wherever we like for the purpose of studying the possibilities. We will choose the position of the diaphragm in contact with the second component, because this renders the discussion as clear and simple as possible. Fig. 70 shows the essential relations which then result, namely that the oblique pencils necessarily pass eccentrically through the first component, so that the aberrational contributions of the latter

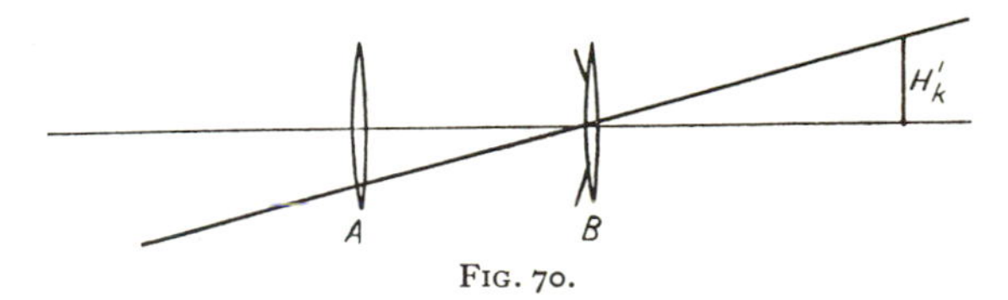

FIG. 70.

become altered from their values for central passage of the oblique pencils to those following from the second fundamental law. This leads to the following important conclusions, which practically settle both the characteristic properties required in the components and the systematic method of designing the system.

(1) As the corrections required include freedom from transverse chromatic aberration (= achromatism of magnification), and as $TchC'_B$ is zero on account of

[61] the central passage of the oblique pencils through the second component, $TchC'^{*}_{A}$ must also be zero in order to give a zero-sum. As for the first component also $TchC'_{A}$ with central passage is zero, we have

$$TchC'^{*}_{A} = LchC'_{A} \cdot Q_{A} \cdot u_{K},$$

and this can only be zero if $LchC'_{A}$ is zero, or if the first component is achromatic. This in turn requires that the second component also is achromatic in order to secure zero longitudinal chromatic aberration at the final focus. Hence the first conclusion :

'In a fully corrected system of two separated thin components, the latter must be individually achromatized.'

(2) If freedom from distortion is to be included in the correction of the system, somewhat similar conclusions apply. The second component will contribute no distortion, on account of the central passage of the oblique pencils through it. Hence $Dist'^{*}_{A}$ must also be zero. But as $Dist'_{A}$ for central passage is zero, we have

$$Dist'^{*}_{A} = 3\,(Ast'_{A} + \tfrac{1}{3}\,Ptz'_{A})\,Q_{A} \cdot u'_{K} + 3\,Coma'_{A}\,Q^{2}_{A} + LA'_{A} \cdot Q^{3}_{A} \cdot u'_{K}$$

in which the first term has a large value on account of the great curvature of field of any thin system with central passage of the oblique pencils, combined with the necessarily finite Q_{A} and u'_{K}. $Dist'^{*}_{A}$ can therefore only be brought to the required zero value if $Coma'_{A}$ and/or LA'_{A} have sufficiently large values to bring this about. A first component free from both coma and spherical aberration is therefore entirely impossible, and as any of these aberrations in the first component must be corrected by the second, we have the conclusion :

'An aplanatic system of two separated thin components can only be rendered free from distortion if the components are *not* aplanatic.'

(3) With reference to curvature of field we will for the present restrict the discussion to the most usual case when both components are positive, as in all doublet microscope objectives and in most photographic objectives. The second component, with central passage of the oblique pencils, will then have the intolerable roundness of field of any positive thin system. Consequently the first component will have to supply an equal *negative* contribution to effect compensation at the final focus. But we have

$$Ast'^{*}_{A} + \tfrac{1}{3}\,Ptz'_{A} = (Ast'_{A} + \tfrac{1}{3}\,Ptz'_{A}) \quad + 2\ Coma'_{A} \cdot Q/u'_{K} + LA'_{A} \cdot Q^{2},$$

where the bracketed term on the right again has the large positive value characterizing central passage of the oblique pencils. The obvious conclusion again is that the left can be brought to the indispensable large negative value only by sufficiently large values of $Coma'_{A}$ and/or LA'_{A}, and as these must be compensated in the second component, we again conclude :

'In an aplanatic system of two positive, thin separated components, a flat field is only attainable if neither component is aplanatic.'

This conclusion becomes less stringent when the second component has negative power, and it can be evaded by making the negative power of the second component approximately equal to the positive power of the first component ; that is the type of the modern fixed telephoto objectives in which it is thus *possible* to [61] attain a flat field with aplanatic components. But our second conclusion as to distortion would then come in with full force, leading to very severe pincushion distortion. The latter defect is in fact peculiar to all telephoto combinations ; but it can be usefully diminished by not insisting on aplanatism of the components.

This discussion thus leads to the result that systems of two thin separated components requiring full correction at the final focus must be built up of strictly achromatic components, with the surprising further requirement that the components must *not* be aplanatic. Hence perfect doublets can only be produced from imperfect thin components!

B. *Systems of two separated thin* unachromatic *components* which are required to be free from transverse chromatic aberration. As the components are not achromatic, the transverse chromatic aberration will vary with the position of the diaphragm or pupils, and the required position of these therefore must enter into the solution instead of being assumed arbitrarily as in the preceding case. Most of the ordinary two-lens eyepieces are of this type, and the four-lens terrestrial eyepiece can also be brought in by treating the two parts, erector and eye-end, separately. The solution of this problem is an extremely useful one for that reason.

In Fig. 71 the lens system to be studied shall consist of the two thin components A and B, with focal lengths f'_A and f'_B respectively, made of glass (or its equivalent

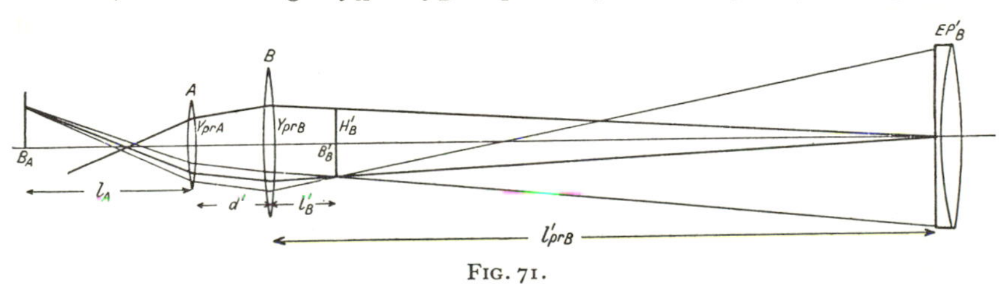

FIG. 71.

in a thin compound lens) of figure of merit V_A and V_B respectively and placed at a separation d'. To fix the ideas, and because the chief application is to eyepieces, the complete course of the rays of an oblique pencil as it would be in an astronomical telescope with 'Ramsden' eyepiece is shown in the lower part of the diagram. The object-glass at the extreme right would focus the light from an extra-axial object-point at a small distance l'_B from the lens B of the eyepiece, and the rays would then pass eccentrically through both lenses of the eyepiece, roughly as sketched, and would eventually emerge into the space to the left. For an observer with normal vision and relaxed accommodation the emerging pencil should consist of parallel rays, and the solution of the eyepiece problem is most usually restricted to this case. But as a hypermetropic eye requires a converging pencil, whilst a myopic eye calls for a diverging one—in severe cases with intersection distances of only a few inches—and as eyepieces are frequently used for projecting an image upon a screen or photographic plate, we will solve for the case when the image is to be projected at a distance l_A from lens A. A negative l_A suitable for actual projection or for a hypermetropic observer is shown in the diagram,

[61] but our solution will of course be equally valid for the positive l_A required by a myopic observer.

As the transverse chromatic aberration to which our present solution is to be limited is an aberration of the principal ray, we require to know only the course of this ray, and this is shown in the upper part of the diagram. For our present purpose the object-glass shown at the right becomes simply the exit pupil of the system to be studied, and we therefore denote its distance from component B by the usual symbol $l'pr_B$, and in accordance with our principle of determining the aberrations at the short conjugate distance we seek the condition which will make total $Tch'^* =$ zero in the image-plane at B'_B.

For our thin components $TchC'$ for central passage of the oblique pencils will be zero. Hence $TchC'^*$ for eccentric oblique pencils will arise solely from the $LchC'$ of the components. Treating these as simple lenses we have the formula of Seidel Aberrations III :

$$LchC' = l_K'^2 \frac{SA^2}{SA_K^2} \cdot c \cdot \delta N.$$

We will transform this by writing

$$c \cdot \delta N = c\,(N-1) \cdot \frac{\delta N}{N-1} = 1/f'V,$$

and we then have, as K is now B,

$$LchC'_A = l_B'^2 \frac{SA_A^2}{SA_B^2} \cdot \frac{1}{f_A \cdot V_A} \quad ; \quad LchC'_B = l_B'^2 \frac{1}{f'_B \cdot V_B} \cdot$$

For the first of these we require the value of SA_A/SA_B. Paraxial rays from B_A will be refracted by component A towards a conjugate at l'_A and will meet component B with an intersection-length $l_B = l'_A - d'$; therefore we have by the usual similar triangles

$$SA_A/SA_B = l'_A/(l'_A - d') = 1/(1 - d'/l'_A),$$

and as by TL (3) $1/l_A = 1/f'_A + 1/l_A$, we obtain finally

$$SA_A/SA_B = 1\Big/(1 - d'/f'_A - d'/l_A).$$

By the second fundamental law we shall have the transverse chromatic contributions of the two thin components

$$TchC'^*_A = LchC'_A \cdot Q_A \cdot u'_B \qquad TchC'^*_B = LchC'_B \cdot Q_B \cdot u'_B$$

with $Q_A = Ypr_A/SA_A$, $Q_B = Ypr_B/SA_B$, and $u'_B = SA_B/l'_B$, whence with the above values of $LchC'_A$ and $LchC'_B$,

$$TchC'^*_A = l'^2_B \frac{SA_A^2}{SA_B^2} \cdot \frac{1}{f'_A V_A} \cdot \frac{Ypr_A}{SA_A} \cdot \frac{SA_B}{l'_B} = l'_B \frac{SA_A}{SA_B} \cdot \frac{1}{f_A V_A} \cdot Ypr_A$$

$$TchC'^*_B = l_B'^2 \cdot \frac{1}{f'_B V_B} \cdot \frac{SA_B}{l'_B} \cdot \frac{Ypr_B}{SA_B} \qquad = l'_B \cdot \frac{1}{f'_B V_B} \cdot Ypr_B.$$

The sum of these two contributions is to be zero, therefore we have the equation [61] of condition

$$0 = l'_B \, . \, Ypr_B \left[\frac{SA_A}{SA_B} \cdot \frac{1}{f'_A V_A} \cdot \frac{Ypr_A}{Ypr_B} + \frac{1}{f'_B V_B} \right].$$

The first factor may safely be treated as always finite, for $l'_B = 0$ would mean that the image fell exactly upon component B, which is a highly undesirable position, as any dust or fine scratches on this lens would be seen sharply with the detail of the useful image ; and Ypr_B being nearly equal to H'_B, necessarily must be finite. Hence our condition can be written

$$0 = \frac{SA_A}{SA_B} \cdot \frac{1}{f'_A V_A} \cdot \frac{Ypr_A}{Ypr_B} + \frac{1}{f'_B V_B},$$

and we see that we only require to know the ratio Ypr_A/Ypr_B and not the separate values. For this ratio, Fig. 71 gives in the usual way, by the convergence of the principal ray,

$$Ypr_A/Ypr_B = l'pr_A/(l'pr_A - d') = (lpr_B + d')/lpr_B = 1 + d'/lpr_B.$$

Then TL (3) gives $\quad 1/lpr_B = 1/l'pr_B - 1/f'_B$

or $\quad Ypr_A/Ypr_B = 1 + d'/l'pr_B - d'/f'_B,$

and if we introduce this value and also the previously determined value of SA_A/SA_B into our equation of condition, the latter becomes

$$0 = \frac{1 + d'/l'pr_B - d'/f'_B}{(1 - d'/f'_A - d'/l_A) f'_A V_A} + \frac{1}{f'_B V_B};$$

and when brought to a common denominator, gives

$$0 = \frac{f'_B V_B + d' \dfrac{f'_B V_B}{l'pr_B} - d' V_B + f'_A V_A - d' V_A - d' \dfrac{f'_A V_A}{l_A}}{\left(1 - \dfrac{d'}{f'_A} - \dfrac{d'}{l_A}\right) f'_A V_A f'_B V_B}.$$

If we now divide numerator and denominator by V_A, the new denominator cannot become infinite in any optically useful case ; hence the numerator must be brought to zero, and if we solve the resulting zero-equation for d' we obtain the final result :

SEIDEL ABERRATIONS VI

$$d' = \frac{f'_A + f'_B \, . \, V_B/V_A}{1 + V_B/V_A + f'_A/l_A - \dfrac{f'_B V_B}{l'pr_B V_A}}.$$

This is a very general solution of the problem of the ordinary eyepieces and will be used in Chapter X, which will be devoted to the study of eyepieces. We will therefore only point out that our general solution gives the ancient text-book formula $d' = \frac{1}{2}(f'_A + f'_B)$, when $V_B = V_A$ and when at the same time l_A and $l'pr_B$ are both infinite. This means that the usual formula is only correct when the two

[61] lenses are made from the same kind of glass, when the telescope is infinitely long and is also focused for a perfectly normal eye with relaxed accommodation.

We must add a definition for V when one of the components is made up of several thin lenses. We introduced V in the modification of the equation for $LchC'$, the original factor $c\,.\,\delta N$ being turned into $1/f'\,.\,V$. If a component consists of several thin lenses, then in accordance with Seidel Aberrations III

$$LchC' = l'^2_K \frac{SA^2}{SA^2_K}\,.\,\Sigma\,(c\,.\,\delta N \text{ of individual lenses}),$$

so that the only change consists in replacing the $c\,.\,\delta N$ of a really simple lens by the sum of all the constituents. We may then, simply as a definition, call that sum $1/f'V$, f' being the focal length of the built-up component, and we shall obtain the correct result from Seidel Aberrations VI if we calculate for such a component

$$V = 1/(f'\,.\,\Sigma c\delta N).$$

This definition should be added to our solution in collecting the formulae.

In concluding this chapter we must once more emphasize that it deals throughout with primary aberrations only, and that many of the most valuable methods are rendered still more inaccurate by involving also the thin lens fiction. The results obtained directly from the numerous methods of solution are for that reason hardly ever close enough to the truth to be turned directly into glass and brass, and most of the profound distrust with which analytical optics meets is due to this limitation not being sufficiently recognized. The results should always be tested and corrected by strict trigonometrical calculations, and the following two chapters are therefore devoted to the principal methods by which this highly important and necessary final operation can be carried out with the maximum of efficiency and in the minimum of total time.

Special Memoranda from Chapter VI

Section [51] on the aperture diaphragm and its images, the pupils, must be thoroughly understood as an indispensable preliminary to the study of the aberrations of oblique pencils.

By introducing the auxiliary optical axis laid through the centre of curvature of any one refracting surface and through the object-point, the oblique pencil admitted by the entrance pupil reveals itself as simply an eccentric part of a larger axial pencil having the auxiliary axis as its axis. The ordinary longitudinal spherical aberration is therefore applicable for the determination of the course of all the refracted rays of the incident oblique pencils ; it immediately leads to the conclusion that, with unimportant exceptions, oblique pencils are subject to very much heavier aberration than the corresponding axial pencil, and that they are likely to display considerable secondary aberration when the axial pencil shows hardly a trace of it. To avoid almost hopeless complications, the detailed investigation is limited to the strictly primary aberrations which are alone effective at very small aperture and field.

Adopting the paraxial image-plane and the linear magnification calculated by the theorem of Lagrange as standards to which the actual imperfect images are to be referred, a principal ray and a D-pair of rays are then traced through a centred lens system, and it is shown that the principal ray fails to arrive at the ideal point in the paraxial focal plane by a small radial displacement called the distortion of the system, and that the final focus of the D-pair displays a longitudinal aberration with reference to the paraxial focal plane and a transverse aberration with reference to the associated principal ray ; these three aberrations involve five constants corresponding to the five Seidel sums, which are proportional to the seriousness of the usual five primary aberrations : spherical aberration, coma, astigmatism, Petzval curvature, and distortion. All these are discussed in section [55], and two primary chromatic aberrations are added in section [56]. The discussion has already been summarized in section [57], and this should be referred to as a condensed account of the principal facts to be remembered. A surprisingly simple method for the numerical calculation of the primary aberrations is worked out, summarized under Seidel Aberrations II, and illustrated by being applied to an ordinary landscape lens.

In section [59] the general solution is specialized for the case of thin systems passed centrally by the oblique pencils, the results being collected as Seidel Aberrations III. It is found that spherical aberration and coma can be calculated by simple G-sums, longitudinal chromatic aberration by another simple equation, that distortion and transverse chromatic aberration are zero, that the Petzval term takes a particularly simple form, and finally, as the most remarkable property of thin systems passed centrally by the oblique pencils, that the astigmatism is absolutely independent of the shape and of the refractive index of the constituent lenses and depends solely on the focal length of the system. On account of this extreme simplicity, it pays to take a certain amount of risk by trying a thin lens solution first, even when the actual thicknesses are fairly large. As a simple and very common example, a two-lens telescope objective is calculated by the G-sums for spherical aberration and coma, so as to be free from both aberrations.

Three fundamental laws of oblique pencils are next formulated. The first puts the three-to-one ratio of tangential to sagittal astigmatism into its most useful form. The second, and by far the most important, law establishes remarkable relations between the known primary aberrations for a particular position of the diaphragm of any centred system and the changed aberrations for any other position of the diaphragm. (See Seidel Aberrations IV.) The third law is restricted to thin systems and emphasizes the fixed and very heavy astigmatism of these systems.

[61] The important conclusions to be drawn from the fundamental laws cannot be usefully condensed into less space than that devoted to them in section [60]; that discussion should therefore be regarded as a part of this summary.

The last section of the chapter generalizes the thin lens equations for the case of systems compounded of a number of thin components at finite separations. The resulting equations will find numerous applications in Part II. Two simple cases are discussed. It is shown first that systems of two separated thin components can only be well corrected with reference to all aberrations if the components are separately achromatized and, strangely, if the components are *not* aplanatic. The second case discussed arises chiefly in eyepieces; an equation of condition is deduced for the correction of the transverse chromatic aberration in systems of two non-achromatic thin components.

Perhaps the most remarkable fact clearly brought out in this chapter is the close interdependence of the five Seidel aberrations, which renders the calculation of the contributions by individual surfaces extremely simple and also leads to the formulation of the important second fundamental law.

CHAPTER VII

THE OPTICAL SINE THEOREM

[62] THE sine theorem was discovered almost simultaneously but quite independently in 1873 by Abbe and Helmholtz, with slight priority of publication for Abbe.† Even with the unnecessarily restricted interpretation hitherto applied to it, it has proved extremely useful in establishing aplanatism, in Abbe's sense of the word, for centred lens systems of every kind. The incomparably wider significance of the famous theorem which is demonstrated in the following pages renders it one of the most remarkable and labour-saving theorems in the whole realm of applied optics.

The sine theorem is an extension of the paraxial theorem of Lagrange to centred systems of any aperture, however large, and the original proof by Helmholtz (Abbe merely *stated* the theorem) was based on this aspect. The defect in the numerous more or less valid proofs of the theorem hitherto offered is that it never was clearly realized that whereas Lagrange's theorem covers *all* the rays of a *thin* pencil of indefinitely small obliquity, the sine theorem only applies to the strictly *sagittal* rays of oblique pencils of *any* aperture. By careful attention to this restriction the sine theorem acquires the vastly extended utility referred to in the opening paragraph.

We will first prove the theorem on the basis of the theory of the primary aberrations in the previous chapter, and will then discuss its range of validity and its limitations by extending the primary theory to higher aberrations as far as is necessary for our present purpose.

In section [53] we proved that the focus of any refracted D-pair of rays falls exactly into the auxiliary optical axis drawn through the intersection point of the incident D-pair, and as the sagittal pair of rays in any given zone of the aperture is merely a special D-pair, the sagittal focus also lies on the auxiliary optical axis. In section [55] we proved further that all the sagittal foci of a zone of given aperture for pencils at all admissible obliquities fall upon a curved surface defining the sagittal curvature of field, and that in first approximation the axial section of this surface is a common parabola with the principal optical axis as its axis, and with the axial focus of the given zone at its pole. If we now refer to Fig. 72, showing a refracting surface with pole at A and centre of curvature at C, and assume that the marginal ray of an incident axial pencil aimed at B_m under angle U, and was trigonometrically traced to B'_m with angle U' after refraction, so that $AB_m = L$, $AB'_m = L'$, whilst the sagittal pair of rays from the same zone aimed at S on the curved sagittal field, then we can draw the auxiliary axis CS and know that the refracted sagittal pair must come to focus on this auxiliary axis at the point S' at

† In 1863 R. Clausius, best known as one of the great pioneers in thermodynamics, gave a differential form of the sine condition in a paper on the concentration of heat by optical means, and pointed out that this differential equation ought to prove valuable in purely optical investigations. As the equation was in terms of squared cosines and ill-adapted to direct optical applications, and as moreover it appeared in a thermodynamical paper, it attracted no attention for several decades. 'Generalized cosine theorems' of similar type have been proposed repeatedly by subsequent investigators, but have never succeeded in replacing the sine theorem.

[62] which the new sagittal curved field cuts the auxiliary axis. We now erect perpendiculars on the principal optical axis at B_m and at B'_m. As these are also tangents to the respective curved sagittal fields, they will fit very closely for some little

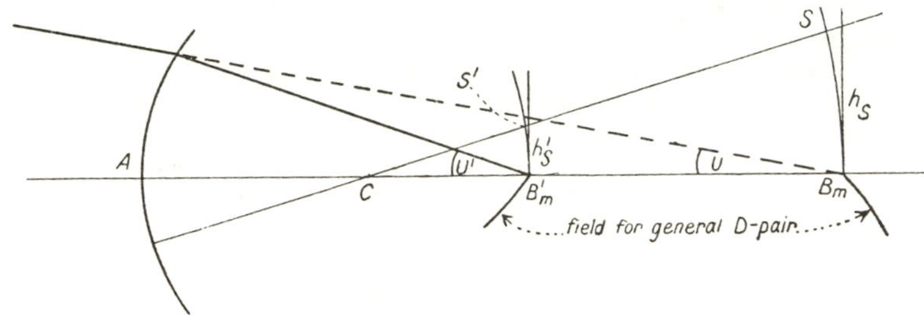

FIG. 72.

distance from the optical axis, and we shall make only a small errror (to be discussed subsequently) by measuring *small* object and image heights h_S and h'_S on these perpendiculars instead of taking the true distances of S and S' from the optical axis. By adopting this simplification we obtain by similar right-angle triangles with a common corner at C: $h'_S/h_S = CB'_m/CB_m = (L'-r)/(L-r)$.

We can then apply the important equation (6) of Chapter I, which by simple transposition gives $(L'-r)/(L-r) = N \,.\, \sin U/N' \,.\, \sin U'$, and substitution of this into the previous equation gives at once the fundamental equation of the optical sine theorem

$$\text{(a)} \qquad h'_S \,.\, N' \,.\, \sin U' = h_S \,.\, N \,.\, \sin U,$$

which will be seen to bear the closest possible resemblance to the corresponding Lagrange equation $h' \,.\, N' \,.\, u' = h \,.\, N \,.\, u$, for the paraxial angles in the latter are simply replaced by the *sines* of the corresponding marginal angles.

Before drawing further conclusions from (a) we will prove that the restriction to sagittal rays is necessary. It was shown in the previous chapter that the curvature of the sagittal field arises from the terms in $H_k'^2$ of the equation

$$LA'_S = g_1 SA^2 + g_3 H_k'^2 + \tfrac{1}{2} g_4 H_k'^2.$$

For any other D-pair of rays we should have

$$LA'_d = g_1 SA^2 + 2\, g_2 l'_k\, Y_k H'_k + g_3 H_k'^2 + \tfrac{1}{2} g_4 H_k'^2,$$

and an additional term therefore appears which is in simple proportion with H'_k instead of growing as its square. The field for such a D-pair would, as a consequence, have a definite slope with reference to the principal optical axis, as is indicated below the axis in Fig. 72, and there would be no similar triangles upon which to found an analogon of (a), no matter how small we chose to make h and h'.

Equation (a) is valid for any one surface. If we call the original object-height h_1 the equation will be

for the first surface: $h'_{S1} N'_1 \sin U'_1 = h_1 N_1 \sin U_1$;

for the second surface: $h'_{S2} N'_2 \sin U'_2 = h_{S2} N_2 \sin U_2$:

and as the image produced by the first surface acts as object for the second, we [62] have $h'_{S1} \equiv h_{S2}$ and also $N'_1 \equiv N_2$, $\sin U'_1 \equiv \sin U_2$; hence the right of the second equation is identical with the left of the first, whence after passage through both surfaces

$$h'_{S2}N'_2 \sin U'_2 = h_1 N_1 \sin U_1.$$

This process can be repeated for the third and any number of following surfaces, with the final result for a centred system of k surfaces

$$\text{(b)} \qquad h'_{Sk}N'_k \sin U'_k = h_1 N_1 \sin U_1,$$

and this is the general equation of the optical sine theorem. It can of course be applied to any ray which has been traced through the system from the given object-point. Usually we trace a paraxial and a marginal ray. If we adopt (b) as the equation for the marginal ray, that for the paraxial ray will merely have paraxial angles instead of the sines, and if we retain the same h_1, but write the final image-height with small s as suffix, it will be the theorem of Lagrange :

$$(\text{b}_p) \qquad h'_{sk}N'_k u'_k = h_1 N_1 u_1.$$

The only use which has hitherto been made of these two equations has been to treat (really inaccurately) h'_{Sk} and h'_{sk} as the true heights of the complete images of the object h_1, in which case h'_{Sk}/h_1 would be the linear magnification produced by the wide-angle pencil, and h'_{sk}/h_1 the corresponding linear magnification by indefinitely thin pencils. It was then argued that if the system were spherically corrected, equality of these two magnifications would denote perfect extra-axial images. As the 'magnifications' are

$$h_{Sk}/h_1 = N_1 \sin U_1/N'_k \sin U'_k \quad \text{and} \quad h'_{sk}/h_1 = N_1 u_1/N'_k u'_k,$$

their equating gives the condition to be fulfilled as

$$N_1 \sin U_1/N'_k \sin U'_k = N_1 u_1/N'_k u'_k, \text{ or simplified } \sin U_1/\sin U'_k = u_1/u'_k,$$

and as this condition applies to sines, this application of our fundamental equations is what is now universally understood as fulfilling the 'Optical Sine Condition'. Even in this form the theorem has saved an enormous amount of labour in the designing of aplanatic systems, as the test for perfection of the extra-axial images consists simply in comparing the ratios of the first and last angular data in the standard system of ray-tracing instead of requiring a complete calculation for an actual oblique pencil. By bearing in mind that our equations apply to sagittal rays only, we shall obtain an absolutely general and remarkably precise determination of the *coma* for any centred lens system, including, when possible, a solution for that position of the diaphragm which will make the coma zero or will adjust it to some prescribed magnitude.

With reference to Fig. 73, the usual trigonometrical ray-tracing will have located the final paraxial and marginal foci of the system at B'_p and B'_m by l'_k and L'_k measured from the last surface of the system at A_k, and the final angles of convergence u'_k and U'_k will also be known. By equations (b) and (b_p) we can then determine the final sagittal foci conjugate to an original object of height h_1 as B'_S at perpendicular distance h'_{Sk} from the marginal focus B'_m and as B'_s at distance h'_{sk} from the paraxial focus B'_p. Remembering that we defined the sagittal coma

[62] as the distance at which the principal ray of an oblique pencil passes above the sagittal focus, and that the sagittal coma is in first approximation proportional to the square of the aperture, we can now conclude that B'_s must lie on the principal

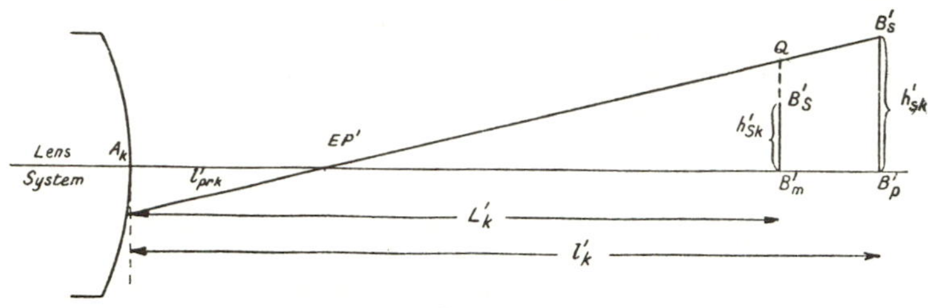

FIG. 73

ray because it represents the sagittal focus for indefinitely small aperture. Hence, if EP' is the exit pupil of our system, we can draw the principal ray as the straight line $EP' - B'_s$, and by the definition of sagittal coma $B'_S Q$ must be the sagittal coma of the marginal zone of the aperture. We therefore have

$$Coma'_S = B'_S Q = B'_m Q - h'_{Sk}.$$

Similar triangles with a common corner at EP' give

$$B'_m Q = h'_{sk}(L'_k - l'pr_k)/(l'_k - l'pr_k),$$

$$\therefore Coma'_S = h'_{sk}(L'_k - l'pr_k)/(l'_k - l'pr_k) - h'_{Sk}.$$

As this equation is true for any small values of the h' and the corresponding h_1, it is more convenient in most applications of the sine theorem to get rid of definite values by forming the *ratio* of $Coma'_S$ to the whole height $B'_m Q$, thus obtaining a pure number independent of linear scale for the measurement of the coma effect. This ratio will be zero when there is no coma ; any finite value of it therefore measures a definite defect of the extra-axial image which we will call the 'Offence against the Sine Condition', with the easily remembered symbol OSC'. Forming the ratio by the last two equations, we thus find

$$OSC' = \frac{Coma'_S}{B'_m Q} = 1 - \frac{h'_{Sk}}{h'_{sk}} \cdot \frac{l'_k - l'pr_k}{L'_k - l'pr_k},$$

or on introducing the value of h'_{Sk}/h'_{sk} by equations (b) and (b_p),

$$\textit{Sine Theorem I:}\quad OSC' = 1 - \left(\frac{\sin U_1}{u_1}\right) \cdot \frac{u'_k}{\sin U'_k} \cdot \frac{l'_k - l'pr_k}{L'_k - l'pr_k}.$$

The first factor $\sin U_1/u_1$ has been enclosed in a mathematically superfluous bracket, because its value will be exactly one and its retention unnecessary whenever we work in accordance with our standard method, beginning the ray-tracing with $u_1 = \sin U_1$. The factor must however be included when the u_1 of the traced paraxial ray is for any reason different from the $\sin U_1$ of an outer ray.

Sine Theorem I enables us to calculate OSC' for a given system with fixed diaphragm. We can turn it into a most valuable analytical solution by solving the equation for $l'pr_k$ so as to determine that location of the exit pupil which will cause a given system to have a prescribed value—most usually zero—of OSC'. A simple transposition gives

$$\frac{l'_k - l'pr_k}{L'_k - l'pr_k} = \frac{1 - OSC'}{\dfrac{\sin U_1}{u_1} \cdot \dfrac{u'_k}{\sin U'_k}} \quad \therefore \quad \frac{l'_k - l'pr_k}{l'_k - L'_k} = \frac{1 - OSC'}{1 - OSC' - \dfrac{\sin U_1}{u_1} \cdot \dfrac{u'_k}{\sin U'_k}},$$

and solution for $l'pr_k$ now produces

$$\textit{Sine Theorem II}: \quad l'pr_k = l'_k - \frac{(l'_k - L'_k)(1 - OSC')}{(1 - OSC') - \left(\dfrac{\sin U_1}{u_1}\right) \cdot \dfrac{u'_k}{\sin U'_k}}.$$

The preceding simple proofs might suggest that the sine theorem is only a primary approximation with an additional inaccuracy due to the neglecting of the sagittal curvature of field. If we now render our proofs more exact by including the higher orders of spherical aberration, we shall find that the sine theorem is greatly superior in precision to the primary coma.

We must note first that the $Coma'_S$ determined by the sine theorem is strictly proportional to the image-height, for we found it $= h'_{sk}(L'_k - l'pr_k)/(l'_k - l'pr_k) - h'_{Sk}$ and found also that both h'_{sk} and h'_{Sk} are strictly proportional to the original object height h_1 and therefore to each other. Consequently we must next inquire what are the types of higher coma which grow directly as H'. We shall obtain these by repeating the deductions of section [53], still for a very small value of object and image height, but with inclusion of the higher orders of spherical aberration; we will include primary, secondary, and tertiary aberration by using the three-term formula

$$LA' = a_2 Y^2 + a_4 Y^4 + a_6 Y^6,$$

remembering that the aperture measure Y may be any quantity which grows directly as the incidence ordinate in the paraxial region. Referring to Fig. 74, which inevitably shows decidedly large values of H and H', but must be interpreted as if H and H' were very small, we measure the semi-aperture in the exit pupil of

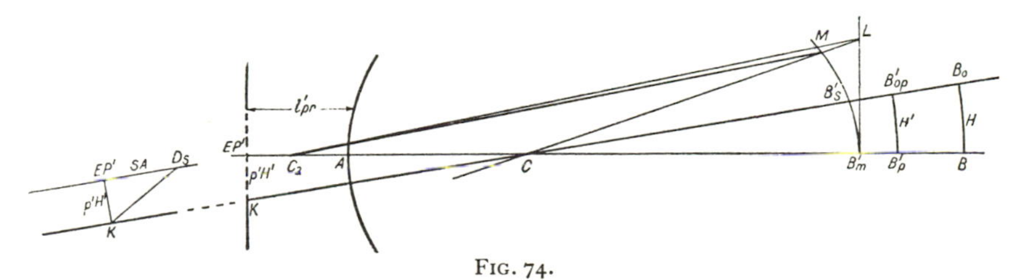

FIG. 74.

the surface under consideration, at EP', and will call it SA. Then we find the new aberrations produced by the surface with pole at A and centre at C by drawing the auxiliary axis CB_o. Whatever the sagittal curvatures of field may be, we can then treat them as insignificant in the close vicinity of the optical axis, so that the

[62] aberration constants a_2, a_4, &c., will be sensibly identical for the axial and for the faintly oblique pencils. Thin pencils traced along the principal and along the auxiliary axis will then locate the paraxial image-points B'_p and $B'o_p$, and for *small* values of H' these will also be sensibly equidistant from C and will also depart to a negligible extent from a perpendicular erected on the principal axis at B'_p. For an axial pencil having semi-aperture SA measured in the exit pupil at EP' there will then result a spherical aberration

$$B'_m B'_p = LA' = a_2 SA^2 + a_4 SA^4 + a_6 SA^6.$$

The auxiliary axis will pass eccentrically through the exit pupil at a distance K–EP' which by similar triangles is proportional to H' and may be put as $p'H'$, p' being a constant. Reference to the end-view of the exit pupil then gives the square of the effective semi-aperture for the sagittal pair of rays as

$$(KD_S)^2 = SA^2 + (p'H')^2,$$

and using this value in the LA' formula we find the spherical aberration of the sagittal pair

$$B'_S B'o_p = LA'_S = \begin{cases} a_2 SA^2 + a_4 SA^4 + a_6 SA^6 \\ +(p'H')^2(a_2 + 2\,a_4 SA^2 + 3\,a_6 SA^4) \\ +(p'H')^4(a_4 + 2\,a_6 SA^2) + (p'H')^6 \,.\, a^6. \end{cases}$$

For the principal ray of the oblique pencil this equation gives, by again regarding the principal ray as the fusion of an infinitely close pair of sagittal rays passing immediately to right and left of EP' in the end-view, or with $SA = 0$,

$$LA'_{pr} = a_2(p'H')^2 + a_4(p'H')^4 + a_6(p'H')^6\,;$$

we can then determine the sagittal coma as the distance at which the principal ray passes above the sagittal focus B'_S exactly as in section [53] as

$$Coma'_S = (LA'_S - LA'_{pr})\,\frac{p'H'}{KB'_{pr}},$$

in which we may—for small H'—take $KB'_{pr} = EP'\,B'_p = l' - l'_{pr}$. Putting in the explicit values of LA'_S and LA'_{pr}, this gives

$$Coma_S = \begin{cases} \dfrac{p'H'}{l' - l'_{pr}}\,(a_2 SA^2 + a_4 SA^4 + a_6 SA^6) \\ + \dfrac{(p'H')^3}{l' - l'_{pr}}\,(2\,a_4 SA^2 + 3\,a_6 SA^4) \\ + \dfrac{(p'H')^5}{l' - l'_{pr}} \cdot 3\,a_6 SA^2. \end{cases}$$

For very small values of H' this will represent a series of extremely rapid convergence, so that only the first line will be of sensible magnitude, and may safely be claimed to determine the sagittal coma near the centre of the field with ample approximation. Simultaneously the only inaccuracy in the tracing of the sagittal foci by the sine theorem, namely the neglecting of the sagittal curvature of the field, will also become indefinitely small, and the sagittal coma deduced from the theorem therefore will also be extremely close to the truth ; hence we may safely conclude that the $Coma'_S$ deduced from the theorem consists of terms of the same form as those in the first line of our last equation, and as OSC' represents $Coma'_S/H'$, we must have [62]

Sine Theorem III: $OSC' = b_2SA^2 + b_4SA^4 + b_6SA^6$, &c.

New letters have been used for the constant coefficients, because they represent the sum of the transferred values of the coefficients at the separate surfaces.

The sine theorem therefore takes full account of all forms of coma which grow directly as H', and this explains the excellent correction of coma in telescope and microscope objectives of every kind when the design is so chosen as to lead to $OSC' = 0$ for the marginal rays. We should however note at once that the latter correction does not necessarily, or even usually, secure zero coma for the aperture as a whole, for just as in a telescope objective with sensible secondary spherical aberration correction of the marginal spherical aberration leaves zonal residues in the intermediate zones, so there are zonal residues of coma when the higher terms of the OSC' series are of sensible magnitude. These zonal residues are of course not a defect inherent in the sine theorem ; as in all other cases, they are the necessary and inevitable consequence of the presence in a system of different orders of aberration of the same type—in the present case coma—which cannot possibly cancel each other for all zones unless they are separately zero, which is an ideal that is very rarely realizable. Moreover, we have another source of imperfection in the case of coma which is not found for the symmetrical axial spherical aberration. The sine theorem deals only with the sagittal coma ; we found in the previous chapter that for other rays the coma effect varies, and we picked out the coma of the tangential rays as leading to an extreme value of precisely three times the sagittal in the case of the primary coma then exclusively considered. With such a fixed ratio as 3 to 1, correction of the primary sagittal coma necessarily implies simultaneous correction of the tangential defect. But it will be shown in Part II that for the secondary simple coma $b_4H'SA^4$ the ratio of tangential to sagittal effect is as 5 to 1, and for the tertiary simple coma $b_6H'SA^6$ the ratio is as 7 to 1. It is easily realized that in the presence of even only primary and secondary coma we may have zero sagittal coma accompanied by a considerable amount of tangential coma, or even positive sagittal by negative tangential ; we may in fact find them in almost any ratio to each other. This complication should be carefully noted, for the trigonometrical tests for coma are very usually limited to tangential rays.

We must finally ask what aberrations of the coma type are neglected in the sine theorem. One answer is supplied by the equation for $Coma'_S$ deduced from Fig. 74, for this gives in its second line a secondary and a tertiary coma growing as the cube of H', and in its last line a tertiary coma growing as the fifth power of H'. As the sine theorem only includes aberrations growing directly as H', these types of coma in the second and third line are missed out. We can easily show that they are missed out owing to the one inaccuracy in the proof of the sine theorem, namely the neglecting of the sagittal curvature of the field. A second auxiliary axis at a much greater angle has been drawn in Fig. 74, to show more clearly that whilst

[62] the sine theorem uses the point L as 'near enough', the true sagittal focus will lie on the curved sagittal field, say at M. In section [55] we determined the distance ML in primary approximation as $= g_3H'^2 + \frac{1}{2}g_4H'^2$, or simply proportional to H'^2. If higher aberrations are included, this simple relation becomes complicated by terms in H'^2SA^2, &c., and other terms in H'^4, but for our present purpose it is sufficient to denote that ML is expressed in terms depending on even powers of H'. At any *one* surface this does not matter very much, because both points lie on the true auxiliary axis. But when we transfer to a new surface with centre, say, at C_2, the correct new auxiliary axis should pass through M whilst we actually lay it through L. This introduces an error in the value of the new h_S equal to ML times (in first and sufficient approximation) the angle between the old and the new auxiliary axes. It is easily seen that this angle is proportional to H', hence the error in the new h_S is proportional to H'^3 and to higher odd powers of H', or is precisely what the neglected higher forms of coma demand.

Summarizing, we have shown that the sine theorem includes *accurately* all forms of coma which grow in direct proportion with H' and with even powers of SA, and that it *ignores* all forms of coma which grow as H'^3, H'^5, &c. We have also shown that OSC' can be represented as a series in even powers of SA, exactly like the longitudinal spherical aberration, and this series may therefore be used to determine the zonal variation.

The two solutions 'Sine Theorem I and II' are applicable in all cases that can arise in practicable centred systems. Solution I at first sight suggests that it would become indefinite or would give wrong results if either l'_k or L'_k should be found infinitely large. But as in that case $l'pr_k$ would have to be finite and in fact fairly small in order to secure a field of finite angular extent, the numerator of $u'_k(l'_k - l'pr_k)/\sin U'_k(L'_k - l'pr_k)$ would become simply $u'_k . l'_k = y_k$ for infinitely large l'_k, and the denominator would become $\sin U'_k . L'_k = Y_k$ for infinitely large L'_k. Hence both are always finite and reasonably small, and can be evaluated even for infinite intersection-lengths. Sine Theorem II would give an ill-determined value of $l'pr_k$ for a very large l'_k if calculated by the standard formula given above, because it would be obtained as the small difference of two big numbers. But by bringing the right-hand side of the standard equation to a common denominator and then simplifying in an obvious way, we obtain two alternative forms :

Sine Theorem II*: $l'pr_k =$

$$= \frac{l'_k - L'_k\left(\frac{u_1}{\sin U_1}\right)\frac{\sin U'_k}{u'_k}(1 - OSC')}{1 - \left(\frac{u_1}{\sin U_1}\right)\frac{\sin U'_k}{u'_k}(1 - OSC')} = \frac{l'_k u'_k - L'_k \sin U'_k\left(\frac{u_1}{\sin U_1}\right)(1 - OSC')}{u'_k - \sin U'_k\left(\frac{u_1}{\sin U_1}\right)(1 - OSC')}.$$

The first of these forms still suffers from the stated drawback, but it has been included because it will in many ordinary cases be found to be more quickly evaluated than the standard equation II. But the second form depends on $l'_k u'_k$ and on $L'_k \sin U'_k$, which at the limit of infinite intersection lengths become simply y_k or Y_k and thus offer no difficulty.

We must finally note once more that our proof of the sine theorem breaks down for a surface at which object and image plane coincide with the centre of curvature. [62] This case will be met in Part II by a proof not depending on the auxiliary optical axis.

APPLICATIONS OF THE SINE THEOREM

[63] We must meet one special case in the general equations Sine Theorem I and II which would seem to render them inapplicable in the very ordinary event of an object-point at infinity, for the factor $\sin U_1/u_1$ would then become 0/0 and apparently indefinite. But if we first consider an object-point at a great distance L_1, we see that then $u_1 = y_1/L_1$ and also $\sin U_1 = Y_1/L_1$, the latter the more accurately the greater L_1; hence the ratio converges towards $\sin U_1/u_1 = Y_1/y_1$, and this is the value which must be used when we treat the object-distance as infinite. A note to this effect should therefore be added to the two equations.

We will begin with the equation I:

Sine Theorem I:
$$\begin{cases} OSC' = 1 - \left(\dfrac{\sin U_1}{u_1}\right) \cdot \dfrac{u'_k}{\sin U'_k} \cdot \dfrac{l'_k - l'pr_k}{L'_k - l'pr_k}. \\ \text{For objects at infinity replace } \sin U_1/u_1 \text{ by } Y_1/y_1. \end{cases}$$

We have already noted that the bracketed factor can be omitted when, in accordance with our usual practice, we begin a calculation with $u_1 = \sin U_1$, or $y_1 = Y_1$ in the case of very distant objects. It is worth remembering also that the final factor becomes exactly one and may be omitted when either $l'_k = L'_k$, that is when there is *exact* spherical correction, or when $l'pr_k$ is infinite. The latter case would demand an original entrance pupil or diaphragm in the anterior principal focal plane of the system, and as such a position of the entrance pupil has occasional advantages either for the actual use or for the design of certain systems, the case is not so improbable as it may appear. In all other cases the final factor alters the value of OSC' very emphatically, and it is this factor which distinguishes our generalized sine theorem from the usual formulation of the 'Sine Condition'. The latter assures freedom from coma only when the system is either strictly corrected for spherical aberration, or else is used with a diaphragm in the anterior focal plane. If our generalized equation gives $OSC' = 0$ the system will be coma-free, no matter how much spherical aberration there may be, when the aperture diaphragm is located so as to yield a final exit pupil at the adopted distance $l'pr_k$.

We will take the landscape lens of the previous chapter as a first test of this claim. For the adopted $Y_1 = y_1 = 0{\cdot}3$, $H'_k = 3$, and $l'pr_k = 0$ we found the primary sagittal coma $= -0{\cdot}00992$, corresponding to

$$OSC' = Coma'_S/H'_k = -0{\cdot}00992/3 = -0{\cdot}00331.$$

The tracing of a paraxial and marginal ray with $y_1 = Y_1 = 0{\cdot}3$ by five-figure logs gave:

$$l'_k = 9{\cdot}8367 \qquad L'_k = 9{\cdot}6713$$
$$\log u'_k = 8{\cdot}49722 \qquad \log \sin U'_k = 8{\cdot}50313.$$

As $l'pr_k = 0$, Sine Theorem I takes the form

$$OSC' = 1 - u'_k \,.\, l'_k/\sin U'_k \,.\, L'_k$$

[63] and we calculate by a few figures :

$$\begin{array}{lll} \log u'_k = 8{\cdot}49722 & \log \sin U'_k = 8{\cdot}50313 & \log u'_k l'_k . = 9{\cdot}49007 \\ +\log l'_k = 0{\cdot}99285 & +\log L'_k = 0{\cdot}98548 & -\log \sin U'_k . L_k = -9{\cdot}48861 \\ \hline \log u'_k . l'_k = 9{\cdot}49007 & \log \sin U'_k . L'_k = 9{\cdot}48861 & \log \text{2nd Term} = 0{\cdot}00146 \end{array}$$

$$\therefore\ OSC' = 1 - 1{\cdot}00337 = -0{\cdot}00337,$$

agreeing with the result by primary aberrations within 0·00006. It will be useful in judging the results of our numerical examples to anticipate that the highest admissible value of OSC' at full aperture of ordinary telescope, microscope, and photographic objectives is $\pm 0{\cdot}00250$. Our first result therefore is in excess of the limit, and the corresponding coma would be distinctly objectionable.

As a second example we will take the spherically corrected Dennis Taylor telescope objective of section [49]. The final trigonometrical results for $y_1 = Y_1 = 0{\cdot}64$ were :

$$l'_k = 9{\cdot}4683 \qquad L'_k = 9{\cdot}4683$$
$$\log u'_k = 8{\cdot}81127 \qquad \log \sin U'_k = 8{\cdot}81346.$$

As $l'_k = L'_k$ and $y_1 = Y_1$, Sine Theorem I becomes reduced to

$$OSC' = 1 - u'_k / \sin U'_k$$

and gives $$OSC' = 1 - 0{\cdot}99497 = +0{\cdot}00503.$$

In section [60] we calculated the coma of this objective from the TL solution by the G-sums, and found for $y_1 = Y_1 = 0{\cdot}64$ and $H'_k = 1$, $Coma'_S = 0{\cdot}00504$; as division by $H'_k = 1$ does not change this result, it is directly comparable with OSC', and shows agreement far closer than we could reasonably expect in view of the low precision of TL aberrations. In this case OSC' amounts to twice our limit, and the objective would be decidedly unsatisfactory if used at the calculated aperture with well-corrected eyepieces. It must however be borne in mind that our objective was calculated for 1·28 aperture at 10 focal length or for $f/7{\cdot}8$. For the $f/15$ ratio usual in astronomical telescopes the aperture for focal length 10 would be only 0·67, and as OSC' changes as the square of the aperture, it would be only $0{\cdot}00503(0{\cdot}67/1{\cdot}28)^2 = 0{\cdot}00138$ at $f/15$; as this is only a little over half our limit, the coma would not be very objectionable.

We will next take an example of the analytical use of the sine theorem by

$$\textit{Sine Theorem II}: \begin{cases} l'pr_k = l'_k - \dfrac{(l'_k - L'_k)(1 - OSC')}{(1 - OSC') - \left(\dfrac{\sin U_1}{u_1}\right) \cdot \dfrac{u'_k}{\sin U'_k}} . \\ \text{For objects at infinity replace } \dfrac{\sin U_1}{u_1} \text{ by } \dfrac{Y_1}{y_1} . \end{cases}$$

The equation gives a definite or reasonable result only when $l'_k - L'_k = LA'$ is finite ; that agrees with our conclusion (2) from the second fundamental law, according to which coma cannot be changed by diaphragm shifts when the spherical aberration is zero.

[63] The OSC' on the right of the equation is that value which it is desired to secure by the diaphragm shift; it will therefore most usually be zero, and the equation then becomes simplified in an obvious way. We will apply it to our landscape lens in order to deduce from the trigonometrical data of the axial pencil that position of the exit pupil which will render the lens free from coma. As this means $OSC' = 0$ and as we also have $y_1 = Y_1$, the equation becomes

$$l'pr_k = l'_k - (l'_k - L'_k)/(1 - u'_k/\sin U'_k).$$

The data in our first example give $LA' = l'_k - L_k = 0{\cdot}1654$ and $u'_k/\sin U'_k = 0{\cdot}98648$, hence

$$l'pr_k = 9{\cdot}8367 - 0{\cdot}1654/0{\cdot}01352$$
$$l'pr_k = 9{\cdot}8367 - 12{\cdot}2338 = -2{\cdot}3971.$$

In section [60] we applied the second fundamental law to the primary aberrations of the same lens, and found that it would be free from primary coma if the diaphragm were so placed as to secure the eccentricity invariant $Q = 1{\cdot}8902$. If we calculate $l'pr_k$ for this value of Q just as it was calculated for $Q = 1{\cdot}3054$ in section [60], we find $l'pr_k = -2{\cdot}3782$, or again a very close agreement with the trigonometrical solution by the sine theorem.

We will apply the third equation of the sine theorem

Sine Theorem III: $\quad OSC' = b_2 SA^2 + b_4 SA^4$, &c.,

to the 2-inch telescope objective in Chapter I, but will use the six-figure results by Mr. J. S. Watkins as in section [31]. The trigonometrical data required are:

for the paraxial ray:	$\log u'_k = 9{\cdot}099262$	$\log l'_k = 0{\cdot}881641$
for the zonal ray:	$\log U'_k = 8{\cdot}950926$	$\log L'_k = 0{\cdot}881321$
for the marginal ray:	$\log U'_k = 9{\cdot}103209$	$\log L'_k = 0{\cdot}881644.$

For the paraxial and marginal ray $y_1 = Y_1$ was exactly 'one', but for the zonal ray Y_1 was taken as $\sqrt{\frac{1}{2}}$ or $\log Y_z = 9{\cdot}849485$.

In telescope objectives it is practically always sufficient to assume the exit pupil in the last surface, or $l'pr_k = 0$, for as the spherical aberration is always small, the small doubt—within the thickness of the objective—as to where the pupil may actually lie, cannot alter the coma to any material extent. We therefore calculate

$$OSC' = 1 - (Y_1/y_1) u'_k l'_k / L'_k \sin U'_k,$$

the bracketed factor requiring inclusion for the zonal OSC'. The results are $OSC'_m = +0{\cdot}009054$, $OSC'_z = 0{\cdot}004272$, and if we measure the semi-aperture on the first surface, it will be exactly 'one' for the marginal and $\sqrt{\frac{1}{2}}$ for the zonal ray, hence

$$b_2 + b_4 = 0{\cdot}009054$$
$$\tfrac{1}{2} b_2 + \tfrac{1}{4} b_4 = 0{\cdot}004272$$

and these give $b_2 = 0{\cdot}008034$, $b_4 = 0{\cdot}001020$.

We can therefore claim that the OSC' of *any* ray traced through this objective will be closely predicted by the equation

$$OSC' = 0{\cdot}008034\, Y^2 + 0{\cdot}001020\, Y^4.$$

[63] We can test this formula in a highly independent way by calculating the primary $Coma'_S$ by Seidel Aberrations II for $H'_k = 1$ and $y_1 = 1$, using the primary SC' values obtained in section [28] to calculate $CC' = SC'.u'_k.i'pr/i'$. The $i'pr$ call for the right-to-left tracing of a principal ray by paraxial formulae and four-figure logs with $l'pr_3 = 0$, $u'_{pr3} = -H'_k/l'_k$ as starting data, giving for the three surfaces

$$\log(i'pr/i'): \quad (1) = 9{\cdot}6065 \quad (2) \doteq 9{\cdot}3937n \quad (3) = 9{\cdot}9672n$$

and with these $CC'_1 = 0{\cdot}008070$ $CC'_2 = 0{\cdot}005855$ $CC'_3 = -0{\cdot}005801$,

or, as the sum, $Coma'_S = +0{\cdot}008124$, and as $H'_k = 1$, this is also the primary OSC'; it will be seen to exceed that deduced from the trigonometrical calculation by only 0·000090.

Exactly as in section [31] for the higher spherical aberration, we may derive two terms of the OSC' formula by using the primary value and the marginal value

$$OSC'_p = 0{\cdot}008124 = b_2 \qquad OSC'_m = 0{\cdot}009054 = b_2 + b_4$$

giving the formula $OSC' = 0{\cdot}008124\, Y^2 + 0{\cdot}000930\, Y^4$,

which looks very different from the previous one, but it gives for the zonal ray at $Y = \sqrt{\tfrac{1}{2}}$: $OSC' = \tfrac{1}{2}(0{\cdot}008124) + \tfrac{1}{4}(0{\cdot}000930) = 0{\cdot}004294$, or wrong by only 0·000022, an insignificant amount. Finally we may use the three values of OSC' to determine a three-term formula, including the tertiary simple coma. We then have

$$OSC'_p = b_2 = 0{\cdot}008124\,; \quad OSC'_m = b_2 + b_4 + b_6 = 0{\cdot}009054$$

$$OSC'_z = \tfrac{1}{2}b_2 + \tfrac{1}{4}b_4 + \tfrac{1}{8}b_6 = 0{\cdot}004272$$

and these give

$$OSC' = 0{\cdot}008124\, Y^2 + 0{\cdot}000750\, Y^4 + 0{\cdot}000180\, Y^6,$$

showing that there is a very sensible amount of tertiary coma in this objective, just as the corresponding discussion of its spherical aberration in [31] disclosed tertiary spherical aberration.

From the practical designer's point of view the most important result in the present case is that this telescope objective has coma to the extent of nearly four times our limit, and would therefore show pronounced deformation of the extra-axial image-points at its full aperture. The latter would have to be reduced to nearly one half—or to $f/8$—to bring OSC' to the highest admissible value.

We will now calculate OSC' for the three bendings of an ordinary cemented achromatic object-glass with crown in front by which the trigonometrical solutions in section [48] were obtained. We found :

for		$c_1=0{\cdot}06,\ c_2=-0{\cdot}41$	$c_1=0{\cdot}16,\ c_2=-0{\cdot}31$	$c_1=0{\cdot}26,\ c_2=-0{\cdot}21$
l'_{3y}	=	9·5664	9·5853	9·6232
$\log u'_{3y}$	=	8·82322	8·81507	8·80600
L'_{3y}	=	9·5414	9·6086	9·5908
$\log \sin U'_{3y}$	=	8·82280	8·81409	8·80896
LA'	=	+0·0250	−0·0233	+0·0324

all for $y_1 = Y_1 = 0{\cdot}64$; we may again assume $l'pr_k = 0$ and have the simple [63] formula : $OSC' = 1 - u_3' l_3' / L_3' \sin U_3'$, which gives

$$OSC' = \quad -0{\cdot}00360 \qquad +0{\cdot}00016 \qquad +0{\cdot}00345$$

We see that the middle bending is very nearly free from coma, but it has a large negative spherical aberration ; the outside bendings have large values (about one-and-a-half times our limit) of OSC' of opposite signs, and are afflicted with heavy positive spherical aberration. As the spherically corrected solutions were located at $c_1 = 0{\cdot}0895$ and at $c_1 = 0{\cdot}224$, both nearer the outside than the middle bendings, and as OSC' shows a nearly uniform rise in value from left to right, we easily conclude that both solutions will have OSC' near our limit. Closer results will be obtained by adding the slightly curved sloping OSC' line to the original graph. This offence against the sine condition is the chief drawback of all cemented objectives of two components when spherical correction is established with the usual varieties of optical glass ; it can only be reduced by using dense barium crown glass in combination with ordinary flint glass, if the cementing is insisted upon. But if a slight air-gap is allowed between the two components, then complete aplanatism is easily established, and the next section will be devoted to that subject, which represents one of the important applications of the sine theorem.

DESIGN OF APLANATIC OBJECT-GLASSES

[64] This section will be limited to the design of object-glasses giving prescribed chromatic, spherical, and coma correction on the assumption that the total thickness is sufficiently moderate to admit of a preliminary solution by the convenient TL methods as a useful first approximation. These TL methods have been fully dealt with in section [40] for the chromatic aberration, in section [49] for the spherical aberration, and in section [59] for coma ; we have therefore only to add the methods of trigonometrical correction of rough designs obtained by these TL solutions.

A. The perfectly safe and systematic procedure consists in first making a complete TL solution by the method exemplified in section [59], in assigning appropriate thicknesses to the components and in some cases to the air-gap, and in then tracing the selected rays trigonometrically through the objective, with correction of the last radius so as to secure exactly the prescribed state of chromatic correction. The final results of this ray-tracing will disclose the residual spherical aberration and, by Sine Theorem I, the residual coma. If both residuals are judged to be sufficiently small as compared with the tolerances, the work will be finished ; but as a rule we shall not be so lucky, and further correction will be necessary.

It will generally be advisable to employ the *strict* trigonometrical method by determining spherical aberration and coma for the extreme marginal ray and by establishing chromatic correction for the $\sqrt{\frac{1}{2}}$ zone, for the objectives will usually be either large astronomical ones or else will have an unusually low ratio of aperture to focal length ; in either case the convenient three-ray method would imply a certain amount of risk. We will take the TL solution of section [59]

[64] as our numerical example, with the thicknesses which have been added to the formula ;

$$
\begin{array}{lll}
r_1 = & 6{\cdot}188 & \\
 & & 0{\cdot}250 \\
r_2 = & -3{\cdot}243 & \\
 & & 0{\cdot}000 \\
r_3 = & -3{\cdot}284 & \\
 & & 0{\cdot}200 \\
(r_4 = & -17{\cdot}39) &
\end{array}
$$

it is permissible to put the axial air-space at zero because the radii are such as to lead to increased air-space towards the margin. The full intended aperture being 1·6, a paraxial and marginal ray were traced through with $y_1 = Y_1 = 0{\cdot}8$. To obtain perfect achromatism r_4 had to be changed to $-17{\cdot}27$, and with this value the *PA* check gave

$$
\begin{array}{llll}
 & l'_4 = 9{\cdot}7656 & & \log u'_4 = 8{\cdot}90365 \\
 & L'_4 = 9{\cdot}7699 & & \log \sin U'_4 = 8{\cdot}90351 \\
\text{or} & LA'_4 = -0{\cdot}0043 & & OSC'_4 = +0{\cdot}00012.
\end{array}
$$

This is a remarkably good result, for the spherical tolerance is $\pm 0{\cdot}0125$, or nearly three times the residual found, whilst with reference to the residual OSC' this is about 1/20th of our limit and would defy detection. Values of OSC' of the order of $\pm 0{\cdot}0001$ are very rarely worthy of notice, because at the large values of H'_k at which they might possibly become visible they will practically always be swamped by astigmatism, curvature of field, and by the forms of higher coma in H'^3 and H'^5 which are not included by the sine theorem. The latter terms in fact frequently call for decidedly higher values of OSC' if the best compromise for a large field taken as a whole is to be achieved.

TL solutions for aplanatic objectives very usually turn out to be closer approximations than similar solutions for cemented objectives, and the realization of coma freedom is nearly always good. In these very common cases the final correction can take the simple form of a very slight bending of the second lens, which will improve the spherical correction without altering the coma correction to any serious extent. This is precisely the method of correction which was employed for the Dennis Taylor objective in section [49], and as glass and focal length are also identical, we can take over the equations $dLA'/dc_3 = 1{\cdot}873 - 14{\cdot}06\, c_3$ and $\delta c_3 = (LA'_{requ.} - LA'_{found})/(dLA'/dc_3)$, but we must carefully note that these equations were worked out for the three-ray method or for $Y_1 = 0{\cdot}64$, whereas our present calculations are for full aperture or $Y_1 = 0{\cdot}80$. As the correcting equations are based on the law of primary spherical aberration, the $LA' = -0{\cdot}0043$ found at full aperture will correspond to that amount times $(0{\cdot}8)^2$ at the 0·8 of the full aperture to which the equations refer, hence we have to introduce

$$LA'_{found} = -0{\cdot}0043 \times (0{\cdot}8)^2 = -0{\cdot}00275.$$

For the present objective c_3 was $= -0{\cdot}3045$, which gives $dLA'/dc_3 = 6{\cdot}154$:

hence $\delta c_3 = (0-(-0{\cdot}00275))/6{\cdot}154 = +0{\cdot}00045$ [64]

giving with original $c_3 = -0{\cdot}3045$

new $c_3 = -0{\cdot}30405$

and by Barlow new $r_3 = -3{\cdot}289$.

As the bending is very slight we can save a new solution for the last radius with perfect safety by applying the same correction $\delta c_4 = +0{\cdot}00045$ to the reciprocal of original $r_4 = -17{\cdot}27$ or original $c_4 = -0{\cdot}05790$, thus determining

$$\text{new } c_4 = -0{\cdot}05745 \quad \text{or} \quad \text{new } r_4 = -17{\cdot}41.$$

A new calculation tracing the rays issuing from the unchanged first lens through this bending of the second lens gives $l_4' = 9{\cdot}7663$, $L_4' = 9{\cdot}7645$ $\therefore$ $LA' = +0{\cdot}0018$, and $OSC_4' = +0{\cdot}00005$. As usually happens, the correction of LA' is slightly overdone, but as it amounts to only 1/7th of the tolerance we can consider the result satisfactory. It may be rendered exact, within the accuracy of a five-figure calculation, by a simple linear interpolation, for as a bending by 0·00045 produced a change in LA' by $0{\cdot}0043+0{\cdot}0018 = 0{\cdot}0061$, we shall secure the desired change by $\delta c_3 = 0{\cdot}00045 \times 0{\cdot}0043/0{\cdot}0061 = 0{\cdot}000317$. We must however remember that the workshop cannot usually produce prescribed radii more closely than to one part in 5,000, even with the best instruments and tools ; hence it is very doubtful whether the last correction would make any genuine difference in the finished objective.

When the first trigonometrical test of a TL solution for an aplanatic objective reveals residuals too serious for the preceding simple method of correction—and this will frequently happen in bold designs—then both components must be bent so as to reduce both LA' and OSC' to the prescribed values. The problem which thus arises, and which we shall later have to solve for even more than two simultaneous conditions, calls for the application of Taylor's theorem for functions of several variables. As this theorem supplies the master-key for the exact solution of nearly all the more intricate optical problems, we will discuss it first in its general algebraical form.

If $f(x, y, z, \&c.)$ is a continuous function of the independent variables x, y, z, &c., then the value of the function for the moderately changed variables $(x+\delta x)$, $(y+\delta y)$, $(z+\delta z)$, &c., will be in first approximation—with which we have to be satisfied in practically all cases—

$$f(x+\delta x,\, y+\delta y,\, z+\delta z,\, \&c.) = f(x,\, y,\, z,\, \&c.) + \delta x\,\frac{\partial f}{\partial x} + \delta y\,\frac{\partial f}{\partial y} + \delta z\,\frac{\partial f}{\partial z}\ \&c.,$$

in which the partial differential coefficients, say for example $\partial f/\partial x$, are determined by treating x as the only variable, y, z, &c., as constant, and by then calculating the numerical value by introducing the known values of x, y, z, &c., into the algebraical expression obtained for $\partial f/\partial x$; similarly for the other partial differential coefficients. We thus secure a linear equation in δx, δy, δz, &c. To be able to solve for the latter, we must have as many different functions $f(x, y, z, \&c.)$, $g(x, y, z, \&c.)$, $h(x, y, z, \&c.)$ as there are variables, and must treat them all in the way shown above for the first of them, thus obtaining as many linear equations as there are unknown δx, δy, &c., when the solution becomes a matter of the simplest routine. An

[64] application to our numerical example will show how simple the method is. Our TL solution was obtained from the equations deduced in [49] and [59] respectively :

$$LA' = 0{\cdot}473 + 1{\cdot}258\,c_2 + 11{\cdot}96\,c_2^2 + 1{\cdot}874\,c_3 - 7{\cdot}03\,c_3^2$$

$$Coma'_S = 0{\cdot}01097 + 0{\cdot}08582\,c_2 - 0{\cdot}05090\,c_3, \text{ for } H'_k = 1, \text{ therefore also } = OSC'_p.$$

Looking upon c_2 as x, on c_3 as y, and treating LA' as $f(x, y)$, OSC'_p as $g(x, y)$, we find the partial differential coefficients :

$$\partial LA'/\partial c_2 = 23{\cdot}92\,c_2 + 1{\cdot}258\,; \qquad \partial LA'/\partial c_3 = -14{\cdot}06\,c_3 + 1{\cdot}874$$

$$\partial OSC'_p/\partial c_2 = 0{\cdot}08582 \qquad ; \qquad \partial OSC'_p/\partial c_3 = -\ 0{\cdot}05090,$$

or on putting $c_2 = -0{\cdot}3084$. $c_3 = -0{\cdot}3045$ of the TL solution into the first two, $\partial LA'/\partial c_2 = -7{\cdot}38 + 1{\cdot}26 = -6{\cdot}12$; $\partial LA'/\partial c_3 = 4{\cdot}28 + 1{\cdot}87 = 6{\cdot}15$ by 10-inch slide-rule. Hence, if we distinguish the aberrations trigonometrically determined for the TL solutions by suffix 'found', and those resulting from a bending of the first lens by δc_2 and of the second lens by δc_3 by suffix 'required', Taylor's theorem gives :

$$LA'_{requ.} = LA'_{found} + \delta c_2(-6{\cdot}12) + \delta c_3(6{\cdot}15)\,;$$

$$OSC'_{requ.} = OSC'_{found} + \delta c_2(0{\cdot}08582) + \delta c_3(-0{\cdot}05090).$$

The 'required' aberrations are both zero. But in putting in numerical values for the 'found' aberrations we must in the present example again remember that the TL solution was worked out for 0·8 of the intended full aperture ; and as both LA' and OSC' grow in primary approximation as the square of the aperture, we must multiply the direct results for full aperture by $(0{\cdot}8)^2$ before putting them into the equations. The latter then become

$$0 = -0{\cdot}00275 - 6{\cdot}12\,\delta c_2 + 6{\cdot}15\,\delta c_3\,; \quad 0 = +0{\cdot}000078 + 0{\cdot}08582\,\delta c_2 - 0{\cdot}05090\,\delta c_3.$$

As the residual aberrations are usually quite small and only known within at most 1 per cent., the slide-rule is the proper tool for the solution ; worries about the decimal place are best avoided by shifting the decimal point in the linear equations to a convenient position. In the present case a shift by two places to the right in the second equation gives :

$$\left.\begin{aligned}(1)\ 0 &= -0{\cdot}00275 - 6{\cdot}12\,\delta c_2 + 6{\cdot}15\,\delta c_3\\ (2)\ 0 &= \ \ 0{\cdot}0078 + 8{\cdot}58\,\delta c_2 - 5{\cdot}09\,\delta c_3\end{aligned}\right\}$$ Multiplication of the first equation by $5{\cdot}09/6{\cdot}15$ then gives :

(3) $0 = -0{\cdot}00228 - 5{\cdot}07\,\delta c_2 + 5{\cdot}09\,\delta c_3$, or by addition of (2) + (3) :

(4) $0 = 0{\cdot}00552 + 3{\cdot}51\,\delta c_2$, $\therefore\ \delta c_2 = -0{\cdot}00552/3{\cdot}51 = -0{\cdot}00157$,

and introducing this into the equation (1), we obtain

(5) $0 = -0{\cdot}00275 + 0{\cdot}00964 + 6{\cdot}15\,\delta c_3$, $\therefore\ \delta c_3 = -0{\cdot}00689/6{\cdot}15 = -0{\cdot}00112$.

It will be noticed that the removal of the residual OSC' of the TL solution simultaneously with the LA' calls for a much larger bending, and one in the opposite direction of that required to remove LA' only by the actually used bending of the second lens.

B. In many cases there will be no difficulty in finding a fairly close approximation to the aplanatic form of a proposed new objective without a complete TL solution ;

we may, for instance, have a graph for cemented objectives of similar type, or an [64] existing solution may merely require modification for a change in the glass data. For all ordinary kinds of glass, the rules for the 'middle bending' given in Chapter V will represent a suitable starting-point. In such cases we can adopt a modified procedure which will usually give the desired final result with less expenditure of time and trouble than method *A*.

If necessary, we begin by testing the adopted approximate prescription trigonometrically with the usual correction of the last radius for chromatic aberration; the result will be values of LA'_{found} and OSC'_{found}, and if these prove too large the components will have to be submitted to the bending process. At that stage we introduce the TL methods by the *G*-coefficients, but as we only require the differential coefficients, the number of terms to be calculated is greatly reduced, which is the advantage secured by this method.

Using the original form of the *G*-sums, the spherical contribution of any one component will be

$$SC' = y^2 l_k'^2 [G_1 c^3 - G_2 c^2 c_1 + G_3 c^2 v_1 + G_4 c c_1^2 - G_5 c c_1 v_1 + G_6 c v_1^2],$$

giving the differential coefficient

$$\partial LA' / \partial c_1 = y^2 l_k'^2 [-G_2 c^2 + 2\, G_4 c c_1 - G_5 c v_1].$$

In our two-lens combinations this will apply to both components, but for the second component the curvature of its first surface will be c_3 and the convergence of the arriving rays $v_3 = v_1 + c_a(N_a - 1)$. The correcting equation by Taylor's theorem therefore becomes

$$LA'_{requ.} = LA'_{found} + \delta c_1 \,.\, y^2 l_b'^2(-G_2^a c_a^2 + 2\, G_4^a c_a c_1 - G_5^a c_a v_1) \\ + \delta c_3 \,.\, y^2 l_b'^2(-G_2^b c_b^2 + 2\, G_4^b c_b c_3 - G_5^b c_b v_3).$$

With regard to OSC' we obtain its TL equation by merely omitting the factor H_k' in the equation for the coma contribution; hence

$$OSC' = y^2(\tfrac{1}{4}\, c_a c_1 G_5^a - c_a v_1 G_7^a - c_a^2 G_8^a) + y^2(\tfrac{1}{4}\, c_b c_3 G_5^b - c_b v_3 G_7^b - c_b^2 G_8^b)$$

and differentiation gives the second correcting equation

$$OSC'_{requ.} = OSC'_{found} + \delta c_1 \,.\, y^2 \,.\, \tfrac{1}{4}\, c_a G_5^a + \delta c_3 \,.\, y^2 \,.\, \tfrac{1}{4}\, c_b G_5^b.$$

There are therefore only eight *G*-terms to be calculated instead of eighteen for a complete TL solution, and for telescope objectives only seven terms, because v_1 is then zero.

When δc_1 and δc_3 come out with rather large values, and this will be the rule rather than the exception, the solution can be greatly improved with very little added trouble by adopting the principle, laid down in section [48. 5] (3), that in the case of LA', which plots as a parabola, the differential coefficient should be worked out for $c_1 + \frac{1}{2}\delta c_1$ and $c_3 + \frac{1}{2}\delta c_3$ and not for c_1 and c_3 of the original form. We shall secure this improvement, not accurately, but with ample approximation, by solving first with the original values of c_1 and c_3 in the differential coefficients of LA', obtaining provisional values of δc_1 and δc_3. We then recalculate the middle

[64] terms in the brackets of the equation for $LA'_{requ.}$ with (original $c+\frac{1}{2}$ provisional δc) and make the final solution, which will usually prove extremely close to the truth.

We will give a decidedly severe test to this solution by seeking the aplanatic flint-in-front form of the objectives in section [48], using the middle bending there adopted for the cemented Steinheil objectives as a starting-point. This bending had $c_1 = 0{\cdot}200$, $c_2 = 0{\cdot}447$, and the collected data give its $LA'_{found} = -0{\cdot}0153$, a decidedly large value. Sine Theorem I gives from the same data $OSC'_{found} = -0{\cdot}00094$, also unpleasantly large. If we now imagine the cement removed, the second component will have of course $c_3 = c_2 = 0{\cdot}447$, and we can calculate the coefficients of δc_1 and δc_3 with $c_1 = 0{\cdot}200$, $c_3 = 0{\cdot}447$, $c_a = -0{\cdot}247$, $c_b = 0{\cdot}470$, $v_1 = 0$, $v_3 = c_a(N_a - 1) = -0{\cdot}247 \times 0{\cdot}6225 = [9{\cdot}1868n]$, $y = 0{\cdot}64$ for the three-ray method, $l'_b = f' = 10$, and with the G-values in [49] and [59], with careful attention to the fact that the order of flint and crown is now reversed. We shall find

$$LA'_{requ.} = 0 = -0{\cdot}0153 \; -6{\cdot}114\,\delta c_1 \; +5{\cdot}985\,\delta c_3$$

$$OSC'_{requ.} = 0 = -0{\cdot}00094 - 0{\cdot}0509\,\delta c_1 + 0{\cdot}0858\,\delta c_3,$$

and these give, by the method fully described in A,

provisional $\delta c_1 = 0{\cdot}0196$; provisional $\delta c_3 = 0{\cdot}0231$;

with which we find

$c_1 + \frac{1}{2}$ prov. $\delta c_1 = 0{\cdot}2098$; $c_3 + \frac{1}{2}$ prov. $\delta c_3 = 0{\cdot}4586$

for the amendment of the original coefficients of δc_1 and δc_3 in the first equation. The equations then become

$$0 = -0{\cdot}0153 \; -6{\cdot}252\,\delta c_1 + 6{\cdot}261\,\delta c_3$$
$$0 = -0{\cdot}00094 - 0{\cdot}0509\,\delta c_1 + 0{\cdot}0858\,\delta c_3$$

and give the final solution

final $\delta c_1 =$	$0{\cdot}0210$	final $\delta c_3 =$	$0{\cdot}0234$,	or with
original $c_1 =$	$0{\cdot}2000$	original $c_3 =$	$0{\cdot}447$	
new $c_1 =$	$0{\cdot}2210$	new $c_3 =$	$0{\cdot}4704$	and with
$c_a =$	$-0{\cdot}247$			
new $c_2 =$	$0{\cdot}4680$,	leading to the prescription :		

$$r_1 = 4{\cdot}529$$
$$0{\cdot}200$$
$$r_2 = 2{\cdot}1368$$
$$0{\cdot}000$$
$$r_3 = 2{\cdot}1258$$
$$0{\cdot}250$$
$$r_4 = ?$$

Trigonometrical calculation by the three-ray method gives $r_4 = 502{\cdot}5$, and the final results by the *PA* check, which is indispensable on account of the exceptionally [64] long radius, are

$$L'_y = 9{\cdot}6584 \qquad l'_y = 9{\cdot}6587 \qquad L'_v = 9{\cdot}6584$$

or $LA' = +0{\cdot}0003.$ Chrom. aberration : Nil.

Sine Theorem I gives $OSC' = +0{\cdot}00005$.
The solution is better than could normally be expected by five-figure logs with angles to the nearest second.

C. The two methods already given will become unsatisfactory when very bold curvatures or considerable thicknesses and separations are introduced into a lens system ; for the contradictions between the TL primary theory and the thick lens practice in presence of heavy higher aberrations become too drastic. A great many of these difficult cases can be met successfully by a complete reversal of the method of using Taylor's theorem as described in *A*. Instead of calculating the partial differential coefficients from an approximate form of the functions and then solving for those increments δx, δy, &c., which satisfy the conditions of the problem, or rather, which would satisfy those conditions if the differential coefficients were exact *and* if in addition the increments were very small, we reduce assumptions to the absolute minimum by treating the functions as quite unknown, merely assuming that they are continuous in the actually used range, and by employing Taylor's theorem purely as an interpolation formula.

Adopting again the form $f(x, y, z)$ as in *A* for the unknown function, we calculate its value for numerical amounts of x, y, and z, judged to lie reasonably close to the solution sought. We then repeat this calculation for as many changed combinations of x, y, z as there are variables ; then Taylor's theorem gives :

$$f(x+\delta x_1, y+\delta y_1, z+\delta z_1) = f(x, y, z)+\delta x_1\frac{\partial f}{\partial x}+\delta y_1\frac{\partial f}{\partial y}+\delta z_1\frac{\partial f}{\partial z}$$

$$f(x+\delta x_2, y+\delta y_2, z+\delta z_2) = f(x, y, z)+\delta x_2\frac{\partial f}{\partial x}+\delta y_2\frac{\partial f}{\partial y}+\delta z_2\frac{\partial f}{\partial z}$$

$$f(x+\delta x_3, y+\delta y_3, z+\delta z_3) = f(x, y, z)+\delta x_3\frac{\partial f}{\partial x}+\delta y_3\frac{\partial f}{\partial y}+\delta z_3\frac{\partial f}{\partial z}$$

and in these three equations the three differential coefficients will be the only unknown quantities, and can be determined from them. The same applies to $g(x,y,z)$ and $h(x, y, z)$, and we then know everything required to apply the solution by *A*. The accuracy of the final solution will be greatly increased if the increments in the successive calculations are chosen on the basis of such experience as may be available so as to effect a closer and closer approach to the desired values in order that the final corrections may be very small.

As a numerical example we will take an $f/4$ aplanatic telescope objective, crown-in-front and identical in glass and focal length with the previous examples. We will, however, effect the aplanatic correction, not by keeping the two components in axial contact with a difference of r_2 and r_3, but by keeping $r_2 = r_3$ and using a small axial air-space instead. This has the advantage that only three different

[64] radii and sets of tools and gauges will be required. As a starting point we will take the form resulting from the rule for middle bendings in [48], but as the clear aperture is now $2\frac{1}{2}$, thicknesses must be increased, and the first calculation will be for this cemented prescription :

$$\begin{array}{llll} c_1 = & 0{\cdot}16 & r_1 = & 6{\cdot}25 \\ & & & \quad 0{\cdot}50 \\ c_2 = & -0{\cdot}31 & r_2 = & -3{\cdot}226 \\ & & & \quad 0{\cdot}30 \\ & & r_3 = & ? \end{array}$$

Calculating through this with $y_1 = Y_1 = 1{\cdot}25$, we find $r_3 = -14{\cdot}34$ by the condition of achromatism, and then final $LA'_I = -0{\cdot}1019$, final $OSC'_I = +0{\cdot}00080$.

The liberties chosen in the present case for the simultaneous correction of LA' and $Coma_S$ are a bending of the whole objective (for as the contact radii are to be equal, both lenses must always be bent by the same amount) and an air-space between the components. But we are not at all bound to use both at once. As the spherical over-correction is by far the most serious defect of our first form, we first try merely an air-space. This will tend towards under-correction because it will lead to a contraction of the cone of rays between the crown and the flint lens, and the latter will produce less negative aberration. A beginner would be almost sure to over-estimate the required separation greatly, and would then be wise to reject the excessive first attempt and substitute a more reasonable one found by rough linear interpolation. We will adopt it as a result of experience that 1/250th of the focal length is nearly always ample, and therefore take next :

$$\begin{array}{ll} r_1 = & 6{\cdot}25 \\ & \quad 0{\cdot}50 \\ r_2 = & -3{\cdot}226 \\ & \quad 0{\cdot}04 \\ r_3 = & -3{\cdot}226 \\ & \quad 0{\cdot}30 \\ r_4 = & ? \end{array}$$

The first calculation remains available up to the determination of sin I_2 and i_2 ; at that point we have to substitute refraction into air for the previous one into flint glass, and find enormous spherical under-correction due to a marginal air-angle of nearly 44°. It is the heavy higher aberration resulting from such severe refraction which causes the correction by the usual TL methods to break down in cases like the present one. Taking the rays through the second lens, we find $r_4 = -16{\cdot}24$ by the condition of achromatism, and then

$$\text{final } LA'_{II} = +0{\cdot}0999, \qquad \text{final } OSC'_{II} = -0{\cdot}00381.$$

As the spherical aberration is very nearly reversed as compared with the first form, we see that an air-space = 0·02 would be almost exactly right ; and as experience shows that a moderate bending of an objective as a whole does not greatly alter either the air-space required for correction in the present example, or the difference of the contact radii in the two previous examples, we decide that the third and final form required for the determination of the differential coefficients shall have $d_2' = 0{\cdot}02$. But we must also bend the objective as a whole in order to determine the effect of bending on the aberrations, and here we must try to improve the coma correction in order that our last form may come out reasonably near the desired perfection. The air-space of 0·04 changed OSC' from $+0{\cdot}00080$ to $-0{\cdot}00381$ or by $-0{\cdot}00461$. By linear interpolation the adopted space 0·02 for the final form would give $OSC' = -0{\cdot}0015$, and we must try to choose such a bending of the whole objective as would tend to remove this. Here again we can draw upon the experience that a bending of a telescope objective of any reasonable type and of any of the ordinary kinds of glass produces a change in OSC' which has the sign of the δc used and which varies only as the square of the *f/ratio* taken as a fraction like $\frac{1}{4}$ for our objective. For the objectives of section [49] calculated at *f/ratio* 1·28/10 = nearly enough 1/8, we found at the end of section [63] that a bending by $\delta c = 0{\cdot}2$ changed OSC' from $-0{\cdot}00360$ to $+0{\cdot}00345$, or by $+0{\cdot}0070$; at *f*/4 this would be increased fourfold or to $+0{\cdot}028$. In our present objective we want to produce a change in OSC' by $+0{\cdot}0015$, therefore by simple proportion

$$\text{required } \delta c = 0{\cdot}2\,\frac{0{\cdot}0015}{0{\cdot}028} = +0{\cdot}011.$$

The second form had $\quad c_1 = +0{\cdot}16 \quad c_2 = c_3 = -0{\cdot}31$;

therefore for the third form $\quad c_1 = +0{\cdot}171 \quad c_2 = c_3 = -0{\cdot}299$,

and we thus have the prescription :

$$\begin{array}{ll} r_1 = 5{\cdot}848 & \\ & 0{\cdot}50 \\ r_2 = -3{\cdot}344 & \\ & 0{\cdot}02 \\ r_3 = -3{\cdot}344 & \\ & 0{\cdot}30 \\ r_4 = ? & \end{array}$$

The trigonometrical test gives $r_4 = -18{\cdot}56$, and with this

$$LA'_{III} = -0{\cdot}0069, \qquad OSC'_{III} = +0{\cdot}00016.$$

We have secured a very good approximation, and in actual practice we should notice at once that the two residuals are practically just of the right magnitude and sign to be corrected simultaneously by a slight increase of the air-space. But as our present purpose is to illustrate completely a general method of extremely wide utility, we will carry out the complete schedule of operations. Remembering that, as compared with form I, form II had $\delta c = 0$, $\delta d_2' = 0{\cdot}04$, and form III had $\delta c = +0{\cdot}011$, $\delta d_2' = 0{\cdot}02$, the application of Taylor's theorem for two variables to LA' and OSC' gives equations of the form

$$\text{changed } LA' = \text{original } LA' + \delta c\,\frac{\partial LA'}{\partial c_1} + \delta d_2'\,\frac{\partial LA'}{\partial d_2'},$$

[64] or with our numerical values, the four equations :

$$0{\cdot}0999 = -0{\cdot}1019 \ldots\ldots\ldots\ldots + 0{\cdot}04 \frac{\partial LA'}{\partial d_2'}$$

$$-0{\cdot}0069 = -0{\cdot}1019 + 0{\cdot}011 \frac{\partial LA'}{\partial c_1} + 0{\cdot}02 \frac{\partial LA'}{\partial d_2'}$$

$$-0{\cdot}00381 = +0{\cdot}00080 \ldots\ldots\ldots\ldots + 0{\cdot}04 \frac{\partial OSC'}{\partial d_2'}$$

$$+0{\cdot}00016 = +0{\cdot}00080 + 0{\cdot}011 \frac{\partial OSC'}{\partial c_1} + 0{\cdot}02 \frac{\partial OSC'}{\partial d_2'}.$$

Calculating by slide-rule, the first and third give directly

$$\frac{\partial LA'}{\partial d_2'} = 5{\cdot}045, \qquad \frac{\partial OSC'}{\partial d_2'} = -0{\cdot}1152,$$

and use of these in the second and fourth determines

$$\frac{\partial LA'}{\partial c_1} = -0{\cdot}536, \qquad \frac{\partial OSC'}{\partial c_1} = +0{\cdot}152.$$

We now use these for the final solution exactly as in sub-section *A*. Starting from form III as by far the closest approximation, and solving for zero values, we have the equations

from LA': $\quad 0 = -0{\cdot}0069 - 0{\cdot}536\,\delta c + 5{\cdot}045\,\delta d_2'$;

from OSC': $\quad 0 = +0{\cdot}00016 + 0{\cdot}152\,\delta c - 0{\cdot}1152\,\delta d_2'$;

and solving these by slide-rule the results are

$$\delta d_2' = +0{\cdot}0014, \qquad \delta c = +0{\cdot}000007.$$

As we anticipated, the bending required is utterly negligible, but as form III had $d_2' = 0{\cdot}02$, the solution for the increment gives for the final solution :

Air-space = 0·0214.

Picking up the third calculation at the air-space and changing the latter to 0·0214, we find $r_4 = -18{\cdot}65$, and with this

final $LA' = -0{\cdot}0004$, $\qquad$ final $OSC' = +0{\cdot}00002$,

an entirely satisfactory solution.

We obtain a remarkable result by tracing a zonal ray at $\sqrt{\frac{1}{2}}$ of the full aperture through our final solution. We find $LZA' = -0{\cdot}0023$, whilst all object-glasses with cemented contact faces give positive zonal aberration. The 2-inch object-glass of 8-inch focus in section [21], which in other respects is closely similar to our present one, gave $LZA' = +0{\cdot}0056$ by six-figure results, or a much larger zonal aberration of opposite sign. Evidently the finite air-space has a specific and powerful effect upon the zonal aberration, and its modest value of 1/500th of the focal length has already over-corrected it. The effect is due to the enormous spherical aberration of the rays emerging from the first component—in our case, with $l_2' = 3{\cdot}8785$ and

$L_2' = 2{\cdot}9841$, no less than $0{\cdot}8944$!—which causes the contraction of the cone of [64] marginal rays to be much larger than that of the paraxial or of any intermediate cone, and diminishes the higher negative aberrations of the third surface. Clearly there may be a particular value of the air-space which, with a suitable small difference between r_2 and r_3, would correct the zonal aberration completely; in our present example the air-space would require diminution, but as the zonal aberration in our final solution amounts to less than 1/3rd of the tolerance, no sensible improvement would result from the introduction of this refinement. But it is worth while to bear the possibility in mind. A small air-space at a deeply curved contact has a similar effect upon the correction of the spherical variation of the chromatic aberration. We found in Chapter IV that in ordinary achromatic objectives, spherically corrected for brightest light, there is spherical under-correction for 'red' and spherical over-correction for 'blue'. This defect becomes serious in photo-visual objectives, and the small air-space found in all well-designed lenses of this type effects its correction. Examples of these higher orders of the correction of aberrations will be given in Part II.

D. We have hitherto effected the aplanatic correction by the use of a 'broken contact', and that is almost the only method in regular use for large objectives. There are a number of objections to the broken contact:

(1) At every glass-air surface about 5 per cent. of the incident light is lost by reflection, whilst at a cemented surface this loss is to all intents and purposes nil. Hence an object-glass with broken contact passes about 10 per cent. less of useful light than a cemented objective. Moreover, some of the reflected light is reflected a second time on reaching preceding surfaces and is then sent towards the useful image; if this doubly reflected light does *not* form a focus close to that of the useful light, it only diminishes contrasts by general faint illumination of the field—the 'brilliance' of the image suffers. But if the doubly reflected light forms a focus close to the focal plane of the useful light, then 'ghost images' result, and these are highly objectionable.

(2) Objectives with broken contact call for workmanship of a very high order if the calculated perfection is to be fully realized. This sensitiveness is due to the enormous aberration in the air-space which has already been referred to. The faintest tilt of either component or a very slight departure from the calculated separation leads to unsymmetrical images or to obvious aberration respectively, and renders objectives with broken contact unsuitable for mass-production.

(3) The narrow and closely confined space between the two components encourages the formation of films of moisture, and not infrequently the growth of fungus or mould.

For these reasons completely cemented aplanatic objectives are highly desirable, especially in small sizes. Previous to the introduction of the valuable barium crown glasses, the only practicable method consisted in the use of triple cemented lenses, a biconvex crown lens cemented between two meniscus flint lenses being nearly always the best type. The additional surface thus introduced provides the required liberty which makes simultaneous correction of spherical aberration and coma possible. The systematic design of these triple objectives is reserved for Part II. We can attain aplanatism with only two cemented lenses by substituting barium

[64] crown of sufficiently high index for ordinary crown, but shall usually have to take advantage of tolerances to a harmless extent. One method of calculating such objectives was given in section [48.5], and we can now realize more clearly why the method was based on the minimum aberration form : this form was shown at the end of section [63] to coincide closely with the form yielding zero coma. We will test the results obtained for a microscope objective in section [48.5] by the sine theorem. For the direct solution, corrected for achromatism by adjustment of the last radius, we found $LA' = -0{\cdot}00374$, or about half the tolerance. The final data give by Sine Theorem I $OSC' = -0{\cdot}00016$, as nearly zero as is of real interest, and this would be a decidedly good solution. It was next found that the removal of the small spherical over-correction called for a very considerable bending ; the solution obtained gave $LA' = +0{\cdot}00065$—quite negligible—but OSC' is found $= +0{\cdot}00450$, nearly twice our limit and hence extremely serious. The bending therefore has completely ruined the objective. In section [50] we applied the more promising method of drawing upon the chromatic tolerance as the most elastic one ; throwing the whole chromatic tolerance upon a shortening of the last radius, we secured reversal of LA' to the entirely harmless value of $+0{\cdot}00079$. For this modification OSC' is found $= -0{\cdot}0016$, an admissible, though not a very creditable, amount. The use of only half the chromatic tolerance then finally suggested would give a satisfactory solution, but by no means the best one.

The sine theorem renders it clear that any considerable departure from the minimum aberration form, either by bending the objective as a whole or by throwing the change of the chromatic correction entirely upon the last surface, leads to the appearance of an objectionable amount of coma. The proper course to be adopted, when the first solution for minimum aberration with perfect achromatism leads to an undesirably large residue of spherical aberration, is therefore to make a complete new solution, first for the values of c_a and c_b which will give the desired focal length with a chromatic aberration equalling one half or even up to the whole of the chromatic tolerance, then for the value of c_2 which gives minimum aberration. Trigonometrical test and adjustment of this complete solution will either prove it satisfactory or will supply the data for the interpolation of the best possible solution. This then is the proper procedure by the method of section [48.5].

There is, however, an alternative method based directly upon the coma correction; it has no notable advantage over the previous method as regards time and trouble involved, but it is a more direct attack of the problem. Having found the values of c_a and c_b, we solve by the TL G-sum equation for zero *coma* of the cemented objective. We will exemplify this method on our microscope objective, but will introduce another change which the previous calculations proved to be permissible : the objective will bear an increased aperture, as with the original aperture of 0·25 the zonal spherical aberration amounted to only 1/8th of the tolerance. We will therefore calculate now for an aperture of 0·30 or initial $y = Y = 0{\cdot}15$.

Recalling that the objective was to be of unit focal length and to work with the TL data $l_1 = -6$, $l_3' = +1{\cdot}2$, we begin by determining the chromatic tolerance $= 1$ wave-length/$\sin^2 U_3'$. With the new semi-aperture 0·15 the TL value of $\sin U_3'$ will be $0{\cdot}15/1{\cdot}2 = 0{\cdot}125$, and as inch-measure is to be used, the wave-

length will be 0·00002, or chromatic tolerance = $0{\cdot}00002/(0{\cdot}125)^2 = 0{\cdot}00128$. [64]
Allowing this full amount in the sense of under-correction, we have to calculate TL, Chr. (4) with $R = 0{\cdot}00128/(1{\cdot}2)^2 = 0{\cdot}00089$, and find for the prescribed unit focal length $c_a = 4{\cdot}483$, $c_b = -2.528$.

In solving for *zero* coma of the thin cemented combination by Seidel Aberrations III we can omit the common factor $H'_k \,.\, SA^2$, and have simply

$$0 = \tfrac{1}{4}\, G_5{}^a c_a c_2 - G_7{}^a c_a v'_2 + G_8{}^a c_a{}^2 + \tfrac{1}{4}\, G_5{}^b c_b c_2 - G_7{}^b c_b v'_2 - G_8{}^b c_b{}^2,$$

by using the alternative G-sum for the first, the original G-sum for the second component. Solved for c_2 the equation gives

$$c_2 = [G_7{}^a c_a v'_2 + G_7{}^b c_b v'_2 - G_8{}^a c_a{}^2 + G_8{}^b c_b{}^2]/(\tfrac{1}{4}\, G_5{}^a c_a + \tfrac{1}{4}\, G_5{}^b c_b).$$

With $v_1 = -1/6$ we find $v'_2 = v_1 + c_a(N_a - 1) = 2{\cdot}408$ and then $c_2 = -3{\cdot}178$, giving by c_a and c_b, $c_1 = 1{\cdot}305$ and $c_3 = -0{\cdot}650$, or the TL prescription for a coma-free lens with chromatic under-correction equal to the tolerance :

$$\begin{array}{ll} r_1 = 0{\cdot}7663 & \\ & 0{\cdot}09 \\ r_2 = -0{\cdot}3147 & \\ & 0{\cdot}04 \\ (r_3 = -1{\cdot}538) & \end{array}$$

The crown thickness has been increased to 0·09 in order to give a sufficient edge thickness with the contemplated increased aperture.

With the initial $l_1 = -6$ and the semi-aperture 0·15, sin U_1 should be about $-0{\cdot}025$, and $U_1 = -1\text{–}25\text{–}46$ was accordingly decided upon. For chromatic under-correction equal to the full tolerance $r_3 = -1{\cdot}356$ was found, and this gave

final $LA' = +0{\cdot}00034$, final $OSC' = -0{\cdot}00062$,

the first utterly insignificant and OSC' less than one quarter of the limit. On allowing only three-quarters of the full chromatic tolerance, with r_1 and r_2 unchanged, the value $r_3 = -1{\cdot}383$ was found, and the usual completion of the calculation gave

final $LA' = -0{\cdot}00063$, final $OSC' = -0{\cdot}00032$.

The tolerance for LA' being found $= \pm 0{\cdot}0048$, it is seen that both residuals are about one-eighth of the permissible amounts.

The primary spherical aberration by Seidel Aberrations IIa for the final form with $r_3 = -1{\cdot}383$ came out at $LA'_p = +0{\cdot}00517$, also a small fraction of the tolerance when discussed as indicative of zonal aberration. The solution may therefore be accepted as an excellent compromise. If the residual OSC' were large enough to deserve attention, it would be easy to give a small positive bending to the complete objective so as to bring OSC' to zero. As the objective lies close to the minimum aberration form, this small bending would not change the spherical correction to any sensible extent.

It would not be difficult to obtain a closer approach than three-quarters of the tolerance to the normal chromatic correction, simply by substituting a barium crown of higher index and/or V-value for the one actually used. This remedy has

[64] not been employed because it has been found in recent years that most of the heavy barium crown glasses are subject to becoming tarnished on long exposure; the glass actually chosen is the most suitable one of those types in Chance's list of 1919 which have not added to their description in the list the warning letter T, which refers to this defect.—On referring back to section [42] (1) it will be realized that departure from the time-honoured visual chromatic correction in the direction of under-correction may actually, in small objectives, prove a desirable improvement. Our final form of the simple microscope objective would merely have minimum focus somewhere in the deep green instead of in the yellow-green.

E. TL equations for OSC′. On account of their importance in approximate solutions and especially in the final differential correction, we will collect the formulae for OSC' of a thin lens or system passed centrally by the oblique pencils, or with $l'pr = 0$. As $OSC' = Coma'_S/H'_k$, the general equations and their differential forms are simply the coma equations of Seidel Aberrations III with omission of the factor H'_k, or

$$OSC' = SA^2 \sum(\text{Coma } G\text{-sums of constituent lenses}),$$

in which

$$G\text{-sum} = \tfrac{1}{4} G_5 c c_1 - G_7 c v_1 - G_8 c^2 = \tfrac{1}{4} G_5 c c_2 - G_7 c v_2' + G_8 c^2,$$

whence $$OSC'_{requ.} = OSC'_{found} + SA^2(\tfrac{1}{4} G_5{}^a \,.\, c_a \,.\, \delta c_a + \tfrac{1}{4} G_5{}^b c_b \delta c_b + \&c.),$$

the amount of bending, δc, of each lens being distinguished by the suffix letter of that lens. The differential formula, whilst very easily calculated by the tabulated G-values, can be rendered still more convenient for many applications by introducing the explicit value of G_5 and the power $1/f'$ of the individual lenses, for we thus find

$$\tfrac{1}{4} G_5 \,.\, c \,.\, \delta c = \tfrac{1}{2} \frac{N+1}{N} (N-1)\, c \,.\, \delta c = \tfrac{1}{2} \frac{N+1}{N} \cdot \frac{1}{f'} \cdot \delta c.$$

Now $(N+1)/N$ varies very little in value for the usual range of refractive indices, for it gives 1·667 for $N = 1·5$ and 1·625 for $N = 1·6$. Hence we shall only introduce an error which is rarely serious in comparison with the inherent inaccuracy of all TL equations, if we adopt a well-chosen mean value of $\frac{1}{2}(N+1)/N$ for all the separate items in the round bracket of the $\delta OSC'$ formula. As achromatic combinations greatly preponderate, and as these nearly always have a higher index for the flint-components than that of the crown-components, 0·86 has been chosen as the average numerical value of $\frac{1}{2}(N+1)/N$, thus leading to the greatly simplified equation

$$\delta OSC' = 0{\cdot}86\, SA^2(\delta c_a/f'_a + \delta c_b/f'_b, \&c.),$$

which for uniform bending by δc of a thin *system* of focal length f' takes the still more simple form

$$\delta OSC' = 0{\cdot}86\, SA^2 \,.\, \delta c/f',$$

or transposed $$\delta c = 1{\cdot}16 f' \,.\, \delta OSC'/SA^2.$$

The formula gives for the third form in subsection C, with $f' = 10$, $\delta OSC' = 0{\cdot}0015$, $SA = 1{\cdot}25$: $\delta c = 0{\cdot}0112$; we used $\delta c = 0{\cdot}011$.

For the final form in subsection D the formula gives as the bending which [64] would remove the small offence $-0{\cdot}00032$, with $f' = 1$, $SA = 0{\cdot}15$, and $\delta OSC' = +0{\cdot}00032$: $\delta c = +0{\cdot}0165$, which may be tested.

The formula is equally applicable to more complicated triple or multiple thin systems, and gives approximations of usually ample closeness.

We will record these equations as :

Sine Theorem IV, for thin systems with central passage of oblique pencils.

$$OSC' = SA^2 \,.\, \sum(\text{Coma-}G\text{-sums of individual lenses})\,;$$

$$\text{Coma-}G\text{-sum} = \tfrac{1}{4}\,G_5cc_1 - G_7cv_1 - G_8c^2 = \tfrac{1}{4}\,G_5cc_2 - G_7cv_2' + G_8c^2\,;$$

$$\delta OSC' = SA^2(\tfrac{1}{4}\,G_5{}^ac_a\delta c_a + \tfrac{1}{4}\,G_5{}^bc_b\delta c_b, \text{ \&c.}).$$

Approximate formulae for uniform bending of any reasonably thin system

$$\delta OSC' = 0{\cdot}86 \,.\, SA^2 \,.\, \delta c/f' \quad \text{and} \quad \delta c = 1{\cdot}16\,f' \,.\, \delta OSC'/SA^2.$$

All the equations become available for $Coma'_S$ by adding the factor H'_k *on the right-hand sides to* SA^2.

F. Tolerances for OSC′ and Coma. We have already met many cases in which complete correction of the important aberrations was impossible, so that some form of compromise had to be adopted. As more aberrations are included in our solutions, so this necessity for compromises will become increased. In order to be able to choose the best, or at any rate a favourable, compromise, it is necessary to know the relative importance or *visibility* of the aberrations between which a balance is to be struck, and it is this aspect which calls for the use of the tolerances which were first introduced and explained in Chapter III. The tolerances stated in that chapter were based on the Rayleigh limit, and determine that value of any aberration at which it begins to have a sensible effect on the quality of the image. For primary coma the corresponding tolerance is

$$\text{Rayleigh limit for } Coma'_S = \tfrac{1}{2}\text{ wave-length}/N' \sin U'_m,$$

an interesting result, for it represents exactly the adopted value of the visual diameter of the spurious disk as recorded in Chapter III. For OSC' we obtain the corresponding tolerance by noting that $OSC' = Coma'_S/H'_k$, hence we have the formulae :

OT(6) Rayleigh limit :

$$Coma'_S = \tfrac{1}{2}\text{ wave}/N' \sin U'_m; \quad OSC' = \tfrac{1}{2}\text{ wave}/N'H'_k \sin U'_m.$$

It should be carefully noted that these tolerances have simply $\sin U'_m$ in the denominator, whilst all the previous ones had the square. This is due to the fact that we measure coma as a transverse aberration, whilst we previously dealt with longitudinal ones. The Rayleigh limit is very severe in the case of coma. The astrographic standard telescopes, for instance, use a field of about 5 inches square, so that the value of H'_k for the corners is about $3{\cdot}5$ inches. The $f/$ ratio is about $f/10$ or $\sin U'_m = 0{\cdot}05$, hence

$$\text{Rayleigh limit for } OSC' = 0{\cdot}00001/(3{\cdot}5 \times 0{\cdot}05) = 0{\cdot}000057,$$

[64] about as close to zero as we could guarantee the result by five-figure logs. For the usual $f/15$ ratio of large astronomical telescopes and H'_k = one inch we find

$$\text{Rayleigh limit for } OSC' = 0{\cdot}00001/0{\cdot}0333 \quad = \quad 0{\cdot}00030.$$

For the Dennis Taylor type we estimated in section [63] $OSC' = 0{\cdot}00138$ at $f/15$, hence, on account of the inverse proportionality of permissible OSC' with H'_k, an objective of this type would reach the Rayleigh limit for coma at $0{\cdot}00030/0{\cdot}00138 = 0{\cdot}217$ inch from the centre of the field.

For the 2-inch objective at $f/4$ calculated in Chapter I we have $\sin U'_m = 0{\cdot}127$, hence for $H'_k = 1''$: permissible $OSC' = 0{\cdot}000079$; we really found in section [63] $OSC' = 0{\cdot}009054$; hence this objective would reach the Rayleigh limit for coma at $0{\cdot}000079/0{\cdot}009054 = 0{\cdot}0087''$ from the centre of the field ; the perfectly defined field would therefore measure only 1/60th of an inch in diameter!

For astrophotographic instruments and for photographic lenses used for the reproduction of maps, &c., we must try to keep within the single, or at any rate the doubled, Rayleigh limit, but shall very often fail to satisfy this severe condition for the outer part of the field. In all other cases we can disregard the Rayleigh limit for coma, and discuss the *visibility* of the coma effect under the actual conditions of the particular instrument. For the very low degrees of coma contemplated by the Rayleigh limit, the actually seen image of a point of light departs only slightly from the usual spurious disk ; it becomes slightly oval, and the light distribution is a little unsymmetrical. When the coma exceeds two or three times the Rayleigh limit, then the deformation of the image becomes pronounced. With a very bright point—or a long exposure in the case of photography—we may see the whole 'coma patch' described in [55] D, which for primary coma measures 3 $Coma'_S$ from tip to tail. But there is very little light in the round end of the patch ; nearly all of it is at the pointed end, and with reasonably favourable illumination, or a properly timed exposure in photography, the *appreciated* image takes the form of a broad arrow, that is, three lines of light at angles of 30° from line to line and meeting at the point of the patch, the central line being much stronger than the two flanking lines, and the linear measurement of this actually seen patch agrees quite closely with the calculated $Coma'_S$. For this reason we have only to discuss at what magnitude our calculated $Coma'_S$ will become easily distinguishable from a mere point. The resolving power of a normal human eye is about one minute of arc, meaning that under the most favourable conditions two points or lines at an angular separation of 1′ may be seen resolved, and also that the diameter of a small patch or the thickness of a line begins to be appreciated when the subtense is 1′ or more. This extreme limit, however, implies a decided strain on the eye, and could not be worked at for any length of time. For comfortable and continued observation a subtense of 2′ or even 3′ must be demanded, and the magnifications chosen by experienced observers for regular work with telescopes or microscopes bear out this conclusion.

On the other hand, the eyepieces of optical instruments afford a field of view which normally varies only from about 30° to 40° or occasionally up to 50°, but really close and critical observations are never made at the extreme limit of the field ; with ordinary photographs very similar angular subtenses come into question, for it is not comfortable to roll the eye through more than 10° or at most 20° from its normal line of sight. If we adopt 10° from the centre of the field as the normal limit of interest, and 20° as the extreme limit attainable without decided discomfort and strain, a coma patch subtending the lowest perceptible angle of 1′ of arc will represent 1/600 and 1/1200 respectively of the whole angle, a patch of 2′ will represent 1/300 and 1/600, one of our adopted maximum of 3′ will amount to 1/200 and 1/400 respectively of the whole angle. As our OSC' determines precisely this ratio of the sagittal coma patch to its distance from the centre of the field, the several fractions just stated *are* the admissible values of OSC' under the respective assumptions. The particular value 0·0025 = 1/400 already anticipated in this section was adopted by the author many years ago as the result of direct experimental tests of microscope objectives specially computed so as to have definitely known coma-values, and has been extensively verified as suitable in actual practice. It *must* however be regarded as a *real limit*, and not as an analogon of the Rayleigh tolerances, which frequently may be exceeded without grave drawbacks. It must also be restricted to eyepieces which always remain in alinement with the optical axis of the object-glass. When eyepieces are fitted on slides which admit decentring, as is frequently the case in astronomical instruments and occasionally in micrometer-microscopes, then the OSC'-limit must be reduced, usually in a drastic manner.

Another exceptional case arises chiefly in photographic reproductions. A definite limit has there to be put to the actual *linear measurement* of the coma patch. For extremely sharp definition 0·001 inch may be adopted, for normal use, as yielding decidedly good definition, we may admit 0·004 inch (0·1 mm.); finally, 0·01 inch is very usually given as a suitable value in photographic books, but this high value corresponds to a very obvious want of sharpness of definition, and it is not advisable to be satisfied with it. The admissible linear measure of $Coma'_S$ is thus defined; and the admissible value of OSC' will be obtained by dividing by the distance H'_k of the point farthest from the centre of the photograph. In photographic 'process'-work H'_k will frequently exceed 10 inches, and if the sharpest definition is demanded, the tolerance for OSC' may become as low as that derived from the Rayleigh limit.

We will collect the formulae as

OT (6) *Tolerances and Limits for Coma and OSC′:*

(1) Rayleigh limit: $Coma'_S = \pm \frac{1}{2}$ wave$/N' \sin U'_m$;

$$OSC' = \pm \tfrac{1}{2}\ \text{wave}/N' H'_k \sin U'_m.$$

(2) For *ordinary* telescopes, microscopes, and photographic objectives, *highest admissible* value of $OSC' = \pm 0{\cdot}0025$.

(3) When the actual linear size of the coma patch has to be considered:

$$\text{Admissible } Coma'_S = \begin{cases} 0{\cdot}001 \text{ inch for extremely sharp definition,} \\ 0{\cdot}004 \text{ inch for good definition,} \\ 0{\cdot}010 \text{ inch for decidedly soft definition.} \end{cases}$$

$$\text{Admissible } OSC' = \text{admissible } Coma'_S / (\text{maximum value of } H'_k).$$

OTHER APPLICATIONS OF THE SINE THEOREM

[65] We have seen in the two preceding sections that the sine theorem supplies extremely simple and accurate methods for the determination and correction of coma in centred systems of every kind. In telescope and microscope objectives coma is almost the only aberration of the oblique pencils which requires attention; but even in these systems an approximate determination of the other Seidel aberrations is frequently desirable, whilst in photographic objectives and in eyepieces these other aberrations are of greater importance than those of the axial pencil. The question arises whether a complete solution by the primary aberration methods can be avoided by suitable simple additions to the calculations required by the sine theorem. In the case of reasonably thin systems an answer is provided by the third fundamental law: for central passage of the oblique pencils we have $Ast'_S = H_k'^2/2f'$; if we adopt $H'_k = 1$, we shall have $OSC' = Coma'_S$ and $Ast'_S = 1/2f'$, together with the Petzval term $Ptz' = \frac{1}{2}$ Petzval sum. Hence, with $Dist' = 0$ for central passage, we know all the Seidel aberrations for $l'pr_k = 0$, and can apply the second law in the manner already exemplified in order to find the best position of the exit pupil. For systems which are too thick to allow the application of the third law we can frequently gain all the information which may be of interest by solving by Sine Theorem II for the position of the exit pupil which yields freedom from coma, and by determining the remaining Seidel aberrations by one of the trigonometrical methods of the following chapter. Time may frequently be saved by adopting one or other of these additions to the calculations required by the sine theorem.

The sine theorem, in which we may include the theorem of Lagrange as merely its paraxial form, also supplies a very convenient method for the determination of the transverse chromatic aberration from the ray-tracing of the axial pencil. With reference to Fig. 75, we proceed exactly as in the application of the sine theorem.

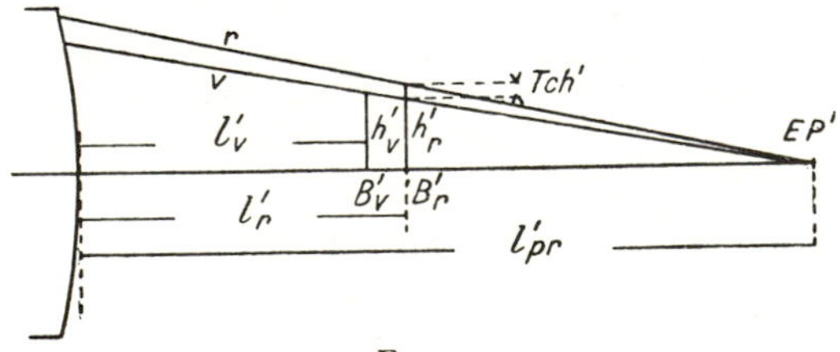

FIG. 75

If paraxial rays in 'red' and 'violet' have been traced through the centred system with initial angle u, and have given the final data l'_r, u'_r and l'_v, u'_v respectively, then the theorem of Lagrange can be applied to find the final image heights h'_r and h'_v corresponding to an original object height h_1. If we preserve the utmost generality by assuming the object in a medium of indices N_r and N_v, and the images in a medium of indices N'_r and N'_v, the theorem will give

$$h'_r = h_1 \frac{u_1 N_r}{u'_r N'_r} \text{ and } h'_v = h_1 \frac{u_1 N_v}{u'_v N'_v}, \quad \therefore \quad \frac{h'_v}{h'_r} = \frac{u'_r N'_r N_v}{u'_v N'_v N_r}.$$

As we are dealing with thin pencils, the upper ends of h'_r and h'_v will be the foci of indefinitely close sagittal rays ; therefore they lie on the principal rays of the respective thin oblique pencils ; the distance between the points at which these principal rays cut through the adopted paraxial image-plane is by definition the transverse chromatic aberration Tch' which we are to determine. We will treat 'red' as the principal colour, and can read off the diagram by similar triangles

$$Tch' = h'_r - h'_v \frac{l'_{pr} - l'_r}{l'_{pr} - l'_v}.$$

As in the case of the sine theorem, it is more convenient and appropriate to adopt the *ratio* Tch'/h'_r as the measure of this chromatic defect ; this ratio represents the constant g_7 of the transverse chromatic aberration, and adopting this symbol we have

$$g_7 = 1 - \frac{h'_v}{h'_r}\frac{l_{pr} - l'_r}{l'_{pr} - l'_v} = 1 - \frac{h_v}{h'_r} \cdot \frac{l'_{pr} - l'_v - (l'_r - l'_v)}{l'_{pr} - l'_v} = 1 - \frac{h'_v}{h'_r} + \frac{h'_v}{h'_r} \cdot \frac{l'_r - l'_v}{l'_{pr} - l'_v}.$$

If we now introduce the previously determined value of h'_v/h'_r, we obtain the final result

$$g_7 = 1 - \frac{u'_r N'_r N_v}{u'_v N'_v N_r} + \frac{u'_r N'_r N_v}{u'_v N'_v N_r} \cdot \frac{l'_r - l'_v}{l'_{pr} - l'_v}.$$

Nearly always we shall have object and image both in air ; as all the indices in the general formula are then exactly 'one', the equation takes the much simpler form

$$g_7 = 1 - \frac{u'_r}{u'_v} + \frac{u'_r}{u'_v} \cdot \frac{l'_r - l'_v}{l'_{pr} - l'_v} = \frac{u'_v - u'_r}{u'_v} + \frac{u'_r}{u'_v} \cdot \frac{l'_r - l'_v}{l'_{pr} - l'_v}$$

and can be easily and quickly calculated by slide-rule, as $u'_v - u'_r$ in the first term and $l'_r - l'_v$ in the second term are always quite small and only known with limited precision. For achromatic systems $l'_r - l'_v$ will of course be zero, so that we then have simply $g_7 = (u'_v - u'_r)/u'_v$; hence for achromatic systems the desirable zero value of g_7 calls for $u'_v = u'_r$, and as the 'red' and 'violet' rays have a common focus, this condition can only be fulfilled if they also have a common point of emergence from the last surface. Hence, for dry systems, reunion of the coloured components of any entering white ray on emergence from the last surface is, strictly speaking, a condition of complete achromatism in both the longitudinal and the transverse sense. In ordinary telescope objectives and in systems built up of separately achromatized and reasonably thin components, the possible departure from fulfilment of this condition is however very small, and it is usually not necessary to pay close attention to it. The defect resulting from non-fulfilment of the condition becomes quite sensible in fairly thick cemented combinations of pronounced meniscus form, such as the ordinary landscape lens, and it becomes serious in the higher powers of microscope objectives in which chromatically over-corrected compound lenses are combined with an uncorrected front lens ; in the latter case the resulting difference of magnification for different colours can be corrected by the use of compensating eyepieces. The reunion of the coloured rays

[65] at the last surface must be borne in mind from the first stage of a new design, and must be closely realized at the final stage, in photographic objectives like the 'Cooke' and the 'Aldis' lens, in which uncorrected lenses at considerable separations occur.

It should be noted that a common point of emergence of the 'red' and 'violet' rays will *not* secure achromatism of magnification—or $g_7 = 0$—when object and/or image lie in media other than air, for if we again assume longitudinal achromatism, the condition for $g_7 = 0$ will then be

$$u'_r N'_r N_v = u'_v N'_v N_r,$$

and will in general demand a difference of the final points of emergence of associated 'red' and 'violet' rays. In oil-immersion microscope objectives this complication always *aggravates* the chromatic difference of magnification, and that is the reason why compensating eyepieces are particularly desirable for these systems.

With regard to tolerances, the close similarity between the transverse chromatic aberration and coma leads to corresponding agreement of the permissible magnitudes of g_7 and of OSC'; but there is the usual difficulty of estimating the range of colours which are sufficiently bright to have to be taken into account. If 'red' and 'violet' correspond to a difference of refractive indices about equal to that from C to F, the limits stated in OT (6), (2), and (3) may be safely adopted with substitution of g_7 for OSC' and of Tch' for $Coma'_S$. The Rayleigh limit is of course meaningless in the case of Tch', as different colours are being compared, so that there is no possibility of mutual interference effects.

The method of determining g_7 and Tch' as described above is a perfectly correct one for the calculation of the true primary transverse chromatic aberration, and will give the same result as that by Seidel Aberrations II within the precision of the underlying ray-tracing. But as we do not calculate paraxial rays in the two colours to be united at a common focus when we carry out our standard methods of optical computation, the above method would require additional ray-tracing before it could be applied. By a small sacrifice of strict mathematical rigour, but in real practice almost invariably with a decided gain in actual precision for the full aperture of any system, we can avoid this difficulty by noting that the deductions drawn from Fig. 75 are equally valid, but for the respective sagittal foci at finite aperture, if we take B'_r and B'_v as the axial foci of the zone through which our usual two coloured rays have been traced. We shall then have by the sine theorem

$$h'_r = h_1 \sin U_1 N_r / \sin U'_r N'_r; \; h'_v = h_1 \sin U_1 N_v / \sin U'_v N'_v.$$

We must however assume that at this finite aperture the coma is nearly the same for the two colours, as otherwise the relation between the two sagittal foci would not imply similar relations between the remaining rays of the two complete pencils. That is the point at which mathematical rigour is sacrificed.

It is easily seen that the resulting formulae will merely have sines instead of paraxial angles, and capital L instead of small l. If we use 'zonal g_7' as a short symbol for the resulting compromise value, we shall therefore have the convenient computing formulae

$$\text{zonal } g_7 = 1 - \frac{N'_r N_v \sin U'_r}{N'_v N_r \sin U'_v} + \frac{N'_r N_v \sin U'_r}{N'_v N_r \sin U'_v} \cdot \frac{L'_r - L_v}{l'_{pr} - L'_v},$$

or for dry systems

$$\text{zonal } g_7 = 1 - \frac{\sin U'_r}{\sin U'_v} + \frac{\sin U'_r}{\sin U'_v} \cdot \frac{L'_r - L'_v}{l'_{pr} - L'_v},$$

the last term on the right being zero in the most usual case when the chromatic longitudinal aberration is corrected. For the aplanatic Steinheil objective, section [64] B, we had perfect longitudinal achromatism, but a small difference between the final U' of the two rays traced through 0·8 of the full intended aperture, namely $U'_y = 3 - 40 - 21$ and $U'_v = 3 - 40 - 17$, whence

$$\text{zonal } g_7 = 1 - \frac{\sin(3 - 40 - 21)}{\sin(3 - 40 - 17)} = -0{\cdot}00030,$$

and meaning that although 'y' and 'v' are focused in the same image-plane they give different sized images, those by 'v' being 0·03 per cent. larger than those by 'y'. Within the small field of a telescope objective this minute difference would of course be utterly invisible, and the fact that this aberration is always of this order of smallness in normal telescope objectives accounts for the universal custom of not troubling about it in these objectives. As has already been stated, it does however become notable and even extremely grave in microscope objectives, in eyepieces, and especially in photographic objectives.

We will collect the principal equations as

Sine Theorem V. Transverse chromatic aberration. To determine the *primary* transverse chromatic aberration, trace an incident white paraxial ray in two colours through the system by the standard paraxial ray-tracing formulae, and calculate

$$g_7 = 1 - \frac{u'_r N'_r N_v}{u'_v N'_v N_r} + \frac{u'_r N'_r N_v}{u'_v N'_v N_r} \cdot \frac{l'_r - l_v}{l'_{pr} - l'_v},$$

or for dry systems :

$$g_7 = \frac{u'_v - u'_r}{u'_v} + \frac{u'_r}{u'_v} \cdot \frac{l'_r - l'_v}{l'_{pr} - l'_v}.$$

To find a usually far more accurate compromise value for the full aperture, trace a zonal ray trigonometrically in two colours, preferably through 0·7071 of the full aperture, and calculate

$$\text{'zonal } g_7\text{'} = 1 - \frac{\sin U'_r \,.\, N'_r \,.\, N_v}{\sin U'_v \,.\, N'_v \,.\, N_r} + \frac{\sin U'_r \,.\, N'_r \,.\, N_v}{\sin U'_v \,.\, N'_v \,.\, N_r} \cdot \frac{L'_r - L'_v}{l'_{pr} - L'_v},$$

or for dry systems :

$$\text{'zonal } g_7\text{'} = 1 - \frac{\sin U'_r}{\sin U'_v} + \frac{\sin U'_r}{\sin U'_v} \cdot \frac{L_r - L'_v}{l'_{pr} - L'_v}.$$

The 'dashed' data refer to the final image space, plain data to the original object space. Tolerances as for OSC' in OT (6).

Historical Notes

When Abbe had discovered the sine condition and had proved its extraordinary effectiveness in freeing microscope objectives from coma, he became interested in the question, how nearly then existing good microscope objectives satisfied it,

[65] although they had of course been produced by trial and error without any knowledge of the condition. He found that all objectives with good definition in the outer part of the usual field closely fulfilled the condition, and he thus secured very strong and highly independent evidence to prove that fulfilment of the condition is *necessary* and *sufficient* for the removal of coma in spherically corrected systems.

Several alternative proposals for an additional condition to be satisfied by a lens system were made before the discovery of the sine theorem. The suggestion by the great Gauss that spherical aberration should be corrected for two different colours has already been mentioned. A totally different suggestion was made by Sir John Herschel : in the case of telescopes it is frequently inconvenient or even impossible to test and adjust them by observation of the very distant objects for which they are intended ; it would be far easier to test them on objects at a comparatively short distance. Therefore the proposal was, so to construct the object-glass with broken contact that it would be simultaneously free from spherical aberration for the very distant objects on which it would eventually be exclusively used *and* for much nearer test-objects convenient in the workshop tests. For telescope objectives this condition may be satisfied sufficiently in the TL preparation by demanding that the spherical G-sum shall be zero for $v_1 = 0$ and that its differential coefficient with reference to v_2' shall also be zero. In this form the 'Herschel Condition' leads to a form of the objective which is rather flatter on the first surface than the form satisfying the sine condition, but the difference is not sufficiently marked to lead to serious coma in telescope objectives of any usual ratio of aperture to focal length. For the glasses used in sections [48], [49], and [59], and for focal length 10, the TL prescriptions are :

Usual Cemented objective	*Satisfying Sine condition.*	*Satisfying Herschel condition.*
$r_1 = 4{\cdot}476$	$r_1 = 6{\cdot}188$	$r_1 = 6{\cdot}807$
$r_2 = -4{\cdot}055$	$r_2 = -3{\cdot}243$	$r_2 = -3{\cdot}095$
$r_3 = +2500$	$r_3 = -3{\cdot}284$	$r_3 = -3{\cdot}132$
	$r_4 = -17{\cdot}39$	$r_4 = -13{\cdot}83$

It will be noticed that the Herschel form is very much closer to that satisfying the sine condition than the usual cemented objective ; yet the coma, even of the latter, can in many cases be tolerated. Substitution of the Herschel condition for the sine condition might therefore be quite justifiable when only comparatively near objects are available for testing and adjusting.

The small difference between objectives satisfying the Herschel and the sine condition respectively has led to occasional mistakes ; Seidel, for instance, was of opinion that the objectives of the great Joseph Fraunhofer (1787–1826) had been designed for freedom from coma, and called his second condition the 'Fraunhofer Condition', because the prescriptions of many Fraunhofer objectives come so close to the form which would produce coma-correction. But in reality all these Fraunhofer prescriptions depart from the form demanded by the sine condition in the direction of the Herschel condition, and as Fraunhofer was a particularly precise and painstaking worker, it seems practically certain that he must have anticipated Herschel's suggestion ; this is the more probable because it is known that Fraunhofer's favourite tests consisted in examining everyday objects, like pages of print, at comparatively short distances. [65]

Special Memoranda from Chapter VII

The sine theorem owes its supreme value and importance to the facts, firstly, that it determines the sagittal coma of a centred system with high accuracy, which becomes practically absolute in that central part of the field within which aberrations proportional to H'^3 may be regarded as insignificant, and secondly, that it does this, however large the clear aperture may be, by a very simple calculation involving only data of the axial pencil. Telescope and microscope objectives utilize a field of at most a few degrees, so that aberrations involving H'^2 or higher powers are comparatively harmless ; hence the sine theorem makes it possible to find out everything that is of real interest in these systems by tracing only rays of the axial pencil. But as the H'^3 coma which the theorem omits would, if large, lead to intolerable zonal variation of the coma in different parts of the field, the theorem gives a good determination of the coma even of eyepieces and photographic objectives, if these are of a type capable of giving good definition throughout their large angular field. As a consequence the equations Sine Theorem I and II will be most extensively employed.

The definition $OSC' = Coma'_S / H'$ must be well remembered, with the added proviso that *strictly* the H' should be measured in the focal plane of the marginal rays and from the optical axis to the point of penetration of the principal ray.

The methods of solution for aplanatic objectives carry the solution of designing problems an important step forward. The extensions of the methods of differential correction should be particularly noted.

For difficult cases method [64] C will be found a very powerful one. It should be carefully studied, as it is capable of extension to include additional aberrations.

The TL formulae 'Sine Theorem IV' and the tolerances for OSC' and coma 'OT (6)' are of wide utility.

The determination of the transverse chromatic aberration by the aid of the sine theorem will be applied to eyepieces in Chapter X.

CHAPTER VIII

TRIGONOMETRICAL TRACING OF OBLIQUE PENCILS

[66] THE only practicable way of determining the exact course of rays is to trace them in strict accordance with the law of refraction, for we have shown that the easily calculated Seidel aberrations are only a first, highly convenient, but usually decidedly rough approximation, whilst the higher aberrations are so numerous and have such complicated coefficients that their direct algebraical determination is utterly out of the question, at any rate for oblique pencils.

A. TANGENTIAL RAYS

We learnt in Chapter VI that the five chief aberrations can be derived from the spherical aberration of the axial pencil, the easily calculated Petzval sum, and three rays of an oblique pencil which pass from the selected extra-axial object-point through the highest, central, and lowest points of the entrance pupil. As all the rays to be traced lie in the plane defined by the extra-axial object-point and the principal optical axis, they remain (for the centred systems which we always assume) in that plane throughout their entire course, and can therefore be determined by the convenient ordinary ray-tracing equations of Chapter I. By this method we obtain the true course of the selected rays, including all higher aberrations, and experience has demonstrated that if we establish the desired state of correction for these rays at the full aperture and for an extra-axial object-point at about 5/6 of the intended full field of view, we have done all that is necessary for the lower types of optical systems, such as eyepieces and ordinary inexpensive photographic objectives, whilst in the case of more ambitious designs like photographic anastigmats, we have provided a sound and necessary foundation for a more complete study.

In order to apply this simplest possible method to a given system, we must determine the usual initial data L and U of the selected rays, and having traced these through, we shall require a few special equations in order to define the aberrations.

The Opening Equations

(1) *For objects assumed at infinity*, which is the usual case for photographic objectives and for eyepieces, when the latter are computed according to the rule previously given, namely towards the short conjugate.

With reference to Fig. 76 (a), let A be the pole of the first refracting surface of the system, EP at distance Lpr the entrance pupil, and SA the semi-aperture of the latter. The numerical value and sign of SA must be that of the Y used for the marginal ray of the corresponding axial pencil, in order to render the oblique aberrations comparable with the axial LA'. An infinitely distant object can only be defined by its direction, and we therefore assume that the angle Upr between the optical axis and the arriving parallel rays is given or has been suitably selected. We shall distinguish the three rays to be traced by suffixes '*a*' for the upper, '*pr*' for the central, and '*b*' for the lower ray respectively. In the present case of parallel entering rays we shall then have $Ua = Ub =$ the given Upr. The convergence angles of the three rays are therefore known. For the initial intersection-lengths [66] of the three rays the diagram gives at sight:

$$(12)\qquad \begin{cases} La = Lpr + SA \text{ cotan } Upr \\ Lpr \text{ given} \\ Lb = Lpr - SA \text{ cotan } Upr. \end{cases}$$

The three rays can now be traced through the system by the standard computing equations of Chapter I.

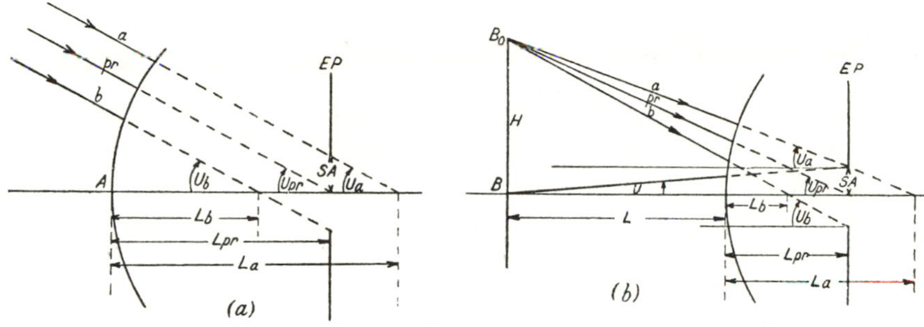

FIG. 76.

(2) *For an object-plane at a finite distance $AB = L$ from the first refracting surface.* With reference to Fig. 76 (b), the extra-axial object-point B_o shall be given and defined by its height H above the axial object-point B. The entrance pupil EP shall again be at a given distance $A - EP = Lpr$ and its semi-aperture shall be SA. The value and sign of SA must agree with the data of the marginal ray of the axial pencil, that is, the U of the marginal ray of the axial pencil and the value of SA must satisfy the relation easily read off the diagram (with due regard to signs)

$$\tan U = SA/(L - Lpr).$$

Hence we must calculate the U for the axial ray-tracing by this equation if SA has been the original datum, or else we must transpose the equation into a solution for SA, if an oblique pencil is to be added to an existing axial ray-tracing based on some other way of fixing U.

Drawing parallels to the optical axis through the upper and lower edges of the entrance pupil, we obtain for the determination of Ua, Upr, and Ub three right-angled triangles with horizontal sides in each case $= B - EP = (Lpr - L)$ and with vertical sides equal to $H - SA$, H, and $H + SA$ respectively: hence the angles are obtainable from the equations:

$$(13)\qquad \begin{cases} \tan Ua = \dfrac{H - SA}{Lpr - L} \\[2ex] \tan Upr = \dfrac{H}{Lpr - L} \\[2ex] \tan Ub = \dfrac{H + SA}{Lpr - L}. \end{cases}$$

We can then determine the L-values exactly as in case (1), but must use the proper angle U for each ray:

$$(14)\qquad \begin{cases} La = Lpr + SA \text{ cotan } Ua \\ Lpr \text{ given} \\ Lb = Lpr - SA \text{ cotan } Ub. \end{cases}$$

The preparatory computation will be recognized as of a very simple type, but it may not be unnecessary to point out that *tangents* of angles largely enter into it. If from force of optical habit the eye should be attracted by the usual sine column, very awkward errors will creep into the calculation.

In straightforward calculations by this method the three rays should be traced simultaneously in parallel columns; radii and refractive indices will then always be the same in each horizontal line where they occur. The L' and U' values found for the three rays at successive surfaces vary in a very puzzling manner, so that it is at times difficult to believe that the rays would meet at least approximately in a common point when produced. Beginners will for that reason find the $3-6-9$ check and the PA check particularly useful and reassuring in work by this method.

The Closing Equations

The final results of the tracing of the three selected rays of the oblique pencil give the three intersection-lengths $L'a$, $L'pr$, and $L'b$ and the corresponding angles $U'a$, $U'pr$, and $U'b$, and these are shown in the rather exaggerated diagram Fig. 77. The marginal ray of the corresponding axial pencil will have given its L' and U', and the paraxial ray, if included, will be defined by l' and u'. We have to determine the aberrations indicated by these results.

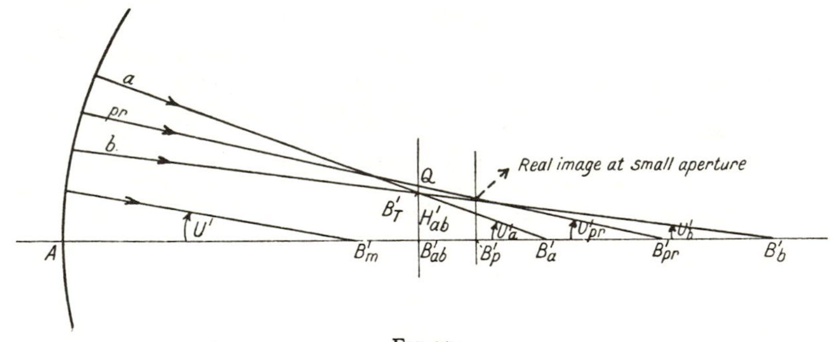

Fig. 77.

The intersection of the rays a and b at B'_T marks the tangential focus of the marginal zone of the oblique pencil; as in all our work on oblique pencils, we shall define this focus in rectangular coordinates referred to the pole A of the last surface and to the optical axis, and we therefore have to determine AB'_{ab}, which we will call L'_{ab}, and $B'_{ab}B'_T$, which we will call H'_{ab}. Right-angled triangles with H'_{ab} as vertical side and B'_a or B'_b as opposite corners give

$$H'_{ab} = (L'_a - L'_{ab}) \tan U'_a = (L'_b - L'_{ab}) \tan U'_b,$$

hence

$$\frac{L'_b - L'_{ab}}{L'_a - L'_{ab}} = \frac{\tan U'_a}{\tan U'_b} \quad \text{or} \quad \frac{L'_b - L'_{ab}}{L'_b - L'_a} = \frac{\tan U'_a}{\tan U'_a - \tan U'_b}.$$

Putting $\tan U'_a - \tan U'_b = \sin (U'_a - U'_b)/\cos U'_a \cos U'_b$, the last equation gives

$$L'_b - L'_{ab} = \frac{(L'_b - L'_a) \sin U'_a}{\sin (U'_a - U'_b)} \cos U'_b.$$

This we use in the second equation for H'_{ab} to determine the latter, and we transpose the equation itself into a solution for L'_{ab}, with the results

$$(15) \qquad \begin{cases} H'_{ab} = \dfrac{(L'_b - L'_a) \sin U'_a}{\sin (U'_a - U'_b)} \,.\, \sin U'_b \\[2ex] L'_{ab} = L'_b - \dfrac{(L'_b - L'_a) \sin U'_a}{\sin (U'_a - U'_b)} \,.\, \cos U'_b. \end{cases}$$

If this solution is repeated with the initial proportion inverted, but otherwise by the same steps, we obtain an alternative pair of equations

$$(15)^* \qquad \begin{cases} H'_{ab} = \dfrac{(L'_b - L'_a) \sin U'_b}{\sin (U'_a - U'_b)} \sin U'_a \\[2ex] L'_{ab} = L'_a - \dfrac{(L'_b - L'_a) \sin U'_b}{\sin (U'_a - U'_b)} \cos U'_a. \end{cases}$$

The two solutions for H'_{ab} are obviously identical, but in the case of L'_{ab} there is a real difference, and the second of (15) will be preferable when U'_a is numerically smaller than U'_b; in the reverse case (15)* will give the more accurate result.

If L'_{ab} is found to differ from the L' of the axial pencil, the difference will be due to curvature of the tangential field; in accordance with the nomenclature adopted in Chapter VI we will adopt the symbol X'_T for this difference of focus, and we can compute it as

$$(16) \qquad X'_T = L'_{ab} - L' \qquad (= B'_m B'_{ab} \text{ of Fig. 77}).$$

A positive value will indicate a 'hollow' field, a negative value will correspond to a 'round' field, according to the generally adopted nomenclature. As was shown in Chapter VI, zero-value of X'_T will in general represent the most desirable state of correction. It should be noted that the curvature of the field can be determined by tracing only three rays, namely the rays '*a*' and '*b*' of the oblique pencil and the corresponding marginal ray of the axial pencil.

By including the principal ray we determine the tangential coma—symbol $Coma'_T$—for this is represented by the distance $B'_T Q$ in Fig. 77, and is shown with a *positive* value inasmuch as the principal ray lies *above* the tangential focus. We calculate it as $B'_{ab}Q - H'_{ab}$; for $B'_{ab}Q$ the right-angled triangle with third corner at B'_{pr} gives $(L'_{pr} - L'_{ab}) \tan U'_{pr}$ as the correct value, hence

$$(17) \qquad Coma'_T = (L'_{pr} - L'_{ab}) \tan U'_{pr} - H'_{ab}.$$

[66] As was shown in Chapter VI, it is almost indispensable to reduce this baneful defect to a practically invisible residue.

When the paraxial ray of the axial pencil is added, we obtain firstly the ordinary axial spherical aberration $LA' = l' - L'$. But as we then can locate the paraxial ideal image-plane, and as we have traced the true principal ray of the oblique pencil, we have also the necessary data for the determination of the true distortion of our lens system. We learnt in Chapter VI that the distortion is always determined for the image formed at very small aperture in the ideal image-plane, and is therefore derived from a comparison of the value of H' by Lagrange's theorem with the height at which the true principal ray pierces the ideal image-plane. The latter has been drawn in Fig. 77 at B'_p, and we can at once read off the height at which the principal ray cuts through it as

$$\text{Actual } H'_{pr} = (L'_{pr} - l') \cdot \tan U'_{pr}.$$

For the ideal location of the image the theorem of Lagrange gives directly, for objects at a finite distance,

$$\text{Ideal } H' = H \cdot N \cdot u/N'u',$$

in which u and u' are the first and last convergence angles of the traced paraxial ray and N and N' the refractive indices of the media surrounding object and image, therefore nearly always both of value 'one' for air. H is, of course, the object-height used for the traced oblique pencil.

In the case of objects assumed at infinite distance the last equation cannot be directly applied, as the mystery product 'infinity times zero' would appear on the right. But as nobody knows even whether there are any objects at infinity, we may be satisfied by working out the product for objects at an indefinitely large finite distance L from the lens system. In that case we have, for a nominal paraxial incidence-height y at the first surface, $u = y/L$, whilst u' is $= y/f'$ by the definition of the equivalent focal length in Chapter I. Consequently the equation becomes

$$\text{Ideal } H' = \frac{N}{N'} \cdot H \cdot f'/L = \frac{N}{N'} \cdot f' \cdot \frac{H}{L}.$$

Now H/L is the tangent of the angle of subtense of the object as seen from the pole of the first refracting surface, and for very distant objects it becomes $= -\tan Upr$ according to the symbols in the opening equations for distant objects. The minus sign is necessary because a positive H and positive Upr correspond to a negative L. Therefore for very distant objects

$$\text{Ideal } H' = -\frac{N}{N'} \cdot f' \cdot \tan U_{pr}:$$

and as we defined the distortion as the distance by which the ideal image lies *above* the actual image, we obtain the computing formulae for distortion :

(18) for near objects :

$$\text{Distortion} = H\frac{Nu}{N'u'} - (L'_{pr} - l') \tan U'_{pr};$$

(18) for very distant objects : [66]

$$\text{Distortion} = -\frac{N}{N'} \cdot f' \cdot \tan U_{pr} - (L'_{pr} - l') \tan U'_{pr}.$$

For systems working in air N/N' will be simply 'one'.

The astigmatism naturally cannot be *determined* by calculating tangential rays only ; but if the obliquity is not excessive and if there are no very deeply curved surfaces in the system, we can estimate the astigmatism by calculating the Petzval curvature and assuming the three-to-one ratio of primary astigmatism for the distances of the tangential and sagittal foci from the Petzval surface according to the primary aberration laws of Chapter VI.

B. NARROW TANGENTIAL AND SAGITTAL FANS OF RAYS

[67] A most valuable addition to the method of the preceding section consists in the direct determination of the foci of the tangential and sagittal rays in the immediate vicinity of the traced principal ray. The difference between these foci gives directly the exact value of the astigmatism at small aperture, whilst comparison of the focus of the close tangential rays with that of the marginal tangential rays gives the true spherical aberration of the tangential fan. We learnt in Chapter VI that in *first* approximation the spherical aberration is uniform throughout the field of view, but when aperture and obliquity become considerable we have to reckon with secondary and even higher aberrations, and one of the most serious of these, especially in photographic lenses, is a spherical aberration term of the form $p \cdot H'^2 \cdot Y^2$, p being a constant, which can reach very grave values for the tangential fan of rays ; it is in fact most usually this aberration which, in otherwise well corrected systems, decides the limit of the field of view beyond which the definition begins to deteriorate very rapidly at full or large aperture.

(1) *Close tangential rays.* With reference to Fig. 78, AP shall be the refracting surface with centre at C, and PB shall be the principal ray of an oblique pencil

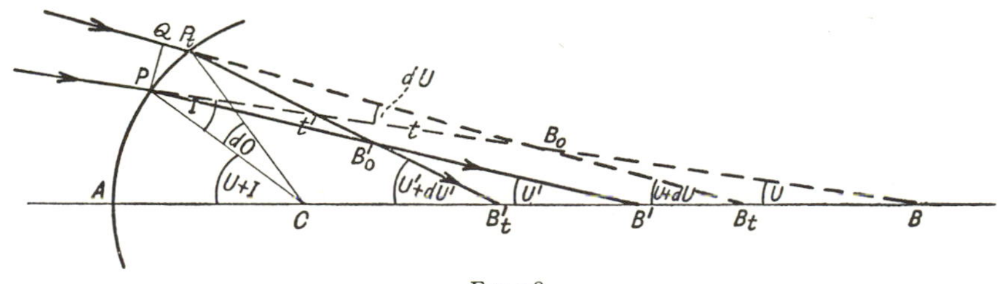

FIG. 78.

which is assumed to have been trigonometrically traced so that all its angles and intersection-lengths are known. To simplify the algebra, we will omit the suffix *pr* in the course of the proof, but we shall restore it in the final computing equations. B_o on the incident principal ray shall be either the original object-point or the tangential focus formed by preceding surfaces. We locate B_o by its distance

[67] $PB_o = t$ from the point of incidence of the principal ray ; this distance is sufficient to fix the position of B_o because we know the course of the principal ray on which it lies. The angles of the principal ray shall be U, I, and, at the centre of curvature, $(U+I)$. We now consider a close tangential ray aiming at B_o and cutting the surface at P_t and the principal optical axis at B_t. It will have slightly different angles of convergence and of incidence, which we will call $(U+dU)$ and $(I+dI)$. If we call the angle PCP_t dO, the angle at the centre of curvature will be, for the close tangential ray, $U+I+dO$, and the universally valid fundamental equation (3) gives

$$U+I+dO = U+dU+I+dI,$$

or by obvious cancellation

(a) $$dO = dU+dI.$$

Up to this point the relations established are exact and valid for any values of dO, dU, and dI. Restriction to very small, strictly differential, values becomes necessary when we seek the relations between r, t and these quantities. Strictly we have *arc* $PP_t = r \,.\, dO$, but if we now take the point P_t extremely close to P, then the arc PP_t will become indistinguishable from the tangent to the refracting surface at P, and we can measure the distance PP_t on this tangent and still claim

$$PP_t = r \,.\, dO$$

as true at the limit of indefinitely small values of dO. If we now draw $PQ \perp PB$, we shall have angle QPP_t equal to I, and therefore

$$PQ = PP_t \,.\, \cos I = r \,.\, \cos I \,.\, dO.$$

The triangle QPB_o next gives $PQ = t \,.\, \tan dU$, but as we have already restricted our investigation to indefinitely small values, we have $PQ = t \,.\, dU$, or on equating the two values of PQ

$$t \,.\, dU = r \,.\, \cos I \,.\, dO.$$

By equation (a) we can replace dU by $(dO - dI)$, giving

$$t \,.\, dO - t \,.\, dI = r \cos I \,.\, dO,$$

which by a simple transposition becomes

(b) $$dI = \frac{t - r \cos I}{t} \,.\, dO.$$

The rays incident at P and P_t will be refracted so as to have angles I', U', $(I'+dI')$ and $(U'+dU')$, but the angle dO at the centre of curvature will necessarily be the same as for the incident rays. The refracted rays will cut each other at some point B'_o, and we call the distance $PB'_o = t'$.

As there are only changes from 'plain' to 'dash', it is clear that a repetition of the above deductions for the incident ray must give for the refracted rays

(b)′ $$dI' = \frac{t' - r \cos I'}{t'} \,.\, dO.$$

To establish a connexion between the equations (b) and (b)′ we have only to notice that $(I+dI)$ and $(I'+dI')$ are the angles of incidence and of refraction for the tangential ray ; they therefore must obey the law of refraction, and the differential form of the latter as given in section [2] gives at once

$$dI \,.\, N \,.\, \cos I = dI' \,.\, N' \,.\, \cos I' :$$

and if we introduce the values of dI and dI' by (b) and (b)′ we obtain

$$N \,.\, \cos I \frac{t - r\cos I}{t} \,.\, dO = N' \,.\, \cos I' \frac{t' - r\cos I'}{t'} \,.\, dO.$$

The quantity dO, which locates the point of incidence of the tangential ray with reference to that of the principal ray, cancels out and proves that the relations between t and t' expressed by the surviving terms hold for all possible values, either positive or negative, of dO, provided of course that the points of incidence P_t are so close to P as to cause only an insensible departure of the arc PP_t from the tangent at P. The existence of a definite tangential focus on the principal ray is thus firmly proved.

A slight transformation of the last equation now leads to

$$N\cos I - \frac{r}{t} N\cos^2 I = N'\cos I' - \frac{r}{t'} N'\cos^2 I'$$

and division throughout by r and a rearrangement of the sequence of terms gives the final solution

$$\text{(c)} \qquad \frac{N'\cos^2 I'}{t'} = \frac{N'\cos I' - N\cos I}{r} + \frac{N\cos^2 I}{t}.$$

(2) *Close sagittal rays.* The sagittal rays in the vicinity of a trigonometrically traced principal ray present a still simpler problem.

With reference to Fig. 79, we again assume that the principal ray incident at P, aiming before refraction at B on the principal optical axis, has been traced and

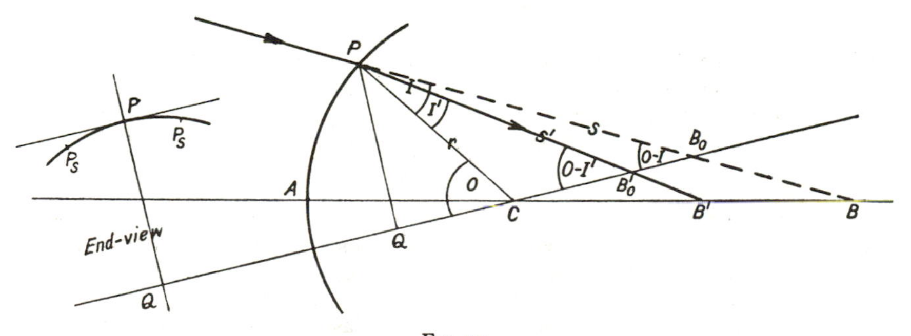

FIG. 79.

that the angles of incidence I and of refraction I' are known. B_o at distance $PB_o = s$ from the point of incidence shall be the point at which the *arriving* close sagittal rays aim. We lay an auxiliary axis through the centre of curvature C and the

[67] object-point B_o and add the end-view of the surface as seen from B_o. The true sagittal rays of the oblique pencil will pass through a straight line laid through P at right angles with the centre line PQ, and the close sagittal rays in the immediate neighbourhood of P will cut the refracting surface at a distance from the auxiliary axis which differs by insensibly small amounts from the distance PQ of the principal ray. Hence these close sagittal rays will have precisely the same longitudinal spherical aberration as the principal ray, and as before refraction they are focused on the principal ray at B_o they must after refraction form a sharp focus at the point B'_o at which the refracted principal ray cuts the auxiliary axis. Therefore PB'_o is the focal distance s' of the refracted fan of close sagittal rays. To determine s' we call the angle between PC and the auxiliary axis O; then we have the angle at $B_o = O - I$ and the angle at $B'_o = O - I'$, and the triangles PCB_o and PCB'_o give by the trigonometrical sine relation

$$\frac{r}{s} = \frac{\sin(O-I)}{\sin O} = \frac{\sin O \cos I - \cos O \sin I}{\sin O} = \cos I - \sin I \,.\, \text{cotan}\, O$$

$$\frac{r}{s'} = \frac{\sin(O-I')}{\sin O} = \frac{\sin O \cos I' - \cos O \sin I'}{\sin O} = \cos I' - \sin I' .\, \text{cotan}\, O.$$

To eliminate the purely auxiliary angle O, we multiply the first equation throughout by $\sin I'$ and the second by $\sin I$, and then subtract the first from the second, with the result

$$\frac{r \sin I}{s'} - \frac{r \sin I'}{s} = \sin I \cos I' - \cos I \sin I'.$$

This gives the final equation by multiplying the terms *alternately* by $N'/r \sin I$ or $N/r \sin I'$, which are equal and therefore freely exchangeable by the law of refraction, and by then throwing the second term of the left to the right.

The solution for close sagittal rays thus obtained is

$$\text{(d)} \qquad \frac{N'}{s'} = \frac{N' \cos I' - N \cos I}{r} + \frac{N}{s}$$

and it will be seen that it differs from equation (c) for close tangential rays only by not having the $\cos^2$ factors.

Both equations also have a close relation with the paraxial equation (6p)**

$$\frac{N'}{l'} = \frac{N'-N}{r} + \frac{N}{l},$$

for they assume that form—apart from the special symbols s and t—when a principal ray happens to pass radially through a surface, because I and I' will then be zero and their cosines will be exactly 'one'.

As equations (c) and (d) contain the same 'astigmatic constant' $(N' \cos I' - N \cos I)/r$, they are particularly convenient for simultaneous calculation of the foci of narrow sagittal and tangential fans. There are absolutely no special cases to be provided for; the equations always give a highly precise result. For plane surfaces the astigmatic constant becomes zero, owing to the infinite value of r, and simplifies the calculation.

As the angles I and I' are necessarily acute and as the cosines of all acute angles are invariably positive, the equations cause very little worry as to signs. The s and t are of course reckoned positive to the right and negative to the left of the surface to which they apply, exactly like the l and L of axial pencils.

In order to adapt the equations for actual numerical work, we must add formulae for the transfer from surface to surface, and instructions for the opening and closing of a complete calculation through any given system.

Opening. With reference to Fig. 80 (*a*), BB_o represents the original object-plane at distance L (negative in the diagram) from the first refracting surface, and B_oP_{pr}

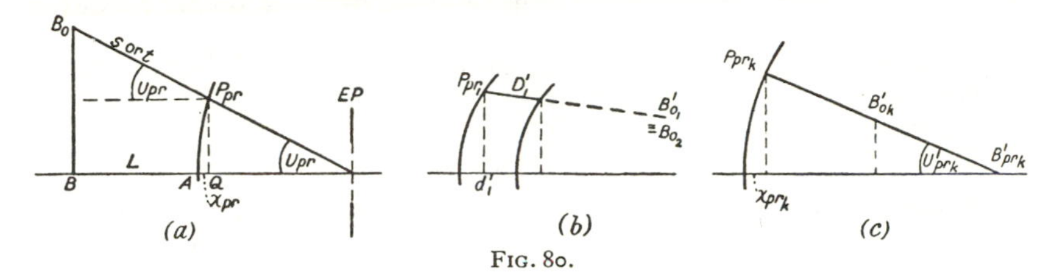

FIG. 80.

is the principal ray of the oblique pencil, at a known angle U_{pr} with the optical axis. Then B_oP_{pr} will be the initial value of both s and t, and will have the sign of L. Projection of B_oP_{pr} upon the optical axis gives the starting values

(e) $$s \text{ or } t = (L - X_{pr}) \sec U_{pr}.$$

The X_{pr} must be calculated from the trigonometrical data of the *principal* ray by one of the formulae (8) in Chapter I. If the original object is assumed infinitely distant, then no opening calculation is required; the equations (c) and (d) become reduced to the astigmatic constant only on their right sides owing to s and t being infinitely large.

Transfer. Referring to Fig. 80 (*b*), a preceding surface distinguished by suffix 1 will have given a focal distance $P_{pr1}B'_{o1}$, whilst for the calculation through the next surface we require $P_{pr2}B'_{o1}$ as the plain s_2 or t_2. Hence the transfer calls for the subtraction of the 'marginal thickness' D'_1, and this has therefore to be calculated for every lens or space by equation (9) of section [14] as

$$D'_j = (d'_j + X_{j+1} - X_j) \sec U'_j,$$

again with the trigonometrical data of the *principal* ray.

Closing. On emergence from the last surface the astigmatic fans will have a final intersection-length on the principal ray $= s'_k$ or t'_k as calculated by equations (d) or (c). In accordance with our general rule we define the final foci in rectangular coordinates referred to the optical axis and to the pole of the last refracting surface. Calling these coordinates l'_t or l'_s and H'_t or H'_s respectively, we can read off the diagram Fig. 80 (*c*):

$$\text{Final } l'_t = t'_k \cos U'pr_k + Xpr_k \qquad \text{Final } H'_t = (L'pr_k - \text{final } l'_t) \tan U'pr_k$$
$$\text{Final } l'_s = s'_k \cos U'pr_k + Xpr_k \qquad \text{Final } H'_s = (L'pr_k - \text{final } l'_s) \tan U'pr_k.$$

As it is known that the final foci lie on the principal ray, it is not often necessary

[67] to calculate the final H' ; our interest is usually limited to the longitudinal differences of these foci, and as the latter are comparable, in aperture of the fans of rays producing them, with the paraxial focus at distance l'_k from the last surface, we shall also derive the curvature of field for small apertures as X'_t = final $l'_t - l'_k$ and X'_s = final $l'_s - l'_k$. Comparison of the final l'_t for the narrow tangential fan with the L'_{ab} of the marginal tangential rays traced by the method of the previous section gives the true spherical aberration of the tangential fan as $l'_t - L'_{ab}$, or rather its projection upon the principal optical axis. As in nearly all optical instruments the focusing movement is along the principal optical axis, it will be realized that the projected value of the oblique longitudinal aberrations is the proper measure of their magnitude.

We will collect the equations in concise form and will assign definitive reference numbers to them ; the temporary reference letters of the present section are added in brackets.

Close Tangential and Sagittal Rays

Preparation. Trace the principal ray trigonometrically through the entire system. Calculate the depth of curvature X_{pr} corresponding to its point of incidence at each surface,

if PA-check has been used : $X_{pr} = PA^2_{pr}/2\,r$

if PA-check has not been used : $X_{pr} = 2r \sin^2 \tfrac{1}{2}(U_{pr} + I_{pr})$.

With these X-values calculate for each lens or space of axial thickness d'_j :

(9) $$D'_j = (d'_j + Xpr_{j+1} - Xpr_j) \sec U'pr_j.$$

Opening. If the original object-plane is at the *finite* distance L from the first surface, calculate

(19) $$\text{initial } s_1 = \text{initial } t_1 = (L - Xpr_1) \sec Upr_1 \quad \text{(e)}.$$

For infinitely distant objects simply use $s_1 = t_1 = \infty$.

Ray-tracing. Calculate for each surface

(20) $$t' = N' \cos^2 I'pr \Big/ \left(\frac{N' \cos I'pr}{r} - \frac{N \cos Ipr}{r} + \frac{N \cos^2 Ipr}{t}\right) \quad \text{(c) transposed;}$$

(21) $$s' = N' \Big/ \left(\frac{N' \cos I'pr}{r} - \frac{N \cos Ipr}{r} + \frac{N}{s}\right) \quad \text{(d) transposed.}$$

Transfer. $t_{j+1} = t'_j - D'_j$; $s_{j+1} = s'_j - D'_j$.

Closing. The rectangular coordinates of the final foci are

(22) $$\begin{cases} \text{Final } l'_t = t'_k \cos U'pr_k + Xpr_k. & \text{Final } H'_t = (L'pr_k - \text{Final } l'_t) \tan U'pr_k. \\ \text{Final } l'_s = s'_k \cos U'pr_k + Xpr_k. & \text{Final } H'_s = (L'pr_k - \text{Final } l'_s) \tan U'pr_k. \end{cases}$$

Conclusions to be Drawn from the Results

(1) The true astigmatism at small apertures, namely primary astigmatism *plus* all higher orders of pure astigmatism, is :

(23) $$\text{True astigmatic difference of focus} = l'_s - l_t.$$

(2) By comparing l'_t and l' with the paraxial intersection-length l'_k we obtain a measure of the curvature of the tangential and sagittal field at the calculated obliquity, namely : [67]

(24) $$X'_t = l'_t - l'_k; \quad X'_s = l'_s - l'_k.$$

(3) If the full aperture tangential rays 'a' and 'b' have also been traced, we obtain the tangential spherical aberration :

(25) $$\text{Tangential } LA' = l'_t - L'_{ab}.$$

The last test should never be omitted in designing photographic objectives of the anastigmatic type, for a large value (either positive or negative) of tangential LA' will prove the system unsuitable for use at large aperture on a field of considerable angular extent.

C. SKEW RAYS

Rays of an oblique pencil which do not come under either of the preceding two sections call for special formulae if they are to be traced accurately in accordance with the law of refraction, for they will in general have a different plane of incidence at every successive surface of even a centred optical system, and this plane has to be determined before the general ray-tracing equations of Chapter I can be applied. [68]

Referring to Fig. 81, which is drawn in the same perspective already extensively introduced in Chapter VI, we assume a spherical refracting surface with pole at A and centre of curvature at C, both of course on the principal optical axis AX of the centred system. We seek formulae by which a ray through *any* point P of the refracting surface can be traced with trigonometrical accuracy. To fix the ideas we

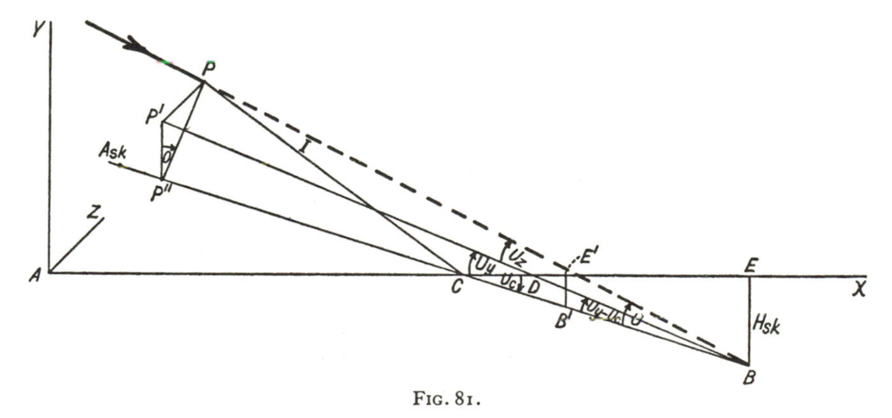

FIG. 81.

will assume P to lie behind the plane of the diagram, but this must not be regarded as a restriction applying to the equations to be deduced. Reserving for separate treatment the case when the ray through P happens to be exactly parallel to the plane of the diagram, there must in all other cases be a definite point B in which the ray cuts the plane of the diagram ; we use this intersection point as the

[68] principal datum by which the ray is defined, and locate B in our usual rectangular coordinates by projection upon the optical axis at E, calling the abscissa $AE : L_{sk}$ and the ordinate $EB : H_{sk}$, the latter being negative in our otherwise 'all-positive' diagram. For the present we assume L_{sk} and H_{sk} as known, either as the coordinates of the given original object-point, or as determined by a ray-tracing through preceding surfaces of the system. As an infinite number of rays could be laid through B in all possible directions, we must next define the particular ray to be traced through point P of the refracting surface. For that purpose we project the ray PB upon the plane of the diagram by dropping the perpendicular PP' and joining P' to B. As point B has already been definitely located by its coordinates L_{sk} and H_{sk}, the projected ray $P'B$ will now be completely defined by the angle U_y—positive if clockwise from the principal optical axis—which it forms with the latter at D, and the ray PB itself is then singled out by the angle U_z between the projected ray $P'B$ and the ray itself. The four coordinates L_{sk}, H_{sk}, U_y and U_z now fix the ray to be traced absolutely, and for the present we assume that all four are known.

If we now draw the auxiliary optical axis BC and produce it until it cuts the refracting surface at the auxiliary pole A_{sk}, then PCB will be the plane of incidence of our ray, $A_{sk}B$ will exactly correspond to the L of the fundamental ray-tracing equations, the angle $CBP = U$ will correspond to the U of the same equations, and angle $CPB = I$ will be the angle of incidence. Finally, CB will be the $(L-r)$ of the fundamental formulae. We have now to find mathematical formulae to define U and CB, and can then apply the standard equations.

For that purpose we first determine the angle U_c between the principal and the auxiliary optical axis, positive if clockwise from principal to auxiliary axis, and the triangle CBE gives at sight the first computing equation

SK (1a) $$\tan U_c = -H_{sk}/(L_{sk}-r),$$

the angle to be taken out as an acute one, with the sign resulting from that of the given quantities.

We next drop a perpendicular $P'P''$ from P' (which necessarily lies in the plane of the diagram) upon the auxiliary optical axis; by joining $P''P$ we then obtain a pyramid $PP'P''B$, with base triangle $PP'P''$ having a right angle at P', and with three triangular faces meeting at B, of which $PP'B$ is right-angled at P' and has the angle U_z at B, whilst face $PP''B$ is right-angled at P'' and has the angle U at B, and face $P'P''B$ is right-angled at P'' and has the angle at $B = U_y - U_c$ because U_y is exterior angle to the triangle DBC.

The angle $PP''P' = O$ of the base triangle measures the *slope* of the incidence plane with reference to the plane of the diagram; the base triangle gives

$$\tan O = PP'/P''P'.$$

We then obtain from the triangle $PP'B$

$$PP' = P'B \,.\, \tan U_z,$$

and from the triangle $P'P''B$

$$P''P' = P'B \,.\, \sin(U_y - U_c);$$

introduction of these values into the equation for tan O then gives the second computing equation [68]

SK (1b) $$\tan O = \tan U_z / \sin (U_y - U_c),$$

and as O is necessarily acute, its sign is definitely determined as that of the right-hand side of the equation.

We can now determine the angle U in a very similar manner. The triangle $PP''B$ gives

$$\sin U = P''P/PB\,;$$

from the triangle $PP'P''$ we then find

$$P''P = PP'/\sin O,$$

and from the triangle $PP'B$

$$PB = PP'/\sin U_z\,;$$

with these the equation for sin U becomes the next computing formula

SK (2) $$\sin U = \sin U_z / \sin O,$$

which again gives a perfectly definite value and sign for the acute angle U.

As it is highly desirable to have a reliable check on the numerical work up to this stage, it is strongly recommended to calculate U a second time by the following highly independent equation : the triangle $PP''B$ gives

$$\tan U = P''P/P''B,$$

and the base triangle gives $P''P = P''P'/\cos O$, whilst triangle $P''P'B$ gives $P''B = P''P'/\tan (U_y - U_c)$. Substitution of these values gives

SK (2*) $$\tan U = \tan (U_y - U_c)/\cos O$$

as a very safe check on SK (1a), (1b), and (2).

When both (2) and (2*) are calculated, we may use for the subsequent calculations whichever value of U appears to be most accurately determined ; usually the sine given by SK (2) will be both more convenient and more accurate.

In order to apply *the usual ray-tracing equations* to the result of the refraction of our ray at the surface, we have only to calculate the distance CB, which represents the $(L-r)$ of our ray with reference to the auxiliary optical axis, as has already been pointed out. The triangle CEB gives at sight the two alternative values of CB stated in the first line of SK (3), and the complete instruction is

SK (3) $$CB = (L_{sk} - r) \sec U_c = -H_{sk}/\sin U_c.$$

$$\sin I = \sin U \,.\, CB/r\,;\ \sin I' = \sin I \,.\, N/N'\,;\ U' = U + I - I'\,;\ CB' = \sin I' \,.\, r/\sin U'.$$

$$PA_{sk}\ \text{Check}: (CB'+r) = (CB+r) \sin U \sec \tfrac{1}{2}(U-I) \operatorname{cosec} U' \cos \tfrac{1}{2}(U'-I')$$

$$L'_{sk} = r + CB' \cos U_c\,;\ H'_{sk} = -CB' \,.\, \sin U_c.$$

The closing equations in the last line follow immediately by inspection of the triangle $CB'E'$ if B' is taken as the intersection point of the refracted ray following from the calculated value of CB'. The whole calculation can be arranged on the standard ray-tracing model with the sole exception that it is convenient to use CB and CB' instead of the full L and L'. As in the tracing of the usual rays proceeding

[68] in the plane of the diagram, the $3-6-9$ check should never be omitted, in order to guard against errors in adding log N/N'; the PA check is always highly advisable and *must* be used in the case of a long radius. In this latter case a further modification is advisable in order to obtain the most accurate value of L'_{sk}: seeing that $CB+r$ in the check-formula represents $L = A_{sk}B$, we can determine this by projection upon the principal optical axis without bringing in the long radius, and we thus obtain the special form:

SK (3) for a long radius:

$$CB+r = (L_{sk} - 2\,r\sin^2\tfrac{1}{2}\,U_c)\sec U_c, \text{ whence } CB = (CB+r)-r.$$

Then second and third line of the standard SK (3) unchanged, but with use of $(CB+r)$ by the present first line in the PA-check. Then the coordinates of B' by

$$L'_{sk} = (CB'+r)\cos U_c + 2\,r\sin^2\tfrac{1}{2}\,U_c\,;\quad H'_{sk} = H_{sk}\,.\,CB'/CB.$$

As the refracted ray remains in the incidence plane defined by the arriving ray, the angle O is the same after as before refraction; the angle U_c between principal and auxiliary axis also remains unchanged; but the angles U'_y and U'_z, which completely define the refracted ray, have to be calculated. As all the relations are exactly the same as for the incident ray, the necessary formulae are immediately obtainable by transposing the opening equations SK (1a), (1b), and (2), and applying the 'dash'. The closing equations are thus found as

SK (4) $\qquad \sin U'_z = \sin O\,.\,\sin U'$;

SK (5) $\qquad \sin(U'_y - U_c) = \tan U'_z/\tan O$; whence

or Check SK (5*) $\quad \tan(U'_y - U_c) = \tan U'\,.\,\cos O \qquad U'_y = (U'_y - U_c) + U_c.$

The four adopted coordinates for the refracted ray being thus determined, we can proceed to the next refracting surface by the obvious transfer equations

SK (6) $\qquad L_{sk1} = L'_{sk} - d'$; $H_{sk1} \equiv H'_{sk}$; $U_{z1} \equiv U'_z$; $U_{y1} \equiv U'_y$;

and these data can be used in the set of equations worked out above to trace the rays through the new surface. A little work may be saved by noting that only *trigonometrical functions* of the angles O and U_z enter into the calculations; hence it is not necessary to extract the angles themselves; it suffices—if the computer has the necessary aptitude—to turn up the log tan first found and to take out the corresponding logs of sin and cos directly without reference to the angle. No special closing equations are required at the last surface, as the final image-point is determined by the standard set of formulae in our usual rectangular coordinates. By reference to Chapter VI it will be easily seen that a ray traced by these skew-ray equations may always (with a centred system) be regarded as one ray of a D-pair, for by symmetry a ray passing through a point P as far in front of the plane of the diagram as our assumed ray is behind that plane must cut this plane in the same point. Each ray-tracing by these equations thus necessarily covers a *pair of rays*, and that is a fairly complete compensation for the greater amount of work required in the calculation.

Only one special case has to be provided for, namely that of a perfectly plane refracting surface. By regarding the plane as part of a sphere of infinite radius, it becomes obvious that in its case the auxiliary optical axis becomes parallel to the [68] principal optical axis ; therefore U_c is zero and does not require a calculation. And as the image-point must fall upon the auxiliary optical axis, it follows as a further

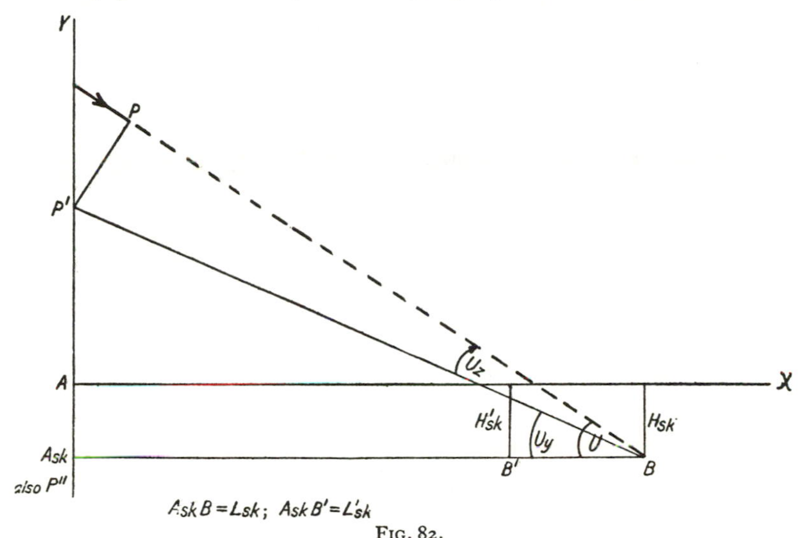

FIG. 82.

simplification that H'_{sk} must always be equal to H_{sk}, so that here also no calculation is required.

As $U_c = 0$, we can at once write down the equations for O and U by putting $U_c = 0$ into the general equations, with the results

SK (1b) for plane : $\tan O = \tan U_z / \sin U_y$.

SK (2) for plane : $\sin U = \sin U_z / \sin O$.

SK (2)* for plane : $\tan U = \tan U_y / \cos O$. (Check)

We can then apply the usual equations for refraction at a plane as given in Chapter I, section 15, and obtain :

SK (3) for plane : $\sin U' = \sin U \, . \, N/N'$

$$L'_{sk} = L_{sk} \, . \tan U \, . \operatorname{cotan} U' = L_{sk} (N'/N) \, . \sec U / \sec U'.$$

Check : $L_{sk} - L'_{sk} = L_{sk} \, . \sec U \, . \sin (U' - U) \, . \operatorname{cosec} U'$

$$H'_{sk} = H_{sk}.$$

By introduction of $U_c = 0$ the general equations (4) and (5) become :

SK (4) for plane : $\sin U'_z = \sin O \sin U'$.

SK (5) for plane : $\sin U'_y = \tan U'_z / \tan O$.

SK (5)* for plane : $\tan U'_y = \tan U' \, . \cos O$. (Check)

The transfer equation SK (6) remains unchanged.

[68]

Opening Equations

We must now add the necessary formulae for the determination of the initial data of a skew ray when the original object-point is defined by our usual coordinates L and H, or by U_{pr} if assumed at infinity, and when the ray is to pass through a prescribed point on the circumference of an entrance pupil of semi-aperture SA at distance L_{pr} from the pole of the first refracting surface.

(1) *For an object-point at a finite distance.* With reference to Fig. 83, the pole of the first refracting surface shall be at A, the given entrance pupil at EP at distance L_{pr} from A, and the object-point B (assumed virtual in the diagram) shall have the

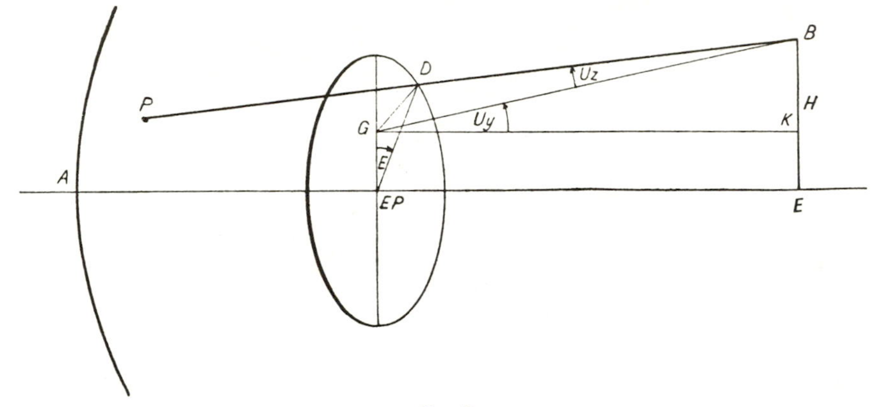

Fig. 83.

coordinates L and H. It is required to define the skew ray passing through point D of the entrance pupil of semi-aperture $EP\text{-}D = SA$. We define the point D by its position angle E with reference to the vertical diameter of the pupil, counting this angle either completely clockwise and positive up to 360°, or positive up to 180° if behind the plane of the diagram, and negative up to − 180° if in front of the plane of the diagram.

We project D upon the XY plane at G, then the right-angled triangle $DG\text{-}EP$ gives

$$DG = SA \,.\, \sin E \quad \text{and} \quad G\text{-}EP = SA \,.\, \cos E:$$

and these formulae will give the correct sign according to the conventions of analytical geometry, whichever way we may choose in measuring the angle E, viz. either right round the clock to 360° or positive and negative to 180°. If we now join GB and draw GK parallel with the principal optical axis, then angle DBG will be the angle U_z and angle BGK will be U_y according to our adopted definitions, the latter angle being negative in the diagram.

We first determine U_y. As $G\text{-}EP = SA \cos E$ and $BE = H$, the triangle BGK gives

$$\tan U_y = (SA \cos E - H)/(L - L_{pr}),$$

the sequence of terms in the numerator being necessary in order to bring out U_y [68] with the required minus sign.

We can then calculate U_z, for in the right-angled triangle BGD we have $DG = SA \sin E$ and the side $GB = (L - L_{pr})/\cos U_y$, hence

$$\tan U_z = SA \sin E \cos U_y/(L - L_{pr}).$$

The other two coordinates of the skew ray require no calculation, for we have

$$L_{sk} = L \quad \text{and} \quad H_{sk} = H.$$

The collected formulae therefore are:

Sk opening for an object-point at finite distance:

$$L_{sk} = L\ ;\ H_{sk} = H\ ;\ \tan U_y = (SA \cos E - H)/(L - L_{pr})$$
$$\tan U_z = SA \sin E \cos U_y/(L - L_{pr}).$$

The skew ray to be traced is nearly always the true sagittal ray which passes through the end of the horizontal diameter of the entrance pupil. The angle E is then exactly 90° for this sagittal ray; hence the third and fourth computing equations simplify to

Sagittal ray: $\tan U_y = -H/(L - L_{pr})\ ;\ \tan U_z = SA \cos U_y/(L - L_{pr}).$

No special cases have to be provided for in this instance of objects at a finite distance, because the radius of curvature of the surface does not enter into the formulae in any way.

(2) *For objects at infinity when the radius is finite.* As the heading implies, we have in this case to exclude the possibility of a *plane* first surface (as frequently found in eyepieces), reserving this for separate treatment. With reference to Fig. 84, the stipulations concerning the entrance pupil remain as before. But the

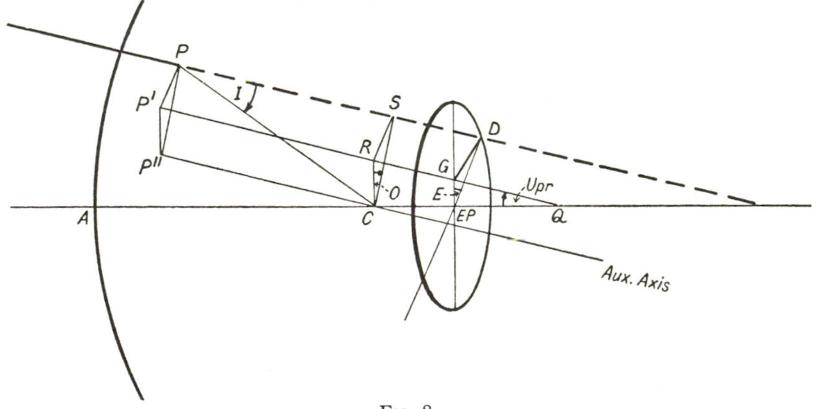

FIG. 84.

infinitely distant object-point can only be defined by its angle U_{pr} from the optical axis, and all the rays received from it will be parallel to each other and to the

[68] auxiliary optical axis laid through the centre of curvature C of the refracting spherical surface.

On account of the parallelism of the ray PD, its projection $P'Q$ and the auxiliary axis $P''C$, we have $PP' = DG = SA \,.\, \sin E$ and we have also $G\text{-}EP = SA \,.\, \cos E$. Q being the point where the projected ray cuts the principal axis, we have angle $GQ\text{-}EP$ equal to the given obliquity U_{pr} of the entering parallel rays, and we can determine the distance from the entrance pupil to Q as

$$EP\text{-}Q = G\text{-}EP \,.\, \text{cotan}\, U_{pr} = SA \cos E \,\text{cotan}\, U_{pr}.$$

If we now draw a perpendicular RC from the centre of curvature upon the projected ray and from R a perpendicular upon the ray itself at S, we obtain a triangle CRS equal to $P''P'P$ and see that the angle O can be obtained from the triangle CRS by

$$\tan O = RS/RC,$$

and that the angle of incidence I is given by the triangle SCP as $\sin I = SC/r$.

As parallels between parallels we have $RS = GD = SA \sin E$. In the triangle RCQ we have side $CQ = C\text{-}EP + EP\text{-}Q$, or with $C\text{-}EP = L_{pr} - r$ and $EP\text{-}Q$ as determined $= SA \cos E \,\text{cotan}\, U_{pr}$,

$$CQ = (L_{pr} - r) + SA \cos E \,\text{cotan}\, U_{pr}.$$

We have next $RC = CQ \,.\, \sin U_{pr}$, or with the preceding,

$$RC = (L_{pr} - r) \sin U_{pr} + SA \cos E \cos U_{pr};$$

therefore

$$\tan O = RS/RC = SA \sin E/[(L_{pr} - r) \sin U_{pr} + SA \cos E \cos U_{pr}]$$

and this is the equation for calculating angle O.

With this value of O we can now find the distance CS required for the determination of $\sin I$, for the triangle CRS gives $CS = RS/\sin O$, whence with the above value of $RS = SA \sin E$:

$$\sin I = SA \,.\, \sin E/r \sin O.$$

For the usual true sagittal ray at $E = 90°$ there is again a simplification, because $\sin E = 1$ and $\cos E = 0$, and we can collect the formulae as follows:

SK opening for objects at infinity and for a finite radius.

$$\text{Tan}\, O = SA \sin E/[(L_{pr} - r) \sin U_{pr} + SA \cos E \cos U_{pr}];$$

$$\sin I = SA \sin E/r \sin O;$$

or for the true sagittal ray ($E = 90°$)

$$\tan O = SA/(L_{pr} - r) \sin U_{pr}; \quad \sin I = SA/r \sin O.$$

With the angle of incidence thus determined, we begin the ray-tracing by Sk(3) with the second equation in the second line of that set of equations and then go straight on in the usual way, excepting that the U_c of the standard formulae becomes replaced by U_{pr} and that in the PA check $r \,.\, \sin I$ takes the place of $(CB + r) \sin U$.

(3) *For objects at infinity when the surface is a plane.* The equations of the

previous subsection cannot be used in this case, because with r infinite O becomes [68] infinitely small and I becomes indeterminate, since the denominator takes the form of infinity times zero. The case is, however, a particularly simple one, for when a bundle of parallel rays meets a plane, all the rays necessarily form equal angles of incidence and therefore also equal angles of refraction ; hence the refracted bundle again consists of parallel rays, and this clearly holds even for a succession of plane refracting surfaces if they are all centred, that is, at right angles with the principal optical axis. With reference to Fig. 85, it will be realized by applying

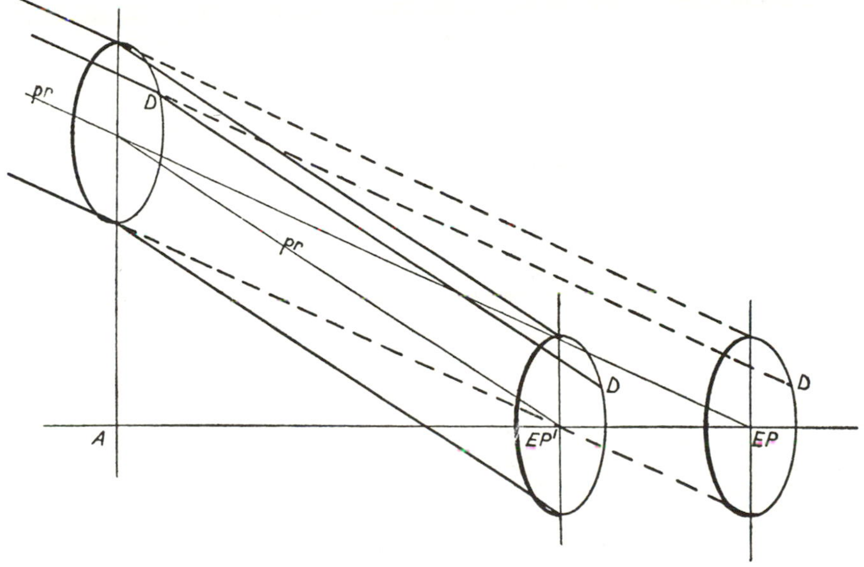

FIG. 85.

elementary principles of solid geometry that an entering parallel bundle fitting the entrance pupil EP will cut the refracting surface in a precisely equal circle, and that the refracted bundle must again have the same cross-section when cut by a plane at right angles with the optical axis, as for instance by the exit pupil EP' of the surface at A. It follows that neither the size of the pupils nor the distribution of rays in them are changed as long as only centred refracting planes are met by parallel rays. Therefore it is not necessary to trace any particular skew ray separately ; it is sufficient to trace the principal ray of the pencil by the ordinary ray-tracing equations up to the first *spherical* surface of the system and then to place the originally prescribed entrance pupil at the point found for the crossing of the axis by the traced principal ray on emergence from the last plane surface. The opening formulae of the preceding subsection are then applied with U_{pr} equal to the U'_{pr} found at the last plane and with the originally prescribed values of the semi-aperture SA and of the position angle E. The case will frequently

[68] occur in the designing of eyepieces, because they very usually have plane external surfaces.

Almost the only skew ray to be traced in regular computing practice is the true sagittal ray at $E = 90°$. By comparing the final sagittal focus found by this ray with the course of the principal ray to which it is referred, and with the focus of the narrow sagittal fan found by equation (22), we obtain results closely corresponding to those obtained by the rays a, pr, and b. We will collect these under the reference

SK (7) *Results from the tracing of the true sagittal ray.*

(a) The distance from the sagittal focus to the principal ray, measured at right angles with the optical axis, determines

$$\text{True } Coma'_S = (L'_{pr} - L'_{sk}) \tan U'_{pr} - H'_{sk}.$$

(b) The difference between L'_{sk} and the corresponding L' of the marginal ray of the axial pencil of the same aperture gives

$$\text{True } X'_S = L'_{sk} - L'.$$

(c) The difference between l'_s of the narrow sagittal fan and the L'_{sk} of the marginal sagittal ray determines the spherical aberration of the sagittal fan as

$$\text{Sagittal } LA' = l'_s - L'_{sk}.$$

(d) The true astigmatic difference of focus between the marginal sagittal and tangential pairs is given by : $L'_{sk} - L'_{ab}$.

By calculating a number of skew rays at suitable intervals of the angle E, the true characteristic focal line may be determined ; it may be stated that in the presence of secondary aberrations the characteristic focal line ceases, as a rule, to be straight and to be bisected by the sagittal focus : it becomes an arc of a common parabola.

D. CLOSE SAGITTAL RAYS

[69] In many cases of simple photographic objectives and of eyepieces it is hardly worth while to apply either the 's' and 't' formulae or to trace skew rays, as the desirable state of correction, and the only one over which we have control in these cases, consists in making the field flat for the tangential rays. As it will nevertheless be desirable to obtain some direct evidence as to the magnitude of the resulting incurable astigmatism, the following surprisingly simple formulae may be used to secure that information with very little labour.

With reference to Fig. 86, PB_s shall be the trigonometrically traced principal ray of an oblique pencil arriving at the spherical refracting surface, and B_s shall be given as the focus of the close sagittal rays of this pencil. As B_s must lie on the principal ray, it is completely defined by the projected distance $AE = l_s$ from the pole of the refracting surface, and we adopt l_s as the quantity determining B_s. If we draw the auxiliary axis B_sC, then the focus of the refracted close sagittal rays must lie on this line in accordance with the proof given for the usual formula for the s-rays, and as this focus must also lie on the trigonometrically determined refracted principal ray, it is completely defined by the intersecting point B_s and can be located by its projected distance $l'_s = AE'$ from the pole of the surface. An equation giving l'_s in terms of l_s and of the usual data of the surface and of the

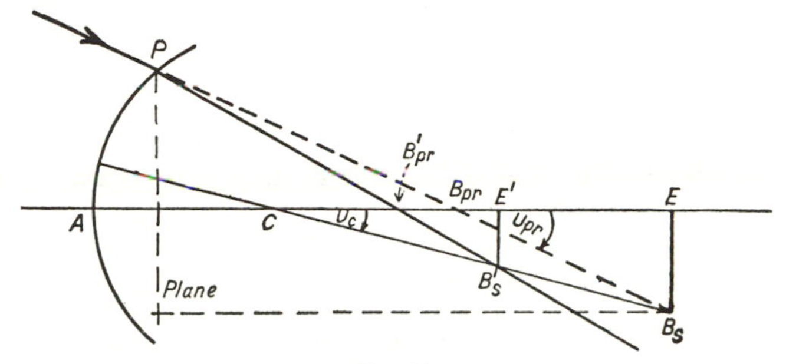

FIG. 86.

principal ray will therefore solve our problem. The known data of the principal ray are $AB_{pr} = L_{pr}$ and $AB'_{pr} = L'_{pr}$, together with the angles of obliquity U_{pr} and U'_{pr}. Taking the incident rays first, we have

$$EB_s = H_s = -B_{pr}E \times \tan U_{pr} = -(l_s - L_{pr}) \tan U_{pr}$$

and with this the triangle CEB_s gives the angle U_c between the principal and the auxiliary axes as

$$\tan U_c = -EB_s/CE = \tan U_{pr}(l_s - L_{pr})/(l_s - r).$$

If l'_s were already known, precisely the same relation would hold for the 'dashed' data; hence l'_s must satisfy the equation

$$\tan U_c = \tan U'_{pr}(l'_s - L'_{pr})/(l'_s - r)$$

and we can obtain an equation in which l'_s is the only unknown quantity by equating the two values of $\tan U_c$, giving

$$\tan U'_{pr}(l'_s - L'_{pr})/(l'_s - r) = \tan U_{pr}(l_s - L_{pr})/(l_s - r)$$

or transposed and slightly modified,

$$\frac{l'_s - L'_{pr}}{l'_s - r} = \frac{l_s - L_{pr}}{(l_s - r)\,\text{cotan}\, U_{pr} \,.\, \tan U'_{pr}}.$$

Applying to this proportion the rule: If $a/b = c/d$, then also $a/(b-a) = c/(d-c)$ we obtain:

$$\frac{l'_s - L'_{pr}}{L'_{pr} - r} = \frac{l_s - L_{pr}}{(l_s - r)\,\text{cotan}\, U_{pr} \tan U'_{pr} - (l_s - L_{pr})} = \frac{1}{\dfrac{l_s - r}{l_s - L_{pr}}\,\text{cotan}\, U_{pr} \tan U'_{pr} - 1}.$$

If we now multiply throughout by the denominator on the left and then transpose the L'_{pr} to the right, we obtain the highly convenient computing formula

(22)* $$l'_s = L'_{pr} + (L'_{pr} - r)\Big/\left(\frac{l_s - r}{l_s - L_{pr}}\,\text{cotan}\, U_{pr} \tan U'_{pr} - 1\right)$$

[69] which requires only half the computing time as compared with the original s and t equations when only the sagittal focus is to be calculated.

Equation (22)* gives a comparatively inaccurate value of l'_s in the rare cases when L'_{pr} is very large compared to l'_s, as the latter will then be found as the small difference of two much larger numbers. To meet this case we invert the starting proportion into

$$\frac{l'_s - r}{l'_s - L'_{pr}} = \frac{l_s - r}{(l_s - L_{pr}) \tan U_{pr} \operatorname{cotan} U'_{pr}}$$

and then turn this into $a/(a-b) = c/(c-d)$. Proceeding otherwise exactly as before, we obtain the alternative formula

$$(22)^* \qquad l'_s = r + (L'_{pr} - r) \Big/ \left(1 - \frac{l_s - L_{pr}}{l_s - r} \tan U_{pr} \operatorname{cotan} U'_{pr}\right).$$

Few cases of a spherical refracting surface are likely to be met with in which one of the two alternative equations will not give a sufficiently exact result.

The plane surface, however, calls for a separate solution, as for it the auxiliary optical axis becomes a parallel to the principal optical axis, as is indicated by broken lines in Fig. 86. It is evident from this that for a plane EB_s becomes $= E'B'_s$, whence we obtain from the triangles $B_{pr}EB_s$ and $B'_{pr}E'B'_s$, resulting for this case

$$(l'_s - L'_{pr}) \tan U'_{pr} = (l_s - L_{pr}) \tan U_{pr} \,;$$

and as for any ray refracted at a plane we have

$$L'_{pr} \tan U'_{pr} = L_{pr} \tan U_{pr},$$

subtraction of this equation from the previous one gives

$$l'_s \tan U'_{pr} = l_s \tan U_{pr}, \qquad \text{or}$$

$$(22)^* \text{ for a plane: } \qquad l'_s = l_s \tan U_{pr} \operatorname{cotan} U'_{pr},$$

which in nearly every case is more quickly and conveniently calculated by the transformation of $\tan U_{pr} . \operatorname{cotan} U'_{pr}$ which was already used in Chapter I, but which holds *only* for plane surfaces, namely

$$(22)^* \text{ for a plane: } \qquad l'_s = l_s(N'/N) . \sec U_{pr}/\sec U'_{pr}.$$

As in the case of either formula the complete factor of l_s will have been already used in the calculation of L'_{pr} from L_{pr}, the work for the finding of the sagittal focus is almost negligible in the case of a plane surface.

Transfer from surface to surface. As the l'_s and l_s are measured on the principal optical axis, the transfer consists simply in deducting the thickness of the lens or space.

No closing equations are required, as the l'_s is directly calculated as our usual abscissa counted from the pole of the refracting surface.

The opening also is excessively simple, for we clearly have: Initial l_s = given distance L of the object-point if the latter is at a finite distance, whilst in the case of objects at infinity the $(l_s - r)/(l_s - L_{pr})$ or its reciprocal in the denominators of the two solutions for l'_s becomes simply 'one', making the opening calculation for [69] *objects at infinity*

$$l'_s = L'_{pr} + (L'_{pr} - r)/(\text{cotan } U_{pr} \tan U'_{pr} - 1)$$

or

$$l'_s = r + (L'_{pr} - r)/(1 - \tan U_{pr} \text{ cotan } U'_{pr}).$$

If with objects at infinity the system begins with one or several planes, these can be simply ignored, as the images will remain at infinity until a spherical surface comes into action.

The equations (22)* are not so precise as the regular s and t formulae of section [67], but they give the desired result with sufficient accuracy and by decidedly less computing work in the cases for which they have been recommended. They break down in one case which is apt to occur in photographic objectives and eye-pieces, namely when the principal ray passes nearly radially through a spherical surface, for as that causes U_{pr} to be nearly equal to U'_{pr} and at the same time L_{pr} and L'_{pr} to be nearly equal to r, the equations approach the indefinite form $l'_s = L'_{pr} + 0/(1 - 1)$. This case is met, without sacrificing the greater convenience of (22)* for the remaining surfaces of the system, by substituting the regular s-method at the one offending surface. It is easily seen by reference to Fig. 86 and to section [67] that we have

$$s = (l_s - X_{pr}) \sec U_{pr} \quad \text{or} \quad \frac{1}{s} = \cos U_{pr}/(l_s - X_{pr}).$$

Introduction of this value of $\frac{1}{s}$ into the regular s equation gives

$$s' = N' \Big/ \left[\frac{N' \cos I'_{pr}}{r} - \frac{N \cos I_{pr}}{r} + \frac{N \cos U_{pr}}{l_s - X_{pr}} \right],$$

and as we also have

$$s' = (l'_s - X_{pr}) \sec U'_{pr} \quad \text{or} \quad l'_s = X_{pr} + s' \cos U'_{pr},$$

we obtain the solution which meets the case of a spherical surface at which nearly radial passage of the principal ray occurs:

$$(22)^{**} \quad l'_s = X_{pr} + N' \cos U'_{pr} \Big/ \left[\frac{N' \cos I'_{pr}}{r} - \frac{N \cos I_{pr}}{r} + \frac{N \cos U_{pr}}{l_s - X_{pr}} \right].$$

This equation, as well as an easily added companion giving l'_t in a corresponding form, could of course be used throughout, but as there would be no gain in computing time over the regular s and t method, it is only recommended for the special case referred to.

Applications of the Exact Equations

[70] The formulae deduced in this chapter should invariably be used in the final test and correction of new designs for the purpose of determining the exact magnitude of the residual aberrations, regardless of whether they are primary, secondary, or still higher. In that application it is the precision and finality of the equations which determines their selection. But the equations are also frequently useful and

[70] at times indispensable in the earlier, analytical, stage of the evolution of new designs, namely when unusually deeply curved surfaces or great obliquity of the extra-axial pencils lead to higher aberrations of such serious magnitude that the ordinarily more convenient primary approximations break down.

A. Example of a final test by the trigonometrical equations.

We will use the ordinary landscape lens of section [60]. By calculating the Seidel aberrations and applying the second fundamental law, we found that this lens should give a flat tangential field with $l_{pr1} = -0{\cdot}997$, and we calculated the other Seidel aberrations for this position of the diaphragm and for the adopted $SA_1 = 0{\cdot}3$ and $H'_k = 3$. As this lens has comparatively gentle curvatures of surfaces and is intended for a field of moderate extent, we should expect good agreement between the primary and the exact total aberrations; we have in fact already shown that there is close agreement as regards the axial spherical aberration, and also, in the previous chapter, that the primary coma is closely confirmed by the sine theorem. We will now trace the characteristic rays of an actual oblique pencil trigonometrically.

As the objects were assumed at infinity and as the value of ideal H'_k used in the primary aberration work was $= 3$, we can calculate the obliquity of the principal ray which, in the absence of distortion, would give this value of ideal H', by the first part of equation (18) for distant objects:

$$\text{Ideal } H' = -f'.\tan U_{pr}, \text{ which gives } \tan U_{pr} = -(\text{Ideal } H')/f',$$

and as $f' = 9{\cdot}5477$, we find $U_{pr1} = -17\text{–}26\text{–}36$, which with $L_{pr1} = -0{\cdot}997$, as fixed by the second fundamental law, defines the principal ray to be traced. For the method of section [66] we then calculate the opening equations (12), and with $SA = 0{\cdot}3$ we find for the three rays a, pr, b, all at $-17\text{–}26\text{–}36$,

$$L_a = -1{\cdot}95177; \qquad L_{pr} = -0{\cdot}997; \qquad L_b = -0{\cdot}04223.$$

Tracing the rays through the three surfaces by the usual routine method, we obtain the final results

$$L'_a = -3{\cdot}07569; \qquad L'_{pr} = -1{\cdot}54350; \qquad L'_b = -0{\cdot}31294;$$
$$U'_a = -12\text{-}56\text{-}1; \qquad U'_{pr} = -14\text{-}34\text{-}59; \qquad U'_b = -16\text{-}19\text{-}52.$$

Equations (15) then give $H'_{ab} = 2{\cdot}93400$; $L'_{ab} = 9{\cdot}7002$.

The marginal ray of the axial pencil having given $L' = 9{\cdot}6713$, we then obtain the residual curvature of the tangential field

(16) $X'_T = +0{\cdot}0289$.....'hollow' instead of zero.

Then by (17) $Coma'_T = 2{\cdot}92523 - 2{\cdot}93400 = -0{\cdot}00877$,

and (18), in which we *made* the first term $= 3{\cdot}0000$ by our choice of U_{pr1}, gives with paraxial $l' = 9{\cdot}8367$ in its second term

(18) Distortion $= 3{\cdot}0000 - 2{\cdot}9607 = +0{\cdot}0393$.....Barrel-Distortion.

The *exact* distortion thus found trigonometrically agrees very well with that given by the Seidel aberrations in section [60], namely $0{\cdot}04060$; the small difference $-0{\cdot}0013$ must be ascribed to secondary distortion growing as H'^5.

The positive value of X'_T similarly is due to higher terms in the curvature of the field which grow as H'^4 or as $SA^2 \,.\, H'^2$ and $Y^2H'^2$; we shall obtain some more evidence in this respect by applying the other trigonometrically exact equations.

With regard to the value found for $Coma'_T = -0{\cdot}00877$, which at first sight appears to clash gravely with the primary $Coma'_S$ found in section [60] as $= -0{\cdot}00307$, we must remember the important three-to-one ratio of primary $Coma'_T$ to primary $Coma'_S$; hence our trigonometrical $Coma'_T = -0{\cdot}00877$ must be compared with primary $Coma'_T = 3$ primary $Coma'_S = -0{\cdot}00921$, and it will be seen that the real disagreement is only $-0{\cdot}00044$, or utterly insignificant, but chiefly to be attributed to secondary coma growing as H'^3.

In the present case all the disagreements between exact and primary values are small compared with the total amounts of the residual aberrations, and we should be entirely justified in treating them as unimportant. When the discrepancies prove too large to be ignored, they must be compensated by a small modification of the design. The graph, Fig. 67, shows at once that the residual hollowness of the field of our landscape lens could be corrected by a small diminution of the distance of the diaphragm, which may be determined either by the graph or by the usual differential correcting method.

We will next apply the formulae for narrow fans of tangential and sagittal rays to the landscape lens; we can use the principal ray already traced for the preceding test. In accordance with the detailed instructions of section [67] we begin by calculating the X_{pr} at the three surfaces, finding successively $-0{\cdot}00869$, $+0{\cdot}00374$, and $-0{\cdot}03108$, and then the 'marginal thickness' D' of the two component lenses measured along the principal ray as $0{\cdot}1149$ and $0{\cdot}2717$ respectively. As the object is assumed at infinity, we have at the first surface $s = t = \infty$, and can begin at once with the working out of equations (20) and (21), finding:

	$s'_1 = -14{\cdot}0398$	$t'_1 = -13{\cdot}7221$
or with	$-D'_1 = -0{\cdot}1149$	$-0{\cdot}1149$
	$s_2 = -14{\cdot}1547$	$t_2 = -13{\cdot}8370.$

The second surface then gives

	$s'_2 = -12{\cdot}7283$	$t'_2 = -12{\cdot}3441$
and with	$-D'_2 = -0{\cdot}2717$	$-0{\cdot}2717$
	$s_3 = -13{\cdot}0000$	$t_3 = -12{\cdot}6158.$

These give for the fans emerging from the third surface

$$\log s'_3 = 1{\cdot}00012 \qquad \log t'_3 = 1{\cdot}00992,$$

and the closing formulae then give

$$\text{final } l'_s = 9{\cdot}6495 \qquad \text{final } l'_t = 9{\cdot}8704$$

as the projected distances of the final foci from the last refracting surface. Comparing these with the paraxial final $l' = 9{\cdot}8367$, with each other, and with the results for a full-aperture pair of tangential rays, we obtain a number of additional data.

70] Equation (23) gives the astigmatic difference of focus at ideal $H' = 3$ as $l'_s - l'_t = -0{\cdot}2209$, whilst the primary aberrations gave $-0{\cdot}2067$ for the same value of H'. The small excess $-0{\cdot}0142$ is secondary pure astigmatism growing as H'^4; being of the same sign as the primary astigmatism, it will cause the defect to grow even more rapidly towards the outer part of the field than would be the case if the secondary astigmatism were absent. In good *anastigmats* the secondary astigmatism has the *opposite* sign of the primary astigmatism, and neutralizes the latter at some particular value of H' which should correspond to a zone close to the margin of the intended full field.

Equation (24) gives the departures of the *s* and *t* fields from the ideal image-plane as $X'_s = l'_s - l' = -0{\cdot}1872$ and $X'_t = l'_t - l' = +0{\cdot}0337$. As the pure primary aberrations gave $X'_S = -0{\cdot}2067$ and $X'_T = 0$, we must attribute the differences to secondary curvature of the field growing as H'^4, which thus amounts to $+0{\cdot}0195$ for the sagittal and to $+0{\cdot}0337$ for the tangential narrow fans at ideal $H' = 3$. The secondary aberration therefore bends both the astigmatic image-surfaces in the 'hollow' direction, the effect being greatest for the tangential fans.

The highly important test by equation (25) gives the true spherical aberration of the full-aperture tangential fan as tangential $LA' = l'_t - L'_{ab} = +0{\cdot}1702$, to be compared with the full-aperture axial $LA' = 0{\cdot}1654$ given earlier. The spherical aberration is thus shown to increase towards the margin of the field, and as the excess grows with H'^2, it would rapidly rise beyond the calculated obliquity. As has already been pointed out, this oblique spherical excess can reach staggering magnitude of either sign in unfavourable types of photographic objectives and of eyepieces, and should therefore invariably be determined. In our present case it is practically harmless in comparison with the heavy general spherical aberration throughout the field.

We will now complete the study of our landscape lens by tracing the extreme sagittal ray at $SA = 0{\cdot}3$ of the selected oblique pencil. As the objects are assumed at infinity, we open by the special formulae of section [68] for this case, and with the initial data $U_{pr1} = -17\text{-}26\text{-}36$, $L_{pr1} = -0{\cdot}997$, and $SA = 0{\cdot}3$, find

$$\log \tan O = 9{\cdot}34147n \quad \text{and} \quad \log \sin I = 9{\cdot}40111,$$

and then by SK(3), with the modifications enumerated for the case of distant objects,

$$L'_{sk} = -13{\cdot}7414 \qquad H'_{sk} = -2{\cdot}57199.$$

SK(4) gives $\log \sin U'_z = 8{\cdot}33016n$ and SK(5) and (5)* give in good agreement $U'_y - U_c = 5\text{-}35\text{-}33$ or $U'_y = -11\text{-}51\text{-}3$. The transfer equations give for the second surface

$$L_{sk} = -13{\cdot}8414,\ H_{sk} = -2{\cdot}57199,\ U_y = -11\text{-}51\text{-}3,\ \log \sin U_z = 8{\cdot}33016n$$

and with these, as a few key-figures,

$$\log \tan U_c = 8{\cdot}95146n,\ \log \tan O = 9{\cdot}26067,\ \log \sin U = 9{\cdot}07662n;$$

then by SK(3) for long radius and by the closing and transfer formulae the co-ordinates of the ray entering the third surface

$$L_{sk} = -12{\cdot}7088,\ H_{sk} = -2{\cdot}44388,\ U_y = -12\text{-}36\text{-}41,\ \log \sin U_z = 8{\cdot}37631n.$$

The standard set of equations applied to the last surface then gives [70]

Final $L'_{sk} = 9{\cdot}4818$; Final $H'_{sk} = 2{\cdot}87144$.

By applying the data already deduced to these results of the tracing of the sagittal ray we find by SK(7) :

(a) True $Coma'_S = -0{\cdot}00304$; (b) True $X'_S = -0{\cdot}1895$;

(c) Sagittal $LA' = 0{\cdot}1677$.

The disagreements between these results and those by the primary aberrations, by the narrow sagittal fan, and by the axial pencil again indicate secondary aberrations. By tabulating all the principal results :

True $Coma'_T = -0{\cdot}00877$; True $X'_T = +0{\cdot}0289$; Tangential $LA' = +0{\cdot}1702$

By [67] $X'_t = +0{\cdot}0337$

Primary $Coma'_T = -0{\cdot}00921$; Primary $X'_T = 0$; Primary $LA' = +0{\cdot}1670$

True axial $LA' = +0{\cdot}1654$

Primary $Coma'_S = -0{\cdot}00307$; Primary $X'_S = -0{\cdot}2067$; Primary $LA' = +0{\cdot}1670$

By [67] $X'_s = -0{\cdot}1872$

True $Coma'_S = -0{\cdot}00304$; True $X'_S = -0{\cdot}1895$; Sagittal $LA' = +0{\cdot}1677$

it becomes evident that the discrepancies due to secondary aberrations are very much more serious in the tangential fan than they are in the sagittal fan. This greater steadiness of the sagittal rays becomes even more pronounced in systems of bolder design with large secondary aberrations, and it would be a most unusual experience to find the judgement formed by the convenient methods of [66] and [67] upset as the result of the tracing of the marginal sagittal ray. The latter rather laborious operation may therefore be safely left till the last stage, when it will be useful as a final confirmation of the whole work.

In order to interpret the results obtained by the exact ray-tracing methods for oblique pencils we have had to anticipate some of the higher aberrations, which are really reserved for Part II. Many of these higher aberrations can be deduced from the symmetry principle as previously applied to centred systems. Thus the curved image surfaces for sagittal and tangential rays must be surfaces of rotation with the optical axis as their axis ; hence X'_S and X'_T must have the same value and sign for $+H'_k$ and for $-H'_k$, and must therefore be represented by series in even powers of H'_k. Similarly coma and distortion must, at any numerical value of H'_k, have either the sign of H'_k or the opposite sign, no matter whether the image-point is taken above or below the optical axis. Their series must therefore be in odd powers of H'_k.—Several of these higher oblique aberrations have also appeared in the discussion of the sine theorem in the previous chapter.

B. Analytical applications of the exact formulae for oblique pencils

(1) The tracing of the usual rays of the axial pencil will give LA' and the coma by the sine theorem. The ray or rays of a different colour will also determine the longitudinal and transverse chromatic aberrations, the latter by Sine Theorem V.

[70] If we then trace trigonometrically a principal ray of suitable obliquity, this gives the distortion by equation (18), whilst *s* and *t* fans along this ray determine astigmatism and curvature of the field. All the aberrations are then known for the chosen position of the diaphragm ; by treating the calculated aberrations as primary, we can then apply the second fundamental law to find the most favourable position of the pupils. Alternatively we can trace *s*-rays by the formulae of section [69] and obtain a less reliable determination of astigmatism and curvature of field by adding the calculation of the Petzval sum.

(2) If instead of using the *s* and *t* formulae, we add rays '*a*' and '*b*' of section [66] to the traced principal ray, we obtain a very accurate determination of $Coma'_T$ and of X'_T at full aperture, but must calculate the Petzval sum in order to obtain the astigmatism. In using the second law we must then assume the required $Coma'_S$ as one-third of $Coma'_T$ in accordance with the relation proved for primary coma.

The solutions obtained by either of these methods will almost invariably prove very much closer approximations than those by the straightforward primary aberration method, and as the amount of work involved is not greatly different, the trigonometrical method will frequently deserve the preference.

(3) A remarkably powerful method for the trigonometrical solution of designing problems results from a modification of the three-ray method of section [66]. It was shown in section [60] that the second fundamental law enables us to find the aberrations of a given system for *any* position of the diaphragm from those for one selected position without any additional ray-tracing, and that we can thus arrive at a general survey of all possible compromises between the oblique aberrations, most instructively by a graph like Fig. 67. The trigonometrical method now to be deduced yields all the data for such a graph.

With reference to Fig. 87, we trace the paraxial rays for the *axial* object- and image-points and *four* equidistant tangential rays from an extra-axial object-point.

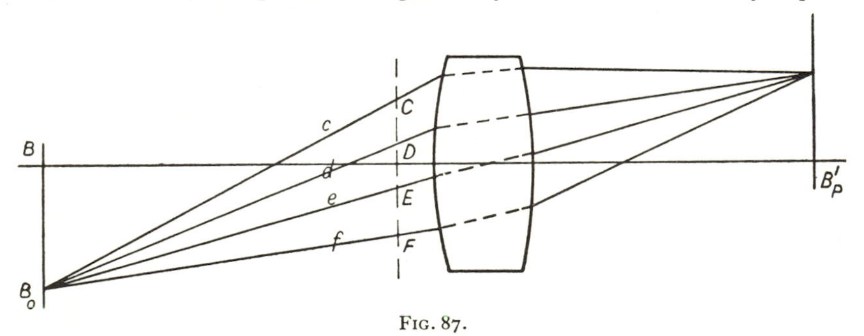

FIG. 87.

To avoid confusion with the three-ray method, we will call these four rays *c*, *d*, *e*, and *f*. It will usually be most favourable to choose the obliquity so as to correspond to about 0·7 of the intended full field. To select the four rays, we make a rough guess at the probable location of the entrance pupil (dotted in the diagram) and symmetrically distribute four points as entering points for the rays. The spacing must be such that all the rays will pass comfortably through the available aperture of the system, with careful remembrance of the vignetting effect in bulky systems ; [70] on the other hand, too close crowding of the rays should be avoided, because the solution depends largely on the intersections of adjacent rays like *c* and *d*, and these would be ill determined if the rays met under very small angles. Treating the ordinates at which the rays cross the adopted pupil position as *SA*-values, the opening formulae of section [66] can be applied to determine the trigonometrical L_c, U_c, &c. of the four selected rays, which can then be traced through the system by the usual routine, giving the 'dashed' final data L'_c, U'_c, &c. We then obtain all the data for the drawing of a graph like Fig. 67 as follows :

(a) We can look upon each one of the four rays as the principal ray of an oblique pencil admitted by a diaphragm or entrance pupil located at the *L*-value of that ray. Using the paraxial data in conjunction with those of each ray, equation (18) will give four values of the distortion, or exactly the right number required for the plotting of the cubic parabola.

(b) We can look upon either *c-d-e* or *d-e-f* as sets of three rays with the middle one of each set as principal. Application of equations (15) or (15*) followed by (17) will give two values of $Coma'_T$ for principal rays crossing the assumed entrance pupil at *D* and at *E* respectively, and as the coma plots as an inclined straight line, this is again exactly the right number for the coma-graph.

(c) To determine the spherical aberration of the system we look upon *d* and *e* as a close pair of tangential rays and upon *c* and *f* as a wide pair of three times the aperture, both referred to the same principal ray (which we need not compute) through the axial point of the assumed entrance pupil. By the second equation of (15) or (15*) we can compute the projected intersection distances of these two pairs as L'_{de} and L'_{cf} respectively. By section [55]F the distance from either of these intersection points to the paraxial image-plane is given by

$$LA'_T = g_1 SA_k^2 + 3\, g_3 H'^2_k + \tfrac{1}{2}\, g_4 H'^2_k$$

in which the first term on the right represents the ordinary LA' of any pencil of the particular aperture. As the other two terms are independent of the aperture, it follows that the difference between L'_{de} and L'_{cf} is equal to the difference of the LA' for the respective apertures, and as the latter are as one to three, the LA' are as one to nine. Hence it follows, if we take the LA'_{de} as our standard, that

$$L'_{de} - L'_{cf} = LA'_{cf} - LA'_{de} = 9\, LA'_{de} - LA'_{de} = 8\, LA'_{de},$$

or

$$LA'_{de} = \tfrac{1}{8}\,(L'_{de} - L'_{cf}),$$

and the other pairs of neighbouring rays will have the same spherical aberration. In the presence of secondary aberrations the LA' thus determined will of course be the tangential LA', which differs from the axial LA' by an amount growing as H'^2. That is the reason why the advice was given to select the obliquity so as to correspond to 0·7 (strictly 0·7071) of the intended field ; the LA' employed will then be the mean value.

(d) We can now calculate the necessary three values of X'_T for the completion of our graph. Looking upon the three possible pairs of neighbouring rays as tangential pairs of aperture $= DE$, we can calculate by the second equation of

[70] (15) or (15*) the projected intersection distances L'_{cd}, L'_{de}, and L'_{ef}. As these are affected by the LA'_{de}, we have to add LA'_{de} to all three to render them comparable with the paraxial l'; hence equation (16) gives for the plotting of the X'_T graph

$$X'_{Tcd} = L'_{cd} + \tfrac{1}{8}(L'_{de} - L'_{cf}) - l'\,;\; X'_{Tde} = L'_{de} + \tfrac{1}{8}(L'_{de} - L'_{cf}) - l'\,;$$

$$X'_{Tef} = L_{ef} + \tfrac{1}{8}(L'_{de} - L'_{cf}) - l'.$$

In practice it will be most convenient to plot the values of the aberrations against the ordinates at which the respective principal rays cut the adopted entrance pupil. The four values of the distortion will then be placed to correspond to points C, D, E, and F. But as the *neighbouring* pairs of computed rays must obviously be referred to a principal ray midway of each pair, the X'_T values have to be correspondingly plotted over the midway points. The two coma-values have to be plotted to correspond to D and E. The solution read off the graph—say for $X'_T = 0$—will be in terms of the position of the principal ray on the adopted entrance pupil. To find the corresponding position of the diaphragm or of the final entrance pupil, we have to lay a ray through B_0 and the particular point of the adopted pupil; the cut of this ray with the optical axis locates the correct final pupil. In the special, but greatly predominating, case when the objects are assumed at infinity, the entering four rays will be parallel and equidistant; in that case we may plot the aberrations against the L-values of the four rays, with the advantage that the graph gives directly the distance of the most favourable pupil from the first surface of the system. This simplification is *not* available in the case of a near object; for the four diverging rays, which then aim at equidistant points in the adopted entrance pupil, will cut the optical axis at unequal intervals, so that the convenient dip-method could not be used in plotting the curves against the L-values.

The method admits of a substantial saving of time in cases like that of ordinary landscape lenses or eyepieces, when the distortion cannot be corrected; for in that case we can trace the axial pencil at the aperture corresponding to the adopted interval between neighbouring oblique rays and obtain the X'_T directly as L'_{de} − axial L', &c., because the spherical aberration will be (in primary approximation) the same for the axial pencil and for the tangential pairs. The calculation of four values of the distortion also drops out. The two chromatic aberrations may be included very easily by tracing the paraxial ray in two colours; the difference $l'_r - l'_v$ gives the longitudinal chromatic aberration directly. The constant g_7 of the transverse chromatic aberration can then be calculated by Sine Theorem V, and by doing this for l'_{pr} respectively $= L'_d$ and $= L'_e$ we obtain two values of $Tch' = g_7 \,.\, H'_k$, which can be plotted on the graph and give the inclined straight line by which Tch' can be read off for any position of the entrance pupil.

As a short example of this valuable four-ray method, we will add a fourth ray to the three already traced through the landscape lens, which easily allows of the addition of another ray at the original interval if we take it below ray b of the three-ray calculation. As the objects are assumed at infinity, the four rays will cut the axis at equal intervals, and the L of the fourth ray therefore follows immediately from that of the other three. The direct results for the four rays, [70] called c, d, e, and f for the present method, will be found as :

	c	d	e	f
Initial L	− 1·95177	− 0·997	− 0·04223	+ 0·91254
Final L'	− 3·07569	− 1·54350	− 0·31294	+ 0·65715
Final U'	− 12-56-1	− 14-34-59	− 16-19-52	− 18-20-51

$$L'_{cd} = 9{\cdot}9865 \qquad L'_{de} = 9{\cdot}4343 \qquad L'_{ef} = 8{\cdot}0157$$
$$L'_{ce} = 9{\cdot}7002 \qquad L'_{df} = 8{\cdot}66745$$
$$L'_{cf} = 9{\cdot}0621.$$

With $LA'_{de} = \frac{1}{8}(L'_{de} - L'_{cf}) = 0{\cdot}0465$ and paraxial $l' = 9{\cdot}8367$ we can then calculate all the aberrations from these trigonometrical results. For the purpose of convenient comparison with the primary aberration results plotted against Q in Fig. 67, the values of Q have been added to the following tabulation ; ray d being the previous principal ray for $l_{pr} = -0{\cdot}997$, its Q-value is that used in Fig. 67, namely $Q = 1{\cdot}3054$. Those for the other three rays then follow at once from the definition of $Q = P_{pr}P_{pr}{}^{*}/SA$, for as our rays pass the entrance pupil at intervals $= SA$, it is evident that the successive Q-values are at exact unit-intervals.

Principal ray at :	C		D		E		F
Q-value :	2·3054		1·3054		0·3054		− 0·6946
Distortion :	+0·0347		+0·0393		+0·0261		− 0·0442
X'_T :		+0·1912		− 0·3610		− 1·7796	
$Coma'_T$:			− 0·0088		− 0·0252		

If these aberrations are plotted against Q to the same scale as Fig. 67 it will be found that the curves for distortion and for X'_T fall very close to those deduced from the primary aberrations by the second fundamental law ; the inclined line for $Coma'_T$ will of course have three times the slope of the former $Coma'_S$ line, on account of the three-to-one relation. This totally independent, purely trigonometrical method of determining the oblique aberrations and their change by shifts of the diaphragm thus supplies a highly satisfactory confirmation of the whole primary theory of these aberrations as deduced in Chapter VI.

C. Limits for astigmatism and curvature of field

In the discussion of the oblique aberrations of a given system, and especially in the selection of a good compromise in that majority of cases when complete correction is not possible, it is again necessary to be able to estimate the real seriousness of given amounts of the various aberrations ; therefore we must make a further addition to our tolerances by fixing suitable limits for astigmatism and curvature of the field. As both these defects are due to longitudinal differences of focus, in the case of pure astigmatism between sagittal and tangential rays, for simple curvature of field (Petzval curvature) between axial and oblique pencils, they both come under the focal range tolerance.

The Rayleigh limit, which must be observed when a system is required to have the full theoretical resolving power, leads to extremely stringent tolerances. For

[70] a system free from astigmatism, the field will have the Petzval curvature measured by our $Ptz' = -X'_{Ptz}$. If we adjust the focal plane midway between the marginal and the axial focus (see Fig. 56 (a)), we shall clearly have X'_{Ptz} = Focal Range = one wave-length/$N' \sin^2 U'_m$ as the Rayleigh limit; for the usual astronomical f-ratio 1/16 we should have $\sin U'_m = 1/32$ or permissible $X'_{Ptz} = 1024$ wave-lengths, which means only half a millimetre. For astigmatism the case is even worse, for as at the S-focus T is out of focus by the whole astigmatic difference of focus = 2 Ast'_S and correspondingly at the T-focus, the Rayleigh limit for Ast'_S is = $\frac{1}{4}$ wave-length/$N' \sin^2 U'_m$, or = 256 wave-lengths at $f/16$; only 1/8 mm. To show how severely this restricts the field which may be used at full resolving power, we will deduce the permissible H'-value for the standard astrographic telescopes. These work at $f/10$, which gives

$$\text{Tolerance on } Ast'_S = 100 \text{ wave-lengths} = 0{\cdot}05 \text{ mm.}$$

As the object-glass may be treated as an aplanatic *thin* one, we have $g_3 = 1/2f'$ or with $f' = 3437$ mm., $g_3 = 1/6874$; and as $Ast'_S = g_3 H'^2$, we have for the determination of the permissible H'

$$\frac{1}{6874} H'^2 = 0{\cdot}05 \quad \therefore \quad H'^2 = 343{\cdot}7, \text{ or } H' = 18{\cdot}5 \text{ mm.}$$

Hence in these telescopes the astigmatism reaches the Rayleigh limit at 18·5 mm. from the centre of the field, whilst the instruments are used for a field of 120 × 120 mm. At the corners of this plate the astigmatism will amount to about twenty times the Rayleigh limit. It is entirely due to the fortunate fact of astigmatism causing a *symmetrical* deformation of the image-points that these plates can be measured within fractions of a second of arc.

This example of an instrument with a field covering only a few degrees will render it clear that in the case of astigmatism we must practically always rely on geometrical estimates of the deterioration of the image, supported as far as possible by direct experimental determinations of the shape and size of images produced in the presence of known amounts of astigmatism. At present it can only be stated that experiments of this kind indicate that in the case of astigmatism the image actually seen or photographed corresponds far more closely to the geometrically deduced dimensions than is the case for coma and especially for spherical aberration. The limits now to be deduced may therefore be accepted as reasonably reliable.

If, with reference to Fig. 88, we had the ideal focal plane of a system at $B'F$ and the sagittal focus for the margin of the field at S, then, supposing first that the tangential rays came to focus somewhere *between* S and F, the best focus would clearly lie midway between S and F, for with that adjustment both the axial pencil and the sagittal fan would be out of focus to the same extent, whilst at any other focal adjustment either one or the other would be more widely diffused. But supposing next that the tangential field had a curvature opposite to that of the sagittal field, then we should have least diffusion midway between S and T. Following this argument to its logical conclusion, we easily see that the best focal adjustment will always be midway between the extreme focal distances, no matter whether these extreme positions correspond to a sagittal, a tangential, or the axial focus, but subject to the assumption that all the pencils are of the same aperture- or convergence-value. The case of pronounced vignetting effect in the oblique pencils must therefore be excluded for special consideration in individual cases. If the convergence angle of the marginal rays is U'_m, the rays aiming at the extreme focal distances will be spread out over a length $= \tan U'_m \times$(difference of extreme focal distances) when received at the *midway* adjustment of the focus, and this length will be the biggest dimension of the resulting diffused images. Hence, with reference to the figures for permissible diffusion selected in the case of the coma tolerance, we may lay down the tolerance for astigmatism and curvature of field as :

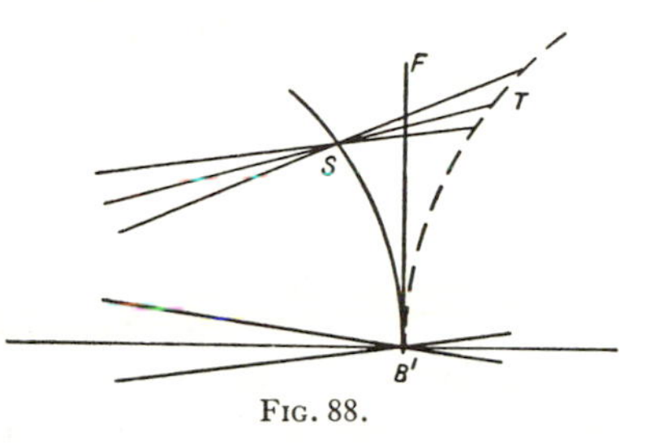

FIG. 88.

OT(7) *Tolerance for astigmatism and curvature of field.*

$$(\text{Extreme difference of focus}) \times \tan U'_m \text{ may be}$$

$$= \begin{cases} 0{\cdot}001'' \text{ for extremely sharp definition,} \\ 0{\cdot}004'' \text{ for good definition,} \\ 0{\cdot}010'' \text{ for decidedly soft definition.} \end{cases}$$

In practice sin U'_m may be used instead of tan U'_m, as their difference is only a very small percentage.

In the case of our landscape lens the biggest difference of focus at full aperture was that between $L'_{ab} = 9{\cdot}7002$ for the extreme tangential rays and 'Final L'_{sk}' $= 9{\cdot}4818$ for the extreme sagittal rays. With sin U'_m nearly enough $= 0{\cdot}0319$ they give $(9{\cdot}7002 - 9{\cdot}4818)\, 0{\cdot}0319 = 0{\cdot}2184 \times 0{\cdot}0319 = 0{\cdot}0070$, or midway between 'good' and 'decidedly soft' definition.

Special Memoranda from Chapter VIII

The ray-tracing formulae of Chapter I are adapted and added to in order to provide convenient equations by which the exact course of the rays of oblique pencils through any centred system can be determined.

In many ordinary cases the usual rays of the axial pencil, together with the three rays *a*, *pr*, and *b* of a selected tangential fan, provide sufficient information by relying for a fuller interpretation of the results upon the Petzval theorem and upon the primary three-to-one ratio of tangential to sagittal coma and astigmatism.

As an alternative the usual rays of the axial pencil can be used to determine spherical aberration and also coma by the sine theorem ; a principal ray trigonometrically traced at a suitably chosen obliquity will then give the distortion, and the *s* and *t* formulae applied to this ray will locate the astigmatic foci for small

[70] aperture. If the marginal tangential rays *a* and *b* are added, this method gives sufficiently full information in the majority of cases.

Skew rays can usually be left for the final stage of the evolution of a new design; they will hardly ever upset the judgement formed by the methods already stated.

The analytical application of the equations, referred to in [70] B, should be borne in mind. Final results will often be reached more quickly in this way than by either strict primary or by TL work.

The tolerances for astigmatism and curvature of field will frequently prove difficult to be complied with. That accounts for the fact that in wide-angle systems spherical and even chromatic correction are frequently sacrificed in order to secure a sufficient reduction of the curvature errors.

CHAPTER IX

GENERAL THEORY OF PERFECT OPTICAL SYSTEMS

IN Chapter I we gave the usual simple formulae for *thin* lenses or systems, which enable us to calculate for objects at any distance the corresponding distance and magnification of the image when nothing more than the focal length of the thin system is known. It had been felt from the earliest days of optics that it ought to be possible to find corresponding general formulae for systems with finite thicknesses and separations, but this problem was never solved until the great Gauss attacked it in 1841. Gauss showed that the paraxial properties of any centred system of spherical surfaces are completely defined by four easily determined planes at right angles with the optical axis, namely the focal planes and the principal planes for the object-space and the image-space respectively. This theory, which is found included in practically all books dealing with optics, is, however, restricted in its full validity to the paraxial region, or as it is usually put, to a thread-like or capillary space immediately surrounding the optical axis, and no legitimate conclusions can be drawn from it as to what may happen when aperture and field are extended to the finite dimensions which are indispensable in actual optical instruments.

About 1873 a less restricted theory was developed by Abbe of Jena, based on purely geometrical conclusions from the rectilinear propagation of light, and leading to highly important deductions concerning the possibilities of correct images at finite aperture and throughout a finite field. In working out this theory Abbe rediscovered a theorem which had been announced about 1857 by Clerk Maxwell, the famous originator of the electromagnetic theory of light. We shall found our theory on this theorem of Clerk Maxwell, and shall also adopt some of Abbe's methods.

The theory will be based on two assumptions, namely a first assumption that in any homogeneous medium the geometrical rays of light are absolutely straight lines, and as this assumption underlies the whole of geometrical optics we may safely accept it ; but we make a second assumption of a very different nature and which we shall subsequently prove to be false, for we also assume that it is possible to construct a perfect or ideal optical system satisfying the following definition :

'A centred optical system is perfect if objects in *any* plane at right angles with its optical axis are depicted sharply and without distortion in a conjugate image-plane also at right angles with the optical axis.'

Accepting this definition, the theorem of Clerk Maxwell states that 'An optical system is perfect for object-planes at *all* distances if it satisfies the above definition of perfection for objects in only *two* different planes at right angles with its optical axis.'

The value of this theorem will be immediately realized, for whilst in its absence it would be a task without end to prove that a given system complied with our definition of perfection for objects at all possible distances, acceptance of the theorem will enable us to prove universal perfection of a given system by testing it for only two different distances.

[71]

Proof of the Theorem of Clerk Maxwell

With reference to Fig. 89 we assume that the lens system lightly indicated at the centre of the diagram depicts objects in the object-plane I sharply and without distortion in the conjugate image-plane I' under a linear magnification M_1', and that objects in the object-plane II are similarly depicted in the conjugate image-plane II' under magnification M_2'. All four given planes shall be at right angles with the optical axis and shall cut it at $B_1\,B_2\,B_1'\,B_2'$ respectively. The distance

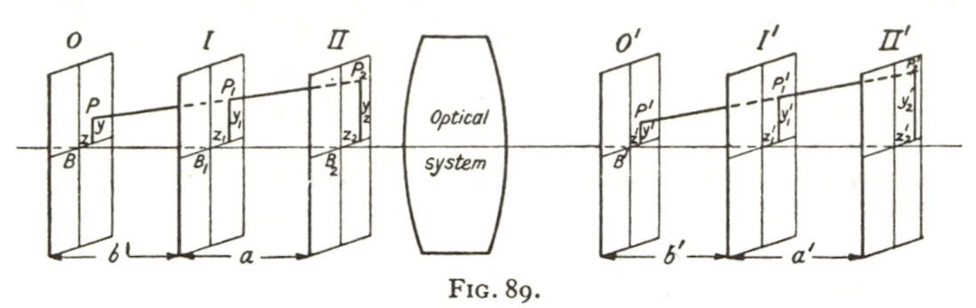

Fig. 89.

from plane I to plane II shall be given as $= a$, positive if II is to the right of I, and the corresponding distance from plane I' to plane II' shall be given as $= a'$. We use the rectangular coordinates of solid geometry with the Y-direction vertical, as in all our two-dimensional diagrams, and the Z-direction counted positive towards the rear. The given magnifications in planes I' and II' then mean that a point P_1 in plane I, defined by the coordinates y_1 and z_1, will be depicted sharply in plane I' at P_1' with the coordinates $y_1' = y_1 M_1'$ and $z_1' = z_1 M_1'$, whilst a point P_2 having the coordinates y_2 and z_2 in plane II will be depicted in the conjugate plane II' with the coordinates $y_2' = y_2 M_2'$ and $z_2' = z_2 M_2'$. The magnifications will of course be subject to our adopted convention, that is, positive for an erect and negative for an inverted image. The diagram represents as nearly as is possible an all-positive case; it may at first sight appear to be incapable of realization, but we shall show a little later that it can be closely realized by a simple arrangement of only two thin lenses.

To prove the theorem we now assume another object-plane O at any distance b from O to I, and consider a 'general' point P with coordinates y and z in this plane as our new object-point. We draw a 'general' ray from P towards the optical system; it shall cut plane I at P_1 with coordinates y_1 and z_1, and plane II at P_2 with coordinates y_2 and z_2; but as light rays are absolutely straight lines we cannot leave y_2 and z_2 in this indefinite general form, because the direction of our ray is completely fixed by the two points P and P_1 through which it is drawn; hence y_2 and z_2 are determined by y, y_1, z, and z_1, and we must express them in terms of these. As the y-coordinate of our ray has changed by $(y_1 - y)$ in the distance b from O to I, it must change by $(y_1 - y)\,(a+b)/b$ in the distance $(a+b)$ from O to II, therefore we have

(a) $$y_2 = y + (y_1 - y)\,(a+b)/b = y_1\,(a+b)/b - y\,.\,a/b.$$

Precisely the same conclusions can be applied to the z-co:ordinate, hence we have also

$$z_2 = z_1\,(a+b)/b - z\,.\,a/b.$$

[71] We can now determine the points P_1' and P_2' in which our ray must pass through the given image-planes I' and II' on emerging from the optical system, for as our ray passes through the point P_1 of plane I we may look upon it as one of the rays sent out by P_1, and as P_1 is sharply depicted in plane I' under magnification M_1', the ray must cut this plane with the coordinates $y_1' = y_1 M_1'$ and $z_1' = z_1 M_1'$, and point P_1' on the emerging ray is thus definitely fixed. In the same way we deduce from the fact that our ray passes through plane II at P_2 with the coordinates y_2 and z_2 the conclusion that it must on emergence from the system pass through the conjugate plane II' with the coordinates $y_2' = y_2 M_2'$ and $z_2' = z_2 M_2'$. Introducing the explicit values of y_2 and z_2 by equation (a) and its companion in terms of z, we thus know that the emerging ray

passes plane I' at $y_1' = y_1 M_1'$ and $z_1' = z_1 M_1'$

and plane II' at $y_2' = [y+(y_1-y)(a+b)/b]\, M_2'$

and $z_2' = [z+(z_1-z)(a+b)/b] M_2'$;

and as the emerging ray again is an absolutely straight line its whole course is definitely determined by these two points through which it passes. Consequently we can calculate the coordinates y' and z' of the point P' at which the emerging ray will cut another image-plane O' placed at right angles with the optical axis at any arbitrary point B' at a distance b' from O' to I', for on account of the unvarying slope of a straight line we must have

$$(y_1'-y')/b' = (y_2'-y_1')/a' \text{ and } (z_1'-z')/b' = (z_2'-z_1')/a',$$

and on account of the close correspondence of all the expressions in terms of y and z respectively, we need only work out the former, as the other can then be written out by analogy. Cross-multiplication gives

$$a'y' = (a'+b')\, y_1' - b'y_2',$$

or on introduction of the above explicit values of y_1' and y_2',

$$a'y' = (a'+b')\, y_1 M_1' - b'\, [y+(y_1-y)(a+b)/b]\, M_2',$$

which is easily put into the form

$$y' = yM_2'ab'/a'b + (y_1/a')\, [M_1'(a'+b') - M_2'b'(a+b)/b],$$

with an accurately corresponding equation in terms of z.

The solution for y' shows that in general the value of y' will vary when y_1 varies, that is, for rays leaving the object-point P at different slopes with reference to the XZ plane ; therefore we shall not obtain a sharp image-point in our arbitrarily chosen plane O'. We could not reasonably expect any other result, for what we have done so far is equivalent to using an optical instrument with a random adjustment of the focus on an object at a given distance ; obviously it would be highly improbable that the object should be in sharp focus. The solution found for y' enables us to find the sharp focus of point P by a simple discussion : The variation of the value of y' with that of y_1 arises from the second term of the last equation, and will be proportional to the magnitude of the factor in the square bracket. If that factor could be brought to zero-value, y' would become inde-

[71] pendent of y_1. The factor is a simple linear function of b'. Therefore there must be one, and only one, value of b' which produces this zero-value, and we find it from·

$$M_1'(a'+b') - M_2'b'(a+b)/b = 0$$

by solving for b', as

G. Th. (1) $$b' = a'M_1'/[M_2'(a+b)/b - M_1'].$$

If we place our image-plane O' into the position defined by this special value of b', the second part of the equation for y' will disappear and we shall have

$$y' = yM_2'ab'/a'b,$$

with a companion by analogy

$$z' = zM_2'ab'/a'b;$$

and as both y' and z' now depend solely on the coordinates of the object-point P and not in any way on the direction of the individual rays sent out by P, we have found the sharp focus conjugate to P and have supplied a first part of the proof of Clerk Maxwell's theorem; for as G. Th. (1) does not contain any y or z values, all points in plane O will be sharply depicted in the conjugate plane defined by that equation. To complete the proof we must show that there is also a fixed and uniform *magnification* in the plane O', and therefore no distortion.

The last equations for y' and z' give by transposition

$$y'/y = z'/z = M_2'ab'/a'b,$$

and as y'/y and z'/z, as ratios of image to object, represent the linear magnification M' in plane O', we have

G. Th. (2) $$M' = M_2'ab'/a'b.$$

The right of this equation does not contain any y or z values; hence the linear magnification is uniform throughout plane O' and our proof of the theorem is complete. We will however put the equation for M' into a more useful form by introducing on the right the value of b'/a' which follows by transposing G. Th. (1) and then slightly simplifying. We thus obtain

G. Th. (3) $$M' = \frac{a}{b} \cdot \frac{M_1'M_2'}{M_2'(a+b)/b - M_1'} = \frac{a \,.\, M_1' \,.\, M_2'}{a \,.\, M_2' + b(M_2' - M_1')}.$$

CHARACTERISTIC DATA OF AN OPTICAL SYSTEM

[72] Having proved Clerk Maxwell's theorem, we can now quickly deduce the general properties of a perfect optical system, as on our assumption that the given system be perfect for two different object distances we must have similar perfection at all distances, and need not discuss that question in each individual case. It should be borne in mind that the conclusions which we shall draw will be true for the full field of the system and at any possible aperture if our assumption proves sound and realizable.

We restrict the entire discussion to cases when the four given planes I, II, I', and II' are all within finite distances of the optical system and when a, a', M_1', and

M_2' have finite values ; but as we admit all possible combinations of signs and of [72] finite magnitude—however large—it will be clear in the light of previous discussions of conceivable infinite values of optical data that the stated restriction is a purely formal one.

(1) *The Principal Focal Planes.* The first or anterior focal plane is defined as that object-plane which yields infinitely distant images in the image-space, and the second or posterior focal plane is that image-plane in which infinitely distant objects in the object-space are depicted. In order to locate these planes we use G. Th. (1) in the slightly modified form resulting when we replace $(a+b)/b$ in the denominator by $(a/b+1)$:

$$b' = \frac{a'M_1'}{\frac{a}{b}M_2' + (M_2' - M_1')}.$$

For the second or posterior focal plane the object distance 'b' must be put $= \infty$, which, with the stipulated finite values of 'a' and M_2', causes the first term in the denominator to vanish. Calling the resulting special value of b' b'_f, we therefore have the distance from the posterior focal plane to the given plane I' determined by

G. Th. (4) $$b'_f = a'M_1'/(M_2' - M_1').$$

For the first or anterior focal plane the image must be formed at infinity, hence it must yield $b' = \infty$. As both a' and M_1' in the numerator of the formula for b' have been stipulated as finite, b' can only become infinite if the denominator of the formula is zero, hence we obtain the requisite value of 'b' from the condition

$$\frac{a}{b}M_2' + (M_2' - M_1') = 0\ ;$$

and if we call the resulting special value of 'b', which determines the distance from the anterior focal plane to the given plane I, b_f, we find by a simple transposition of the last equation

G. Th. (5) $$b_f = -aM_2'/(M_2' - M_1').$$

As the equations for b_f and for b'_f both contain $(M_2' - M_1')$ in the denominator, both distances become infinite when M_2' is $= M_1'$. This is a highly important special case, which we reserve for a subsequent separate discussion. For the remainder of the present general investigation we will therefore exclude the case when $M_2' = M_1'$.

(2) *The Gaussian Principal Planes.* These are the two conjugate planes for which the linear magnification M' is exactly equal to 'one'. Any object in one of these planes is therefore depicted erect and natural size in the other. For that reason these two particular conjugate planes are frequently called the 'planes of unit-magnification'. We locate them most conveniently by solving our fundamental equations for the values of b and b' which will yield *any* prescribed linear magnification M'. The second form of G. Th. (3)

$$M' = \frac{aM_1'M_2'}{aM_2' + b\,(M_2' - M_1')}$$

[72] gives by a simple transposition

$$\text{G. Th. (6)} \qquad b = \frac{aM_2'(M_1' - M')}{M'(M_2' - M_1')},$$

and by this equation we could calculate the position of an object-plane which would yield images magnified M' times. To find the corresponding value of b' we transpose G. Th. (2) into the form

$$\frac{b'}{b} = \frac{a'}{a} \cdot \frac{M'}{M_2'}$$

and multiply G. Th. (6) by this equation, with the immediate result

$$\text{G. Th. (7)} \qquad b' = \frac{a'(M_1' - M')}{M_2' - M_1'}.$$

We shall use these general solutions a little later for other important deductions. For our present purpose of locating the Gaussian principal planes we have merely to introduce $M' = +1$. We will call the distance from the first or anterior principal plane to plane I b_{pp}, and the distance from the second or posterior principal plane to plane I' b'_{pp}, then we have

$$\text{G. Th. (8)} \qquad b_{pp} = aM_2' \frac{M_1' - 1}{M_2' - M_1'}; \; b'_{pp} = a' \frac{M_1' - 1}{M_2' - M_1'}.$$

(3) *The Equivalent Focal Lengths.* We define the first or anterior equivalent focal length as the distance from the first principal plane to the first focal plane, and correspondingly we define the second or posterior equivalent focal length as the distance from the second principal plane to the second focal plane. In accordance with the principles of our nomenclature, we will assign to the two focal lengths the symbols f and f' respectively. As b_{pp} gives the distance from the first principal plane to plane I, and b_f gives the distance from the first focal plane to plane I, we have

$$f = b_{pp} - b_f = aM_2' \frac{M_1' - 1}{M_2' - M_1'} + a \frac{M_2'}{M_2' - M_1'} = \frac{aM_1'M_2'}{M_2' - M_1'}$$

and correspondingly

$$f' = b'_{pp} - b'_f = a' \frac{M_1' - 1}{M_2' - M_1'} - a' \frac{M_1'}{M_2' - M_1'} = - \frac{a'}{M_2' - M_1'}.$$

We will collect these solutions as

$$\text{G. Th. (9)} \qquad f = \frac{aM_1'M_2'}{M_2' - M_1'}; \; f' = - \frac{a'}{M_2' - M_1'}.$$

(4) The equations deduced in the preceding three subsections lend themselves to numerous applications both in the calculation of optical systems and in the experimental determination of the characteristic data. Comparison of the final equations for the two equivalent focal lengths with those for the various b-values discloses such close relations as to suggest the possibility of simplification of the

earlier equations by introduction of the values of f or f'. Beginning with b'_f we [72] easily find

$$b'_f = a'M'_1/(M'_2 - M'_1) = -f'M'_1.$$

Similarly the equation for b_f gives

$$b_f = -aM'_2/(M'_2 - M'_1) = -aM'_1 M'_2/(M'_2 - M'_1)\, M'_1 = -f/M'_1.$$

If we next examine the general solution G. Th. (6) for the value of b which will lead to a prescribed magnification M' we can transform this as follows:

$$b = \frac{aM'_2\,(M'_1 - M')}{M'\,(M'_2 - M'_1)} = \frac{aM'_1 M'_2}{M'\,(M'_2 - M'_1)} - \frac{aM'_2}{M'_2 - M'_1} = \frac{f}{M'} + b_f = \frac{f}{M'} - \frac{f}{M'_1}.$$

Treating G. Th. (7) in a corresponding manner we find

$$b' = \frac{a'\,(M'_1 - M')}{M'_2 - M'_1} = \frac{a'M'_1}{M'_2 - M'_1} - \frac{a'M'}{M'_2 - M'_1} = b'_f + f'M' = f'M' - f'M'_1,$$

and as b_{pp} and b'_{pp} are merely the special values of b and b' for $M' = 1$, we have

$$b_{pp} = f - \frac{f}{M'_1};\ b'_{pp} = f' - f'M'_1.$$

The b-values, whilst greatly simplified, are still counted from the various object- and image-planes to plane I and plane I' respectively, and as these two planes have no particular interest attached to them, and moreover are not directly identifiable by some visible material mark, a further transformation is desirable in order to refer objects and images to such a permanent reference point. In most cases, and especially in actual calculations, it is most convenient to measure distances in the object-space from the pole of the first surface of the system, and distances in the image-space from the pole of the last surface. But any other starting points will answer our purpose, as long as they are permanent and sufficiently definite. In microscope objectives distances on the side of the object are usually measured from the end of the mount, and distances on the side of the eyepiece from the shoulder of the standard objective screw. On photographic objectives the diaphragm slot is probably the most favoured reference mark. Whatever choice may be made, we will call the distances of the two given object-planes l_I and l_{II}, and the conjugate distances in the image-space l'_I and l'_{II}, then we shall have $a = l_{II} - l_I$ and $a' = l'_{II} - l'_I$. If we use l and l', with suitable suffix when necessary, for the distances of other object- and image-planes, the previously determined b-values will be defined by $b = l_I - l$ and by $b' = l'_I - l'$, and the required distances will therefore be $l = l_I - b$ and $l' = l'_I - b'$ These general formulae give the position of the two principal focal planes by

$$l_f = l_I - b_f = l_I + f/M'_1 \quad \text{and} \quad l'_f = l'_I - b'_f = l'_I + f'M'_1;$$

they locate the two principal planes at

$$l_{pp} = l_I - b_{pp} = l_I - f + f/M'_1 \quad = l_f - f,$$
$$l'_{pp} = l'_I - b'_{pp} = l' - f' + f'M'_1 = l'_f - f',$$

[72] and the two conjugate planes which yield a linear magnification M' in the image-plane are determined by

$$l = l_I - b = l_I - \frac{f}{M'} + \frac{f}{M_1'} = l_f - f/M'$$

$$l' = l_I' - b' = l_I' - f'M' + f'M_1' = l_f' - f'M'.$$

We will collect these final equations as follows :

G. Th. (10).

An optical system having been found to give a linear magnification M_1' with conjugate distances (measured from any suitable permanent reference points) l_I and l_I' and magnification M_2' with conjugate distances l_{II} and l_{II}', calculate $a = l_{II} - l_I$ and $a' = l_{II}' - l_I'$; then the equivalent focal lengths are

$$f = a \, . \, M_1'M_2'/(M_2' - M_1') \quad \text{and} \quad f' = -a'/(M_2' - M_1'),$$

the principal focal planes are located by

$$l_f = l_I + f/M_1' \qquad \text{and} \qquad l_f' = l_I' + f'M_1',$$

the Gaussian principal planes are determined by

$$l_{pp} = l_f - f \qquad \text{and} \qquad l_{pp}' = l_f' - f',$$

and any prescribed magnification M' will be obtained with

$$l = l_f - f/M' \qquad \text{and} \qquad l' = l_f' - f'M'.$$

Note : It is obviously arbitrary which of the given pairs of planes we call I; therefore we may calculate l_f and l_f' either by the above equations or by the alternative ones

$$l_f = l_{II} + f/M_2' \qquad \text{and} \qquad l_f' = l_{II}' + f'M_2';$$

calculations can occasionally be considerably shortened by using these alternative solutions.

Example. The lens system shown in the centre of Fig. 90 gave a linear magnification $M_1' = \frac{2}{3}$ when a test object was placed at $l_I = -3\frac{1}{2}$ from the first surface, and the sharp image was found at $l_I' = 2\frac{2}{3}$ from the last surface. On placing the test object at $l_{II} = -2\frac{3}{4}$ the sharp image was formed at $l_{II}' = 3\frac{1}{3}$ and its magnification was $M_2' = 1\frac{1}{3}$.

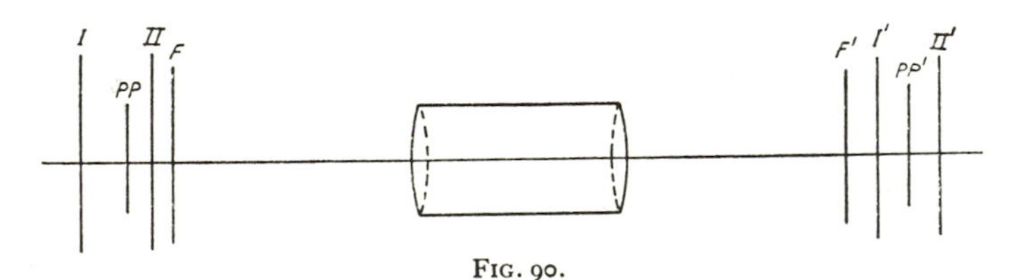

FIG. 90.

The data give

$$a = l_{II} - l_I = -2\frac{3}{4} - (-3\frac{1}{2}) = +\frac{3}{4}; \; a' = l_{II}' - l_I' = 3\frac{1}{3} - 2\frac{2}{3} = +\frac{2}{3}.$$

With these and $M'_1 = \frac{2}{3}$, $M'_2 = 1\frac{1}{3}$, we then find [72]

$$f = aM'_1M'_2/(M'_2 - M'_1) = \tfrac{3}{4} \times \tfrac{2}{3} \times 1\tfrac{1}{3}/(1\tfrac{1}{3} - \tfrac{2}{3}) = +1$$
$$f' = -a'/(M'_2 - M'_1) \qquad = -\tfrac{2}{3}/\tfrac{2}{3} \qquad\qquad = -1.$$

As the second, posterior, or usual focal length comes out with a negative sign, the system has properties corresponding to those of a concave simple lens, and would be called a negative system.

We next find

$$l_f = l_I + f/M'_1 = -3\tfrac{1}{2} + 1/\tfrac{2}{3} = -3\tfrac{1}{2} + 1\tfrac{1}{2} = -2$$
$$l'_f = l'_I + f'M'_1 = 2\tfrac{2}{3} - 1 \times \tfrac{2}{3} = 2\tfrac{2}{3} - \tfrac{2}{3} \quad = +2.$$

The two focal planes therefore lie at these distances from the first and last surface respectively.

For the principal planes we obtain

$$l_{pp} = l_f - f = -2 - 1 = -3\;;\; l'_{pp} = l'_f - f' = 2 - (-1) = +3.$$

If we wanted to know at which conjugate distances this system would produce a linear magnification $M' = -2$, that is, an inverted image of twice natural size, we should find by the general solution

$$l = l_f - f/M' = -2 - 1/(-2) \quad = -2 + \tfrac{1}{2} = -1\tfrac{1}{2}$$
$$l' = l'_f - f'M' = 2 - (-1)(-2) = 2 - 2 \quad = 0.$$

The object would have to be placed at distance $1\frac{1}{2}$ to the left of the first surface, and the sharp image would be found *on* the last surface.

It will be found highly instructive to calculate l and l' for a series of positive and negative values of M' and to plot the distances against M'; in the interpretation of the resulting graphs it should be remembered that positive values of l signify a virtual object, whilst negative values of l' correspond to virtual images.

As far as the paraxial region is concerned, this rather unusual system, which closely corresponds to the diagram upon which we founded our theory, is completely realized by two thin convex lenses of focal length = 1 placed at a separation = 3. We can prove this by the thin lens equations of section [23]:

$$\text{TL (3)} \qquad \frac{1}{l'} = \frac{1}{f'} + \frac{1}{l} \quad \text{and TL (4)} \quad \frac{h'}{h} = m' = \frac{l'}{l}.$$

For our given plane I at $-3\frac{1}{2}$ from the first surface we shall have for the first thin lens $l_a = -3\frac{1}{2}$ and $f'_a = +1$, hence $\frac{1}{l'_a} = \frac{1}{1} - \frac{1}{3\frac{1}{2}} = 1 - \frac{2}{7} = \frac{5}{7}$ or $l'_a = 1{\cdot}4$, and then the magnification $h'_a/h_a = l'_a/l_a = 1{\cdot}4/(-3{\cdot}5) = -0{\cdot}4$. The image formed by the first lens will act as object for the second, hence, with $d' = 3$, $l_b = l'_a - d' = 1{\cdot}4 - 3 = -1{\cdot}6$, and the second lens will give $\frac{1}{l'_b} = \frac{1}{f'_b} + \frac{1}{l_b} = 1 - \frac{1}{1{\cdot}6} = \frac{0{\cdot}6}{1{\cdot}6}$ or $l'_b = \frac{1{\cdot}6}{0{\cdot}6} = \frac{8}{3} = 2\frac{2}{3}$, which agrees with the l'_I used in our example. The magnification

[72] produced by the second lens alone will be

$$h'_b/h_b = l'_b/l_b = 2\tfrac{2}{3}/(-1{\cdot}6) = -8/4{\cdot}8, \text{ or } h'_b = -\frac{h_b}{0{\cdot}6},$$

and as by the calculation for the first lens we have $h_b = h'_a = -0{\cdot}4\, h_a$ we find $h'_b = -\dfrac{-0{\cdot}4\, h_a}{0{\cdot}6} = +\tfrac{2}{3}\, h_a$, or, in the symbols of our solution by G Th (10) $M'_1 = +\tfrac{2}{3}$, exactly the value which we used. The other pairs of planes corresponding to finite values of M' may be verified in the same way. For infinitely distant objects the first lens would give $l'_a = 1$; therefore $l_b = l'_a - d' = 1 - 3 = -2$. Then for the second lens $\dfrac{1}{l'_b} = \dfrac{1}{f'_b} + \dfrac{1}{l_b} = 1 + \dfrac{1}{-2} = 1 - \tfrac{1}{2} = \tfrac{1}{2}$ or $l'_b = 2$, again in agreement with the l'_f by our general solution.

It is easily realized that, whatever our final decision may be as to the validity of our initial assumption that perfect optical systems be possible at any aperture and for any diameter of field, the assumption is certainly justified for the paraxial region, that is, for a very small aperture and a very small field, merely provided that the system is a centred one; for we proved in Chapter I that any centred system yields a geometrically sharp image by paraxial rays of an axial object-point at any distance, and moreover that, by the theorem of Lagrange, the system also gives sharp images of the surrounding field, with a perfectly definite and uniform linear magnification. The theorem of Clerk Maxwell therefore is applicable in the paraxial region, and our simple logical deductions from it, culminating in G Th (10), must then be true. Hence this set of equations rests on a full and satisfactory proof of the Gaussian theory, and we may safely apply it to any centred lens system as far as the usual limitation to the paraxial region is concerned. We will therefore discuss suitable methods by which the Gaussian constants of any given system can be conveniently determined with the desirable degree of precision.

(A) The most generally useful and also the quickest method for the optical designer consists nearly always in tracing paraxial rays from two sufficiently widely separated axial object-points. One such tracing—from the actual object-distance for which the projected system is intended—forms the primary basis of the usual designing routine and is therefore ready for use. But in practically the only cases in which knowledge of all the Gaussian constants is of value, namely, those of photographic objectives and of eyepieces, a second ray-tracing, namely, that of the principal ray of an oblique pencil, is also part of the regular routine, so that as a rule no additional ray-tracing of any kind will be required. In any case, the results of the two ray-tracings supply all the data necessary for the calculation of G Th (10). As a short example we will take the simple photographic objective dealt with in section [58]. The paraxial result for the object-point at infinity was, for nominal $y = 0{\cdot}300$, final $u' = 0{\cdot}031421$, final $l' = 9{\cdot}8367$. As the object was at infinity, the final focus is the principal focus, and if we adopt the pole of the last surface as the point from which we measure all the l', in accordance with our universal computing practice, we have at once $l'_f = \text{final } l'_1 = 9{\cdot}8367$. We also—with object at infinity *only*—can determine f' at once by the formula of Chapter I, $f' = y/\text{final } u'$, which gives the second or posterior focal length as $f' = 9{\cdot}5477$, and with this G Th (10) [72] gives $l'_{pp} = l'_f - f' = 9{\cdot}8367 - 9{\cdot}5477 = +0{\cdot}2890$. The second or posterior principal plane therefore lies at 0·2890 to the right of the last surface. All the Gaussian data for the image space are thus determined. To obtain the corresponding data for the object space we use the calculations for the principal ray, applying the suffix II to these in accordance with the symbolism of G Th (10). We can then extract from section [58] $l_{II} = l_{pr1} = +0{\cdot}249$; $l'_I = l'_{pr3} = 0$. To obtain the linear magnification for the point II' we extract $u_{pr1} = -0{\cdot}31427$ and $u'_{pr3} = [9{\cdot}4843_n] = -0{\cdot}30500$, and obtain by the theorem of Lagrange $M'_2 = u_{pr1}/u'_{pr3} = [0{\cdot}0130] = 1{\cdot}0304$. To complete this example, we must, on account of $l_I = \infty$, anticipate a subsequent proof that for any centred system working in air we must have $f = -f'$, or in the present case $f = -9{\cdot}5477$. This gives, with the alternative formula for l_f,

$$l_f = l_{II} + f/M'_2 = 0{\cdot}249 + \frac{-9{\cdot}5477}{1{\cdot}0304} = 0{\cdot}249 - 9{\cdot}266 = -9{\cdot}017,$$

and then the location of the first or anterior principal plane

$$l_{pp} = l_f - f = -9{\cdot}017 + 9{\cdot}548 = +0{\cdot}531.$$

In systems calculated for objects at a finite distance it is of course always possible to calculate all the equations of G Th (10) directly without drawing upon the fact, as yet unproved by us, that $f = -f'$, and this relation may then be used as a check, for if f should come out $\gtrless -f'$ an error is sure to be in the work.

(B) Various graphical methods have been worked out by which the Gaussian constants can be obtained on the drawing-board without, or with very little, actual numerical work. It is impossible to obtain even the accuracy of four-figure log work by these methods, which, moreover, lead to a confusing tangle of lines in a great many cases, and for that reason are laborious and greatly subject to error. A properly trained designer is therefore not likely to employ these methods.

As a matter of fact, the Gaussian constants do not enter very largely into the work of designing properly corrected optical systems of considerable aperture and field, for the simple reason that such systems can only be designed for *one* particular pair of conjugate planes and become imperfect for any other pair. There are, however, two simple cases which frequently occur in the preparatory stages of working out new designs, for which it pays to work out complete algebraical solutions:

(C) *The Gaussian constants for systems of two thin lenses* at a finite separation. As the general solution by G Th (10) is valid for any two pairs of conjugate points, we can secure an elegant and simple solution by a judicious choice. The two lenses shall have focal lengths f'_a and f'_b, and shall be placed at a separation d'. The constants for the object space shall be counted from the first lens, those for the image space from the second lens. With reference to the diagram, Fig. 91, we choose the first pair of conjugates so that the image falls upon the thin second lens, or in the symbols of G Th (10) so that $l'_I = 0$. Being treated as thin the second lens will have no effect on either the location or the linear magnification of this image, and

[72] we can deduce these data from the first lens only. The universal TL formula $1/l' = 1/f' + 1/l$ then gives, as $l = l_I$ and $l' = d'$,

$$\frac{1}{d'} = \frac{1}{f'_a} + \frac{1}{l_I}, \text{ or } \frac{1}{l_I} = \frac{1}{d'} - \frac{1}{f'_a} = \frac{f'_a - d'}{d'.f'_a}, \text{ or } l_I = \frac{d'.f'_a}{f'_a - d'}$$

and by the TL magnification formula $m' = l'/l$ we find

$$M'_1 = d'/l_I = (f'_a - d')/f'_a.$$

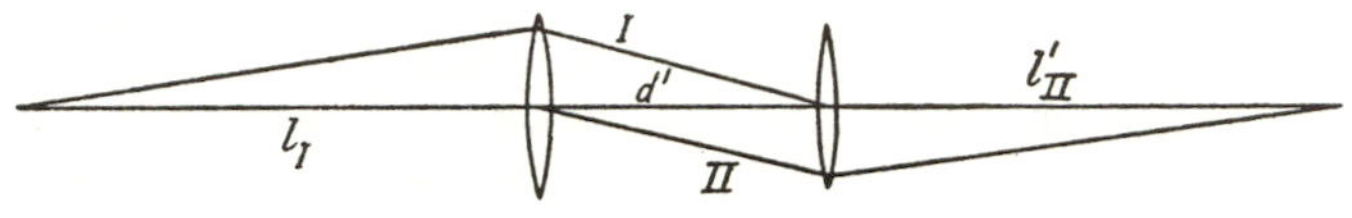

FIG. 91.

As the second pair of conjugates we choose that which has the object in contact with the first lens, or $l_{II} = 0$, and as the entire refraction effect will now be due to the second lens, with $l = -d'$, the TL formula gives

$$\frac{1}{l'_{II}} = \frac{1}{f'_b} + \frac{1}{l} = \frac{1}{f'_b} - \frac{1}{d'} = \frac{d' - f'_b}{d'.f'_b} \therefore l'_{II} = \frac{d'f'_b}{d' - f'_b} \text{ and } M'_2 = \frac{l'_{II}}{-d'} = -\frac{f'_b}{d' - f'_b}.$$

We now have all the data required for the working out of G. Th. (10), namely,

$$l'_I = 0;\ l_I = d'f'_a/(f'_a - d');\ l_{II} = 0;\ l'_{II} = d'f'_b/(d' - f'_b)$$

$$M'_1 = (f'_a - d')/f'_a;\ M'_2 = -f'_b/(d' - f'_b);$$

and these give $\quad a = l_{II} - l_I = -l_I = -d'f'_a/(f'_a - d');$

$$a' = l'_{II} - l'_I = l'_{II} = d'f'_b/(d' - f'_b).$$

The working out of the equivalent focal lengths becomes simplest if we invert the formulae of G. Th. (10). The reciprocal of the equation for f' then gives

$$\frac{1}{f'} = \frac{M'_1 - M'_2}{a'} = \left[\frac{f'_a - d'}{f'_a} + \frac{f'_b}{d' - f'_b}\right] \cdot \frac{d' - f'_b}{d'.f'_b} = \frac{(f'_a - d')(d' - f'_b)}{d'f'_a f'_b} + \frac{1}{d'}.$$

Multiplying out the first numerator and then bringing the second term to the common denominator gives

$$\frac{1}{f'} = \frac{-f'_a.f'_b + d'(f'_a + f'_b) - d'^2 + f'_a f'_b}{d'f'_a f'_b} = \frac{f'_a + f'_b - d'}{f'_a f'_b} = \frac{1}{f'_a} + \frac{1}{f'_b} - \frac{d'}{f'_a f'_b}$$

in agreement with the result in Chapter I.

Treated in the same way, the reciprocal of the equation for f gives

$$\frac{1}{f} = \frac{1}{a.M'_1} - \frac{1}{a.M'_2} = \frac{1}{\dfrac{-d'f'_a}{f'_a - d'} \cdot \dfrac{f'_a - d'}{f'_a}} - \frac{1}{\dfrac{-d'f'_a}{f'_a - d'} \cdot \dfrac{-f'_b}{d' - f'_b}} =$$

$$-\frac{1}{d'} - \frac{(f'_a - d')(d' - f'_b)}{d'f'_a f_b}$$

and, as the last result has the same magnitude (but the opposite sign) as the corresponding value of $1/f'$, we need not carry the transformations to a finish, but can [72] at once conclude that $1/f = -1/f'$ or also $f = -f'$; another confirmation of our assertion—as yet unproved—that systems working in air always have the two focal lengths equal but of opposite sign. As the second focal length f' is the one almost invariably stated and employed, we will in the remainder of the work replace f by $-f'$ whenever it occurs. For l'_f, usually called the 'back focus' of a system, we employ the equation $l'_f = l'_I + f'M'_1$, because l'_I is zero, and find

$$l'_f = f'.\frac{f'_a - d'}{f'_a} = f' - \frac{d'f'}{f'_a}.$$

For l_f, the distance of the anterior principal focus from the first lens, we use the alternative equation $l_f = l_{II} + f/M'_2$, because in our case $l_{II} = 0$, and find, by using $f = -f'$,

$$l_f = -f'.\frac{d' - f'_b}{-f'_b} = -f' + \frac{d'f'}{f'_b}.$$

The principal planes are then located by the equations of G Th (10) at

$$l_{pp} = l_f + f' = \frac{d'f'}{f'_b} \quad ; \quad l'_{pp} = l'_f - f' = -\frac{d'f'}{f'_a}.$$

This completes the determination of all the usual Gaussian constants; but considerable interest is attached in the present case to the distance separating the two principal planes, which we will call positive when PP' is to the right of PP. If in Fig. 91 we assume PP to the left of lens 'a' and therefore with a negative l_{pp}, and PP' to the right of lens 'b', we read off

$$PP \rightarrow PP' = d' + l'_{pp} - l_{pp} = d' - d'f'(1/f'_a + 1/f'_b) = d'f'(1/f' - 1/f'_a - 1/f'_b).$$

By transposing the second form of our equation for $1/f'$ it is found that the term in the last bracket represents $-d'/f'_a f'_b$, hence

$$PP \rightarrow PP' = -d'^2.\frac{f'}{f'_a \cdot f'_b}.$$

The collected results now are:

Gaussian constants for two thin lenses of focal lengths f'_a and f'_b placed at a separation $= d'$:

$$\frac{1}{f'} = -\frac{1}{f} = \frac{1}{f'_a} + \frac{1}{f'_b} - \frac{d'}{f'_a \cdot f'_b} = \frac{f'_a + f'_b - d'}{f'_a \cdot f'_b};$$

$$l'_f = f' - \frac{d'f'}{f'_a};\ l_f = -f' + \frac{d'f'}{f'_b};\ l'_{pp} = -\frac{d'f'}{f'_a};\ l_{pp} = \frac{d'f'}{f'_b};$$

$$PP \rightarrow PP' = -d'^2.f'/f'_a f'_b.$$

l_f and l_{pp} are measured from the first lens, l'_f and l'_{pp} from the second lens. For the determination of the conjugate distances l and l' which realize any prescribed magnification M', the formulae

G Th (10) $$l = l_f + f'/M' \text{ and } l' = l'_f - f'M'$$

[72] are then immediately available, and these may of course be transposed to find the magnification corresponding to any given value of l or l' by the equations

$$M' = f'/(l - l_f) = (l'_f - l')/f'.$$

Transpositions of the other equations similarly will supply solutions of a great variety of problems which are apt to arise in the course of designing new systems.

To illustrate the extreme convenience of this solution we will apply it to the system which has already been shown to realize our first example, namely, two thin lenses with $f'_a = f'_b = 1$ and $d' = 3$. Our solution gives

$$\frac{1}{f'} = -\frac{1}{f} = \frac{f'_a + f'_b - d'}{f'_a \cdot f'_b} = \frac{1+1-3}{1 \times 1} = -1 \therefore f' = -1, f = +1,$$

$$l'_f = f' - \frac{d'f'}{f'_a} = -1 - \frac{3\times(-1)}{1} = -1+3 = +2\,;\ l'_{pp} = -\frac{d'f'}{f'_a} = +3$$

$$l_f = -f' + \frac{d'f'}{f'_b} = +1 + \frac{3\times(-1)}{1} = 1-3 = -2\,;\ l_{pp} = \frac{d'f'}{f'_b} = -3$$

$$PP \rightarrow PP' = -d'^2 . f'/f'_a f'_b = -3^2(-1)/1\times 1 = 9.$$

To secure $M' = \frac{2}{3}$ (as for our planes I and I'), the solution gives

$$l = l_f + f'/M' = -2 + -1/\tfrac{2}{3} = -2 - 1\tfrac{1}{2} = -3\tfrac{1}{2}$$
$$l' = l'_f - f'M' = 2 - (-1) . \tfrac{2}{3} = 2 + \tfrac{2}{3} = 2\tfrac{2}{3}$$

and similar confirmations of all the other earlier results.

As many important types of optical instrument can in first approximation be treated as composed of two thin separated components, our solution for combinations of this type will repay a brief discussion. There are three main classes, for both components may have the properties of a convex lens, or both may have the properties of a concave lens, or one may be convex whilst the other is concave. In the first class f'_a and f'_b will both be positive, in the second class both negative, in the third one positive and the other negative ; d' will, of course, always be positive.

First class. In the equation determining the power of the combination $1/f' = [(f'_a + f'_b) - d']/f'_a f'_b$, $f'_a + f'_b$ in the numerator and $f'_a . f'_b$ in the denominator will be necessarily positive, and the d' in the numerator will be really subtractive. Hence the combined power will have a maximum value $= (f'_a + f'_b)/f'_a f'_b = 1/f'_a + 1/f'_b$ with a negligibly small separation ; increased separation will cause the power of the combination to sink (and therefore f' itself to grow), and when d' reaches the value $= f'_a + f'_b$ the power of the combination becomes zero and its equivalent focal length infinite. If d' grows still further, the numerator of the equation, and with it $1/f'$ and f' become negative and the system acquires the properties of an ordinary concave lens, with a focal length which becomes shorter and shorter with increasing d'. The system used as a numerical example is of this last type. We shall see more clearly, subsequently, that the forms with $d' < (f'_a + f'_b)$ are essentially object-glasses, magnifying glasses, or ordinary eyepieces, the form with $d' = (f'_a + f'_b)$ represents the simplest type of astronomical or Keplerian telescope, whilst forms with $d' > (f'_a + f'_b)$ have the properties of a compound micro- [72] scope. In the case of $d' = (f'_a + f'_b)$, l'_f, l_f, l'_{pp}, and l_{pp} all become infinite, and the solutions for conjugate distances become useless. The equation for $PP \to PP'$—usually called the 'interstitium'—shows that this is necessarily negative as long as f' of the combination is positive; hence in combinations of the eyepiece type the principal planes are always 'crossed', the 'anterior' one lying to the right of the 'posterior' one. For the telescope case this distance also becomes infinite, and for the compound microscope type (f' negative) it has positive values as in our numerical example.

Second class. When f'_a and f'_b are both negative, both parts of the numerator in the $1/f'$ equation are negative, whilst the denominator is positive. Hence the power of these combinations is always negative and increases steadily with increasing separation. The interstitium is always positive, l'_{pp} always negative, l_{pp} always positive; hence both principal planes are invariably virtual, but also invariably in the natural order suggested by the terms anterior or posterior. There are no important representatives of this class among the usual optical instruments.

Third class. In a combination of a positive with a negative component the denominator of the $1/f'$ equation will be negative, which we must carefully bear in mind. The class must be subdivided according to whether the concave component has the longer or the shorter focal length.

(a) When the negative component has a longer focal length than the positive component, then $f'_a + f'_b$ will be negative; hence both parts of the numerator in the $1/f'$ equation will be negative, and on account of the negative denominator $1/f'$ will always be positive and will grow in value with increasing separation. The equivalent focal length is therefore at a maximum when the two lenses are in contact, and diminishes steadily with increasing separation. The principal plane for the space adjacent to the concave component is always virtual, that for the space adjacent to the convex element always real. The interstitium is always positive. Dialyte telescope objectives of moderate separation and a few photographic objectives are of this type.

(b) When the negative component has the shorter focal length, $f'_a + f'_b$ will be positive, and as the second numerator term $-d'$ in the $1/f'$ equation as well as the denominator are negative, the combination will have negative power for *small values of d'*. Compound eyepieces for Galilean telescopes have occasionally been made to this type, which is, however, rather unimportant. When the separation d' is increased, the negative power of this combination diminishes, and reaches zero when $d' = f'_a + f'_b$. This, therefore, is another case when the equivalent focal length, and with it the other Gaussian constants, becomes infinite and useless for solutions. The combination in this state of adjustment represents the simplest form of the Galilean telescope. When d' is still further increased the numerator of the $1/f'$ equation becomes negative, and therefore $1/f'$ assumes positive values which rise as d' is further increased. This also is an important form, for practically all telephoto lenses are of this type. The principal plane on the side of the convex lens is real and at a considerable distance from the lens, that on the side of the concave lens is virtual; the shortening of l'_f which results is the chief *raison d'être* of the telephoto type.

[72] (D) *The Gaussian constants for a thick lens.* We easily obtain a solution of this problem by the same choice of two pairs of conjugates as in the previous case. The lens (Fig. 92) shall have radii r_1 and r_2, thickness d', and refractive index N.

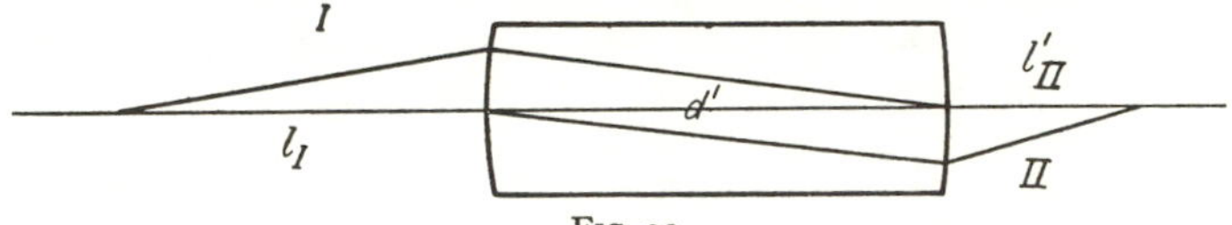

FIG. 92.

To find the conjugate distances we must in this case apply equation (6p)** of Chapter I for refraction at any one refracting surface : $N'/l' = (N' - N)/r + N/l$, and for the magnification (10p)* : $m' = Nl'/N'l$.

For the first pair of conjugates, producing an image *on* the second surface, we apply these formulae to the first surface, for which the N of the formulae $= 1$ for air, whilst the N' of the formulae is the lens-index. As the left-hand conjugate is to be called l_I and the right-hand conjugate is to be $= d'$, we find

$$N/d' = (N-1)/r_1 + 1/l_I \therefore 1/l_I = (Nr_1 - (N-1)d')/d'r_1 \quad \text{or}$$
$$l_I = d'r_1/(Nr_1 - (N-1)d'),$$

whilst l'_I, counted from the second surface, will be $l'_I = 0$. The magnification formula gives $M'_1 = d'/Nl_I$.

For the second pair of conjugates N' in the standard formulae $= 1$ for air, whilst N is the lens-index ; $l = -d'$ and l' is to be called l'_{II}, hence (6p)** gives for the second pair

$$1/l'_{II} = -(N-1)/r_2 - N/d' \quad \text{or} \quad l'_{II} = -d'r_2/(Nr_2 + (N-1)\,d')$$

to be coupled with $l_{II} = 0.$

The magnification formula gives $M'_2 = -N\,.\,l'_{II}/d'$;

G Th (10) then gives $a = l_{II} - l_I = -l_I$ because $l_{II} = 0$,

and $a' = l'_{II} - l'_I = l'_{II}$ because $l'_I = 0$.

It is again simplest to work out the reciprocals of the two focal lengths. For the second focal length f' we find

$$\frac{1}{f'} = \frac{M'_1}{a'} - \frac{M'_2}{a'} = \frac{d'/Nl_I}{l'_{II}} - \frac{-Nl'_{II}/d'}{l'_{II}} = \frac{d'}{Nl_I l'_{II}} + \frac{N}{d'}.$$

For the first focal length we obtain

$$\frac{1}{f} = \frac{1}{aM'_1} - \frac{1}{aM'_2} = \frac{1}{-l_I d'/Nl_I} - \frac{1}{-l_I(-Nl'_{II}/d')} = -\frac{N}{d'} - \frac{d'}{Nl_I l'_{II}}.$$

Clearly we again have $1/f = -1/f'$; hence we only work out $1/f'$ by putting in the explicit values of l_I and l'_{II}.

$$\frac{1}{f'} = \frac{N}{d'} + \frac{d'}{N} \cdot \frac{Nr_1 - (N-1)d'}{d'r_1} \cdot \frac{Nr_2 + (N-1)d'}{-d'r_2} =$$
$$\frac{N}{d'} - \frac{N^2 r_1 r_2 - N(N-1)r_2 d' + N(N-1)r_1 d' - (N-1)^2 d'^2}{Nd'r_1 r_2}.$$

If the first fraction is now brought to the denominator of the second fraction [72] it is seen to cancel the first term of the latter. The surviving terms have $(N-1)$ as a common factor, and, dividing out, we obtain the final fairly simple formula

$$\frac{1}{f'} = (N-1)\left[\frac{1}{r_1} - \frac{1}{r_2} + \frac{N-1}{N} \cdot \frac{d'}{r_1 r_2}\right] = -\frac{1}{f}.$$

It should be noticed that it gives the usual formula for the power of a simple *thin* lens on putting $d' = 0$.

For l'_f we use the regular formula of G Th (10), because $l'_I = 0$, and find

$$l'_f = f'M'_1 = f' \cdot \frac{d'}{N} \cdot \frac{1}{l_I} = f' \cdot \frac{d'}{N} \cdot \frac{Nr_1 - (N-1)d'}{d'r_1} = f' - \frac{N-1}{N} \cdot \frac{d'f'}{r_1},$$

which gives
$$l'_{pp} = l'_f - f' = -\frac{N-1}{N} \cdot \frac{d'f'}{r_1}.$$

For l_f we use the alternative formula $= l_{II} + f/M'_2$, because $l_{II} = 0$, and substituting also $-f'$ for the f, we find

$$l_f = -f'/M'_2 = f' \cdot \frac{d'}{N} \cdot \frac{1}{l'_{II}} = f' \cdot \frac{d'}{N} \cdot \frac{Nr_2 + (N-1)d'}{-d'r_2} = -f' - \frac{N-1}{N} \cdot \frac{d'f'}{r_2},$$

and
$$l_{pp} = l_f - f = l_f + f' = -\frac{N-1}{N} \cdot \frac{d'f'}{r_2}.$$

The distance between the two principal planes is of no great interest in this case, especially because it does not yield a very compact expression. The collected equations are:

Gaussian constants of a thick lens in air, r_1 and r_2 being the radii of curvature of the two surfaces, d' the thickness and N the refractive index of the lens. l, l_f, and l_{pp} are referred to the pole of the first surface, l', l'_f, and l'_{pp} to that of the second surface.

$$\frac{1}{f'} = -\frac{1}{f} = (N-1)\left[\frac{1}{r_1} - \frac{1}{r_2} + \frac{N-1}{N} \cdot \frac{d'}{r_1 r_2}\right]$$

$$l'_f = f' - \frac{N-1}{N} \cdot \frac{d'f'}{r_1}; \quad l'_{pp} = -\frac{N-1}{N} \cdot \frac{d'f'}{r_1};$$

$$l_f = -f' - \frac{N-1}{N} \cdot \frac{d'f'}{r_2}; \quad l_{pp} = -\frac{N-1}{N} \cdot \frac{d'f'}{r_2}$$

$$l = l_f + f'/M'; \qquad l' = l'_f - f'M'$$

$$M' = f'/(l - l_f); \qquad M' = (l'_f - l')/f'.$$

The discussion of these equations yields quite a number of interesting conclusions:

(1) If one surface is plane, or one radius infinite, the third term in the square bracket of the equation for $1/f'$ becomes zero; hence the power and equivalent

[72] focal length of planoconvex or planoconcave lenses is quite independent of their thickness. As the equations for the location of the principal planes have the *other* radius in the denominator, the principal plane on the side of the curved surface coincides with the pole of the surface, whilst the other principal plane is always virtual. The focal distance measured on the curved side is therefore the true equivalent focal length.

(2) In a biconvex lens r_1 is positive, r_2 is negative. It follows that $1/r_1 - 1/r_2$ in the square bracket of the equation for $1/f'$ is necessarily positive, whilst the last term is negative on account of r_2 in its denominator. Hence a biconvex lens has a maximum of power, or shortest equivalent focal length, for $d' = 0$. The power sinks steadily as d' is increased, and when d' reaches the value $= \frac{N}{N-1}(r_1 - r_2)$ the square bracket becomes equal to zero and the equivalent focal length as well as all the other Gaussian constants become infinite. The lens—really a glass rod of considerable length—then has the properties of a Keplerian telescope. If the already extravagant thickness grows still further, the lens acquires negative properties and its effects are similar to those of a compound microscope. Both principal planes are virtual if f' is positive, both become real when f' reaches negative values through great thickness.

(3) A biconcave lens has r_1 negative, r_2 positive, with the result that all three terms in the square bracket of the equation for $1/f'$ are permanently negative. Hence a biconcave lens is always negative, but its negative power increases steadily with increasing thickness. Both principal planes are invariably virtual.

In meniscus lenses r_1 and r_2 have the same sign, and the third term in the square bracket of the equation of $1/f'$ is therefore invariably positive and increases in value with increasing thickness. We must, however, distinguish two cases :

(4) In a meniscus in which the concave surface has the longer radius, or the smaller curvature, $1/r_1 - 1/r_2$ is positive no matter which way the lens is placed. Hence for this type the square bracket is invariably positive but increases in numerical value with d'. The power of the lens therefore also increases steadily with d'. The principal plane on the concave side is virtual, that on the convex side is real.

(5) In a meniscus in which the concave surface has the shorter radius, $1/r_1 - 1/r_2$ is negative, whilst the term depending on d' remains positive. Hence this type has negative power for sufficiently low values of d'. The power sinks to zero and all the other Gaussian constants reach infinity when d' reaches $\frac{N}{N-1}(r_1 - r_2)$. The thick meniscus is then the simplest possible Galilean telescope and has occasionally been made for actual use at very low magnification. For still greater thickness the power becomes positive and the properties of the thick meniscus are then similar to those of the telephoto type.

(6) An interesting special case is that of the 'monocentric' lens—that is, a lens in which the centres of curvature of the two surfaces coincide. With due regard to signs, this means that $d' = r_1 - r_2$ and there are two possible types, for the lens may be a thick biconvex with the common centre of curvature inside the lens, or it

may be a meniscus with the common centre outside the lens. For either type the [72] formula for $1/f'$ becomes simplified on account of $d' = r_1 - r_2$, for this gives

$$\frac{1}{f'} = (N-1)\left[\frac{1}{r_1} - \frac{1}{r_2} + \frac{N-1}{N}\,\frac{r_1 - r_2}{r_1 r_2}\right] = (N-1)\left[\frac{1}{r_1} - \frac{1}{r_2} - \frac{N-1}{N}\left(\frac{1}{r_1} - \frac{1}{r_2}\right)\right] = \frac{N-1}{N}\left(\frac{1}{r_1} - \frac{1}{r_2}\right).$$

The power of a monocentric lens is therefore $1/N$th of that of a *thin* lens with the same radii and is positive for the biconvex, negative for the meniscus type. If we put the special values of $1/f'$ and of d' into the formula for l'_{pp}, we find

$$l'_{pp} = -\frac{N-1}{N} \cdot \frac{r_1 - r_2}{r_1} \cdot \frac{1}{\frac{N-1}{N}\left(\frac{1}{r_1} - \frac{1}{r_2}\right)} = -\frac{r_1 - r_2}{r_1\left(\frac{1}{r_1} - \frac{1}{r_2}\right)} = \frac{-r_2(r_1 - r_2)}{r_2 - r_1} = r_2,$$

and l_{pp} is in the same way found $= r_1$.

This means that the two principal planes coincide at the common centre of curvature. The simplest case is that of a sphere or glass ball ; this has $r_2 = -r_1$ or $d' = 2\,r_1$ and gives for a sphere :

$$\frac{1}{f'} = 2\,\frac{N-1}{Nr_1}; \quad l'_{pp} = -r_1; \quad l_{pp} = r_1.$$

The peculiarity of monocentric lenses is that they have no definite optical axis because there is only one centre of curvature.

Bending rule for thick lenses. The convenient TL rule, that a change of c_1 and c_2 of a lens by the same amount does not alter the power, is not strictly correct for thick lenses. Our new exact equations for thick lenses enable us to extract the true rule. If we introduce our usual c_1 for $1/r_1$ and c_2 for $1/r_2$ into the exact equation for the power of a thick lens, it becomes

$$1/f' = (N-1)\,(c_1 - c_2 + \frac{N-1}{N} \cdot d' \,.\, c_1 \,.\, c_2).$$

When f' and d' are treated as fixed, prescribed data, this equation can be solved to give the value of c_2 which will yield the prescribed f' with the prescribed d' for any chosen value of c_1, or vice versa. Solved for c_1, the equation gives

$$c_1 = \left(\frac{1}{f'(N-1)} + c_2\right) \Big/ \left(\frac{N-1}{N}\,d' c_2 + 1\right);$$

and solved for c_2 it gives

$$c_2 = \left(c_1 - \frac{1}{f'(N-1)}\right) \Big/ \left(1 - \frac{N-1}{N}\,d' c_1\right).$$

Like the other equations for a thick lens, these are perfectly exact paraxial formulae and may be worked out with the full accuracy which the precision of the data may justify. The equations are a universal solution of the bending problem for a simple lens ; as for $d' = 0$ the denominators become simply 'one', the

[72] numerators alone give the solution for an infinitely thin lens. We will apply the solution for c_2 to the biconvex lens of Chapter I, adopting the form there used: $r_1 = 10{\cdot}000$, $d_1' = 0{\cdot}600$, $r_2 = -5{\cdot}000$, $N = 1{\cdot}518$, as a standard for the focal length to be maintained in all the bendings. As the bending equations contain $1/f'(N-1)$, we calculate this by transposing the $1/f'$ equation into:

$$1/f'(N-1) = c_1 - c_2 + \frac{N-1}{N}\, d' c_1 c_2,$$

in which for our standard form $c_1 = 0{\cdot}1$, $c_2 = -0{\cdot}2$, $d' = 0{\cdot}6$, and $N = 1{\cdot}518$, hence

$$1/f'(N-1) = 0{\cdot}3 - \frac{0{\cdot}518}{1{\cdot}518} \times 0{\cdot}6 \times 0{\cdot}1 \times 0{\cdot}2 = 0{\cdot}295905.$$

Choosing the bendings $c_1 = -0{\cdot}1$, 0, 0·1, 0·2, 0·3, 0·4 as used in section [30] and calculating also the denominator $= 1 - \frac{N-1}{N} \cdot d' . c_1$ for these values, we find:

for c_1	=	−0·1	0	0·1	0·2	0·3	0·4
Numerator = correct c_2 for *thin* lenses	=	−0·395905	−0·295905	−0·195905	−0·095905	+0·004095	+0·104095
Denominator	=	1·02047	1	0·97953	0·95905	0·93858	0·91910
Ratio = correct c_2 for lenses 0·6 thick	=	−0·387963	−0·295905	−0·200000	−0·100000	+0·004362	+0·113258
By TL-rule, inexact c_2	=	−0·4	−0·3	−0·2	−0·1	0	+0·1

The agreement between correct c_2 for $d' = 0{\cdot}6$ and the c_2 by the TL rule for $c_1 = 0{\cdot}1$ is of course due to our choosing that thick form as our standard, and the agreement must be repeated for $c_1 = 0{\cdot}2$ because that is the same lens merely turned round. The agreement between correct c_2 for a thin lens and correct c_2 for the thick lens at $c_1 = 0$ is necessary because we proved that the equivalent focal length of a plano-convex lens is independent of thickness. But in the remaining cases the disagreements will be seen to attain considerable magnitude. It therefore pays in many cases, when comparatively thick lenses are used, to calculate the bendings by the exact formulae. The thin lens values in the first line differ from those in the last line by a constant amount; that is simply due to our having taken a biconvex *thick* lens as our standard for focal length. All the forms in lines one and three have the exact equivalent focal length of the lens which was the subject of the first trigonometrical example in Chapter I.

CONJUGATE ANGLES OF CONVERGENCE.

[73] In the first and again in the seventh chapter we found that a very close relation exists between the magnification for two conjugate points and the angles which the rays passing between these points make with the optical axis before and after passage through the optical system. It will therefore be of great interest to look for a corresponding relation by the method of the present chapter. To obtain this relation in its most general form we return to the first diagram of this chapter, but as we have already worked out all the principal consequences of our initial assumptions, we have only to refer to the general pair of conjugate planes O and O' and to the given pair I and I', which are therefore alone shown in Fig. 93. A ray from

the axial point B of plane O drawn in the plane of the diagram to point P_1 at [73] ordinate y_1 in plane I will form an angle with the optical axis defined by $\tan U = -y_1/b$. Because this ray passes through B in the object space it must pass through

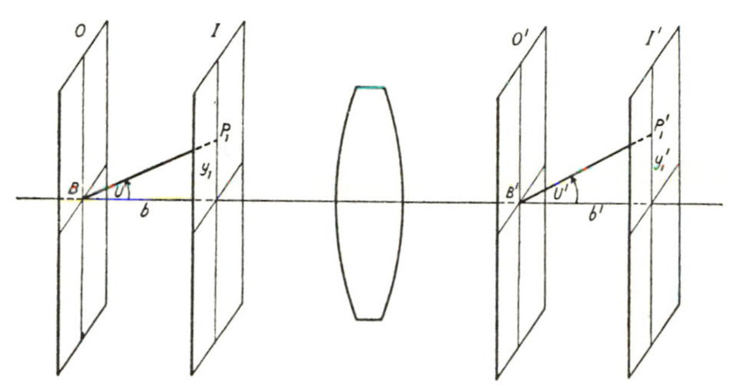

FIG. 93.

the conjugate point B' in the image space ; and because it passes through point P_1 of plane I it must pass through the conjugate point P_1' in plane I' at $y_1' = M_1' \cdot y_1$; therefore

$$\tan U' = -y_1'/b' = -y_1 \,.\, M_1'/b'.$$

A minus sign has to be added on the right of both equations in order to cause the angles to agree in sign with our *optical* sign conventions. A mere division of the equations for $\tan U$ and $\tan U'$ now gives the desired relation of conjugate angles of convergence in the form

$$\tan U/\tan U' = b'/bM_1'.$$

We modify this by introducing for b and b' the values found in the proof of G Th (10), namely

$$b = f/M' - f/M_1' = f(M_1' - M')/M'M_1' \quad \therefore \quad bM_1' = f(M_1' - M')/M'$$
$$b' = f'.M' - f'.M_1' = -f'(M_1' - M')$$

and obtain the result

$$\text{(a)} \qquad \frac{\tan U}{\tan U'} = -\frac{f'}{f} \cdot M'$$

as a necessary consequence of our assumption that the optical system be perfect for all object-distances, or, by the theorem of Clerk Maxwell, at merely two different object-distances. As f' and f have perfectly fixed values for such a system and as M' also has a definite value for each pair of conjugate planes, the result means that the *tangents* of conjugate convergence angles must be in a constant ratio for all rays that can pass from an axial object-point to the conjugate image-point.

In Chapter VII we proved the optical sine condition, and Chapter VIII supplied proofs of the extraordinary precision with which fulfilment of that condition leads

[73] to freedom from coma, perhaps the most objectionable of all aberrations. By the sine condition we must have

(b) $$M' = \frac{h'_{Sk}}{h_1} = \frac{N \sin U}{N' \sin U'},$$

or a constant ratio of the *sines* of the same convergence angles for which our general theory demands a constant ratio of the tangents. As tan U/tan U' may be put into the form (sin U/sin U') (cos U'/cos U), it is seen at once that these two conditions clash hopelessly excepting when $U = \pm U'$, which is a highly exceptional special case. Hence either the sine condition or our general theory must be unsound. The former is based on the true law of refraction and on the ray-tracing methods following from that law, and both its proof and its practical test by actual calculations supply abundant evidence of its absolute soundness. Our general theory in its present form, in which no limit is imposed on either aperture or field, is a purely geometrical deduction from two assumptions: Firstly, the rectilinear propagation of light, which we may safely regard as amply established and therefore sound; but we assumed secondly that it be possible to correct an optical system so as to be perfect for two different object-distances. We did not supply the slightest proof that this was possible or compatible with the laws of reflection and refraction. Neither could we find any optical instrument which had the postulated property at considerable aperture and diameter of field, no matter where we might search for it. Hence we are driven to the inexorable conclusion that our second assumption must be unrealizable, or that

'It is impossible to correct an optical system of considerable aperture and field for freedom from all aberrations at two different distances of the objects.'

We can, however, make some comforting additions to this depressing conclusion. In the first place our conclusion does not apply at all to optical systems required for use with only one object-distance; perfection may therefore be attainable in that case. In the second place the clash between the constant tangent ratio demanded by the general theory and the constant sine ratio required by the sine condition is only slight as long as the angles U and/or U' do not exceed a few degrees; from this point of view it is highly significant that the only known lens systems which come near universal utility are certain photographic objectives of low aperture—maximum about $f/6$—in which the maximum value of U or U' is less than five degrees. Thirdly, we must remember that there are always certain tolerances as regards aberrations, and as we may safely assume that the worsening of the definition, as we go away from the planes for which correction has been established, will be a continuous and not a sudden process, there must in all cases be a useful, though possibly quite small, *range* of object and image distances within which the aberrations are not highly objectionable, and within that curtailed range our general theory will still be applicable at full aperture with sufficiently close approximation.

Having proved that a perfect optical system in strict accordance with our original stringent definition of perfection is absolutely impossible, we must henceforth adopt the restriction which we have found to be necessary, namely, that the conjugate angles U and U' must be so small for all object distances to be considered that the difference between tan and sine is practically negligible. In all strictness this would mean that the angles must be indefinitely small, but on account of tolerances we need not go quite to that limit but may admit values up to a few degrees. The two conditions (a) and (b) will then be compatible ; writing them in paraxial angles they will be [73]

$$\text{(a)}\ \frac{u}{u'} = -\frac{f'}{f} \cdot M' ; \qquad \text{(b)}\ \frac{u}{u'} = \frac{N'}{N} \cdot M' ;$$

and equating the two values of u/u' we obtain a highly important result :

G Th (11)
$$\frac{f'}{f} = -\frac{N'}{N}.$$

This gives us in absolutely general form, but restricted to small aperture, the ratio of the two focal lengths of any centred optical system as equal to the negative ratio of the refractive indices of the media in which the respective focal points are located. It follows that

(1) The two focal lengths of any centred lens system always have opposite signs.

(2) If object- and image-space are in media of the same refractive index (nearly always air) the two focal lengths are numerically equal.

(3) If object- and image-space have different refractive indices, then the two focal lengths are in the negative ratio of the indices of the media in which they are measured.

We have had numerous confirmations of the first two conclusions and one of the third conclusion in Chapter I, namely, for the two focal lengths of a single refracting surface. We now have the proof that these isolated cases are due to a universally valid law.

[74] The perfectly definite relation which we have now proved to exist between the two focal lengths of any lens system enables us to make some very valuable additions to the relations between object and image.

(1) By G Th (10) we have for the principal planes

$$l_{pp} = l_f - f ; \qquad l'_{pp} = l'_f - f',$$

and for the conjugate distances at which a prescribed magnification M' is produced,

$$l = l_f - f/M', \qquad l' = l'_f - f'.M',$$

and all the l and l' are measured from some fixed but arbitrary mark or marks on the lens system. This is perfectly convenient in experimentally determining the constants of a given system, but we will now show that we obtain a very important result if we make a special choice of the reference points, consisting in measuring all distances in the object-space from the first principal plane and all distances in the image-space from the second principal plane. That obviously means $l_{pp} = 0$ and $l'_{pp} = 0$, or by using the above values of the general solution,

$$l_{pp} = l_f - f = 0 \ \therefore\ l_f = f, \quad \text{and} \quad l'_{pp} = l'_f - f' = 0 \ \therefore\ l'_f = f'.$$

[74] If we now put these special values of l_f and l'_f into the general solution for the conjugate distances yielding magnification M', we find as the distances from the respective principal planes :

$$l = f - f/M' = -f(1-M')/M' \, ; \; l' = f' - f'.M' = f'(1-M') \; \therefore \frac{l'}{l} = -\frac{f'}{f} \cdot M'.$$

These formulae contain absolutely nothing that would distinguish one system of focal lengths f and f' from any other system with the same focal lengths. Hence we have proved that

> 'If the conjugate distances are measured from the respective principal planes, then the properties of a lens system are absolutely defined by its two focal lengths, quite irrespective of the construction or bulk of the system.'

In the most usual case of systems working in air we shall have $f = -f'$, and can write the above results

$$l = f'(1-M')/M' \, ; \quad l' = f'(1-M') \, ; \quad \frac{l'}{l} = M' \, ;$$

comparison with TL (4) of section [23] will show that these equations agree exactly with those there given for a thin lens, and we thus have a confirmation of our conclusion. This result is of great value in the designing of new systems, for it enables us to proceed from the usual first thin lens scheme to components of any construction. All we have to do is to design the thick components so as to have an equivalent focal length equal to that of the thin lens prototypes and then to space them so that the distance from the original object to the first principal plane of the first component is equal to l_a of the first thin lens, then to place the second component so that the distance from the second principal plane of the first component to the first principal plane of the second component is equal to d'_a in the thin lens scheme, and so right through the system. The final image will be at a distance from the second principal plane of the last component exactly equal to the l'_k in the thin lens scheme, and will have exactly the same linear magnification as in the thin lens scheme. The total distance from original object to final image will, however, be changed by the sum of the interstitia $PP \rightarrow PP'$ of the several thick components, and may therefore be greater or smaller than the corresponding distance in the thin lens scheme, but usually only to a moderate extent. From the practical point of view this is one of the most important conclusions in the general theory.

To avoid confusion with the general solution we will adopt the symbols X_{pp} and X'_{pp} for object- and image-distances measured from the principal planes, and record the formulae as

G Th (10)* $X_{pp} = -f(1-M')/M' \, ; \; X'_{pp} = f'(1-M') \, ; \; X'_{pp}/X_{pp} = -M'.f'/f.$

(2) Another special choice of the reference points for conjugate distances is due to Abbe, and leads to still simpler equations. We measure the conjugate distances from the respective principal *focal* points, which evidently means $l_f = 0$ and $l'_f = 0$; the formulae for conjugate distances in G Th (10) then become

$$l = -f/M', \quad l' = -f'.M', \; \therefore \; l.l' = f.f' = constant.$$

We will introduce the special symbols X_f and X'_f for object- and image-distances referred to the principal focal planes of lens-systems, and will record the Abbe equations as [74]

G Th (10)** $\quad X_f = -f/M'\ ;\quad X'_f = -f'.M'\ ;\quad X_f\,.\,X'_f = f\,.f'.$

They clearly bring out the same important fact as G Th (10)*, namely, that the properties of a lens system are completely defined by the two focal lengths. They have the further advantage of being easily remembered and extremely convenient in solving all the little preliminary problems of optical design. Thus supposing we wanted to produce a linear magnification $M' = -5$ with a lens or system working in air and having $f' = 8$, whence $f = -8$, then the first equation tells us at once that the object will have to be placed at $X_f = 8/-5 = -1{\cdot}6$, therefore by this amount to the left of the anterior focus, and the second equation gives $X'_f = -8 \times -5 = +40$; therefore the image will be formed at 40 units of length to the right of the posterior focus. The last equation gives the same result by transposition: $X'_f = f\,.f'/X_f = -64/-1{\cdot}6 = +40$.

Theoretically the two last forms of solution of the problem of image formation also bring out important facts in the clearest possible manner. They obviously will give real solutions for any conceivable value of M' between $-\infty$ and $+\infty$, no matter what the signs of f and f' may be. Therefore any prescribed linear magnification, and either erect or inverted images, may be produced by either positive or negative systems (corresponding to convex or concave thin lenses). The prevalent notion that concave lenses always give erect and minified images, and that convex lenses give either erect magnified and virtual images or else inverted real images, is only true when we limit ourselves to the examination of real, material objects *and* to *thin* lenses or systems. When we include virtual objects in the form of real images projected by suitable optical means, and/or thick lenses or systems, such as the simple negative system already introduced as an example, then the above conclusion that every lens system of finite focal length can produce any value and sign of linear magnification is easily verified. It is of the utmost importance that the optical designer should thoroughly familiarize himself with this greatly extended range of possibilities, as the solution of apparently formidable problems may become quite simple by using a lens or system in the less well-known range of its possible applications.

The Abbe equations supply the proof of another general property of all lens systems by the last equation: The product $X_f\,.\,X'_f$ has the constant value $= f\,.f'$, and as f and f' always have opposite signs, the product $X_f\,.\,X'_f$ is always negative. That means, firstly, that if an object lies to the left of the anterior principal focus ($\therefore X_f$ negative) then the image must lie to the right of the posterior focus so as to make X'_f positive, and of course vice versa. And as the product $X_f\,.\,X'_f$ is of constant value, a diminution of the numerical value of X_f must lead to an increase in the numerical value of X'_f. Together with the opposite signs this leads to the remarkable conclusion that in any lens system a longitudinal shift of the object in a certain direction necessarily leads to a shift of the image in the *same* direction. The only exception to this rule occurs when the object crosses the focal plane; the image then performs the *salto mortale* from positive into negative infinity.

[74] The exception is therefore limited to the one possibility of a discontinuity in the associated movements of object and image. As soon as this critical point has been passed the regular rule again applies.

(3) *Relations between object and image when measurements by the angle of subtense are used.*

It has been repeatedly pointed out in earlier sections that in many cases, and especially in that of distant objects, it is convenient or even necessary to introduce the angle of subtense instead of the linear measure. The general theory enables us to deduce a number of highly interesting and important results with reference to these cases.

(a) The most generally useful case is that in which the angle of subtense is taken at one of the principal focal points of the system. With reference to Fig. 94, we will first consider an object-point at any distance from the anterior focus F which,

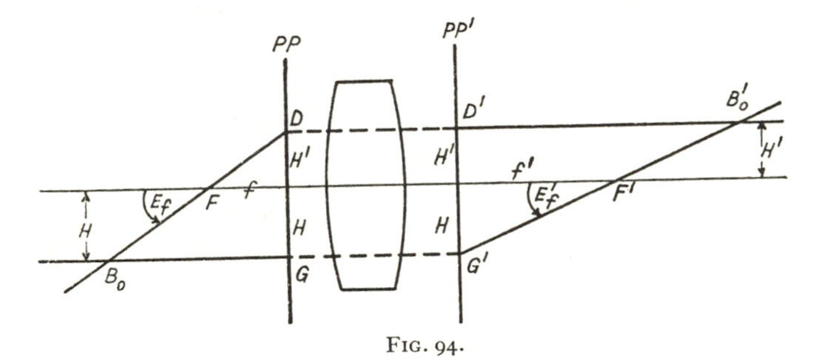

FIG. 94.

as seen from that point, appears under an angle E_f from the optical axis. We trace the ray from the object-point through the anterior focus. It will cut the first principal plane at a definite point D; as there is unit magnification for the principal planes, the emerging ray must pass the second principal plane at the same height as it passed the first one, and as the ray came from the anterior focus it must also emerge parallel to the optical axis. Hence it will everywhere in the entire image-space be at the distance H' from the optical axis at which it cut the first principal plane, and as the sharp image of our object-point must lie on this ray, the image-height must be $= H'$, no matter at what distance the sharp image may be found, or whether the image is real or virtual. We obtain this absolutely fixed value of H' from the right-angled triangle having the anterior focal length f (negative in the diagram) as horizontal side as

$$H' = f \,.\, \tan E_f,$$

and this will give the correct sign of H' if angle E_f is subjected to our optical sign convention according to which it is negative in the diagram. As this relation is unconditionally correct—excepting the immediate neighbourhood of F, it supplies another highly convenient method of determining the equivalent focal length by transposing the last equation. The smaller the distance of the object of known angular subtense, the more accuracy is required in placing the anterior focus F

at the point from which the subtense was measured, hence it is convenient to select objects at a considerable distance. If the latter amounts to several hundred times the focal length to be measured then a rough guess as to the location of F will be sufficient.

There is an important corollary to the above proposition. Supposing that we placed an object of linear size H' in the space to the right of the system and examined it through the latter with the eye (or the entrance pupil of an auxiliary observing instrument) at F, then we should see an image subtending the angle E_f. This deduction also can be readily adapted to the measurement of the equivalent focal length by a spectrometer or theodolite ; but we shall find it most useful as the true key to a proper appreciation of the magnification produced by magnifying glasses and by compound microscopes.

The application of the same reasoning to the second principal focus F' obviously leads to the corresponding result

$$H = f' \tan E'_f,$$

and to precisely analogous interpretations.

Incidentally our last diagram shows the well-known graphical method of completely locating the image B'_o of a given extra-axial object-point B_o by two rays, of which one passes parallel to the optical axis, the other through the principal focus. In the practice of optical design this method plays no part, as it is both quicker and more accurate to compute conjugate distances and magnifications by the Abbe equations.

(b) The subtense equations in conjunction with the Abbe equations enable us to solve the problem of the equivalent focal length and of the location of the principal focal points of a *combination of two systems* of known constants in a perfectly general and very elegant way.

With reference to Fig. 95, the two systems to be combined shall be A and B,

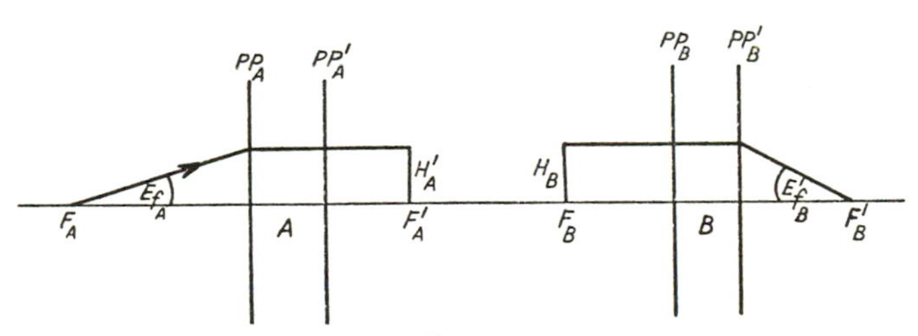

FIG. 95.

all their data being assumed to be known, as well as the distance $F'_A F_B$ from the second principal focus of system A to the first principal focus of system B. It must of course also be stipulated that the medium to the right of the first system is identical with that to the left of the second system, but the outside media may be different. To find the anterior equivalent focal length of the combination we assume a very distant object-point at angle E_{fA} from the optical axis. It will be

[74] depicted by system A alone in its posterior focal plane, and by the subtense equation the height of the image will be

$$H'_A = f_A \,.\, \tan E_{fA}.$$

This image will act as object for system B and we can apply the Abbe equations G Th (10)** to find the final image. For system B we have by the first equation transposed $M'_B = -f_B/X_{fB}$ as the magnification under which H'_A will be depicted at the final focus. X_{fB} is the distance $F_B F'_A$ or $-F'_A F_B$, whence $M'_B = f_B/F'_A F_B$ or the final image height

$$H' = H'_A \,.\, M'_B = f_A \,.\, f_B \,.\, \tan E_{fA}/F'_A F_B.$$

As the subtense equation is universally valid, it must also apply to the complete combination and to its focal lengths f and f'. Hence we must also have

$$H' = f \,.\, \tan E_{fA},$$

and, equating the two values of H', we find

$$f = f_A \,.\, f_B/F'_A F_B.$$

We can also locate the posterior principal focus of the complete combination with reference to the known posterior focus F'_B of the second component, for the third Abbe equation gives this as

$$X'_{fB} = f_B \,.\, f'_B / X_{fB} = -f_B \,.\, f'_B / F'_A F_B.$$

To find the posterior focal length and the anterior focal point of the complete combination we apply the same processes to a right-to-left pencil. For a very distant object-point at any angle E'_{fB} the second system will give an image in its anterior focal plane of height

$$H_B = f'_B \,.\, \tan E'_{fB}.$$

This image will act as object for system A and will produce the final image H. The second Abbe equation will apply and will give $M' = H_B/H = X'_{fA}/-f'_A$, and as X'_{fA} is $= F'_A F_B$, we obtain

$$H_B/H = F'_A F_B/-f'_A \quad \text{or} \quad H = -H_B \,.\, f'_A/F'_A F_B\,;$$

or on introducing the above value of H_B,

$$H = -f'_A f'_B \,.\, \tan E'_{fB}/F'_A F_B.$$

On the other hand the subtense equation applied to the complete system gives

$$H = f' \tan E'_{fB},$$

and on putting this value of H into the previous equation we obtain the final result

$$f' = -f'_A \,.\, f'_B / F'_A F_B.$$

Finally we obtain the distance of the anterior focus of the complete system from the anterior focus of system A by the third Abbe equation, which with $X'_{fA} = F'_A F_B$ gives

$$X_{fA} = f_A \,.\, f'_A / X'_{fA} = f_A \,.\, f'_A / F'_A F_B,$$

and all necessary data of the complete system are now known. These equations include our previous solution for two *thin* lenses at a finite separation d', namely,

$f' = f'_a \cdot f'_b / (f'_a + f'_b - d')$, as a special case, for it is easily seen that for thin lenses [74] we have $d' = f'_a + F'_a F_b + f'_b$, and if this is substituted in the old solution the new one results. The new and perfectly general solution finds its most important application in the theory of the compound microscope; the distance $F'_A F_B$ is then called the 'optical tube-length'.

(c) We obtain a remarkable result which is of the greatest utility in the determination of the magnifying effect of telescopes, if for a system of two separated components we measure the angular subtense of the object at the anterior focal point of the first component, and the angular subtense of the final image at the posterior focus of the second component. With reference to Fig. 96, the two

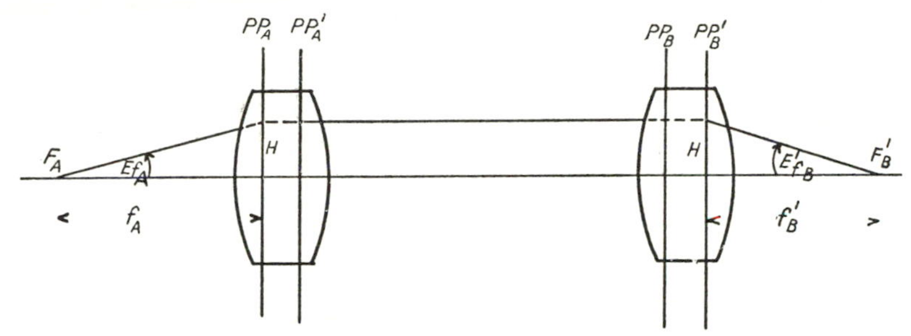

FIG. 96.

systems are indicated by their principal planes and their external focal points. If an extra-axial object-point at any distance from the system appears, as seen from F_A, under an angle E_{fA} (negative or counter-clockwise in the diagram) with the optical axis, then the ray passing from the object-point through F_A will cut the anterior principal plane of system A at a height

$$H = f_A \cdot \tan E_{fA},$$

and will leave system A as a parallel to the optical axis still at distance H from it. Refraction at system B will direct it towards the posterior focus F'_B of this system, and as it met its anterior principal plane PP_B at height H, it must also leave the posterior principal plane PP'_B at that height. Hence the angle E'_{fB} under which it meets the optical axis on emergence is determined by

$$\tan E'_{fB} = H/f'_B,$$

and if we introduce the value of H by the previous equation and transpose, we obtain the remarkable result

$$\frac{\tan E'_{fB}}{\tan E_{fA}} = \frac{f_A}{f'_B}.$$

The result *is* remarkable because f_A and f'_B are fixed data of the two systems and because the separation between the two systems does not enter into the equation in any way; hence there is an absolutely fixed ratio, for such a combination of two systems at any separation, between the subtense of the image seen from F'_B and the subtense of the object as seen from F_A.

[74] Leaving the application of the equations deduced in this sub-section to the following section, we will collect the formulae and assign references to them :

G Th (12) $H = f' . \tan E'_f$; $H' = f . \tan E_f$.

These give the relations between object and image when one is measured by angular subtense at the focal point, the other in linear measure.

G Th (13) $f = f_A . f_B / F'_A F_B$; $X_{fA} = f_A . f'_A / F'_A F_B$

$f' = -f'_A . f'_B / F'_A F_B$; $X'_{fB} = -f_B . f'_B / F'_A F_B$.

These determine the two focal lengths and the two focal points of a combination of two systems A and B when the separation is measured by the optical tube-length $F'_A F_B$.

G Th (14) $\tan E'_{fB} / \tan E_{fA} = f_A / f'_B$.

This fixes the ratio of the subtense of the image seen from the posterior focal point of system B to the subtense of the object as seen from the anterior focus of system A.

MEASURES OF MAGNIFICATION

[75] The term 'magnification' is very commonly used in a very loose and misleading way ; the equations which we have deduced from the general theory of perfect optical systems enable us to make the necessary sharp distinctions and to impart a perfectly precise and definite meaning to the terms which we shall use.

A. Linear Magnification

This term can be legitimately applied only in the case of projected real images. If we project an image of a real object by an optical system upon a suitable focusing screen or on a photographic plate, then we can measure both object and image in inches or millimetres or any other unit of length, and by forming the ratio $H'/H = m'$ obtain the linear magnification as a perfectly definite pure number with a simple and precise significance. All cases of magnification which we have hitherto discussed have been of this type. By *calculation* we can ascertain the actual linear dimensions of virtual images with equal facility and precision on assuming definite values for the distance of the object ; but virtual images are only of use for visual observation, and in these focusing for the most comfortable vision by the individual eye forms a regular part. As the single eye ('monocular vision') has very little appreciation of distance, the location of the virtual image produced by the instrument may vary, at any rate for young observers with normal vision, from a few inches away to infinity, with proportionate variation in the true 'linear' magnification. For that reason there is absolutely no sense in speaking of linear magnification in the case of direct visual observations with optical instruments.

B. Angular Magnification

As was pointed out in section [64] the normal human eye has a limit of resolving power of approximately one minute of arc and it follows that fine detail invisible to the naked eye must be raised to at least that subtense if it is to be visible through an optical instrument. Given good definition of the image the effectiveness of an instrument therefore depends on its power of increasing the angular subtense of the objects to be examined by it, and that leads us to the recognition of 'angular magnification' as the essential property of optical instruments for visual use. The more the subtense of objects is magnified, the more additional detail will be brought within the resolving power of the unaided eye. Hence we must in the case of visual instruments determine the angular magnification, and the number frequently stated as the 'linear magnification' of a microscope or even of a telescope employed in a given observation really represents the angular magnification. As the *tangent* of the angle of subtense appears in all our formulae for perfect optical systems, we will define the angular magnification as the ratio of the tangent of the angle of subtense (taken from the optical axis as the fixed reference line) of the image presented to the eye by the instrument, to the tangent of the angle under which the object would be seen by the naked eye; and to distinguish angular magnification from linear magnification in our symbols we will use MA and MA' for the angular magnification. [75]

(1) *Telescopes.* We will define a telescope as an optical instrument which magnifies the angle of subtense of objects, usually at a considerable distance, when it is not possible, convenient, or permissible to obtain improved vision by diminishing that distance. As a telescope always (at any rate in practice) consists of two systems called object-glass and eyepiece respectively, we can apply G Th (14) to it; but as telescopes are always used in air, we can replace f_A by $-f'_A$ in order to have the usual equivalent focal length of the object-glass in the formula. As E'_{fB} in the equation represents the angle of subtense of the image as seen from the posterior focal point of the eyepiece and E_{fA} the angle of subtense of the object as seen from the anterior focus of the object-glass, we have, subject to adoption of these respective viewing points,

G Th (14)* $$MA' = \frac{\tan E'_{fB}}{\tan E_{fA}} = -\frac{f'_A}{f'_B}$$

as the angular magnification of any telescope. The equation will be recognized as the standard formula given by all textbooks, but it is really a much more general solution of the problem, for whilst the textbook formula is restricted to a properly focused telescope applied to very distant objects, our equation was shown to hold no matter what the separation between the two systems might be. Subject to the restriction as to the two viewing points our equation is therefore applicable to every conceivable case regardless of the distance of the object and of the consequent adjustment of the focus.

As the object-glass always has the properties of a convex thin lens, f'_A is always positive. Hence we obtain a negative MA' or inverted images if the eyepiece also has a positive equivalent focal length, and we obtain a positive MA' or erect images if the eyepiece is negative.

(2) *Microscopes.* A microscope may be defined as an optical instrument which magnifies the angle of subtense of usually very small objects when the distance of the object is under complete control. This control over the distance of the object brings in another distinction between telescope and microscope, for it calls for the

[75] adoption of a standard distance at which the naked eye view is supposed to be taken. In the case of small objects we use our accommodation to obtain the closest possible view of them and the standard distance is therefore called the least distance of distinct vision. This varies even in the case of normal sight according to age, and as we must have a definite value, 10 inches or 250 mm. has been adopted as the 'conventional' least distance of distinct vision ; it is about right for people with normal sight and 30–40 years of age, but is too large for younger or myopic people, too small for older or hypermetropic people. For that reason the angular magnifications assigned to microscopes are not so definite in their significance as those of telescopes.

Whatever the construction of the microscope may be, the general theory of perfect optical systems will assign to it a perfectly definite, positive or negative equivalent focal length f'. Therefore we can use the first equation of G Th (12) to find the angle of subtense E'_f of the image conjugate to a small object of height H observed through the instrument by

$$H = f' . \tan E'_f.$$

On the other hand this small object when examined by the naked eye at the conventional distance of distinct vision, for which we will introduce the symbol Dv, will subtend an angle E_v determined by

$$\tan E_v = H/Dv,$$

and if we now introduce H by the previous equation and transpose, we find the angular magnification

G Th (12)* $$MA' = \frac{\tan E'_f}{\tan E_v} = \frac{Dv}{f'}$$

or equal to the distance of distinct vision divided by the equivalent focal length of the complete microscope. We must, however, note that this is unconditionally true only if the eye is placed at the posterior focal point of the complete instrument.

The equation applies to any common magnifying glass, and although in that case there is considerable latitude in the distance between the lens and the pupil of the eye of the observer, so that the actual angular magnification may vary by a considerable percentage, it is hardly worth while to discuss this variation seriously, as magnifying glasses are habitually used in a haphazard manner with complete disregard of theoretical refinements.

In the case of the compound microscope the equivalent focal length is always very short—rarely more than half an inch—and the eye cannot be moved away from the posterior focus of the instrument by more than a small fraction of an inch without losing the view of the complete field ; hence in this case the variation of the angular magnification which might result from the possible variation in the position of the eye is practically negligible. We may therefore accept G Th (12)* as a sufficiently exact solution in all cases. For the compound microscope a transformation of the equation is desirable, because the instrument is made up of two separate parts called the objective and the eyepiece, and the focal lengths of these are usually the known data. We therefore introduce by G Th (13)

$$f' = -f'_A . f'_B / F'_A F_B$$

and obtain the convenient equation : [75]

G Th (12)** $$MA' = -\frac{Dv \times F'_A F_B}{f'_A \cdot f'_B} \quad \text{(compound microscope)}.$$

As this equation may be regarded as determining the magnifying power by the product of two fractional factors, of which the one having f'_A in its denominator may be assigned to the object-glass and the other to the eyepiece, and as this division into factors is in fact convenient because there are usually several objectives and several eyepieces which yield a whole number of different combinations with varying magnifications, the question has been frequently raised and answered in the two possible ways as to which of the two numerator terms should be assigned to objective and eyepiece respectively. As the objective really does produce a real inverted image in the lower focal plane of the eyepiece it seems a matter of common sense to assign to the objective the share $-F'_A F_B/f'_A$, and to the eyepiece, which acts as a simple magnifying glass, the share Dv/f'_B. The great J. J. Lister proposed the opposite distribution of the numerator terms in a paper of 1841, which, however, was not published until 1913, and this distribution was actually introduced by Abbe when the Zeiss apochromatic objectives were put upon the market, Dv/f'_A being called the 'initial magnification' of the objective and $F'_A F_B/f'_B$ the 'angular magnification' of the eyepiece, the latter having the resulting number engraved upon it. Whilst not very convenient in practical microscopy, the Abbe resolution of the total angular magnification of a microscope into factors is decidedly useful in the designing of eyepieces for compound microscopes, for it will be shown in the following chapter that an eyepiece of given construction yields its best results as regards freedom from coma and flatness of field for one particular value of the angular magnification as defined by Abbe.

C. Range of Useful Magnification

It is evident from the equations for both telescopes and microscopes that we could produce any desired angular magnification by a suitable choice of the focal lengths of objective and eyepiece and, in the case of the microscope, of tube-length. The question then arises as to the range of magnifications which will be really useful in the case of a given instrument. That question can only be answered on the basis of physical optics.

It was shown in Chapter III that a telescope of clear aperture $= A$ inches has an extreme limit of angular resolving power $= 4{\cdot}5$ seconds of arc$/A$. On the other hand we learnt in the previous chapter that the limit of resolving power of the human eye is normally one minute of arc, but that for comfortable and long continued observation a minimum angular separation of two or three minutes is desirable. Clearly the angular magnification should be such as to raise the subtense of the actual objects to the values required by the eye. This gives :

For visual subtense.	*Condition for MA'*	*Solved for MA'*
$= 1'$	$\frac{4''.5}{A} \cdot MA' = 60''$	$MA' = 13\frac{1}{3}\,A$
$= 2'$	$= 120''$	$= 26\frac{2}{3}\,A$
$= 3'$	$= 180''$	$= 40\,A$
$= 4\frac{1}{2}'$	$= 270''$	$= 60\,A$

[75] and may be stated in words as follows :

> To render the finest detail accessible to a given telescope just visible (with considerable effort) to a normal eye an angular magnification is required = $13\frac{1}{3}$ per inch of clear aperture.
>
> Comfortable visibility calls for angular magnifications of $26\frac{2}{3}$ to 40 times the aperture in inches.
>
> Occasionally, but rarely, it may pay to use as much as 60 times the aperture in inches.

Thirty times the aperture in inches may be taken as a fair average of the magnifications employed by good observers in work calling for the full resolving power of astronomical telescopes. But in the observation of faint objects like comets, nebulae, and the outer planets it is necessary to employ very much lower magnifications down to only four times the aperture in inches ; for as the image of an extended object is produced by the light which the object itself sends through the object-glass and therefore contains a constant amount of light, the image will become fainter and fainter as increasing magnification spreads that light over a greater area. With astronomical telescopes having only fairly high magnifications it is a common experience to see extensive faint nebulosities quite well in the 'finder' owing to its low magnification, when nothing whatever can be detected in the big telescope.

There are corresponding limits to the really useful magnification of microscopes. For objects of known linear size we found in Chapter III :

$$\text{Least resolvable separation} = \tfrac{1}{2}\,\text{wavelength}/NA.$$

At the distance of distinct vision Dv this will subtend an angle = $\frac{1}{2}$ wavelength/$Dv.NA$, and the angular magnification of the microscope must be such as to raise the angle of subtense to the value of 1, 2, or 3 minutes of arc, or, in our present radian-measure, to 0·000291 and multiples. Hence we have to choose the magnification by

$$MA' \times \tfrac{1}{2}\,\text{wavelength}/Dv.NA = 0{\cdot}000291 \text{ or multiples,}$$

or on putting $\frac{1}{2}$ wavelength = 0·000011″ and Dv at the conventional 10 inches,

$$MA' = \frac{0{\cdot}00291}{0{\cdot}000011}\,.\,NA = 265 \times NA \text{ or multiples.}$$

We therefore conclude :

> To render the finest detail resolvable by a microscope objective of given numerical aperture NA just visible, the magnification must be = $265 \times NA$.
>
> For comfortable observation this figure must be raised to from $530 \times NA$ to $800 \times NA$.
>
> Definition will become unpleasantly soft for magnifications of $1200 \times NA$ or more.

For the usual 1/6th inch objective of 0·75 NA these rules give magnifications from a minimum of 200 to a maximum of 900 ; for an oil-immersion objective of 1·30 NA the limits would be 350 and 1560.

[75] As in the case of the telescope, magnifications below the above minimum value may occasionally be employed for very dark objects or when the full resolving power is not required.

As actual observations are nearly always made in light composed of a considerable range of colours and therefore of wavelengths, and as, moreover, there are very decided individual variations in the acuity of vision, the figures given must not be regarded as exact, but merely as average values subject to a doubt of at least 20 per cent. in either direction.

D. Diameter of the Emerging Pencils

There are interesting and important relations between the diameter of the individual image-forming pencils and the aperture and magnification of the telescope or microscope.

For the telescope we obtain these relations in their most instructive form by working out the linear magnification for objects at any distance by means of the Abbe equations, G Th (10)**. With reference to Fig. 97 (a), which shows a

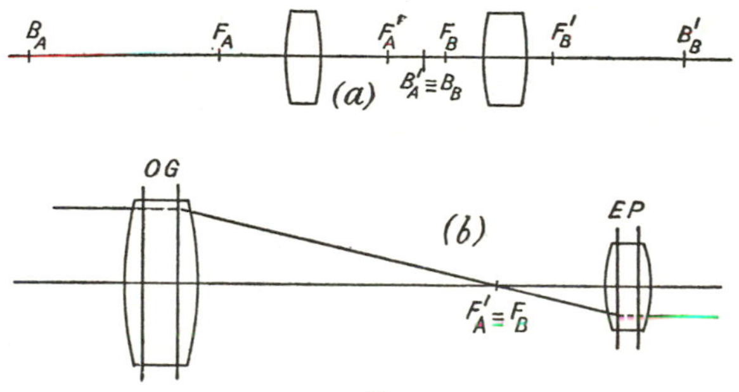

Fig. 97.

generalized telescope with *any* separation between the object-glass and eyepiece, the Abbe equations give for an object at any distance $F_A B_A = X_{fA}$ the conjugate distance $F'_A B'_A = X'_{fA}$, and the linear magnification M'_A by the object-glass only as

$$X'f_A = f_A \cdot f'_A / Xf_A \text{ and } M'_A = -f_A / Xf_A.$$

This image produced by the object-glass acts as object for the eyepiece, with $Xf_B = F_B B'_A = X'f_A - F'_A F_B$, and the Abbe equations give for the effect produced by the eyepiece alone

$$X'f_B = f_B \cdot f'_B / Xf_B = f_B \cdot f'_B / [f_A \cdot f'_A / Xf_A - F'_A F_B];$$

$$M'_B = -f_B / Xf_B = -f_B / [f_A \cdot f'_A / Xf_A - F'_A F_B].$$

But as the object presented to the eyepiece is a rendering of the original object at B_A under magnification M'_A, the total linear magnification of the final image at B'_B is

$$M' = M'_A \cdot M'_B = f_A \cdot f_B / [f_A \cdot f'_A - Xf_A \cdot F'_A F_B].$$

[75] These solutions for $X'f_B$ and M' cover every conceivable case of a combination of two centred systems A and B, whatever the indices of the three media, to the left of A, between A and B, and to the right of B, may be. But as these three media are invariably air in the telescopic systems which we are discussing, we can specialize the equations without any real loss of generality by putting $f_A = -f'_A$ and $f_B = -f'_B$, thus introducing the usual focal lengths. We thus find

$$M' = -f'_A \cdot f'_B / [f'^2_A + Xf_A \cdot F'_A F_B];$$
$$X'f_B = f'^2_B / [f'^2_A / Xf_A + F'_A F_B] = Xf_A \cdot f'^2_B / [f'^2_A + Xf_A \cdot F'_A F_B]$$

for any combination of two centred systems working in air. It is obvious that both equations become simplified when $F'_A F_B = 0$; that is the case of a telescope accurately focused so as to give very distant images of very distant objects, and is the only case usually discussed under the name of a telescopic system. The equations then become

$$M' = -f'_A f_B / f'^2_A = -f'_B / f'_A; \; X'f_B = Xf_A \, (f'_B / f'_A)^2 = Xf_A \, . \, M'^2.$$

The first of these equations proves that a strictly telescopic system has a perfectly fixed linear magnification for objects at all finite distances; it is formally necessary to exclude infinitely distant objects, because the simplification of the equations resulted from putting $F'_A F_B = 0$ in the product $Xf_A \, . \, F'_A F_B$; for $Xf_A = \infty$ this product assumes the indefinable value $\infty \times 0$. Excluding this infinitely improbable special case, we see that the strictly telescopic system accurately realizes the case $M'_1 = M'_2 = M'$ which we had to exclude from the general discussion in the earlier part of this chapter, because all the usual Gaussian constants became infinite. Fig. 97 (b) shows that this is the necessary consequence of the coincidence of F'_A with F_B. We are now learning that this case readily yields to the methods of the general theory when the telescopic system is split up into 'object-glass' and 'eyepiece'.

We obtain a startling result by comparing our determination of the linear magnification of a telescopic system with the angular magnification previously deduced, for we then have

$$M' = -(f'_B / f'_A), \quad MA' = -(f'_A / f'_B),$$

so that there is a reciprocal relation between the two measures of magnification. Hence an accurately focused telescope magnifying MA' = say 100 times, in the accepted meaning of the term, yields an image which, when measured with the foot-rule, would prove to be to a scale of 1/100th of the linear measurements of the objects. At first sight this appears to be an absolutely hopeless contradiction, but it reveals itself as a neat and perfectly harmless paradox when we include the distances of object and image in the discussion. By the equation for $X'f_B = Xf_A \, . \, M'^2$ the image, although linearly minified M' (say 1/100th) times, is brought nearer to the eye to the extent of M'^2 (say 1/10,000th) times. The angular subtense of the small image is therefore $1/M'$ (say 100) times as large as that of the object, and the value of the angular magnification is fully confirmed.

The reciprocal relation of MA' and M' finds an important practical application in the determination of the actual magnifying power of a given telescope. It is [75] inconvenient and requires costly instruments to determine MA' by measuring the angular subtense of a suitable distant object and of its image. But the linear magnification is easily determined. As it is constant for the whole object-space, we can treat the clear aperture of the object-glass as an object of known or easily measured size. An image of this clear aperture appears as a bright little circle, known as the Ramsden disk of the telescope, a little beyond the eyepiece, and can be measured by a micrometer scale placed at the focus of a Ramsden eyepiece, the whole of this small instrument being known as a Ramsden Dynamometer. Then the ratio : Actual clear aperture of OG/diameter of the Ramsden disk, is the correct value of MA'.

The diameter of the Ramsden disk has a very direct bearing on the brightness and definition of the images seen through the telescope. Calling the clear aperture of the object-glass A and the diameter of the Ramsden disk RD, we have just proved that $RD = A/MA'$. In subsection C we proved that the finest detail resolvable by a telescope would subtend an angle of 1′ and be just visible to a normal eye if $MA' = 13\frac{1}{3}\,A$, A being measured in inches. If we put this value of MA' into the equation for RD, we find the value

$$RD = A/13\tfrac{1}{3}A = \tfrac{3}{40} \text{ inch}$$

as the diameter of the Ramsden disk with which a normal eye can just see all the detail which any given telescope can resolve. Under these conditions the image obviously would appear to the eye as an extremely sharp one. It is evident that the other values of MA' selected in C as giving subtenses of the finest resolvable detail of 2, 3, and $4\frac{1}{2}$ minutes would give Ramsden disks of respectively 3/80, 1/40, and 1/60 inch with progressive loss of definition ; for as the eye would not find any detail barely resolvable by it—as in the case first considered—and would see points and lines rendered as soft dots or streaks several minutes in diameter or width, the whole image would appear softer and vaguer than any naked-eye view of natural objects. The important and remarkable fact to be remembered is that the diameter of the Ramsden disk is an absolutely dependable indication of the definition to be expected from a well-corrected instrument : $RD = 3/40$ inch or 2 mm. is the smallest diameter of the emerging pencils which will yield definition perfectly satisfying to a normal eye. With smaller diameters there will be progressive softening of the definition. It follows that a comparison of the definition yielded by telescopes of different apertures is only fair if eyepieces are chosen which give the same diameter of Ramsden disk on the several instruments. Another interesting aspect of these relations is that the connexion between clear aperture and angular resolving power applies equally to the naked eye ; hence the degree of definition obtained with the naked eye when looking through apertures of 2 mm. or less is exactly the same as that given by a telescope with a Ramsden disk of similar diameters.

If we now briefly examine the case of the generalized telescopic system, that is the case when owing to nearness of the telescopic objects there is a finite value of $F'_A F_B$, then the general formulae for M' and for $X'f_B$ will still apply, but will give varying values for M' when the object-distance Xf_A is changed. But we again

[75] obtain a simplification when $Xf_A = 0$, that is for an object placed into the anterior focal plane of the object-glass. For this case we obtain

$$M' = -(f'_B/f'_A)\,;\; X'f_B = 0\,;$$

and as we found for this case also $MA' = -(f'_A/f'_B)$ on condition that the subtense of the telescopic object is taken at the anterior focus of the object-glass, the reciprocal relation of M' and MA' remains in force. Hence MA' may be determined as the reciprocal of the minification of the image yielded by an object in the anterior focal plane of the object-glass. This may be measured by the Ramsden dynamometer : we place a scale into the anterior focal plane of the object-glass and measure its image. If the adjustment of the telescope is not very different from that for distant objects, then a rough estimate of the location of F_A will be sufficient. The relations between definition and diameter of the Ramsden disk remain exactly the same as for strictly telescopic systems.

To determine the same relations in the case of the microscope we regard the latter, in accordance with the general theory, as equivalent to a simple lens of a definite focal length f'. We can then determine the diameter of the emerging pencils in the posterior focal plane of the instrument, with which the Ramsden disk coincides as nearly as is of interest, as follows :

With reference to Fig. 98, which is drawn so as to correspond to the case of a compound microscope, the axial object-point B shall be depicted at B', at distance

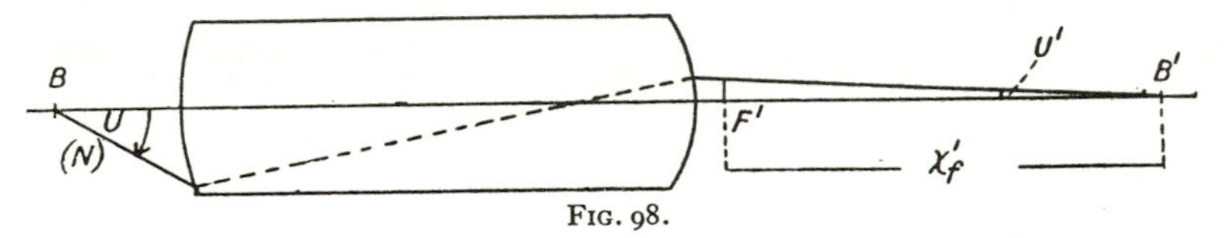

FIG. 98.

X'_f from the posterior principal focus F'. A real image is assumed in the diagram ; for the more usual virtual image B' would merely shift to a position to the left of F' and to a negative value of X'_f. By the general theory the linear magnification at B' will be given by

$$M' = -X'_f/f'.$$

We can obtain an alternative solution for M' by the ratio of the convergence angles U and U', but as the angle U may reach 70° or more, we should obtain a fallacious result by the tangent ratio of the general theory ; the microscope must be free from coma and we must therefore employ the Sine Condition. If, in order to include immersion objectives, we assume that the object lies in a medium of index N, the sine condition gives

$$M' = N\,.\,\sin U/\sin U' = NA/\sin U'.$$

U' is always a very small angle, so that the difference between its sine and its tangent is negligible. We obtain the latter by the right-angled triangle with $F'B'$ as base and $\frac{1}{2}\,RD$, the semi-diameter of the emerging pencil, as vertical side, as $= RD/2\,X'_f$, whence

$$M' = NA/\sin U' = NA/(RD/2\,X'_f) = X'_f\,.\,NA/RD$$

and if we now equate the two values of M' we find [75]

$$-X'_f/f' = 2\, X'_f\, .\, NA/RD \quad \text{or } RD = -2 f'\, .\, NA.$$

We now eliminate f' and introduce the true angular magnification of the microscope by transposing G Th (12)* into $f' = Dv/MA'$, and then putting this value of f' into the equation for RD, with the result

$$RD = -2\, Dv\, .\, NA/MA'.$$

As Dv is fixed by convention, this is an exact relation between the diameter of the Ramsden disk, the numerical aperture, and the angular magnification of the microscope, and any one of these three variables can be determined when the other two have been measured. Thus we can find the angular magnification, in close analogy with the telescope, by measuring RD and NA, or if we have no convenient means for a direct determination of NA, we can find its value by measuring RD and MA'.

For inch measure we have $Dv = 10$, hence

$$RD = -20\, NA/MA'.$$

In subsection C we fixed the value of MA' at which the finest resolvable detail will be seen under an angle of 1′ at $MA' = 265\, NA$. If we put this value into the equation for RD, we find the diameter of the Ramsden disk as

$$RD = -20/265 = -3/39\tfrac{3}{4} \text{ inch};$$

it has been put as a fraction with 3 in the numerator to show the almost absolute agreement with the 3/40 found under the same conditions for the telescope. It will be accepted without proof that we should find corresponding agreement between telescope and microscope at other degrees of definition of the image.

THE NODAL POINTS

[76] We proved in section [73] that the angles of convergence U and U' of all rays passing from an axial object-point B to its conjugate image-point B' satisfy the equation (a) of that section

$$\tan U/\tan U' = -M'.f'/f$$

if M' is the linear magnification at B' and if f and f' are the two focal lengths of the centred system.

If we now suppose that there be an extra-axial object-point anywhere on the ray passing through B under angle U, then, because every ray from an object-point is necessarily a geometrical locus for the image of that point, the image of the supposed extra-axial object-point must lie on the emerging ray passing through B' under angle U'. In accordance with our definition of the telescopic angular magnification, the latter is therefore $MA' = \tan U'/\tan U$ at point B', where the linear magnification is M'. If we now replace $\tan U/\tan U'$ in equation (a) by its exact equivalent $1/MA'$, we obtain the result

$$1/MA' = -M'.f'/f,$$

which becomes simply $1/MA' = M'$ if we assume that the system is one working in air, because we then have $f = -f'$. This is an extremely interesting result,

[76] because it proves that the reciprocal relation between linear and telescopic angular magnification, which we proved in the previous section for the telescope only, is an absolutely general property of any centred system, with the mere addition of a correcting factor $-f'/f = N'/N$ in the case of systems separating a medium of index N from another of index N'.

The nodal points, which are occasionally useful in special applications of the general theory, are merely that pair of conjugate points B and B' for which the telescopic angular magnification is exactly 'plus one', so that all objects appearing under any given angle U as seen from the first nodal point B will yield images which will appear under the same angle and in the same clock-sense when viewed from the second nodal point B'.

The condition satisfied by the nodal points is therefore tan U = tan U' or tan U/tan $U' = +1$, and if we introduce this value into equation (a) we obtain the equation of condition

$$M' = -f/f'.$$

For systems working in air—or in the same medium other than air on *both* sides—we have $f/f' = -1$ or $M' = 1$; hence for such systems the nodal points coincide with the respective principal planes, seeing that the latter are *defined* by the condition $M' = +1$. For systems with a difference of the refractive indices for the object- and image-space respectively, our equation of condition enables us to locate the nodal points by the Abbe equations $X_f = -f/M'$ and $X'_f = -f'.M'$, which give the immediate result

$$\text{First Nodal Point at } X_f = f'$$

$$\text{Second Nodal Point at } X'_f = f.$$

As the principal planes lie at $X_f = -f$ and at $X'_f = -f'$, we see, bearing in mind the opposite signs of f and f', that the nodal points always lie in the same direction as the principal planes with reference to the respective focal points, but are shifted by the difference between the numerical values of the two focal lengths. The interstitium between the two nodal points is always the same in magnitude and sign as that between the two principal planes.

Some Additional Deductions

[77] A. Although this chapter has been strictly limited to the consideration of object- and image-planes at right angles with the optical axis, it is easily proved by a slight modification of the method of proof employed in section [71] that perfect optical system would depict straight lines and planes in any position whatsoever as straight lines and planes in the image-space, forming definite angles with the XY, XZ, and YZ planes. This property is utilized in practical photography in order to obtain a sharp rendering of house fronts and similar essentially flat objects, which stand at an angle other than 90° with the optical axis, by adjusting the focusing screen to the proper tilt with reference to the optical axis by the so-called swing movements of the camera. The sharp images thus obtainable are, however, not geometrically similar to the objects in the tilted object-plane; they show the well known perspective of all pictures of this type.

B. Our theory of perfect optical systems is restricted throughout to the consideration of the *sharply focused* images produced by a given *perfectly fixed* combination of centred surfaces and intervening media. The theory therefore cannot be legitimately applied to the case of small 'fixed focus' cameras, or to the eyes of very old people, which are in fact fixed focus cameras. Arrangements of this type, if sharply focused for some intermediate distance, give diffused images of more distant objects which are relatively too large, and they give diffused images of nearer objects which are relatively too small, in either case by comparison with the size of the sharp image which would be obtained on shortening or lengthening the camera or eyeball to the proper extent. Our theory is also inapplicable to systems, of which the eye of *not* very old people is the chief example, in which a sharp focus of objects at varying distances is obtained by altering the strength of a constituent lens. Such systems show a falsification of the size of the image in the same sense as fixed focus arrangements. It follows that the theory cannot be applied legitimately to the human eye at any period from the cradle to the grave. Serious errors are the result of ignoring this fact. Thus it is stated in a book on the theory of optical instruments that the objectionable minifying effect of concave spectacle lenses could be reversed by applying the lenses very close to the cornea, and very cheerful but utterly fallacious values are calculated for the 'magnification' obtained by concave 'contact' spectacles. Under the actual conditions such an effect could only be secured by meniscus lenses of impracticable thickness and depth of curvature, practically a modification of the solid glass Galilean telescope mentioned in [72] D, 5. [77]

C. The general theory enables us to make some additions to the theory of achromatism. In tacit anticipation of the impossibility of a perfect optical system, we only discussed seriously achromatism for one pair of conjugate distances and found that two conditions were then involved which we formulated under the names of longitudinal and transverse achromatism. It is self-evident by the Abbe equations that complete achromatism for all conjugate distances would demand coincidence of the 'red' and 'violet' focal points, and equality of the 'red' and 'violet' f as well as f'. For a dry system the anterior and posterior focal lengths in either colour would automatically be of equal magnitude and opposite sign. Hence only three achromatic conditions would have to be fulfilled to secure complete achromatism at all conjugate distances. It appears highly doubtful whether all three conditions could be closely satisfied with lenses of the inevitable finite thickness; there are, however, many photographic objectives in which the departure from stable achromatism is unimportant in comparison with the variation of the spherical aberrations for changes in the conjugate distances. But for immersion systems there is a definite and serious impossibility of stable achromatism, for assuming the immersion medium at the left and air at the right, we shall have $f_r = -N_r \cdot f'_r$ in red and $f_v = -N_v \cdot f'_v$ in violet; and as for all known media we have $N_v > N_r$, it follows at once that we cannot possibly have $f_r = f_v$ when $f'_r = f'_v$; hence the second condition of stable achromatism cannot be fulfilled.

D. On account of the fundamental importance of the proof that a perfect optical system of finite aperture is impossible, it is worth pointing out that the primary

[77] aberrations dealt with in Chapter VI lead to the same conclusion. Taking as an example the formula for the coma contribution of any one surface,

$$CC' = \tfrac{1}{2}(N'-N)(N'/N)\,l'u'i'i'_{pr}\,(i'-u)/N'_k u'_k,$$

we can put this entirely into terms of the constant data (radii, thicknesses, and refractive indices) of the system and of the initial l_1 and nominal y_1; for by the paraxial relations we can express l'_1 in terms of l_1, then also $l_2 = l'_1 - d'_1$, and so right through; it does not affect our arguments that fearfully complicated expressions must result for the later surfaces. With these l-values we then obtain $y_2 = y_1 (l'_1 - d'_1)/l'_1$, &c., also in terms of y_1 and l_1 only. We next have any $u = y/l$ also in terms of y_1 and l_1, and from the u we obtain the other angles by the paraxial ray-tracing equations. The i'_{pr} will be constant if we take all the extra-axial object-points on the traced principal ray. It follows that each CC' and therefore also the total final $coma'_S$ is a very complicated but rational function of l_1 only; the same applies to spherical aberration, astigmatism, and longitudinal chromatic aberration. As it is easily realized that the majority of the coefficients of the powers of l_1 must be finite, our proof of the impossibility of zero aberration for all object distances is then complete, for algebra knows no rational function of one variable with finite coefficients which is zero for all possible real values of that variable. Such a function may have a number of roots for which it is zero, but it will have finite values for intermediate values of the variable. The Petzval term has to be excluded from these arguments, because it is totally independent of intersection-lengths. And distortion together with transverse chromatic aberration have been excluded also, because in either case a highly specialized condition exists which makes the particular aberration zero for all distances.

E. The impossibility of a perfect optical system of finite aperture also explains the obstinacy with which different methods of determining the Gaussian constants of optical systems persist in disagreeing; for as all the methods involve measurements for two different pairs of conjugates if all the constants are to be measured, they invariably become affected by aberrations and can only give compromise values. To minimize this difficulty, that method should be selected in each individual case which gives the really required constants for the approximate range of conjugates within which the particular system gives sufficiently good definition.

Special Memoranda from Chapter IX

It was proved in Chapter I that any centred system of spherical surfaces has the property of depicting a very small object located at any distance, but at right angles with the optical axis, in a conjugate normal plane with a definite linear magnification, on condition that the aperture also is very small and that monochromatic light is used. Practically all the subsequent work was devoted to the problem of extending this desirable property to objects and apertures of finite size and to light of mixed wavelengths or colours.

Hence it appears at first sight a reasonably modest assumption that an optical system might be corrected so that extended objects in only two different normal planes of the object-space would be depicted sharply and without distortion in [77] two conjugate normal image-planes, and this, together with the rectilinear propagation of light, is the only assumption underlying the general theory as given in this chapter.

It is then shown, as a strict, inevitable, and logical consequence of the two assumptions, that such a system would be equally perfect for object-planes at all possible distances, and that location and magnification of the conjugate images would be in accordance with the theory given by Gauss for the 'threadlike space surrounding the optical axis', but not for objects and apertures of finite size.

The proof, however, rests solely on the two assumptions and on strict geometry; the laws of refraction and of reflection do not enter into it in any way, as no ray-tracing is required. Hence there is a possibility that the result obtained may be incompatible with fundamental properties of light, and this proves to be the case when the law of the ratio of conjugate angles of convergence is deduced from the geometrical general theory. The latter demands a constant ratio of the tangents of these angles, whereas the inexorable sine theorem demands a constant ratio of their sines. The two ratios are compatible only for the two pairs of conjugate points for which $U' = \pm U$. For all other pairs our supposed perfect system would, by the sine theorem, suffer from coma and would therefore be seriously imperfect. That leads to the conclusion that one of our assumptions in the general theory must be wrong, and as the rectilinear propagation of light may be regarded as amply established, we are driven to the only alternative, namely, that it is absolutely impossible to correct a system of finite aperture for more than one pair of conjugate planes. This conclusion is confirmed at the end of the chapter by the equations for those primary aberrations which affect the definition of individual image-points. The utility of systems corrected for a particular pair of conjugates when tried for other pairs therefore depends solely on tolerances; there may be a considerable useful range of distances for systems of small angles of convergence, but there can be only a decidedly small range when the angles of convergence become large. These conclusions are in full agreement with practical experience.

Although the validity of the general theory becomes greatly restricted by the limitation to small apertures and a small field, the equations deduced in this chapter find numberless applications in the working out of new designs. The equations numbered G Th (10) and higher are the most useful ones. Those relating to simple thick lenses, to combinations of two separated lenses or systems, and to magnification and its limits, should be particularly borne in mind.

CHAPTER X

ORDINARY EYEPIECES

78] THE eyepieces usually found in telescopes and microscopes, namely, the 'Huygenian', the 'Ramsden', and the four-lens 'terrestrial' eyepiece are with rare exceptions made up of ordinary planoconvex crown glass lenses, but are nevertheless capable of yielding perfectly satisfactory images when the best use is made of the restricted liberties for varying the design. The great majority of eyepieces found with even the most expensive instruments fall decidedly short of the best possible correction, and could be considerably improved without any addition to their cost. The methods developed in Chapters VI–VIII for the oblique aberrations, together with the general theory of Chapter IX, enable us to build up an entirely systematic procedure in the design of these simple but highly important and interesting systems, by which the best possible forms can be quickly and accurately determined.

We may profitably begin with a few general observations on the chief characteristics of the two principal types and on certain important requirements. Fig. 99 shows above the optical axis a half section of a normal Huygenian eyepiece, below the axis a similar view of a Ramsden eyepiece. In the former both planoconvex

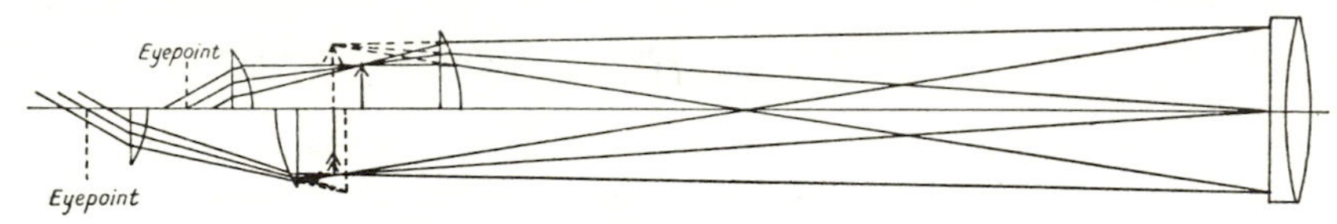

FIG. 99.

lenses turn their flat sides towards the eye of the observer ; the 'field lens' next the telescope or microscope objective intercepts the converging rays before they have come to a focus and produces a diminished real image between the two lenses or within the eyepiece, and this image is then viewed by the eye as magnified by the 'eye lens'. In the Ramsden eyepiece the objective produces its real image of the objects under observation at some little distance in front of the field lens, which is placed with its flat side towards the objective. The field lens produces a slightly magnified virtual image, and this is seen further magnified by the eye lens, the latter being placed with its flat side next the eye, as in the Huygenian. The chief advantage of the Huygenian eyepiece is that it can be completely freed from transverse chromatic aberration, and that it is therefore quite free from false one-sided colour-bands in the outer part of the field ; hence it is greatly preferred for purely visual observations not involving fine micrometrical measurements. For the latter it is unsuitable, as the micrometer would have to be inside the eyepiece in contact with the real image, which is not only highly inconvenient but also gravely objectionable because the real image between the two lenses is a decidedly imperfect one, suffering from considerable amounts of distortion and transverse chromatic aberration. The Ramsden eyepiece is suitable for use with micrometers of every description, as the eyepiece is quite clear of the plane of the direct image produced [78] by the objective. The only drawback to the ordinary Ramsden eyepiece is that it cannot be freed from a very considerable residue of transverse chromatic aberration, and that as a consequence it shows unpleasant unsymmetrical false colour in the outer part of the field. With regard to *all other* aberrations the Ramsden eyepiece is very decidedly superior (if properly designed) to the Huygenian, and should be used exclusively on instruments in which the images are monochromatic, such as spectroscopes and certain types of polarimeter. A very important detail in all eyepieces for visual observations is the 'eye-point' or 'Ramsden disk', which marks the position where an image of the objective is formed and where all the light admitted by the latter is concentrated within the smallest possible circle. The pupil of the eye of the observer must be placed very close to this point (whence the name eye-point) if the whole field is to be seen simultaneously and comfortably, and this is only possible if the eye-point is far enough from the eye lens to leave room for eyelids and especially eyelashes. Five or six millimetres clear distance is about the minimum, and eight millimetres represent probably the best distance, for if the distance exceeds the latter amount the right position for the eye is difficult to find and to maintain, especially in night observations. That is the reason why low-power eyepieces with an unusually distant eye-point are fitted with a projecting 'cap' with an opening of some 8–10 mm. within about 8 mm. of the eye-point.

The course of an oblique pencil through a four-lens terrestrial eyepiece was shown—in the reverse direction as compared with our present diagram—in Fig. 46 (h) of Chapter VI, which should be referred to. This eyepiece is best regarded as a compound microscope; the 'erector' represents the object-glass of this microscope, and in type is a projecting Ramsden eyepiece; the 'eye end' is an ordinary Huygenian eyepiece. As the erector, although of the type of the Ramsden eyepiece, is not required for direct observation, the 'virtual' position of its eye-point, which would render the latter inaccessible to the eye, does not matter in any way. Under these conditions the Ramsden eyepiece *can* be freed from transverse chromatic aberration, and there is for that reason no difficulty in completely correcting the terrestrial eyepiece in this important respect.

In all these eyepieces the placing of the planoconvex lenses, with reference to whether the convex face is on the right or the left side, is dictated by the resulting oblique aberrations, more especially by astigmatism and curvature of field. It was shown in Chapter VI that a positive system required to cover a considerable angle of field cannot be satisfactory if there is any positive or 'under-corrected' astigmatism in the final image, and that the best compromise calls for a fairly considerable amount of negative astigmatism. We can quickly determine the sign of the astigmatism produced at the separate surfaces by discussing the equation for astigmatism in

Seidel Aberrations II (5) $AC' = \frac{1}{2}(N'-N)(N'/N) \,.\, l' \,.\, u' \,.\, i'^2_{pr}\,(i'-u)/N'_k u'^2_k$,

remembering that u, l', u', and i' refer to the *axial* pencil, whilst i'_{pr} is the angle of emergence of the principal ray of an oblique pencil. It is only necessary to discuss the sign of $(N'-N) \,.\, l' \,.\, u'\,(i'-u)$ as the remaining terms are unconditionally positive, and $l' \,.\, u'$ may be most easily discussed by remembering that it represents the incidence height y of the paraxial ray. Taking the eye lens of any of the three

[78] types as an example and discussing it in our usual computing direction, we see that the arriving rays of the axial pencil will be nearly parallel to the optical axis if the eyepiece is properly focused for a reasonably normal eye, hence in the orthodox position with plane next the eye (left-hand of Fig. 100) u and u' at the

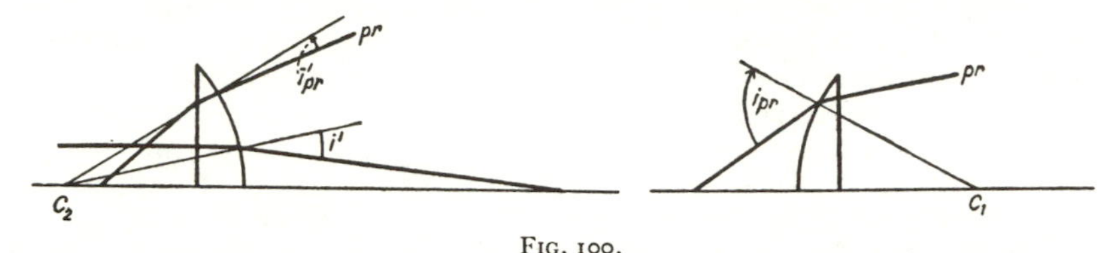

FIG. 100.

plane will be extremely small and its astigmatic contribution will be quite unimportant. At the curved surface i' will be negative, u very small, hence $(i'-u)$ negative; $l'u'=y$ is positive, but $N'-N$ (Air index − Glass index) is negative. Hence AC' is positive and unfavourable. But the tracing of the principal ray shows that in this position of the eye lens i'_{pr} will be small and AC', being multiplied by i'^2_{pr}, will also be small. In the reverse position (right-hand of Fig. 100) u for the curved first surface would be very small, but i and i' considerable and positive, and as $l'u'$ and $N'-N$ are both positive, AC' of the curved surface will again be positive. But in this position of the eye lens i'_{pr} obviously would be very large, making AC' also extremely large, and in fact quite unmanageable. Moreover the plane now makes a further and not unimportant contribution to the positive astigmatism, and we see that the orthodox position of planoconvex eye lenses is abundantly justified. If the field lenses are studied in the same way it will be found that in the orthodox position the plane face gives a moderate positive AC', whilst the curved face gives a considerable *negative* AC', and thus becomes capable of flattening the field. If the field lens is reversed, both faces give positive astigmatism and flattening of the field would be utterly out of the question. We can thus safely conclude:

(1) With planoconvex lenses flattening of the field is only possible if the lenses are placed into the orthodox positions.

(2) At the eye lens the oblique pencils must pass nearly (but as a rule not quite) radially through the convex surface in order to keep the inevitable positive astigmatism as low as possible.

(3) At the curved face of the field lens a large value of i'_{pr} is desirable in addition to a reasonably large incidence height $l'u'=y$; for the latter reason the anterior image-plane must not lie very close to the field lens, because this would render the negative astigmatism too small.

It may at once be added, as the result of considerable practical experience, that departures from the simple planoconvex form of eyepiece lenses do not open any very alluring prospects. Properly tested and corrected solutions in which the forms of the two components are included as variables never lead to any great departure from the planoconvex form, and as for these cheap eyepieces the latter form has many advantages, we will limit our studies to it.

The primary aberration methods of Chapter VI are not very suitable for the [78] systematic design of simple eyepieces, for on account of the heavy aberrations and wide separation of the components the approximation is very unsatisfactory and has to be followed by a large amount of trigonometrical correction. We can, however, reach a satisfactory solution quickly and easily by first using Seidel Aberrations VI to find the separation which would give freedom from transverse chromatic aberration for thin components, and by then relying on the extremely valuable conclusion (5) in the discussion of the second fundamental law, section [60]. All these eyepieces are spherically under-corrected, and we have seen that flatness of the field depends on the negative astigmatism of the curved face of the field lens, which is seldom sufficiently large to overdo the correction of curvature of the field. Now conclusion (5) tells us that a spherically under-corrected positive system will have a minimum of astigmatism in the mathematical sense when its coma is zero ; hence we shall obtain the utmost flattening of the field (with a remote possibility of a moderately hollow tangential field) if we solve for that position of the exit pupil which will secure zero value of the coma, which latter in itself is a highly desirable state of correction. The easiest way of determining this position of the exit pupil, and at the same time a highly exact way, is to trace a paraxial and a marginal ray of the *axial* pencil and to apply the sine condition ; this therefore will be our royal road in the design of simple eyepieces. Direct trigonometrical tracing of an oblique pencil through an eyepiece arrived at by this simple method will usually merely confirm the result obtained by the sine condition ; but as the trigonometrical test also gives the actual measure of the residual curvature of field and of the distortion, it should not be omitted, at any rate not until considerable experience has been gained.

We will complete this preliminary general discussion by a few numerical examples of the results obtainable from Seidel Aberrations VI

$$d' = (f'_A + f'_B \cdot V_B/V_A)\Big/[1 + V_B/V_A + f'_A/l_A - f'_B V_B/l'pr_B V_A].$$

We may now define f'_A as the focal length of the eye lens, f'_B as the focal length of the field lens, d' as that separation which will secure TL achromatism of magnification when the real or virtual final image is formed at a respectively negative or positive distance l_A from the eye lens, and when the object-glass, or more strictly the iris, is at distance $l'pr_B$ from the field lens ; we may describe $l'pr_B$ conveniently, though a little inaccurately, as the tube-length of the instrument. We will choose $f'_A = 1$ and $f'_B = 2$—a very usual ratio in Huygenian eyepieces—and will in each case also calculate the equivalent focal length of the complete eyepiece as

$$f' = f'_A \cdot f'_B/(f'_A + f'_B - d').$$

(1) For the usual and ancient text-book assumption of $V_A = V_B$, $l_A = l'pr_B = \infty$, the equation becomes simply $d' = \frac{1}{2}(f'_A + f'_B)$ and gives $d' = 1\frac{1}{2}$, $f' = 1 \times 2/(1 + 2 - 1\frac{1}{2}) = 2/1\frac{1}{2} = 1\frac{1}{3}$.

(2) If we retain $V_A = V_B$ and the final image distance $l_A = \infty$, but, thinking of inch-scale, put $l'pr_B = 8$ as an average microscope tube-length, we find

$$d' = (1 + 2)/(1 + 1 - \tfrac{2}{8}) = 3/1\tfrac{3}{4} = 1\tfrac{5}{7}, \text{ nearly } 0{\cdot}2 \text{ longer than by (1)};$$

$$f' = 2/(1 + 2 - 1\tfrac{5}{7}) = 2/1\tfrac{2}{7} = 1\tfrac{5}{9}, \text{ more than } 0{\cdot}2 \text{ longer than by (1)}.$$

[78] (3) Putting $V_A = V_B$, $l'pr_B = \infty$ (for a very long telescope), but $l_A = \pm 8$, the upper sign—for inch-scale—suiting an observer with $-5\,D$ of myopia, the lower sign either projection of a real image at 8 inches from the eye lens or a case of $+5\,D$ hypermetropia, we find $d' = (1+2)/(1+1\pm\frac{1}{8})$, $f' = 2/(3-d')$ or

$$\text{for myopia} \quad d' = 24/17 = 1{\cdot}41 \quad f' = 2/1{\cdot}59 = 1{\cdot}26\,;$$

$$\text{for projection}\; d' = 24/15 = 1{\cdot}60 \quad f' = 2/1{\cdot}40 = 1{\cdot}43.$$

(4) For microscope eyepieces it can be advantageous to make the eye lens of crown glass and the field lens of flint glass. Taking $V_A = 60$, $V_B = 36$, or $V_B/V_A = 0{\cdot}600$, l_A for normal vision $= \infty$, but $l'pr_B = 8$, we find

$$d' = (1+2\times 0{\cdot}6)/(1+0{\cdot}6-0{\cdot}6\times\tfrac{2}{8}) = 2{\cdot}2/1{\cdot}45 = 1{\cdot}52$$

$$f' = 1\times 2/(1+2-1{\cdot}52) = 2/1{\cdot}48 = 1{\cdot}35,$$

or very nearly the same values as in case (1) for an eyepiece of two lenses of the same glass but adjusted for a long telescope. With lenses of the same focal length a fractional value of V_B/V_A diminishes both separation and equivalent focal length.

It will now be realized that the additional terms which distinguish Seidel Aberrations VI from the ancient simple formula are not by any means negligible in their effect on the solutions.

HUYGENIAN EYEPIECES

[79] We will adopt the restriction to planoconvex lenses and for the present the further restriction that both lenses shall be made of the same kind of ordinary crown glass. As any optical design can be carried out to any linear scale without altering its corrections, we may then, for the purpose of a systematic study, adopt a convenient focal length or net curvature for the eye lens, for that will merely fix the scale of the computed specimens. Having done this, we have two liberties left for the purpose of correcting aberrations, namely, the *ratio* of the focal length of the field lens to that of the adopted eye lens, and the *separation* between the two lenses; we have already decided that transverse chromatic aberration and coma are to be the two aberrations which call for attention. As coma is a highly objectionable defect, and as in the case of simple eyepieces it also aggravates the curvature of the field, we will insist on complete correction. But with regard to achromatism of magnification there is a considerable amount of tolerance, and in some cases a certain amount of over- or under-correction can be directly beneficial, as for instance in the microscope, where the high-power objectives call for eyepieces with over-correction of the transverse chromatic aberration. Accordingly our systematic method will consist in trying each ratio of f'_B/f'_A at three evenly progressing separations in the neighbourhood of the value pointed out by Seidel Aberrations VI, in determining for each of these separations the value of the tube-length which by the sine condition will give zero coma, and in then calculating the corresponding value of the transverse chromatic aberration. Tabulation, or (more instructively) plotting of the results, will then enable us to pick out the best possible eyepiece for any given case. We shall find that the ratio f'_B/f'_A has to be about $2{\cdot}3$ for high astronomical magnifications, and sinks to $1{\cdot}4$ or even less for the lowest microscopic magnifica-

tions at short tube-lengths. For the actual numerical example we will select the [79] ratio 2, which we shall find most suitable for a fairly long range of medium magnifications. Similar calculations for ratios 2·3, 1·7, and 1·4 will cover the whole ordinary range of magnifications. For the calculations it will be most convenient to select $r_2 = -1$ for the eye lens, which leads to $r_4 = -2$ for the field lens of the example, and to $r_4 = -2{\cdot}3,\ -1{\cdot}7,\ -1{\cdot}4$ for the supplementary calculations. Eyepiece lenses are usually made with a clear diameter fully equal to one half of their focal length, and this requires an axial thickness of about 0·15 of the radius of the curved face of planoconvex lenses. For the example this gives 0·15 thickness of the eye lens, 0·30 thickness of the field lens. It will be advisable to retain the latter thickness for all the other ratios, because with a given eye lens the diameter of the field lens required to cover the usual angle of field does not diminish in proportion to the focal length of the field lens; it is more nearly constant. As the refractive index for brightest light we will adopt $1{\cdot}517 = N_y$, and as the index for 'nominal violet' we will choose $N_v = 1{\cdot}526$; these are fairly representative of ordinary cheap crown glass.

The first step consists in selecting suitable separations by applying Seidel Aberrations VI. For the usual assumption that the observer adjusts the focus for a very distant virtual image we have $l_A = \infty$, and as our two lenses are of the same glass the equation reduces itself to

$$d' = (f'_A + f'_B)/(2 - f'_B/l'pr_B);$$

but this immediately presents a problem, because l'_{prB}, the tube-length, is to be determined by the zero-coma condition and is therefore not yet known. Its influence on d', however, is not very large, and as the ratio 2 suits fairly high magnifications we will for this ratio treat $l'pr_B$ as infinite, thus coming back to the text-book formula. For the focal length of a planoconvex lens with curved side on the right we have $f' = -r_2/(N-1)$, or in our example

$$f'_A = 1/0{\cdot}517 = 1{\cdot}934 \qquad f'_B = 2/0{\cdot}517 = 3{\cdot}868,$$

giving

$$d' = \tfrac{1}{2}(1{\cdot}934 + 3{\cdot}868) = 2{\cdot}90.$$

This TL result really calls for a reduction on account of the considerable thickness of our lenses, for d' ought to be taken as the distance between the second principal plane of the eye lens and the first principal plane of the field lens. By the thick-lens equations of the preceding chapter the reduction will be found to amount to about 0·2, making the proper separation in the actual eyepiece about 2·7. We will incorporate the reduction in our selection of three values of d' for actual trigonometrical test by choosing $d' = 2{\cdot}6$, 2·9, 3·2. It is this selection of separations for the other ratios which will be facilitated by the results for the ratio 2.

We are now ready for the main part of the work, which consists in finding the values of $l'pr_B$ which will secure freedom from coma with each of the selected separations, and in then determining the value of the transverse chromatic aberration.

To be able to apply the sine condition we must trace a paraxial and a 'marginal'

[79] ray through the eyepiece; we must therefore select a suitable initial $y = Y$. As in the actual use of an instrument the eyepiece receives the rays concentrated by the object-glass at its focus, it is evident that eyepiece and objective always work at the same ratio of aperture to focal length, or the same f-ratio or f-number, as it is usually put. We therefore ought to calculate our eyepieces for the f-ratio of the object-glasses with which they are to be used, but this varies from about $f/4$ for the objectives of prismatic binoculars through practically all intermediate values to about $f/15$ for the usual astronomical object-glasses, and still further to as little as $f/40$ for oil-immersion microscope objectives when the latter are considered from the eyepiece end. Fortunately, higher aberrations are small in the axial pencil of eyepieces, and as a consequence the results obtained do not vary much with the calculated aperture. As the sine condition will give higher precision of its results for a fairly large aperture we will choose (for our standard $r_2 = -1$) $y = Y = 0{\cdot}2$, or one-fifth of the radius of the curved side of the eye lens. This will give f-numbers for our eyepieces ranging from six upwards, and not too far from those for which the eyepieces may at times be used.

We therefore trace a paraxial and a marginal ray with the initial Y thus selected completely through the eyepiece, and in order to obtain also the chromatic aberrations, we add to these two rays in brightest light a third ray, namely, a paraxial ray in 'nominal violet'. It would really be more strictly correct to trace two coloured rays, say C and F, just as in the exact achromatization of object-glasses; but in the case of ordinary eyepieces this would be an entirely extravagant refinement. We shall nearly always begin with a pencil of rays strictly parallel to the optical axis as the normal case for observation with relaxed accommodation of a normal eye; but it should be remembered that if the eyepiece is to be used for projection, or is to be specially designed for an observer with highly abnormal vision, then the calculation must begin with the proper value of l_A, and the latter must, if short, be roughly included in the preliminary solution for the separation d'. With the usual parallel rays we have the advantage that these will go normally through the plane first surface of the eye lens, so that the ray-tracing begins at the second surface.

The tracing of the three rays will give the final results l'_y, l'_v, L'_y, and the angles u'_y, u'_v, and U'_y, and the former supply important information; for $l'_y - l'_v$ will be the longitudinal chromatic aberration of the eyepiece and $l'_y - L'_y$ will be its longitudinal spherical aberration at the calculated aperture. These are the defects which cannot be corrected in the eyepiece, and which, if sufficiently serious in comparison with the respective tolerances, must be compensated by a corresponding over-correction of the object-glass if the instrument as a whole is to be really satisfactory. The data also give the equivalent focal length of the eyepiece—provided that the calculation was begun with $l_a = \infty$ – as $f'_{EP} = y/u'_y$.

We then apply Sine Theorem II to determine the tube-length at which the eyepiece will be free from coma. A decidedly satisfactory result would be obtained by solving for $OSC' = 0$, for this would give complete coma correction in the central part of the field where critical observations are always made. But in the outer part of the field coma would reappear and would usually attain undesirable magnitude in the marginal region owing to the presence in all eyepieces of the secondary coma growing as H'^3, which has already been discussed in previous sections. In eyepieces this 'cubic' coma is always negative and does not vary very seriously in different types, provided that the f-ratio is the same. Hence it is possible to reduce the residual coma of eyepieces on the zonal principle by solving for a suitable positive value of OSC' instead of zero. There will then be slight positive coma in the central region of the field, zero value for a particular zone of the field, and negative coma beyond this zone, but the highest numerical value will be greatly reduced so as usually to be harmless if the coma-free zone is skilfully chosen. Experience has led to the conclusion that the coma-free zone will be located at about 12°–16° from the centre of the field if Sine Theorem II is solved for $OSC' = 0{\cdot}2 \sin^2 U'_y$.

As this is merely a rough empirical formula, it may be calculated by slide-rule or by three-figure logs.

With the small change in symbols adopted in the present chapter, Sine Theorem II takes the form

$$l'pr_B = l'_y - \frac{(l'_y - L'_y)(1 - OSC')}{(1 - OSC') - u'_y/\sin U'_y}.$$

We can render this still more convenient for eyepiece calculations by noting that $l'pr_B$ is the distance from the field lens to the object-glass, whilst l'_y is the distance from the field lens to the focal point of the eyepiece; hence in the case of a telescope $l'pr_B - l'_y$ is the focal length of the object-glass with which the eyepiece will yield the desired coma correction, and we will introduce the symbol f'_O for this quantity. In the case of the compound microscope the exit pupil of the microscope objective always lies fairly close to the posterior focus of the objective, because the illumination is supplied either by a fairly distant window or lamp, or else by the iris-aperture of a substage-condenser, which should appear as a distant object as seen through the condenser. Hence, in the microscope the distance $l'pr_B - l'_y = f'_O$ may be treated as sufficiently nearly identical with the $F'_A F_B$ of our equations for the magnification of microscopes, and thus again represents a fundamental datum. With a convenient adjustment of signs, the equation to be calculated thus becomes

EP (1)
$$\begin{cases} f'_O = (l'_y - L'_y)(1 - OSC')\Big/(u'_y/\sin U'_y - (1 - OSC')) \\ OSC' = 0{\cdot}2 \sin^2 U'_y, \text{ numerical factor empirical.} \end{cases}$$

We then use the f'_O found by this solution in the equation for g_7 under Sine Theorem V in order to determine the state of correction of the eyepiece with reference to transverse chromatic aberration, or 'chromatic variation of the magnification', as it is more usually called in eyepieces. In the equation for g_7 we slightly change the original symbols by replacing suffix r by the present y and transform the denominator of the final term thus:

$$l'pr_B - l'_v = l'pr_B - l'_y + (l'_y - l'_v) = f'_O + (l'_y - l'_v);$$

the equation to be computed then becomes

EP (2)
$$g_7 = \frac{u'_v - u'_y}{u'_v} + \frac{u'_y}{u'_v}\,\frac{l'_y - l'_v}{f'_O + (l'_y - l'_v)}.$$

[79] The most convenient number by which the results of such a calculation may be recorded is the angular magnification $MA' = -f'_O/f'_{EP}$ which each eyepiece would give when applied to a telescope objective of the calculated focal length. The advantage secured by using MA' is that as a ratio of two lengths MA' is a pure number and independent of the scale to which the eyepiece may eventually be actually made.

We will now apply these extremely convenient formulae to our selected example with the prescription :

$$r_1 = \infty$$
$$0{\cdot}15$$
$$r_2 = -1{\cdot}000$$
Air-space, 2·6, 2·9, and 3·2.
$$r_3 = \infty$$
$$0{\cdot}30$$
$$r_4 = -2{\cdot}000$$
Semi-aperture $y = Y = 0{\cdot}200$.
$$N_y = 1{\cdot}517\,; \quad N_v = 1{\cdot}526.$$

Two paraxial rays are to be traced right through in 'yellow' and 'violet' respectively, and one marginal ray in 'yellow' only. Good four-figure log work is really sufficient for *practical* purposes, but five-figure results are more dependable when the theory is to be tested. The five-figure results are :

For an air-space	=	2·6	2·9	3·2
Final l'_y	=	−1·11172	−1·66411	−2·35425
Final l'_v	=	−1·17143	−1·74382	−2·46511
Final L'_y	=	−1·27101	−1·86379	−2·61067
Chrom. Aberration $l'_y - l'_v$	=	+0·05971	+0·07971	+0·11086
Spher. Aberration $l'_y - L'_y$	=	+0·15929	+0·19968	+0·25642
Final log u'_y	=	8·90482	8·85914	8·80808
Final log u'_v	=	8·90540	8·85808	8·80498
Final log sin U'_y	=	8·90464	8·85914	8·80427
$f'_{EP} = y/u'_y$	=	2·49006	2·76624	3·11136
For Coma-compromise :				
$OSC' = 0{\cdot}2 \sin^2 U'_y$	=	0·00129	0·00104	0·000812
f'_O by EP (1)	=	93·427	38·437	26·627
$MA' = -f'_O/f'_{EP}$	=	−37·520	−13·895	−8·5580
g_7 by EP (2)	=	+ 0·00196	− 0·00038	−0·00299

A very large amount of important information can be extracted from these results.

With regard to the longitudinal chromatic aberration $l'_y - l'_v$ we will take the eyepiece with an air-space = 2·9 for the detailed discussion. It has $l'_y - l'_v$ = [79] 0·080, with a focal length = 2·766, and we know that this aberration is in first, and for eyepieces sufficient, approximation independent of the aperture, hence the numerical value will not change whatever the f/ratio of the instrument may be. The value, however, will change with the scale to which the eyepiece is made, and as the average Huygenian eyepiece may be taken as having a focal length = about 0·7 inch, we may assign $l'_y - l'_v = 0{\cdot}080 \times 0{\cdot}7/2{\cdot}766 = 0{\cdot}020$ inch to this more normal eyepiece. It will be noticed that this is just the amount which we assumed for the chromatic over-correction of an object-glass in section [40], example (2), when we found that such a departure from perfect achromatism profoundly changed the prescription for an object-glass of given glasses and focal length. From the point of view of chromatic tolerance the $l'_y - l'_v$ of our eyepiece is also very serious, for 1 wavelength/$\sin^2 U'_y$, with $\log \sin U'_y = 8{\cdot}85914$, gives the tolerance $= \pm 0{\cdot}0038''$, so that the chromatic aberration of the eyepiece amounts to over five times the Rayleigh limit at the calculated aperture corresponding to an f-number = 2·766/0·4 = 7, and would remain above the Rayleigh limit even at the f/15 of the usual astronomical object-glass. It will thus be realized that the longitudinal chromatic aberration of Huygenian eyepieces is by no means negligible, even when allowance is made for the less definite or more elastic nature of chromatic tolerances. A remarkable and little known fact may be added: The longitudinal chromatic aberration of all Huygenian eyepieces is *greater* than that of a simple thin lens of the same glass and focal length. The glass of our eyepieces has $V = (N-1)/\delta N = 0{\cdot}517/0{\cdot}009 = 57$, and in accordance with TL, Chr. (2) the chromatic aberration at the principal focus of a thin simple lens of 0·7″ focus would be $= f'/V = 0{\cdot}7/57 = 0{\cdot}0123''$. Our eyepiece has a chromatic aberration $= 0{\cdot}020''$, or as much as a thin lens with $V = 0{\cdot}7/0{\cdot}020 = 35$. Hence the chromatic aberration of Huygenian eyepieces made from *crown* glass is as large as that of simple lenses made from fairly dense *flint* glass.

We will now discuss the longitudinal spherical aberration of the same eyepiece. To the computed scale and at f/7 it had $LA' = 0{\cdot}19968$ with $f'_{EP} = 2{\cdot}766$. Change of scale for $f'_{EP} = 0{\cdot}7$ then gives $LA' = 0{\cdot}19968 \times 0{\cdot}7/2{\cdot}766 = 0{\cdot}0506$, still for f/7. As the spherical tolerance is four times the chromatic tolerance, which latter was worked out above as $= \pm 0{\cdot}0038''$, we have spherical tolerance $= \pm 0{\cdot}0152$; hence at f/7 our eyepiece has a spherical aberration amounting to more than three times the Rayleigh limit, and if the eyepiece were applied to a spherically corrected object-glass the performance of the complete instrument would be gravely imperfect at f/7. There will, however, be a far more rapid improvement in this case when the aperture is reduced, because the actual spherical aberration will diminish as the square of the aperture, whilst the tolerance—on account of $\sin^2 U'_y$ in the denominator—will increase inversely as the square of the aperture. At f/15 the spherical aberration will therefore be $= 0{\cdot}0506(7/15)^2 = 0{\cdot}011$, and the tolerance will be $= 0{\cdot}0152(15/7)^2 = \pm 0{\cdot}069$. For the usual astronomical telescopes and for the microscope the spherical aberration of normal Huygenian eyepieces may therefore safely be treated as negligible; but 'finders' and other small telescopes are frequently made with f/7 or still larger aperture, and then the spherical aberration of the eyepiece calls for careful consideration.

[79] With reference to the oblique aberrations, the f'_O calculated by EP (1) represents that 'tube-length', or more strictly that focal length of a thin telescope object-glass, with which each eyepiece will give a good compromise correction for coma, namely, a small amount of positive coma in the central part of the field, zero coma at about 12° to 16° from the centre of the field, and rapidly growing negative coma in the outer field beyond the coma-free zone. As these eyepieces all have considerable spherical under-correction, they will also have the flattest possible field at the same tube-lengths. The g_7 calculated by EP (2) then gives the constant of the primary transverse chromatic aberration of each eyepiece at the calculated value of f'_O, and we see that this comes out positive for the eyepiece with 2·6 separation, nearly zero with 2·9 separation, and with a considerable negative value for the 3·2 separation. In accordance with our sign convention for g_7, this means that if these eyepieces were used in the computing direction—that is, projecting images of distant objects situated to the left of the eye lens, to be viewed through the field lens from a distance $= f'_O + l'_y$—then the eyepiece with 2·6 separation would give 'yellow' images larger than the corresponding 'violet' images, the eyepiece with 2·9 separation would be almost perfectly achromatic as regards magnification, and the eyepiece with 3·2 separation would give 'yellow' images smaller than the 'violet' ones. For the majority of instruments we require eyepieces with the most complete achromatism of magnification that is attainable. In accordance with the close analogy between transverse chromatic aberration and coma which we have repeatedly noticed, the best correction for achromatism of magnification for the field as a whole calls for a compromise, because there is a secondary form of the transverse chromatic aberration which grows as H'^3 and which is invariably negative in otherwise well-corrected eyepieces. Hence we should allow a little positive transverse chromatic aberration in the central part of the field to balance the negative higher aberration at a suitable distance from the centre ; experience suggests that this zone of perfect achromatism of magnification will lie at about 12°–16° from the optical axis if we select that eyepiece which has $g_7 = +0{\cdot}001$, and we will adopt this compromise.

Plotting the values of g_7 found for the three calculated forms of the eyepiece as ordinates against the respective separations as abscissae, they will be found to fall nearly into a straight line, but the small curvature should be allowed for by drawing the graph on the dip-principle according to section [47]. The resulting curve then shows that g_7 will have the desirable value $= +0{\cdot}001$ for a separation or air-space $=$ 2·725 ; hence, an eyepiece of our chosen glass and specification will have the best possible correction of coma *and* of transverse chromatic aberration when the lenses are mounted with the stated separation and when the eyepiece is used to produce the angular magnification found by EP (1) and by $MA' = -f'_O/f'_{EP}$. In practice it is not necessary to calculate MA'; it is simpler and more instructive to add to the g_7-graph another curve found by plotting the calculated values of $1/MA'$ for the three trial separations. The reciprocal value should be used in preference to MA' itself, because it gives a better curve in all cases and avoids an actual discontinuity in cases (sure to arise when the 2·3 ratio is tried) when f'_O and MA' come out with reversed sign for one of the trial separations, for such a change of sign takes place via infinity. In our present case this added curve shows at once that the separation $= 2{\cdot}725$ calls for $1/MA' = -0{\cdot}047$, or for MA' itself $= -21{\cdot}3$. [79] Yet another curve may be added to the graph to represent f'_{EP}; this gives for the solution $f'_{EP} = 2{\cdot}60$.

The direct result of our simple calculations is that the ratio 2 of the focal lengths of field lens and eye lens gives simultaneous correction of coma and of transverse chromatic aberration when the eyepiece is adjusted for an angular magnification $= -21{\cdot}3$. To obtain an equally perfect correction for any other magnification, we should, strictly, have to vary the ratio of the focal lengths. There is, however, a very considerable tolerance. Our graph covers a long range of magnifications and gives for each chosen value of MA' that value of the air-space which will yield an eyepiece free from coma and therefore also with the flattest possible field. The only important defect in eyepieces read off the graph for magnifications differing from the optimum value $= -21{\cdot}3$ therefore consists in a certain amount of transverse chromatic aberration, and the tolerance for this aberration will decide the range of magnifications for which the ratio 2 may be safely used. In accordance with the discussions in sections [64] and [65], the defect will be practically invisible if g_7 is within 0·001 of the optimum, or in our case between 0·002 and zero; for the image of a point would at the limits of this range be drawn out into a spectrum subtending an angle of one minute of arc at $16\frac{2}{3}°$ from the centre of the field. The graph shows that $g_7 = +0{\cdot}002$ corresponds to $MA' = -38$ and that $g_7 = 0$ is reached at $MA' = -15$; hence the ratio 2 may certainly be used with perfect safety for magnifications from 15 to 38 times. In section [65] we fixed the *limiting* value of g_7 at the much higher figure $\pm 0{\cdot}0025$ based on extensive experience. That would bring all eyepieces with g_7 between $+0{\cdot}0035$ and $-0{\cdot}0015$ within the permissible range, and by moderate extrapolation the graph fixes the corresponding range of permissible magnifications as 11 to infinite, so that the ratio 2 may be expected to answer quite well for all astronomical magnifications. It should, however, be borne in mind that it is highly inadvisable to use up the full tolerance to cover imperfections of a design; therefore the ratio 2 should, at most, be adopted for the 15 to 38 range corresponding to a practically invisible transverse chromatic aberration. It should also be clearly realized that the tolerances here discussed refer to eyepieces of ratio 2 which have been accurately adjusted for coma-freedom at the magnification with which they are to be actually used. For an eyepiece in a fixed mount the permissible range of magnifications is very much smaller, as both coma and transverse chromatic aberration will change, and will change at a greater rate than in the case which we have discussed, namely, when the separation is carefully adjusted.

As the ratio 2 only covers magnifications between 15 and 38, whilst astronomical telescopes usually magnify over 100 times and microscope eyepieces have angular magnifications of less than 12 times, it will be seen that the other ratios advised above must be similarly investigated in order to enable the designer to choose the best possible proportions for every purpose. Each ratio must be tried for three separations or air-spaces, and the only additional instruction required refers to the selection of suitable values. Experience has shown that the air-space does not vary very much for different ratios if the same eye lens (in our case with $r_2 = -1$) is retained. Hence we may begin for both ratio 2·3 and for 1·7 with an air-space of

[79] 2·6, near which the best solution for ratio 2 was found. This means that we can take over the original calculation for ratio 2 up to the fourth surface, and that we merely have to recalculate through the latter with radius $-2{\cdot}3$ or $-1{\cdot}7$ instead of the original $-2{\cdot}0$. The closing calculation by EP (1) and EP (2) will give a certain definite value of g_7 for this first trial. We then again profit from previous experience which teaches that g_7 for any given ratio always diminishes when the air-space is increased. Hence we try next an increase of the air-space by 0·3, if g_7 for the first separation was found larger than $+0{\cdot}001$, and we try a diminution by 0·3 if the first separation gave too small a value of g_7 or perchance a negative value. The two values of g_7 then available will clearly indicate the tendency of the g_7-graph, and the final separation can be chosen as another 0·3 step either upward or downward. It will be seen that an almost absurdly small amount of calculation will give complete graphs like that described for ratio 2. If low-power eyepieces for the microscope are to be covered, ratio 1·4 will have to be added; this ratio will also be required for the eye ends of four-lens terrestial eyepieces. To complete the work, we take from the graphs the best possible eyepiece of each ratio in the manner exemplified for ratio 2 and construct a final graph by plotting the values of $1/MA'$, of the air-space, and of f'_{EP} of the respective best forms as ordinates against the ratios of f'_B/f'_A as abscissae. This final graph will give directly for any required angular magnification the equivalent focal length, the requisite air-space, and the ratio f'_B/f'_A for an eyepiece having $r_2 = -1$ and $r_4 = (-\text{ratio})$. If the eyepiece actually required has a different focal length from that read off the graph, it is only necessary to change all radii, thicknesses, and separations in the proportion of required f' to f' from graph. If the longitudinal spherical and chromatic aberrations of the directly calculated forms are added to all the graphs, then the final one will also give this important information, which will be required if an object-glass is to be designed to correct the longitudinal aberrations of the eyepiece with which it is to be fitted.

The quick and simple method just described will be found to produce the best possible Huygenian eyepiece with as close an approximation as is of practical interest. But as the method does not give any information as to the distortion or as to astigmatism and curvature of the field, it must be supplemented by direct trigonometrical calculation of an oblique pencil in order to determine the magnitude of these aberrations. It will also be realized that the two empirical allowances for OSC' and for paraxial g_7 can only be derived from such direct and exact calculations.

The eyepiece to be trigonometrically tested should first be submitted to the method already given in order to find f'_{EP}, f'_O, and g_7 as well as the location of the axial focal points. The trigonometrical work begins with the tracing of the principal ray of the selected oblique pencil right-to-left from the pupil fixed by f'_O through the eyepiece to its intersection with the optical axis at the eye-point. The initial intersection-length of this principal ray will be $l'pr_B = f'_O + l'_y$. The initial U-value will depend on the zone of the field for which the trigonometrically exact aberrations are to be determined. It has already been suggested that a pencil reaching the eye under an angle of about 15° from the optical axis appears appropriate. Hence the angle Upr_A under which the principal ray traced right-to-left finally emerges should be selected, probably as $= -15^\circ$; but, as $U'pr_B$ is required for the opening of the calculation, this must be calculated to correspond to the chosen Upr_A. In accordance with G Th (14) the relation for a perfect telescopic system is, in our present symbols, [79]

$$\tan U'pr_B/\tan Upr_A = -f'_{EP}/f'_O,$$

and a ray traced with the initial U'_{prB} given by this equation would therefore emerge under the angle Upr_A if the eyepiece were free from distortion. Hence, if we trace the principal ray with the initial $U'pr_B$ so determined and find a final Upr_A differing from that used in working out the formula, the difference will be indicative of distortion and the latter can be determined.

The starting data for the principal ray therefore are

$$L'pr_B = f'_O + l'_y; \quad \tan U'pr_B = -\tan\,(\text{ideal } Upr_A) \,.\, f'_{EP}/f'_O,$$

'ideal' being added to the selected value of Upr_A (usually -15°) because it will only be actually realized by the traced ray if there is total absence of distortion. The results of the ray tracing will be Lpr_A and (actual Upr_A), and these supply important information.

Lpr_A is the distance from the plane face of the eye lens to the eye-point; we learnt that observations with an instrument become uncomfortable on account of the irritating contact of the eyelashes with the eye lens if this distance is less than 5 or 6 mm., and it is for this reason that Lpr_A is an important quantity.

With reference to the difference between ideal and actual Upr_A it is evident that if the sharp image presented to the eye lay at a projected distance L' from the eye-point, then the ideal image-height would be $= -L'.\tan$ (ideal Upr_A), whilst we should actually see an image of height $= -L' \tan$ (actual Upr_A); hence the distortion as defined in Chapters VI and VIII would be $= -L'$ (tan (ideal Upr_A) $-$ tan (actual Upr_A)); but there would be little sense in thus defining the distortion of an eyepiece, because the observer has no means of determining L' which depends on the accommodation of his eye. This difficulty is best met by stating the distortion as the percentage which the linear distortion, as stated above, bears to the ideal image height. This percentage is obviously independent of L' and is given by

EP (3) $$\text{Percentage distortion} = 100\,.\left(1 - \frac{\tan\,(\text{actual } Upr_A)}{\tan\,(\text{ideal } Upr_A)}\right).$$

As the linear distortion grows with H^3, it follows that the percentage distortion obtained by dividing the former by H grows with H^2, or as the *square* of the distance from the centre of the field; this distance may be most conveniently expressed by tan Upr_A. As before, a positive value of the percentage distortion indicates barrel distortion, a negative value pincushion distortion. The latter predominates in eyepieces and can reach over 5 per cent. at $Upr_A = 15^\circ$ in unfavourable forms.

By tracing the principal ray also in 'violet', we can secure a direct determination of the transverse chromatic aberration for the selected zone of the field. Distinguishing the two values of Upr_A thus obtained for 'yellow' and 'violet' by the

[79] symbols Upr_y and Upr_v, we can apply the same reasoning already used in discussing the distortion, and, calling the resulting value of Tch/H 'marginal g_7', we obtain the formula

EP (4) $$\text{marginal } g_7 = 1 - \frac{\tan Upr_v}{\tan Upr_y}.$$

In accordance with our sign convention for g_7 a positive value will mean that a star image is drawn out into a spectrum with violet pointing towards the optical axis and red away from the axis, or, in terms of magnification, an eyepiece with positive 'marginal g_7' will, in the selected zone of the field, have higher magnification in 'yellow' than in 'violet'. In comparing the marginal g_7 found by right-to-left calculation with the paraxial g_7 obtained by the left-to-right simple method it must be remembered that the exchange of object and image which is implied by calculations in opposite direction carries with it a reversal of sign ; hence the sign obtained for 'marginal g_7' must be reversed before comparing it with the g_7 by EP (2).

To complete the trigonometrical test, we finally add rays '*a*' and '*b*' by section [66], using the Lpr_A and Upr_A found for the 'yellow' principal ray together with SA = (Y of the axial pencil) in the opening equations. The distortion by (18) need not be calculated, as it has already been determined, but we may note that, if it were calculated and expressed as a percentage, it would come out with the opposite sign of that by EP (3), again on account of the exchange of object and image.

If the eyepiece submitted to these trigonometrical tests has been selected by the simple method first described, with the allowances for OSC' and g_7 as advised, then very little coma and transverse chromatic aberration will be found, and comparison with the respective tolerances will prove these residuals to be practically negligible. The field will nearly always prove round, but as it has been flattened to the utmost possible extent by correcting the coma (second fundamental law) the residual must be accepted as incurable if plano-convex simple lenses are to be adhered to. Huygenian eyepieces usually have about −3 per cent. of pincushion distortion at 15° obliquity; but this also must be accepted, as it can only be reduced at the cost of greatly aggravated curvature of the field.

We will apply the trigonometrical test to the best possible eyepiece with f'_B/f'_A = 2, for which the graph gave 2·725 as the necessary air-space. The simple method gives for the three axial rays $L'_y = -1{\cdot}50252$, $l'_y = -1{\cdot}32792$, $l'_v = -1{\cdot}39488$, and with the empirical OSC' allowance = 0·00118, EP (1) gives $f'_O = 58{\cdot}787$ and EP (2) gives $g_7 = +0{\cdot}00100$, exactly as intended. The equivalent focal length comes out = 2·5981, also very close to the 2·60 deduced from the graph. The angular magnification f'_O/f'_{EP} is found by the direct calculation = −22·627, whereas the graph gave −21·3 ; this apparently serious disagreement is due to the fact that f'_O is found by small differences of big numbers and therefore with a considerable uncertainty ; but as these differences are small because the aberrations are small, the uncertainty of f'_O and of MA' has only a very small and practically negligible effect upon the perfection of the final result. The uncertainty is only a small fraction of the tolerance.

Accepting the directly calculated f'_O, we next find the starting data for the right-to-left principal ray with ideal $Upr_A = -15°$: [79]

$$L'pr_B = f'_O + l'_y = 58{\cdot}787 - 1{\cdot}328 = 57{\cdot}459 \,;$$

and tan $U'pr_B = -$ tan (ideal Upr_A) . f'_{EP}/f'_O gives $U'pr_B = 0$–40–43. The tracing of this ray in the colours y and v then gives $Lpr_{Ay} = -0{\cdot}57338$, $Upr_y = -15$–26–49, $Upr_v = -15$–26–59.

Lpr_A being the clear distance of the eye-point, we see that if our eyepiece were made to centimetre scale it would just about have the distance of the eye-point (5 to 6 mm.) which was stated above as the least distance compatible with comfortable observation. As the focal length would then be 2·6 cm., or practically one inch, we conclude that Huygenian eyepieces of less than one inch focal length are uncomfortable. This is undoubtedly true, but the type is habitually used by astronomers down to less than one-quarter inch focal length.

The actual Upr_A being nearly half a degree in excess of the ideal Upr_A, we recognize serious pincushion distortion, which is determined by EP (3) as $-3{\cdot}13$ per cent. Although distortion of this order of magnitude is usual in otherwise well-corrected Huygenian eyepieces, it is curious how rarely it is recognized by observers.

The correction of the transverse chromatic aberration is practically perfect, for no human eye could appreciate the dispersion of only ten seconds of arc for a colour-interval corresponding to that between C and F. Equation EP (4) gives marginal $g_7 = -0{\cdot}00021$, or one twelfth of the normal tolerance. We may note, however, that if we had calculated the marginal g_7 in the opposite direction so as to be directly comparable with the paraxial g_7 by EP (2), it would have been $+0{\cdot}00021$, indicating that our allowance of $+0{\cdot}001$ for higher aberration was too large and that $+0{\cdot}0008$ would have been just right for the calculated zone.

For the pair of tangential rays 'a' and 'b' equations (12) give $L_a = -1{\cdot}29716$ and $L_b = +0{\cdot}15040$ as the initial intersection-lengths, with obliquity $= Upr_y = -15$–26–49 for both. The left-to-right tracing of these rays through the entire eyepiece gives the final results $L'_a = 6{\cdot}4347$, $U'pr_a = 5$–0–27, $L'_b = -10{\cdot}7266$, $U'pr_b = -3$–35–14, and the closing equations 15*, 16, and 17 then lead to the final results :

$$X'_T = -0{\cdot}0543, \quad Coma'_T = -0{\cdot}00129, \quad H'_{ab} = 0{\cdot}70021.$$

The negative sign of X'_T indicates roundness of the tangential field, but at the focal length $= 2{\cdot}6$ of our eyepiece this would easily be neutralized by the accommodation of the eye and need not be regarded as serious. As, owing to the practically complete freedom from coma, our eyepiece has the nearest approach possible to a flat field, by the second fundamental law, this residual roundness also must be regarded as incurable. If we calculate $X'_{Ptz} = -\frac{1}{2} H'^2 . \Sigma(N-1)c/N$, we find with H' (nearly enough) equal to the above H'_{ab}, $N = 1{\cdot}517$, and $c_A = 1$, $c_B = 0{\cdot}5$: $X'_{Ptz} = -\frac{1}{2}(0{\cdot}7)^2 \times 0{\cdot}517 \times 1{\cdot}5/1{\cdot}517 = -0{\cdot}125$; hence the roundness of the field would amount to a difference of focus $= -0{\cdot}125$ in the absence of astigmatism. As we found $X'_T = -0{\cdot}0543$ we see that there is a considerable amount of negative astigmatism in our eyepiece and that the greater part of the Petzval curvature of

[79] the field is corrected by it. The state of correction as regards astigmatism is therefore decidedly favourable.

With regard to residual coma, the tangential amount $-0{\cdot}00129$ found by the ray-tracing may be assumed to correspond nearly enough to one-third of that amount of sagittal coma, or to $Coma'_S = -0{\cdot}00043$. As we found H' as $= 0{\cdot}7$, this coma residual corresponds to

$$OSC' = Coma'_S / H' = -0{\cdot}00043/0{\cdot}7 = -0{\cdot}0006\dot{1}.$$

It is therefore less than one-quarter of our limiting value for admissible OSC' and quite harmless.

A closer discussion of the residual coma, however, is highly instructive. We deliberately solved for eyepieces with a considerable positive value of OSC' as calculated from the data of the axial pencil; in the present example the allowance was $OSC' = +0{\cdot}00118$, which at $H' = 0{\cdot}7$ would give $Coma'_S$ by Sine Condition $= 0{\cdot}00118 \times 0{\cdot}7 = +0{\cdot}00083$. The direct trigonometrical calculation indicates an actual negative $Coma'_S = -0{\cdot}00043$. As has been repeatedly pointed out, the sine condition only takes account of those forms of sagittal coma which grow in strict proportion with H', and the disagreement usually found by trigonometrical determinations of the coma in pencils of finite obliquity is due to the presence of secondary and possibly still higher forms of coma, which grow as the cube of H' and are included in the trigonometrically determined coma. Now we know by the OSC' allowance that in our eyepiece the coma proportional to H' is $= 0{\cdot}00118\,H'$ $= 0{\cdot}00083$ for the calculated oblique pencil. The trigonometrically estimated total coma of this pencil has been found as $= -0{\cdot}00043$; hence the coma proportional to H'^3 must amount to $-0{\cdot}00083 - 0{\cdot}00043 = -0{\cdot}00126$ at the calculated obliquity corresponding to $H' = 0{\cdot}7$. The first and practically important conclusion to be drawn is that our OSC' allowance was not large enough to produce zero coma at the calculated obliquity; it should have been one and a half times as large; hence $OSC' = 0{\cdot}3 \sin^2 U'_y$ would be a more suitable empirical allowance for eyepieces of the present type if it were desired to have freedom from coma at 15° obliquity.

The conclusions drawn enable us to build up a formula by which the coma in any part of the field can be closely predicted. We have already stated that the part proportional to H' is $= 0{\cdot}00118\,H'$. The part growing as H'^3 has been estimated as $= -0{\cdot}00126$ at $H' = 0{\cdot}7$, hence this part for any other value of H' will, by the proportionality to the cube, be

$$-0{\cdot}00126\,H'^3/0{\cdot}7^3 = -0{\cdot}00367\,H'^3.$$

We can therefore claim that the coma for any value of H' will be fairly closely given by the formula:

$$\text{Total } Coma'_S = 0{\cdot}00118\,H' - 0{\cdot}00367\,H'^3.$$

This formula gives:

for H' =	0·2	0·4	0·5	0·6	0·7	0·8
$0{\cdot}00118\,H'$ =	0·00024	0·00047	0·00059	0·00071	0·00083	0·00094
$-0{\cdot}00367\,H'^3$ =	−0·00003	−0·00024	−0·00046	−0·00078	−0·00126	−0·00188
Total $Coma'_S$ =	0·00021	0·00023	0·00013	−0·00007	−0·00043	−0·00094

It will be seen that the compensation of the negative cubic coma by the deliberately introduced positive ordinary coma greatly reduces the net amounts, and therefore leads to a valuable improvement in the coma correction. It will also be easily realized that an increase in the positive allowance would extend this improvement to still higher values of H', but at the expense of larger residuals in the more central parts of the field. The allowance actually made produces zero coma at $H' = \sqrt{(0{\cdot}00118/0{\cdot}00367)} = 0{\cdot}57$, corresponding to $Upr_A =$ about 12°. [79]

The whole discussion had to be safeguarded by the steady employment of the word 'estimated' instead of 'determined', because the case of coma is in reality more complicated than has been assumed, for there are really three different types of secondary coma instead of the one included in the estimates. But in ordinary eyepieces the term included is by far the largest, and the above simple method will therefore give fairly close results in their case. A more complete analysis must be reserved for Part II.

RAMSDEN EYEPIECES

[80] In the ordinary Ramsden eyepiece, of two planoconvex lenses made from the same kind of glass, the eye lens obviously must have a focal length which is larger than the airspace in order to secure the accessible focal plane, or positive l'_B, which is the chief advantage and indeed the *raison d'être* of the type. On the other hand we shall find that the field lens practically always has to have a longer focal length than the eye lens in order to secure freedom from coma and flatness of field. Now the condition of achromatism of magnification demands as a minimum an airspace $d' = \frac{1}{2}(f'_A + f'_B)$, or, as f'_A must be smaller than f'_B, an airspace necessarily larger than f'_A, which contradicts the demand for an accessible or *real* focal plane; and as the latter demand cannot be evaded, it follows that ordinary Ramsden eyepieces cannot be freed from transverse chromatic aberration.

Evidently we shall reduce this unavoidable defect to the utmost practicable extent by fixing the airspace at the greatest proper fraction of f'_A which appears reasonably safe, bearing in mind that the thickness of the field lens must also be allowed for. If it were safe to assume that the observer always adjusted the focus for a very distant virtual image, then $d' = \frac{7}{8} f'_A$ would be admissible. But as myopic people usually object to wearing their proper spectacles when observing, and as young observers have a very prevalent though pernicious habit of working with a close virtual image calling for a considerable strain on the accommodation, it is not advisable to exceed $d' = \frac{3}{4} f'_A$, and we will for our example adopt a value yet a little smaller, namely, $d' = 0{\cdot}7 f'_A$. Still lower values down to $\frac{1}{2} f'_A$ are occasionally met with, but they lead to a considerable and quite unjustifiable aggravation of the transverse chromatic aberration.

The airspace being thus fixed, there is only one liberty left for varying the correction of a Ramsden eyepiece, namely, the ratio of the focal lengths of eye lens and field lens; consequently only one aberrational condition can be satisfied, and as this we choose freedom from coma because it carries with it the greatest possible flattening of the field by the second fundamental law. The calculation is carried out by the methods and formulae of the preceding section, but is much simpler because there is no variation in the airspace. As the cubic coma is again

[80] negative and comparable in magnitude with that of Huygenian eyepieces, the OSC' allowance suggested in EP (1) will give a greatly improved solution by the simple method. The useful range of values of the ratio f'_B/f'_A is much shorter than in the Huygenian type, the limits being about 1 and 1·3.

As an example we will take a Ramsden eyepiece with the eye lens used for the Huygenian, so that we shall have $f'_A = 1{\cdot}9341$ and $d' = 0{\cdot}7 \times f'_A = 1{\cdot}35387$ by ordinary multiplication. We may note that although we shall use this value with all the stated decimals in the calculations, it should be rounded off to three or even only two decimal places in a workshop prescription in accordance with the precision reasonably attainable in the brasswork. Choosing $f'_B/f'_A = 1{\cdot}25$, we shall have the specification

$$\begin{array}{ll} r_1 = \infty & \\ & d'_1 = 0{\cdot}15 \\ r_2 = -1{\cdot}000 & \\ & \text{Air } d'_2 = 1{\cdot}35387 \\ r_3 = +1{\cdot}250 & \\ & d'_3 = 0{\cdot}19 \\ r_4 = \infty & \end{array}$$

As in the Ramsden eyepiece the field lens increases the convergence produced by the eye lens whilst the convergence is diminished by the field lens of a Huygenian, the previously used semi-aperture would lead to an absurdly high f-ratio. For that reason we adopt $y = Y = SA$ of the oblique pencil $= 0{\cdot}12$ instead of the previous 0·2. The tracing of the three rays of axial pencils gives

$$L'_y = 0{\cdot}32107, \quad l'_y = 0{\cdot}34267, \quad l'_v = 0{\cdot}32027, \quad f'_{EP} = 1{\cdot}5599,$$

whence $$LA' = l'_y - L'_y = 0{\cdot}02160, \quad Lch' = l'_y - l'_v = 0{\cdot}02240.$$

The f-number $f'_{EP}/2Y$ is $1{\cdot}56/0{\cdot}24 = 6{\cdot}50$, almost exactly identical with that of our completely tested Huygenian eyepiece which had $2{\cdot}60/0{\cdot}4 = 6{\cdot}50$. The two eyepieces are therefore directly comparable as regards the ratio of aperture to focal length. For the Huygenian we found $LA' = 0{\cdot}17460$; but if this eyepiece were reduced in scale so as to secure $f'_{EP} = 1{\cdot}56$ instead of the calculated 2·60, then the spherical aberration, as a linear quantity, would also shrink in the same proportion; hence we have

$$LA' \text{ of Huygenian with } f'_{EP} = 1{\cdot}56, \text{ at } f/6{\cdot}5 : 0{\cdot}1746 \times 1{\cdot}56/2{\cdot}60 = 0{\cdot}1048.$$

As the Ramsden under the same conditions gives $LA' = 0{\cdot}0216$ we see that it has an enormous advantage over the Huygenian by having only about 1/5th of its spherical aberration.

There is a corresponding but smaller advantage in the longitudinal chromatic aberration. At its calculated focal length the Lch' of the Huygenian was $= 0{\cdot}06696$, which at the focal length of the Ramsden would become $= 0{\cdot}06696 \times 1{\cdot}56/2{\cdot}60 = 0{\cdot}0402$. As the Ramsden has $Lch' = 0{\cdot}0224$, its chromatic aberration is only 0·56 of that of a Huygenian of the same focal length.

It is well worth while to show that our airspace $= 0{\cdot}7 f'_A$ could not be greatly increased without risk of the field lens being brought into contact with the micrometer device by myopic observers or by young people straining their accommodation. At a low $f/$ratio objects at the calculated distance $l'_y = 0{\cdot}343$ from the field lens would be seen through the eyepiece as infinitely distant images, and an experienced observer with normal sight would choose this adjustment. We can then calculate the distance of the image which would result if the object were placed into contact with the field lens by the third form of G Th (10)** : $X_f \,.\, X'_f = f.f' = -f'^2$; for X'_f, being the distance from the posterior focus to the object, will be $= -l'_y = -0{\cdot}343$, and we know $f' = f'_{EP} = 1{\cdot}56$. Hence $X_f = -f'^2/X'_f = (1{\cdot}56)^2/0{\cdot}343 = 7{\cdot}1$, meaning that the virtual image would be located at 7·1 units of length to the right of the anterior focus, which for the present purpose we may identify sufficiently nearly with the eye-point of the eyepiece. If the eyepiece were made to a scale of inches—and 1·5 inches focal length would be quite normal on the *microscope*—the distance would be 7·1 inches ; for an observer with $-5\frac{1}{2}\,D$ of myopia this would be the longest distance at which he could see sharply without spectacles, and young observers with normal sight could easily accommodate for it. Hence in either case there might be trouble and possible damage. But as the vast majority of Ramsden eyepieces have $f'_{EP} < 1{\cdot}56''$, the risk is small and we may regard $d'_2 = 0{\cdot}7\,f'_A$ as quite safe in nearly all cases. [80]

For the calculation of EP (1) the OSC' allowance is found $= +0{\cdot}001185$ and gives $f'_O = 33{\cdot}488$, whence $MA' = -21{\cdot}468$, very close to that of our Huygenian specimen. EP (2) then gives paraxial $g_7 = +0{\cdot}01004$, and discloses the one rather grave defect of the Ramsden eyepiece, for this value of g_7 represents four times our adopted tolerance. This defect will be more closely discussed subsequently.

Proceeding to the study of an oblique pencil at finite angles, we again adopt ideal $Upr_A = -15°$ and find the starting data for the corresponding right-to-left principal ray by the previously given equations

$$L'pr_B = f'_O + l'_y = 33{\cdot}488 + 0{\cdot}343 = 33{\cdot}831 :$$

$$\tan U'pr_B = -\tan\,(\text{ideal } Upr_A)\,.\,f'_{EP}/f'_O,\ \text{giving } U'pr_B = 0\text{-}42\text{-}54.$$

Traced with these initial values in '*y*' and '*v*', the ray gives

$$Lpr_{Ay} = -0{\cdot}50197\ ; \quad Upr_y = -15\text{-}12\text{-}34\ ; \quad Upr_v = -15\text{-}20\text{-}33.$$

The distance of the eye-point $= Lpr_A$ is less than in the Huygenian, which gave $-0{\cdot}57338$; but as the latter had $f'_{EP} = 2{\cdot}60$ against 1·56 for the Ramsden, we must make a reduction for scale and find for a Huygenian of the focal length of our Ramsden

$$\text{reduced } Lpr_A = -0{\cdot}57338 \times 1{\cdot}56/2{\cdot}60 = -0{\cdot}344.$$

Hence at the same focal length the clear distance of the eye-point of a Ramsden eyepiece is practically one and a half times that of a Huygenian, and the former again scores a valuable advantage. The angles of the principal ray give

by EP (3) percentage distortion $= -1{\cdot}46$ per cent. ;
by EP (4) marginal $g_7 = -0{\cdot}00918$.

The distortion, which is independent of scale, is less than half that of the

[80] Huygenian, but g_7, although reduced compared with its paraxial value, remains very high.

For the completion of a tangential fan by the left-to-right tracing of rays 'a' and 'b' parallel to the principal ray, the standard formulae give $L_a = -0{\cdot}94336$, $L_b = -0{\cdot}06058$, and on emergence from the field lens

$$L'_a = 5{\cdot}2937,\ U'_a = 4\text{-}49\text{-}11\ ;\quad L'_b = -6{\cdot}7004,\ U'_b = -3\text{-}24\text{-}5.$$

These give the final results

$$H'_{ab} = 0{\cdot}41813\ ;\quad Coma'_T = -0{\cdot}00011\ ;\quad X'_T = +0{\cdot}0138.$$

The sagittal coma inferred by the three-to-one ratio would be $-0{\cdot}00004$ and utterly negligible; but as the OSC' allowance would produce ordinary coma proportional to H' equal to $0{\cdot}001185 \times 0{\cdot}418 = +0{\cdot}00049$, we conclude that the H'^3-coma amounts to $-0{\cdot}00053$ and that the allowance suggested in EP (1) is very nearly right for the present eyepiece, but a trifle too small for ideal $Upr_A = -15°$; the coma would be zero at a fraction of a degree below 15°. The magnitude of the residual coma in all zones of the field may be estimated in the manner fully described in the discussion of the specimen Huygenian eyepiece.

The positive value of X'_T shows that in our Ramsden eyepiece the negative astigmatism is sufficiently large to lead to a *hollow* tangential field. But as the Petzval sum of this eyepiece is $= 1{\cdot}8 \times 0{\cdot}517/1{\cdot}517 = 0{\cdot}614$, we have

$$X'_{Ptz} = -\tfrac{1}{2} g_4 H'^2 = -0{\cdot}307(0{\cdot}418)^2 = -0{\cdot}054,$$

and as, according to the discussion in section [55] G, X'_T may be allowed to reach the amount $-\frac{1}{2} X'_{Ptz} = +0{\cdot}027$, we see that the slight hollowness of the tangential field of our eyepiece is well inside the range of adjustment which we concluded to be favourable. Moreover, we must remember that any ordinary telescope objective has the pronounced positive astigmatism and roundness of field peculiar to thin systems passed centrally by the oblique pencils. A rough estimate shows that the roundness of the tangential field of an ordinary telescope objective of $f'_O = 33{\cdot}5$ would compensate the greater part of the hollowness of field in our eyepiece. We may therefore safely claim that our Ramsden eyepiece is superior to our Huygenian specimen as regards the adjustment of astigmatism and curvature of field.

The Ramsden eyepiece has another advantage of considerable practical importance; it is very much less sensitive than the Huygenian to departures from the exact value of f'_O calculated by our method. We can easily prove this by the second fundamental law of oblique pencils deduced and discussed in section [60], for as the f'_O determines the location of the exit pupil of the eyepiece in its reversed position, changes in f'_O amount to a shift of the pupil and therefore come under this important law. The changes in all the aberrations might be calculated by the standard method exemplified in [60], but in the present case we can shorten the work very considerably by utilizing the conclusions drawn in section [60]:

Our eyepieces, when chosen by the methods given above, are as nearly coma-free as is practicable; hence astigmatism and curvature of field are at a *minimum* for the calculated value of f'_O, and will only change to a minute extent for quite considerable changes in f'_O; therefore we need not pay any attention to astigmatism. And as our corrected eyepieces also have a very small value of X'_T, the distortion will be at a 'stationary' value and can be omitted. [80]

This leaves Coma as the only aberration which will change quickly when f'_O is changed, and therefore as the predominating cause of the diminished efficiency, which becomes obvious when an eyepiece well corrected for a particular value of f'_O is used with a decidedly different value. Now we have learnt that the sine theorem provides by far the quickest, and at the same time a highly precise, way of determining coma by its equation I:

$$OSC' = 1 - \left(\frac{\sin U_1}{u_1}\right) \cdot \frac{u'_k}{\sin U'_k} \cdot \frac{l'_k - l'pr_k}{L'_k - l'pr_k}$$

which can be worked out for any possible value of the pupil distance $l'pr_k$. We will specialize the equation for our present eyepiece problems. By adding and subtracting L'_k in the numerator of the last fraction, we obtain

$$l'_k - l'pr_k = L'_k - l'pr_k + l'_k - L'_k, \text{ giving } \frac{l'_k - l'pr_k}{L'_k - l'pr_k} = 1 + \frac{l'_k - L'_k}{L'_k - l'pr_k};$$

and if in the last denominator we similarly add and subtract l'_k, it becomes $l'_k - l'pr_k + L'_k - l'_k = -f'_O - (l'_k - L'_k)$ by the definition of our f'_O; hence

Sine Theorem I for Eyepieces :

$$OSC' = 1 - \left(\frac{\sin U_1}{u_1}\right)\frac{u'_k}{\sin U'_k} + \left(\frac{\sin U_1}{u_1}\right) \cdot \frac{u'_k}{\sin U'_k} \cdot \frac{l'_k - L'_k}{f'_O + (l'_k - L'_k)},$$

which gives the part depending on f'_O separately by the last of its three terms. As we always calculate the axial pencils with initial $u_1 = \sin U_1$ or $y_1 = Y_1$, the bracketed factor may be safely omitted, and as the whole last term is always small because the $LA' = l'_k - L'_k$ is invariably a very small fraction of the tube-length f'_O, it will practically always be safe to neglect also the small departure from 'one' of the factor $u'_k/\sin U'_k$ and the small change in f'_O by the addition of $(l'_k - L'_k)$. Therefore we may in nearly all cases estimate the change in coma resulting from a change in f'_O by the highly convenient simplified equation :

Eyepieces only :

$$OSC' = 1 - \frac{u'_k}{\sin U'_k} + \frac{l'_k - L'_k}{f'_O}.$$

The first two terms obviously represent the value of OSC' for $f'_O = \infty$, or for very long instruments. The final term now shows most clearly that its effect is directly proportional to the amount of spherical aberration and also to the reciprocal of f'_O. The latter result once more demonstrates the persistence with which aberrational changes in optical systems tend to be proportional to the reciprocals of radii and intersection-lengths, and not to these quantities themselves.

As our specimen Huygenian and Ramsden eyepieces both work at $f/6{\cdot}5$ and suit the same angular magnification, they will have (nearly enough) the same value of f'_O when made to the same focal length. And as it was shown above that under these conditions their longitudinal spherical aberrations were as 5 to 1, we can

[80] render our introductory statement as to the advantage of the Ramsden more precise by stating that our Huygenian eyepiece will be five times as sensitive to changes of tube-length as our Ramsden; this ratio will be found to be roughly maintained for all ordinarily useful modifications of the f'_B/f'_A ratio.

We will now work out a few numerical values of the change in OSC' of our Ramsden eyepiece which results from changes in f'_O. We know that the eyepiece will have the coma correction which we have accepted as the most desirable compromise when $f'_O = 33{\cdot}5$. With the calculated $l'_k - L'_k = 0{\cdot}02160$ the variable last term in the OSC' formula will then be $= 0{\cdot}0216/33{\cdot}5 = 0{\cdot}00064$. If we used the eyepiece on an instrument of huge focal length, the last term would be zero and we should depart from the most desirable value of OSC' by 0·00064; as our adopted limit for admissible OSC' is $\pm 0{\cdot}0025$, the amount found would be only one quarter of the admissible maximum, and we may safely conclude that our eyepiece would give good results for any conceivable increase in f'_O. If, on the other hand, we used the eyepiece with $f'_O =$ half the best value, or $= 16{\cdot}75$, then the final term would become doubled to 0·00128 and OSC' would again be wrong to the extent of 0·00064, which we have just shown to be comparatively harmless. Hence we may safely say that our eyepiece will show very little variation in its corrections for the whole range, $16{\cdot}75 < f'_O < \infty$, corresponding, with $f'_{EP} = 1{\cdot}56$, to a range of angular magnifications from $-10{\cdot}8$ to ∞, and thus covering almost every ordinary application of Ramsden eyepieces. For our specimen of a Huygenian eyepiece, when made to the scale giving 1·56 focal length, we should have, as shown above, $l'_k - L'_k = 0{\cdot}1048$ and $f'_O = 58{\cdot}8 \times 1{\cdot}56/2{\cdot}60 = 35{\cdot}3$, hence the final term at the tube-length giving the most desirable correction would be $0{\cdot}1048/35{\cdot}3 = 0{\cdot}00297$. If this eyepiece were used on a very long instrument, the last term would be zero instead of the desirable 0·00297, or wrong by more than the full OSC' limit. Hence this eyepiece would not be satisfactory at high magnifications. If we adopt as small enough the change in the last term by $\pm 0{\cdot}00064$ as admitted for the Ramsden, the last term might be allowed to vary between 0·00233 and 0·00361. By a simple transposition of the last term the corresponding values of f'_O are found as $= 0{\cdot}1048/0{\cdot}00233 = 45{\cdot}0$ and $= 0{\cdot}1048/0{\cdot}00361 = 29{\cdot}0$, or a range of admissible magnifications from $-29{\cdot}0/1{\cdot}56 = -18{\cdot}6$ to $-45/1{\cdot}56 = -28{\cdot}8$. Magnifications outside this range would call for an adjustment of the separation or, better, for a different ratio f'_B/f'_A. All the calculations are done by slide-rule because it gives results of ample precision in a minimum of time whenever questions of tolerance are concerned, simply by reason of the inevitable uncertainty of the tolerance itself.

Although we have shown that a Ramsden eyepiece of given construction covers a very considerable range of magnifications on account of its relatively small spherical aberration, it is advisable or even necessary in the case of very high or very low values of MA' to vary the ratio f'_B/f_A. It will be found that—with our selected eye lens having $r_2 = -1$—all probable requirements will be covered if the simple solution is repeated for $r_3 = 1{\cdot}15$ and $r_3 = 1{\cdot}05$. The resulting graph will give curves with such moderate departure from the straight line that extrapolation on the dip-principle will be quite safe to the small extent required to cover infinite MA'.

B. The 'Achromatized' Ramsden Eyepiece

The one outstanding defect of the ordinary Ramsden eyepiece, its high value of the transverse chromatic aberration, can be reduced to comparatively harmless magnitude by replacing the simple planoconvex eye lens by an approximately achromatic cemented combination. The effectiveness of this change is easily shown by our formula for achromatism of magnification, which for the usual case, when $l_A = \infty$, is

$$d' = (f'_A + f'_B \cdot V_B/V_A) \Big/ (1 + V_B/V_A - f'_B \cdot V_B/L'pr_B V_A).$$

If we solve this equation for V_B/V_A, it will give us directly that ratio of the V-values of the two components which would produce achromatism of magnification with any given values of f'_A, f'_B, d', and $L'pr_B$. The solution is

$$\frac{V_B}{V_A} = -\frac{f'_A - d'}{f'_B \cdot \dfrac{L'pr_B + d'}{L'pr_B} - d'}.$$

There is practically no exception to the rule that in a well-corrected eyepiece f'_B is greater than d', and as the correcting factor $(L'pr_B + d')/L'pr_B$ is obviously greater than 'one', we may safely claim that the denominator on the right is always finite and positive. The numerator $(f'_A - d')$ will be negative for eyepieces of the Huygenian type, and the general minus-sign on the right will then lead to a positive value of V_B/V_A, which can be realized by simple lenses as the V-numbers of individual glasses are all positive. If we choose $f'_A = d'$, then the right of the equation will become zero and achromatism of magnification will accordingly demand an infinite value of V_A, which means a fully achromatized eye lens. This was the case realized by the first eyepiece of this kind as constructed many decades ago by the German optician, Kellner, whose name is still associated with this type. Its objectionable features are that it cannot be used on micrometers, that any dust or imperfection of the field lens becomes painfully clearly visible, and that it is practically impossible to secure sufficiently good correction of other important aberrations for low f-numbers such as are now prevalent; at the modest $f/15$ of astronomical object-glasses and still higher $f/$numbers of microscopes, this last objection becomes less pronounced. If we now proceed to the case of the Ramsden type, we have $f'_A > d'$ and achromatism of magnification calls for a negative value of V_B/V_A, which means a chromatically over-corrected eye lens. It is almost impossible to satisfy this condition without a pernicious aggravation of other aberrations of the eyepiece on account of the deep curvatures which result, more especially at the contact surface. For that reason nearly all eyepieces of the achromatized Ramsden type have a very appreciable remnant of transverse chromatic aberration in the sense of the simple lens Ramsden.

Fig. 101 shows the usual type and renders it obvious that the principal rays of the oblique pencils cross the contact surface under severe angles, which are further increased for the rays 'a' of the corresponding tangential fans. Angles of incidence up to and exceeding 50° are quite common in the marginal pencils. The most ective way of reducing the resulting higher aberrations to manageable magnitude onsists in making the value of $N' - N$ at the contact very small by using dense barium crown with rather light flint ; *this combination of glasses is the chief*

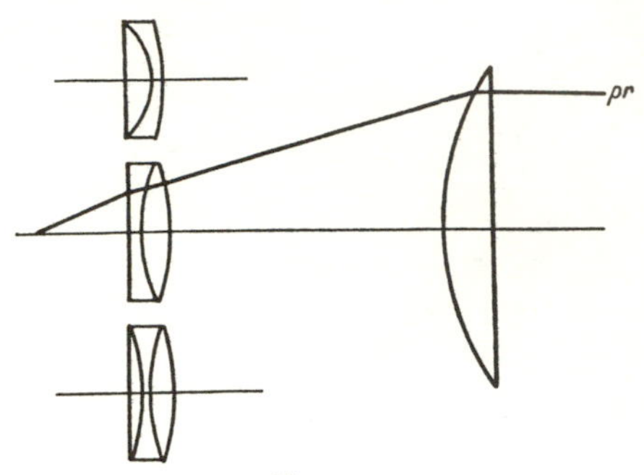

FIG. 101

secret of success. The reversed sequence of crown and flint shown above the usual eye lens at first sight suggests that it might be more favourable ; but the extremely deep contact surface which results will be found to defeat this scheme. The triple cemented form shown below the usual eye lens would be more favourable, but as it would also cost more it must be reserved for desperate cases. Even in these compounded eye lenses the externally planoconvex form can usually be adhered to, but it should be remembered that a moderate bending may secure a small improvement. As the theoretically indicated heavy chromatic over-correction of the eye lens is rendered inadvisable by the serious zonal spherical aberrations which would accompany it, and as small changes in the chromatic correction of the eye lens produce quite unimportant alterations in all the principal aberrations, it would be stupid pedantry to expend much time in establishing *precisely* some arbitrarily chosen type of achromatism ; hence we may simply solve for the usual TL achromatism and accept the resulting total curvatures as final. The residual g_7 is then kept as low as possible by choosing the highest value of the air-space which appears reasonably safe for possible actually or wilfully myopic observers.

We will choose as a numerical example an eyepiece of the usual type with the glasses :

Eye lens Flint : Chance 1034 ; $Nd = 1{\cdot}6041$; $Nf - Nc = 0{\cdot}01599$; $V = 37{\cdot}8$.
Eye lens Crown : Chance 9002 ; $Nd = 1{\cdot}5744$; $Nf - Nc = 0{\cdot}00995$; $V = 57{\cdot}7$.
Field lens : Ordinary crown as in all previous examples.

Choosing $f'_A = 1$, we find by the usual TL, Chr. solution for visual achromatism $c_a = -3{\cdot}143$, $c_b = +5{\cdot}050$. As in the previous cases, the strength of the field lens determines the f'_O for which the eyepiece will be free from coma ; r_4 and f'_O change in the same sense, but not in strict proportion. Choosing r_4 on the basis of previous experience—which the novice would have to buy with a few systematic [8 trials—and the airspace as $\frac{3}{4}$ of the adopted f'_A, the prescription is :

$$
\begin{array}{ll}
r_1 = \infty & \\
 & 0{\cdot}050 \\
r_2 = +0{\cdot}318 & \\
 & 0{\cdot}150 \\
r_3 = -0{\cdot}524 & \\
 & \text{Air } 0{\cdot}750 \\
r_4 = +0{\cdot}700 & \\
 & 0{\cdot}120 \\
r_5 = \infty &
\end{array}
$$

The thicknesses must be chosen on the basis of a scale drawing, with decided generosity in the case of the eye lens crown, because these eyepieces are usually made for a large angular field. Even more than the adopted value would be necessary if the eyepiece were to be used on a prismatic binocular with the usual field of quite 50° and an object-glass working at $f/4$, and it would *then* probably also be advisable to *diminish* yet further the small difference in the refractive indices of eye lens crown and flint.

The calculations were carried out with the usual indices $N_y = N_d + 0{\cdot}188\ (N_f - N_c)$ and $N_v = N_y + (N_f - N_c)$ for the eye lens and with semi-aperture $y = Y = SA = 0{\cdot}05$. The principal results for the three axial rays were

$$L'_y = 0{\cdot}12216\ ;\ l'_y = 0{\cdot}12934\ ;\ l'_v = 0{\cdot}12893\ ;\ f'_{EP} = 0{\cdot}8351,$$

$$\text{giving } LA' = 0{\cdot}00718 \quad l'_r - l'_v = 0{\cdot}00041.$$

The LA' appears very small compared with the 0·02160 found for our simple Ramsden eyepiece. But as the latter had $f'_{EP} = 1{\cdot}56$ and worked at $f/6{\cdot}5$, whilst our achromatized form has respectively 0·8351 and $f/8{\cdot}35$, the LA' found for the latter requires multiplying by 1·56/0·8351 to correct for scale and by $(8{\cdot}35/6{\cdot}5)^2$ to correct for f-ratio, hence for a really fair comparison :

$$\text{Corrected } LA' = 0{\cdot}00718\ (1{\cdot}56/0{\cdot}835)\ (8{\cdot}35/6{\cdot}5)^2 = 0{\cdot}0222 \text{ by slide-rule.}$$

The achromatized form, therefore, is actually a little worse than the simple form as regards spherical aberration. It has, however, a decided advantage in chromatic correction, for the 0·00041 found directly requires correction for scale only. As these eyepieces are chiefly used when a large field of view is required, the empirical OSC' allowance in EP (1) was increased to $0{\cdot}3 \sin^2 U'_m = 0{\cdot}001075$ so as to increase the previously contemplated diameter of the coma-free zone. The results were

$$f'_O = 5{\cdot}98 \text{ by } EP\,(1) \text{ and paraxial } g_7 = 0{\cdot}00433 \text{ by } EP\,(2),$$

the former giving $MA' = -5{\cdot}98/0{\cdot}8351 = -7{\cdot}16$, about midway between the two most usual binocular magnifications, 6 and 8.

o] Ideal $Upr_A = -20°$ was chosen for the trigonometrically traced tangential fan ; the results were

$$\left.\begin{array}{ll}\text{Barrel distortion} & = +0{\cdot}9 \text{ per cent.} \\ \text{Marginal } g_7 & = -0{\cdot}00367\end{array}\right\} \text{at } 20° \text{ from optical axis ;}$$

$$\left.\begin{array}{ll}Coma'_T & = +0{\cdot}00022 \\ X'_T & = +0{\cdot}032\end{array}\right\} \text{at } H' = 0{\cdot}297.$$

The distortion is of the opposite sense compared with that of the simple lens eyepieces, but is very much smaller, for as the percentage distortion grows as H'^2, the Huygenian eyepiece would have, at 20°, approximately $-3{\cdot}13 \times (20/15)^2 = -5{\cdot}6$ per cent., and the simple Ramsden would have $-1{\cdot}46 \times (20/15)^2 = -2{\cdot}6$ per cent. ; our eyepiece with achromatized eye lens, with only + 0·9 per cent., is therefore decidedly superior to the simpler forms in this respect.

The marginal g_7 shows the usual diminution as compared with the paraxial value ; it would not be objected to by most observers, for the reason given in the concluding section.

The residual coma is utterly insignificant ; but the positive value shows that the *OSC'* allowance was a little too large.

The value of X'_T signifies a decidedly hollow tangential field, and as the Petzval sum of our eyepiece gives $X'_{Ptz} = -0{\cdot}053$, the hollowness of field actually exceeds to a small extent the highest value which we justified, chiefly for wide-angle photographic objectives, in section [55] G.

We must, however, remember that our eyepiece is suitable for use with an object-glass of only 5·98 focal length. Such an object-glass will have a tangential curvature of field determined by the general equation $X'_T = -H'^2 (3\,g_3 + \frac{1}{2}\,g_4)$. For any reasonably thin telescope objective we shall have for g_3 the value $1/2\,f'$, and g_4 for such thin objectives never departs far from the value $0{\cdot}7/f'$. Hence the X'_T of any normal telescope objective may be estimated, with an error probably not exceeding 5 per cent., as

$$X'_T = -H'^2 (1\tfrac{1}{2}/f' + 0{\cdot}35/f') = -1{\cdot}85\,H'^2/f'.$$

In our eyepiece problems this has to be worked out with the H'_{ab} found for the trigonometrically traced oblique pencil and with the value of f'_O. As our eyepiece gave $H'_{ab} = 0{\cdot}297$ and $f'_O = 5{\cdot}98$, we find the X'_T value of the objective as

$$X'_T = -1{\cdot}85\,(0{\cdot}297)^2/5{\cdot}98 = -0{\cdot}027.$$

This value applies to the objective in our usual computing direction ; if it were reversed right for left, so as to fit our eyepiece in the computing direction adopted for the latter, the sign of the X'_T of the objective would also be reversed, showing that the value $+0{\cdot}032$ of the eyepiece would very nearly agree with the $+0{\cdot}027$ of the objective, and would produce an almost perfectly flat tangential field for the complete instrument. No estimate of similar simplicity is possible in the case of a microscope, but as microscope objectives nearly always have a decidedly round field, the positive X'_T of our eyepiece would be beneficial in the great majority of cases. It will be seen, and should be impressed upon the memory, that at low magnifications the curvature of field of the objective has an important influence. [80]

The achromatized Ramsden is the favourite eyepiece for prismatic binoculars; we must therefore study the effect of the prisms, which in these instruments take the place of the 'erector' in the old-fashioned four-lens terrestrial eyepiece. As the optical axis of the lenses of the complete telescope must pass at right angles through the refracting surfaces of the prisms in order to retain the advantages and symmetry of a *centred* system, and as the reflections at the plane internal surfaces introduce no aberrations of any kind, the aberrational effect of the prisms is simply that of planoparallel glass plates at right angles with the optical axis and of thickness equal to the path of the optical axis within each prism. The most usual size of binocular prisms has a hypotenuse face $1\frac{3}{4}$ inches long, and, as this is also the length of the path of the optical axis within the prism, we may assume that two plates, each $1\frac{3}{4}$ inches thick, are equivalent to the two erecting prisms. Moreover, we showed in section [15] that a longitudinal shift of a planoparallel plate does not have any effect on the co-ordinates of the emerging rays; hence we may reduce our problem to its simplest form by calculating the effect of a single plate $3\frac{1}{2}$ inches thick, placed in contact with the last surface of our eyepiece, assuming inch-scale for the latter. In fair accordance with usual practice we will take as the indices of the prism glass $N_y = 1{\cdot}560$ and $N_v = 1{\cdot}570$. Tracing the three rays of the axial pencils through this glass block by the usual formulae for plane surfaces, we find on emergence

$$L'_y = -2{\cdot}11910 \quad l'_y = -2{\cdot}11427 \quad l'_v = -2{\cdot}10034,$$
$$\text{giving } LA' = +0{\cdot}00483\;;\; l'_y - l'_v = -0{\cdot}01393.$$

The spherical aberration is greatly reduced from the previous 0·00718, and the chromatic aberration is reversed into a fairly considerable over-correction. Applying EP (1) and EP (2) in the usual way, we find

$$f'_O = 4{\cdot}025\;;\; \text{paraxial } g_7 = +0{\cdot}00080.$$

Hence our eyepiece, together with the erecting prisms, will have almost perfect achromatism of magnification, but it now gives freedom from coma with an objective of 4·025″ focal length. This objective would have to be designed for fulfilment of the sine condition, for a spherical over-correction at $f/8{\cdot}35$ of $-0{\cdot}00483$, and for a chromatic under-correction of $+0{\cdot}01393$, and a very well corrected telescope would result. But as the instrument would have a Ramsden disk only 0·1 inch in diameter and also an unpopular magnification of about 4·8 times, the evolution of a commercially useful design would have to begin with a study of the eyepiece at 0·15″ to 0·20″ diameter of the entrance pupil, and r_4 would have to be lengthened, probably to about 0·75, so as to secure freedom from coma for eyepiece plus prisms at one of the usual magnifications, such as 6 or 8 times. The smallest safe size of the prisms would also have to be determined, because compactness and light weight are the chief attractions of the prismatic telescope. These desiderata account for the low f-numbers, down to $f/4$, at which these instruments work.

[81] Before concluding this chapter on ordinary eyepieces we must add some decidedly important general remarks.

[81] (1) All the eyepieces which we have discussed consist of two widely separated convex lenses or components. All systems of this type suffer from the grave drawback that their Petzval sum is much larger than that of a thin single or achromatic lens of the same focal length. For our typical Huygenian eyepiece we used above its Petzval sum $= 0{\cdot}517 \times 1{\cdot}5/1{\cdot}517 = 0{\cdot}511$. Any ordinary achromatic lens of the same $f' = 2{\cdot}60$ would, by the convenient formula also given above, have a Petzval sum $= 0{\cdot}7/f'$, or in our case $= 0{\cdot}7/2{\cdot}60 = 0{\cdot}269$, practically half that of the Huygenian eyepiece. For the two other forms a similar disadvantage will be found. It *is* a disadvantage because it was shown in section [55] G that the amount of negative astigmatism required to produce any selected type of flattened field is directly proportional to the Petzval sum. Hence all these eyepieces have either a more curved field, or greater negative astigmatism, than a fairly compact achromatic combination. Some eyepieces of the latter type will be dealt with in Part II. We may add that bulky photographic objectives, like the Petzval portrait lens and the old rectilinear objectives, suffer from the same drawback of aggravated Petzval curvature.

(2) Very important conclusions as to the most desirable location of the eye-point can be drawn from certain peculiarities of naked-eye vision. Whilst the human eye has an exceptionally large field of view, extending to more than 90° from its axis in the unobstructed outward direction, and to about 45° or more in the inward and upward directions where the field is obstructed by nose and eyebrows, its really *sharply defined* field is surprisingly small, measuring only a fraction of one degree in angular diameter. This may and should be verified by *fixing* the eye on, say, a chosen letter of this page, when it will be noticed that even the immediately surrounding letters are less distinct, and that those appearing a few degrees away from the chosen letter cannot be identified at all ; experience merely suggests that they also are 'print'. It must be added that this experiment requires stern discipline and grim determination to keep the eye *fixed* ; for, whilst few are clearly aware of the fact, everybody has a subconscious knowledge of the restricted sharp field of the eye and by long habit rolls the eye or turns the head in the direction of anything attracting attention, with the result that sharp vision is *then* obtained. As this action becomes purely automatic from early childhood, the true state of gross imperfection in the extra-axial regions of the field of our eye is rarely realized.

(3) In the light of the preceding explanations we will now first discuss the case of vision through fairly large lenses not restricted by diaphragms. All kinds of spectacle lenses, low-power magnifying glasses, and the lenses through which stereoscopic pictures or small-scale single photographs are examined, are examples of this type. If, with reference to Fig. 102 (*a*), we assume that the eye is directed towards the axial part of the field of view, this part of the field will be seen by 'direct vision' with the full acuity of the eye, and we must therefore demand good correction of the axial pencils delivered by the lens. The extra-axial parts of the field will be seen by 'indirect vision' obtained by oblique pencils admitted by the pupil of the eye, and we thus secure a general though rather vague impression of the whole field of view, sufficient to show the mutual relations of the various objects, but not sufficient to discern delicate detail. Hence, assuming that the lens

is reasonably fit to be used for the particular purpose, it is highly improbable that [81] the aberrations of the oblique pencils admitted by the pupil would be large enough sensibly to aggravate the bad definition which is peculiar to the outer parts of the

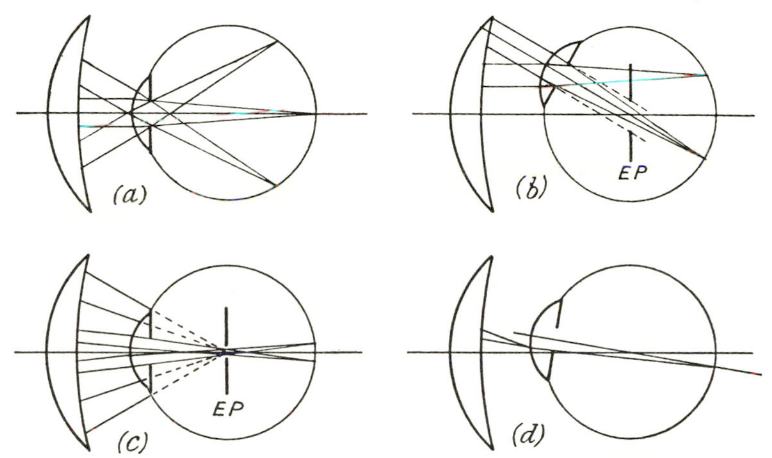

FIG. 102

retina. But if anything dimly perceived in the extra-axial field attracts the attention, the eye is immediately rolled in its socket so as to secure 'direct' vision of that part of the field, and as this is now seen at the highest acuity of vision, we must demand the best possible correction of the oblique pencil admitted along the axis of the rolled eye ; we need not worry about any loss of definition in the rest of the field (including the previous centre), because the outer parts of the retina are quite incapable of appreciating it. It is easily seen by Fig. 102 (*b*) that the realization of the best possible direct vision throughout the field of the lens demands that the lens should have its oblique aberrations corrected for an exit pupil of the lens at the centre of rotation of the eyeball of the observer. This centre lies about 13 mm. ($\frac{1}{2}$ inch) behind the vertex of the cornea ; allowing a reasonable addition for eyelids and lashes, we thus arrive at the conclusion that lenses of the type here referred to should be designed for an 'eye-point' at a clear distance of 18–22 mm. from the last surface. As such lenses are almost invariably of simple form and not aplanatic, they are strongly subject to changes of the oblique aberrations by the second fundamental law; it is therefore important to locate the eye-point correctly. Moreover, it must be remembered that different observers will choose slightly different positions of the eye, so that the actual distance of the eye-point may become longer or shorter by several mm. than that assumed by the designer. These variations will have the least effect—again by the second fundamental law—*if the coma is made zero for the mean distance*; for the coma will then be small throughout a considerable range of l'_{pr}, and the astigmatism and curvature of field will be near minimum, or maximum in the case of concave spectacle lenses, and will be practically constant throughout the range.

[81] (4) The reasoning employed in the preceding subsection would be equally valid in all cases of visual observations if the attainment of the sharpest possible 'direct' vision throughout the field of view, by rolling the eyeball round its stationary centre, were the *only* important consideration. But we have learnt that eyepieces are severely diaphragmed by the limited aperture of the object-glass, and that, as a consequence, the emerging individual pencils are rarely more than two millimetres in diameter, and may be even less than half a millimetre. Fig. 102 (*c*) shows that under these conditions and direct vision along the axis the pencils from the outer part of the field could not enter the pupil of the eye, and that only a vignetted central part of the field would be available for observation. If the eye is rolled—and the eye is rarely still for even a few seconds—other parts of the field come into view, but only at the expense of a corresponding loss of the previously visible part ; the whole field can never be seen simultaneously. Owing to the usual rapid, though small, movements of the eyeball—rendered necessary by the minuteness of the really sharp field—there results a perpetual fluttering about of the visible field which is intolerably worrying and distracting. That is the reason why we must, in the case of normal eyepieces, depart from the optimum defined in (3) by choosing the design so that the Ramsden disk of the eyepiece becomes nearly coincident with the pupil of the eye, thus establishing the arrangement of the oblique pencils shown in Fig. 102 (*a*). The whole field is now visible when looking along the optical axis, and remains visible as long as the eye does not roam beyond the usual ten or twenty degrees in diameter to which normal sharp observations are restricted. But even when the eyeball is turned so much that the Ramsden disk ceases to fall entirely within the circle of the pupil, early experience in peeping through narrow holes or cracks causes us to do automatically exactly what is necessary to recover full vision, namely, to move the eye bodily to the right in order to see more of the left side of the object, and vice versa. In the case of Fig. 102 (*c*) the required movement of the eye is in the opposite sense and contrary to all our early experience ; that is what renders this case almost maddening.

The penalty attached to coincidence of the Ramsden disk with the pupil is, that if the eye is merely rolled round its centre from the correct position for the sharpest axial vision, all the pencils enter through an eccentric part of the pupil ; as the eye is not corrected either for spherical or for chromatic aberration, a certain amount of coma, astigmatism, and transverse chromatic aberration results. If this is sufficient to disturb clear direct vision, instinct again causes us to apply the right remedy : a slight transverse shift of the whole eye.

(5) The last paragraph leads directly to important conclusions with reference to achromatism of magnification in eyepieces, for as eccentric entry of a pencil imparts transverse chromatic aberration to it, it must also be possible to correct moderate amounts of this aberration left uncorrected in the eyepiece by receiving the pencils through a suitably chosen eccentric part of the pupil. We will now show that this possibility does in fact bear an important part in observations with eyepieces having residues of transverse chromatic aberration.

As far as its chromatic aberration is concerned, we may replace the complicated optical apparatus of the eye by a single air-into-water refraction at a convex surface (replacing the cornea) of about 5·7 mm. radius. Whilst not by any means exact the results will be near enough for the purpose of an approximate estimate. If we [81] trace a ray parallel to the optical axis and at $Y = 1$ mm. into the water † for C-light, and then an F-ray coinciding in the water with the C-ray outward, we shall find an angle of $3'$ of arc between the two rays in the air, in the sense shown with gross exaggeration in Fig. 102 (*d*). It follows that an eccentricity of the entering pencils of only 1 mm. will correct an angular transverse chromatic aberration of $3'$, and this small eccentricity will practically always be easily realizable, as under proper observing conditions the pupil will be at least 3 to 4 mm. in clear aperture. But at night the pupil opens to 8 and more mm. aperture, and an eccentricity of 3 mm., corresponding to $9'$ of angular chromatic aberration, would be quite practicable. For our Ramsden eyepiece we found $Upr_y - Upr_v$, *its* angular chromatic aberration, equal to $8'$ at $15°$ from the centre of the field, hence the whole of this could be compensated at night by a suitable decentration of the eye!

Fig 102 (*d*) shows that in the normal case, when the centre of rotation of the eyeball is kept on the optical axis, the eccentric entry of the pencils resulting from rolling of the eye is such as to suit an 'under-corrected' eyepiece like the Ramsden. As the pupil lies about 3 mm. behind the cornea or 10 mm. from the centre of rotation of the eye, an eccentricity of 1 mm. will be produced by a turn of the eye through 1/10 radian or $5\frac{3}{4}°$. As we calculated the corresponding C to F dispersion of the principal ray for this eccentricity as $3'$, the corresponding g_7-value is $3/5\frac{3}{4} \times 60 = 0{\cdot}0087$, very nearly that of our Ramsden eyepiece. Hence the want of achromatism of the human eye happens to correct the transverse chromatic aberration of the ordinary Ramsden eyepiece throughout that central part of the field where nearly all accurate observing is done, and for which it is not necessary to shift the eye transversely! This explains the otherwise puzzling fact that unsophisticated observers are usually under the impression that the Ramsden eyepiece is perfectly corrected for a considerable angular field surrounding the axis and then goes off rather suddenly towards the extreme edge. It also explains why an eyepiece perfectly corrected for transverse chromatic aberration is apt to give the false impression of 'over-correction' in the more central part of the field.

It does not appear advisable, however, to conclude that eyepieces ought to be corrected for a positive paraxial g_7 of a magnitude approaching that of our Ramsden eyepiece, for the marginal parts of the field, especially with the large angles of 50° and even 60° which are becoming quite common, would display at least approximately the true state of correction and would lead to condemnation. But we may safely say that a moderate *positive* value of paraxial g_7, up to perhaps 0·004, is decidedly *preferable* to a negative residual.

Special Memoranda from Chapter X

As all ordinary types of eyepieces are spherically under-corrected for the axial pencil, their oblique aberrations vary when the aperture diaphragm is shifted along the optical axis, and there can be only one position which gives the best possible state of correction. In all ordinary instruments the object-glass acts as aperture diaphragm for the eyepiece: hence there will be a particular 'tube-length' at which

† Indices of water: $N_c = 1{\cdot}3314$; $N_d = 1{\cdot}3332$; $N_f = 1{\cdot}3374$; whence $V = 55{\cdot}5$.

[81] any given eyepiece gives its best results. The systematic method of solution given in this chapter is based on these facts.

By applying the sine theorem to the results of the axial raytracing, that tube-length is determined which ensures freedom from coma ; on account of the positive spherical aberration this freedom from coma carries with it the utmost possible flattening of the field, and as there is little risk in the ordinary eyepieces of reaching an objectionably hollow field, the sine theorem alone is sufficient to select the best tube-length for any given eyepiece. In the Huygenian type there is a second liberty which can be used to secure also achromatism of magnification, again by the sine theorem.

The stratagem employed of actually solving for small positive amounts of coma and transverse chromatic aberration in order to balance higher aberrations in a suitably chosen zone of the field on the basis of previous experience, should be carefully noted, as empirical allowances of this kind will accelerate the solving of many other problems which have to be worked out at frequent intervals with comparatively small departures from a known type.

The concluding remarks in section [81] on the effects of idiosyncracies of the human eye on the design of eyepieces may be profitably followed up by reading books on physiological and ophthalmic optics.

APPENDIX

FUNCTIONS OF N

N	G_1	G_2	G_3	G_4	G_5	G_6	G_7	G_8
1·43	·440	·830	1·137	·516	1·461	·946	·580	·307
4	·456	·854	1·170	·526	1·491	·966	·593	·317
5	·473	·878	1·204	·535	1·521	·985	·605	·326
6	·490	·902	1·237	·545	1·550	1·005	·618	·336
7	·508	·926	1·271	·555	1·579	1·025	·630	·345
8	·526	·950	1·306	·564	1·609	1·044	·642	·355
9	·544	·975	1·340	·574	1·638	1·064	·654	·365
1·50	·562	1·000	1·375	·583	1·667	1·083	·667	·375
1	·581	1·025	1·410	·593	1·696	1·103	·679	·385
2	·601	1·050	1·446	·602	1·724	1·122	·691	·395
3	·620	1·076	1·481	·611	1·753	1·141	·703	·405
4	·640	1·102	1·517	·621	1·781	1·161	·715	·416
5	·661	1·128	1·554	·630	1·810	1·180	·727	·426
6	·681	1·154	1·590	·639	1·838	1·199	·739	·437
7	·702	1·180	1·627	·648	1·866	1·218	·752	·447
8	·724	1·206	1·665	·657	1·894	1·237	·764	·458
9	·746	1·233	1·702	·666	1·922	1·256	·776	·469
1·60	·768	1·260	1·740	·675	1·950	1·275	·788	·480
1	·791	1·287	1·778	·684	1·978	1·294	·799	·491
2	·814	1·314	1·817	·693	2·005	1·313	·811	·502
3	·837	1·342	1·855	·702	2·033	1·332	·823	·513
4	·861	1·370	1·894	·710	2·060	1·350	·835	·525
5	·885	1·398	1·934	·719	2·088	1·369	·847	·536
6	·909	1·426	1·973	·728	2·115	1·388	·859	·548
7	·934	1·454	2·013	·736	2·142	1·406	·871	·559
8	·960	1·482	2·054	·745	2·170	1·425	·882	·571
9	·985	1·511	2·094	·753	2·197	1·443	·894	·583
1·70	1·012	1·540	2·135	·762	2·224	1·462	·906	·595
1	1·038	1·569	2·176	·770	2·250	1·480	·918	·607
2	1·065	1·598	2·218	·779	2·277	1·499	·929	·619
3	1·092	1·628	2·259	·787	2·304	1·517	·941	·631
4	1·120	1·658	2·301	·795	2·331	1·535	·953	·644
5	1·148	1·688	2·344	·804	2·357	1·554	·964	·656
6	1·177	1·718	2·386	·812	2·384	1·572	·976	·669

LOGARITHMS OF FUNCTIONS OF N

N	log G_1	log G_2	log G_3	log G_4	log G_5	log G_6	log G_7	log G_8
1·43	9·6431	9·9190	0·0559	9·7124	0·1648	9·9758	9·7637	9·4878
4	·6591	·9313	·0683	·7206	·1735	·9848	·7729	·5008
5	·6749	·9432	·0805	·7286	·1820	9·9936	·7819	·5136
6	·6904	·9550	·0925	·7365	·1904	0·0022	·7907	·5261
7	·7057	·9666	·1043	·7441	·1985	·0106	·7992	·5384
8	·7207	·9779	·1158	·7515	·2065	·0188	·8076	·5505
9	·7355	9·9890	·1272	·7588	·2142	·0269	·8159	·5624
1·50	9·7501	0·0000	0·1383	9·7659	0·2218	0·0348	9·8239	9·5740
1	·7645	·0108	·1493	·7729	·2293	·0425	·8318	·5855
2	·7787	·0214	·1600	·7797	·2366	·0500	·8395	·5968
3	·7926	·0318	·1707	·7863	·2437	·0574	·8471	·6079
4	·8064	·0420	·1811	·7928	·2507	·0647	·8545	·6189
5	·8200	·0521	·1914	·7992	·2576	·0718	·8618	·6297
6	·8334	·0621	·2015	·8055	·2643	·0788	·8689	·6403
7	·8466	·0718	·2115	·8116	·2709	·0857	·8759	·6507
8	·8597	·0815	·2213	·8176	·2774	·0924	·8828	·6611
9	·8726	·0910	·2310	·8235	·2838	·0990	·8896	·6712
1·60	9·8854	0·1004	0·2405	9·8293	0·2900	0·1055	9·8963	9·6812
1	·8980	·1096	·2500	·8350	·2962	·1119	·9028	·6911
2	·9104	·1187	·2593	·8406	·3022	·1182	·9092	·7009
3	·9227	·1277	·2684	·8460	·3081	·1243	·9155	·7105
4	·9348	·1366	·2775	·8514	·3140	·1304	·9217	·7200
5	·9468	·1454	·2864	·8567	·3197	·1364	·9279	·7294
6	·9587	·1540	·2952	·8619	·3253	·1423	·9339	·7386
7	·9705	·1625	·3039	·8670	·3309	·1480	·9398	·7478
8	·9821	·1710	·3125	·8720	·3364	·1537	·9457	·7568
9	9·9936	·1793	·3210	·8770	·3417	·1594	·9514	·7657
1·70	0·0050	0·1875	0·3294	9·8818	0·3470	0·1649	9·9571	9·7745
1	·0162	·1957	·3377	·8866	·3523	·1703	·9627	·7832
2	·0274	·2037	·3459	·8913	·3574	·1757	·9682	·7918
3	·0384	·2116	·3540	·8960	·3625	·1810	·9736	·8003
4	·0493	·2195	·3620	·9005	·3675	·1862	·9789	·8088
5	·0601	·2272	·3699	·9050	·3724	·1913	·9842	·8171
1·76	0·0708	0·2349	0·3777	9·9095	0·3772	0·1964	9·9894	9·8253

INDEX

A CATALOG OF SELECTED

DOVER BOOKS

IN SCIENCE AND MATHEMATICS

RELATIVITY, THERMODYNAMICS AND COSMOLOGY, Richard C. Tolman. Landmark study extends thermodynamics to special, general relativity; also applications of relativistic mechanics, thermodynamics to cosmological models. 501pp. 5⅜ × 8½. 65383-8 Pa. $11.95

APPLIED ANALYSIS, Cornelius Lanczos. Classic work on analysis and design of finite processes for approximating solution of analytical problems. Algebraic equations, matrices, harmonic analysis, quadrature methods, much more. 559pp. 5⅜ × 8½. 65656-X Pa. $11.95

SPECIAL RELATIVITY FOR PHYSICISTS, G. Stephenson and C.W. Kilmister. Concise elegant account for nonspecialists. Lorentz transformation, optical and dynamical applications, more. Bibliography. 108pp. 5⅜ × 8½. 65519-9 Pa. $3.95

INTRODUCTION TO ANALYSIS, Maxwell Rosenlicht. Unusually clear, accessible coverage of set theory, real number system, metric spaces, continuous functions, Riemann integration, multiple integrals, more. Wide range of problems. Undergraduate level. Bibliography. 254pp. 5⅜ × 8½. 65038-3 Pa. $7.00

INTRODUCTION TO QUANTUM MECHANICS With Applications to Chemistry, Linus Pauling & E. Bright Wilson, Jr. Classic undergraduate text by Nobel Prize winner applies quantum mechanics to chemical and physical problems. Numerous tables and figures enhance the text. Chapter bibliographies. Appendices. Index. 468pp. 5⅜ × 8½. 64871-0 Pa. $9.95

ASYMPTOTIC EXPANSIONS OF INTEGRALS, Norman Bleistein & Richard A. Handelsman. Best introduction to important field with applications in a variety of scientific disciplines. New preface. Problems. Diagrams. Tables. Bibliography. Index. 448pp. 5⅜ × 8½. 65082-0 Pa. $10.95

MATHEMATICS APPLIED TO CONTINUUM MECHANICS, Lee A. Segel. Analyzes models of fluid flow and solid deformation. For upper-level math, science and engineering students. 608pp. 5⅜ × 8½. 65369-2 Pa. $12.95

ELEMENTS OF REAL ANALYSIS, David A. Sprecher. Classic text covers fundamental concepts, real number system, point sets, functions of a real variable, Fourier series, much more. Over 500 exercises. 352pp. 5⅜ × 8½. 65385-4 Pa. $8.95

PHYSICAL PRINCIPLES OF THE QUANTUM THEORY, Werner Heisenberg. Nobel Laureate discusses quantum theory, uncertainty, wave mechanics, work of Dirac, Schroedinger, Compton, Wilson, Einstein, etc. 184pp. 5⅜ × 8½. 60113-7 Pa. $4.95

INTRODUCTORY REAL ANALYSIS, A.N. Kolmogorov, S.V. Fomin. Translated by Richard A. Silverman. Self-contained, evenly paced introduction to real and functional analysis. Some 350 problems. 403pp. 5⅜ × 8½. 61226-0 Pa. $7.95

PROBLEMS AND SOLUTIONS IN QUANTUM CHEMISTRY AND PHYSICS, Charles S. Johnson, Jr. and Lee G. Pedersen. Unusually varied problems, detailed solutions in coverage of quantum mechanics, wave mechanics, angular momentum, molecular spectroscopy, scattering theory, more. 280 problems plus 139 supplementary exercises. 430pp. 6½ × 9¼. 65236-X Pa. $10.95

THE ELECTROMAGNETIC FIELD, Albert Shadowitz. Comprehensive undergraduate text covers basics of electric and magnetic fields, builds up to electromagnetic theory. Also related topics, including relativity. Over 900 problems. 768pp. 5⅜ × 8¼. 65660-8 Pa. $15.95

FOURIER SERIES, Georgi P. Tolstov. Translated by Richard A. Silverman. A valuable addition to the literature on the subject, moving clearly from subject to subject and theorem to theorem. 107 problems, answers. 336pp. 5⅜ × 8½. 63317-9 Pa. $7.95

THEORY OF ELECTROMAGNETIC WAVE PROPAGATION, Charles Herach Papas. Graduate-level study discusses the Maxwell field equations, radiation from wire antennas, the Doppler effect and more. xiii + 244pp. 5⅜ × 8½. 65678-0 Pa. $6.95

DISTRIBUTION THEORY AND TRANSFORM ANALYSIS: An Introduction to Generalized Functions, with Applications, A.H. Zemanian. Provides basics of distribution theory, describes generalized Fourier and Laplace transformations. Numerous problems. 384pp. 5⅜ × 8½. 65479-6 Pa. $8.95

THE PHYSICS OF WAVES, William C. Elmore and Mark A. Heald. Unique overview of classical wave theory. Acoustics, optics, electromagnetic radiation, more. Ideal as classroom text or for self-study. Problems. 477pp. 5⅜ × 8½. 64926-1 Pa. $10.95

CALCULUS OF VARIATIONS WITH APPLICATIONS, George M. Ewing. Applications-oriented introduction to variational theory develops insight and promotes understanding of specialized books, research papers. Suitable for advanced undergraduate/graduate students as primary, supplementary text. 352pp. 5⅜ × 8½. 64856-7 Pa. $8.50

A TREATISE ON ELECTRICITY AND MAGNETISM, James Clerk Maxwell. Important foundation work of modern physics. Brings to final form Maxwell's theory of electromagnetism and rigorously derives his general equations of field theory. 1,084pp. 5⅜ × 8½. 60636-8, 60637-6 Pa., Two-vol. set $19.00

AN INTRODUCTION TO THE CALCULUS OF VARIATIONS, Charles Fox. Graduate-level text covers variations of an integral, isoperimetrical problems, least action, special relativity, approximations, more. References. 279pp. 5⅜ × 8½. 65499-0 Pa. $6.95

HYDRODYNAMIC AND HYDROMAGNETIC STABILITY, S. Chandrasekhar. Lucid examination of the Rayleigh-Benard problem; clear coverage of the theory of instabilities causing convection. 704pp. 5⅜ × 8¼. 64071-X Pa. $12.95

CALCULUS OF VARIATIONS, Robert Weinstock. Basic introduction covering isoperimetric problems, theory of elasticity, quantum mechanics, electrostatics, etc. Exercises throughout. 326pp. 5⅜ × 8½. 63069-2 Pa. $7.95

DYNAMICS OF FLUIDS IN POROUS MEDIA, Jacob Bear. For advanced students of ground water hydrology, soil mechanics and physics, drainage and irrigation engineering and more. 335 illustrations. Exercises, with answers. 784pp. 6⅛ × 9¼. 65675-6 Pa. $19.95

NUMERICAL METHODS FOR SCIENTISTS AND ENGINEERS, Richard Hamming. Classic text stresses frequency approach in coverage of algorithms, polynomial approximation, Fourier approximation, exponential approximation, other topics. Revised and enlarged 2nd edition. 721pp. 5⅜ × 8½.
65241-6 Pa. $14.95

THEORETICAL SOLID STATE PHYSICS, Vol. I: Perfect Lattices in Equilibrium; Vol. II: Non-Equilibrium and Disorder, William Jones and Norman H. March. Monumental reference work covers fundamental theory of equilibrium properties of perfect crystalline solids, non-equilibrium properties, defects and disordered systems. Appendices. Problems. Preface. Diagrams. Index. Bibliography. Total of 1,301pp. 5⅜ × 8½. Two volumes. Vol. I 65015-4 Pa. $12.95
Vol. II 65016-2 Pa. $12.95

OPTIMIZATION THEORY WITH APPLICATIONS, Donald A. Pierre. Broad-spectrum approach to important topic. Classical theory of minima and maxima, calculus of variations, simplex technique and linear programming, more. Many problems, examples. 640pp. 5⅜ × 8½. 65205-X Pa. $12.95

THE MODERN THEORY OF SOLIDS, Frederick Seitz. First inexpensive edition of classic work on theory of ionic crystals, free-electron theory of metals and semiconductors, molecular binding, much more. 736pp. 5⅜ × 8½.
65482-6 Pa. $14.95

ESSAYS ON THE THEORY OF NUMBERS, Richard Dedekind. Two classic essays by great German mathematician: on the theory of irrational numbers; and on transfinite numbers and properties of natural numbers. 115pp. 5⅜ × 8½.
21010-3 Pa. $4.95

THE FUNCTIONS OF MATHEMATICAL PHYSICS, Harry Hochstadt. Comprehensive treatment of orthogonal polynomials, hypergeometric functions, Hill's equation, much more. Bibliography. Index. 322pp. 5⅜ × 8½. 65214-9 Pa. $8.95

NUMBER THEORY AND ITS HISTORY, Oystein Ore. Unusually clear, accessible introduction covers counting, properties of numbers, prime numbers, much more. Bibliography. 380pp. 5⅜ × 8½. 65620-9 Pa. $8.95

THE VARIATIONAL PRINCIPLES OF MECHANICS, Cornelius Lanzcos. Graduate level coverage of calculus of variations, equations of motion, relativistic mechanics, more. First inexpensive paperbound edition of classic treatise. Index. Bibliography. 418pp. 5⅜ × 8½. 65067-7 Pa. $10.95

MATHEMATICAL TABLES AND FORMULAS, Robert D. Carmichael and Edwin R. Smith. Logarithms, sines, tangents, trig functions, powers, roots, reciprocals, exponential and hyperbolic functions, formulas and theorems. 269pp. 5⅜ × 8½. 60111-0 Pa. $5.95

THEORETICAL PHYSICS, Georg Joos, with Ira M. Freeman. Classic overview covers essential math, mechanics, electromagnetic theory, thermodynamics, quantum mechanics, nuclear physics, other topics. First paperback edition. xxiii + 885pp. 5⅜ × 8½. 65227-0 Pa. $17.95

ORDINARY DIFFERENTIAL EQUATIONS, Morris Tenenbaum and Harry Pollard. Exhaustive survey of ordinary differential equations for undergraduates in mathematics, engineering, science. Thorough analysis of theorems. Diagrams. Bibliography. Index. 818pp. 5⅜ × 8½. 64940-7 Pa. $15.95

STATISTICAL MECHANICS: Principles and Applications, Terrell L. Hill. Standard text covers fundamentals of statistical mechanics, applications to fluctuation theory, imperfect gases, distribution functions, more. 448pp. 5⅜ × 8½. 65390-0 Pa. $9.95

ORDINARY DIFFERENTIAL EQUATIONS AND STABILITY THEORY: An Introduction, David A. Sánchez. Brief, modern treatment. Linear equation, stability theory for autonomous and nonautonomous systems, etc. 164pp. 5⅜ × 8¼. 63828-6 Pa. $4.95

THIRTY YEARS THAT SHOOK PHYSICS: The Story of Quantum Theory, George Gamow. Lucid, accessible introduction to influential theory of energy and matter. Careful explanations of Dirac's anti-particles, Bohr's model of the atom, much more. 12 plates. Numerous drawings. 240pp. 5⅜ × 8½. 24895-X Pa. $5.95

ORDINARY DIFFERENTIAL EQUATIONS, I.G. Petrovski. Covers basic concepts, some differential equations and such aspects of the general theory as Euler lines, Arzel's theorem, Peano's existence theorem, Osgood's uniqueness theorem, more. 45 figures. Problems. Bibliography. Index. xi + 232pp. 5⅜ × 8½. 64683-1 Pa. $6.00

GREAT EXPERIMENTS IN PHYSICS: Firsthand Accounts from Galileo to Einstein, edited by Morris H. Shamos. 25 crucial discoveries: Newton's laws of motion, Chadwick's study of the neutron, Hertz on electromagnetic waves, more. Original accounts clearly annotated. 370pp. 5⅜ × 8½. 25346-5 Pa. $8.95

INTRODUCTION TO PARTIAL DIFFERENTIAL EQUATIONS WITH APPLICATIONS, E.C. Zachmanoglou and Dale W. Thoe. Essentials of partial differential equations applied to common problems in engineering and the physical sciences. Problems and answers. 416pp. 5⅜ × 8½. 65251-3 Pa. $9.95

BURNHAM'S CELESTIAL HANDBOOK, Robert Burnham, Jr. Thorough guide to the stars beyond our solar system. Exhaustive treatment. Alphabetical by constellation: Andromeda to Cetus in Vol. 1; Chamaeleon to Orion in Vol. 2; and Pavo to Vulpecula in Vol. 3. Hundreds of illustrations. Index in Vol. 3. 2,000pp. 6⅛ × 9¼. 23567-X, 23568-8, 23673-0 Pa., Three-vol. set $38.85

ASYMPTOTIC EXPANSIONS FOR ORDINARY DIFFERENTIAL EQUATIONS, Wolfgang Wasow. Outstanding text covers asymptotic power series, Jordan's canonical form, turning point problems, singular perturbations, much more. Problems. 384pp. 5⅜ × 8½. 65456-7 Pa. $8.95

AMATEUR ASTRONOMER'S HANDBOOK, J.B. Sidgwick. Timeless, comprehensive coverage of telescopes, mirrors, lenses, mountings, telescope drives, micrometers, spectroscopes, more. 189 illustrations. 576pp. 5⅜ × 8¼. 24034-7 Pa. $8.95

CATALOG OF DOVER BOOKS